Management of EMS

Bruce E. Evans
Jeff T. Dyar

Pearson
Upper Saddle River, New Jersey

Library of Congress Cataloging-in-Publication Data:

Evans, Bruce E.
  Management of EMS / Bruce E. Evans, Jeff T. Dyar.
    p. cm.
  Includes index.
  ISBN-13: 978-0-13-232432-8
  ISBN-10: 0-13-232432-6
  1. Emergency medical services—Management. I. Dyar, Jeff T. II. Title.
  RA645.5.E93 2010
    362.18068—dc22

                                                    2008047912

Publisher: Julie Levin Alexander
Publisher's Assistant: Regina Bruno
Senior Acquisitions Editor: Stephen Smith
Associate Editor: Monica Moosang
Development Editor: Jo Cepeda
Editorial Assistant: Heather Luciano/
    Patricia Linard
Director of Marketing: Karen Allman
Executive Marketing Manger: Katrin Beacom
Marketing Specialist: Michael Sirinides
Marketing Assistant: Lauren Castellano
Managing Production Editor: Patrick Walsh
Production Liaison: Julie Li
Production Editor: Lisa S. Garboski, bookworks
    editorial services

Media Product Manger: Amy Peltier
Media Project Manager: Lorena Cerisano
Manufacturing Manager: Ilene Sanford
Manufacturing Buyer: Pat Brown
Senior Designer Coordinator: Chris Weigand
Cover Designer: Chris Weigand
Director, Image Research Center:
Manager, Rights and Permissions: Zina Arabia
Manager, Visual Research: Beth Brenzel
Manager, Cover Visual Research and Permissions:
    Karen Sanatar
Image Permission Coordinator: Vicki Menanteaux
Composition: Aptara®, Inc.
Printing and Binding: Edwards Brothers
Cover Printer: Phoenix Color Corporation

Notice on Gender Usage

The English language has historically given preference to the male gender. Among many words, the pronouns "he" and "his" are commonly used to describe both genders. Society evolves faster than language, and the male pronouns still predominate our speech. The authors have made great effort to treat the two genders equally, recognizing that a significant number of EMS providers are female. However, in some instances, male pronouns may be used to describe both male and females solely for the purpose of brevity. This is not intended to offend any readers.

Pearson Education Ltd.
Pearson Education Singapore Pte. Ltd.
Pearson Education Canada, Ltd.
Pearson Education—Japan

Pearson Education Australia Pty. Limited
Pearson Education North Asia Ltd.
Pearson Educación de Mexico, S.A. de C.V.
Pearson Education Malaysia Pte. Ltd.

Prentice Hall
is an imprint of

www.pearsonhighered.com

10 9 8 7 6 5 4 3 2 1
ISBN 13: 978-0-13-232432-8
ISBN 10:   0-13-232432-6

Dedication

This book is dedicated to my wife Debora and son Oliver for their patience and understanding as I wrote this material and to my mother, Sandra Laidlaw for putting me to work on the hometown ambulance and steering me to a great career.

To my paramedic mentors at Mary Greeley Medical Center, especially Lee Thomas, Scott Guetzko and the late Larry Rossman for giving me the technical skills; the staff of Life Flight Des Moines and retired flight nurse Cathy O'Brien, who taught me compassionate care; and the crew from Henderson Fire Station 94 "B" shift (Robbie, Sean, Chad, Jimmy and Brian) for building my leadership skills and being a band of brothers.

Bruce E. Evans

I would like to thank my family and colleagues for understanding and supporting my globetrotting ways and responsibilities. I am grateful to my wife Val and sons Joel and Mason for teaching me patience and giving me energy and direction for the future. I thank my mentors—James O. Page for the spark and the fire storm we call EMS, shepherding its first 30 years, and giving me great hope in my times of greatest need; Ronny Coleman for having the perseverance to shape the fire service and mentoring some of the finest leaders I know; Alan Brunicini for teaching me that humor and humility are essential and enduring qualities.

And to my brothers and sisters, who respond to the needs of our nation at any moment, for any need, without question . . . you are my heroes.

Jeff T. Dyar

Contents

Foreword

Being asked to write the foreword for this *EMS Management* book is a great honor for me. Jeff Dyar and Bruce Evans have been leaders in the field of EMS for a long time. I have had the pleasure of knowing both authors for many years, during which time I have had one-on-one conversations, read some of their writings, and listened to their lectures. I also worked with Jeff at the National Fire Academy while I served as the acting chief operating officer of the United States Fire Administration. As for my relationship with Bruce, I had a birds-eye view: while I served as president of the IAFC, he was making significant contributions to the EMS section. While not all of the interactions were about EMS, there were plenty that indicated to me the competence of both Bruce and Jeff in this emerging field. (Yes, emerging, as EMS as we know it today had its genesis in the fire service in the 1970s, just when these two were cutting their teeth!)

The information provided in this text comes from two individuals that are extremely qualified. They have years of meaningful experience and have taken advantage of their access to many in the profession to expand their knowledge base. Both have a tremendous passion for EMS and have made it their business to become leaders who are recognized throughout the country. They are both disciples of the late (and, I might add, great) Jim Page, the father of modern, fire-based EMS. Having known Jim, I have no doubt he would have approved of the lessons presented in this book.

So many EMS managers have learned the business every way possible—through experience, seminars, and workshops and from mentors. There is finally one source that captures all the elements of EMS management. There is information on leadership, interagency relations, legal issues, ethics, costing, fleet management, special operations, quality management, personnel management, and much more. It is all in here. This will be a great reference source and an outstanding text for higher education.

In closing, I reiterate my tremendous respect for Jeff and Bruce—as professionals with solid EMS backgrounds but more importantly as "top-shelf" people. Their knowledge base is real, obtained from the field and numerous other sources. They both know their stuff and make a great team. The lessons are delivered to the reader in a way that makes sense and is easily understood. We in the profession are fortunate to have Jeff and Bruce share their knowledge. There is no doubt that this text solidifies their status as "legends" in EMS.

Richard A. Marinucci

Fire Chief, Farmington Hills, MI, Fire Department since 1984
President of the International Association of Fire Chiefs, 1997–98
Chief Operating Officer of the United States Fire Administration, 1999

Preface

If you have picked up this text, you have taken one of the most important steps to serving your community and your organization, the first being called to serve as an EMS provider. EMS managers are handed a badge only to learn this aspect of the profession on the job. Ideally, this text will be a resource for the reader to hit the ground running, as the work of EMS management is much harder than patient care. This text provides information on how to build a system that serves the public and employees and on enabling a management team to ensure the public trust.

OUR DEVELOPMENT STORY

The idea for this text came after the passing of James O. Page. While many great managers and a few exceptional leaders in EMS remain, we will never be so lucky as to have Jim's knowledge and, more importantly, his wisdom. As with many professions, the mechanics of the trade are often never captured in the written word. It was our intention to collect the common knowledge and experience—from those with whom we have interacted in the EMS community and from those who have attended the National Fire Academy's EMS programs—and present all here in this text.

ORGANIZATION OF THIS TEXT

Chapter 1 describes the landscape of EMS, the contributors, organizations, and the history of the EMS systems. It is there that this book differs from other EMS management texts in that it is more about "how to" than "what is." Strategic Planning for EMS in Chapter 2 provides core principles that lead into Chapter 3: Manager to Leader. Many agencies have good managers without real leadership, and Chapter 3 clearly points out the distinc-

tions between the two. Chapter 4 emphasizes the need for injury prevention to be just as important for EMS organizations as is response. Chapter 5 prepares the reader for interacting with customers from the perspective of public trust and servant leadership.

Risk Management and Safety (Chapter 6), EMS Human Resources Management (Chapter 7), and Management of EMS Education (Chapter 8) could be bound together to educate a manager on the core principles of the management of people, as each of these content areas ensures a stable, healthy, and competent workforce. Chapter 9 encompasses the financial aspects of EMS, not only from a billing perspective but also from a complete perspective of where to find and how to manage money. Because reimbursement strategies change so rapidly, readers should visit the Web site that accompanies this text for more up-to-date information on maximizing reimbursement.

Chapter 10 reflects the principles of the physician organizations ACEP and NAEMSP and instructs the EMS organization on how to fully integrate a physician into an EMS organization. Chapter 11 covers Fleet Management, which is often a weak point in many of our operations. The chapter reflects some of the best practices from the private ambulance industry and the International City/County Management Association. In Chapter 12, Career Development and Staff Focus, the content reflects on the work of the Centers for Public Safety Excellence, the National Fire Academy, and the International Association of Fire Chiefs to advance the EMS profession.

Chapter 13 is about quality management and reflects some of the most current processes to improve services. Chapter 14, Incident Management, and Chapter 15, Interagency Relations and

Operations, go hand in hand, illustrating the National Incident Management System specific to EMS and emphasizing how we must rely on and be a resource to our neighbors. Chapter 16, Data Collection and EMS Research, is designed for EMS personnel to engage in research. Legal aspects and labor relations are covered in Chapter 17, which encourages a pro-labor perspective and an understanding that EMS operations are a collaborative process that is always ongoing. Often forgotten yet one of the most important aspects of EMS is the communication center. Chapter 18 is a guide to how an EMS manager interfaces and provides leadership in the communication center. Finally, Chapter 19 provides detailed coverage of how EMS leadership manages, prepares, and responds to special operations.

FEATURES

Accreditation Standards: For EMS leadership to achieve a level of excellence, standards for those in the organization must be set. Each chapter of the text begins by identifying the Committee on Ambulance Accreditation Standards (CAAS) accreditation standards covered in the chapter.

Chapter Objectives: Objectives are identified at the beginning of each chapter and outline the material the reader should understand upon completion of the chapter.

Key Terms: Key terms are listed at the beginning of each chapter and are bold upon introduction in the chapter. They are defined in the comprehensive glossary at the end of the book.

Point of View: Every chapter starts with a personal reflection from EMS leaders in the field. These unique success stories are tied to the content of each chapter.

Figures and Tables: Figures and tables are presented throughout the text to highlight and illustrate important concepts. Checklists, assessments, and models to incorporate in your EMS organization are provided. In addition, samples from fire and EMS agencies around the United States and the United Kingdom are imbedded in this book.

Chapter Summaries: Chapter summaries recap the key content discussed in the chapter.

Research and Review Questions: These questions are designed to promote active learning. Students are required to draw on the knowledge presented in the chapter to answer the questions, as well as to conduct outside research.

References: A complete list of references appears at the end of each chapter. Students are encouraged to review references of interest for further study.

ROAD MAP/HOW TO USE THIS TEXT

This text is designed to be used as an anchor for a course in the Management of Emergency Medical Services at the National Fire Academy. It is intended to be used as a primary text for the Fire and Emergency Service Higher Education (FESHE) EMS Operations and Administration course and as a reference for other courses in the FESHE EMS curriculum. While there is enough content to use this text for all or part of every course in an EMS management curriculum, it is important to understand that this is a comprehensive text and the material is designed to be a resource for frontline EMS supervisors to use as they progress up the chain to leadership in the organization. Many of the chapters can be bundled to accommodate one particular course. Each chapter can be used for management as continuing education in a particular aspect of an EMS operation.

TEACHING PACKAGE

The following resources can be downloaded from Pearson's Instructor Resource Center:

PowerPoint Slides: (0-13-232434-2) provide you with the tools you need for a dynamic presentation.

Instructor's Resource Manual: (0-13-232433-4) includes lecture outlines, chapter quizzes with answer keys, and handouts to further enhance chapter discussions.

Within the Instructor's Resource Manual tab you will also find the following:

Additional Accreditation Standards: The federal government's Malcolm Baldrige Health Care Criteria for Performance Excellence are included here with page references to the text.

Your Web Connection: Organized by chapter, these web links expand upon topics and organizations discussed throughout the textbook.

Acknowledgments

The detail in this book would not be possible without the help of the nation's EMS community. Special thanks to Lori Moore and the International Association of Firefighters for their assistance; the Harvard Business Press for the release of material related to coaching and leadership; and ESRI press, Bradshaw Consulting, and Apperson for their technical expertise related to technology, education, and GIS. Special thanks to Theresa McCallion and Cyndee Rose for their review and input to the customer service and marketing chapter and to Tom Bay for his insight and contributions on leadership. We'd like to thank the National Academies of Emergency Dispatch for their review of the material on communication centers. In addition the richness of this book is possible by the contributions of those whose personal triumphs are shared at the start of each chapter.

We would also like to thank the following reviewers:

David S. Becker
Sanford-Brown College
Fenton, MO

Chris Cannon
EMT-P Instructor
Cowley College, Winfield, KS

Chris G. Caulkins, MPH, FF, EMT-P
Century College Paramedic Program
White Bear Lake, MN

Deb Crager
Whidbey Island EMS
Coupeville, WA

Steve Dargan, Senior Coordinator
Washington County Emergency Medical Services Office
Hillsboro, OR

Theresa DeVito, RN, EMT-P, M.Ed, EMS I
EMS Education Coordinator
Capital Community College
Hartford, CT

Mike Giannini, EMT-P
Battalion Chief/Emergency Medical Officer
Marin County Fire Department
Woodacre, CA

James L. Jenkins, Jr. BA, NREMT-P
Tuckahoe Volunteer Rescue Squad
Mechanicsville, VA

Darren Lacroix
Faculty Instructor, Emergency Medical Services Program
Del Mar College
Corpus Christi, TX

Jeffrey Lindsey, PhD
Fire Chief (Ret) Estero Fire Rescue
Assistant Professor, George Washington University
Washington, DC

Ryan Maloney, BS, EMT-P
EMS Educator
University of New Mexico Health Sciences Center
Emergency Medical Services Academy
Albuquerque, NM

Captain William R. Montrie EMT-P
Springfield Township Fire
Holland, Ohio
EMM Certification Coordinator
Owens Community College
Perrysburg, Ohio

David Sarazin, M.Ed., NREMT-P
Captain, Duluth Fire Department
Program Director
Lake Superior College ERTC
Duluth, MN

Geoffrey L. Shapiro
The George Washington University
Emergency Health Services Program
Washington, DC

Bradley Van Ert
Captain/EMS Coordinator
Downey Fire Department
Downey, CA

About the Authors

Bruce Evans, MPA, NREMT-P, has been in EMS for over 25 years and currently serves as the EMS chief for the North Las Vegas Fire Department in Nevada. Bruce has been teaching at the College of Southern Nevada for over 20 years, first in EMS and then as the program coordinator for the Fire Technology program. Bruce also has been an adjunct instructor for over a decade at the National Fire Academy in the EMS, Hazardous Materials, and Fire and Life Safety programs and he helped initiate the Fire Service Higher Education project for EMS management at the National Fire Academy. Bruce is a member of the International Association of Fire Chiefs and serves as the liaison to the National Association of EMS Physician. He remains a member of the International Association of Fire Fighters and the National Association of EMS Educators. He is also involved with the National Association of EMTs, where he serves as the chair of the Health and Safety Committee.

With the encouragement of his parents, who were EMT-Intermediates with the West Des Moines EMS in West Des Moines, Iowa, Bruce began his career in EMS as summer employment while he was attending college. As an EMT-Intermediate he served as a flight attendant on the helicopter with Life Flight out of Methodist Medical Center in Des Moines, Iowa. After completing paramedic training at Mary Greeley Medical Center in Ames, Iowa, Bruce was recruited to Mercy Ambulance in Las Vegas and later served as the corporate recruiter before leaving for the fire service. Bruce formerly worked for the Henderson, Nevada, Fire Department and served as an EMS and fire suppression captain, assisting in starting the paramedic program and achieving some of the highest levels of accreditation an organization can achieve in the fire-based EMS.

In addition to authoring this text, he writes the bimonthly column "EMS Viewpoints" in *Fire Chief* magazine and is on the editorial board of *JEMS* magazine. He speaks frequently at state, regional, and national conferences.

Jeff T. Dyar, B. S., NREMT, started his career as a volunteer EMS provider in a small rural town 36 years ago. As with many idealistic and motivated young persons at the time, he was consumed and captured forever by the television show *Emergency*. Not long after becoming an EMT, he enlisted and served as a combat medic during the Vietnam war. Jeff describes himself as first a teacher and then a provider and has held positions in public, private, military, and federal capacities focusing on EMS. Known as the architect of EMS management programs at the National Fire Academy, he spent nearly 12 years in the post of program chair for EMS in Emmitsburg, Maryland. He has also responded with FEMA to dozens of national emergencies and achieved the level of chief of operations at the National Interagency Operation Center in Washington, D.C.

Over the years, James O. Page mentored and "tormented" him to bring leadership and meaning to national EMS. Advocating partnerships and breaking down walls is his forte, and many a student attending courses at NFA remembers his saying, "What opportunity does this situation or problem offer you?" In 1999 Jeff received the James O. Page Award for contributions to advancing EMS on a national level, given by the EMS Section of the International Association of Fire Chiefs.

Jeff and his wife Val landed back in a small rural town in Colorado, where he serves on the board of directors of a fire protection district providing ALS-level EMS services to a 258-square-mile response area.

Government Structure and Structure and EMS

1 CHAPTER

Accreditation Criteria

CAAS

101.01 Legal Organization: Full disclosure of the agency ownership is required (pp. 11–12).

101.02 Organizational Structure: Documentation of the organizational structure is required so that lines of responsibility and authority can be clearly delineated (pp. 12–14).

Objectives

After reading this chapter the student should be able to:

1.1 Identify six key historical events and key figures that have impacted the progress of emergency medical services (EMS), then discuss their collective influence (pp. 1–6).

1.2 Identify four key federal, state, and local legislative events that have formed the emergency medical services, then discuss their collective influence (pp. 4–5).

1.3 Identify federal, state, and local EMS system components and discuss their interrelationship (pp. 8–14).

1.4 Identify the components of an EMS system (pp. 5–6).

1.5 Explain the various National Highway Traffic Safety Administration EMS agendas for EMS-related topics (pp. 4, 6, 8–10, 19).

1.6 Identify the various trade organizations that support EMS activities (pp. 14–22).

Key Terms

501(C)(3) organizations (p. 14)

Accidental Death and Disability: The Neglected Disease of Modern Society (p. 4)

Centers for Medicare and Medicaid Services (CMS) (p. 9)

computer-based testing (CBT) (p. 20)

DOT national standardized curricula (p. 9)

Emergency Medical Services Systems Act of 1973

(Public Law 93-154) (p. 5)

EMS Agenda for the Future (p. 6)

Federal Interagency Committee on EMS (FICEMS) (p. 8)

Highway Safety Act of 1966 (p. 4)

Malcolm Baldrige National Quality Award (p. 10)

medical authority (p. 12)

National Fire Academy (NFA) (p. 6)

National Highway Traffic Safety Administration (NHTSA) (p. 4)

National Registry of Emergency Medical Technicians (NREMT) (p. 20)

pre-EMS (p. 5)

Robert Wood Johnson Foundation (p. 11)

Star of Life (p. 4)

Strategic National Stockpile (SNS) (p. 9)

Wedworth-Townsend Act (p. 12)

Point of View

It has been said that a career in EMS can be challenging as well as rewarding. As managers and leaders, we are often presented with the challenge of separating ourselves from the direct patient care that drew us to this great profession. Now our activities are typically focused on developing policy and possibly directing other EMS professionals. As a wise mentor once said, "You are no longer one of the horses racing to the fire pulling the hose cart. Now you are setting atop the hose cart and driving the team." Following this sage advice has been invaluable [to me] as a manager and leader of others.

A career in EMS has the potential of providing rewards many times over. Maybe you are someone who got started with a rural volunteer fire department or worked on an EMS squad in a major metropolitan city and progressed to director of a state EMS office. Along the way, you are sure to cross paths with some of the great movers and shakers of this industry. In doing so, managers also encounter numerous opportunities to effect change and influence others. The future development of our profession rests in the hands of visionary leaders and managers and those who are open to change.

Fergus Laughridge, *Director Emergency Medical and Trauma Systems, Nevada State Health Division*

■ THE BEGINNING OF EMS

JAMES O. PAGE

A book on EMS management would be remiss without identifying James O. Page as the father of modern emergency medical services (EMS) (Figure 1.1). Though the exact origins of EMS are widely debated, he was the light that guided EMS into the realm of being a profession and a specialty. Jim often remarked that it does not matter who the provider is, just as long as he is doing a good job. To ensure that the next generation of providers can do the best job possible,

today's managers and leaders must continue to improve the system not only based on its history but also with a vision of the future (Figure 1.2).

THE HISTORY OF PREHOSPITAL EMERGENCY MEDICINE

Prior to the 1960s the victim of a medical emergency in the United States did not receive much medical assistance other than transportation to the hospital. Ambulance services until that time offered little in the way of medical care, and ambulances themselves were often staffed only with a driver. The first records of the use of

FIGURE 1.1 ■ James O. Page (1936–2004), the "Father of Modern EMS."
(Tom Page/Tom Page Photography)

ambulances date back to the Crusades, and the first prehospital care was provided during the Napoleonic Wars in the late 18th and early 19th centuries. The Union Army started America's first ambulance service in 1862, when specially designed wagons were used to transport wounded soldiers from a "medical-aid station" near battle lines to one of several field hospitals outside the range of artillery fire. Following the Civil War, ambulance services were started in many cities, typically by hospitals. This type of system—hospitals providing a vehicle to transport seriously ill or injured patients—continued in urban areas of the country through the 1930s, while rural areas retained the age-old system of local doctors making house calls.

During World War II, the loss of personnel to the war effort caused many hospitals to turn their ambulance services over to volunteer groups and agencies capable of operating this type of motor vehicle. Typically, the service was added to the role of the fire department, police department, or funeral home. In some places local citizens joined together to provide the service, forming independent ambulance corps or rescue squads. Ambulance services continued to be little more than "horizontal-taxi" operations, because medical training for the ambulance crews was rudimentary, if there was any medical training at all.

Then during the 1960s, the medical community began focusing on the problem of heart attacks.

FIGURE 1.2 ■ The Future of EMS.

- ◆ Extend basic and advanced first-aid training to a greater number of the lay population.
- ◆ Prepare a nationally acceptable text, training aids, and courses of instruction for rescue squad personnel, police officers, firefighters, and ambulance attendants.
- ◆ Implement recent traffic legislation to ensure completely adequate standards for ambulance equipment and supplies and for qualifications and supervision of ambulance attendants.
- ◆ Adopt state-level policies and regulations pertaining to EMS.
- ◆ Adopt ways and means of providing EMS by local levels of government that are applicable to the conditions of the locality.
- ◆ Initiate pilot programs to determine the efficacy of providing physician-staffed ambulances for care at the site of injury and during transportation.
- ◆ Initiate pilot projects to evaluate helicopter services to sparsely populated areas.
- ◆ Delineate radio frequencies suitable for voice communication between ambulances and hospitals.
- ◆ Actively explore the feasibility of designating a single nationwide phone number to summon help for medical emergencies.

FIGURE 1.3 ■ Summary of the Recommendations of the White Paper.

Technical developments led to increased medical knowledge, including new concepts of electrical defibrillation and closed-chest pulmonary resuscitation. While these innovations were reducing the death rate of patients who reached the hospital, the effects of these breakthroughs were still limited because most heart-attack patients died at the scene or en route to the hospital. Consequently, emphasis began to be placed on upgrading ambulance technicians nationwide to advanced first aid and cardiopulmonary resuscitation (CPR) training from the American Red Cross.

The medical community continued to pursue a goal of reducing the overall mortality rate, and emergency medical care began to receive added attention throughout the country. The long-range effects of quality emergency medical care on mortality were discussed in a 1966 government white paper from the National Academy of Sciences (NAS) report, *Accidental Death and Disability: The Neglected Disease of Modern Society,* which documented widespread deficiencies in trauma care. The report concluded that the average response time of most mobile emergency teams was over 40 minutes, and the crews of these teams were inadequately equipped and

trained. It was common for emergency patients to be transported to the hospital in vehicles operated by mortuary services, and few hospitals had emergency wards staffed by doctors. This study, along with others, was the stimulus for the development of prehospital EMS, although it would require nearly a decade for this transformation to take shape. The NAS report revealed that the average American had a greater chance of surviving in the combat zones of Korea or Vietnam than on the nation's highways. This was the catalyst that created public support for the creation of the EMS system as we know it today. The recommendations of the white paper are summarized in Figure 1.3.

Also in 1966, Congress passed legislation enabling the creation of the **National Highway Traffic Safety Administration (NHTSA),** setting the stage for the first federal standards in EMS. Primarily as a result of concern over deaths and injuries due to vehicle crashes, the **Highway Safety Act of 1966** addressed EMS issues by developing specifications covering ambulance attendant training, equipment requirements, and the design of the emergency vehicle itself. A series of national training courses was created and implemented in many communities across the country. Those are the courses on which the national standard curricula of today are built.

In 1972 the U.S. Department of Transportation (DOT) and NHTSA adopted a national EMS symbol. Just as physicians have the caduceus and pharmacists have the mortar and pestle, EMS has the **Star of Life** (Figure 1.4), a symbol whose use is encouraged by both the American Medical Association

FIGURE 1.4 ■ Star of Life.

(AMA) and the U.S. Department of Health and Human Services (DHHS). The symbol's six-barred cross represents the six-system function of the EMS. Those six functions include detection, reporting, response, on-scene care, care in transit, and transfer to definitive care. The staff in the center of the symbol represents medicine and healing. Approved uses of the Star of Life can be found in the DOT publication "Star of Life Emergency Medical Care Symbol," DOT HS 808-721.

In the mid-1980s much of the funding for EMS programs was reduced or eliminated by the federal government, and several federal EMS programs were slashed or eliminated. As a result, EMS systems became more dependent on state and local funding. Many of the differences in levels of service that persist today—differing training standards and the lack of comprehensive 911 coverage are examples—can be traced to the decline in federal funding.

Since the mid-1960s and particularly since the mid-1980s, the number of fire departments providing EMS as a significant part of their service has risen. Many departments already providing EMS have expanded the level of service they provide or have started providing transport services in addition to first response. The demand for increased services may be one reason, as may a decrease in the number of fires due to increased fire prevention and protection efforts.

In the years since 1966, the efforts of EMS providers at all levels have helped make our EMS system the most advanced in the world. But despite its remarkable progress, the EMS system still faces the challenge of declining support for state and federal EMS programs.

FEDERAL INITIATIVES

The **Emergency Medical Services Systems Act of 1973 (Public Law 93-154)** designated federal funding for improved EMS across the nation through the development of regional EMS systems. This act was the first federal or national initiative on EMS and outlined 15 specific components of an EMS system. The EMS Act produced the development of the first set of DOT national curricula for EMTs and paramedics. Individual states soon began developing and implementing medical-practice acts that authorized EMS activities and defined the skills and abilities of various levels of EMS providers.

■ THE EMS SYSTEM

EMS TODAY

The Highway Safety Act of 1966 authorized the DOT to fund communication and education for EMS services, as well as purchases of ambulances and equipment. Congress enacted the EMS Systems Act of 1973, which funded and authorized the Department of Health, Education, and Welfare to develop EMS systems throughout the country. As a result more than 300 regional EMS management entities have been designated to develop a systematic approach to EMS care, training and empowering paramedics to deliver prehospital patient care. Figure 1.5 represents Public Law 93-154, which identified the 15 components essential to an EMS system.

At first the EMS design proved to be deficient in many respects, including medical direction and accountability, prevention, rehabilitation, financing, and operational and patient-care protocols. But EMS continued to be refined in the 1980s and 1990s, even after most federal funding had ended. Today, prehospital care has evolved into an indispensable element of public health.

EMS once offered only the most fundamental resources. Now, however, it encompasses professional, occupational, and lay disciplines that include prehospital-care personnel, physicians, nurses, emergency medical dispatchers, directors and administrators, first responders, and other allied-health personnel. The lay public or the **pre-EMS** contingent is also an integral part of the EMS system. Without the prompt

1. Communications.
2. Training.
3. Manpower.
4. Mutual aid.
5. Transportation.
6. Accessibility.
7. Facilities.
8. Critical-care units.
9. Transfer of care.
10. Consumer participation.
11. Public education.
12. Public-safety agencies.
13. Standard medical records.
14. Independent review and evaluation.
15. Disaster linkage.

FIGURE 1.5 ■ Original Components of an EMS System (1973).

notification and key interventions by lay public and lay professionals, the outcomes of EMS services could dramatically change.

Today, the effectiveness of EMS is a function of the coordination among many autonomous and interdependent professional disciplines. It is a comprehensive, coordinated arrangement of resources and functions provided in a timely, staged manner to targeted medical emergencies—regardless of their cause or the patient's ability to pay—and in a way that minimizes the physical and emotional impact on the patient.

THE *EMS AGENDA FOR THE FUTURE*

In 1996 the DOT and NHTSA updated the view of EMS. It was a significant step in integrating EMS into the health-care system and designing a more community-based approach to EMS. They published their recommendations in a document called the ***EMS Agenda for the Future***. That document changes the focus of national EMS activities from the 15 components of the EMS Systems Act of 1973 to 14 attributes or functions (Figure 1.6). All of those attributes or functions are meant to be integrated into the system design and throughout the phases of an EMS response.

As the EMS system continues to be modified for the future, it also must continue to be in line with the U.S. health-care system. The EMS systems of the future will look longitudinally at the patient—from initial detection of the emergency through to in-hospital care, rehabilitation, and follow-up. In other words, the effectiveness of EMS will be measured by the impact it has on the movement of the patient through the health-care system. Figure 1.7 indicates the key benchmarks of a health-care system with EMS integration.

Present and Future Resources

Facilities, agencies, and other aspects of health care are considered key resources in partnership with EMS. For example, hospitals, which provide an important link to EMS, act as receiving facilities, provide medical control, and offer educational opportunities. Along with private, government, and other nonprofit services, hospitals also are a resource for transportation, providing the traditional ground ambulance as well as air- and water-response vehicles. In addition, fire services, rescue squads, law enforcement, and third-service agencies are the local (municipal city or county) or state resources that are involved with EMS.

Other resources include the following:

- Regional, state, and national EMS professional organizations are resources that provide advocacy, education, and lobbying efforts. The regional and state EMS regulatory agencies provide direction, regulation, enforcement, and funding on occasion to the EMS system. Educational programs operated by associations, grant agencies, or for-profit training centers also contribute to the educational requirements of the system.
- Federal EMS agencies, such as the DOT, the **National Fire Academy (NFA),** and the Centers for Disease Control (CDC), set national and strategic agendas and provide grants and funding for EMS activities. In addition, agencies such as the American Society for Testing and Materials (ASTM) and the National Fire Protection Association (NFPA) build consensus for operational standards and then publish those standards for local and state EMS systems to use.
- No system can operate without funding or some other source of input. Funding mechanisms for EMS systems include federal, state, and local agencies. Dedicated revenue sources and reimbursement mechanisms such as Medicare, Medicaid, and private insurance provide the bulk of financial resources for EMS. On occasion, private-sector or foundation grant monies can be another common source of funding for EMS.

---

1. Integration of health-care services.
2. EMS research.
3. Legislation and regulation.
4. System finance.
5. Human resources.
6. Medical direction.
7. Education systems.
8. Public education.
9. Prevention.
10. Public access.
11. Communication systems.
12. Clinical care.
13. Information systems.
14. Evaluation.

FIGURE 1.6 ■ Fourteen Attributes of EMS.

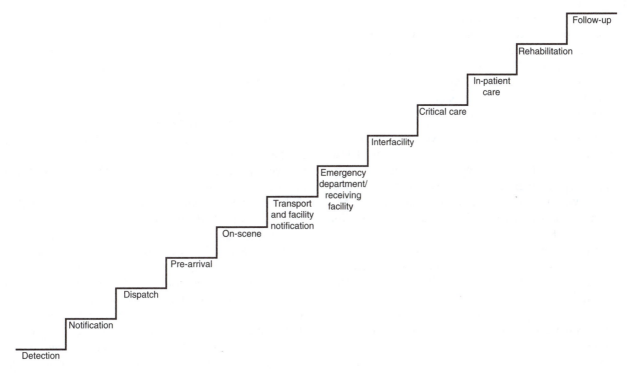

FIGURE 1.7 ■ Integrated EMS and Health-Care System.

Present and Future Functions

One function of an EMS system includes an organization and management structure that has the authority to direct, develop, and allocate resources. The authority needs to develop and implement local regional and state infrastructure; for EMS this refers to radio systems, facilities, ambulances, and computer or other record management systems.

In addition, an EMS system should incorporate the following functions:

- An EMS system should not operate without medical direction. Medical direction occurs off line and on line or directly. It also takes the form of protocols that direct transport, treatment, and triage decisions.
- Management of human resources and the education of career and volunteer members are ongoing functions of EMS. These include recruitment, retention, and training and continuing education.
- Access to EMS's communication system must address detection, recognition, and pre-arrival instructions for an emergency. Communication functions also include interagency, mutual aid, and disaster communication.
- EMS also has the responsibility to transport patients. This is accomplished by air, ground, or water trans-

portation to hospitals or other medical facilities, such as acute-care facilities or rehabilitation centers.

- Another function of EMS is quality improvement (QI) and quality assessment (QA) based on evaluation and data collection of EMS activities. QI/QA activities include the collection of run records, which are linked to hospital records in order to measure patient outcomes. A local trauma center or trauma system needs to link EMS data to the trauma registry. Other specialty data collection includes cardiac-arrest save rates, pediatric databases, and toxicology information. An ongoing review of performance of all medical and operational systems is a management function that requires data collection and analysis.
- Public information and education is an important, yet underdelivered, function in most EMS systems. Topics for public education should involve prevention, system access, system education, education of politicians, and citizen CPR and first-aid training.
- Managers must carve out time for disaster planning and education. This should include integrated disaster planning, mutual aid, and mass-casualty training and exercises. Management also should establish disaster caches and triage systems.

◆ Research on EMS system design, including medical and operational procedures, should be studied. Such research should focus on measurements of the effectiveness of EMS medical interventions, and include data on interventions for special-needs and handicapped patients, as well as trauma and medical patients, pediatric patients, poisonings, and mental-health patients.

The impact of the popular television shows *Emergency* and *Rescue 9-1-1* cannot be overlooked. These shows brought EMS into living rooms across the country in a way that increased both awareness and expectations. Without these shows emphasizing how EMS can and should be delivered, the public demand for high-quality EMS might not have existed. The public now uses television shows, news broadcasts, and home videos as benchmarks and evaluation tools to ensure that the services they receive (including EMS) meet their expectations.

Today, trained personnel in the field routinely perform specialized advanced life support procedures at the scene and en route to specialized medical facilities. Most states have minimum certification requirements and staffing levels for ALS and BLS ambulances, and EMT-level certification is recognized nationally as the minimum level of EMS training for ambulance services. Every ambulance has some type of communication capability with area hospitals. In some cases, ambulances have multiple modes of voice and data communication that are able to send 12-lead ECGs from the field over the Internet to hospitals. Ambulances must meet federal specifications, national standards, and state requirements for design, patient-care areas, storage space, medical equipment, and vehicle safety.

■ GOVERNMENT AND EMS

THE FEDERAL GOVERNMENT

The Department of Homeland Security (DHS) is becoming more and more influential in EMS. In 2005 the Safe, Accountable, Flexible, Efficient Transportation Equity Act (SAFETEA) was signed into law; it authorized a federal oversight committee for EMS. The **Federal Interagency Committee on EMS (FICEMS)** is located under the aegis of NHTSA. The previous FICEMS committee had been located under the U.S. Fire Administration. FICEMS has a congressional

mandate to report to Congress yearly on emerging issues in EMS. FICEMS comprises government representatives from different agencies that have statutory or legal responsibility for some aspect of EMS (Figure 1.8).

The DOT and the NHTSA

The lead federal agency on EMS will continue to be the DOT. NHTSA has elevated EMS to a division status within the administration. NHTSA's EMS Office leads the national EMS efforts, and NHTSA has produced the *EMS Agenda for the Future*, a strategic plan for the nation's EMS systems.

In addition to the *EMS Agenda for the Future*, there are several EMS agendas specific to key components of the national plan. NHTSA collaborated with the National Association of EMS Educators to build the *EMS Education Agenda for the Future*. The *EMS Scope of Practice* was the project recently completed that looked at the practice of EMS and categorized the skills and knowledge set for each of the designated titles of EMS providers. NHTSA produced an *EMS Quality Improvement Manual*, one of the first comprehensive documents instructing EMS providers on quality-improvement initiatives. NHTSA provides a two-day seminar instructing providers on how to complete QI/QA activities in their departments. The *EMS Research Agenda for the Future* and the *Consensus Statement on the Role in Injury Prevention* are two other documents that NHTSA facilitated with a collaborative agency or trade group.

Department of Transportation
National Highway Traffic Safety Administration
Department of Homeland Security
U.S. Fire Administration
Department of Health and Human Services
Maternal and Child Health Bureau
Health Resources and Services Administration
Coordinating Center for Environmental Health and Injury Prevention
Centers for Disease Control and Prevention
Centers for Medicare and Medicaid Services, Indian Health Services
Assistant Secretary for Public Health Emergency Preparedness
Department of Defense
Office of the Assistant Secretary of Defense Health Affairs
Federal Communications Commission
Wireless Telecommunications Bureau

FIGURE 1.8 ■ FICEMS Government Members.

NHTSA provides grant monies through each state's highway safety office. The statewide EMS technical assessment process helps states identify system development needs and strategies based on the *EMS Agenda for the Future*. NHTSA provides monies to help states meet the challenges based on the assessments. Monies and application for those funds are made through the state's highway safety office and the regional NHTSA office.

NHTSA also is responsible for creating **DOT national standardized curricula**—core content covering knowledge, skills, abilities, and attitudes—for the EMS positions of paramedic, advanced EMT, emergency medical technician, emergency medical responder, EMS injury prevention specialist, and EMS instructor. Many of these curricula are created with grant money from NHTSA and delegated to trade groups, associations, or educational institutions to complete using a collaborative approach. NHTSA also facilitated the development of the KKK standard for the construction and manufacturing of an ambulance.

Office of Domestic Preparedness

DHS is the parent organization to the Office of Domestic Preparedness (ODP). DHS has two key federal agencies that provide some EMS training and resources—the Federal Emergency Management Agency (FEMA) and the National Fire Academy (NFA).

The national emergency training center that houses the NFA is located in Emmitsburg, Maryland. It has several EMS programs and resources. Located within the Response Branch of the National Fire Academy, the EMS programs target the nation's fire services as part of the NFA's mission that encompasses rescue. The EMS programs at the NFA focus on EMS management, EMS special operations, EMS Community Risk Reduction, and EMS safety and incident command. Figure 1.9 lists the educational offerings at the NFA.

R-150	Management of EMS
R-151	Advanced Leadership Issues in EMS
R-152	EMS Special Operations
R-247	ALS Response to Hazardous Materials Incidents
R-154	Advanced Safety Operations and Management
R-149	Management of Community Health Risks

FIGURE 1.9 ◼ National Fire Academy EMS Courses. These are on-campus offerings, requiring attendance at the NFA in Emmitsburg, Maryland.

◆	ICS for EMS
◆	Emergency Response to Terrorism/Tactical Considerations in EMS
◆	NIMS ICS/EMS
◆	Infection Control for Emergency Response Personnel
◆	Health and Safety Officer
◆	Incident Safety Officer

FIGURE 1.10 ◼ Direct-Delivery EMS Courses.

In addition to the on-campus courses at the National Fire Academy, two-day direct-delivery courses can be held locally. Several EMS courses and courses related to fire and EMS operations are applicable for EMS managers and leaders. Figure 1.10 lists some of the outreach courses offered by the National Fire Academy.

At the U.S. Fire Administration, some research and funding activities encompass EMS. The National Fire Incident Reporting System (NFIRS) has a small section that collects EMS data. Work is often assigned to private consultants who serve as subject-matter experts.

DHS is facilitating the development of a national EMS credentialing system that will allow incident commanders to credential EMS workers accurately who respond across state lines. The national credentialing system will document minimum qualifications, certifications, training, and education and allow EMS providers to be plugged into the national incident management system.

Health and Human Services

The Department of Health and Human Services (DHHS) has several agencies that contribute significantly to national EMS activities. The CDC has been taking on more of an EMS role in government. It conducts injury-prevention programs, does surveillance, and maintains the Web-based Injury Statistics Query and Reporting System (WISQARS) database. The CDC also provides education on infectious disease, promotes standards for bloodborne pathogens, and maintains the **Strategic National Stockpile (SNS)** that would dispense pallets of medication in the event of a terrorism incident or biological terrorist event. The **Centers for Medicare and Medicaid Services (CMS)** reviews and sets reimbursement schedules, standards for billing, and privacy regulations for ambulance services. CMS also designates hospital-destination information and works jointly with the Joint Commission on the Accreditation of Healthcare Organizations (JCAHO) to certify hospitals to be able to bill for Medicare and Medicaid.

The Health Resources and Services Administration (HRSA) has three partnerships with the NHTSA EMS efforts. A special effort is being made to increase the level of pediatric care and education of EMS providers. A national clearinghouse for EMS material related to the care of children and adolescents is maintained by the EMS for Children program funded by Congress through HRSA and the Maternal and Child Health Bureau. HRSA also works closely with NHTSA to improve the nation's trauma system.

The federal Office of Rural Health Policy (ORHP) partnered with NHTSA to create the *Rural EMS Agenda for the Future*. The ORHP provides funding for demonstration projects and assists agencies with meeting their needs in the rural EMS associations.

Federal Communications Commission

The Federal Communications Commission (FCC) has had a long history of helping EMS serve their communities. Many of the telemetry units, field radios, and base-station frequencies, including the telephone services of 911, are coordinated and facilitated through the FCC. In addition, grants to public-safety agencies and funding from private sources and the automotive industry have funded new intelligent communication devices such as OnStar®, which initiate a public-safety call when there has been an emergency.

The Wireless Telecommunication Bureau within the FCC oversees the development of public-safety radio systems and wireless frequencies for paperless systems. The primary interface with EMS is in the planning and regulation of 800- and 700-megahertz radio systems. The FCC helped develop the 800-megahertz and trunked communication systems. In an attempt to promote interoperability among emergency responders, in 2005 the FCC designated the entire bandwidth in the 700-megahertz range for public safety.

The FCC also helps develop telemetry, dispatch software, and the 911 telephone system. It has led the way in creating cellular phone technology that can triangulate the cell phone and route the call to the closest public safety call center. A related emerging issue is the use of voice-over Internet protocol (VOIP), by which phone service is routed through the Internet service, which often complicates communication to the local 911 public safety answering system.

Department of Commerce

The Department of Commerce (DOC) is an unlikely partner in EMS. However, the DOC manages the national quality award called the **Malcolm Baldrige National Quality Award**. The Baldrige National Quality Program is a public-private partnership to improve the performance of national organizations. The program was developed by the late Secretary of Commerce Malcolm Baldrige and has, for the last two decades, provided global leadership training in the performance of excellence. Every year the president of the United States gives only three awards for each category—health care, small business, manufacturing, and education. The fact that there are so few awards gives them a prominent status and helps reinforce the need for continuing improvements in performance in a marketplace that becomes more competitive every day.

The Baldrige Health Care Criteria for Performance Excellence provide a framework for improvement without being dictatorial. Organizations are encouraged to develop creative and flexible approaches aligned with organizational needs and to demonstrate cause-and-effect linkages between these approaches and their results. The criteria are inclusive, providing an integrated framework that addresses all the factors that characterize an organization, its operations, and outcomes. The criteria focus on common requirements, rather than procedures, tools, or techniques, as compared to other improvement efforts. Insurance Services Organization (ISO), Committee on Accreditation of Ambulance Services (CAAS), Centers for Public Safety Excellence (CFI) accreditation may be integrated into the organization's performance-management system and included as part of the response to criteria requirements. The Baldrige Health Care Criteria for Performance Excellence are adaptable and can be used by large or small EMS agencies, regardless of their affiliation.

The Baldrige criteria are on the leading edge of validated management practice and are updated based on strategy-driven performance practices and stake-holder input, and they accommodate important organizational needs and practices.

The Baldrige criteria are a set of critical factors that drive organizational success. Approximately 100 questions are grouped into an organizational profile, with seven categories; the criteria include key influences on how an EMS organization operates and the challenges that face the organization. The seven health-care categories are:

* Leadership.
* Strategic planning.
* Focus on patients, other customers, and markets.
* Measurement, analysis, and knowledge management.
* Staff focus.
* Process management.
* Organizational performance results.

The seven categories are organized into items, and the items are broken down into areas that contain questions to which the organization will need to respond. A sample of the Baldrige criteria is in Figure 1.11 and denotes the broad category of a standard. Each criterion has several key components, or descriptions of what to examine in each of the standards.

If your EMS agency makes a decision to pursue the Baldrige criteria, the first step is to educate the organization on the principles of performance. This in-cludes building a common mission and vocabulary. Conduct a one-day self-assessment of your organization's strengths and weaknesses to help facilitate an action plan. Apply for a state, local, or regional quality award to get detailed feedback on the performance of the organization.

STATE ORGANIZATIONS AND ACTIVITIES

State Initiatives

In the 1960s the U.S. Department of Health and Human Services allocated $16 million for EMS demonstration projects in five different states. The **Robert Wood Johnson Foundation,** a private foundation, provided $15 million for 44 projects in 32 states. The first paramedic programs evolved in Miami, Florida; Hollywood, North Carolina; Columbus, Ohio; Los Angeles County, California; and Seattle, Washington. Dr. Eugene Nagel from the University of Miami Medical School held the first paramedic school at the University of Miami and called these graduates "Physician Extenders." By March of 1967 these paramedics were transmitting heart rhythms to Jackson Memorial Hospital, with an ECG unit that weighed a combined 54 pounds.

Though the first volunteer rescue squad that actually provided first aid was established in Roanoke, Virginia, in 1928, it was not until 1969 in North Carolina

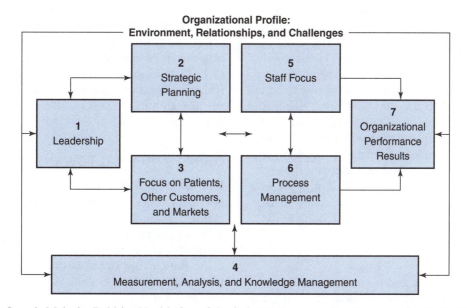

FIGURE 1.11 ■ Sample Malcolm Baldrige Health-Care Criteria for Performance Excellence Applicable to EMS.

that a volunteer rescue squad received the training and equipment to function as paramedics.

In 1967 Dr. Michael Criley and Dr. James Lewis commenced a pilot program at Harbor General Hospital in Los Angeles that consisted of 18 firefighters—12 from the Los Angeles County Fire Department and six from the Los Angeles City Fire Department. On September 12, 1969, the firefighters began an intensive 180 hours of training that included classroom, laboratory, and clinical instruction under the tutelage of critical-care unit nurse Carol Bebout. By December of 1969 they were ready to go, but they had no legal authority to do so.

A member of the Los Angeles County Board of Supervisors, Kenny Hahn (who became known as the "father" of the paramedic program), pushed legislation through the board and presented it to state senator James Wedworth and state assemblyman Larry Townsend. Both houses of the state legislature approved legislation that gave legal authority for paramedics to perform, and on July 14, 1970, Governor Ronald Reagan signed the **Wedworth-Townsend Act** into law, effective immediately. This law provided the first legal definition of paramedics and listed them as "mobile intensive care paramedics." Los Angeles County and City paramedics began service shortly thereafter. Within five years, 29 other states had established titles.

At the same time, in Seattle, Washington, Dr. Leonard Cobb set out to design a citywide system for handling out-of-hospital cardiac arrest. Physicians and nine specially trained Seattle firefighters began the Medic One system in the "Heartmobile," responding to cardiac arrest. They focused on a system approach to cardiac arrest that incorporated bystander CPR. Through this and other successful programs in Cincinnati, Ohio, and Jacksonville, Florida, it was determined that, with proper training, paramedics could perform the same procedures that physicians would if the physicians were on the scene. Despite the success of these individual pilot projects, there was not yet any leadership or direction from the federal level to create a comprehensive EMS system.

State EMS Offices

Each state in the union and most of the U.S. territories have state-chartered departments of EMS. The Association of State EMS Officials is an active group of administrators on the national stage. Most of the EMS grants that are funneled from the federal government often are given as block grants to the states and are distributed to local providers through the state department of EMS. In some states, the EMS divisions are located in the department of health; in others, they are under the umbrella of the department of public safety, which often includes law enforcement and/or the state fire marshal office. State EMS directors often advise the state legislatures on the medical-practice acts, certification requirements, educational standards, and funding appropriations.

State offices of highway traffic safety provide funding and education to EMS agencies. Many states have various mechanisms to fund EMS. For example, Maryland funds its EMS and trauma systems with a $5 surcharge on vehicle registrations. That money provides EMS helicopter support, trauma-center funding, and EMS education. Some state EMS offices only provide medical oversight, while others have a very comprehensive education and program-support system. In Delaware, for instance, all of the advanced life support (ALS) services are provided by state-run ALS vehicles and paramedics.

In some states the state fire marshal's office provides EMS support by providing equipment, training for the incident command system (ICS), and extrication or rescue training. The U.S. Fire Administration also makes money available to local fire departments for EMS-related operational needs. Under the fire act, grant provisions have been made to allow the U.S. Fire Administration to fund EMS and safety-related topics. Each of the state fire marshal offices and in some states the Department of Health or EMS Division will provide educational grants to EMS organizations. State Medicaid offices fund and pay for ambulance transportation for indigent and low-income people.

COUNTY AND LOCAL GOVERNMENTS

EMS Medical Authority

For most EMS providers, a state, county, or local **medical authority** provides oversight and accountability to the public. There are a variety of medical-authority models. In some areas the medical authority is the state, and it oversees the health-related activities for all services on a statewide basis. Large counties and some major metropolitan areas or cities have their own health departments that oversee EMS under the guidance of state law. In some jurisdictions the state health departments or medical-authority delegates, or in some locations legislative mandates, place the accountability for EMS oversight in city or county health departments. A delegated model works best when budgets and the demand for oversight are disproportionate from one geographic area to the next. In some states there are requirements and regulations imposed by state, local, or regional health authorities. Whether they are fire, private, law-enforcement, volunteer, or third-service providers, individual providers have medical directors who provide agency-specific direction.

The role of the medical authority is to grant certification, recertification, and decertification of prehospital personnel. Most government authorities utilize medical advisory boards to assist in governing or advising an EMS system in these areas. In many jurisdictions the medical authority has a method in place for appeals and a procedure to grant privileges and with due process to remove or suspend personnel from providing care.

The medical authority is responsible for establishing, implementing, revising, and authorizing system-wide protocols, policies, and procedures for all medical-related activities. This includes the dispatch, treatment, and transport of the patient. It also includes establishing the criteria for the priority and type of resource to be sent to an incident. In addition, the EMS medical authority establishes destination criteria that typically include trauma-center, pediatric-center, and burn-center destination criteria. Destination criteria also are being developed and implemented for stroke, cardiac, and obstetrics as hospitals focus on becoming centers of excellence.

A key role of the medical authority is to ensure competency of personnel who provide patient care, including physicians, nurses, paramedics, EMTs, and emergency medical responders. Most EMS authorities handle the certification, background checks, and licensure of EMS providers and agencies. The medical authority also is responsible for establishing the training levels and minimum hours required for certification, including that of air and ground ambulances.

A key function of the medical authority is to ensure that quality-improvement activities are being conducted in the organization. Most medical or EMS authorities or services that maintain or seek accreditation must conduct quality-improvement activities. The

health authority conducts several functions, and its roles vary in each organization.

EMS regulations are approved by the local, county, or state board of health. Boards of health can be elected or appointed and most often are composed of representatives of the jurisdictions. In some cases a nonprofit or quasi-government structure is formed to oversee EMS. One example is the Regional Emergency Medical Services Council of New York City, Inc., a nonprofit, tax-exempt corporation whose function is to improve EMS services in New York City. The Regional Emergency Medical Services Council of New York City was formed under Article 30 of the Public Health Law by the State of New York. In Denver, Colorado, the mayor appoints the board of directors to Denver Health, the county organization that oversees all aspects of community health, including EMS in the city and county of Denver. In most rural states, such as Wyoming, the state department of EMS oversees all EMS, despite the presence of local health districts in many communities. In San Diego, California, the Division of Emergency Medical Services is a division of the Health and Human Service Agency's Public Health Services. It is the "local EMS agency" as defined in California law. The San Diego EMS agency conducts similar oversight of that city's EMS. The common functions of government medical authority are shown in Figure 1.12.

Other Government Interactions

An EMS agency may be involved with only operational issues and have little to do with the financial or business aspects of ambulance services. Often the city or county business license or contracting office oversees franchise or contractual agreements with ambulance providers. The business license department often also monitors compliance with taxes, response times, and franchise fee collections. Some municipal EMS agencies and fire departments have the city or county business or utilities offices conduct the billing

---

- ◆ Develop protocols and monitor compliance.
- ◆ Establish and enforce training standards.
- ◆ Approve continuing education.
- ◆ Regulate policies.
- ◆ Enforce standards of care.
- ◆ Provide medical direction.

FIGURE 1.12 ■ Functions of Government Medical Authority.

operations to eliminate having regular employees handle cash, checks, or transfer of monies in unsecured locations. Governments often contract billing services to private entities that are experienced in ambulance financing. This is done to increase the collection rates and shift the risk when dealing with complex federal regulations.

Structure of Local EMS Organizations

The community EMS provider can be a single organization, or EMS services may be provided through the combined efforts of multiple organizations. The most common service in the country is fire department–based, which means that a response-and-patient-transportation system uses cross-trained, dual-role firefighters. A variation on the fire-based EMS is the EMS service that operates under fire department oversight, which provides a response-and-patient-transportation system using EMS personnel who are not cross-trained as fire-suppression personnel.

Public single-role EMS systems that are separate from the fire department are a "third service." These systems may use fire department resources to supplement or complement an EMS response. Many systems in the United States are serviced by private ambulance-provider systems in which EMS is run by a corporation or business interest without public-safety oversight. Like the public single-role EMS systems that are separate from the fire department, private ambulance systems often utilize fire department resources to complement or supplement EMS responses.

Some small business owners still operate ambulance services. However, most private ambulance systems are run by large corporations with franchise agreements or contractual services with communities. Additional provider types are police, wilderness, military, law enforcement, nonprofit volunteer, and hospital-based services.

Regardless of the provider type, participant roles and responsibilities need to be organized to ensure that every part of the system contributes to the effectiveness of the system as a whole.

Division or Bureau. Some legal issues have defined the development of EMS in a fire-based or government organization. In some large metropolitan areas, EMS in the fire departments is a separate division, and EMTs and paramedics are government employees managed by the fire department, yet they are not firefighters.

Often EMS units are on separate schedules and are not on 24-hour shifts. This structure has many different legal issues defining compensation and benefits; one important issue is that overtime must be paid for a fire-based EMS worker who spends more than 40 hours in a given week on EMS-related work. In addition the chain of command in this type of EMS is often not through the fire officer; those EMS providers report to other managers who oversee the entire EMS system and are not able to effectively service EMS employees due to the span of control.

In the division or bureau model, EMS is operating in a more autonomous role. When EMS is organized as a support function, EMS personnel report up a normal chain of command in the fire-operations division. This design often leaves EMS managers with little or no authority to handle operational issues unless a solid culture of collaboration is embraced by the organization.

Other Forms of Management. Rural and volunteer EMS organizations are, at times, organized under special taxing districts and governed by fire boards. The authority of these administrative boards to set special tax rates for the delivery of emergency services is defined by state law. The members of the fire boards are elected public officials and have the fiduciary responsibility to oversee the funding and management of local fire departments on behalf of the local citizens.

There continue to be several forms of quasi-government or charitable organizational structures to operate ambulance services. For example, **501(c)(3) organizations** are nonprofit corporations that operate in the public interest, as recognized by the Internal Revenue Service. Common forms of nonprofit EMS organizations are ambulance clubs and volunteer EMS associations. Several college campuses operate ambulance services as nonprofit EMS organizations.

■ PROFESSIONAL ORGANIZATIONS

MANAGEMENT ORGANIZATIONS

The International Association of Fire Chiefs (IAFC) maintains a strong EMS presence on the national stage. EMS has been elevated to section-level status within the IAFC, and the IAFC EMS section is routinely present at almost every collaborative initiative brought forth by the federal government. The EMS section conducts an annual conference called Fire Rescue Med that focuses on EMS managers and administrators. The EMS section responds to national media stories, as it did in 2005 with a series of *USA Today* articles on EMS in the United States.

Representatives from IAFC have been part of the National Scope of Practice project, the IAFF Benchmarking initiative, the CMS ambulance collection procedures, and the EMS core content group. The IAFC EMS section has produced position papers on SARS, smallpox, and the avian bird flu. It annually gives away two awards: the James O. Page Award for EMS Leadership and the Medtronic/Physio-Control Heart Safe Community Award. EMS section members provide key input to the National Fire Protection Association on two standards—NFPA 450 on Emergency Medical Services and NFPA 1710 on Deployment Standards for Fire and Emergency Service.

The International City/County Management Association (ICMA) is the leading organization for information on local government management. Its members include city managers, county managers, and other appointed officials such as fire and EMS chiefs. ICMA works hard to create excellence in local government by developing, researching, and demonstrating best practices. It has developed ride-along guidelines, paramedic student internship programs, fire/EMS merger guidelines, consolidation proposals, and MCI response plans. ICMA continuously produces documents on EMS for city and county managers and showcases excellence in government at the annual Best Practices Conference.

UNIONS AND LABOR GROUPS

The International Association of Firefighters (IAFF) has made major contributions to EMS in the last two decades (Figure 1.13). The IAFF EMS division is a leader in using geographic information system (GIS) mapping to improve EMS funding and response times. The IAFF has developed a comprehensive quality improvement program, assisted in system design and compensation packages, and championed employee rights.

Specifically, the IAFF EMS division addresses the concerns of local affiliates on various components of fire departments and EMS system operations, including staffing, deployment, transportation, equipment, communications, record keeping, public education, injury prevention, and quality assurance. It can provide

FIGURE 1.13 ■ The International Association of Firefighters (IAFF) Logo.
(Reprinted with permission.)

customized assistance to local affiliates including the processing of daily requests for printed or video materials; writing, evaluating, or responding to requests for proposals; and developing strategies to compete with private providers.

The IAFF created and maintains a library of EMS resources and assists the federal Department of Education in developing fire-based EMS educational materials. The IAFF plans and conducts a biennial "EMS in the Fire Service" conference. The EMS division acts in coordination with other IAFF departments to educate federal, state, and local government agencies on EMS-related issues. Additionally, the IAFF department of EMS works cooperatively to seek clarification of laws pertaining to the public provision of EMS and seeks state and national policy to protect fire-based systems.

The IAFF collects and maintains information covering various types of fire departments and EMS systems. The IAFF plans for trends in the industry that impact firefighter wages, working conditions, pensions, and job security. This information will be used in conjunction with customized technical assistance. Survey results and data are maintained in IAFF databases for researching and responding to local affiliate requests. These databases include the contract database, the economic database, the Death & Injury (D&I) Survey database, and the Phoenix Survey.

The American Federation of State, County, and Municipal Employees (AFSCME) represents several groups of EMS workers that are private ambulance, nonprofit EMS agencies, and non-fire-based government EMS providers.

The International Association of EMTs and Paramedics (IAEP) has a powerful voice in Washington, D.C., with a dedicated legislative team that works full time to safeguard the jobs and bargaining rights of EMS professionals. IAEP has shaped legislative policy that directly benefits members. The IAEP has regional offices around the country and provides support for collective bargaining and organizing EMS workers.

The International Brotherhood of Teamsters represents EMS workers in several states and brings the benefits of a large union to EMS. Many of the larger private ambulance companies have organized under the Teamsters. The Teamster pension system and insurance services have often been some of the best benefits in the industry.

TRADE AND ADVOCACY GROUPS

The federal government has provided the funding and seed money for EMS, but the work and development of EMS is placed on the shoulders of groups of EMS personnel, who affiliate with trade groups, associations, nonprofit organizations, and trade or labor groups. The contributions of nongovernmental organizations have driven the movement of EMS forward over the last two decades. Federal agencies that direct EMS consult and partner with many of these organizations.

The American Red Cross is not a government agency. Its authority to provide disaster relief was formalized when in 1905 the Red Cross was chartered by Congress. The charter is not only a grant of power but also an imposition of duties and obligations to the nation, to disaster victims, and to the people who generously support its work with their donations.

Red Cross disaster relief focuses on meeting people's immediate disaster-related emergency needs. When a disaster threatens or strikes, the Red Cross provides shelter, food, and health and mental-health services to address basic human needs. In addition to these services, the core of Red Cross disaster relief is the assistance given to those individuals and families affected by disaster to enable them to resume their normal daily activities independently. The Red Cross also feeds emergency workers, handles inquiries from concerned family members outside the disaster area, provides blood and blood products to disaster victims, and helps those affected by disaster to access other available resources. The American Red Cross is one of the only nonfederal agencies assigned a primary role in the National Response Plan for an emergency support function.

The American Heart Association (AHA) and the American Stroke Association (ASA) both have signif-

icant resources for EMS. The AHA sets guidelines for the care of cardiovascular emergencies and conducts research into emergency medicine as it relates to the care and treatment of heart problems, stroke, and other cardiovascular diseases. A scientific assembly meets regularly to discuss and set recommendations based on evidence-based medicine. AHA provides core emergency medicine education by providing certification courses in advanced cardiac life support, pediatric advanced life support, CPR training, and neonatal resuscitation. The ASA now has a training package for assessing stroke victims and advocates treatment guidelines. Both organizations provide preventive health care and activities aimed at reducing heart disease and stroke.

The American Ambulance Association (AAA) is a trade association that seeks to advance the business interest of the private-ambulance industry. The AAA focuses on emergency and nonemergency transportation and ways to help private ambulances deliver better service and patient care while maximizing business profits. The AAA was formed in 1979 and serves not only as the voice of the private-ambulance industry but also as a lobbying organization to the federal government and state governments. The AAA provides model agreements and contracts for municipal agencies to obtain services from various private-ambulance providers.

PHYSICIAN GROUPS

The American College of Emergency Physicians (ACEP) has long been a dedicated and effective advocate on behalf of EMS (Figure 1.14). ACEP's advocacy focuses on the delivery of quality emergency medical care, including emergency-department overcrowding, liability reform, and physician reimbursement. ACEP conducts the world's largest emergency

FIGURE 1.14 ■ The American College of Emergency Physicians (ACEP) Logo. *(Reprinted with permission.)*

medicine conference and has a wide variety of publications, training materials, and audiovisual products; the *Airway Cam Video Series* is one example of the quality of educational programs the organization offers. ACEP also publishes the EMS Quality Improvement Manual.

In 1984 the National Association of EMS Physicians (NAEMSP) was formed as an organization of physicians and allied health professionals to provide leadership and foster excellence in out-of-hospital emergency medical services (Figure 1.15). NAEMSP promotes an annual meeting, produces quarterly publications, and creates products to serve its members. NAEMSP conducts the National EMS Medical Directors Course and provides position papers on out-of-hospital care, including protocol development. It also provides position papers, sample protocols, and sample contracts for physician medical directors, as well as memoranda of understanding.

The American College of Surgeons (ACS) is a nonprofit scientific and educational association of surgeons. Founded in 1913, ACS improves the care of the surgical patient by setting high standards for surgical education and practice. Examples of activities conducted by ACS include educational programs such as the Clinical Congress and standard-setting programs in trauma care. The organization published some of the first EMS training books and continues to serve EMS by improving trauma and disaster care and research.

AEROMEDICAL ASSOCIATIONS

Established in 1980, the Association of Air Medical Services (AAMS) is an international association that serves providers of air and surface medical transport systems. The association is a voluntary nonprofit organization that helps maintain a standard of performance reflecting safe operations and efficient, high-quality patient care. The AAMS is built on the idea that representation from a variety of medical-transport services and businesses can be brought together to share information, collectively resolve problems, and provide leadership in the medical-transport community.

The Air and Surface Transport Nurses Association (ASTNA) was renamed in 1998 from the National Flight Nurses Association that was founded in 1980. ASTNA is a nonprofit member organization whose mission is to advance the practice of transport nursing and enhance the quality of patient care. ASTNA has committees that focus the association's work on issues that involve ground, maternal and high-risk obstetrics patient, military, pediatric, neonatal, fixed-wing, and rotor-wing transport. ASTNA provides clear and decisive leadership for the unique and distinct professional specialty of transport nursing.

The International Flight Paramedic Association (IFPA) represents an alliance of medical professionals from every specialty involved in critical care transport. IFPA members are involved in transporting critical-care patients by airplane, helicopter, and ground ambulance. The majority of IFPA members are flight paramedics, but they welcome paramedics involved in ground transport. IFPA helps develop standards for the air medical profession and contributes to the *Air Medical Journal* associates committee, which provides guidance to the publishers of the air medical transport industry's professional journals. The IFPA develops standards for critical-care paramedic education and the establishment of a nationally recognized credentialing process.

The Air Medical Physician Association (AMPA) is an association of physicians and EMS professionals involved in air medical transport who are committed to promoting safe and efficacious patient transportation through quality medical direction, research, education, leadership, and collaboration. AMPA seeks to attract as members all physicians with an interest in air medical transport. Its mission is to offer opportunities to collectively study the impact of air transport upon patients and to share expertise so that patients may receive the best care possible in the safest operating environment. AMPA collaborates with other medical and aviation disciplines in air medical transport.

STANDARDS-SETTING ORGANIZATIONS

The American Society for Testing and Materials (ASTM) was formed in 1898 by a group of engineers and scientists to address railroad-industry accidents. ASTM International is a voluntary standards-development organization for technical standards for materials, products, systems, and services. ASTM standards are categorized as medical-equipment specifications, test methods, practices, guidelines, and terminology. The ASTM Committee F30 on EMS was formed in 1984 and meets twice a year. ASTM International standards have an important role in setting standards for search and rescue, medical equipment, EMS personnel training and education, dispatch procedures.

The mission of the National Fire Protection Association is to reduce the worldwide burden of fire and other hazards on the quality of life by providing and advocating consensus codes and standards, training, and education. NFPA membership comprises more than 50,000 individuals from around the world and more than 80 national trade and professional organizations. Established in 1896, NFPA serves as the world's leading advocate of fire prevention and is an authoritative source on public safety. NFPA's focus is on true consensus, as the EMS documents built by consensus have earned accreditation from the American National Standards Institute (ANSI). NFPA publishes two EMS documents. The first publication, *NFPA 473: Standards for Competencies for EMS Personnel Responding to Hazardous Materials Incidents*, identifies the levels of compe-

tence required of EMS personnel who respond to hazardous-materials incidents. It specifically covers requirements to reduce accidents, exposure, and injuries for basic life support and advanced life support personnel in the prehospital setting. The second publication, *NFPA 450: Guide for Emergency Medical Services and Systems*, is a document that is designed to assist individuals, agencies, organizations, or systems as well as those interested or involved in EMS system design.

The purpose of the National Association of State EMS Officials (NASEMSO) is to promote the national network of coordinated and accountable state, regional, and local EMS systems (Figure 1.16). NASEMSO was formed in 2005 when the National Association of State EMS Directors merged with the National Association of State EMS Training Coordinators. Other organizations exist within NASEMSO—the National Council of State EMS Data Managers, the State Trauma Managers Council, the State Medical Directors Council, and the State EMS-C Council. NASEMSO is the lead national organization for EMS, a respected voice for national EMS policy with comprehensive concern and commitment for the development of effective, integrated, community-based, universal, and consistent EMS systems. NASEMSO supports its members in providing vision and leadership in the development and improvement of EMS systems and national EMS policy.

The National Volunteer Fire Council (NVFC) is a nonprofit membership association representing the interests of the volunteer fire, EMS, and rescue services. The NVFC serves as the information source regarding legislation, standards, and regulatory issues.

ACCREDITATION AGENCIES WITH EMS INTEREST

The Commission on Accreditation of Allied Health Education Programs (CAAHEP) accredits programs

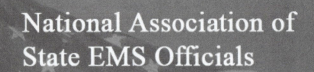

FIGURE 1.16 ■ The National Association of State EMS Officials (NASEMSO) Logo. *(Reprinted with permission.)*

upon the recommendation of the Committee on Accreditation of Educational Programs for the Emergency Medical Services Professions (CoAEMSP). The mission of CoAEMSP, under the umbrella of the parent accrediting body of CAAHEP, is continuously to improve the quality of EMS education through accreditation and recognition services for EMS professions. CAAHEP maintains and establishes minimum standards of quality used in accrediting programs that prepare individuals to enter the EMS professions. The accreditation standards constitute the minimum requirements to which an accredited program is held accountable.

In 1990 the Commission on Accreditation of Ambulance Services (CAAS) was created to champion the cause of accreditation of EMS services with a set of quality standards. Many of the standards came from a collaborative process that brought together the AAA, NAEMSP, IAFF, IAFC, and NASEMD. CAAS conducts its process by asking the EMS agency to do a self-study of its organization based on the accreditation standards. Then up to three other EMS professionals validate the process by conducting a site visit. The standards are very comprehensive, and a fee schedule for the site visit is on a three-tiered scale according to the size of the department. CAAS is an entirely voluntary program. Accreditation is an accepted standard placed into many ambulance contracts with government agencies. In some city ordinances ambulance providers must be accredited. If an EMS provider is subject to encroachment by other providers, CAAS accreditation can protect the public from lower standards of care in the community by competitors and leads to better patient care, improved performance, and more cost-effective operations.

PROVIDER ASSOCIATIONS

The National Association of EMS Educators (NAEMSE) promotes EMS education, development, and delivery of educational resources and advocates research and lifelong learning. NAEMSE holds an annual symposium in conjunction with CoAEMSP and provides educational rollouts throughout the year. NAEMSE provides some model curricula, textbook development for EMS instructors, and input into national EMS efforts from NHTSA and EMS organizations (Figure 1.17). NAEMSE publishes the newsletters *Domain[3]* and *Educator Update*.

FIGURE 1.17 ■ The National Association of EMS Educators (NAEMSE) Logo.
(Reprinted with permission.)

The Emergency Nurses Association (ENA) is a national association of professional nurses dedicated to the advancement of emergency nursing. ENA promotes the profession through advocacy, publishes the *Journal of Emergency Nursing*, and provides emergency nursing curricula. ENA sponsors a continuing education program and conducts three standardized courses: the Trauma Nursing Core Course (TNCC), a 16- to 20-hour course that integrates trauma-nursing knowledge and psychomotor skills; the Course in Advanced Trauma Nursing II (CATN-II), which reinforces the concepts central to trauma nursing by evaluating ongoing clinical conditions; and the Emergency Nursing Pediatric Course (ENPC), a 16-hour course with six psychomotor skill stations encompassing pediatric emergency-nursing knowledge and psychomotor skills. ENA arranges an annual meeting with a general assembly, a scientific assembly, and a leadership challenge. ENA is organized into state and local chapters that provide effective networking opportunities. ENA also endorses the certified emergency nurse (CEN) and the certified flight nurse (CFRN) credentials.

The International Tactical EMS Association (ITEMS), in cooperation with other law-enforcement and emergency medical associations, was established to maintain an organized network of tactical officers, tactical medical providers, law-enforcement agencies, and other interested parties within the special weapons and tactics community. This network gathers tactical emergency medical support–related information that can be disseminated to would-be and established tactical medical providers and their teams. The association's position as a central source of information contributes to a reduction in the morbidity and mortality associated with tactical training and special operations. ITEMS publishes a quarterly magazine called the *Journal of Tactical Emergency Medical Support*.

Formed in 1997, the National Public Safety Telecommunication Council (NPSTC) is a federation of organizations representing public safety telecommunications. NPSTC was originally formed to encourage and facilitate implementation of the findings and recommendations of the Public Safety Wireless Advisory Committee (PSWAC). This committee was established in 1994 by the FCC and National Telecommunications and Information Administration (NTIA) to evaluate the wireless communications needs of local, tribal, state, and federal public-safety agencies through the year 2010; identify problems; and recommend possible solutions. NPSTC has since taken on additional responsibilities, including implementing the recommendations of the FCC Public Safety National Coordination Committee (NCC) and the support and development of the Computer Assisted Pre-coordination and Resource Database (CAPRAD) System for a 700-MHz spectrum to assist the Regional Planning Committees (RPCs). NPSTC develops and makes recommendations to appropriate governmental bodies regarding public-safety communications issues and policies that promote greater interoperability and cooperation between local, state, and federal agencies.

The National Association of Emergency Medical Technicians (NAEMT) represents and serves EMS personnel through advocacy, educational programs, and research. EMS strategic plans, policy papers, conferences, and position papers are produced by NAEMT. It is the primary sponsor of the EMS Expo, one of the largest EMS conferences in the country. NAEMT's administrator division is engaged on a national level with the development of EMS management policies and activities. NAEMT also sponsors an EMS provider to the Harvard Public Safety Fellowship.

The **National Registry of Emergency Medical Technicians (NREMT)** has for years provided standardized testing for EMTs and paramedics. This national certification includes a written or didactic test and several practical-skill stations. The certification is accepted in most states as a way to gain reciprocity between state, county, and city certification and licensure. The NREMT maintains a two-recertification cycle that requires skills verification and a standardized set of content areas for refresher training. The registry also maintains a large database called the LEEDS project that contains information on paramedics and EMTs in a variety of different categories.

In January 2007 the National Registry moved to **computer-based testing (CBT)** through Pearson VUE, a vendor that provides standardized testing for several medical disciplines and physicians. This requires a student to enter a testing facility that is monitored for security and to take a standardized test on a computer. After the candidate completes the exam, his scores are transmitted electronically and encrypted to the EMS authority or training institution. CBT offers increased flexibility, security, and easier registration. It also offers computer-adaptive testing (CAT), a process in which the person taking the test is assessed by the computer, which estimates the person's ability and selects the next set of questions. As the test progresses through each set of questions, the ability of the program to estimate the person's knowledge base becomes more accurate.

LOBBY ORGANIZATIONS

Advocates for EMS (AEMS) is a nonprofit EMS lobbying organization. It was founded on October 22, 2002, and is dedicated to educating decision makers in Washington on issues affecting EMS providers. AEMS supports all types of EMS, whether fire based, volunteer, third-service, public, or private, by monitoring and influencing legislation and regulatory activity involving EMS and raising awareness among lawmakers on issues of importance to EMS.

AEMS uses both congressional funding requests (appropriating legislation) and authorizations (non-appropriating legislation) to increase support for EMS. AEMS works closely with DHHS, DHS, and the DOT and is active in promoting EMS issues in Washington, working with the Federal Interagency Committee on EMS and other federal and professional organizations.

Fire Service-based EMS Advocates is a coalition that came together in 2007 to educate policy makers at the local, state, and federal level about the important role of fire service–based EMS in local communities. It is a coalition of IAFF, IAFC, NVFC, NFPA, and the Congressional Fire Service Institute based in Washington, D.C. Its mission is to help validate the use of fire-based EMS as the most efficient use of tax dollars for EMS spending. The coalition is responding to the Institute of Medicine's reports on EMS and in the future will play a significant role in legislation and funding of EMS.

VOLUNTEERS

The primary representative of fire, EMS, and rescue volunteers is the National Volunteer Fire Council (NVFC). The NVFC represents the issues of volunteer fire and EMS agencies across the country. It lobbies for legislation to support the volunteers and many rural fire and EMS issues. The NVFC is comprised of 49 state firefighter organizations that take fire and EMS issues to the federal agencies and Congress.

CHAPTER REVIEW

Summary

The responsibility of managing an EMS system often takes on a local and internal focus at the expense of understanding the larger picture, which involves outside agencies and government sources. Since the business of management involves relationships, an effective manager must cultivate contacts at all levels, including the highest levels of government. Both government and nongovernment agencies created the parameters for which EMS systems are accountable. Research, standards, regulation, and education emanate from those outside sources, and it is in the best interest of all EMS organizations to understand and utilize the resources they offer.

Research and Review Questions

1. Contrast the five original recommendations of the 1966 report entitled *Accidental Death and Disability: The Neglected Disease of Modern Society* with the current state of EMS today. The report was done by the National Academy of Sciences (NAS), which was at the time of the report called the National Research Council's National Science Foundation.
2. Research one of the following agencies or organizations: FEMA, NHTSA, IAFC EMS Section, or IAFF EMS section. During your research, determine the following: agency/organization mission as it relates to EMS, the agency's constituent group, annual budget as it relates to EMS, and products or services offered to its constituents. In addition, how can this organization assist your EMS agency?
3. Research the following agencies, and identify eight training opportunities available for EMS personnel: NHTSA, FEMA, IAFF, IAFC, ODP, NFA, and AAA.
4. Research the EMS systems in Washington, D.C.; Denver, Colorado; Kansas City, Kansas; St. Louis, Missouri; and Oklahoma City, Oklahoma. Identify which EMS delivery system is being used in each of the cities: fire department–based; fire department–based oversight; public single-role EMS systems; or private provider.

References

American College of Emergency Physicians. (1993). Guidelines for trauma care systems. *Annals of Emergency Medicine, 22,* 1079–1100.

National Academy of Sciences, National Research Council. (1966). *Accidental death and disability: The neglected disease of modern society.* Washington, DC: The National Academies Press.

National Highway Traffic Safety Administration and Maternal and Child Health Bureau. (1996). *EMS agenda for the future* (DOT Publication No. HS 808-441). Washington, DC: U.S. Government Printing Office.

National Highway Traffic Safety Administration, U.S. Department of Transportation. (1995). *Star of Life emergency medical care symbol background specifications and criteria* (DOT Publication No. HS 808-721). Washington, DC: NHTSA Office of Injury Control.

National Institutes of Standards and Technology, U.S. Department of Commerce. (2007). *Health care criteria for performance excellence.* Gaithersburg, MD: Baldrige National Quality Program.

Page, J. O. (1978). *Emergency medical services* (2nd ed.). Boston: National Fire Protection Association.

Page, J. O. (1979). *The paramedics.* Morristown, NJ: Backdraft Publications.

Roush, W. R. (1994). *Principles of EMS systems* (2nd ed.). Dallas, TX: American College of Emergency Physicians.

Stewart, R. D. (1989). The history of emergency medical services. In A. E. Kuehl, *National Association of EMS Physicians EMS medical directors handbook*. St. Louis, MO: Mosby.

Strategic Planning for EMS

Accreditation Criteria

CAAS

103.02   Strategic Planning: The agency shall have a process in place for short- and long-range strategic planning (pp. 29–31).

Objectives

After reading this chapter the student should be able to:

2.1   Define strategic planning and the time frames that are incorporated in strategic planning (pp. 29–31).

2.2   Define the contingent planning model for strategic planning (pp. 29–30).

2.3   Use a 15-point plan to create components of a strategic plan (pp. 30–31).

2.4   Demonstrate how to prioritize strategic planning concepts and items (pp. 31–32).

2.5   Define the strategic planning items—and their purposes—that would be included in a fire or EMS plan (pp. 25–29).

2.6   Explain and incorporate a SWOT analysis into the planning cycle for a strategic plan (pp. 34–35).

2.7   Discuss the various considerations for calculating the cost of EMS services (pp. 37–39).

2.8   Define or identify the stakeholders in EMS (p. 31).

2.9   Identify the management tools used to conduct project planning (pp. 41–43).

2.10 Describe the primary components of an EMS budget (pp. 40–41).

2.11 Identify proactive approaches to EMS funding (pp. 43–44).

Key Terms

Applied Strategic Planning
  Model (p. 30)
average cost curve (p. 37)
balanced score card (BSC)
  (p. 29)
brainstorming (p. 34)
business plan (p. 43)
capital equipment cost
  (p. 38)
commitment planning
  (p. 29)
cost shifting (p. 45)
critical path method (CPM)
  (p. 41)
economy of scale (p. 37)
executive summary (p. 25)

Fire Department Strategic
  Planning Model (p. 30)
fixed expenses (p. 38)
government budget (p. 40)
incremental planning
  approach (p. 30)
inspirational leadership
  (p. 30)
line-item budget (p. 40)
master plans (p. 31)
mission statement (p. 31)
operational plans (p. 31)
Oval Mapping Technique
  (p. 36)
performance-based budget
  (p. 40)

Program Evaluation and
  Review Technique
  (PERT) (p. 42)
quality planning (p. 45)
rational strategic planning
  (p. 29)
stakeholders (p. 25)
strategic map (p. 25)
strategic planning (p. 24)
SWOT analysis (p. 25)
values statements (p. 33)
variable expenses (p. 38)
vision statement (p. 31)
zero-based budget (p. 41)

Point of View

Motivational speaker Anthony Robbins professes, "If you don't have a plan for success, someone has a plan for your failure." This reads as a hard-line and extreme statement, but there is arguably no truer statement ever made. For EMS leaders this comes in the form of losing out during the budget cycle, failure to be recognized as members of the medical community, and an inability to create meaningful change. Strategic planning is the process of getting your organization from there to here. It is the ability to envision future actions, quantify your organizational goals (budget), and determine the appropriate time for implementation.

Many EMS organizations do not plan strategically, because they are constrained to surviving the present. They fail to realize their obligation to create an organizational road map. Their leaders often know the cost of everything but the value of nothing. Effective strategic planning is the balancing of vision, mission, and realities.

For example, an EMS system received a federal grant to develop an employee health and wellness program. This included a full physical examination (with blood work and cardiac stress testing), a fitness evaluation, and employee education. This department counted on receiving a follow-up grant to ensure a continuation of the program. Unfortunately, the follow-up grant did not materialize, and there were no alternative plans. The EMS chief was criticized by the town administration and labor groups for using the grant and this one-time effort as self-promotion.

Harold Cohen, PhD, FACHE, NREMT-P
SPC/TriData

■ WHAT IS STRATEGIC PLANNING?

The most important task an EMS manager and leader can do is to conduct planning. An EMS leader must look at EMS operations from a strategic perspective; that is, management must ensure that equipment, systems, and personnel are matched with the community's EMS needs. The goal of strategic planning is not just the plan itself; rather, strategic planning is a planning method or how the planning will be done.

Strategic planning refers to the entire set of processes and behaviors that an organization uses to identify, prioritize, focus, and schedule action in order for it to remain viable in the future. Strategic planning contributes to organizational leadership by providing a vision and direction. City managers, county administrators, and other government officials expect EMS leaders to understand and be able to create a strategic plan. It is a necessity in the private-ambulance industry and in agencies that survive on limited funding. Strategic planning can be used to solve department problems, improve overall efficiency, and establish priorities.

Before actually beginning the mechanics of a strategic plan, it is important to create a **strategic map.** A strategic map enables EMS managers to define and communicate the cause-and-effect relationship between monies spent on EMS and their EMS organization's value to the community. Strategic maps apply the planning process to the organization, using four perspectives: *financial*, *customer*, *process*, and *learning and growth*. Mapping these strategic themes is well suited to public-safety organizations, which have limited political freedom to experiment with structural changes.

EMS leadership can make this planning process more successful by assigning one manager to each perspective, or theme. This tactic can help avoid the lack of clarity that often occurs at the beginning of any planning process, as well as encourage each manager's personal commitment. Organizing the plan by perspective, or theme, keeps each manager focused on a single critical aspect of the plan, which has the benefit of eliminating the potential for gaps. This tactic also is a powerful tool for implementing and monitoring the progress of the organization.

■ OVERVIEW OF A STRATEGIC PLAN————

A strategic plan is a product that is developed by the coalition of **stakeholders** to move an organization forward or to define its service delivery and development over a period of time. This plan forms the framework of the actions that will be taken in order to achieve the vision that was expressed at the beginning of the planning process.

A *written* strategic plan has eight sections: **executive summary,** definition of the problem, assessment of critical factors, intervention strategies, stakeholders, organizational objectives, budget, and plan evaluation.

- *Executive summary.*   Generally written at the end of the development process, the executive summary appears first in the written strategic plan. It identifies the

key issues related to the plan. This section serves as a hook to the reader and needs to entice the individual to read further. Failure to write a compelling executive summary may doom the plan to not being read further. The executive summary should contain the program mission and vision statements, both of which are discussed later in this chapter.

- *Problem definition.*   This takes the information identified in earlier processes and crafts it into a clear and concise definition of the issue or problem. It should include the data that demonstrate the issue, the impacts of the issues, and the potential outcomes of interventions. The problem definition should identify clearly the issue and its effects upon the target audience.

- *Assessment of critical factors.*   This is an assessment of the three critical factors of project management—time, costs, and performance. The assessment also should include the identification of the dominant and the weakest factors or project constraints.

- *Intervention strategies.*   This list is created using the information that was obtained through research and problem-identification efforts. The plan should identify both the potential and the selected intervention strategies that reflect the mission, legal, and community responsibilities for which the agency is chartered.

- *Stakeholders.*   The stakeholders in the coalition should be identified, listed, and discussed. Include their vested interests as potential adverse effects of the project. The plan also should address their anticipated levels of involvement in the project.

- *Organizational objectives.*   An overview of the strategies and measurable objectives that have been identified, as well as their time frames for completion, should be written into the strategic plan. This section should include:
  – Overall strategies and milestones.
  – Objectives and corresponding tasks.
  – Financial and resource requirements.
  – Responsibilities.

- *Budget.*   Each component of the proposed budget needs to have a **SWOT analysis,** a tool used to evaluate the Strengths, Weaknesses, Opportunities, and Threats involved in a project or business venture. The summary of the SWOT analysis should specifically include how strengths will be tapped into to correct weaknesses, and how opportunities will be explored while simultaneously avoiding threats.

- *Plan evaluation.*   The last section of a strategic plan is to identify how, when, and what will be evaluated. The plan is a living document, and it will need monitoring, as well as periodic and frequent review and adjustment. Consequently, the plan evaluation must include a description of the monitoring and review processes that will be used for each budget item or component of the strategic plan. It also must include a discussion of the means that will be used to monitor performance levels and trigger points for comprehensive review, revision, completion, and even "bail-out" points.

A strategic plan is of tremendous value to EMS leadership when explaining to personal and political leaders where the agency needs to go. It helps in budgeting and to provide vision. See Figure 2.1 for excerpts from a fire-based EMS organization's strategic plan. It shows an executive summary, a description of the planning process, strategic focus areas, and details of components of one of the focus areas.

DATE: August 24, 2006
TO: All Members
FROM: Chief Bruce W. Edwards
SUBJECT: Department of EMS Strategic Plan, City of Virginia Beach

I am pleased to present the Virginia Beach Department of Emergency Medical Services' Strategic Plan for 2006–2011. This document is the product of several months of research and review at many levels both inside and outside the department. The plan is divided into five focus areas: Our People, Our Programs, Our Facilities, Our Operations and Our Equipment. A set of goals and objectives is provided for each of these focus areas. Timeframes for achievement and lead division are identified for each objective.

The challenge now is to convert the goals and objectives into outcomes. Even though we have just completed two years of rapid growth and change, we cannot rest on our laurels. The City's population and call demands continue to grow. We face technical challenges as well. The new American Heart Association's Guidelines alone represent a major shift in how we care for thousands of our patients. Fortunately, we are in an ideal position to move forward into the future. Our leadership positions are fully staffed. Rescue Council and volunteer rescue squad officers are actively engaged in the organizational development process. Our career medic program is running smoothly with no major adjustments immediately expected. We should now be able to shift from our daily mode of thinking into future thinking using the Strategic Plan as our guide. Working together, we will make Virginia Beach a premiere EMS organization for others to emulate. . . .

Planning Process

Phase I: Committee Formulation
The Strategic Planning committee was convened in the fall of 2005. The core committee consisted of Deputy Chief Kiley and the chiefs of each of the department's four divisions. Throughout the process, participation was solicited from all levels of the organization. Monthly meetings were held along with a number of ad hoc sessions and virtual meetings online.

Phase II: Methodology
The committee started with a review of strategic plans from public safety agencies around the country. It was immediately apparent that no standard model existed for a strategic plan. The plans reviewed varied widely in both format and methodology. In some cases, the actual goals of the organization were lost amidst a complex process of voting, cumbersome numbering systems and vague objectives. The committee felt that a simplified format was both the easiest to read and was most likely to be a useful planning document. The following steps were established for the project:

1. Review the Mission, Vision, and Values of the Department. Revise as necessary.
2. Perform a Horizon Scan (SWOT).
3. Establish Strategic Focus Areas.
4. Establish Key Goals for each Focus Area.
5. Develop Objectives to achieve each goal.
6. Establish a Timeframe and Lead Division for each Objective.
7. Conduct final review at the senior staff level.
8. Publish the Strategic Plan after approval.

Phase III: Current Environment Assessment
The Committee conducted an analysis of strengths, weaknesses, opportunities, and threats (SWOT) impacting the department. Stakeholders from throughout the organization participated in facilitated SWOT brainstorming sessions. Groups surveyed included administrative staff, career medics, volunteers, and senior officers. Issues were identified and discussed at length. While each group had some unique concerns, there were a significant number of common issues across the department. A particular concern was the need to produce a realistic plan that would serve stakeholder needs. The Committee identified and evaluated many groups that had a vested interest in both the planning process and outcomes. Dialogue with partner agencies was conducted to identify future external issues that

FIGURE 2.1 ■ Four Excerpts from a Strategic Plan.
(Reprinted with permission of Virginia Beach EMS.)

might impact the Department of EMS. Some inquiries were conducted formally via interviews while other inputs were drawn from ongoing inter-agency exchanges. Stakeholder inputs came from the following City of Virginia Beach departments:

Fire	Police
Parks and Recreation	Public Works
Risk Management	COMIT

Contact was also made with the following outside agencies:

Oceanfront Lifeguards	Physician Groups
Commercial Ambulance Services	Tidewater EMS Council
Military Fire Departments	Virginia Office of EMS

The Committee was also able to draw information from an ongoing organizational development initiative. An outside consultant had been brought onboard to look at the department's organizational structure and management techniques. While this project was separate from the Strategic Plan, there were a number of insights gained from the survey process that translated into desired goals and objectives.

Phase IV: Plan Development
The actual planning began with the validation of our Mission Statement. This is the basis for the department's existence. A review was also made of our desired Vision and Values. While EMS has traditionally used the City of Virginia Beach's Vision and Values, the committee opted to add one more value to highlight our commitment to volunteerism as a core philosophy.

The Committee developed a number of broad goals associated with solving a variety of critical issues. When these goals were compared against the inputs from the SWOT analysis and partner dialogue, it was clear that the more important/valuable goals could be grouped into distinct areas. Five "Focus Areas" were ultimately named in order to provide practical targets for planning. This became the basic framework for the plan.

Within each focus area, the Committee assigned goals. Under each goal, specific objectives, lead divisions, and a timeframe were identified. After much internal review, the draft focus areas and goals were forwarded across the organization via mass email. Literally every member, career or volunteer, with an email address was given the opportunity to comment. Copies were also sent to the Virginia Office of EMS, Tidewater EMS Council, Virginia Beach Department 10 of Management Services, and all partner hospitals. Feedback was incorporated into the plan or filed for later use in strategy development. The last external review came in late May 2006 when copies were presented to the senior leaders of our ERS partners in the Police and Fire Departments.

Phase V: Plan Strategy
Individual objectives each have a time frame assigned in terms of continuous, short, medium, and long-range achievement. For purposes of this plan, these time frames are defined as follows:

- Continuous: Daily focus of the organization.
- Short: Six months to two years.
- Medium: Three to four years.
- Long: Five years or more.

The majority of the objectives fell within the short and medium windows. This is not unexpected due to the volume of recent changes experienced by the department.

A lead division has been identified for each objective. The responsible officers will work with the appropriate stakeholders to create strategies and work plans related to each objective. The committee gathered a sizable amount of member input to assist with this step during plan development and ongoing organizational development sessions.

It is anticipated that the Strategic Plan will receive an annual progress review at the Chief and Deputy Chief level. It should serve as the template for major projects and budgetary program proposals for the next three to four years.

It is recommended that the next Strategic Planning Committee be convened in 2010 to prepare for 2011 and beyond. . . .

Strategic Focus Areas

The Department of EMS intends to "focus" its attention and resources on five primary areas. Within these major areas are numerous subcategories of activities and roles sharing a common theme within that Focus Area. Main goals with primary objectives have been developed and are presented here. A projected timeframe and lead division is provided with each objective.

Our People

The most critical asset the department has is our people. Without field providers, we cannot respond to calls. Without administrative staff, we cannot plan or support the organization. Without training staff, we cannot keep our medics sharp. Constant focus on recruiting and retention is required to maintain adequate staffing levels to meet service demands.

(continued)

Our Programs A number of programs and initiatives are required to support both the department and the community at large. Some of these provide direct service delivery while others support the ongoing evolution of the department. We need to focus on programs and initiatives that will maintain the strength and abilities of the organization for years to come.

Our Facilities Our facilities house our people and equipment. We need to assess their condition, renovate where feasible, and plan for new structures that will support future operations.

Our Operations High quality emergency medical care relies on timely response by well-trained, well-led, and well-equipped personnel. This is where we bring our resources, policies and practices together to deliver service to the community.

Our Equipment Medics need the right tools to save lives. The department will invest in the latest technology and act as good stewards for its equipment. . . .

Focus Area = Our Equipment

Goal I: Utilize the Latest Technology

Objectives:	Timeframe	Lead Division
1. Integrate new glucometers	S	Regs/Train/Ops
2. Evaluate 12-lead cardiac monitor systems	S	Regs/Train
3. Identify next generation monitor/defibrillators	M	Regs
4. Integrate electronic reporting systems	L	Regs
5. Provide GPS mapping systems onboard response vehicles	L	Ops
6. Establish a program to identify technologies and integration into other health-based systems	M	Ops

Goal II: Improve Fleet Management

Objectives:	Timeframe	Lead Division
1. Identify long-range special operations vehicle needs	M	Ops
2. Evaluate alternative supervisor vehicles	M	Ops
3. Review current approach to vehicle maintenance tracking/management	M	Ops

Goal III: Establish Durable Medical Equipment (DME) Management Plan

Objectives:	Timeframe	Lead Division
1. Establish DME plan for public access and police AEDs	M	Regs
2. Establish DME plan for monitor/defibrillators	S	Regs
3. Establish major medical equipment maintenance plan	M	Regs

Goal IV: Standardize Response Vehicles and Equipment

Objectives:	Timeframe	Lead Division
1. Provide standardized equipment on all zone cars	S	Ops
2. Identify a model ambulance design	L	Ops
3. Identify model equipment locations in response vehicles	M	Ops
4. Provide standard equipment on all response vehicles	S	Ops/Regs
5. Engage regional partners to standardize regional supplies and pharmaceuticals	M	Regs

FIGURE 2.1 ■ Continued.

BALANCED SCORECARD

The **balanced score card (BSC)** is based on the principle that what gets measured gets done, and that financial performance is not the only measure of an agency's success. The EMS manager also must measure performance with respect to customers, the public, and processes.

There are two key components of the BSC. The first is "what gets measured gets done." The second is that financial measures are not sufficient to manage the organization. The BSC creates an instrument panel of indicators from all aspects of the organization's activities. It identifies measures from the perspectives of financial performance, customers, internal business processes, clinical competencies, patient outcomes, and personnel growth and learning. The expectation is that when these measures are linked to the organization's strategy, managers, staff, and employees will naturally adopt appropriate behaviors to achieve the EMS agency's goals.

CONTINGENCY MODEL OF STRATEGIC PLANNING

Most strategic plans use a rational approach to planning incorporated into a multi-year budget cycle. The planning approach that will be taken is contingent upon a choice made by analyzing a model that includes four boxes, a vertical axis, and a horizontal axis (Figure 2.2).

The vertical axis looks at the complexities of technology and the uncertainty of the environment. Some examples of technology that will have a tremendous impact on EMS planning are GIS, modeling, and mapping. Technology and simulation help reduce uncertainties in the environment, such as the fluctuations in hospital diversion, overcrowding in the emergency rooms, disasters, and economic or political events.

The horizontal axis of the model deals with how clear and how well defined and communicated the organization's goals are. The horizontal axis also identifies whether or not the labor unit, stakeholders, and leadership agree on the goals or if the goals are unstable and subject to change.

Using strategic leadership, an organization's situation should dictate the technique that makes sense. The first box identifies **rational strategic planning.** This approach typically takes the following steps: plan, act, and evaluate. This is often useful for organizations that have stable EMS demands and clear organizational goals.

The **commitment planning** box signals an approach to planning in which EMS managers and leadership are not in agreement or the goals are not clear. Commitment planning, in the second box, is a search to find common ground through a labor-management

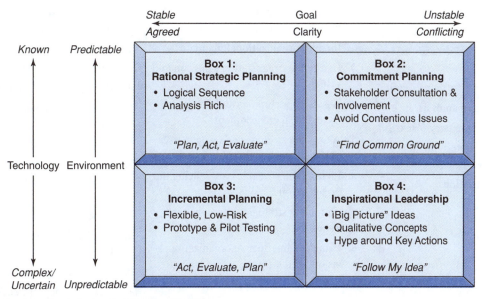

FIGURE 2.2 ■ Jim MacKay's Strategic Planning Matrix.
(Reprinted with permission of Berkeley Consulting Group)

partnership or by stakeholder consensus building. The focus is on getting buy-in on operational issues.

Incremental planning, the third box, is for complex and unpredictable environments with identified and agreed-upon goals. An **incremental planning approach** does not lead to a master strategic plan and focuses on initiatives that are flexible, reversible, and bite size. Incremental planning requires innovation, pilot projects, and experimentation. Part of that process involves scenario planning and experimenting with possible future programs.

The fourth box requires **inspirational leadership** for complex, unpredictable environments with unstable and conflicting goals. This requires a charismatic leadership and new approaches to service delivery at critical times; for example, peak-load ambulance needs during an EMS system's high-volume periods.

New approaches to EMS service delivery are often done with little evidence-based medicine. The science behind most of what EMS does is weak and unsupported and comes more from bold ideas or anecdotal experience and needs to be validated. This will be more important in the future as reimbursement becomes tied to outcomes.

PLANNING PROCESSES

Several strategic-planning processes are used in government. The stakeholder-management approaches, competitive analysis, and strategic-issues management are some of the systems used for strategic planning in quasi-government structures. Three models for strategic planning prevail in EMS. The first is the National Fire Academy's model, created by Dr. J. William Pfeiffer, a process called the **Applied Strategic Planning Model.** A second, more widely used model is from John Bryson's work on strategic planning for nonprofit and government organizations, commonly known as strategic planning. A very popular and well-accepted process is Mark Wallace's **Fire Department Strategic Planning Model,** which has a 12-step process that merges the Fire Academy model and the Bryson nonprofit model. Table 2.1 describes the three common processes and their differences.

Strategic planning is completed in several cycles that overlap each other to create a comprehensive plan for the organization (Figure 2.3). One year's plans incorporate the budget and operational considerations for the immediate year, which would start in the begin-

---

TABLE 2.1 ■ Comparison of Most Common Public-Safety Strategic Plans

NFA Applied Strategic Planning Model	*Bryson Strategic Plan*	*Wallace's Fire Department Strategic Planning Model*
1. Plan to plan.	1. Initiate and agree on a strategic planning process.	1. Identify the department's values.
2. Formulate mission statement.	2. Identify organizational mandates.	2. Plan to plan strategically.
3. Perform strategic business modeling.	3. Clarify the organization's mission and values.	3. Select and recommend a strategic planning process.
4. Conduct a performance audit.	4. Assess opportunities and threats of the external environment.	4. Identify the department's mandates.
5. Perform a gap analysis.	5. Assess the strengths and weaknesses of the internal environment.	5. Develop the mission of the department.
6. Prepare contingency plans.	6. Identify the strategic issues facing the organization.	6. Create a philosophy of operations.
7. Create integrated functional plans.	7. Formulate strategies to manage the issues.	7. Assess the opportunities and threats of the external environment.
8. Implement actions.	8. Establish an effective organizational vision for the future.	8. Assess the opportunities and threats of the internal environment.
		9. Identify the strategic issues of the department.
		10. Create strategies for strategic issues.
		11. Create the department's ideal future.
		12. Conduct operational planning from a strategic perspective.

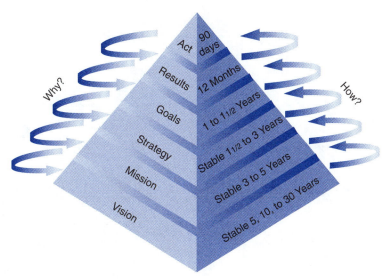

FIGURE 2.3 ■ Strategic Planning Cycles.
(Reprinted with permission of Myrna Associates, Inc.)

ning of the fiscal year. **Operational plans** are short term and generally range from 3 to 5 years. **Master plans** are either 10- or 20-year plans. For EMS operations it is often difficult to plan more than 10 years out due to the constant changes in the health-care system. When developing a master plan, EMS leadership should take into consideration population trends, tax base, planned development, and trends in the EMS industry.

In order to ensure a successful strategic-planning process and the development of a good strategic plan, several key characteristics must be in place. The first is to identify the stakeholders that can help to clearly define a mission and vision of success and to define strategies and objectives. A stakeholder is any individual or organization that has a vested interest in the implementation or outcome of the plan. These stakeholders will form the organization's coalition for success, and their involvement from the start is essential.

Stakeholders can be either "internal" or "external" to the EMS manager's particular organization. Since health issues affect many operating organizations within the community, the external stakeholders generally will be numerous in comparison to the internal stakeholders. Depending upon the organization's actual composition and structure, internal and external stakeholders are defined as:

+ *Internal:* system managers, chief officers, financial managers, planning, information and technology (IT) employees, public-education personnel, and public-information officers.

+ *External:* hospital administrators, insurance company representatives, advocacy groups, homeowner associations, community and service organizations, risk managers, emergency-department staff, public-health officials, elected officials, constitutional officers, subject-matter experts, and grant consultants.

It is best to identify the stakeholders in an EMS system and display them graphically, as in Figure 2.4. It becomes obvious from this list that the project quickly can become larger than any single EMS organization or provider. Therefore it is essential to have the ability to recruit the assistance and support of the key stakeholders in order to build a successful coalition. The EMS manager should work diligently to identify all the possible stakeholders in the community. There is nothing worse than getting into the middle of a project only to come to the realization that a key member of the support system is missing. Not only do these situations potentially slow the planning and implementation process, the political impacts of hurt feelings can be hard to overcome.

CLEARLY DEFINED MISSION AND VISION

In addition to the key stakeholders and the EMS manager's ability to build a coalition, a plan will only work if its mission is defined clearly. Often, a **vision statement** sets the stage for a **mission statement.** A vision statement gives an EMS leader a chance to set the direction of the EMS division and the organization. A vision statement provides a futuristic look at, and

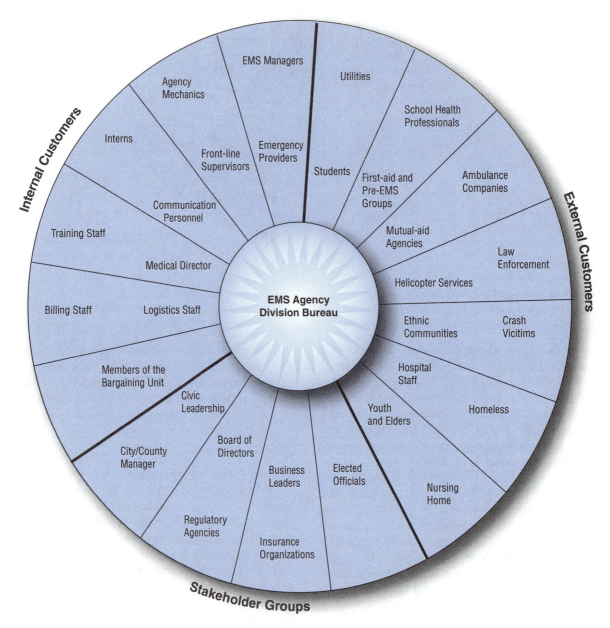

FIGURE 2.4 ■ EMS Stakeholders.

broad guidance for, the EMS agency or system. In simple terms the vision statement helps to ensure everyone is going in the same direction. Each objective and action plan that is subsequently devised is consistent with the vision statement, thus ensuring constancy of purpose and compatibility of actions.

While development of the vision statement is directed by the leaders of the state, regional, or local EMS system, the system or agency "players" should be deeply involved in the development process. People are more likely to help implement what they help to develop. Typically, a vision statement would be a short, motivational description of the EMS system's ideal condition. The vision statement can serve as motivation for those involved in the system and can be a steady guide through the numerous changes necessary to achieve a quality system. When constructing vision statements, be aware of attitudes and actions that kill an organization's vision. Often fire and EMS organizations fail in developing their vision statements due to tradition, stereotypes, fatigue, and short-term thinking. A vision statement should build loyalty through involvement, commitment, and a need for change.

Often a vision statement is supplemented by a values statement. **Values statements** should be positive and inspiring and provide a focus to the organization. A vision statement communicates the leader's beliefs, values, and outcomes as they relate to the organization's mission and vision statement. Values statements establish the base for the organization's culture. Often it is helpful to establish core values for an organization. Core values are used as a basis for decision making. They help attract people to the organization who reflect the core values of the department, agency, or company.

After the EMS manager has identified the issue and the target audience, it is important to be able to articulate the organization's mission, as well as its vision of success, concisely. The mission statement can

be developed properly only after the EMS manager has identified the problem and established the beliefs of the organization and its leaders.

The mission statement defines a focal point. When an organization or project starts to run off into new and uncharted territory, a mission statement provides a direction that allows the EMS manager's team to realign to a common purpose. It defines the reason that the organization, committee, or group exists or what action it is expected to take. Working outside the mission tends to become counterproductive. Simply put, a mission statement should address what the organization is going to do and for whom the organization is doing it.

Many people hold different definitions as to how mission and vision statements differ. A mission statement will be described as a statement that answers the questions, "What does success look like?" and "What measurable and visible observations indicate that success had been achieved?" A mission statement is a short statement, not more than two sentences. An example of a department's vision, mission, and values statement is shown in Figure 2.5.

Our Vision

* We will be a global leader in emergency services.
* We will be a diverse workforce that provides quality fire and life safety through proactive and innovative training, education, code enforcement, risk assessment, and community involvement.
* We will be vigilant, brave, and prepared.

Our Mission

* To advance the City of North Las Vegas vision by providing dedicated emergency and community services in a professional manner.

Our Values

Noble:	We will possess all the characteristics and qualities of professional fire and life safety emergency service providers.
Leadership:	We believe in positive leadership with vision toward the future. We will mentor and empower ethical leaders throughout the organization. We will conduct ourselves as leaders in the community.
Vigilant:	We will diligently watch over ourselves, our family, and our community.
Family:	We will remember: what affects one, affects all.
Diligent:	We will enthusiastically complete all tasks safely and with detailed perseverance. We promote fiscal responsibility and accountability.
Professionalism:	We will remain skilled, knowledgeable, and ready to serve. We believe training and education are the foundation of professionalism.
Respect:	We will treat others with compassion, in a dignified and courteous manner.
Integrity:	We will conduct ourselves in a way that brings honor and respect to our profession. We believe that maintaining public, personal, and professional trust is paramount.
Diversity:	We will embrace diversity and foster a workforce reflective of our community.
Excellence:	We will continually strive for performance that surpasses all expectations. We believe that a quick and safe response, training, education, and preparedness are the keys to excellence.

FIGURE 2.5 ■ North Las Vegas Fire EMS Vision, Mission, and Values Statement. *(Reprinted with permission.)*

1. Define and agree upon the objective.
2. Brainstorm ideas and have an agreed-upon time limit.
3. Categorize, condense, combine, and refine.
4. Assess and analyze effects or results.
5. Prioritize options and rank list them as appropriate.
6. Agree upon actions and time frames.
7. Control and monitor follow-up.

FIGURE 2.6 ■ The Brainstorming Process.

BRAINSTORMING

When engaged in any strategic-planning session or work group, EMS managers should capitalize on the opportunity to solicit new ideas from employees by **brainstorming.** Brainstorming can be one of the first processes in creating a strategic plan. It is a method of shared problem solving in which all members of a group spontaneously contribute ideas.

Brainstorming is often conducted at an organizational retreat to create ideas, solve problems, and motivate and develop teams by making their ideas part of the organization's operations.

Brainstorming needs to be structured, and it should employ a process led by a facilitator familiar with managing the process, people, their involvement, and their sensitivities (Figure 2.6). Most facilitators recognize the importance of using an easel or marker board to ensure that all participants can see what is happening and can improve on each other's ideas. Brainstorming can be a powerful tool and can often yield results that produce services or processes that improve on an organization's existing services or processes.

Ensure that everyone understands the goal of the brainstorming session by making the objective simple and specifying the time allotment. When managing the ideas, do not discount any ideas. Record every suggestion, to ensure that every person feels comfortable participating. Make headings and lists to combine ideas by condensing or refining those ideas. The facilitator should combine or include weaker suggestions with other topics to avoid dismissing or rejecting ideas. Prioritize and develop the list of ideas.

■ SWOT ANALYSIS

There are several types of brainstorming models. One of the most popular is used by the National Fire Academy and is called the SWOT analysis, which was intro-duced earlier in this chapter. Other models include the McKinsey Seven-S Framework, which evaluates the criteria for a successful company. The Action-Centered Leadership model is a brainstorming technique used to evaluate the management of people. The Delegation model of brainstorming evaluates the success of task delegation and staff development through delegation.

USING A SWOT ANALYSIS

EMS managers are often asked to solve problems, and often people bring solutions or ask for assistance to bring an idea on line. The SWOT analysis is a resource-analysis process that identifies strengths and weaknesses and examines the opportunities and threats the program or idea will face. A SWOT analysis examines key characteristics or organizations (or, as in the case of a coalition such as stakeholders, the collective characteristics) and an agency's ability to complete the project effectively and achieve the desired outcome.

INTERNAL CHARACTERISTICS: STRENGTHS AND WEAKNESSES

When we think about the strengths and weaknesses of our own organization, we are thinking about its internal characteristics. For example, highly trained employees, high levels of motivation, available internal financial resources, and individuals who are particularly knowledgeable in the topic might all be examples of internal strengths. Internal weaknesses might include resistance to change, a lack of understanding of the issues related to the project, and lack of financial resources. When assessing internal characteristics, look for ways to tap into the strengths and correct weaknesses as they relate to the project. Some helpful questions necessary to gauge strengths and weaknesses include the following:

Strengths:

- What advantages does the organization have?
- What does the organization or its resources do well?
- What are the relevant resources to which you have access?
- What do other people see as the strengths in such a program?

Weaknesses:

- What current operations could be improved?
- What is the program or organization doing badly?
- What should you avoid?

EXTERNAL CHARACTERISTICS: OPPORTUNITIES AND THREATS

As with an internal audit, look externally to the organization or coalition to identify opportunities that can serve to enable your collective success with the project. Opportunities such as highly positive public opinion, financial opportunities (such as grants, foundations, and alternate funding sources), support of elected officials, changes in laws or regulations that would increase flexibility, and even the interest of insurance agencies should be considered.

On the other hand, threats are external characteristics that might impede a coalition's ability to succeed. Lack of public empathy, education levels and decreasing reading levels, competition for available public funds, and a changing or restrictive regulatory environment are all possible threat considerations. When assessing characteristics external to the project, pursue opportunities while avoiding or minimizing threats.

Therefore, the goal of the SWOT analysis is to develop an assessment that is designed to use strengths and correct weaknesses while you pursue opportunities for success and avoid or minimize threats.

Following are the components and questions to consider when generating answers about opportunities and threats for a SWOT analysis:

Opportunities:

- Where are the good opportunities for the program?
- What are the trends in the field?
- What are the changes in technology and markets on both a broad and a narrow scale?
- What are the changes in government policy related to your field?
- What are the changes in social patterns, populations, industry changes, and so on?
- What are the local, state, and national events?

Threats:

- What obstacles do you face?
- What are other agencies doing?
- Are the required specifications for the job, products, or services changing?
- Is changing technology threatening your position?
- Could the weakness seriously threaten the agency or the operations?

A SWOT analysis is a very powerful tool when applied to programs and processes that an EMS organization is considering. Often programs are put in place without much thought or because the program is popular, only to have the process incur an unforeseen expense or hurdle. As an example, Figure 2.7 shows the application of a SWOT analysis for a tactical EMS program. Tactical EMS programs can be difficult political and logistically complex operations.

STRENGTHS	WEAKNESSES
• Reduced officer injury/disability/death. • Increased effectiveness of tactical team. • Operational security. • Reduced need on local EMS and fire, freeing unit for response in local area. • Preparation for school shooting and acts of violence.	• Fire or EMS assigned to law enforcement command. • Equipment specific to tactical EMS expensive. • Baseline training infrequent. • Overtime and training cost. • Scheduling. • Post-certification needed for carrying weapons.
OPPORTUNITIES	**THREATS**
• Joint training with other area law enforcement. • Joint training with tactical training businesses. • Highlight as best practices. • Utilization for dignitary protection. • Migration of training to fire and EMS crews not assigned to tactical operations.	• Injury to fire and EMS personnel. • Fire and EMS seen as law enforcement. • Cooperative use of other tactical team with metro. • Budget expense, overtime. • Weapons of mass destruction. • Methamphetamine labs.

FIGURE 2.7 ■ SWOT Analysis.

A = Audience: Who will do it?

B = Behavior: What must be done?

C = Conditions: By when, where, and how?

D = Degree: What is the quantity and quality expected?

An objective does not have to follow this order. However, it should include the four elements.

FIGURE 2.8 ■ The ABCD Method.

CLEARLY DEFINED GOALS AND OBJECTIVES

A major component of a successful plan is that of clearly defined strategies, goals, and tasks: What overall strategies or interventions will be used? What goals need to be achieved in each? What tasks will be required to achieve these intermediate goals? A brainstorming session answers these questions. The results of the ideas need to be transferred into a means of continuous monitoring and periodic review in order to guide EMS managers in the success of the project. Organizational goals are written to express what the agency needs to accomplish in the given short-term and long-term budget cycles. In some situations goals evolve from problems. A commonly used method for creating goals in EMS is the SMART method.

The SMART method defines goals as:

- *Specific.* The goals must be concise and to the point without generalities and ambiguities.
- *Measurable.* Goals must be able to determine success or failure, and actual results must be recordable.
- *Attainable.* The goal must be within reach of the organization.
- *Realistic.* The goal must be realistic or part of the overall mission and appropriate for the organization.
- *Tangible.* The accomplishment of specific goals or objectives must be confirmed. It must be visible and have value.

Once the goals of the agency have been completed, a more specific set of objectives is needed to

- Local.
- State.
- Federal.
- Contract.
- Consensus standards.

FIGURE 2.9 ■ Prioritizing Objectives.

link to the goals of the organization. Goals and objectives are two distinct parts of a strategic plan, much like tactics and strategies. An objective is a specific description of an expected outcome to be attained over an identified period of time. When writing objectives, an ABCD format can be used (Figure 2.8).

Objectives should define what the organization intends to accomplish as specifically as possible. It can be likened to an initial focus or detailed assessment. One finds the problem with the patient; the others identify the specific steps in detail. Objectives are often prioritized by level of authority, as described in Figure 2.9. Objectives are often unfunded and made a priority based on whether they are funding or legal requirements.

The **Oval Mapping Technique** is one method used to help prioritize objectives. This technique requires that EMS managers apply a collaborative approach to setting the organizational goals that will be used in the strategic plan. Key goals or aspects of the EMS leader's vision of the organization are placed on a wall. With an EMS leader or management representative in the lead, each participant is given several oval Post-it notes. Oval mapping starts with an open discussion, with each participant writing down one of his issues on each Post-it note. Then each note is placed on the wall near the specific goal, and each group discusses and links the issues and establishes priorities.

Oval mapping is a process that helps stakeholders and the EMS teams contribute to the organization's mission. It helps the EMS leader identify issues with the strategic goals and develop an action plan to manage those issues.

■ COSTING OUT THE PLANNING PROCESS—

Several factors must be considered to determine the cost of operating and the strategic goals of an EMS system. First, consider the type of service levels that will be established and maintained. This includes direct services to the public, such as public education, preventive programs, and special events. Functional services include internal needs such as physicals, employee assistance, uniform allowances, and office support. The first step is to identify the workload indicators.

ECONOMIES OF SCALE

Most economic measurements identify that for an EMS system to be efficient and supported it requires a population base of 200,000 people. If the population served is fewer than 200,000, clinical and response-time reliability will suffer if there are insufficient subsidies to offset the economies of scale. System designs that have multiple providers, or those that split the nonemergency and emergency business between two different providers, can also have a negative impact on economic factors.

Economy of scale is an increase in production efficiency as the number of goods or services being produced increases. Typically an agency that achieves economies of scale lowers the average cost per unit through increased production, since fixed costs are shared over an increased number of goods. There are two types of economies of scale: external and internal. External economies happen when the cost per unit hour depends on the size of the industry, not the agency. Internal economies occur when the cost per unit depends on the size of the individual department, agency, or company.

An example of an economy of scale is awarding nonemergency and emergency services to a single provider in a community. In theory this allows Medicare and private-insurance patients in the nonemergency market to subsidize, or help pay for, the more expensive emergency work. Economies of scale, or the concept that nonemergency work can subsidize the more expensive emergency work, resulted in the proliferation of the public-utility model as an ambulance-system design in the 1980s. Many systems cover multiple jurisdictions with one ambulance; this is another example of jurisdictions using an economy of scale.

Poor economies of scale have both financial and operational impacts. When there is a low call volume, the financial impact and fixed cost of operating ambulances and unit hours move upward. EMS leadership must then look to reduce the unit-hour cost by decreasing the fixed cost, human-resource cost, training, and vehicle maintenance. Operationally, the impact of poor economies of scale appears as a reduced unit-hour utilization that affects response times.

On the other hand, economies of scale are maximized when a single provider exclusively serves a population of 1.2 million people. This is derived by looking at the **average cost curve.** The average cost curve is equal to total cost divided by the number or quantity of goods or services produced (Cost-C/Quantity-Q). This also can be assessed by measuring average variable costs by dividing costs such as operating expenses by Q. Average fixed costs (total fixed costs divided by Q) is used for capital or long-term investments of infrastructure or vehicles. Most agencies use a combination or merging of these costs to create the average cost.

Average costs may be dependent on the time period considered (increasing emergency services may be expensive or impossible in the short term, for example). Average costs affect the supply curve and are fundamental components of supply units and demand for emergency services. The curve continues to decline until it reaches the population of 1.2 million.

Providers that service large populations in noncontiguous communities, such as EMS districts that cover multiple communities or fire districts that cover large geographical regions, are able to maximize economies of scale. Taking advantage of economies of scale helps agencies fund specialty teams and fleet programs and reduce medical-supplies costs by buying in bulk.

WORKLOAD INDICATORS

Formulating the cost of goods and services starts with a review of the workload indicators—the key components of the production or services that the EMS organization provides. Figure 2.10 shows workload indicators for EMS. Looking at past workload indicators helps organizations prepare and determine budgets.

- ◆ Population served.
- ◆ Total responses.
- ◆ Rescue/EMS calls.
- ◆ Transports.
- ◆ Training contact hours.
- ◆ Special events.
- ◆ Standby for police, fire, or other agency support.
- ◆ Special operation calls.
- ◆ Response times, including call processing, turnout, and at the patient's side.
- ◆ Responses by unit.

FIGURE 2.10 ■ Workload Indicators for EMS.

CAPITAL EXPENSES

Capital expenses are defined as the costs of any property, furniture, fixtures, machinery, and any item worth more than $5,000. They include the cost of new or replacement vehicles and medical equipment, the building of new stations, and the upgrading of engines, ambulances, and computer systems. Most agencies require that a depreciation schedule be developed to compare costs with private companies. Many of the depreciation schedules for capital expenses will be 5-, 10-, or 20-year schedules that indicate replacement or rehabilitation.

Capital expenses are often large portions of a government or EMS budget. Most EMS systems may not get a significant amount of capital unless replacement schedules are due or an expansion of service is being implemented.

OPERATING EXPENSES

Operational, or maintenance, expenses should be divided into fixed and variable expenses. **Fixed expenses** are those aspects of the service that continue to be the same regardless of the level of activity. Fixed expenses include salaries, depreciation, preventive vehicle maintenance, and training. **Variable expenses** change with the level of activity and include unscheduled vehicle maintenance, station repair, medical supplies, linen, fuel, office supplies and reports, and accounting and legal expenses.

Training cost is a fixed cost and is often mandated by a regulatory agency, labor contract, or state law. Calculating training cost involves determining the total number of hours required for all aspects of the job for all personnel. This includes mandated EMS re-certification training, OSHA compliance training, and in-service training on new equipment. This includes the support services that are needed to maintain or deliver the training, such as a media technician to create instructional programs, an analyst, or clerical support for recordkeeping and photocopying.

FACILITY COST

A depreciation schedule should be used to determine the yearly and hourly cost of the agency's facilities. For example, if a new building or station is constructed at a cost of $750,000 and has a life expectancy of 30 years, the depreciation schedule per year is $62,500 or approximately $5,200 per month. A further breakdown of $175 per day and $7.25 per hour can assist in billing for the use of the facility. For example, a fire station is used by a wilderness fire team for a based or camp assignment. Federal regulations allow the site to be reimbursed for the use of that facility. A worksheet for facilities is shown in Figure 2.11.

CAPITAL EQUIPMENT COST

A similar depreciation schedule can be applied to **capital equipment cost.** Consider the cost of a new ambulance at $160,000 with a life expectancy of five years. The ambulance has an annual depreciation cost of $32,000 or $2,666.67 per month. Adding the cost of medical equipment, vehicle maintenance, insurance, fuel, and oxygen will give the agency a total cost of the

Name of Program:		Amount of Funds:			
Title of Position, Purpose of Item, Type of Equipment*	Project Time (FTE)	QTY	Salary, Rental, or Unit Cost	Budget	
*Equipment is defined as any individual line item with a unit (per each) cost of $2000 or more.					

FIGURE 2.11 ■ Facilities Worksheet.

- FICA.
- Retirement.
- Life and health insurance.
- Worker compensation.
- Disability insurance.
- Clothing allowance.
- Specialty pay.
- Educational incentives.

FIGURE 2.12 ■ Common Costs Associated with Dual-role, Cross-trained Firefighter Paramedics.

$54,000	Base salary.
$20,000	Benefits (usually 40% to 42% of base salary).
$5400	Paramedic incentive @ 10%.
$3000	Overtime for medical continuing education.
$2700	Retirement contribution.
$2500	EMS conference fees (optional).

FIGURE 2.13 ■ Sample Firefighter Paramedic Classifications.

vehicle. Since fuel, oil, and oxygen are variable expenses, three separate calculations should be made to be applied to vehicles that have low, medium, and high call volume.

PERSONNEL COST

Determining personnel cost requires more than just an hourly wage. Each pay scale, base salary, certification level, and incentive-pay category needs to be calculated. Each rank will also have several other costs associated with the calculation of the true cost of an employee (Figure 2.12).

Many systems have a pay-for-performance or incentive-pay program. The type and level of compensation assigned to a variety of incentives needs to be tracked. A budget item for that level of compensation should be calculated based on the triggers for incentives and the past practices.

Numerous systems exist by which the plan can be developed. However, the actual series of steps used does not matter nearly as much as the completed project and plan. The following process is a simplified method that can be adapted as needed. Line items or specific budget requests need to be placed for human-resource costs for EMS personnel. Pay differentials or incentive pay should be in the budget. In strategic planning, costs can be substantially different in systems in which paramedics and advanced EMTs receive premium pay as a percentage of their base salary as compared to having a classification as a paramedic.

Classification usually requires that the salary cost be applied to retirement, and the employer makes a contribution to the retirement plan (Figure 2.13). Premium pay, in most cases, is not figured into the retirement contributions. If recertification training is not conducted on duty, then overtime wages must be projected for the certification cycle. The Fair Labor Standards Act (FLSA) requires the employer to pay to maintain certifications needed for the job performance. Additional incentive monies may be allocated in a budget to reward paramedics and EMTs with seminar or conference attendance.

PROBLEM RESEARCH TECHNIQUES

Gathering information on the needs related to local EMS issues and operations is a lengthy task. Such information gathering is required in order to research community problems that the organization's strategic plan must address. For an EMS manager this means spending time meeting with individuals, interviewing EMS providers, looking through reports, and visiting with those affected by the issue. It is a very personal process.

This gathering of information requires the EMS provider to make contact with a variety of people from the community, including clergy, community leaders, members of the target audiences, other local health-care providers, social-service workers, emergency-services personnel, and so on. These contacts then become a library of information on the people and the problem. The local union, labor groups, and employees need to be part of the process to establish the strategic plan. Using a labor-management process in a collaborative approach to develop a strategic plan can avoid delays in the approval process and bring valuable resources that provide an insight from performing providers.

During the information gathering the EMS manager should attempt to answer six questions about the problem. Those questions are:

- Who is involved or affected by the item or budget request?
- What is the nature of the problem solved or service provided by the item or request?
- When will the item operate, and what will be the duration or life span?
- Where will the item be located, housed, or operated?
- How does the item impact the mission of the organization?
- Why does the item or request need to exist?

It would seem that the EMS manager is more of an investigative reporter than a strategic planner. These questions are very similar to those used by the media to get to the heart of a story, but in this case the questions are being used to get to the heart of a problem.

WHO IS INVOLVED WITH THE PROBLEM OR BUDGET ITEM?

This question seeks to identify the population group or groups directly affected by the problem or budget item. This is known as the primary target audience. Another group identified by this question is the secondary target audience. This is the group that has the ability to influence the risk factors associated with an EMS issue or budget item. For example, the primary target audience for children who are injured in bicycle crashes is the children riding the bicycles. The secondary target audience is the children's parents.

EMS managers must identify the nature of the problem or budget item. This question explains the specific problem being researched. In the bicycle injury scenario, for example, this might be the specific physiological damage done to the body by the injury as well as the process that leads to the harm. The budget item would be a line item that is designated to create or operate an injury-prevention program targeted at helmet use in children.

EMS managers need to identify when a budget item needs to be on line. This provides information on what time of day, what day of the week, what month of the year, and so on, that the problem or issue occurs. The goal is for a pattern to emerge, which would provide additional insight into why and how the issue or problem occurs. In the example of the bicycle-injury program, it may be one yearly event or an ongoing process.

Budget items need to identify where an item or operation is to occur. This identifies the physical location of the problem or issue. It may also provide insight into any sociocultural locations of the issue or problem. For example, activities for a bicycle-helmet program may be targeted to a specific neighborhood or a skateboard park.

EMS managers need to determine how a budget item is specified or what kind of budget will be applied. This is the heart of the matter. The answer to this question is all about understanding the chain of events leading to the problem or issue and is essential for developing appropriate interventions. EMS managers can establish an ongoing program where the budget item is recurring with adjustments; some can be budgeted under a sunset clause, meaning that the program is funded for only a certain cycle or given time frame.

The need for this item must be justified. This is always a tough task, but the root causes of the problem are identified with this information. The answer to this question is a result of examining all the other information provided through the questions. The answer should reflect the big picture of the issue and make an impact. Often, this is done using a cost-benefit analysis.

Once an initial summary and matrix has been designed and a solution drafted, EMS management needs to solicit feedback from the stakeholders. It is important to include those affected by the management decision in the process of how the work or project will be implemented. Organizations that do this have greater support and employee buy-in of the organizational processes. Governments can use a variety of budgets, as shown in Figure 2.14.

A **government budget** is a legal document that is often passed by the legislature, city council, fire or EMS board, or county commission and approved by the chief executive. For example, only certain types of revenue may be imposed and collected. Property tax is frequently the basis for local revenues, while sales tax may be the basis for state revenues, and income tax and corporate tax are the bases for national revenues. The two basic elements of a budget are revenues and expenses.

Government budgets have an economic, a political, and a technical basis. Unlike a pure economic budget, they are not entirely designed to allocate scarce resources for the best economic use. They also have a political basis, wherein different interests push and pull in an attempt to obtain benefits and avoid burdens. The technical element is the forecast of the likely levels of revenues and expenses.

A **line-item budget** is simple and provides an appropriation that is itemized on separate lines in the budget. A **performance-based budget** uses statements of missions, goals, and objectives to explain why the money is being spent. It is a way to allocate resources to achieve specific objectives based on program goals

---

- ◆ Line-item budget.
- ◆ Performance-based budget.
- ◆ Zero-based budget.

FIGURE 2.14 ■ Types of Budgets.

and measured results. A **zero-based budget** is a technique of planning and decision making. It reverses the working process of traditional budgeting. In traditional incremental budgeting, department managers need to justify only increases over the previous year's budget. This means that what has been already spent is automatically sanctioned.

There are three critical factors organized as a triangle—performance, time, and cost—with the scope of the project in the middle of the triangle. Performance levels are one of the primary concerns; reviewing performance shows how cost or time reductions can increase or decrease the performance of a budgeted program. As the cost moves more to the right, the performance decreases or increases.

The time component is a dependent factor when time is short and resources are in higher demand. When time is stretched out, people lose interest in the project, so when a program begins to lose momentum, a concept called "crashing" can be applied to a project. Crashing involves applying a significant, sudden increase in resources to accelerate the time it takes to complete a project. This increases the cost significantly and may not affect performance.

Cost is based upon a relationship with performance and will increase as performance requirements do (e.g., unit hours, response-time standards). If more resources are required to meet response times, cost will increase. The inverse is also true: if response times or calls decline, cost will decrease.

All of these are interrelated factors and define the scope of a budget item or project. If one of the components changes, it impacts the components of the system.

PROJECT AND PROGRAM MANAGEMENT

Most goals, budget items, and operations in a strategic plan are broken down into programs or projects. The EMS manager needs to manage these projects and programs through a systematic approach. Projects need to be defined, and stakeholders need to be identified. EMS managers must then analyze, implement, evaluate, and readjust the program or project.

■ TOOLS FOR STRATEGIC PLANNING

Several tools are available to assist an EMS manager with projects considered to be part of the strategic plan. These strategic-planning tools help monitor projects and programs to allow EMS managers to analyze and readjust. Many of these tools are used in manufacturing, government contracting, and the service industry.

CRITICAL PATH METHOD

A popular strategic-planning tool created by Dupont Chemical to facilitate the restarting of chemical plants has become widely accepted as a way to get critical and time-sensitive processes on line. The **critical path method (CPM)** provides a graphic view of projects and predicts the time required to complete a project by showing which activities are critical to maintaining the program or the construction of the physical plant. The CPM shows each activity as a node or circle within a network. The beginning or ending of an event or activity is depicted as lines between the nodes.

The CPM has a six-step process that can be used to develop a project or program. The first step is to specify the individual activities by making a list of all the activities in the project or program. This list will have further details applied to it in a later process. The sequence of how those activities are put together is important, as some activities must be done first or in sequence. Any subtask needs to be added to the role of the main activity. A network is then drawn connecting all of the nodes or activities. A time frame to complete each activity can be estimated using previous times, and a single number is used to ensure the completion time. No provisions are made for a range of time in order to stay focused on the completion of the budget cycle.

The critical path is then identified—it is the path of longest duration through the network. The activities on the critical path cannot be delayed without delaying the entire program or project. The critical path requires that several time issues be identified. First, identify the earliest start time (ES) at which the activity can start, given that its precedent activity must be completed first. The earliest finish time (EF) is the earliest start time for the activity, plus the time required to complete the activity. The latest finish time (LF) is the latest time at which the activity can be completed without delaying the project. And finally, the latest start time (LS), which is equal to the latest finish time minus the time required to complete the activity.

The critical path is the path through the project network in which none of the activities have slack and the path is ES = LS and EF = LF for all the activities on the path. A delay in the critical path delays the entire project. If the project or program needs to be accelerated, it is necessary to reduce the total time required for the activities in the critical path.

PROGRAM EVALUATION AND REVIEW TECHNIQUE

The **Program Evaluation and Review Technique (PERT)** is a way to organize activities that have more random completion times. PERT is generally used for complex, strategic projects that may take years to develop and bring on line. It incorporates and builds on the CPM and can reduce cost and time to complete a project. One of the main differences is that it takes into consideration variations in the time it takes to complete a task. Figure 2.15 is an example of a PERT chart used for the construction of a new fire or EMS station. It is excellent for construction projects and vehicle acquisitions.

When making a PERT project, an *activity* is a task that must be performed and an *event* is a milestone marking the completion of one or more activities. Before an activity can begin, all of the precursor activities must be completed. The PERT diagram contains circles that are called nodes. These nodes start small and gradually increase, with the ending node being the largest. The nodes are placed originally in increments of 10, which allows an EMS manager to add nodes if the original plan needs more detail or a task needs to be added. The activities are labeled with letters and the expected times required to complete the activity.

PERT is useful in strategic planning because it helps identify project completion time. PERT identifies the critical-path activities that can delay a project or directly impact the completion time. PERT has some limitations, in that activity times are often estimates and this often causes underestimation of the expected completion time. PERT has a six-step process, which is displayed in Figure 2.16.

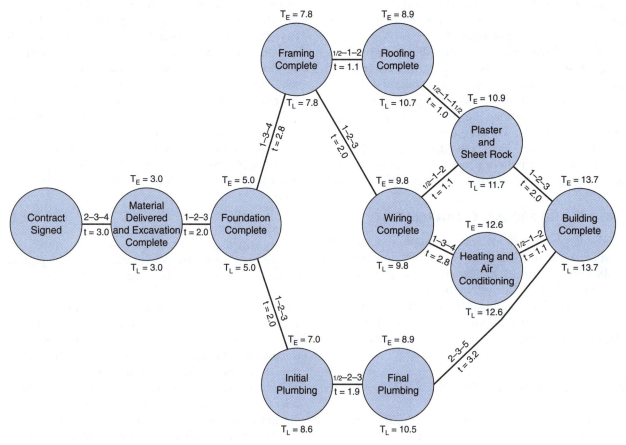

FIGURE 2.15 ■ PERT Chart for Constructing an EMS Facility or Fire Station.

1. Identify the specific activities and milestones.
2. Determine the proper sequence of the activities.
3. Construct a network diagram.
4. Estimate the time required for each activity.
5. Determine the critical path.
6. Update the PERT chart as the project progresses.

FIGURE 2.16 ■ The PERT Planning Process.

PERT charts can be created with any project-management software. And managers should learn how to use a software program that can create the charts with simple data input from staff members or other personnel working on projects. PERT charts are excellent tools for specification projects, such as ordering ambulances, building training programs, or specialty services like tactical medic programs.

To create a more detailed schedule, a manager or project leader may want to build a Gantt chart, a matrix that lists on the vertical axis the entire task and on the horizontal axis the time the task will take. Tasks that are linked can be connected in the project and expressed in different time units. Figure 2.17 illustrates a Gantt chart used to create a plan for the acquisition of a new ambulance.

■ STRATEGIC PLAN AS A LIVING DOCUMENT

Two different types of change are proactive change and reactive change. Proactive change occurs when managerial decisions are made to try new things in anticipation of changes in the environment (i.e., planned change). Reactive change involves decisions about change that are made in response to crisis events in an organization.

Following are the five types of planned change:

♦ Strategic change is change involving the strategic direction of an organization in terms of products, services, markets, and distribution.

♦ Technological change involves altering the organization's equipment, engineering processes, research techniques, or production methods.

♦ Structural change focuses on increasing organizational effectiveness through modifications to the existing organizational structure.

♦ Personnel change focuses on increasing organizational effectiveness by upgrading the skills and knowledge of employees.

♦ Cultural change affects the set of shared values, attitudes, beliefs, and assumptions that serve to shape the behaviors and guide the actions of all employees.

Organizations are much like human systems and comprise interdependent elements that are in sequence. Efforts to change one segment of the service or organization are likely to affect other segments. Any effort to change only one area without concern for other parts of the system is apt to result in failure. An intervention involves special actions taken within the organization to bring about specific changes. There are a number of different interventions (such as surveys, team building, job enrichment, process consultation, teamwork training, or sensitivity training) that can be initiated by EMS managers with employees or with the assistance of consultants. A common error made by EMS leaders is to discount the ideas and suggestions of the field personnel when aspects of the service need to change.

Whatever combinations of interventions are used, a collaborative approach with labor and management is the best method available for aiding organizational change. Culture always emerges as one of the most common barriers to change. Because the actions of employees have been programmed over a long period of time, employees tend to get acclimated to particular ways of doing things and will resist change, which, of course, makes changing the culture difficult. Often, because culture is so pervasive, changes must occur to support changes in other parts of the organization.

A learning organization is one that has woven together a continuous and enhanced capacity to learn, adapt, and change its culture. It is one in which change and improvement are institutionalized to help the company more effectively and efficiently meet its day-to-day challenges as well as its long-term goals. Given the times, opportunities, and employee expectations in today's EMS organizational environment, employee-centered management—with participation of management and labor in the pursuit of the mission statement—seems to be the preferred management model. A leader can only lead if followers follow, and a significant part of this relationship is allowing the followers to actively participate.

BUSINESS PLAN

A **business plan** is an accountability program for the current budget year and sets the stage for multi-year budgets and operational targets. Several commercially

ID	Task Name	Duration	Start	Finish
1	Assemble Committee	28 days	Mon 12/3/07	Wed 1/9/08
2	Review Budget and Needs	3 days	Tue 12/4/07	Thu 12/6/07
3	Review KKK Standards	3 days	Wed 12/5/07	Fri 12/7/07
4	Determine Specifications	1 day	Mon 12/3/07	Mon 12/3/07
5	Determine Chassis	0 days	Mon 12/3/07	Mon 12/3/07
6	Type of Ambulance	1 day	Tue 12/4/07	Tue 12/4/07
7	Brand Eligibility	1 day	Wed 12/5/07	Wed 12/5/07
8	Interior Layout	2 days	Thu 12/6/07	Fri 12/7/07
9	Cabinets	1 day	Mon 12/10/07	Mon 12/10/07
10	Color Scheme and Paint	0 days	Mon 12/10/07	Mon 12/10/07
11	Warming Equipment	1 day	Tue 12/11/07	Tue 12/11/07
12	Electrical Options	1 day	Wed 12/12/07	Wed 12/12/07
13	Factory Upgrades	1 day	Thu 12/13/07	Thu 12/13/07
14	Climate Controls	1 day	Fri 12/14/07	Fri 12/14/07
15	Turnout Storage	1 day	Mon 12/17/07	Mon 12/17/07
16	Submit Initial Spec	1 day?	Tue 12/18/07	Tue 12/18/07
17	Open for Bids	7 days	Wed 12/19/07	Thu 12/27/07
18	Bid Review and Award	9 days	Fri 12/28/07	Wed 1/9/08
19	Initial Pre-Build Meeting	0 days	Wed 1/9/08	Wed 1/9/08
20	Site Visit Mid Build	1 day	Wed 1/30/08	Wed 1/30/08
21	Water Test	0 days	Wed 1/30/08	Wed 1/30/08
22	Ambient Temperature Test	0 days	Wed 1/30/08	Wed 1/30/08
23	Sound Test	0 days	Wed 1/30/08	Wed 1/30/08
24	Order Gurney and Delivery	35 days	Tue 12/25/07	Mon 2/11/08
25	Order MDT and Radios	35 days	Thu 12/13/07	Wed 1/30/08
26	Acceptance Visit	0 days	Wed 1/30/08	Wed 1/30/08
27	Quality Test	0 days	Wed 1/30/08	Wed 1/30/08
28	Road Test	0 days	Wed 1/30/08	Wed 1/30/08
29	Workmanship Test	0 days	Wed 1/30/08	Wed 1/30/08
30	Transport to Destination	7 days	Thu 1/31/08	Fri 2/8/08

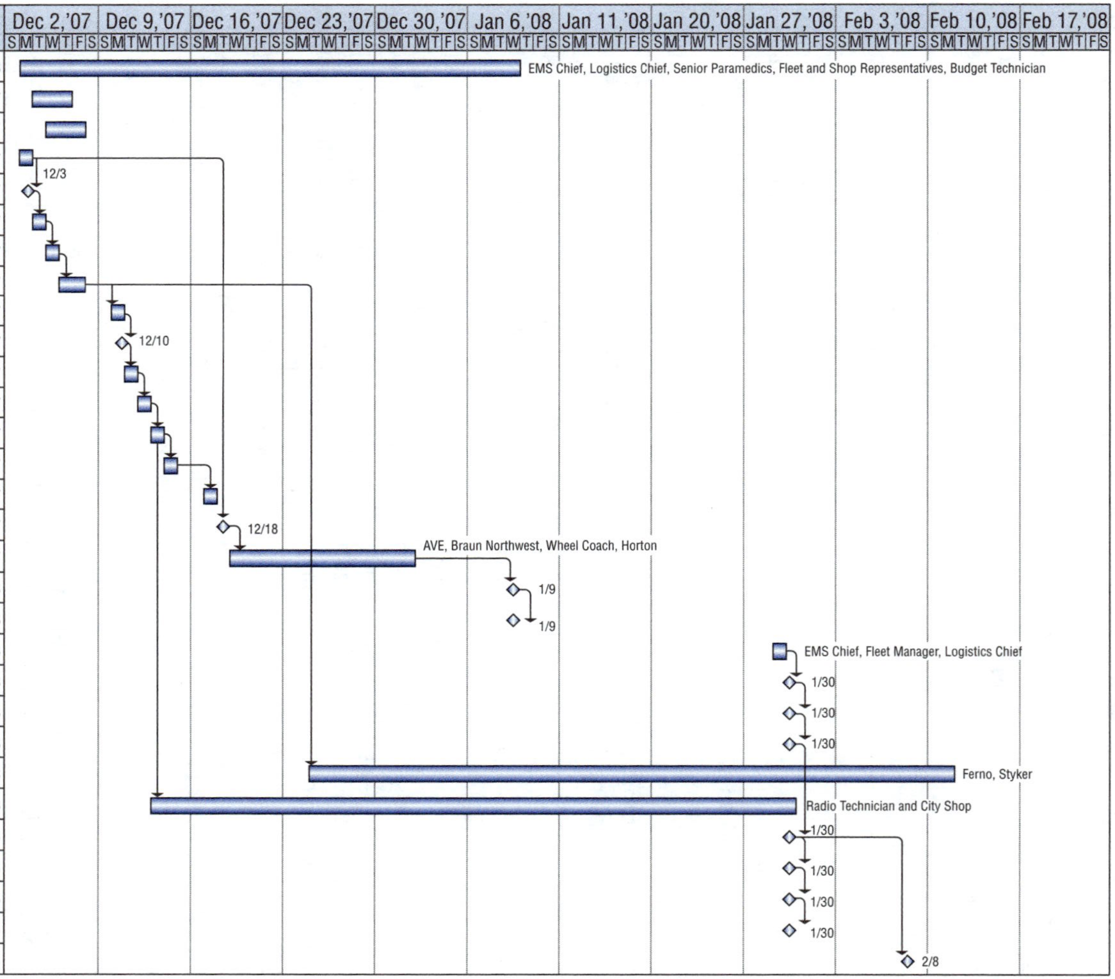

FIGURE 2.17 ■ Gantt Chart Used to Create a Plan for the Acquisition of a New Ambulance.

available software programs provide templates for business plans, and while government budgeting is slightly different from budgeting in private industry, the principles are the same.

Creating a business plan is the process of arriving at a document that outlines how the organization will achieve its objectives in conjunction with the fiscal constraints set by the budget process. An EMS business plan is a joint action taken by an administrative body such as a government or board of directors and the EMS agency or department that outlines the major task, performance level, and associated cost. Business plans generally have four distinct sections: the description of the business, marketing, finances, and management.

The benefits of conducting a business plan include keeping EMS leaders engaged in the process of critical thinking and ensuring political priorities are in line with the EMS organization's goals. A business plan acts as a compunction tool to local governments, citizens, and EMS workers (Figure 2.18). Business plans also help set the guidelines for performance measurement. If the department's business plan is approved, then the budgetary approval follows shortly after.

A formalized and approved business plan needs to be reviewed at regular intervals. A business analysis or audit compares the EMS system performance to recognized industry benchmarks and the business plan. A business audit includes a review of financial performance and a global analysis of the prevailing rates and costs. Business audits also look for opportunities for **cost shifting** to direct the cost of service to other providers or health-care agencies. For example, supplies are restocked at the hospital, and costs are shifted to the hospital or the patient's insurance or hospital bill. Since ambulance-transport reimbursement no longer allows for itemization, shifting those costs to the hospital—which can bill for individual items such as medications—is a good business practice.

- ◆ Executive summary.
- ◆ History, mission, and goals.
- ◆ Analysis of services.
- ◆ Information technology.
- ◆ Human-resource planning.
- ◆ Asset management.
- ◆ Finances.
- ◆ Performance measurement.
- ◆ Marketing and public information.
- ◆ Appendix.

FIGURE 2.18 ■ Components of a Government Business Plan.

Any review of the business plan needs to measure and ensure adequate resources are in reserve. Any business plan needs to ensure that the funding is matched to resources to support the ongoing EMS activities. Any business plan should also be able to identify expansion and growth needs in the service.

Marketing is an important element of the business plan. It has a strategic function, which is to convince city or county managers and the public of the value and effectiveness of EMS services. A specific market strategic plan takes into account an analysis of the agency and the external environment. Figure 2.19 shows a process by which a strategic marketing plan is carried out in an agency. It involves analysis, strategy, and implementation.

QUALITY PLAN

Quality planning is the successful design or redesign of a system to perform to the quality standards expected by patients or other stakeholders. In the quest for continual improvement, strategic planning can be closely linked with quality planning and combined into a single organization- or agency-wide planning process. It is called *strategic quality planning*.

Strategic quality planning is simply an organized method of determining where an EMS system or organization wants to be and how it plans to get there. Strategic quality planning is an integral and ongoing part of the system. It involves the careful integration of all components of the EMS system, including clinical performance, financial support, legal authority, personnel management, education and training, and data collection and analysis. Individual components are mutually interdependent; planning and evaluation of one component cannot occur in isolation from the others.

EMS systems involve many different organizations and individuals with separate authorities, management, and governing bodies, each of which may have its own strategic quality-planning process. EMS often involves organizations and individuals not traditionally viewed as health-care providers (such as law-enforcement personnel, dispatchers, and the general public). Yet it is because of the diversity of the organizations involved that strategic quality planning is imperative to the overall improvement and smooth functioning of the entire EMS system.

The activities of each level (local, state, and regional) of EMS are different but complementary. Strategic quality planning, as well as the entire quality improvement (QI) process itself, should occur at

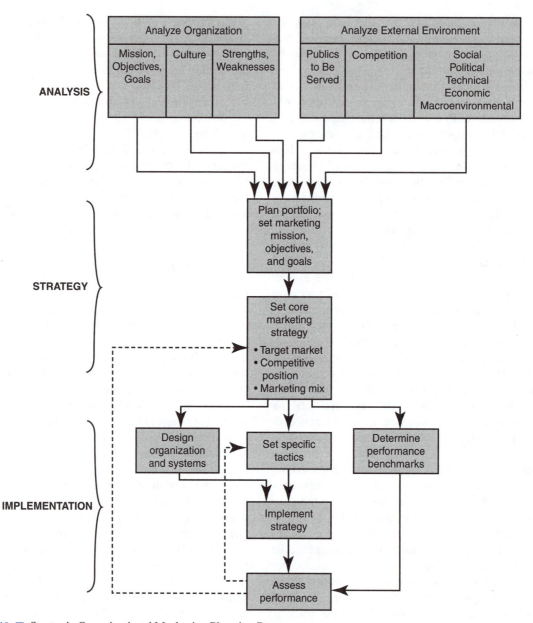

FIGURE 2.19 ■ Strategic Organizational Marketing Planning Process.
(Andreasen, Alan; Kotler, Philip; Strategic Marketing for Nonprofit Organizations, *6th, ©2003. Electronically reproduced by permission of Pearson Education, Inc., Upper Saddle River, New Jersey.)*

all levels. Just as the state EMS system must support the local EMS system, these activities should be compatible with an overall statewide vision and mission.

Strategic Quality Planning at the State and Regional Levels

Strategic quality planning for EMS at the state level differs from strategic planning at the local or agency level. Unlike the process in the private sector, there are no competitive or entrepreneurial demands placed on a government agency. Instead the state EMS agency is charged with creating or maintaining public policy and meeting the needs of different types of citizens, as well as the patient, hospitals, and health-care providers.

The state EMS agency is typically charged with designing or in most cases redesigning systems of care that will lead to optimal patient outcomes. The state EMS agency is also charged with identifying demographic and economic issues that will impact the delivery of EMS care and planning to meet these changing needs. Whether it involves the design of a trauma system from the ground up or the redesign of a system that has evolved over several decades, strategic quality planning can help the state EMS agency better meet the needs of the EMS community and those they serve.

The state EMS strategic quality plan serves as the road map for achieving quality improvement in EMS for the entire state. One key to successful strategic quality planning at the state level is to involve all those individuals and organizations that will be affected by the plan. A strategic quality plan that lacks input and buy-in from those affected will only gather dust on the shelf.

In order for a state EMS plan to be a useful tool in measuring performance across the state and improving service to the patient and other stakeholders, it must be all of the following:

- Monitored and revised frequently according to the results obtained and in response to the changing health-care environment in the state.
- Simple, easy to use, and applicable to both rural and urban settings.
- Achieved through consensus building throughout the state with local and regional EMS agencies, hospitals, EMS providers, patients, and other stakeholders.
- Conveyed to all members of all involved organizations.
- Easily and quickly changed to reflect new demands and changing conditions.

State and Regional Vision

A vision statement provides a futuristic look at and broad guidance for the state or regional EMS agencies or system. In simple terms the vision statement helps ensure that everyone is going in the same direction. Each objective and action plan that is subsequently devised is consistent with the vision statement, thus ensuring constancy of purpose and compatibility of actions. A state EMS vision statement can help unify EMS agencies. The components of mutual- and auto-aid agreements and disaster-response agencies are derived from a clear sense of purpose identified by state and regional EMS leadership.

While development of the vision statement is directed by the leaders of the state, regional, or local EMS system, the system or agency "players" should be deeply involved in the development process. People are more likely to help implement what they help to develop. Typically, a vision statement would be a short, motivational description of the EMS system's ideal condition. The vision statement can serve as motivation for those involved in the system, and it can be a steady guide through the numerous changes necessary to achieve a quality system. The state's vision statement is broad, encompassing the entire statewide EMS system. The regional system's vision statement should be unique to that system and consistent with the state vision statement.

Strategic Quality Planning Structure

The right strategic quality planning structure is based upon the organizational characteristics of the EMS system. The structure must account for the fact that strategic quality planning is an ongoing process, based on the principles of quality improvement and involving the EMS system's organizational, financial, and clinical aspects. Strategic quality planning is not something new and different that requires a separate system or a separate process; instead, strategic quality planning is a process that drives all planning and all quality-improvement efforts. The state or regional strategic quality planning structure must take into consideration existing planning and quality mechanisms, including state EMS advisory councils, local EMS councils, health-care advisory councils, and other specific statutory or regulatory requirements.

CHAPTER REVIEW

Summary

Strategic planning is a critical process for successful EMS agencies. Effective strategic planning is built upon four key principles: development of a vision, strategic planning, a business plan, and quality planning. These principles, and the philosophy they represent, are the foundation for effectively defining the issues and problems affecting EMS and understanding the target audiences. EMS managers should take time to review the principles and consider how they apply to themselves, their organization, and their community.

EMS issues are seldom simple. Rather, they are complex interactions among the characteristics of the person or group involved, the problem, the physical environment where the person or group lives or works, and the sociocultural environment of the person or group. Effective assessment of the people and the problem requires unraveling this relationship into distinct, understandable parts. The information from an effective assessment allows the EMS man-

ager to hit the target bull's-eye with the intervention arrows.

The assessment of the people and the problem is a basic, straightforward process. The EMS manager can accomplish it successfully if there is a system in place that relies on concrete performance indicators. Assessment of the people and the problem is a personal process; it requires meeting with the members of the group, health-care providers, social-service providers, and others who work with, or are part of, the people and the problem. It cannot be done simply by reading reports and reviewing data; there must be face-to-face contact.

Research and Review Questions

1. Identify and provide a sample budget for a project or service-delivery item in your jurisdiction. Identify what kind of budget system is used.
2. Create a goal and plan for one project within your scope of practice, and design it to be integrated into the strategic plan.
3. Write a short-term plan for adding a peak-load rescue to an EMS system using a firefighter-and-paramedic team to be deployed in the system during your agency's highest call volume of the day. You may not deploy this on a 24-hour basis.
4. Conduct a SWOT analysis on the potential for a bicycle-medic program in your community.
5. List the components of the PERT planning process, and apply the implementation of a new station with an EMS vehicle.
6. List the components of a government business plan, and apply it to a request by a neighboring community of 20,000 people to take over the emergency response.

References

Andreasen, A. R., & Kotler, P. (2003). *Strategic marketing for nonprofit organizations* (6th ed.). Upper Saddle River, NJ: Prentice Hall.

Bryson, J. (1995). *Strategic planning for public and nonprofit organizations* (Rev. ed.). San Francisco: Jossey-Bass Publishers.

Kaplan, R. S., & Norton, D. P. (2006, March). How to implement a new strategy without disrupting your organization. *Harvard Business Review, 100*.

MacKay, J. M. (2004). *Does strategic planning still fit in the 2000s?* Toronto, Ontario, Canada: The Berkeley Consulting Group.

Wallace, M. (1998). *Fire department strategic planning: Creating future excellence by fire.* Saddlebrook, NJ: Engineering Books and Videos. A Division of Penwell.

Manager to Leader

Accreditation Criteria

CAAS

103.03 Management Development: The agency shall demonstrate its commitment to the ongoing development of its leadership (pp. 60–72).

Objectives

After reading this chapter the student should be able to:

3.1 Define the difference between management and leadership (p. 61).
3.2 Identify the skills needed to be an EMS manager (pp. 53–54).
3.3 Identify the management processes and sentinel events in EMS operations (p. 55).
3.4 Develop and define leadership activities and roles (pp. 60–64).
3.5 Understand how to create a vision and values statement (pp. 61–62).
3.6 Apply the concept of values to organizational leadership (p. 51).
3.7 Understand and apply the principles of mentoring and coaching to EMS workers (pp. 68–72).

Key Terms

coaching (p. 57)
culture (p. 51)
hygiene factors (p. 58)
leader emergence (p. 61)

leadership (p. 50)
management (p. 50)
mentoring (p. 71)
motivating factors (p. 58)

power (p. 52)
sentinel event (p. 55)
theory X (p. 58)
theory Y (p. 58)

Point of View

Managers who are examining whether or not they have what it takes to become good leaders must thoroughly examine their ability to stand back and look at issues and people in a broad sense. Do you have the ability to remove yourself from the emotion or politics that may be involved in an issue or decision? The best analogy to this question relates to a cardiac monitor and how many "leads" you have the ability to examine. A good manager looks at a situation and examines it with a more straightforward approach, a four-lead view. He or she can identify those issues that have the ability to cause immediate problems. This manager, however, often isn't able to see the whole picture and identify hidden issues or perceive future problems that might result from the decision.

A good leader has the ability to not only look at the underlying "rhythm" of a situation but also to dig deeper to see what else might lie beneath and affect an organization as a result of his or her decision. A good leader is able to take a 12-lead view of the issue and hopefully make decisions that not only have a positive effect on the immediate situation but also on future issues that may result from the decision.

A good leader also has to master two basic principles. Look at issues as if they were a glass half full, not half empty. Life is way too short to approach issues with a negative view. Finally, surround yourself with good, trustworthy people. I have been blessed with a great family, great administrative staff, and great managers who make my job so much easier and have given me the ability to excel in my career. You can't do it alone.

Jeffery Dumermuth
Chief, West Des Moines (Iowa) EMS

■ ORIGINS OF MANAGEMENT

The concept of managing goes back to Stone Age times for humans. There was always a set of given tasks that must be performed for primitive humans to survive. The gathering of wood for heating, lighting, and cooking; foraging for food; hunting wild animals; and planting and harvesting crops are just a few of those priorities. What if the priorities were out of order? Maybe they lit the fire before getting any food, or they planted crops in the fall. The outcomes of a few mistakes could have a catastrophic effect on the plight of the group, which is not unlike an emergency service organization. What has just been described is the nature of work, which is defined as the division of labor to achieve an outcome. The overall direction of that process—prioritizing jobs and applying the right combination of resources and talents to achieve an acceptable outcome—is what we call **management.** The manager in the cave paintings could always be identified as the person with the biggest club.

In the past, management typically sets rules, policies, and procedures, and employees followed them. Today many organizations seek to take full advantage of their employees' talents and abilities and to make the best use of everyone's time. It makes sense for those who best understand work processes and improvement opportunities to make the decisions. Therein lies the challenge, because we typically do not prepare new supervisors or operations officers for their new duties, which require a completely different skill set, as well as management and leadership abilities. It is often observed that even the most clinically competent paramedic can make the worst supervisor.

Management is a function that focuses on the delivery of good emergency medical services and patient care. The processes necessary to meet the needs of the organization should be managed by a team. Different people can take different parts of the EMS management function. For example, one team member may do a budget, another may monitor quality, and team members may take turns facilitating meetings. If the organization's training is good, then even the weakest manager can survive.

It is important to shift thinking toward reinforcing the team approach—a process that is not always reinforced in clinical EMS training. So why do we need managers and leaders? The challenge being placed on EMS in general will make the job tougher in the future. Managers and leaders are needed to ensure that EMS personnel remain motivated and efficient.

The EMS mission statement will identify the organization's purpose and help you answer the question, "What business are we in?" Certainly EMS is in the business of saving lives and mitigating injury and illness. So what is the big deal? Everyone understands this concept. The challenge is in seeing it from a common viewpoint so that everyone on the team can pursue the same vision. Vision is "seeing something that other people don't see." The difficulty is getting the organization to see the vision, understand it, and implement it. This may sound simple, but only about 5% of all emergency-service organizations get to this point.

■ THE ROLE OF MANAGEMENT

First and foremost, management must understand the mission of the organization. A mission statement is important because it conveys the organization's guiding principles or objectives to employees and customers. It is the basis for the organization's actions. It also galvanizes decision making so that there is consistency, whether the decisions are those of employees or managers. It is necessary for management to align with the mission (Figure 3.1). EMS managers and leaders need to integrate the organization's mission into the values and culture of the organization. An important task of managers and leaders is to define themselves by developing a personal mission statement.

VALUES STATEMENTS

In addition to the mission statement, it is helpful for an EMS agency to list and explain the organization's beliefs by creating a values statement—a written description of the beliefs, principles, and ethical guidelines that direct an EMS organization's planning and operations. The values statement should reflect what the organization and its leadership believe and what they are trying to do. The values statement helps promote the culture of the organization and creates a positive work environment. An EMS organization's values should not change; however, its culture and operations must constantly change. It is the responsibility of the managers or leaders to work to make sure their organization is aligned with its values and vision.

CULTURE

Culture can be a pattern of behavior that helps remove uncertainty for employees and provides a con-

Step 1:	*Identify past successes.* Spend some time identifying four or five examples where you have had personal success in recent years. These successes could be at work, in your community, at home, and so on. Write them down. Then, try to identify a common theme—or themes—to these examples.
Step 2:	*Identify core values.* Develop a list of attributes that you believe identify who you are and what your priorities are. The list can be as long as you need it to be. Once your list is complete, see if you can narrow your values to five or six of the most important ones. Then, finally, see if you can choose the one value that is most important to you.
Step 3:	*Identify contributions.* Make a list of the ways you could make a difference. In an ideal situation, how could you contribute best to the world in general, your family, your employer or future employers, your friends, and your community?
Step 4:	*Identify goals.* Spend some time thinking about your priorities in life and the goals you have for yourself. Then make a list of your personal goals, perhaps in the short term (up to three years) and the long term (beyond three years).
Step 5:	*Write your mission statement.* Base it on the first four steps and the better understanding of yourself that you achieved.

FIGURE 3.1 ■ How to Build a Personal Mission Statement. *(Original article written by Randall S. Hansen, Ph.D. Complete article can be found at: http://www.quintcareers.com/creating_personal_mission-statements.html. Copyright: Quintessential Careers.)*

text for doing the organization's work. Culture provides employees with cues that guide them in performing their functions. These cues often originate with top management and pervade all parts of the organization. Thus culture influences employee behavior by giving cues as to what is, and is not, appropriate behavior. The influence of culture on all employees throughout the organization will affect the organization's overall performance. The values, beliefs, and assumptions that underlie the organization's culture, coupled with the strength of that culture, will determine whether or not an organization will successfully continue to meet the challenges posed by the environment.

Culture is a pattern of artifacts, behaviors, values, beliefs, and underlying assumptions that is developed by a given group of people as they learn to cope with internal and external problems of survival and prosperity. In other words, the members of the organization have a common set of values and beliefs that enable them to come together to communicate and work together effectively. Their culture is rooted in the community in which the agency serves.

Culture, as a body of learned behaviors, is specific to a given agency or EMS organization and acts rather like a template (i.e., it has predictable form and content), shaping behavior and consciousness within the agency from generation to generation. Culture resides in all learned behavior; in some organizations, it shapes its template or consciousness prior to behavior as well (that is, a "cultural template" can be formed in an individual's initial EMS training).

This shaping of a cultural EMS template and body of learned behaviors is based on a system of meaning, a set of relationships between one group of variables (such as words, behaviors, physical symbols, and so on) and the meanings that are attached to them. Relationships in meaning systems are arbitrary: there is no particular reason why the word "bus" should refer to an ambulance, for example. However, when an organization agrees on certain relationships between a certain class of variable (like words or behaviors) and meanings, a system of meaning is established. Language is perhaps the most formal of human meaning systems. At the same time we all know what it means to wink at someone or to give someone "the finger." This suggests that human behavior, like language, can be a part of a complex and established system of meaning or an organizational culture.

Leaders and managers often set the stage for culture by establishing the mission, vision, and goals that give the organization direction. Given their position of authority, managers are often able to influence the organization's culture; that is, they have the ability to ensure that the founder's vision is passed on to future generations (e.g., they may be responsible for socializing new employees). Managers can also use their **power** to change organizational culture by creating new mission statements, new visions, and new goals. Managers must often do just that in response to changing economic conditions. Because the culture can facilitate or inhibit the organization's performance, the manager's role in shaping and transmitting the culture is an important one.

Often management and leadership styles are dependent on understanding the organization's culture. A strong cultural environment has both advantages (pros) and disadvantages (cons). The pros and cons of strong culture are as follows:

Pro: Strong culture implies uniformity or consensus among all organizational members. As a culture matures, it becomes stronger and leads to greater uniformity of behavior. Other pros of strong culture are improved communication and diminished need for bureaucratic rules and autocratic controls.

Con: Strong culture may also encourage stagnation, narrow-mindedness, and the "not invented here" or "we never do it that way" syndrome, in which employees are uninterested in, or feel threatened by, external ideas. These characteristics can lead to the death of an organization. In "groupthink" managers feel the environment cannot sustain debate and is intolerant of other ideas.

Because they are intimately familiar with the processes by which they get their work done, employees often understand decision-making situations better than their managers. Thus, in an organization that seeks to take full advantage of its employee experience, knowledge, and abilities, it makes sense to delegate decision making to those who understand the particular problem or improvement opportunity best.

It is critical that everyone involved have a clear sense of the organization's processes and mission, as well as how each employee's work fits in with that of others in achieving the organization's mission. A simple test of this concept is to ask an organizational member from the lowest seniority pool, "What is the mission statement, and how do you apply it to your customers?" If the answer is anything but clear and concise, there is indeed an organizational values void that must be addressed, because this person faces customer-service opportunities on a daily basis.

MANAGING POWER

The line between management and staff is drawn by the projection of power. Power is defined as the ability to get work done. Power is a necessary thing in organizations and society, although many times it has a negative connotation. Abuse of power and position is the most likely cause for this negativity, and potential abuse should be monitored at all times in an organization. In EMS there can be many types of power. It is the application of this power that helps the organization achieve its mission. In the simplest of terms, it is power that runs every engine. An EMS manager can obtain power by one of the five methods described in Figure 3.2. For an EMS manager to be successful and prepared to move up to a leadership role, it is important to develop some aspect of each type of power.

1. *Reward power* is provided by the organization, giving the manager access to various rewards. It operates when employees believe their manager will reward them for compliance and for performing various tasks.
2. *Coercive power* is provided by the organization, giving the manager the ability to deliver punishment. It operates when employees believe that a manager can deliver punishments when they do not conform to the manager's directives.
3. *Legitimate power* is provided by the organization, giving the manager certain rights and authorities. It is evidenced when employees believe that managerial positions automatically afford managers certain rights and authorities.
4. *Expert power* is gained through the manager's own mastery of technical information, experience, and specialized skills. It operates on the employee beliefs that the manager has specialized knowledge or skills to accomplish goals and objectives.
5. *Referent power* is gained through the manager's own ability to prove he is trustworthy and likable, such that subordinates want to identify with the manager. It is a function of the degree to which the manager's employees identify with and respect the manager.

FIGURE 3.2 ■ The Five Bases of Power. *(Source: The National Fire Academy.)*

It is assumed that the manager or leader will be inclined to use power for win–win solutions and to use interpersonal skills effectively in applying that power. A manager's or leader's effectiveness is based on the cooperation of others; power minimizes this dependency. The unique thing about power is that it must be given away to get it back. Effective use of power has evolved in modern management to something called empowerment. Empowerment means giving people responsibility for their task or work and including with it the ability to make change and to do it their way. It involves ownership of the quality and content of the work.

■ BASIC MANAGEMENT SKILLS

EMS managers usually consist of people who have experience in the fire service, EMS, or health-care field. Often they are people who have worked their way up through the agency. An EMS manager knows how the EMS system works and has good technical knowledge, but frequently, fire and EMS services make the mistake of promoting their best EMTs and paramedics into management, only to find out they lack the skills or the control over results that they had as clinicians.

EMS providers should inventory their skill sets before making the decision to move into the management role. A specific set of basic skills is needed for frontline EMS managers that enables a person to manage the processes in EMS operations.

If a broad group of members in a given EMS organization were questioned about their performance, their answers would probably not include comments such as, "I'm here to do the minimum job I can today" or "I hope to have a mediocre career and retire from the department at the lowest level" or "I've learned all I can, and now I just want to coast for the next 20 years." However, if we were to analyze any organization, chances are there would be members who had outcomes that match those statements.

So how does one accomplish excellence and success in one's management and leadership role and not fall prey to these conditions? There are seven essentials that will help ensure success in a manager-leader position (Figure 3.3). Mastery of all seven—keen perception, broad-based knowledge, critical thinking, mature judgment, clear communication, common sense, and ethical conduct—is necessary to be an effective leader.

Keen perception is the ability to act or become aware by using one's senses. This means that EMS managers are able to take in information around them and act on it accordingly. Many times, we switch off surrounding inputs and available information. One example is being focused on a project or activity to the point of screening out all other available information. To be aware of opportunities and threats, the EMS manager must be skilled at sensing the environment, whether it is political, personal, or organizational in nature.

Broad-based knowledge means having a wide base of information and knowing how to apply it. After we learn and master a task or vocation, what is it that causes us to seek out continuing information about it? Intellectual curiosity is a trait of successful leaders and managers. The ability of leaders and managers to continue to learn about their area of expertise is critical to maintaining currency and effectiveness. Without continued exploration, leaders and managers become stagnant and eventually revert to defending antiquated knowledge bases, ideas, and practices. Learning is a lifelong process that is driven by curiosity and the desire to be the best in one's area of expertise.

Critical thinking is the application of logical concepts to the analysis of everyday reasoning and

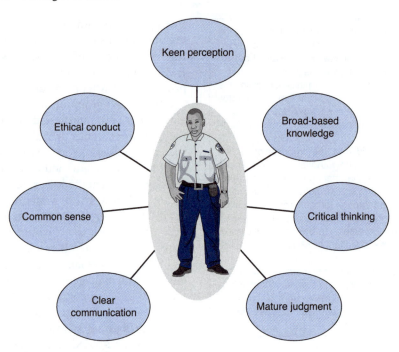

FIGURE 3.3 ■ Basic EMS Management Skills.

problem solving. Critical thinking combines common sense and problem solving and applies them to common situations. This is the intersection of knowledge and reasoning applied to desired outcomes. Critical thinking is a learned process and is something that must be developed and practiced over time.

Mature judgment is the ability to use experience-based decision making. As we accumulate different experiences, we develop a context for making decisions. In other words, we may have had positive or negative experiences, and we make judgments and decisions based on these experiences. Maturity is a difficult thing to quantify and infuse into one's talent base. Elements of mature judgment include control of emotions (rational decisions), fact-based arguments, and well-thought-out solutions.

Clear communication is the ability to transfer to and clarify information for an individual or group. This element is the most difficult and time-intensive task of a manager and leader. It is also the most valuable tool one can master; without it nothing can be achieved in an organization. Mastering clear communication involves the use of many media, including conversation, written correspondence, values statements, standing orders, and opportunities to discuss and clarify the information contained in those media. The human

element of communication is fraught with challenges that include language, jargon, trust, and ability. Successful leaders will endeavor to communicate often and embrace opportunities to clarify their messages.

Common sense is making decisions based on inherent good judgment. The challenge with common sense is that what is common to one person is alien to another. Measurements of common sense consist of reasonable and prudent decisions and actions. Cultural norms such as organizational values and acceptable practices also define common sense. Organizations frequently attempt to define these parameters through operating procedures and personnel rules. Common sense is highly individual and intuitive and incorporates mature judgment, critical thinking, and a good knowledge base.

Ethical conduct is how a person behaves based on accepted principles of right and wrong. Ethical conduct is most often shaped by an individual's lifelong experience and guided by organization norms, community, religion, and family. This is where an individual distinguishes himself with character. The six attitudes or behaviors that define character are trustworthiness, respect, responsibility, justice and fairness, caring and civil virtue, and citizenship. Because of the complexity of defining organizational ethical conduct, there are always significant differences of opinion in evaluating personal and professional conduct. Individuals often have their own definition of the six attributes of character. Although simplified to the extreme, the term "doing the right thing" may be the best way to define and describe ethical conduct.

■ MANAGEMENT PROCESSES

Managers focus their efforts on process. The most important aspect of managing process is to identify the events that signal problems within it. A signal that identifies the need for management to take action is called a **sentinel event.**

In day-to-day operations a sentinel event is an undesirable event or phenomenon that triggers the need for further analysis and investigation. It is an unexpected occurrence that can result in a serious injury that specifically includes loss of limb or function. A list of sentinel events appears in Figure 3.4. The phrase "or risk thereof" includes any process or operational variation that can result in an adverse outcome resulting in a death of a patient or economic loss to the organization. A process ensuring that sentinel events are reported within five business days should be implemented and should be the main focus of management's supervision of EMS field operations. A thorough root-cause analysis should occur within 30 days of the sentinel event. Following the review, a plan for implementation to reduce risk should follow.

WHAT EMPLOYEES WANT

What is the importance of using the power of management to encourage personnel involvement in the organization? Why do employees need to be good followers? Goal attainment and teamwork require

people who can think but also follow direction. Strong leadership empowers followers to meet a common goal and carry out a mission. But the commitment to the goal should also allow the employees or followers to challenge a leader's decisions. This empowerment is not doled out indiscriminately or without thought. Followers have a significant obligation to follow a chain of command to maintain their right to challenge a leader. A leader will acknowledge the difference of opinion, accept constructive criticism, and recognize that differences of opinion will crop up within any high-performance teams present in fire and EMS systems. One of the core tenets of followership is that the authority of the leader is preserved and protected unless the leader is incapacitated.

Good followers understand that stress is an operational distracter. Good followership requires that the leader minimize stress; he should do so by communicating the mission and ensuring high levels of training in preparation for the task. The physical and mental condition of any team member is critical to the team's success. The attributes of good followership skills are listed in Figure 3.5. Attitude, memory limits, and behavioral tendencies play a significant role in followership as well. The components of attitudes, frames of mind, prejudices, and interests all play a role in the actions and interactions of people. Strong leadership employs guarded speech and commitment to the task and mission without attitude interference.

Leaders and managers spend a lot of time trying to figure out the successful traits necessary to get good followership from members in an organization. It makes sense to ask the followers these questions, since they are the ones best suited for the questions. Over the last

---

- BLS and ALS Response >12 minutes.
- Any arrest en route from the scene.
- Any defibrillation or pacing.
- Any vehicle failure with patient embarked.
- Procedural errors.
- All patient complaints against EMS personnel, organization, or system.
- Multiple responses to the same location within 24 hours.
- All utilization of physical restraints on patients.
- All medical air evacuations from the scene.
- Any unexpected diversion from an intended medical receiving facility.

FIGURE 3.4 ■ List of Sentinel Events.

- Respect authority.
- Be safe.
- Keep your fellow followers and leaders safe.
- Accept that authority comes with responsibility.
- Know the limits of your own authority.
- Desire to make the leader succeed.
- Possess good communication skills.
- Develop and maintain a positive learning attitude.
- Keep ego in check.
- Demand clear assignments.
- Establish an assertiveness-authority balance.
- Accept direction and information as needed.
- Publicly acknowledge mistakes.
- Report status of work.
- Be flexible.

FIGURE 3.5 ■ Followership Skills.

decade EMS leadership has done this very exercise in fire- and emergency-service organizations. Amazingly, the desire expressed by EMS members parallels the thinking of members in private-sector organizations. The four leadership traits identified by members of emergency-service organizations as most important are honesty, consistency, vision, and competency.

When assessing honesty ask one person what it means to be honest and he will give responses such as "telling the truth," "accurate information," and "authentic dialog." A group will not give a clear and concise idea of what honesty is. One description of honesty is that trust is the commodity that honesty is traded in. When someone trusts the source of information, he is likely to believe he is receiving an honest message. The complication of honesty is more in the perception than the reality of the message being sent (from leader to follower). Perception often trumps reality when a follower judges actions and information from leaders. If the follower has doubts about previous encounters with a leader, then the truth surrounding that leader's message will always be in doubt in future encounters. This is a "catch-22" situation, since the follower is always suspicious of messages coming from that source. The solution to this dilemma is creating a dialog (a two-way communication) in which each party tests, clarifies, and authenticates each other's message, preferably face to face.

An example of multiple messages (or interpretations by followers) is a paper-based communication instrument, or a written memo—either on paper or via e-mail—sent out to all department members. The opportunity for interpretation of this written message is as large as the membership; in other words 100 members can (and will) interpret the message in the memo, potentially, 100 different ways. The basic reason is complex, but it lies in the limitation of a one-way communication (leader to follower) with no opportunity to test, clarify, or authenticate the message. Very little is known of the intent or expectation of the written message without validating it with the original source. If followers were to be asked if this memo was "honest" or "truthful," the response would be confusion, since they have no way of asking clarifying questions or understanding the intent of the message. (This is how a member determines the truth or honesty of the message.)

Honesty revolves around trust, which is established on a continual basis on behalf of the leader and the follower. Trust is difficult to establish between peo-

ple and represents a lot of time, effort, and many encounters between individuals. Trust that has been established over time and with much effort can be destroyed by a single event of distrust by either party. The truth is that honesty between leaders and followers is a difficult, long-term, and time-intensive endeavor. It is also a critical component in effective communications and organizational performance, as judged by followers.

When establishing consistency, ask a follower what is most frustrating in his interactions with a leader or the organization. He will most likely cite a lack of consistency. Once again, the perception of consistency is in the eye of the beholder—ask the leader about consistency in department communications, and he will complain, "We've constantly sent out memos and e-mails, and posted information on bulletin boards, but they just don't read them." The follower responds, "When we can't understand the overload of information, which often conflicts with other communications, how are we supposed to act?"

Consistency comprises such things as predictable behavior, repetitive practices (such as written and verbal communications), and honesty (authentic dialog). Consistency can also be a trap. Overreliance on constancy leads to stagnation, and a leader must look to improve the organization constantly. The rates of change in the world today require updates and the revisiting of procedures. Change is the only constant in EMS.

SUPERVISOR-SUBORDINATE COMMUNICATION

The job of a supervisor is to support subordinates so they may complete their jobs. Assuming subordinates have the necessary knowledge, skills, and abilities to do their jobs, it is up to the supervisor to empower them in pursuit of their tasks. This is a reasonable definition of a team-centered supervisor-subordinate relationship. Unfortunately, some supervisors believe that only they know "the right way" to approach a task. If a subordinate fails to "guess" what the manager is thinking, the subordinate must deal with the wrath of the supervisor.

Employees and their coworkers are in a unique position to problem solve and find reasonable solutions. Supervisors should listen carefully to them and intervene only when there are inconsistencies with acceptable organizational values or practices. A supervisor must exercise patience and tolerance for a subordinate's decision. With practice the supervisor can provide gentle corrections and **coaching** to improve the decisions a subordinate makes and build on the skills of that employee for the benefit of not only the employee but also the organization.

SUPPORTIVE WORK ENVIRONMENT

By declaring that each member of the organization has a responsibility to his job, empowerment to pursue it, and accountability to the outcomes, the organization can build confident and competent members. A key component is tolerance of others' ideas and approaches that may differ from the supervisor's. Subordinates must understand that this is not a license to abuse the power of their position or betray the values and beliefs of the organization. Accountability is a concept that all members of the organization must understand. The best organizations realize that making mistakes is a part of the learning experience of all its members. When members make mistakes after good-faith efforts, they should be coached and encouraged to understand how calculated risk taking and failure are part of what is expected of them.

DELEGATION OF TASKS

Delegation must be taught to most members, and accountability is an important aspect of delegation. When delegating to a subordinate, refer to this checklist for success:

- Does the member have the knowledge, skills, and abilities and the responsibility to pursue the assignment?
- Is the assignment properly defined, in writing, and has there been an opportunity for a face-to-face meeting to clarify any aspects of the assignment?
- Have the resources available to complete the assignment been well defined and acknowledged by both parties?
- Have a timeline and the final product been agreed upon by both parties?

When delegation fails, it is generally traceable to these items. There is nothing more frustrating to an individual or team that has been delegated a task than to produce a product that is received with "well, that's not what I was looking for." If the supervisor checks in frequently, without interfering with the process, this will benefit both parties. Good communication between supervisor and subordinate will result in small "midcourse corrections" rather than an unsuitable end product or outcome. Remember, it is the supervisor's job to provide the necessary support to the subordinate to accomplish the mission.

PERFORMANCE APPRAISAL AND COACHING

A typical scenario for an annual performance appraisal goes something like this: The subordinate arrives for a one-on-one with the supervisor, and both parties are nervous. The supervisor explains the outcome of the appraisal form as good, bad, or indifferent and sends the employee on his way. If there was an item of poor performance, there is a good chance this is the first time the subordinate has heard about it. The supervisor sends the subordinate on his way with a "just do better in the future" directive.

There are many reasons for that type of experience: poor appraisal processes and "one-size-fits-all" municipal employee appraisal processes. EMS leaders and supervisors are sometimes forced to use performance-appraisal systems and instruments that do not fit EMS personnel. There may be little that can be done to overcome this obstacle. It may be best to define a system that will work for EMS members, outside municipal constraints, while still meeting the requirements of management.

First and foremost, performance appraisal is an ongoing, continuous process. If an EMS manager were to be confronted with a single meeting, once a year, only to find out that he is exceptionally good or bad, that manager would have a difficult time adjusting his performance. Imagine training a new recruit and only giving him feedback and coaching once a year. Supervisors must give and receive constant feedback to subordinates if they want to influence behavior and performance. This does not mean that hounding and constant negative criticism are the answer. People like to be treated with respect, courtesy, and professionalism. Most individuals will respond with the same traits if they are team players. Imagine being yelled at like a dog every time one makes a mistake or fails to please the "master." It does not work very well for dogs, and it does not work very well for people.

By interpreting the values and beliefs of the organization, the supervisor does much to ensure subordinate performance. By defining the job tasks and team performance standards and using constructive criticism to coach within those parameters, the supervisor gives the employee a good idea of what to expect. Every new member of the team—whether a new rookie or a transferred, established member—needs an orientation to the team and shown what is expected of him. Imagine having a group of eight-year-old children arrive in the station and expecting them to know how to behave, how to perform on the team, without any direction or orientation. Every new member of the team should meet with the supervisor and other team members to get oriented.

The supervisor sets the expectations and performance measures. When this is not defined for the team (or by the team), anything goes as far as behavior or performance. The supervisor in this situation will spend the majority of his time chasing the individuals of the team and trying to "herd the cats" back into line. So how do leaders or supervisors reinforce behavior and performance in the organization? First, they must model the standard of behavior; without doing so, they cannot expect the organization to do so. Second, they must recognize organizational excellence through the membership, by rewarding members that do it right. The higher profile the reward is, the more dramatic the effect on the rest of the organization.

■ MOTIVATION AND TEAM BUILDING

FOUNDATIONS OF MOTIVATION

An established foundational concept in motivation is that a relationship exists between satisfaction and performance. This was established in the 1950s by Fredrick Hertzberg and is known as the two-factor theory of motivation, the cornerstone of modern management's motivational efforts. Hertzberg found that there are factors that affect the employee's attitude about his job or work and that factors leading to job satisfaction differ from those that lead to dissatisfaction.

The two-factor theory in Table 3.1 displays one set of factors called **hygiene factors** that, when present, can keep employees from being dissatisfied but cannot cause them to be satisfied with the work. For instance, employees expect reasonable department policy, good working conditions and reliable equipment, and reasonable relationships with their managers and EMS peers. If these factors are not present, employees will be dissatisfied; however, the presence of these factors alone will not create satisfaction. The presence of satisfiers, or **motivating factors,** will cause workers to be satisfied with the job; the absence of these factors does not cause dissatisfaction.

Two management approaches are derived from a classical assumption about management's approach to supervising employees. While the science to support Macgregor's **theory X** and **theory Y** concepts of management is weak, they remain part of the foundation of management. Theory X in early human relations was the belief that employees dislike work and that managers must coerce, control, and direct them. This is often demonstrated by an EMS supervisor who micromanages crews, instructing them on procedures they perform well or having his hands in the action because he believes he must do this in order to have it done "right." The management philosophy of theory Y in human relations assumes that employees accept work as natural and that managers do not need to use exclusively external controls.

Everyone wants to feel fulfilled in his life. Everyone has the natural desire to do something he can feel proud of. Abraham Maslow developed a pyramid model of human needs that starts with the basic physi-

---

TABLE 3.1 ■ Differences Between Motivation and Hygiene Factors

Hygiene Factors	Motivation Factors
Department policy and administration	Opportunity for achievement
Supervision	Opportunity for recognition
Relationship with supervisors	Work itself
Relationship with EMS peers	Responsibility
Working conditions	Advancement
Safety and reliable equipment	Opportunity to have a say on how the work is done

(*Robbins, Stephen P., Decenzo, David A.;* Fundamentals of Management, *5th,* © 2005. *Electronically reproduced by permission of Pearson Education, Inc., Upper Saddle River, New Jersey.*)

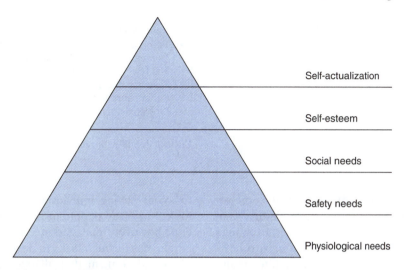

FIGURE 3.6 ■ Maslow's Hierarchy of Needs.
(Maslow, Abraham H.; Frager, Robert D. (Editor); Fadiman, James (Editor); Motivation and Personality, 3rd, ©1987. Electronically reproduced by permission of Pearson Education, Inc., Upper Saddle River, New Jersey.)

cal needs we all have—food and water—and moves up to satisfying needs for safety, then social and ego needs, and finally needs that actualize ourselves in some significant way. Maslow claims that people are motivated to move up the pyramid only when the lower needs have been satisfied. It must also be remembered that once a level is reached, it is not necessarily permanent. We can move up and down in the pyramid based on our life situations at any given time. This process is known as Maslow's Hierarchy of Needs (Figure 3.6).

Motivation is an internal drive; the individual must find motivation within himself. Managers cannot really motivate anyone; they can only create an environment in which the person wants to find motivation. Current management theory on motivating workers, known as the human-resources model, encourages use of the participative leadership style to create an environment in which all workers can contribute to the limits of their ability. Full worker participation is encouraged on all important matters, and worker self-direction and self-control is continually broadened. Basically, managers need to make full use of each member's "untapped" human resources.

Motivational and inspirational experiences improve employee attitudes, confidence, and performance. Good leadership demands good people-motivation skills and the use of inspirational techniques. Motivated people perform better; for example, people working on the same crew or people competing in teams learn about each other, and they communicate better to see

each other in a new light. When an EMS manager breaks down barriers, misunderstandings, prejudices, insecurities, divisions, territories, and hierarchies, he begins to build teams. People become motivated to achieve and do better when they have experienced the feeling of success and achievement, regardless of context. This is why team-building activities and employee-development activities that are team based have such a powerful motivational effect.

TEAM BUILDING AND PARTICIPATION

Teams are excellent mechanisms for improving the quality of EMS operations. Modern management principles recognize the value of participatory, multi-functional, and multi-disciplinary teams. Teams are formed to solve specific problems on a temporary basis, or they can be permanent. Types of teams common to EMS are displayed in Figure 3.7.

- Quality improvement team.
- Safety team.
- EMS week activities team.
- Workplace standards team.
- Special operations team.
- Equipment specification team.
- Recruiting/hiring and orientation teams.
- Health and wellness.
- Training and education.

FIGURE 3.7 ■ Examples of EMS Teams.

- ◆ Collaboratively establish objectives.
- ◆ Reinforce a participatory style.
- ◆ Focus and provide resources.
- ◆ Organize meetings.
- ◆ Organize team members.
- ◆ Define and enforce rules.
- ◆ Hold members accountable.
- ◆ Establish timelines.

FIGURE 3.8 ■ Team Checklist.

- ◆ Team members share leadership.
- ◆ Teams develop their own scope of work.
- ◆ The team commits to time allotted to do the work.
- ◆ The team produces tangible work.
- ◆ Team members are mutually accountable.
- ◆ Performance is measured on the team's product.
- ◆ Problems are discussed and resolved by the team.

FIGURE 3.9 ■ Characteristics of Effective Teams.

An EMS team is a group of people who come together to achieve an organization's agenda or a specific, common objective and in the process exchange experiences and ideas while respecting each other's contributions. EMS managers need to plan for team-oriented manager activities. Planning can be accomplished with the team checklist shown in Figure 3.8. Team members should be committed to participating in the process by effectively communicating with each other. Often team members have different duties and functions within the organization. And different people, depending on the problem that needs to be solved or the task that needs to be completed, form a quality team. Teams should not exceed 8 to 10 members, and their assignment should be voluntary to be effective.

The EMS manager will be required to supervise, organize, and evaluate the use of teams, whether the teams comprise EMS crews or are more complex, focusing on a process within the organization. There are several ways to ensure the EMS supervisor and staff can become a strong team. Supervisors or leaders of a work team must ensure clear objectives and the actions needed to accomplish those tasks in order to succeed. A trend in labor relations is for these objectives to be defined by a labor-management partnership. This type of participatory style reinforces buy-in, and supervisory personnel need to listen to all ideas and acknowledge different points of view. The EMS manager works for the team, and it is important to identify the needs of each of the team members and find ways in which they can enhance their participation and contributions, as listed in Figure 3.9.

Team meetings need to be conducted according to an agenda to keep focused and make efficient use of personnel's time. Agendas should be set ahead of time and should consider the input of any member of the team who has a discussion item to include. Roles and responsibilities need to be defined by the team leader

to avoid having individuals become frustrated. It is logical to place people in positions that suit their talents; however, managers need to consider mentoring employees through tasks to develop new skill sets. Explain and confirm that all team members know the rules and their individual responsibilities. Responsibility and accountability can be maintained by open communication and progress reports. Finally, it is important to establish a timeline for the team project and expectations of the organization when there will be a budget impact or change in operations.

■ LEADERSHIP

Leadership is the process of guiding others toward accomplishing goals. Management is the rational assessment of a situation; the development of goals and strategies; and the design, organization, direction, and control of the activities required to attain the goals. Leadership involves vision, motivation, and empathy; management is more detached and analytical. Leaders know the organization's mission. Leaders share information, help people advance in their careers, and create a work environment that develops people. It is important for leaders to communicate by asking employees what they think and communicating the organization's priorities. A 2005 poll by Fitch and Associates identified five common core competencies of a successful leader: listening skills, adaptability, decision making, team playing, and good communication skills.

The International City/County Management Association has identified the future leadership challenges for public government. Transferring those ideas to EMS managers, the future will require EMS leaders to engage in community building. This means that EMS will need to play a part in building a sense of community and a value to public service. EMS leaders will be required to blend technology with efficiency in

TABLE 3.2 ■ Differences Between Leaders and Managers

Subject	Leader	Manager
Essence	Change	Stability
Focus	Leading people	Managing work
Have	Followers	Subordinates
Seeks	Vision	Objectives
Detail	Sets direction	Plans detail
Power	Personal charisma	Formal authority
Appeal	Heart	Head
Energy	Passion	Control
Dynamic	Proactive	Reactive
Persuasion	Sell	Tell
Style	Transform	Transact
Exchange	Excitement for work	Money for work
Risk	Takes risk	Minimizes risk
Rules	Breaks rules	Makes rules
Conflict	Uses conflict	Avoids conflict
Direction	New roads	Existing roads
Blame	Takes blame	Blames others

- Vision.
- Values.
- Leadership traits.
- Ensure members know what is expected of them.
- Establish/maintain high performance expectations.
- Let members know where they stand.

FIGURE 3.10 ■ Elements of Effective Leadership. (*Source:* The Art of Leadership *by Tom Bay. Reprinted with permission.*)

order to modernize the organization. EMS leaders need to bridge gaps implied in the challenges of leadership. That includes bringing the community, hospitals, regulatory agencies, citizens, and EMS professionals together.

COMMONALITIES AND DIFFERENCES

Leader emergence is a process in which peers recognize the leadership potential of a group member and allow that person to influence their futures. Four factors appear to be associated with leader emergence: higher participation rates, higher expertise and contribution to group problem solving, more polished social skills, and strong power motives. The EMS manager has no doubt witnessed informal leaders, those without position or rank, influencing leaderless or weakly led processes. Leadership requires shifting perspective and behaviors. Leadership requires focus on people versus processes. Table 3.2 indicates the differences between leaders and managers.

LEADERSHIP DEFINED

Leaders must let vision, strategies, goals, and values be guideposts for action and behavior rather than attempting to control others. Leadership is about building and energizing relationships. A leader is someone whom people will naturally follow by their own choice, compared to a manager who must be obeyed. People will follow a leader based on hope of success, trust, excitement about a project, or the opportunity to improve. A leader's main function is to choose a team, build trust, motivate team members, and coach employees. The glue that holds this activity together is vision, and vision unites people. The primary role of an EMS leader is to serve as a facilitator and build collaborative relations. This is done by identifying and removing barriers so that others can get their work done and find meaning in that work. The basic elements of effective leadership are listed in Figure 3.10.

Leaders stand out by being different and questioning assumption. They seek the truth, make decisions based on facts and experience, and desire innovation. Employees are often more loyal to leaders than to a manager. Being observant and sensitive creates loyalty. Leaders develop confidence and trust within the team by taking blame when things go wrong, celebrating achievements, and giving credit for achievement.

ESTABLISHING VISION FOR THE ORGANIZATION

Many management scientists often identify vision as the single most defining difference between leaders and managers and leaders and followers. A test described by a major metropolitan fire chief best describes vision and its interpretation by the followers in the organization. This test consists of finding the newest member of an "on-the-street" response unit and asking him to describe the mission of the organization, the vision of the leadership, and how he applies those principles in service to the citizens and the community. If the "rookie" can pass this test successfully, then it is reasonable to assume the followers understand the leader's vision.

Often leaders and managers find that most new members, as well as older members, are not able to pass this test. Vision implies items such as: Where are we going, and how do we get there? What is the follower's role in this process, and how does the leadership support this effort? When it comes to leadership skills in the organization, vision works. It inspires people to achieve what they know they can achieve by concentrating on what is possible.

EMS administrators and officers should set the direction and create a patient focus with clear and visible values and high expectations. The directions, values, and expectations should balance the needs of all stakeholders. Those stakeholders need to include both external and internal customers.

EMS leaders should ensure the creation of strategies, systems, and methods for achieving excellence in prehospital care, stimulating innovation in the delivery of care, and building knowledge and capabilities within the EMS system. The values and strategies should help guide all activities and decisions of the organization. EMS leaders should inspire and motivate their entire staff and should encourage all staff members to contribute, develop and learn, be innovative, and be creative. This is best done by a collaborative approach between the EMS leadership and labor.

While goals and objectives need to incorporate a labor-management process, the vision of the organization should be articulated by the very top of the organization's EMS leadership or senior staff. When it comes to establishing a collaborative vision, an EMS leader should ensure the components shown in Figure 3.11.

EMS leaders should be responsible for the organization's actions and performance. The management team should be responsible ultimately to all stakeholders for the ethics, vision, actions, and performance of the organization and its senior EMS managers. EMS leaders should serve as role models through their ethical behavior and their personal involvement in planning, communications, coaching, development of future leaders, review of organizational perform-

ance, and staff recognition. As role models they can reinforce ethics, values, and expectations while building leadership, commitment, and initiative throughout the organization.

■ LEADERSHIP STYLES

Leadership is the process of influencing others toward the achievement of the organization's goals. Successful organizations have dynamic and effective leaders who respond to changing needs. Successful leaders often change styles according to the situation. Situational leadership is based on flexibility, diagnosis, and communication. Leaders should change their leadership style to fit the needs of their people based on assessing the needs of the people and the organization.

Leaders generally have four choices of style to get the work completed: situational, coaching, supportive, and delegation. In all situations or approaches, the leader sets goals, monitors progress, observes performance, and provides feedback. The four choices are:

- *Situational leadership.* In this leadership style a leader can choose to direct, support, coach, or delegate. When a leader directs, the leader tells the follower what, how, where, and when about the task. The leader closely supervises the work without dictating. The follower has no involvement with the decision making.
- *Coaching.* In a coaching leadership style, the leader continues to give specific directions but also offers reassurance and encouragement. The leader explains decisions and solicits suggestions and input.
- *Supportive leadership.* In the supportive leadership style, the leader no longer needs to give a lot of direction. The process helps the follower build self-confidence and motivation through encouragement and praise. The leader helps a person reach his own solutions by asking questions that expand thinking. This is shared responsibility.
- *Delegated leadership.* A leader who delegates no longer needs to give very much direction or support at all. The leader turns over responsibility for decision making and problem solving to the follower.

Another concept of leadership was developed by Richard Ross and Tom Isgar, who related leadership styles to high-performance teams. EMS and fire-service crews fit the criteria of high-performance teams. In some situations the scenario may call for more task-

---

- ◆ Feasible.
- ◆ Desirable.
- ◆ Focused.
- ◆ Flexible.
- ◆ Communicable.
- ◆ United.

FIGURE 3.11 ■ The Mechanics of Creating a Vision.

oriented versus process-oriented leadership. Four more categories of leaders are identified in Ross and Isgar's work: innovators, achievers, organizers, and facilitators:

- Innovators are those who are great at brainstorming, stimulating thinking, creating a risk-free atmosphere, and not relying on time.
- Achievers are leaders who are goal oriented and often overcommit to many projects; they feed on momentum. Achievers are action oriented and celebrate a finished project.
- Organizers like to ensure excellent preparation and follow the steps with resources that are well defined. For the organizer time efficiency is critical to the organization's plan.
- Facilitators like good processes and consensus building; they create clear expectations for the roles and purposes of each team member and design how that happens.

COMPETENCY

Eldridge Cleaver, an author and prominent civil rights leader, said, "What we're saying today is that you're either part of the solution, or you're part of the problem." The competent professional thinks hard about what he is doing and must be able to move back and forth between realities and the abstract theories that give meaning to situations.

People and organizations are looking for problem solvers. But competency is a rather tricky trait to pin on a leader, since it resides in two different realms. They are:

- *Personal.* The leader's success is based on his having the knowledge, skill, and ability to do the job. Competence involves his combining skills, abilities, values, and personal strengths to perform the task of a leader.
- *Perceivable.* The leader's perceived ability is judged by the rest of the organization. There are leaders who have very little knowledge of the technical aspects of jobs in their organizations, yet they are able to command the respect of those who perform very technical aspects of certain jobs in the organizations.

Competency is rooted in knowledge and analytical skills. EMS is a good base because prehospital medicine provides a background that requires a person to diagnose, apply knowledge, and evaluate. Yet the best medics often fail at leadership tasks that are nonmedical. Leadership builds on medical skills by engaging the complexity, ambiguity, and uniqueness of a condition.

Leadership investigates by asking questions until a decision-making framework emerges. For example, the leader of an EMS organization may not have the highest medical training but may enjoy the respect of the members because of his leadership style. Competency weighs heavily on followers, and they will reflect on the ability of leaders to do their job and will discuss it often.

An important item missing from this description is experience, which is a difficult thing to judge. The competencies of an entry-level firefighter or EMT differ greatly from those of the leader of the organization, a fact often lost on the members. Competencies are also heavily embedded in interpersonal communication and dynamics. A member who is a tremendous technician in his area of expertise but cannot communicate with leaders, members, or the public is of limited benefit to the organization. Communicating well, building teams to execute the mission, and describing a vision for the organization is the work of the leader.

COMMUNICATION

Communication should focus on understanding the other person. Mutual respect develops when people see skills and attributes in others that they did not know existed. It is important that an EMS leader identify the beliefs, behaviors, and characteristics of good participative or team-centered communication.

For example, positive- and negative-performance feedback that is specific and sensitively presented can help employees improve their performance. Feedback is unlikely to bring improvement if it seems that the negative feedback is aimed at the employee as a person, rather than at his specific areas of deficient performance—feedback that seems aimed at the person is likely to cause the employee to react defensively and to dismiss any other information that could have led to improved performance.

Several aspects of communication need to be refined among EMS managers and supervisors. Managers should develop communication with coworkers, manage electronic communication, and practice written communication. Clear, concise, accurate, and complete communication is vital in organizations. Only when employees know what is going on can they work intelligently and with a clear sense of purpose. Communication, therefore, is a manager's number-one challenge in improving the organization's efficiency and effectiveness. Figure 3.12 lists some common challenges as they relate to communications in EMS organizations.

♦ **Information overload.** This occurs when more information is received than can be handled. Selective attention is employed, resulting in the tendency to lose, ignore, or misinterpret much of the information. Managers can address this problem by prioritizing their tasks and the information needed to accomplish them. This is especially important with information formats such as e-mail, in which recipients can receive hundreds of messages each day.

♦ **Frame-of-reference differences.** These reflect the differences in educational, developmental, intellectual, cultural, and work-related experiences. To the extent that these differences are large and the sender of the message is not sufficiently attentive to them, frame-of-reference differences can contribute to the miscommunication of a message. Managers can minimize the impact of frame-of-reference differences by seeking information about the receiver's background and taking that information into account when transmitting a message (i.e., know your audience).

♦ **Status differences.** These pose a barrier to communication when individuals of perceived lower status are threatened by the status difference and become hesitant to engage in clear and accurate two-way communication. Managers can minimize status differences by speaking with employees on their own "turf" and by making sure that conversations in the manager's office do not take place across the desk.

♦ **Poor listening skills.** This problem poses yet another barrier, as when managers listen to respond, rather than listen to listen. Managers can remove this barrier by practicing active listening skills and creating an environment that facilitates listening (e.g., not interrupting, paying attention to nonverbal behaviors, encouraging the speaker to be open, interpreting and paraphrasing the sender's message, and so on).

FIGURE 3.12 ■ Communication Challenges.

The speed and efficiency of communication in the organization is helped by technology. This in turn results in an increase in productivity. Additionally, technology, such as voice mail, online video communication, and e-mail, makes communication more convenient (e.g., an employee can receive messages after hours and access them when away from the office).

Communication technology tends to be impersonal, which many people dislike. It also allows more information to be transmitted quicker. This makes the time needed to make a decision shorter and allows the manager to be more efficient. However, quicker communication can contribute to information overload. Information overload, in turn, causes information to be lost, ignored, or miscommunicated. Abuses of such systems (for instance, sending personal messages on e-mail) can prevent them from being effective in that the transmission of important information is delayed or unable to be transmitted.

When managers experience information overload, they begin to attend to information selectively, noting the information they believe to be important while ignoring the rest (that is, they filter out the information they determine to be least important or least in keeping with their expectations). The result is that much information gets lost, ignored, misinterpreted, or underinterpreted.

MANAGING E-MAIL

E-mail and wireless communications are part of an EMS manager's or leader's daily routine. When the supervisor-to-paramedic ratio is high or there is significant distance between crews, e-mail communication may be the primary form of daily communication. It is easy for an EMS manager to become a slave to the e-mail system, and often the momentum of the day is destroyed by excessive e-mail. In an attempt to help people process e-mail communications, Dr. Francine Galliore has written an article entitled "Seven Habits of Highly Effective E-mail Communicators."

Among Dr. Galliore's recommendations are the following:

♦ Write something meaningful in the e-mail subject line; never leave it blank. Subject lines—such as "Would like to discuss quality project" or "Following up on last week's meeting"—will help the receiver distinguish your e-mail from junk and infectious communiqués.

♦ Use proper grammar and punctuation in the body of the message. Even brief messages should contain complete sentences. Otherwise, you risk having the recipient think you do not know how to use a computer. Also, do not capitalize all the letters, because the receiver may think you are angry or that he is being yelled at.

♦ Get to the point early, preferably within the first three sentences. The best method is to use only one sentence to state the point. Examples: "I need more information about the project before I make a decision" or "Your presentation inspired me to make changes in the operation."

♦ Keep e-mail messages discreet. Avoid adding people to an e-mail list without their permission or request. And never share confidential, private, or potentially damaging information via e-mail. If the EMS manager has something sensitive to say, whether it is business or personal, he should do so by phone or in person.

- E-mail can be used to request a meeting, but it should not be used to lay out a case. Doing so may risk both being misinterpreted and potential liability.
- Be respectful of people's time and expertise. A respectful letter of inquiry follows this template: Dear [Mr., Ms., Dr.]___: I heard about you from___. [This is who I am:]___. I may be interested in using your services. Briefly, this is my situation:___. Sincerely yours,___.
- Use e-mail for quick congratulations. Although all our lives are busy, we should not forget to send a note of appreciation for a job well done. Modern quality-improvement theory promotes frequent acknowledgment of a job well done, and e-mail is a great way to send a quick "Great job on that call" or "Your customer service was very impressive. Thank you!" or "I read about your chart—excellent documentation!" The EMS manager will be well remembered for taking the time to do this.

COMMUNICATING WITH PEOPLE

Work-related and medical care–related training commonly concentrate on process, rules, structure, and logic. However, leadership really is about people skills, or "soft" skills, and attributes like self-motivation, confidence, initiative, empathy, and creativity. A good leader strengthens motivation and learning by good coaching and mentoring. Good communication seeks to find ways to improve the conditions of others by knowing what they are about and what their goals are. Often it is about finding resources that fit needs communicated to them by the members of their team.

Networking

A key communication skill for a leader is how well he develops a professional network. Networking is building cooperative relationships with people who can, or might someday, assist you in reaching a goal. A network is usually an informal, unstructured support system that enhances one's ability to use power effectively. Networks can include communication that is developed with people who:

- Have expertise in a variety of areas.
- Have power.
- Have access to vital information.
- Control necessary resources.
- Have access to decision makers.

Effectiveness

If leaders or managers are not communicating effectively, the organization is operating neither effectively nor as a cohesive team. So they periodically must check the effectiveness of their communication with team members, as well as the cohesiveness of the team. The simple act of asking a team or crew for their feedback is an opportunity for leaders and managers to strengthen the team and their own techniques. Figure 3.13 is a tool used in wildland fire management that applies to any emergency team. The instruction and assessment tool measures the basics of crew cohesion.

Leadership as an Art

It is often said that leadership is an art. The skills and abilities are based in the humanities, yet the extent to which those skills and abilities are applied by leaders is an art. Figure 3.14 lists some ideas thought to enhance the art of leadership. A person must have certain traits to make him a leader. Ideally, a leader displays enthusiasm, initiative, and integrity. The ABCDE model of leadership also provides a guide to developing leadership traits. In this model the elements of professionalism are crafted into leadership:

A—Attitude.
B—Behavior.
C—Communication.
D—Demeanor.
E—Ethics.

A leader must have an attitude that is contagious and must see problems as opportunity. A leader's behavior must reflect self-discipline, emotional control, moderation, and discretion. A leader must be able to communicate and listen. A leader's demeanor must be poised and professional so that, as a figurehead of an organization, he portrays confidence and self-discipline. A leader must be ethical in his behavior and beliefs. He also must be a lifelong learner and have a desire for self-improvement.

CONFLICT RESOLUTION

Solving conflicts is an extremely important leadership or management trait. Conflict almost always arises between organizations with several different viewpoints. It is important to have the skills and a process to solve conflict inside and outside the organization. Figure 3.15 presents some basic rules about conflict resolution. Conflict resolution is frequently used by the management team and EMS leaders. Working through conflict and finding a win-win out-

Crew Cohesion Assessment Tool

Purpose: In terms of crew cohesion, each crew is different, and even the same crew with the same people will vary in the level of cohesion from time to time. The Crew Cohesion Assessment is designed to provide a tool to measure crew or team behaviors as they relate to cohesion.

Crew cohesion is no mystery. The factors that make crews and teams cohesive are well known, documented through the centuries in both literature and research. This tool describes behaviors that are grouped into seven general categories representing characteristics of cohesive groups. Although not all-inclusive, the list can provide a place to start in determining the strengths and weakness of your crew or team in relation to team cohesion.

The interpersonal dynamic of teams or crews changes constantly. What was true last year may not be true today. A leader's responsibility is to continually monitor and assess the health and well-being of the crew and its members. This tool can be used independently by a single crew leader, by a leadership team, or with the crewmembers.

Instructions: Following are the recommended steps for using the survey tool:

Step 1: Determine the time frame to which you will confine your assessment. Although you can use the tool to assess past or present behavior, for the most useful results you should limit your assessment time to a short window of recent time (i.e. the past 4 weeks). Do not mix the distant past and the present because it may blur the focus of the results.

Step 2: Rate the *behavior* of the crewmembers and leaders during the assessment period, using a scale of 1 to 5 as follows:

Score	Description
1 = never observed	Behavior is not seen except in very rare cases or only with a single person—indicates potential serious weakness that should be addressed immediately.
2 = rarely observed	Behavior is seldom seen or exhibited by only a few crewmembers or leaders—indicates a weakness the improvement plan should address.
3 = sometimes observed	Behavior is seen unpredictably or by only some of the crewmembers—indicating average performance. Improvement plan dependent upon scope and criticality of issue.
4 = often observed	Behavior is evident with most crewmembers most of the time—indicates a definite strength to be preserved. Further improvement is likely limited to working with a few people or specific operations.
5 = always observed	Behavior is seen in all crewmembers and leaders nearly all the time—indicates exceptional strength. Improvement plan should include actions to maintain this strength for the future.

Try to be as honest and pragmatic as possible. Sometimes your initial reaction is the best response you can provide.

Step 3: If you are assessing the crew with others, compare your assessment, discussing areas where you have differing perceptions. Discuss which items are the most important to your crew and its mission. Note items that rate unusually high or low for later discussion.

Step 4: Build a plan of action to improve upon the identified weaknesses, and plan ways to sustain the strengths. If crewmembers were not involved with the assessment process, consider sharing your perceptions with the crew and including them in the planning.

Step 5: File your assessment and set a date for your next assessment to determine your progress against your planned goals.

LEADERSHIP TOOLBOX REFERENCE
Crew Cohesion Assessment
Courtesy of Mission-Centered Solutions

FIGURE 3.13 ■ Crew Cohesion Assessment Tool.
(Reprinted with permission of Mission-Centered Solutions)

Communication

Crew leaders communicate intent clearly and crewmembers understand the intent of orders that are given.	1 2 3 4 5
Crewmembers are willing to bring up problems, successes, or issues with the leadership team.	1 2 3 4 5
Crewmembers communicate well with the other crewmembers.	1 2 3 4 5

Conflict

Crewmembers are willing to address conflicts with others when they occur.	1 2 3 4 5
Crewmembers focus on what is right, not who is right, when resolving conflict or other problems.	1 2 3 4 5
Conflicts between individuals on the crew are short-lived and do not persist over extended periods of time.	1 2 3 4 5

Trust

The crew environment allows and encourages all crewmembers to be heard.	1 2 3 4 5
Crewmembers complete assignments without excessive supervision.	1 2 3 4 5
Crewmembers implement decisions of the leadership team without delay.	1 2 3 4 5
Crewmembers are willing to experiment on new ideas without risking embarrassment.	1 2 3 4 5

Teamwork

Crewmembers show ownership in the crew's accomplishments or failures.	1 2 3 4 5
Crewmembers show a strong consciousness of the history, tradition, and lore of the crew.	1 2 3 4 5
Crewmembers anticipate the needs of others and act in anticipation of those needs, especially during high tempo operations.	1 2 3 4 5
The intensity of the work is determined more by the crewmembers than by direction from leaders.	1 2 3 4 5

Effectiveness

Crew remains focused on the quality of the service they provide, even when others do not hold them directly accountable for that service.	1 2 3 4 5
Crew works well with other resources to accomplish the mission.	1 2 3 4 5
Crew actively discusses situation awareness when planning and conducting operations.	1 2 3 4 5
Crewmembers can pause an action to clarify their situation awareness or voice concerns.	1 2 3 4 5
Crew recognizes and successfully transitions between high-stress and low-stress conditions.	1 2 3 4 5

Leadership

Crew has a complete set of standard operating procedures or crew handbook that is well understood and used.	1 2 3 4 5
Leaders demonstrate and adhere to a consistent set of values and standards.	1 2 3 4 5
Leaders conduct themselves in an ethical manner.	1 2 3 4 5
The leadership team speaks and acts with one voice and mind.	1 2 3 4 5

Learning

Crew debriefs daily as part of the standard operating routine.	1 2 3 4 5
Crew conducts impromptu on-the-job training events to build crew skill sets.	1 2 3 4 5
Crew conducts training to discuss lessons from other fires or other operations.	1 2 3 4 5
Training is conducted by more than just one or two crew leaders.	1 2 3 4 5
Crew leaders understand their responsibility to mentor crewmembers.	1 2 3 4 5

LEADERSHIP TOOLBOX REFERENCE
Crew Cohesion Assessment
Courtesy of Mission-Centered Solutions

FIGURE 3.13 ■ Continued.

- ◆ Active listening.
- ◆ Think large.
- ◆ Suspend judgment.
- ◆ Follow through.
- ◆ Communicate.
- ◆ Lead and practice followership.
- ◆ Seek to serve.
- ◆ Cross-train.
- ◆ Show you care.
- ◆ Authenticate yourself ("walk the talk").

FIGURE 3.14 ■ The Art of Leadership.
(Source: The Art of Leadership *by Tom Bay. Reprinted with permission.)*

1. Avoid jumping to conclusions. (Fully analyze the problem.)
2. Investigate and describe the facts. (What is the situation? What is the context of the conflict? Who is involved? Who are the stakeholders? What is each person's point of view?)
3. Define the parameters of the conflict. (Where is the difference of opinion? Get each person's viewpoint and analyze viewpoints.)
4. Diagnose the conflict and identify the power issues. (What led up to the conflict? What is each party's issues? What are the power issues?)
5. Consider the alternatives and find a solution.

FIGURE 3.15 ■ Five Steps to Conflict Resolution.

come, or finding constructive alternatives, is paramount in keeping motivation and morale in an EMS team.

Several new ways of solving conflict have emerged in the last decade. Leaders need to understand when they are at an impasse in a conflict-resolution scenario. An alternative dispute-resolution system is a process employed by the leader that uses facilitation and arbitration by a third party. The idea is that the third party brings a clear perspective to a conflict and sees both sides without bias. It is important for a leader to identify this process as a resource in the leader's toolbox.

ETHICS AND THE LEADER

At some point in his career, an EMS leader or manager should study the application of ethics in both the medical and operational aspects of the organization. A leader defines and clarifies standards, values, and ethics. Managers demonstrate a commitment to those ethics and expect a similar commitment by others. A leader should ensure the organization supports and rewards ethical behavior and ethical solutions to problems.

Ethical leaders model ethical behavior and attempt to balance personal ethics with the organization's ethics. A leader must understand the impact of decisions on others and operate with integrity, honesty, and courage. Those decisions require that the leader consider how those decisions affect an EMS leader's professional life, relations with the public or society, relations with employees, and relationships with coworkers. The approaches to ethics from a positive point of view are guided by the leader's ethical compass or conscience.

■ SUCCESSION PLANNING

Succession planning is not a common phenomenon in fire and EMS organizations. Often the choosing of the next generation of managers and leaders is left to political processes or patronage. More and more, organizations are realizing the need to plan who will fill key vacancies in the future. Succession planning requires that an organization take an assessment of current talent and project the needs of fire and EMS operations. This may include looking at skills, education, and desires of people in the organization. This requires an honest look at what talent and experience needs to be brought in from outside the organization to meet its needs.

Once an assessment of needs and an inventory of current talent and interested personnel are completed, current management and leadership need to be matched to employees wishing to promote. The matching between a mentor and a protégé will be to facilitate coaching, education, and mentoring opportunities for employees to help them build skills and abilities for future positions in the organization.

COACHING

When people are asked who the people are who most influenced their lives, most answer that second to a parent is usually a high school or college coach. Coaching is defined as an activity through which managers work with subordinates to foster skill development, impart knowledge, and instill values and behaviors that will help them achieve organizational goals and prepare them for a more challenging assignment or performance. There has been a significant increase in personal coaches; in fact, an entire profession has developed around the coaching of executives. The increasing number of these services is a testament to the desire for people to achieve more and improve on themselves.

Coaching benefits an EMS organization in several ways. Coaching often is identified as part of union performance-improvement programs. It is often a first step or an informal process that deals with behavior or performance issues when an employee is in violation of standard operating guidelines (SOGs) or the organization's standards. Coaching can increase performance and develop employee skills and abilities.

Coaching is a difficult process to master. The skills needed to coach a family member's sports team transfer only slightly to the process of workforce coaching. Failure in the coaching process can occur; several factors can create a coaching failure. A common source of failure is a lack of commitment by the EMS officer or the employee to maintain the relationship long enough to complete the goal of the coaching activities. It is important that EMS managers engaged in coaching projects clearly define and have sensible expectations of the relationship. Other common errors made by EMS managers and employees being coached include defensiveness, passiveness, or lack of risk taking. Real leadership involves taking a risk by trying new ideas, opening new lines of communication, and establishing relationships.

Coaching an employee requires that EMS supervisors involve others in the process. This is best accomplished by using 360-degree evaluation tools and performance appraisals. An EMS officer may be coaching a paramedic who is not directly in the chain of command and is supervised by a station captain, lieutenant, or field supervisor. It is important to involve others in feedback on the coaching or development process.

It is important to select good coaches and preceptors in EMS. Often this is the transition from manager to leader and frequently is a paramedic's first step into management. Training and coaching new employees, new supervisors, and committees or teams is an essential skill to build in the organization. When selecting good coaches, identify and choose those who are lifelong students of EMS, with those attributes shown in Figure 3.16. Look for paramedics and EMTs who have a constant desire to improve and learn. Find the latest medical procedures, training techniques, and anything else that will gain an advantage for the patient and the trainee.

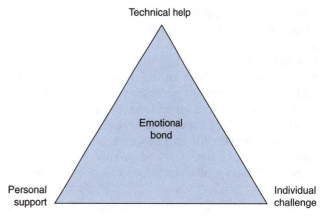

FIGURE 3.16 ◼ Characteristics of a Good Coach. *(Reprinted by permission of Harvard Business School Press. From* Coaching and Mentoring: Harvard Business Essentials. *Richard Luecke and Herminia Ibarra. Boston, MA 2004, p. 3. Copyright © 2004 by the Harvard Business School Publishing Corporation; all rights reserved.)* Source: Susan Alvey, with permission.

Good organizational skills are a must; many coaching opportunities will have corresponding paperwork, agendas, and orchestrated meetings. For example, the first meeting of a quality-improvement committee should have an agenda, be coordinated with members' schedules, and have minutes forwarded to upper-level EMS leadership.

Coaches need critical-thinking and problem-solving skills. People skills are a must; good coaches are patient and have respect for others, an ability to relate, and common sense. Great coaches also have discipline, and discipline brings structure to an event, a call, or management activity.

Observation skills are important for EMS managers and leaders. They should be able to read a call or medical situation like a book and make adjustments in the progression of the call. Good EMS coaches have a feeling or sense about a situation, based on their experience, that governs their decision making. This allows the EMS manager to conduct efficient operations, meetings, or communication.

Last, each new EMS manager or leader develops his own philosophy about how he approaches his job and career and how he represents his agency.

Fostering an environment that promotes coaching, as an approach to developing people and solving performance problems, is a more positive approach to labor units. The majority of the people who come to work in EMS arrive with the intent to do the right thing, as do most people who join a team. The success of the team, or the end results of the organization's processes, is a reflection of the effectiveness of coaches, mentors, and leadership within an organization. Contrast that with an organization that focuses on failures and does not provide a chance for more experiences or for educated personnel to guide, teach, or coach those lacking experience or skills.

It is important for an EMS manager or leader to monitor coaching activities and promote an environment that fosters good coaches. Leadership should be on the lookout for influences that can cause a failure in the coaching process. Signs of failure include the following:

* Lack of commitment.
* Unrealistic expectations.
* Defensiveness.
* Playing a passive role.
* Playing it safe.
* Failing to involve others.

For a leader it may be helpful to assess the coaching skills of EMS managers and leadership. Figure 3.17

Harvard ManageMentor — COACHING TOOLS
A Coach's Self-Evaluation Checklist

The questions below relate to the skills and qualities needed to be an effective coach. Use this tool to evaluate your own effectiveness as a coach.

Question	Yes	No
1. Do you show interest in career development, not just short-term performance?		
2. Do you provide both support and autonomy?		
3. Do you set high yet attainable goals?		
4. Do you serve as a role model?		
5. Do you communicate business strategies and expected behaviors as a basis for establishing objectives?		
6. Do you work with the individual you are coaching to generate alternative approaches or solutions which you can consider together?		
7. Before giving feedback, do you observe carefully, and without bias, the individual you are coaching?		
8. Do you separate observations from judgments or assumptions?		
9. Do you test your theories about a person's behavior before acting on them?		
10. Are you careful to avoid using your own performance as a yardstick to measure others?		
11. Do you focus your attention and avoid distractions when someone is talking to you?		
12. Do you paraphrase or use some other method to clarify what is being said in a discussion?		
13. Do you use relaxed body language and verbal cues to encourage a speaker during conversations?		
14. Do you use open-ended questions to promote sharing of ideas and information?		
15. Do you give specific feedback?		
16. Do you give timely feedback?		
17. Do you give feedback that focuses on behavior and its consequences (rather than on vague judgments)?		
18. Do you give positive as well as negative feedback?		
19. Do you try to reach agreement on desired goals and outcomes rather than simply dictate them?		
20. Do you try to prepare for coaching discussions in advance?		
21. Do you always follow up on a coaching discussion to make sure progress is proceeding as planned?		
TOTALS		

When you have these characteristics and use these strategies, people trust you and turn to you for both professional and personal support.
*If you answered "**yes**" to most of these questions, you are probably an effective coach.*
*If you answered "**no**" to some or many of these questions, you may want to consider how you can further develop your coaching skills.*

FIGURE 3.17 ■ Coach's Self-Evaluation Checklist.
(Reprinted by permission of Harvard Business School Press.
From Coaching and Mentoring: Harvard Business Essentials.
Richard Luecke and Herminia Ibarra. Boston, MA 2004, p. 140.
Copyright © 2004 by the Harvard Business School Publishing Corporation; all rights reserved.)

offers a sample of a coaching skills assessment from the Harvard Business Press. It is incumbent on the leadership to ensure senior EMS personnel are supportive and that they remove any barriers to effective coaching. EMS leadership should look for ways to identify people who can overcome these common failures in coaching.

MENTORING

Mentorship, like coaching, is a way of developing people. **Mentoring** is about guiding others in their personal quest for growth through learning. Mentors are trusted guides who offer advice and open doors to learning opportunities when possible and appropriate. Mentoring, then, is the offering of advice, information, and guidance by a person with useful experiences, skills, or expertise for another individual's personal and professional growth. The paramedic internship or field shift is often a common mentorship relationship between a preceptor or field training officer and a new paramedic. That relationship can be applied to potential leaders and managers in an EMS system. The same model for teaching paramedic skills and decision making in the field using competencies, demonstration, and immersion in real scenarios can be used to prepare managers and leaders.

A mentor relationship is initiated by the employee seeking mentorship and thus requires him to be a self-starter.

The scope of mentoring is much larger than that of coaching. Coaching is a component of mentoring, but mentoring involves sponsorship, in which an EMS officer opens a relationship between the employee and the organization. Mentors support the employees they mentor by acting as a buffer between the organization and the employee and the errors he may make in his development process. EMS officers conducting mentoring need to continue to challenge the employees to broaden their capabilities. This increase in skills is coupled with the mentor's finding opportunities for the employee to conduct operations or complete projects that become visible to management.

Mentors provide counseling and act as role models. The mentor provides a lasting friendship and an acceptance of the employee. The mentor should not be in the employee's direct chain of command, unlike coaches, who are often the employee's direct supervisor.

1. Sets high standards.
2. Is willing to invest time and effort.
3. Devises experiences to develop those mentored.
4. Is respectful of members of organization.
5. Demonstrates good people skills.
6. Understands teaching techniques.
7. Has access to information.
8. Is open and honest with thoughts.
9. Has good chemistry.
10. Is connected to the organization.

FIGURE 3.18 ■ Ten Traits of a Mentor.
(Reprinted by permission of Harvard Business School Press. From Coaching and Mentoring: Harvard Business Essentials. *Richard Luecke and Herminia Ibarra. Boston, MA 2004, pp. 100–101. Copyright © 2004 by the Harvard Business School Publishing Corporation; all rights reserved.)*

Coaching involves the job tasks and skills, but mentoring is about an employee's career.

Successful mentorship can be defined by some observed behaviors, such as those in Figure 3.18. First there is enthusiasm and satisfaction on the part of both the mentor and the mentored. There is observable personal bonding between the two and observed real learning. The mentored has an increase in self-awareness and self-confidence in his role within the organization and the environment in which he is operating.

Rarely do mentorships last more than a few years with one individual. The person being mentored needs to seek out new relationships and develop a diversified network of mentors.

Mentoring and coaching are two separate and distinct processes that need to be incorporated into a succession plan for any organization. Coaching is an ongoing process necessary to be performed almost daily by EMS managers. EMS managers and supervisors coach subordinates routinely on performance career development and on patient care. Table 3.3 shows the differences between coaching and mentoring.

Leaders and supervisors in an ideal succession plan should select three people to mentor. Mentorship, while not always a formal or sanctioned process in an EMS organization, is an invaluable way to develop the next generation of EMS leaders and supervisors. Both processes—coaching and mentoring—need to be managed and supported by the organization.

TABLE 3.3 ■ Differences Between Coaching and Mentoring

	Coach	*Mentor*
Goals	Corrective behavior, improve skills	Support personal growth
Initiative	Coach directs and instructs	Mentor in charge of learning
Volunteer	Subordinate must accept coaching	Mentor and mentored participate voluntarily
Focus	Immediate problems	Long-term development
Roles	Heavy on telling, with feedback	Heavy on listening, making suggestions, and connections
Duration	Usually on the short-term or as-needed basis	Long term
Relationship	Coach is the coachee's boss	Mentor should not be in the mentoree's chain of command

(Reprinted by permission of Harvard Business School Press.
From Coaching and Mentoring: Harvard Business Essentials.
Richard Luecke and Herminia Ibarra. Boston, MA 2004, p. 79.
Copyright © 2004 by the Harvard Business School Publishing Corporation; all rights reserved.)

■ LEADERSHIP FAILURES

The Denver patient-abuse scandal, the San Francisco fire/EMS conflict, and the *USA Today* description of the Washington, D.C., EMS system highlight systems that failed to deliver a good team effort and quality service. These incidents reflect the leadership of the organization, and while they are isolated incidents played up by the media, they demonstrate what can happen. Why does leadership fail in EMS? A shift in focus away from the core values or the mission statement of the organization can cause a leadership failure.

Leaders can lose sight of what is important. Often new managers get caught up in the details and become too action oriented. How often has a frontline supervisor been involved in patient care when he should have been coaching the employee through a difficult call and not becoming immersed in the details? This is a current issue with paramedic-certified officers who do not exercise efficient scene management because they micromanage. This is known as role conflict and role ambiguity.

Poor communication can be the result of a lack of focus, and often a leader will hide confusion and uncertainty in poor communication. Clarity and feedback are the tools of effective communication. Without effective communication leadership is bound to fail. Consider the EMS manager's role as a supervisor who is mentoring an experienced paramedic for the implementation of a public-relations program for "EMS week." The window comes and goes, with a comprehensive program arriving a week late. In such a situation was the supervisor absolutely clear about the deadline?

Leadership involves taking risks, and risk aversion or fear of failure can lead to actual failure in organization.

An ethics slip can result in leadership failure when the leader produces a discrepancy between what he does and who he is. When an EMS supervisor rationalizes away the ethics of the organizational processes and the dignity of the worker as a means to achieve results, integrity slides down a slippery slope. When EMS managers confuse manipulation with what they perceive as leadership, they become people pleasers in order to ease the guilt over the lapse of integrity. It is the act of doing what is popular versus what is right that defines the loss of integrity in the management process and in the leadership of people. An EMS manager can test for this by asking his staff if they think they have ever felt used or taken advantage of. Poor self-management can lead to a failure in an EMS supervisor's ability to lead.

The position of an EMS leader is exhausting and yet rewarding. An EMS leader needs the capacity to be stimulated by emotion and interpersonal issues, rather than to be worn down or exhausted by them. Frontline EMS supervisors frequently work in conditions that impact physical, psychological, emotional, and spiritual needs. The success of an EMS leader requires the ability to absorb the emotional strains of the uncertainty of interpersonal conflict, responsibility, and other stressors.

A leader must plan and follow a regimen that attempts to achieve a state of spiritual, physical, and emotional well-being. Finally, leadership can fail when it is perceived that leaders have lost their love for what motivated them in the first place. (See Figure 3.19.)

Dan Netski

Dan Netski was an EMS field supervisor at Mercy Ambulance in Las Vegas, Nevada, for many years. During his time of service, the organization was undergoing tremendous growth and the company serviced a very unique population that could stress even the best paramedics. Mercy was an organization that was often a paramedic's first full-time job. As a supervisor Dan was well respected by his crews and always kept his cool under pressure. Dan had weathered a high-volume system for more than 20 years, yet he still had a focus on patient care and customer service that was as sharp as the first day he signed on after returning from Vietnam as a navy corpsman.

After a particularly long shift filled with frustration, I asked him how he kept so upbeat about this job, considering the abuse the industry and urban system puts on a paramedic. He opened his wallet and showed me a picture of his sons on a powerboat. He said, "On my days off, I go to the lake and drive this boat as fast as I can across the lake. I get away from this place and enjoy my hobbies and family, and that gives me the strength to come back and give 100% to the crews and the company."

His point was well taken: You have to have something else that restores your energy, because the system and the job will

take it from you. Dan was an exceptional supervisor, as was evident by the hundreds of people from many agencies (many traveled great distances) who were at his funeral.

FIGURE 3.19 ■ Dan Netski, EMS Captain Nevada Test Site Fire Department. His success and tenure in EMS were the result of a "balanced life."
(Printed with permission of On Assignment Studios.)

CHAPTER REVIEW

Summary

Managing and leading require two separate sets of skills. Managers use formal and rational methods, as compared to a leader who stirs emotions and passion among personnel. There is a clear delineation between a manager who should focus on the processes and leaders who should focus on the people of the organization. The two separate sets of skills require a person to master the skills of a manager, gain the knowledge of a coach, and serve as a mentor to become a leader. Managers and leaders can be trained. The skills and abilities needed to successfully serve as an EMS manager or leader require focus on skills, practice, and feedback to perform well. Success involves bringing the right people together and then fostering a climate that promotes trusting relationships, loyalty, and commitment.

Research and Review

1. Describe the role of values and beliefs of organizational members and how they affect the definition of the culture of the organization. Design a process to galvanize the beliefs of the members and the organization that will result in a common culture.

2. What is the role of the leader in mission-statement development, and what tools does he have at his disposal?

3. When considering the different types of power as they relate to the leader, which type is best utilized to gain member confidence and followership in a participative management setting? Why?

4. As the fire service and EMS progressed from paramilitary organizations to participatory, team-based organizations, how have the uses of power changed?

5. The four leadership traits of honesty, consistency, vision, and competency can influence followership. When subordinates are mentored for these traits, what are some of the effects on organizational culture?

6. Theories X and Y define differing management styles. Which style is most beneficial in a team-centered environment? How can you transform traditional managers into that particular style?

7. Describe ways to hold team members accountable without diminishing their enthusiasm or micromanaging their duties.

8. Discuss the difference between coaching and mentoring. Describe a situation that best utilizes each activity.

References

Bay, T. (2000). *Look within or do without*. Franklin Lakes, NJ: Career Press.

Blanchard, K., & Shula, D. (1995). *Everyone's a coach*. New York: Harper Business Press.

Certo, S. C. (1994). *Supervision: Quality and diversity through leadership*. Homewood, IL: Richard Irwin, Inc., and Austen Press.

Collins, J. C. (1996). Aligning action and values. *Leader to Leader 1*:19–24.

Greyson, D., & Larson, K. (2000). How to make the most of the coaching relationship for the person being coached. In M. Goldsmith, L. Lyons, & A. Freas (Eds.), *Coaching for leadership* (pp. 121–129). San Francisco: Jossey-Bass/Pfeiffer.

Hargrove, R. (1999). *Masterful coaching: Extraordinary results by impacting people and the way they think and work together*. San Francisco: Jossey-Bass/Pfeiffer.

Hargrove, R. (2004). *Harvard business essentials: Coaching and mentoring*. Boston: Harvard Business School Press.

Hargrove, R. (2004). *Harvard business review on developing leaders*. Boston: Harvard Business School Press.

Linsky, M., & Heifetz, R. (2002). *Leadership on the line*. Boston: Harvard Business Press.

Maccoby, M. (2000). Understanding the difference between management and leadership. *Research Technology Management 43(1)*: 57–59.

Mink, O. G., Owen, K. Q., & Mink, B. P. (1993). *Developing high performance people: The art of coaching*. Reading, MA: Addison-Wesley Publishing.

Newkirk, W. L. (1984). *Managing emergency medical services: Principles and practices*. Reston, VA: Reston Publishing Company.

Robbins, S. P., & Decenzo, D. A. (2005). *Fundamentals of management: Essential applications*. Upper Saddle River, NJ: Pearson Prentice Hall.

Sanborn, M. H. (1992). *Teambuilt: Making teamwork work*. New York: Mastermedia Limited, Mark Cook Associates.

Injury Prevention and EMS

4 CHAPTER

Accreditation Criteria

CAAS

105.01.02   Community Education: Through community education initiatives, the agency shall be actively involved in informing the public of out-of-hospital care, health promotion, and injury prevention (pp. 86–91).

Objectives

After reading this chapter the student should be able to:

Key Terms

attributable risk
(p. 81)

behavior-modification
approach (p. 90)

disease-management
coordinator (p. 79)

E code (p. 86)

educational approach
(p. 89)

educational interventions
(p. 86)

EMS for Children (EMS-C)
(p. 96)

enforcement attempts
(p. 87)

environmental interven-
tions (p. 86)

epidemiology
(p. 78)

Haddon Matrix
(p. 83)

"Healthy People 2010"
(p. 78)

International Classification
of Diseases (ICD-9)
(p. 86)

National Center for Injury
Prevention and
Control (NCIPC)
(p. 95)

National Institute of Occupational Safety and
Health (NIOSH) (p. 95)

National Training Institute
(NTI) (p. 78)

N code (p. 86)

persuasion approach (p. 89)

social-influence approach
(p. 90)

three Es of prevention
(p. 86)

Web-Based Injury Statistics
Query and Reporting
System (WISQARS)
(p. 79)

Point of View

Injury prevention has long been a topic of disdain for many EMS providers, yet the benefits of active participation are immense. After years of looking back at preventable injuries and death, we find that injury prevention is an activity we become impassioned about. Let me give you an example of what a little effort can accomplish.

In a particular neighborhood the incidence of auto-pedestrian deaths far exceeds any other neighborhood anywhere. Year after year children died because they ran out into traffic on a busy highway. After a series of several deaths within a few short months, the neighborhood gathered to discuss a solution. Ideas included more adult supervision, teaching the children the hazards of the traffic at a younger age, more police presence, and the list went on. Finally one of the parents suggested that a fence be placed around the playground. You see, the reason there were so many fatalities in that area was that this was the playground beside the main highway. The prevention activity of a simple chain-link fence prevented the children (and their toys) from leaving the playground and being injured or killed by the passing traffic. In the ensuing five years, not a single child was injured.

Another example: After his son was killed in a motorcycle crash, Randy O. tried to figure out why this had happened. As a paramedic, he began to think about how he could make sure no one ever had to go through the tragedy he and his family was going through. The number of children drowning in home swimming pools caught his attention. He further realized that most of these deaths were preventable. Within months Randy began talking with parents, homeowner associations, legislators, and anyone else who would listen. After hundreds of meetings and even more training sessions, the city, the county, and the state passed laws to mandate safety enclosure fences with self-closing gates around all swimming pools, private and commercial. Over the ensuing year the death rate dropped nearly by half and continued to drop for the next several years. It is hard to determine exactly how many children this one injury-prevention activity saved, but it is conservatively stated to be over one hundred children in the subsequent 10 years.

Robert (Bob) Waddell II, EMT-P (Ret.)
Cheyenne, Wyoming

■ INTRODUCTION

Consider the following scenario:

> Mrs. Smith was trying to find her way from her bedroom to the bathroom at 3:00 A.M. in her comfortable suburban home. Sleepy and trying to navigate a poorly lit hall, unsteady due to the diuretic blood-pressure medicine she took before going to bed, Mrs. Smith tripped on a rug and fell, sustaining a fracture of her left hip. She was able to access 911 from the hall phone, and an engine company arrived within four minutes of the call. They carefully placed Mrs. Smith on a scoop stretcher, padding both sides around the hip. Soon an ambulance arrived and gave Mrs. Smith 250 cc of normal saline through an intravenous line and a 5-mg dose of morphine sulfate before transport. Mrs. Smith was seen in the emergency department and went directly to surgery for a total hip repair. She spent several days in the hospital and was discharged to a rehabilitation hospital. She never completely regained her movement. She passed away from pneumonia after being bedridden for 18 months. The total cost for Mrs. Smith's event from start to finish was recorded by her HMO and calculated at $650,000. The loss to her husband of 50 years, children, and grandchildren cannot be measured.

But in truth this event never happened. Mrs. Smith's HMO had given her local fire department a grant to implement the National Fire Protection Association's (NFPA) "Remembering When" fall-prevention program after EMS managers identified several falls among the elderly in Mrs. Smith's neighborhood. An engine company visited the Smiths' house and installed donated nightlights in the hallway, instructed Mrs. Smith to remove her rugs from the slick floor, and provided her with a referral to a nonprofit agency that installed a handrail next to the bathroom. They talked to Mr. and Mrs. Smith about the height of their bed and slippers with improper soles. The time on task for the engine company was only 20 minutes in one afternoon. However, they were not awakened after midnight for Mrs. Smith's call that shift. Mrs. Smith never fell. Instead the ambulance was available for an asthmatic child who required ALS.

HISTORY OF INJURY PREVENTION

Injury is still the leading cause of death and disability in the United States. According to the Centers for Disease Control and Prevention (CDC), unintentional injury in the United States is the leading cause of death of children ages 1 through 14. Intentional injuries are also a leading cause of death. Homicide and suicide are the second and third leading causes of death for 15- to 24-year-olds. Suicide is the third leading cause of death for 10- to 14-year-olds. Homicide is the fourth leading cause of death for 1- to 4-year-olds and 5- to 9-year-olds. More than 1000 children ages 1 through 9 die each year from intentional injuries, with abuse as a probable cause. For every injury death, there are 45 hospitalizations and 1300 emergency-department visits.

This text recognizes that EMS has only briefly made progress in implementing injury prevention into the mainstream of EMS operations. As health-care costs continue to increase, there remains a need to put more money toward preventing injury to avoid a more expensive response. This requires that EMS professionals and managers possess the competencies necessary to integrate injury prevention into EMS operations. Who better than EMS and fire personnel to promote injury prevention, since EMS providers are on the front lines facing the results of unsafe acts and providing care to the victims?

Historically, EMS has sold itself as the guardian angel of the masses, ready to respond to every life-threatening event, bringing people back from the dead. The challenge with this approach is in the assumption that investing solely in a reactive mechanism is the best use of EMS. For the past 35 years, EMS has spent approximately 80% of its time, money, and expectations in advanced life support "upgrades" to provide high-end services such as cardiac and trauma resuscitation, promising to resuscitate expired victims. The problem is that EMS has not increased survival rates from cardiac and trauma arrests, no matter how many dollars, training programs, or paraprofessionals are thrown at the problem. Since it is difficult to qualify and quantify "near miss" episodes, we are left with life-and-death statistics to make the case for the profession.

In January 2003 the State and Territorial Injury Prevention Directors Association (STIPDA), in a partnership with the National Association of State EMS Directors (NASEMSD) and the National Highway Traffic Safety Administration (NHTSA), started to assess EMS activities related to injury prevention. This information has been used to set up a multi-agency injury-prevention grant program. Those involved form the National Training Initiative for Injury and Violence Prevention (Figure 4.1).

- ◆ National Association of Injury Control Research Centers.
- ◆ National Highway Traffic Safety Administration.
- ◆ Centers for Disease Control and Prevention.
- ◆ Maternal and Child Health Bureau.
- ◆ National Center for Injury Prevention and Control.
- ◆ Indian Health Services.
- ◆ State and Territorial Injury Prevention Directors Association.

FIGURE 4.1 ■ National Training Initiative (NTI) for Injury and Violence Prevention.

INJURY-PREVENTION PROGRAMS

The National Training Initiative for Injury and Violence Prevention has developed a draft set of core competencies, or essential skills and knowledge, that is the backbone of injury- and violence-prevention programs. Basic to any program is a fundamental understanding of injury as a public-health problem. Being able to describe injury as a public-health problem is a core competency an EMS manager should have.

The first criterion for defining a public-health problem is that it affects the health of a community or population and that it can be prevented by applying health-promotion and disease-prevention principles. The four objectives developed by the **National Training Institute (NTI)** are outlined in Figure 4.2.

Are injuries preventable or are they accidents? They are often referred to as car accidents, accidental drownings, and accidental falls. But injuries are not accidents. Not only is the word *accident* vague, but it also suggests a lack of predictability. It suggests that we have no idea why this injury occurred; it just did, because of bad luck. There's an **epidemiology**, or pattern, to injuries that make injuries foreseeable and preventable. Once that is established, EMS can begin to design interventions to prevent injuries from occurring.

Law enforcement has implemented a prevention model by using the Drug Abuse Resistance Education (DARE) program and community policing. The long-term cost savings to EMS, criminal-justice, and social-services systems by keeping one child off drugs pays for the activities. This is similar to the way fire prevention has prevented countless deaths with smoke- and carbon monoxide–detector education and community outreach.

The American College of Emergency Physicians (ACEP) has a list of injury-prevention resources for medical directors. ACEP has long focused on violence as a disease that an injury-prevention model could impact. ACEP's focus on interpersonal violence has made firearm violence a major focus. Physician resources can include emergency-department doctors, epidemiologists, and public-health and primary-care doctors.

NHTSA has a long history of funding highway traffic safety grants, drunk driving prevention programs, bicycle safety programs, and a variety of other transportation-related programs. Often NHTSA's activities reach the daily operations of EMS with money and activities. The Maternal and Child Health Bureau has supplied money for child safety seats and trained emergency responders to inspect and install car seats properly. The NFPA has expanded from its traditional fire-protection emphasis into health-related prevention topics. The National Association of Emergency Medical Technicians (NAEMT), through prehospital trauma life support (PHTLS), has one of the most comprehensive lists of resources, including Internet links and tools for implementing injury-prevention activities related to trauma.

A national initiative entitled **"Healthy People 2010"** has collected the national agendas of National Training Initiative (NTI) agencies to identify injury-prevention targets. "Healthy People 2010" is an upgrade to a document with a similar name dated 10 years prior. The "Healthy People 2010" benchmarks offer EMS leadership goals to work toward. For example, approximately 35 deaths per 100,000 people were caused by unintentional injuries in 1998, and "Healthy People 2010" seeks to reduce those numbers by targeting specific injuries.

The "Healthy People 2010" initiative is a national injury-prevention strategy to which fire and EMS services can compare their community to a national target for injury reduction. An example of the objectives is reducing deaths caused by unintentional injuries to no more than 29.3 per 100,000 people from the 34.7 per 100,000 in the 1987 targets. Another benchmark that "Healthy People 2010" seeks to achieve is to reduce the average 131 hospital

- ◆ Define and describe concepts of intentionality.
- ◆ Explain how injuries are preventable.
- ◆ Describe general approaches to prevention.
- ◆ Demonstrate how conceptual models are used to describe multiple risk factors.

FIGURE 4.2 ■ National Training Initiative (NTI) Objectives for Injury Prevention.

TABLE 4.1 ■ Healthy People 2010 Target and Baseline for Hip Fractures

Objective	Reduction in Hip Fractures	1998 Baseline	2010 Target
		Rate per 100,000	
15-28a.	Females age 65 years and older	1055.8	416
15-28b.	Males age 65 years and older	592.7	474

emergency-department visits per 1000 people caused by injury in 1997 to 45 per 100,000. Several other baselines and targets are listed in the document. The 1998 statistics of 1.6 drownings per 100,000 people is targeted to be reduced to 0.9 drownings per 100,000 people by 2010.

An example of how EMS can establish an effective program is to look at the baseline of 4.7 deaths per 100,000 people caused by falls in 1998 (age adjusted to the year 2000 standard population). The target is to get that number down to 3.0 deaths per 100,000 people. Fall-related injuries have been subcategorized to identify hip fractures; the strategy is shown in Table 4.1. If a fire or EMS service's community has not reached the 2010 target for hip fractures, the potential exists to partner with an insurance company, health maintenance organization, or community group to reduce the hip fractures.

■ DATABASES AND INJURY PREVENTION——

In most cases EMS providers have an idea of the major sources of death and disability in their community. When it comes to budgeting and prioritizing injury prevention, it could be a specific group at risk; for example, teens and gun violence in an inner-city community or ATV crashes in a rural community. To assist the EMS manager with identifying statistically the leading causes of morbidity and mortality, the CDC has a comprehensive database known as **Web-Based Injury Statistics Query and Reporting System (WISQARS)**. WISQARS is an interactive database system that provides customized reports of injury-related data and can query information by state and cause of death and injury (Figure 4.3).

Many state EMS systems or departments of highway traffic safety have state numbers of highway or crash deaths based on road miles. Deaths and injuries are also reported by county and, with new geographic information system (GIS) software, can be plotted in a map grid or at specific global position system (GPS) locations. Many health maintenance organizations (HMOs) track longitudinal costs and event costs for specific diseases such as asthma, diabetes, and injuries of the elderly. Even the basic information in the National Fire Incident Reporting System (NFIRS) data under the EMS screen can help create a baseline for creating an injury-prevention program.

Most HMOs employ a **disease-management coordinator** for each disease process to examine how to reduce medical costs and provide cost-effective activities such as injury prevention. Disease-management coordinators can capitate money, which means that HMOs can award a specified amount of money to an EMS agency to care for a certain type of patient. The principle is that a given amount of money is provided each year for the transportation of a medical condition. If an EMS agency can prevent that condition from requiring transport by injury prevention, the agency keeps the money.

Injury-prevention surveillance systems are used to identify the greatest risk to the public and track the progress of EMS activities as they relate to injury prevention. EMS agencies should have the ability to collect some statistics and measures of the cost of an injury-prevention program. Figure 4.4 is a 10-step process to employ in monitoring EMS injury-related activities.

Fire and EMS managers who oversee injury-prevention activities should monitor the programs for impact. A lot of effort goes into delivering and implementing injury-prevention activities, and a program

10 Leading Causes of Death by Age Group, United States – 2003

Rank	<1	1-4	5-9	10-14	15-24	25-34	35-44	45-54	55-64	65+	Total
					Age Groups						
1	Congenital Anomalies 5,621	Unintentional Injury 1,717	Unintentional Injury 1,096	Unintentional Injury 1,522	Unintentional Injury 15,272	Unintentional Injury 12,541	Unintentional Injury 16,766	Malignant Neoplasms 49,843	Malignant Neoplasms 95,692	Heart Disease 563,390	Heart Disease 685,089
2	Short Gestation 4,849	Congenital Anomalies 541	Malignant Neoplasms 516	Malignant Neoplasms 560	Homicide 5,368	Suicide 5,065	Malignant Neoplasms 15,509	Heart Disease 37,732	Heart Disease 65,060	Malignant Neoplasms 388,911	Malignant Neoplasms 556,902
3	SIDS 2,162	Malignant Neoplasms 392	Congenital Anomalies 180	Suicide 244	Suicide 3,988	Homicide 4,516	Heart Disease 13,600	Unintentional Injury 15,837	Chronic Low. Respiratory Disease 12,077	Cerebro-vascular 138,134	Cerebro-vascular 157,689
4	Maternal Pregnancy Comp. 1,710	Homicide 376	Homicide 122	Congenital Anomalies 206	Malignant Neoplasms 1,651	Malignant Neoplasms 3,741	Suicide 6,602	Liver Disease 7,466	Diabetes Mellitus 10,731	Chronic Low. Respiratory Disease 109,139	Chronic Low. Respiratory Disease 126,382
5	Placenta Cord Membranes 1,099	Heart Disease 186	Heart Disease 104	Homicide 202	Heart Disease 1,133	Heart Disease 3,250	HIV 5,340	Suicide 6,481	Cerebro-vascular 9,946	Alzheimer's Disease 62,814	Unintentional Injury 109,277
6	Unintentional Injury 945	Influenza & Pneumonia 163	Influenza & Pneumonia 75	Heart Disease 160	Congenital Anomalies 451	HIV 1,588	Homicide 3.110	Cerebro-vascular 6,127	Unintentional Injury 9,170	Influenza & Pneumonia 57,670	Diabetes Mellitus 74,219
7	Respiratory Distress 831	Septicemia 85	Septicemia 39	Chronic Low. Respiratory Disease 81	Influenza & Pneumonia 224	Diabetes Mellitus 657	Liver Disease 3,020	Diabetes Mellitus 5,658	Liver Disease 6,428	Diabetes Mellitus 54,919	Influenza & Pneumonia 65,163
8	Bacterial Sepsis 772	Perinatal Period 79	Benign Neoplasms 38	Influenza & Pneumonia 72	Cerebro-vascular 221	Cerebro-vascular 583	Cerebro-vascular 2,460	HIV 4,442	Suicide 3,843	Nephritis 35,254	Alzheimer's Disease 63,457
9	Neonatal Hemorrhage 649	Chronic Low. Respiratory Disease 55	Chronic Low. Respiratory Disease 37	Benign Neoplasms 41	Chronic Low. Respiratory Disease 191	Congenital Anomalies 426	Diabetes Mellitus 2,049	Chronic Low. Respiratory Disease 3,537	Nephritis 3,806	Unintentional Injury 34,335	Nephritis 42,453
10	Circulatory System Disease 591	Benign Neoplasms 51	Cerebro-vascular 29	Cerebro-vascular 40	HIV 178	Influenza & Pneumonia 373	Influenza & Pneumonia 992	Viral Hepatitis 2,259	Septicemia 3,651	Septicemia 26,445	Septicemia 34,069

Source: National Vital Statistics System, National Center for Health Statistics, CDC.
Produced by: Office of Statistics and Programming, National Center for Injury Prevention and Control, CDC.

FIGURE 4.3 ■ Example of WISQARS Data on Injuries.

1. Define the objectives for the injury-surveillance system.
2. Form a data committee (hospitals, HMOs, news media, and youth groups).
3. Identify existing data sources.
4. Determine the strengths and limitations of each potential data source.
5. Conduct a preliminary data analysis.
6. Reevaluate objectives for the surveillance system based on steps 3 through 5.
7. Consider linking information from existing data sources (e.g., CDC, state highway traffic safety).
8. Perform validation studies to evaluate the injury-surveillance system.
9. Develop a dissemination plan for sharing data.
10. Tie surveillance to actions and funding.

FIGURE 4.4 ■ A 10-Step Process for EMS Injury Surveillance.

that does not make an impact should be redesigned or discontinued. To defend against challenges, the proposed or existing programs need to be drawn from a cost-benefit or cost-effectiveness viewpoint.

The opportunity—and the future of EMS—is to adjust from being a primarily reactive emergency-response service to operating as an injury-prevention service. No one will argue that the "emergency response" aspect of EMS will not go away. However, prevention activities also offer a community-wide approach to healthy living. A question often arises in EMS organizations concerning their role in community health. Perhaps the best description of this relationship is that EMS is a public-service agency that provides public-health services. To embrace this concept we must accept responsibility to provide health services beyond emergency response to the citizens.

The CDC states that every dollar spent on prevention results in many times that cost in savings on health-care and rehabilitation services. Here are some examples:

Every dollar spent on:

- Bicycle helmets saves $29 in medical costs.
- Child safety seats saves $29 in medical costs.
- Smoke detectors saves $65 in medical costs.
- Counseling by pediatricians to prevent injuries saves $10 in medical costs.
- Poison-control-center services saves $7 in medical costs.

The issue for the future is how EMS spends its time, money, and expectations in managing community-health risks. Clearly, federal and state funding is squarely behind proactive and preventive activities and strategies.

■ SCIENCE OF INJURY PREVENTION

Epidemiology is the study of distribution and determinants of health-related states and events in specified populations and the application of this study to control health problems. Injuries are not accidents if they can be prevented by examining the pattern of occurrence and providing an action to stop that pattern. Epidemiology can focus on description or analysis. In the analysis phase EMS managers should attempt to identify **attributable risk**. Attributable risk assesses the ways in which a particular type of injury is associated with a particular exposure within a population. Following are some examples:

- Failure to properly restrain motor-vehicle occupants results in injuries in a motor-vehicle crash. Was it an accident that seat belts were not used? With seat belt use the injury was preventable.
- The lack of a fence around a person's swimming pool results in a child's drowning. Is that predictable and preventable?
- Parents who were not taught appropriate parenting and anger-management skills abuse their child, which results in injuries. With proper education and skills, this is preventable.

Evaluating the developmental age of a child also helps to predict injuries. In fact, age strongly influences injury rates and patterns. Understanding how assists in the design of age-appropriate interventions. Following are developmental factors and characteristics that increase an infant's and a toddler's risk of injury:

- Rapid and unexpected motor development.
- Drive for autonomy.
- Need to explore environment.
- Cannot control impulses.
- Cannot understand consequences of behavior.
- Unable to comprehend causality or assess danger and consider multiple aspects and risks of a situation.

Based on these factors, predictable injury patterns emerge for this population. Common injuries seen in this age group are assault, suffocation, drowning, falls, pedestrian injuries, poisonings, and burns. A curious infant will explore his or her environment, find a pill bottle, and then try to open it and eat the pills. Teaching infants and toddlers not to touch the bottle will not be sufficiently effective. The proper prevention strategy is to prevent the toddler from getting the bottle or from getting the pills out of the bottle, as is accomplished through child safety caps.

School-age children (5 to 12 years of age) have acquired some of the skills needed to understand risks and consequences. However, they may perform risk-taking behaviors to achieve social acceptance and as a result of their impulsiveness and belief in their own invulnerability. They have difficulty fully understanding causal relationships and like to challenge rules and convince adults that they are capable. They also have inadequate perception of sound, movement, distance, and speed.

Teenagers (children 15 to 18 years old) can understand risk and consequences and have a greater understanding about how these interrelate. However, the drive for autonomy and independence plus peer pressure make them susceptible to risk taking and impulsiveness.

PUBLIC-HEALTH MODEL

Since injury is a public-health problem, a public-health approach to injury prevention is an effective way to proceed. Looking at causes and solutions in the community as a whole, EMS managers need to focus on prevention activities. Using a public-health approach also requires looking at the multiple causes of an injury and devising multiple solutions based on the many causes. It means considering the individual, physical, and social environments and the characteristics

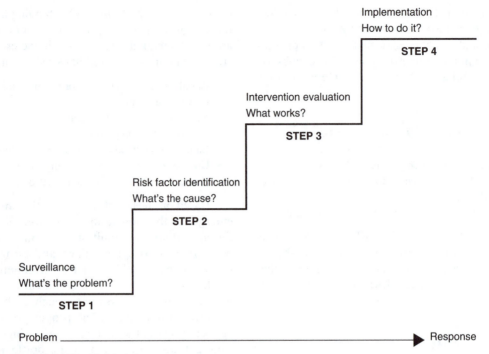

FIGURE 4.5 ■ The Public-Health Model.

of the products with which people come into contact. The CDC has a four-step model (Figure 4.5) that parallels a fire-service perspective, moving from problem identification to response.

Using conceptual models for injury prevention will assist in moving from recognizing a community injury problem to acting to prevent the problem. For example, monitor the incidences of injury by collecting data on what injuries occurred, whether they were nonfatal or fatal injuries, and the costs associated with the injuries. Once it is known what injuries are occurring, identify the risk factors for that injury and determine how and where to intervene. Risk factors can be social, genetic (such as in the ways infants react to stimuli), or environmental. Once the risk factors are identified (Figure 4.6), the

information can be used to determine where, how, and when to intervene. Finally, evaluate the intervention to determine if it affected the incidence of the injury.

Using the public health model, Figure 4.7 displays an analysis of a problem in which children are being injured in falls from playground equipment.

EPIDEMIOLOGICAL MODEL

The epidemiological model looks at the environment, the host or human, and the agent or vector, and how the three interrelate (Figure 4.8). This model is central to the evaluation of public-health problems, uncovering multiple causes and devising multiple solutions to prevent injuries.

General Model for Injury Control

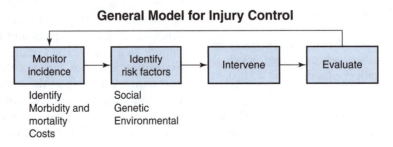

FIGURE 4.6 ■ Simple Model for Injury Prevention.

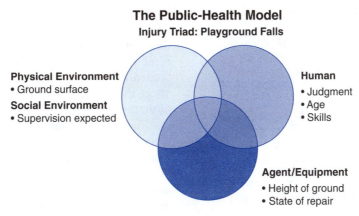

FIGURE 4.7 ■ The Public-Health Model Applied to Playground Falls.

The host is the person (or people) injured. The agent deals with the products people use, from cars to pills to cleaning supplies. Changing the agent factors requires changing design issues; for example, vehicle designs and construction methods in the design of pools or playground equipment. Packaging methods are another example of design changes, such as the modifications that have been made in pill bottles with child safety caps.

The environment is the physical and social environment in which people live, play, and work. Examples of physical-environment factors are found in how highways are built and whether or not homes are built with fire escapes. The social environment refers to the informal practices and formal laws that influence how our society functions and how it encourages or discourages certain behaviors. Some examples of social-environmental factors that would influence injury rates include policies for serving alcohol, legal driving age, concealed-weapons laws, and worker-safety policies.

What could be causing these injuries, and where could we intervene to lessen or prevent them? Consider the factors related to children and their development that could contribute to the injury:

- *Age.* Some behaviors can be traced to the predictable developmental stages of a growing child, such as how the child will respond to stimuli.
- *Judgment.* Knowing what is and is not dangerous and how to avoid danger (both toddlers and teens will explore and take risks as part of normal development).
- *Skills, knowledge, and coordination.* These may be poor and can lead to frequent falls.

Factors related to equipment (such as playground equipment) are height off the ground, state of repair, surface shape, and materials with which it is made. Environmental factors are key in determining the kinds of injuries and how often they occur. For example, a child may be on a slide that is too high off the ground or the surface below the equipment is so hard that energy is not dissipated when a fall occurs.

THE HADDON MATRIX

The **Haddon Matrix** is another important conceptual model for developing injury-prevention strategies (Figure 4.9). This model was developed by Dr. William

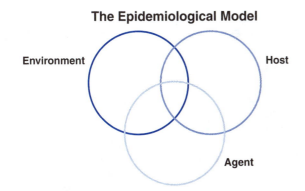

The Epidemiological Model

FIGURE 4.8 ■ The Epidemiological Model of Injury Prevention.

- ◆ Human factors.
- ◆ Agent or vehicle.
- ◆ Physical environment.
- ◆ Sociocultural environment.

FIGURE 4.9 ■ Haddon's Matrix Factors.

Haddon's Matrix: Phases of Injury Prevention

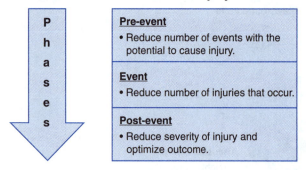

Pre-event
• Reduce number of events with the potential to cause injury.

Event
• Reduce number of injuries that occur.

Post-event
• Reduce severity of injury and optimize outcome.

FIGURE 4.10 ■ Haddon's Matrix: Phases of Injury Prevention.

Haddon, who was the first administrator of what is now NHTSA. Haddon took the public-health model's host, agent, vehicle, physical environment, and sociocultural environment a step further by looking at injuries in terms of casual and contributing factors.

Haddon believed that injuries occur with a certain time sequence, consisting of pre-event, event, and post-event phases (Figure 4.10). The buildup of uncontrolled energy is released in the pre-event phase, energy is transferred in the event phase, and factors about the state of the person, agent, or environment affect what the energy does in the post-event phase. The value of the Haddon Matrix is that it points out different areas in which interventions can be applied.

Interventions in the pre-event phase are designed to reduce the number of events with the potential to cause injury. Interventions in the event phase do not stop the event, but they reduce the number of injuries that occur as a result of the event.

And finally, interventions in the post-event phase do not stop the event or the injury from occurring, but they reduce the severity of injury and optimize the outcome to the injured party. The Haddon analysis is intended to focus on prevention, and interventions that affect the post-event phase would need to be accomplished primarily prior to an injury event's occurring.

The goal in using Haddon's Matrix is to identify major modifiable factors that lead to unhealthy outcomes. Look at the matrix itself. It is divided into a series of cells. Each of the cells represents an opportunity to think through the prevention possibilities and avoid devoting all of one's attention to only one or two cells on the matrix, such as pre-event education to the host. Figure 4.11 displays the use of the matrix in motor vehicle crashes.

Conceptual models are used in injury prevention to describe multiple factors contributing to the underlying injury or violence. The use of the Haddon Matrix can visually convey complex issues to politicians and employees who are funding or trying to impact the injury factors.

ACCESS, INTERPRET, USE, AND PRESENT INJURY DATA

EMS managers need to have a core competency in interpreting data. Some specific tasks EMS managers and leaders may be required to do are to:

♦ Explain the importance of data for use in program planning and evaluation, advocate for injury prevention, and understand a given community's unique needs.
♦ Be able to describe and understand key sources of data at the national, state, and community level and describe their strengths and weaknesses.

Phase	Host (Human)	Vector (Vehicle)	Physical Environment	Cultural Environment
Pre-event	Alcohol. Experience. Judgment.	Brake status. Tires.	Night. Rain.	Acceptance of drinking and driving.
Event	No seat belt.	No air bag. Hardness of surfaces.	Tree too close to road. No guardrail.	Speed limits. Enforcement of seat-belt laws.
Post-event	Physical condition.	Fuel system integrity. Cell phone.	Distance of emergency response.	Support for trauma systems. Training level of EMS personnel.

FIGURE 4.11 ■ Example of Haddon's Matrix Using a Motor-vehicle Crash.

Access to good injury data makes it possible to identify and understand the local community injury problem. It also helps to prioritize a list of problems and determine how large specific problems are. It enables the EMS manager to note patterns of when and how injuries occur and to monitor trends to determine if injuries have increased or decreased over time. Data can help assess emerging injury issues and identify behavioral and environmental risk factors so the EMS manager can design, implement, and evaluate an effective injury-prevention program.

Consider how we might use data to help define an injury problem. Data can help answer the who, how, where, what, and when questions that are important to selecting a cause of injury to address, a target population, and other issues related to the problem that are useful in choosing an appropriate intervention. The key questions are:

- Who is being injured (age group and ethnicity)? This will assist in determining the target age group for the intervention, as well as the target ethnicity, gender, and community.
- How is the person being injured? What is the cause, for example, suicide with firearms or suicide by poisoning? This is the mechanism of injury.

- Where are the injuries taking place? If the injury is drowning, is it a drowning in a pool or an open body of water?
- What about the circumstances under which injuries occur? In the case of a burn, what is causing the burn? Space heaters? Cigarettes? This looks at the causes of exposure and how they resulted in an injury.
- How serious are the injuries? How many are fatal? These questions help us determine the severity of the cause of injury.
- When are the injuries occurring? This evaluates time trends, rather than the time of day. If there was a drop in traumatic brain injuries (TBI) in 1995, look at what else happened at that time; for example, was a helmet law passed?

■ FRAMING AN INJURY PROBLEM

To understand the nature of an injury problem, an EMS manager needs to assess both fatal and nonfatal injuries. The use of an injury pyramid shows that deaths are just the tip of the iceberg: the farther down on the pyramid, the more injuries there are (Figure 4.12). This is another way to display information and send a message about a particular injury

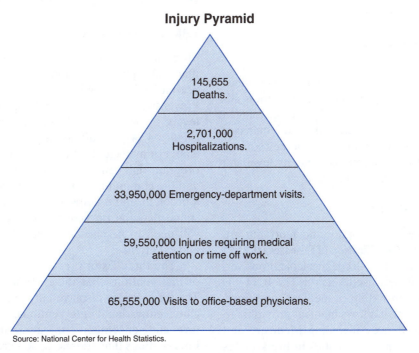

Injury Pyramid

145,655 Deaths.

2,701,000 Hospitalizations.

33,950,000 Emergency-department visits.

59,550,000 Injuries requiring medical attention or time off work.

65,555,000 Visits to office-based physicians.

Source: National Center for Health Statistics.

FIGURE 4.12 ■ Injury Pyramid.

pattern. Far more people suffer nonfatal injuries than fatal injuries, and the cost associated with the survivors often has a larger impact on the system than fatalities.

FINDING DATA TO DEFINE THE INJURY PROBLEM

The primary sources of data on deaths are vital statistics and coroner or medical-examiner data. Information from these two sources is sent to the CDC for analysis and tabulation. Child-death-review teams, Fatal Accident Reporting System (FARS), and law enforcement are also sources of mortality data. Soon the National EMS Information System (NEMSIS) will collect data from local EMS agencies and pass it to state EMS offices, who will in turn feed the data into federal EMS databases.

To get a complete picture of a problem, it is important to look at morbidity or nonfatal injury data. Because of its complex nature, data on nonfatal injuries can be difficult to obtain, and their coding and completeness can vary widely. Reporting of nonfatal injuries is not consistent. An example is someone who falls at home, is injured, but does not seek medical care. Sources of morbidity data in your community may include:

- Trauma registries.
- Emergency-department records.
- Hospital-discharge data.
- Emergency medical services.
- Poison control center.

EMS leadership will need to explain the importance of data for use in program planning and evaluation, advocating for injury prevention, and understanding a given community's unique needs. An EMS manager should be able to describe and understand key sources of data at the national, state, and community level and describe their strengths and weaknesses.

FEDERAL COST TRACKING

Many of the sources of morbidity and mortality are based on **International Classification of Diseases (ICD-9)** code. The next version, ICD-10, is available; however, ICD-9 is the most widely used version. The ICD-9 is the world classification system of medical conditions, diseases, and injuries that allows medical providers to bill government insurances, including Medicare and Medicaid. Injury-prevention activities in fire services and EMS can utilize ICD-9 codes to determine the

| N codes | 800–829 | Fractures |
| E codes | 880–888, 885 | Fall on the same level |

FIGURE 4.13 ■ Sample ICD-9 Codes for a Hip Fracture.

nature of an injury and how it occurred. Information on the nature of an injury is in the **N code**, and information on the external cause is captured in the **E code**. E codes are required on all death certificates, and they provide common categories for physicians, paramedics, and insurers. Figure 4.13 shows a common series of E and N codes for ICD-9 that involves an injury from a fall that results in a fractured hip.

Now that the injury-prevention problem is known, the EMS manager needs to identify and choose an injury-prevention strategy to employ. The ability to identify and choose strategies must be accompanied by the ability to design and implement injury- and violence-prevention activities. To develop this competency the EMS manager will demonstrate the use of conceptual models for identifying intervention opportunities. Exercises provide opportunities for interventions; the EMS manager will need to choose and justify an intervention and outline the steps for developing an implementation plan.

■ INTERVENTIONS

THREE STRATEGIES

There are three models that can be used to develop multifaceted injury-prevention strategies. The **three Es of prevention** in the Haddon Matrix detail the Spectrum of Prevention, which talks about the different community levels at which the EMS manager can intervene (Figure 4.14). Identifying a problem moves the EMS leader or manager to the next step, which is designing a program that can impact the outcome for the patient and the community. There are often many solutions to a problem, and it helps to define the option to make a sound decision on actions to be taken. Interventions will need to be chosen based on resources, finances, and experience.

The three Es of prevention look at the key types of interventions that the EMS manager can employ. The three Es are displayed in Figure 4.14, and they include **educational interventions**, enforcement activities and enactment of laws, and **environmental interventions** that address the environment or product engineering.

The Three Es of Prevention

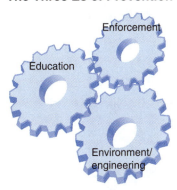

FIGURE 4.14 ■ The Three Es of Prevention.

Educational interventions are probably familiar to most people. They are generally the easiest ones to implement. Educational interventions, however, may not be the most effective in reducing the incidence or severity of an injury. Generally, they are most effective when used with strategies identified in the other Es, such as a policy change. Educational interventions attempt to initiate behavioral changes by informing a target group about potential hazards, explaining risks, and persuading people to adopt safer behavior. They also can be used to inform policy makers about issues. Education will not always cause individuals to alter their behavior, but it can make them more receptive to additional injury-prevention strategies. Education also can be used as part of social-norm change.

Enforcement attempts reduce dangerous behaviors through legislation and enforcement of that legislation. Legislation can target behaviors by individuals, manufacturers, and local governments. Examples of state and local injury laws include:

- Regulating the color and speed capacity of school buses.
- Mandating that buildings be constructed to meet codes and standards.
- Requiring reporting of child abuse.
- Restricting the sale or giving of alcohol to minors.

These laws would be useless if they were not enforced, and education plays a key role in informing people of their responsibilities under these laws.

Environmental interventions make changes to the environment or product design to protect everyone automatically. These are sometimes called passive or automatic interventions, because they require no work on the part of the individual. Interventions that do not require direct action by the individual are usually more effective. Examples of environmental interventions include child-resistant pill bottles, changing the type of windows installed or installing locks, and providing soft surfaces for playgrounds.

The most effective prevention strategies combine tactics from each of the three categories. The following example demonstrates the use of the three Es with seat-belt laws:

- *Enforcement/enactment:* Policy change and consequences by law enforcement.
- *Education:* Inform people of the new law.
- *Environment/engineering:* Mandate seat belts in design of cars.

The three Es are used as a way to approach the different types of interventions. EMS managers should remember to consider developmental stages when designing their interventions. For example, education may not be appropriate for toddlers.

THE SPECTRUM OF PREVENTION

The Spectrum of Prevention expands on the three Es by describing seven levels at which prevention activities can occur (Figure 4.15). Often prevention consists of distributing brochures or doing community education, but the Spectrum goes beyond those activities. Since injury problems are often complex, the best solutions are usually comprehensive and address all levels of the Spectrum. The levels are:

- *Level 1, strengthening individual knowledge and skills.* An example is training expectant mothers on correct use of car seats.
- *Level 2, promoting community education.* An example of a communitywide program is the "April Pools Day" held at swimming pools in conjunction with mass media.
- *Level 3, training EMS providers.* EMS managers can train bicycle sales staff on the risks of head injuries and guidelines for properly fitting bike helmets.
- *Level 4, fostering coalitions and networks.* An example is the "Safe Routes to School" program, which was formed to address high rates of childhood pedestrian injuries.
- *Level 5, changing organizational practices.* An example is a day-care center that adopts new rules for medication storage.

Levels of the Spectrum of Prevention

1. Strengthening individual knowledge and skills.

2. Promoting community education.

3. Training providers.

4. Fostering coalitions and networks.

5. Changing organizational practices.

6. Mobilizing communities and neighborhoods.

7. Influencing policy and legislation.

FIGURE 4.15 ■ Levels of the Spectrum of Prevention.

- *Level 6, mobilizing communities and neighborhoods.* We see this implementation strategy in the national "Neighborhood Watch" program and the "Exit Drills in the Home" project.
- *Level 7, influencing policy and legislation.* An example is a coalition that petitions the city council to upgrade playground surfaces.

To apply a public-health approach to injury prevention, look at the different locations where an intervention would be most effective. If while using the Haddon Matrix you identified the problem as a lack of pool barriers and a potential intervention is to install pool barriers, the Spectrum will help determine at what levels to work to implement the solution. To say "install pool barriers" is one thing, but how do you convince current homeowners to install them? It may be possible to create a coalition that can do educational outreach to consumers. Primary-care providers, for instance, can discuss this issue with their patients.

■ **PUTTING IT INTO ACTION**

INJURY-PREVENTION LEADERSHIP

Injury prevention requires solid leadership and, much like the success that is enjoyed by fire-prevention public-education specialists, the role of the EMS injury-prevention specialist needs to be defined and have appropriate human resources. Often injury-prevention activities require a tremendous effort in networking, planning, meetings, and preparation that can monopolize an EMS manager's time.

Another important human resource is to partner with a local university or college that has a health-education program. A permanent internship program that utilizes college students in the last year of their studies can be a win-win for all parties. It affords the student a positive experience with EMS services, and it brings the latest information and practices in injury prevention and community health to the doorstep. Often these programs run continuously and require an EMS manager responsible for injury prevention to establish goals and objectives each semester for the student, to write a midterm exam, and provide a final evaluation on the student's ability to meet those goals. This is an excellent opportunity for the student to be mentored by EMS leadership. Most internships pay the intern a stipend or hourly wage and require 20 hours per week of time and usually award a student three to six college credits. Most positions do not offer benefits and are outside the labor contract.

IMPLEMENTING STRATEGIES AT THE COMMUNITY LEVEL

In order for injury-prevention efforts to be successful, concurrent changes often must be taken together by individuals and agencies. A systems approach to injury prevention involves changes at the individual, agency, and environmental levels. If EMS managers think of their community as one big system, these levels are all part of the community and therefore must be part of the change.

This systematic, step-by-step approach is outlined in Figure 4.16. Each step will not necessarily happen in the order shown. The key is to remain flexible. The first

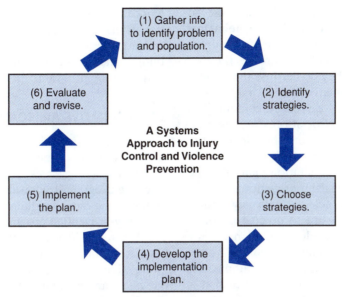

FIGURE 4.16 ■ Sample of a Systematic Approach to Injury Prevention.
(Reprinted with permission of UNC Injury Prevention Research Center, the University of North Carolina at Chapel Hill.)

step in a systematic approach to injury prevention is to learn as much as you can about the problem. Gather data on who is affected, what affects them, where they are affected, why they are affected, and when they are affected. Any solution used has to be done within the larger context of the community. Find out more about what resources are already available in the community and what barriers exist in the community.

Once the data have been gathered, select the injury priorities. This will help define prevention strategies, identify resources and potential partners, and clarify objectives for evaluation. If there are specific priorities and narrowly focused efforts, chances are that intervention will be more successful. The questions to ask when selecting priorities are, What are the injury problems? Is there already an effective intervention strategy? Is the community interested in doing something about this problem?

The EMS manager should not be afraid to start small. Even small successes are still successes, and they will help bring additional partners and resources into the injury-prevention program. Once the problem has been identified and the focus narrowed, choose the intervention strategies. Use the Haddon Matrix, the three Es, and the systems approach to assist in the process.

To select strategies start by looking at the typical scenario of an injury problem, or how the injuries are occurring. If playground injuries are a problem, what types of injuries are they? If most injuries are falls, what are characteristics of the playground equipment and environment that could be contributing to these fall injuries? Pick a proven strategy for the first effort to ensure a greater likelihood of success. Successes help establish credibility, gather resources, and get more partners on board.

There is a tendency for people to think they have to, or they want to, develop something brand new. First and foremost, explore what is available. Talk to people who are working in the field and find out what exists. Existing products, models, and solutions may be more cost effective and specifically tailored to shorten and improve the task.

There are four general approaches to induce behavioral change in injury prevention: education, persuasion, behavior modification, and social influence. The **educational approach** is based on the belief that individuals will do the right thing if they understand why and know how to carry it out. The **persuasion approach** subscribes to the idea that people are motivated after a careful argument and the triggering of their motivational hot buttons. It requires the person listening to the message to adopt the message's philosophy. In the **behavior-modification approach**, the influence of thoughts and feelings is minimized; that is, it stresses the theory that people do what they do because they find it rewarding. The **social-influence approach** advocates a campaign that influences the entire community's collective behavior. It uses social pressures to promote conforming to safe behaviors, showing that the safe behavior is preferred.

Next, develop an implementation plan or scope of work to guide and focus the prevention efforts. An evaluation should be included with any injury-prevention activity. Make sure to involve partners and other agencies in the development plan so they have buy-in from the beginning. Too many groups decide to give away bike helmets, for example, but do not think about why they are doing it or if it is the appropriate intervention. The implementation plan includes:

- ◆ *Goals and objectives.* What you are ultimately trying to do and how you are going to do it.
- ◆ *Activities.* Who is doing what and how it is going to be done.
- ◆ *Evaluation measures.* How you are going to track what you are doing to see if it is successful.
- ◆ *Resources assessment.* What resources are needed.
- ◆ *Timeline.* A time frame for completion of each step.

It may be difficult to measure outcomes, so it is important to have appropriate evaluation measures in the plan. An example of an outcome is to lower fall-injury rates by 15%.

FUNDING STRATEGIES

Injury-prevention initiatives cost money. The start-up money for an injury-prevention program can come from grants or line items in a fire or EMS budget. A budget for administrative costs and money for program costs or the delivery of the program should be delineated in a line item. Any injury-prevention programs need a plan for sustaining the program activities. Money for programming can come from fees for service provided by EMS activities. Fines and surcharges, such as a percentage of vehicle registration or speeding tickets, can be earmarked for injury-prevention activities. The cost of human resources can be offset by pro bono work, work done by light duty or injured fire and EMS

workers, and volunteers. Sponsorships and in-kind donations from business can fund specialty projects or sustain programs. Whatever the program is, a budget and sources of funding must be identified to continue the program.

COLLABORATIVE APPROACHES

When implementing injury-prevention programs and activities, the EMS manager should look to collaborate with other agencies or community partners to deliver the program. The diversity of forming a coalition or task force can strengthen the impact and delivery of an injury-prevention program.

When considering a collaborative approach to injury prevention, pick one problem to work on. Injury prevention should reflect a local problem that is prevalent and preventable. Formulate a plan using the FEMA five-step process shown in Figure 4.17 and identify a list of people or organizations that have a compelling reason to get involved. The FEMA five-step process is designed to help build coalitions for public injury-prevention education.

Some of the people who potentially will get involved are the lead agency or a fire or EMS unit affected in the problem area. Using skateboard injuries as an example problem, get fire and EMS crews who cover a skateboard park involved with injury-prevention activities. Look for communities to assist within the service area. For example, Hispanic groups may help with injury-prevention activities that benefit people riding in the beds of pickup trucks.

Finally, identify survivors or the people at risk to help support and promote the program. Identify a leader or an interested person to be personally and professionally committed to the program. The leader should seek advice and ideas. It is important to create a list of potential coalition contacts or resources. Figure 4.18 is a sample form for maintaining contact information of key resources and people to help ensure collaborative effort in injury prevention.

Injury-prevention programs make an excellent opportunity to partner with local colleges and universities. Under the direction of an EMS division, college interns can be utilized in health promotion, education, and sociology programs to meet their independent study or internship degree requirements. Often this requires EMS leadership to conduct or fill out college evaluation forms for the student to receive college credit.

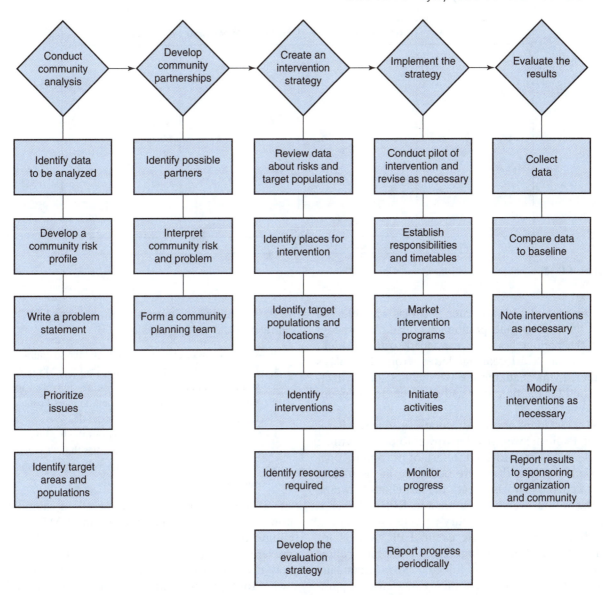

FIGURE 4.17 ■ The FEMA Five-Step Public Education Planning Process on Public Fire and Injury Prevention.

■ BEST PRACTICES IN INJURY PREVENTION—

SAFE COMMUNITIES

"Safe Communities" is a unique approach to a comprehensive injury-control coalition that includes community-based representation by citizens, law enforcement, education, business, civic groups, and other government agencies. The coalition uses the best-practices model incorporating new strategies. NHTSA provides a complete start-up kit for "Safe Communi-

ties" that has a step-by-step process for implementing the program.

THE EPIC MEDICS STORY

On May 8, 1996, a San Diego County paramedic unit responded to a call to assist a two-year-old boy named Nicholas Rosecrans. The boy drowned after he wandered away from a day-care center into the unfenced pool of the house next door. The paramedics' resuscitative efforts were successful only to the point of return

Area/Expertise	Name/Contact Info	Member Responsible	Outcome

PIER Injury Prevention Curriculum, Module V: Strategies for Safety Coalitions.

FIGURE 4.18 ■ Coalition Recruitment and Contact Planning Sheet.

of a pulse, and 12 hours later Nicholas was released from life support and pronounced dead. A few days later the paramedics who responded to the incident received a heartbreaking letter from Nicholas's mother, Lynn, thanking the paramedics for their efforts and for the time they had given her to say good-bye to her son.

This incident and its aftermath so touched Paramedic Paul Maxwell that he vowed to do everything in his power to try to end the rash of preventable pediatric drownings that were occurring in his district and to make Nicholas's death the last child-drowning call he would ever have to answer. EPIC Medics, a local paramedic-based injury-prevention initiative, was born from this tragic event. EPIC stands for Eliminate Preventable Injuries of Children. Fellow paramedic Josh Krimston joined Paul in organizing other paramedics to utilize their experience as paramedics to turn tragedy into action. Together they founded EPIC Medics.

Paul Maxwell contacted the San Diego Safe Kids Coalition. The coordinator of the coalition, Roxanne Hoffman, immediately recognized the value of using a paramedic as a spokesperson to influence public policy and deliver the prevention message. "Our paramedic should know. . . he sees it every day," she would say to the press. Together, they used EMS data, the media, and many other strategies to help influence the passage of AB3305, a law that requires barriers surrounding all new pool construction in the state of California.

EPIC Medics is an example of a local nonprofit organization of paramedics, firefighters, EMTs, and others who work toward eliminating or reducing preventable childhood injuries. The acronym EPIC also represents the group's mission:

E—Eliminate or reduce preventable childhood injuries.
P—Provide educational opportunities for EMS personnel to enhance emergency care delivered to sick and injured children and their families. Protect EMS personnel from injury or death.
I—Influence public opinion in favor of the health and safety of children in the communities that we serve.
C—Coalition building to mobilize other groups and individuals who share our goals.

This grassroots effort is an excellent example of how a small group of paramedics and EMTs can facilitate best-practice local community-based injury-prevention efforts. The Nicholas Rosecrans Award is now an annual award that showcases the best-practices award for injury prevention. The award is designed to recognize emergency responders who demonstrate leadership, commitment, and innovation in preventing injuries.

FRISCO FIRE SAFETY TOWN PROJECT

The Frisco Fire Safety Town Project in Frisco, Texas, is an example of a collaborative effort designed to address safety concerns for all age groups. Its broad focus and comprehensive facility make an effective, innovative solution for a wide variety of safety issues in the community. The Frisco Fire Safety Town concept provides life-size learning and experience (Figure 4.19). Its goal is to create a model community complete with streets, stoplights, and businesses that allow for participants to

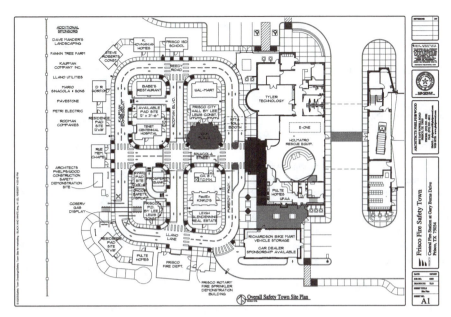

FIGURE 4.19 ■ Frisco Fire Safety Town Schematic.
(Reprinted with permission of Frisco [TX] Fire Department)

be actual drivers and pedestrians while receiving hands-on training in motor and bicycle safety (Figure 4.20).

Model businesses and sponsored buildings can accommodate safety exhibits and interactive simulators to enhance the educational experience. For example, the Frisco Fire Safety Town has an exhibit on tornado safety that uses a simulated home with an animated tornado. It allows students to tour a home and take action during a tornado warning and to learn where in a home to seek shelter.

FIGURE 4.20 ■ Safety Town.
(Reprinted with permission of Frisco [TX] Fire Department)

Case Study: Bicycle Helmet Program for Elementary School Children

The following [is] an example program developed and implemented by the Harborview Injury Prevention and Research Center (HIPRC). . . . Please contact the HIPRC if you want more information and educational materials. . . . Also note that the statistics are no longer current.

Step One: Gather and Analyze Data

Here is a sample of the type of information that was used to support the need for the program.

◆ According to the National Electronic Injury Surveillance System (NEISS), there were over 500,000 emergency room visits per year due to bicycle crashes. Nearly 400,000 of these were children. Bicycle injuries were the main recreational cause of emergency room visits and the second cause of hospital admissions.
◆ According to the Washington State Department of Health, from 1985–1989 there were 40 children (1–19 years of age) who died from bicycle-related injuries and 322 were hospitalized over the same time period.
◆ Nationally about 2% of children wore helmets in 1986.

Step Two: Select Target Injuries and Population

According to the above information as well as other data, children were at greater risk for injury and death due to bicycle crashes. While the 10–14 year olds were at somewhat higher risk, the 5–9 year old group was targeted because the age group was known to be more amenable to parental influence and had spent a shorter time riding a bike bareheaded.

Step Three: Determine Intervention Strategies

A few of the ideas developed in brainstorming for strategies included:

◆ Educational strategies: Do not let children ride in streets. Do not let children ride after dark. Train children to ride properly.
◆ Technological strategies: Barriers to separate bicyclists from motor vehicles. Helmets resistant to impact.
◆ Legislative strategies: Stop producing bikes. Outlaw sale of bikes to children under 15 years old.

Of the above strategies, the most effective, easily accomplished and realistic was to promote the use of a helmet to prevent head injury. Because the HIPRC is a research institute, there were resources available as well as the desire to examine factors of the potential program in greater detail. It was decided that answers to the following questions could help shape the program further:

1. How effective are helmets in reducing injury to the head?
2. Why do not parents buy their children helmets?
3. Why do not children wear bicycle helmets?

Step Four: Develop an Implementation Plan

To address question #1 above, a research study looked at people coming into five Seattle-area emergency rooms for bicycle-related head injuries. The study also surveyed people who did not come into the ERs because they had been wearing a helmet that worked. The results of the study showed that helmets reduce the risk of serious injury to the head by 85% and to the brain by 88%. Therefore, helmets were proven to be a very effective prevention intervention.

A survey of 1000 third graders and their parents answered questions #2 and #3. The most common reason parents gave for not having bought a helmet for their child was they had never thought about it, the second reason was cost and the third was they did not think their child would wear one. The children responded that helmets were uncomfortable and they worried about what their friends would think.

Based on these results the specific program goals and objectives were:

◆ Increase helmet purchase by parents: increase awareness of problem and solution, reduce financial barrier: discount coupons, bulk buy programs, subsidies; focus on helmet use only; not on riding behavior.
◆ Increase helmet use by children: focus on 5–9 year olds who are susceptible to parent pressures and have less history of bareheaded riding; increase pressure to use—parent modeling, more colors and styles, stickers, etc.; increase availability in department stores, toy stores, etc.; link helmet wearing with fun activities—"hero" modeling and incentives; require use when possible.

FIGURE 4.21 ■ Case Study: Bicycle Helmet Program.

The Frisco Fire Safety Town is available for all area schoolchildren, community groups, businesses, and visitors from across the region. The classrooms can be used by the public to utilize information that is up to date, comprehensive, and in a usable format for injury prevention. Professional instruction from firefighters, police officers, and health-care professionals is provided for accurate and age-appropriate training.

Several safety towns are now being developed across the country. In Hagerstown, Maryland, a five-acre Safety Village systematically educates every elementary school student in the county on pedestrian, fire, and bicycle safety. A safety town or village serves as a linchpin for community safety efforts and provides an opportunity to establish injury prevention as a continuous and necessary part of public-safety operations.

SEATTLE BIKE HELMET PROGRAM

The Harborview Medical Center Injury Prevention and Research Center initiated a program to reduce the number of injuries and deaths in the King County area from bicycle falls and crashes. The Harborview project used a 10-step process with an outcome-and-process evaluation to assess impact. Figure 4.21 shows a sample of a well-documented program on

Step Five: Identify, Select and Commit Community Agencies to Implement the Plan

The lead agency in the Seattle/King County helmet program was the HIPRC. In other counties across the state it was, in most instances, the local health department for the county. The designated coordinator for the Seattle program was a staff person at the HIPRC and a nurse or health educator in the local health departments. Enthusiastic and committed coalition members included: Health Departments; State Medical Association; bicycle organizations/clubs; health maintenance organizations, hospitals and clinics; helmet manufacturers and distributors; advertising and marketing representatives; representatives of TV, radio, or print media.

Step Six: Develop an Action Plan

A multifaceted approach was developed to involve the entire community which required the contribution of all coalition members: physician advice; media—victim stories, Public Service Announcements, repeated articles in the press, TV and radio talk shows, press conferences; informational materials—posters, brochures, flyers, videos, etc.; manufacturers and distributors—discount coupons, bulk purchase, more helmet choices; PTAs, youth groups, schools, health fairs, rodeos and other community events.

Step Seven: Orient and Train Agencies/Individuals Implementing the Intervention Plan

Initial coalition meetings were the most appropriate time to orient people to the prevention program and the reasons for its implementation. The problem was described both nationally and locally through references to data. The program plan was outlined.

Step Eight: Implement the Program

The coordinators of programs in communities across Washington gathered local data and planned the program goal and objectives prior to the first meeting. Coalition meetings started early in January. Some coalitions met once a month, others met more frequently. Activities took place in spring, summer and fall. During the late fall and early winter coalition members could rest but the coordinators were planning for the next season's program and activities.

Step Nine: Monitor and Support the Program

Meetings were held and memos and information were sent to coalition members. Telephone contact was frequent by the coordinator. Small decisions were made by the coordinator, larger ones brought to meetings for coalition consensus. Coalition members were encouraged to share stories of activities and keep each other aware of opportunities for making helmets visible.

Step Ten: Evaluate and Revise the Program

Program evaluation:

- Process evaluation. Some of the measurements used included: number of parent groups reached; number of educational brochures and flyers distributed; number of discount coupons for helmets distributed and the number redeemed; number of helmets donated; number of low-cost programs: discount programs, bulk purchase plans.
- Outcome evaluation. Observations of helmet use also took place: observation by the HIPRC of the number of children wearing helmets in the early spring before activities began and again in the fall when activities were over for the bicycling season showed an increase from 5.5% to 33% over five years of program implementation in the Seattle/King County area.

Experimental research: A survey was done by the HIPRC in 1986 through King County pediatricians' offices which revealed the helmet use by children to be at 1%. A year later, when a few program activities had already started, a study compared the use of helmets by children in Seattle with the use by children in Portland, Oregon, where there was no helmet promotion program. The usage in Seattle was 5.5% compared to 1.0% in Portland. After two seasons of the program in Seattle the usage had risen to 15.7%, while in Portland it was 2.9%.

The following surveillance mechanisms were used to detect changes in bicycle head injury trends: county vital statistics; hospital discharge records; State Patrol data registry at the regional trauma center.

FIGURE 4.21 ■ Continued

how to reduce bicycle injuries and head trauma related to bicycle injuries.

RESOURCES

The lead federal agency for injury prevention is the CDC and the **National Center for Injury Prevention and Control (NCIPC)**. The NCIPC provides services through three branches: the division of unintentional injury; the violence prevention division; and the acute care, rehabilitation research, and disability prevention division. The NCIPC coordinates injury-prevention activities and provides an infrastructure for injury-prevention programs. The goal is to provide lower morbidity and mortality on nonoccupational injuries.

Occupational injuries are under the jurisdiction of the **National Institute of Occupational Safety and Health (NIOSH)**. NIOSH is also under the management of the CDC and is charged with supporting the Occupational Safety and Health Administration (OSHA) with research and scientific analysis of occupational injury. NIOSH investigates workplace hazards and deaths and supports a university-based educational research center. The National Center for Health Statistics (NCHS) focuses on monitoring U.S. health data, collecting and analyzing mortality and illness.

- ◆ National Association of State EMS Directors.
- ◆ Emergency Medical Services for Children.
- ◆ National Highway Traffic Safety Administration.
- ◆ Centers for Disease Control.
- ◆ Local governments.
- ◆ Nonprofit groups and patient-advocacy groups.
- ◆ National charities and foundations.

FIGURE 4.22 ■ Agencies Involved in Injury Prevention.

The Maternal and Child Health Bureau (MCHB) and the Health Resources and Services Administration (HRSA) provide services to low-income and high-risk populations through grants. The MCHB supports the Traumatic Brain Injury (TBI) program and the **EMS for Children (EMS-C)** program to improve the way EMS serves children in emergency situations. The EMS-C program conducts several injury-prevention programs and training efforts. The National Institutes of Health (NIH) conducts medical, social, and behavioral research on injuries. Figure 4.22 displays a more comprehensive list of federal agencies involved with injury prevention.

EVALUATING INJURY-PREVENTION PROGRAMS

A formative evaluation can be used to refine a program's implementation before full-scale implementation (Figure 4.23). This generally includes pretesting, pilot studies, and focus groups to determine the likely response of target audiences and program developers. With formative evaluation, information is collected and fed back to program developers early in the planning process in order to enhance the program and maximize chances of success.

A process evaluation can document the degree to which an injury-prevention program is being implemented as designed. Instead of looking at program results, a process evaluation looks at whether program activities and their delivery are being carried out as planned. A process evaluation looks at what was actually done; where and when it was done; how often it was done; and by whom, with whom, and for whom it was done.

An outcome evaluation looks at whether or not the objectives of a particular injury-prevention program have been achieved. It documents a program's effectiveness in producing expected outcomes related to the program goals. For example, has there been a change in the number or severity of injuries the injury-prevention program was trying to prevent? This may include a particular risk

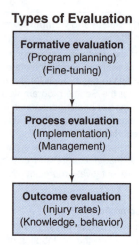

Types of Evaluation

FIGURE 4.23 ■ Types of Evaluations.

factor, knowledge, attitude, or policy. A helmet campaign may measure an increase in the number of children wearing bike helmets. The implementation plan should include some combination of each of the evaluation tools.

TIME TO IMPLEMENT

After the partners have had time to review and comment on the implementation plan, it is time to implement. When starting the plan, think about who needs to be involved and how they might be pulled together. Look for partners who are credible, have a similar mission, and also have the organizational structure to carry out the plan. Look beyond the obvious, and include the broader community. To pull these groups together, think about forming a coalition or task forces that can help carry out implementation, identify and solicit resources, and solve problems that arise along the way. Consider working with an existing coalition, rather than developing a new one.

How can the EMS manager ensure the activities will continue to be carried out after the manager has left or after initial funding ends? Building injury-prevention activities into what people are already doing, rather than giving them a whole new program to implement, will create the opportunity for something to continue over a long period of time. An example is having fire and EMS personnel inspect playground equipment after conducting a fire drill in a school.

There should also be an intervention protocol that defines the roles and responsibilities of each person and agency involved. This will make it easier for people to follow through with their assigned tasks. Even if there is a coalition, a single person will be the catalyst that will make things happen. Make sure that training is provided to agencies and individuals who will conduct the intervention.

Finally, EMS managers should evaluate their programs and revise them if necessary. Remember to design evaluation as part of the implementation plan before beginning the intervention. Evaluations help determine if the program is actually preventing the injury it is meant to prevent. Evaluations also enable the EMS manager to modify the intervention to increase its chances of success, help convince funding agencies and the community of the value of the program, and help make it more effective. The key questions in the evaluation are, What is the program designed to do? Is the program doing what it was designed to do? Is the intervention making a difference?

CHAPTER REVIEW

Summary

Investing in injury-prevention efforts give EMS an opportunity to shift from a reactive to a proactive nature. The investment pays for itself several times over and allows for perhaps the most important mission of EMS: saving lives. The old saying, "pay me now or pay me later" is applicable here. The question to ponder is how to best invest our limited time, money, effort, and expectations in reducing risk and preventing injury. Preventing death and injury is much more defendable than attempting to resuscitate the dead.

EMS professionals should be actively involved in safety and injury-prevention programs. Programs should be implemented that address the population at risk in the community. Get the entire department involved, and find individuals who have passion or a connection to the people in need. Ensure integrity by conducting business in a way that demonstrates the belief or by being a mentor to those who are attempting to deliver the program. Educate the public, and encourage widespread dissemination of the information and public advocacy.

Research and Review Questions

1. Using Haddon's Matrix, develop an abbreviated plan for the prevention of one of the following: gun safety for ages 4 to 7, motorcycle-helmet use for 18-year-old youth, and drown-proofing at a reservoir.
2. Describe the public-health model for injury prevention, and explain a method to implement it in a typical emergency-response organization.
3. Identify one local source of funding and one federal source of funding for a prevention program on bicycle safety and use of helmets.
4. Using the WISQARS database from CDC, identify the leading causes of injury in the county in which you reside or work.
5. Create a strategic plan for your agency's injury-prevention activities with a 5-year and a 10-year timeline.

References

Barker, W. H. (1998). Prevention of disability in older persons. In R. B. Wallace (Ed.), *Public health and preventive medicine* (14th ed., p. 1063). Stamford, CT: Appleton & Lange.

Chrisoffel, T., & Gallagher, S. C. (1999). *Injury prevention and public health, practical knowledge, skills, and strategies*. Gaithersburg, MD: Aspen Publications.

Haddon, W. J. Jr., & Baker, S. P. (1981). Injury control. In D. Clark & B. MacMahon (Eds.), *Preventive and community medicine*. Boston: Little Brown & Co.

Institute of Medicine. (1997). *Improving health in the community*. Washington, DC: National Academy Press, 48–56.

Institute of Medicine. (1999). *Reducing the burden of injury*. Washington, DC: National Academy Press.

National Association of Injury Control Research Centers and State and Territorial Injury Prevention Directors Association. (2001). *Survey of safe USA participants*. Chapel Hill: Joint Committee on Infrastructure Development and University of South Carolina.

National Association of Injury Control Research Centers and State and Territorial Injury Prevention Directors Association. (2001). *Training survey of health departments*. Chapel Hill: Joint Committee on Infrastructure Development and University of South Carolina.

National Association of Injury Control Research Centers and State and Territorial Injury Prevention Directors Association. (2004, September). *National Training Initiative for Injury and Violence Prevention: Draft core competencies*. Chapel Hill: Joint Committee on Infrastructure Development and University of South Carolina.

National Center for Health Statistics. (1999). *Healthy people 2000 review, 1998-99*. Hyattsville, MD: Health and Human Services.

National Center for Injury Prevention and Control. (2007). *Unintentional injury fact sheet: Falls and hip fractures in the elderly*. Atlanta, GA: Centers for Disease Control.

Powell, P., Sneed, M., & Hall, R. (1997). *Fire and life safety educator* (2nd ed.) Stillwater: International Fire Service Training Association and Oklahoma State University.

U.S. Department of Health and Human Services. (2000). *Healthy people 2010: Understanding and improving health* (Public Health Service publication No. 017-001-00550-9). Washington, DC: Author.

World Health Organization. (2003, May 20). Traumatic injury 2000, Press Release, "The Global Burden of Injury" & Appendices, *The Injury Chart Book*. New York: Author.

Customer Service and Marketing

Accreditation Criteria

CAAS

105   Community Relations and Public Affairs: Due to the high visibility and unique expertise of EMS agencies, there exists a responsibility to keep the public well informed about out-of-hospital care and related health issues. These agencies must maintain a respected high profile to enhance out-of-hospital care in their communities (pp. 116–120).

105.01   Community Education, Health Promotion, and Injury Prevention: The agency shall have established programs designed to inform the public of out-of-hospital care, health promotion, and injury prevention (pp. 112–113).

105.02   Community Relations: The agency shall have practices in place to strengthen its image within the community (pp. 113–116).

105.03   Media Relations: The service shall have established methods to promote positive media relations (pp. 107–112).

Objectives

After reading this chapter the student should be able to:

5.1 Understand the principles of customer service (pp. 100–102).
5.2 Create a customer-service program for an EMS organization (pp. 102–108).
5.3 Understand and apply marketing concepts for any EMS agency (pp. 116–120).
5.4 Identify image-building activities to be conducted by EMS agencies (pp. 113–114).
5.5 Analyze and modify customer-service programs from EMS industry standards (pp. 117–120).
5.6 Understand the branding process for EMS agencies (p. 113).

Key Terms

benchmarking (p. 102)	EMS complaint (p. 120)	public information, educa-
brand (p. 113)	EMS Week (p. 114)	tion, and relations
branding (p. 113)	external customer (p. 118)	(PIER) (p. 103)
call avoidance (p. 103)	identity (p. 113)	public service announce-
customer complaint	internal customer	ment (PSA) (p. 107)
(p. 120)	(p. 118)	speaker's bureau (p. 114)

Point of View

While marketing is defined in many ways, it boils down to a social and managerial process by which individuals and groups obtain what they need and want through creating, offering, and exchanging products of value. In most EMS organizations, marketing is a poorly understood concept. Since most EMS organizations do not generate demand for their services directly with the customer, marketing is often achieved in an indirect way. In many instances, the objective of EMS marketing is to develop a positive reputation for the EMS service in the community, so the public can feel confident that those services will be reliable and professionally delivered in their time of need. Marketing also involves the process of identifying service needs in the community and selling those services to stakeholders and customers within the local health care community.

Customer service is an essential process to maintain an organization's reputation and serves as a primary tool to improve the quality of services. Many EMS customer service programs are nar-rowly focused on the resolution of complaints and the monitoring of clinical care. An ideal customer service program not only addresses these issues, but also ensures that employees present a professional, polite, and caring image to the community both during and aside from patient care encounters. EMS services are often judged by nonpatient interactions with fellow providers/customers in the health care and public safety communities. It is, therefore, important that EMS organizations look to identify and cultivate employees with good customer interaction skills in addition to clinical care competence. Customer service standards must be cultivated by management to include positive interactions at all times with the community at large, the driving public, bystanders, and family and friends of patients who are often the first to either recognize good service, or to file a complaint.

Sean Caffrey, MBA, NREMT-P
Summit County Ambulance Service
Frisco, Colorado

■ INTRODUCTION TO CUSTOMER SERVICE—

On August 30, 2004, the Henderson Fire Department in Henderson, Nevada, was called to check the welfare of a 78-year-old man. Upon arrival the fire crew realized that the elderly man was not physically hurt. However, they found that he had become a shut-in after his wife left him and he had forgotten to eat for two days. Instead of returning to the station, the crew cooked the man his dinner, retrieved his mail, and made several phone calls to family members. The crew stayed with him until a volunteer from a local community-action team called the Trauma Intervention Program (TIP) arrived to assist in finding family and additional help for the man. The crew spent approximately two hours at the incident.

The real story is not that the crew showed compassion and kindness, but it is about the response by the Henderson Fire Department management when they learned of an award that TIP wanted to give to the crew for their actions. The deputy chief authorized overtime compensation to cover the crew that attended the evening awards banquet at the Four Seasons Hotel in Las Vegas. The event was held from 6 to

10 P.M., but the chief authorized compensation for an entire 24-hour shift. Furthermore, the deputy chief asked each of the crew members to bring their wives and paid $1000 for the table that evening as part of the fundraiser for the community-sponsored program.

When thanked for the generosity of the fire administration, the chief commented, "We preach customer service, and when one of our crews does something that reflects it and someone notices, we're going to back it up." Many fire and EMS services claim to focus on customer service, but this example illustrates a real commitment to it. Imagine the impact this has on EMS and fire crews when management demonstrates this level of commitment.

CUSTOMER RELATIONS

Within any EMS system the majority of managers and street responders spend a significant amount of time in training to improve their response capability. Technical-skills training concentrates on clinical patient care in dangerous and often life-threatening situations—as it should—and is reinforced through continuing-education programs. Unfortunately, it appears that little time is spent in learning about the need for or means of providing for the personal and emotional needs of patients who are the system's customers. Without this important aspect of total patient care, EMS responders can end up providing a patient with excellent clinical care and still be accused of mistreatment by the customer or the customer's family. Helpful, positive people are sued less often than negative, grumpy people. This is a further incentive to provide positive attitudes and courteous customer service.

A displeased customer will tell 9 to 10 people about an unhappy experience, even more if the problem is serious. One in six of those unhappy customers will tell up to 30 people about poor service. The same customer will tell five people if a problem is handled adequately. One-third of the customers who experience service problems never register a complaint because the effort is not worth the trouble.

EMS managers must select EMS workers capable of providing genuine care. The Health Occupations Basic Entrance Test (HOBET) and the Health Occupations Aptitude Examination are standardized tests used to assess the applicant's ability to be successful in the health-care professions. Myers-Briggs is a standardized personality-trait and communication-trait as-

sessment. It identifies 16 different personality types. The standard Myers-Briggs Type Indicator (MBTI) exam identifies the ESTP (extrovert, sensing, thinking, and perceiving) or ISTP (introvert, sensing, thinking, and perceiving) communicators as most suitable to be paramedics and firefighters.

Responders see themselves going out on an ever-increasing number of calls, spending long hours on the street or in the back of an ambulance or rescue vehicle, and if patient accusations of mistreatment are true, there is not much gratitude expressed for the lives they save. It is a fact that negative members' attitudes do occur, and there are causes that sustain them: violence against EMS responders, abuse from intoxicated individuals, personal problems, stress, and too many responses to nonemergency calls. The task for an EMS manager is to provide the system commitment, leadership, support, and training necessary to change negative attitudes into positive ones and integrate customer-service standards and routines into the delivery process. Implementing customer service is easier if managers effectively convey the compelling reasons for members to interact using courtesy and customer concern, in addition to providing the best patient care, during every response. In an article that describes the STAR CARE program, Mike Taigman and Thom Dick refer to the three Cs of customer care as courtesy, competence, and customer sense. Some agencies add "concern" to their customer-care lineup. STAR CARE was one of the first organized customer-service programs in EMS (Figure 5.1).

As is the case in most scenarios, EMS managers are not present when conflict (defined as some form of inappropriate or misunderstood behavior) occurs between an EMS member and a patient/customer. The member may think everything went well on a call, but the customer may perceive things differently. These perceptions may stem from an unflattering or derogatory comment made "out of earshot." Comments regarding the patient should be held confidential. Simply failing to manifest concern and common courtesy toward the patient can also lead to feelings of ill will. While EMS responders may feel these perceptions are not true, the facts indicate that each customer views reality differently and each has his own expectations regarding customer service.

There are tools available that EMS managers can use to assist in the development of local programs regarding customer service. The use of consultants to bring the best of other industries' customer-service

Safe	Were my actions **safe** for me, for my colleagues, for other professionals, and for the public?
Team-based	Were my actions taken with due regard for the opinions and feelings of my **coworkers**, including those from other agencies?
Attentive to human needs	Did I treat my patient as a **person**? Did I keep him warm? Was I gentle? Did I use his name throughout the call? Did I tell him what to expect in advance? Did I treat his family and/or relatives with similar respect?
Respectful	Did I act toward my patient, my colleagues, my first responders, the hospital staff, and the public with the kind of **respect** that I would have wanted to receive myself?
Customer accountable	If I were face to face right now with the **customers** I dealt with on this response, could I look them in the eye and say, "I did my very best for you"?
Appropriate	Was my care **appropriate medically**, professionally, legally, and practically—considering the circumstances I faced?
Reasonable	Did my actions **make sense**? Would a reasonable colleague with my experience have acted similarly under the same circumstances?
Ethical	Were my actions **fair and honest** in every way? Are my answers to these questions also fair and honest?

FIGURE 5.1 ■ STAR CARE Customer Service Program Checklist. Use it to analyze almost any patient-care issue you might encounter. Go through the list in order from top to bottom. Ask yourself if your care meets each criterion. If it does, chances are that you can defend your actions in almost any forum.
(Reprinted with permission of Mike Taigman and Thom Dick.)

ideas is an effective way to introduce new ideas to fire and EMS services. In 2006 MiRos Services, based in Elkridge, Maryland, developed a program and standards of service specifically for fire and EMS organizations. In an effort to take customer service into the next generation, the Dekalb County Fire and Rescue Services in Atlanta, Georgia, has coined the term *customer care* versus customer service. Its customer-care plan is displayed in Figure 5.2. The key to developing a customer-care plan for internal and external customers is building it with performance-measurement strategies for ongoing management.

Fitch & Associates of Kansas City, Missouri, developed a program called "Service First" that looks at all the different interaction points that occur from the beginning to the end of a response. The challenge for

EMS managers is to develop and maintain a work environment that supports the EMS members during good and bad times and also reflects an organizational concern for its customers. Managers must establish a system or program that begins with clear guidelines, incorporates tracking and response for each customer-service complaint, and recognizes a job well done.

EMS SERVICE BENCHMARKING

EMS service managers need to learn how to benchmark customer service as it relates to their organization. **Benchmarking** is a means of determining which public and private EMS agencies are doing the best job of customer service and complaint resolution and then using that information to improve their own agency's performance. EMS managers should study the gap between their organization and those who are the best in the business to help them determine where their agency is on the customer-service continuum. A process for creating a customer-care standard as a service benchmark is shown in Figure 5.3.

One pivotal component of any customer-service system is to hire and train the best for service-related jobs. EMS agencies should look for a variety of character traits, skills, and experience for customer-service jobs. Some examples of those behaviors and skills include problem-solving ability, skill in handling stressful situations, multitasking, a strong desire to help, compassion, and effective communication skills.

Some organizations build a job rotation or designate an acting or mentored position into the career ladder. This process allows each person to learn almost every aspect of the organization and provides tremendous experience in problem solving. Training in assertiveness and communicating across cultures is an important component of hiring the best. Another approach is to implement a customer-service team with representatives from every aspect of the EMS delivery.

EMS leadership must understand that patient complaints are more often the result of procedures and policies that do not meet patient or employee needs. Baldrige-winning agencies have a single focus: the customers. Every EMT, paramedic, communication operator, billing specialist, and EMS supervisor recognizes that it is a job requirement to focus on customer care. One strategy is **call avoidance**, which determines how to prevent customers from being dissatisfied. When employees understand that leadership is focusing on doing a good job for the customer rather than placing blame, fear and resistance will decrease and employees will be

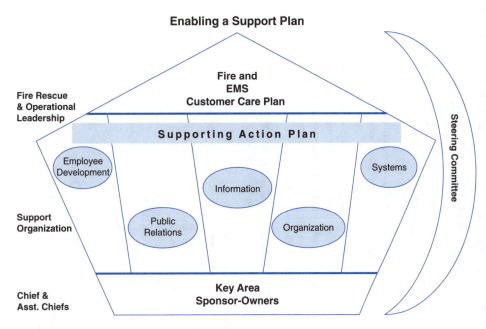

FIGURE 5.2 ■ Dekalb County Fire Customer Care Program.
(Reprinted with permission of Dekalb County Fire Customer Care Program.)

empowered to help find and fix problems. Figure 5.4 illustrates the fire and EMS leadership customer-service policy from the West Midlands Fire Service in England.

PUBLIC INFORMATION, EDUCATION, AND RELATIONS

Dealing with the public and the public's perception of EMS is an ongoing affair. **Public information, education,** **and relations (PIER)** must be a priority for EMS system managers, supervisors, and responders because they are topics that impact all levels within the system. For the system manager PIER determines how the system is perceived as a public-service entity and, at budget time, if it is worth the cost. A good example of public service is ensuring that the community knows how to access EMS. Publicizing the U.S. Fire Administration's

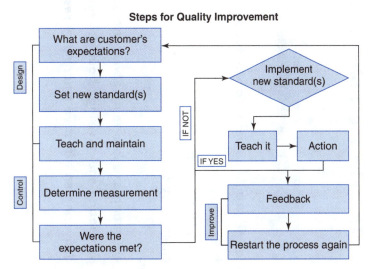

FIGURE 5.3 ■ Customer Service Benchmarking Process.
(Reprinted with permission of MiRos Services.)

Our commitment to you

Our Vision Statement

"To make West Midlands safer"

Our Mission Statement

"We will provide a value-for-money service to reduce:

- fire casualties
- the effects of fire
- emergency calls

by:

- educating and informing about the danger of fire
- promoting and advising about risk reduction
- delivering an effective emergency response."

You're at the heart of our service

Customer Care Policy

We are proud of the services we provide to the community and you, the customer. You are our top priority

Our commitment to you is to continuously improve and to provide a caring, efficient, and cost-effective fire service.

We have a range of standards that we aim to achieve. We will consistently check to ensure we meet our standards and supply a professional service to you. Any areas where we need to improve will be identified and addressed as necessary.

Customer Care Standard 1
Telephone Inquiries

a. **General (Nonemergencies)**
- We will endeavour to answer calls within 7 rings (where possible) in a courteous and professional manner.
- We will answer the call and give our location and name.
- If we have to transfer calls, we will endeavour to, where possible:
  - ensure the caller is connected to the person he or she requires.
  - give the caller the correct telephone number.
  - take a message if the contact is not available and ensure it is given to the person concerned as soon as possible.

We will not:
- intentionally ignore a ringing telephone.

b. **In Operational/Emergency Conditions**
- We will obtain a call back number (where possible) at the start of the call and explain that we will call back when the emergency has been dealt with.

FIGURE 5.4 ■ West Midlands Fire Customer Policy.
(Reprinted with permission of West Midlands Fire Service)

You're at the heart of our service

Customer Care Standard 2
Written Communication

a. When you write to us:
- we will acknowledge receipt of your correspondence by telephone wherever possible and send a written reply within 10 working days.
- we will let you know when you can expect a full reply. If we cannot reply within 10 working days, we will let you know why.

b. When we write to you, we will:
- include thanks for the correspondence.
- state who is dealing with the issue.
- state what is being done.
- give a name and contact number for any queries.
- act on any stated proposals and give a timescale for these.

Customer Care Standard 3
Face-to-Face Communications

All visitors will:
- be greeted in a courteous and friendly manner.
- be given the name of the person greeting them.
- be given any information requested: at the time of the request.
- be given an explanation of why information is not available at the time of the visit, and when it will be available.

Please remember, we are an emergency service and at times, although we will always strive to do so, we may not be able to meet these standards.

You're at the heart of our service

Customer Care Standard 4
Response Standards for E-mails

We will:
- acknowledge e-mail inquiries within 2 working days, where possible, and reply in full within 7 working days.
- let you know when you can expect a full reply. When we cannot reply within 7 working days, we will let you know why.

Customer Care Standard 5
Comments, Compliments, and Complaints (CCC) Procedure

When you think we can improve:
- contact us by telephone, letter, or e-mail. (See page 4 for details of contact numbers.)

When things go well:
- please let us know by contacting us by telephone, letter, or e-mail.

When things go wrong we will:
- receive your complaint at any fire station or fire service premises in the West Midlands by telephone, letter, e-mail, or in person.
- acknowledge your complaint within 2 working days; this may be by telephone call, personal visit, or a letter.
- contact you within 10 days and inform you of the outcome of the investigation of your complaint.
- give an explanation for any delay in the investigation and a new deadline for the outcome of the complaint.

If you feel your complaint has not been resolved satisfactorily you may also contact your local councillor, the local Government Ombudsman, or seek independent advice from the Citizen's Advice Bureau or a solicitor.

FIGURE 5.4 ■ Continued

pamphlet, "Make the Right Call," is a good way to get that message across.

For EMS managers it is important to motivate field personnel to present a good public image while offering tours for school children and conducting community blood-pressure screenings. It is also important to ensure that the appearance of a paramedic unit or ambulance throughout a shift presents a positive, professional image of the service. EMS crews should present themselves with clean uniforms and outwardly caring behavior, in addition to an expected cool, calm, and collected clinical competence. In the end the public-relations effort of all EMS personnel collectively determines the strength of the system's public image and reputation.

If a public relations/information program is going to have a positive impact, EMS managers and leadership must allocate time to train everyone involved. The best time to work on creating the desired public image is *before* a negative incident occurs because a negative incident has the potential to erode public confidence and perception. Consequently, all EMS personnel must be prepared to support a positive image of EMS at every occasion. If the public has a negative perception or image of an EMS system, it generates negative comments and an increased number of complaints. As mentioned earlier, EMS managers must be prepared to respond to comments or resolve complaints in order to reestablish a positive image.

The public image of EMS is largely based upon news-media coverage. Every EMS response can be considered a public event, and the media have a right to cover it. EMS managers and supervisors must train responders to deal with aggressive and sometimes overbearing news media without compromising patient care. Legal media requests should be complied with whenever possible; however, everyone must pay due regard to the issue of protecting patient confidentiality. The alternative—preventing the media from doing its job—has legal ramifications and will often make the headlines that neither boost a unit's public image nor improve EMS and media relations. On a positive note, the media can assist in enhancing the public image of EMS by making time or space available upon request for public-service announcements, press releases, and messages regarding EMS.

EMS managers and leadership must never lose sight of the fact that what they do to promote and enhance the public and community image of EMS translates into system support. New recruits or volunteers, support at budget time, support by community and agency officials when dealing with complaints, and better press may manifest in system support. EMS responders must learn that management may shape a system's image, but it is the responder's direct interface with the public, day by day on the street, that is the basis for a system's public image. In Figure 5.5 paramedics from

FIGURE 5.5 ■ The Durango Fire Rescue removes a stump from a section of river.
(Photo reprinted with permission of Durango Fire Rescue)

the Durango Fire & Rescue Authority removed a dangerous stump in the middle of a popular section of white water in the Animas River in Colorado. The snag posed a danger to rafters and kayakers, and the fire department was the only agency willing to remove it and make this section of river safer.

The Importance of Public Information

Providing members of the public with necessary safety information actually serves three goals: public service, making our job easier and sometimes safer, and marketing. In these days of budget cuts, the reality is that we must often fight for what we once took for granted. All life-safety education programs include self-promotion tactics. Educating the public about the value of our service allows us to serve the public better while furthering the mission of this public service. PIER programs can be an extremely effective tool in achieving most of those goals. Positive public opinion often can be the determining factor in budget and contracting issues. The key is coordinated effort, based on high-level goals. PIER strategies are closely linked to customer service and marketing/networking programs. Care must be taken to ensure that efforts are coordinated for maximum effectiveness.

When developing PIER programs, it is important to keep in mind several key elements that will improve the chances of a successful campaign (Table 5.1): First, messages that are consistent, easily understood, and disseminated on a regular basis are more likely to be noticed. One effective technique for enhancing message communication is the use of symbols, such as logos, art, and themes. Symbols simplify complex messages and enhance the appearance of consistency.

Second, material should be of high quality and presented in an effective manner. Everything the organization prints and distributes is a public-relations tool. It either improves the organization's image or undermines it. Simple things such as letterhead design, stationery quality, letter format, grammar, and spelling send significant messages about the organization's attention to detail.

Third, because there is a variety of EMS customers (young, old, married, single, and so on), public-relations messages should be sent in a variety of ways. For example, the use of video games recently has been a successful recruitment tool for the military via the MP3 player and iPod. Media should be chosen based on marketplace research and knowledge.

One of the best places to obtain market demographics is from the local newspaper. Compare newspaper and

TABLE 5.1 ■ The PIER Process

1. Adopt a vision.	Make PIER a high priority within the organization.
2. Analyze the system.	Analyze your current PIER/customer service/marketing strategies. Are they coordinated? Why or why not? How effective are they? Are you achieving your goals? Are the goals good? Use interviews, surveys, questionnaires, etc.
3. Establish policies.	Set an overall strategy based on your research and goals. The strategy should reach across programs and systems.
4. Direct implementation.	Select appropriate members, delegate, and make them responsible. Ensure the coordination of all system functions.
5. Monitor effectiveness.	Public surveys are useful. Randomly observe programs as they are presented. Is the presentation professional? Are the presentations consistent (as delivered)?
6. Revise appropriately.	Generate solutions to refine/improve the approach and continue to monitor.

magazine rates by readership and cost per thousand. It may be worth paying more to get a message targeted directly at the appropriate audience by radio, television, or print. The Sunday newspaper is a good choice for small communities. A **public-service announcement (PSA)** is an effective way to use radio and television. PSAs are typically run free on radio and television stations, but they tend to run during nonpeak hours when fewer audience members are listening or watching.

Each ambulance and fire truck should be considered a mobile billboard. Vehicles should display a clean, professional image. Often the passenger step and cab of an EMS vehicle gives the patient's family the first impression of the service. A dirty crew area or unclean step or passenger-seating area can damage the image of what would otherwise be excellent patient care.

A Fire Education Model for Injury Prevention

Many fire and EMS organizations have applied a five-step process of public education and fire prevention with success. Fire prevention has an approach that can be used in EMS and injury-prevention activities. Both approaches seek to change public behavior and influence the community to have fewer dangerous situations or injury-generating activities.

The five-step planning process for developing fire and life-safety public-education programs provides a detailed, step-by-step approach to launching a program plan (refer back to Figure 4.17). Following this process really addresses the problem and avoids doing prevention activities that may not have any significant impact.

Interacting with the Media

Interacting with the media can intimidate and frustrate EMS managers, in part because they often think they have little control over the situation. Too often the only time managers deal with the media is in times of crisis or during an incident (at which time they have many other tasks). In fact, the media can be a useful leadership tool, if leaders spend the time and effort on doing their homework.

It is important for EMS managers to avoid thinking of the media as adversarial. Fire or EMS officials should take care not to approach the media with an attitude that conveys suspicion or that can be interpreted as stonewalling. Media professionals have a job to do, and that job is to inform the public without bias. They want to complete their assignments and go home to their families, just as fire and EMS personnel want to do. Building a trust relationship with the reporter assigned to city government is a good idea. When EMS managers earn the reporter's confidence, the reporter will seek out the EMS manager, knowing he will get a straight answer, even if the answer is, "I don't know yet, but I will let you know when I do."

When working with the media, be sure not to speak ill of the competition, other agencies; instead, focus on their merits, always providing straight and truthful information. Any attempt to deceive the media will ruin the possibility of forming a trust relationship, and therefore, the EMS manager will not be a trusted source for information, which may make the EMS manager and his agency a target in the future.

It is important to contact the media directly when significant changes and improvements within the EMS system take place or are slated. Contact the media often and early. Be proactive rather than reactive. EMS managers have the responsibility of educating the media on the various aspects of the EMS system and its operations. It is routine to thank the media for their reporting on the issues or the story. Government

- Present the ground rules.
- Prepare as thoroughly as possible.
- Remain calm, and be professional.
- Be honest and straightforward.
- Present the organization in the best light possible.
- Offer to research issues that are unknown—and then do it.
- Speak in "sound bites," concisely and in an organized manner.
- Avoid off-the-record comments!
- Speak to the interviewer, rather than the cameras.
- Manage the interview, and bring it to a close after answering all reasonable questions.
- Provide handouts that include a paragraph describing the system, stations, and population services.
- Never ask to review a story before it airs or is printed; instead ask reporters to read back your quote for *your* benefit.

FIGURE 5.6 ■ Tips for Handling Media Interviews.

cannot take out ads to tell our side of the story, so the media are often a free promotional activity. Figure 5.6 offers some tips for working with the media.

Handling a Crisis

A *crisis* is an unstable event, critical time, or state of affairs in which a decisive change is impending and a possibility exists of a highly undesirable outcome. EMS managers should be prepared to handle them. However, despite all planning and efforts, a crisis that is potentially damaging to the agency's credibility—such as an EMS vehicle collision, lawsuits, dispatch errors, embarrassing human-resources issues, and bad patient outcomes—can occur. When a crisis is escalating or when insufficient information is available, an encounter with the media can be surprising. During these times it is important to get a public-information officer in place to call for a press conference.

Figure 5.7 lists common events that may require immediate intervention by an EMS manager or EMS leadership to communicate with the press. With proper preparation, planning, and execution, it is possible to counteract or minimize an image-damaging outcome. Identify potential scenarios along with guidelines for management and publicity. Consider using role-play techniques to train staff to deal with EMS-related media events. Do not permit the staff to berate the media during training exercises. If such behavior is modeled and tolerated in practice sessions, it will likely reappear during high-stress and crisis situations.

Media Mitigation

Before a crisis happens it is important to establish good media relations. The news media are not your friends.

- Emergency-vehicle crash or collision.
- Catastrophic natural events.
- Human-resources issues.
- Criminal events.
- Bad patient outcomes.
- Mass-casualty events.
- Lawsuits involving products, events, or services.
- Dispatch errors.
- Injuries in the workplace.
- Rate hikes or disruption of service.
- Reduction of service, such as hospital closures.

FIGURE 5.7 ■ Crisis Events Internal and External to the Organization.

They are professionals with a job to do. EMS managers should never assume that a congenial relationship with the media will keep bad things out of the spotlight. However, it might provide the benefit of the doubt.

A crisis is not the time to avoid the media, and it is certainly not the time to begin a relationship. So collect contact information, such as phone numbers to the news desk, the fax machine, home or cell-phone numbers, and after-hours numbers or e-mail addresses from key reporters and news agencies and place those in a fact sheet (Figure 5.8). It is important for the EMS leader of an agency to meet the people of the key news agencies, such as the news director, producers, writers, reporters, photographers, and the assignment-desk editor. Learn how a newsroom is set up and staffed, when the deadlines are, and how to format and submit materials for print, radio, TV, and official online news agencies. Likewise, have the media visit your agency on a routine basis to cover stories of accomplishment and human interest. Prepare a media kit for your agency that includes a description of your agency, the population served, square miles covered, the number of stations, and staff contact information with phone numbers for key personnel.

When an event happens, it is important to tell the media the truth. Covering up, avoiding, or outright lying may seem a reasonable choice at the time, but the consequences can be extreme. Imagine the media or a reporter discovering and broadcasting the real story. Situations will tend to be dragged out as the various facts are uncovered and reported. The event may be reported for a longer period if it is "sweeps week" and if the event has been reported by the local news to a network news center.

Being dishonest in communications can damage an agency's credibility. Present one effective voice, and choose a credible spokesperson in times of crisis. This is typically the EMS organization's leader or frontline supervisor. This leader comes equipped with the automatic credibility of his or her position and, hopefully, credibility and reputation earned during extensive networking and public-relations efforts. After the event, be sure to send updated media releases. Schedule news conferences, and make sure that the facts are correct before they are released.

When establishing a designated media area, keep in mind that an event receiving nationwide coverage requires space for satellite trucks and numerous news vehicles. If the EMS agency is handling the media attendance, remember food, water, mints, a smoking area, and aspirin. The site should be set up with phones, faxes, a copier, and computers with e-mail and Internet access. The ability to monitor television and radio transmissions can keep an EMS manager ahead of the story. Many news helicopters and mobile command posts have the ability to link communications to enable the command team the advantage of real-time information and the luxury of reconnaissance from the air with telephoto lenses.

Types of Media Available to EMS

An EMS manager needs to be familiar with several types of media, including press releases, briefings, and conferences. The most common are press releases, which have a standard format and serve as the official statement of the incident. Figure 5.9 is an example of a simple press release. Press releases should come from one source, and facts should always be double-checked. Press releases should be distributed at every press conference; they should be on official letterhead with 1½ or double line spacing, be limited to one page, and include a contact name and phone number. The title should be in all capital letters, centered, and underlined.

When writing material for dissemination to the public, remember that most of America reads at the sixth-grade level, so keep terms simple and avoid technical or "insider" terminology. Press releases should cover the who, what, when, where, why, and how surrounding an event.

Press conferences should be called if there is a need to be the first to announce, to clarify recent news published with a discrepancy, or to improve the agency's image by providing more detail to an ongoing event. Provide media agencies with as much advance notice as possible, which will help keep them from embellishing the story or collecting information from unofficial sources. Begin scheduled press conferences on time. The EMS manager conducting the

---

EXAMPLE MEDIA FACT FORM
Print/Newspaper

DATE_____

NAME OF PUBLICATION _____

ADDRESS _____

_____

TELEPHONE ( )_____ AFTER HOURS # ( ) _____

DESCRIPTION OF PUBLICATION _____

DISTRIBUTION: ☐ DAILY   ☐ WEEKLY   ☐ MONTHLY   ☐ OTHER_____

CIRCULATION _____   COST _____

DEADLINES: NEWS_____   (TIME) _____

FEATURES _____   PSA'S _____

HOLIDAYS _____   WEEKENDS _____

GEOGRAPHIC COVERAGE AREA _____

DEMOGRAPHICS/READER PROFILE _____

_____

KEY PERSONNEL:	NAME	TELEPHONE
EDITOR	_____	_____
CITY/LOCAL EDITOR	_____	_____
FEATURES EDITOR	_____	_____
LOCAL REPORTER	_____	_____
POLICE REPORTER	_____	_____
MEDICAL REPORTER	_____	_____
AD DEPARTMENT	_____	_____
OTHER	_____	_____

FIGURE 5.8 ■ Sample Media Fact Forms for Print/Newspaper and Radio/Television.

event may need to have notes or prepared comments available. Make sure the room is big enough for everyone and that there is a podium, water, and easy access. Some other EMS agencies use the Internet, e-mail, recorded telephone lines, and local TV and radio. Situations that require evacuation or extreme action by the pubic should be announced using the emergency-alert system, public-address systems, and Reverse 911 systems to deliver targeted messages to the public.

Interviews

On occasion an EMS manager or leader is authorized or requested to give an interview. Phone calls from media sources should be returned promptly, because these professionals are working with tight deadlines and should not be put in a position to report that their attempts to contact received "no response." In many cases, coaching a paramedic or EMT through an interview can save the agency and the employee from embarrassing remarks.

EXAMPLE MEDIA FACT FORM
Radio/Television

DATE_____

STATION (Call Letters/Frequency/Channel)_____

ADDRESS _____

_____

TELEPHONE ( )_____ BUSINESS HOURS_____

DESCRIPTION OF PUBLICATION _____

DISTRIBUTION: ❒ DAILY   ❒ WEEKLY   ❒ MONTHLY   ❒ OTHER_____

DEADLINES: NEWS_____   (TIME) _____

               COMMUNITY CALENDAR _____ PSA'S _____

               FEATURE PROGRAMS _____

               HOLIDAYS/WEEKENDS _____

BROADCAST AREA _____

DEMOGRAPHICS/AUDIENCE PROFILE _____

_____

PSA PROFILE _____

_____

KEY PERSONNEL:	NAME	TELEPHONE
STATION MANAGER	_____	_____
PROGRAM DIRECTOR	_____	_____
NEWS DIRECTOR	_____	_____
ASSIGNMENT EDITOR	_____	_____
PSA EDITOR	_____	_____
LOCAL REPORTER	_____	_____
NEWS HOTLINE	_____	_____
OTHER	_____	_____

FIGURE 5.8 ■ Continued

It is a reporter's responsibility to identify himself and state the purpose of the interview. According to etiquette, the reporter should specify a date, time, location, and the expected duration of the interview. It is reasonable to expect the reporter to listen and allow the interviewee time to respond. Ideally, a reporter will forward a prepared set of questions in advance to provide the interviewee time to formulate reasonable and intelligent answers; it is appropriate to request this list. EMS managers should avoid interviews in their offices, and they should dress professionally, stick to the facts, or use prepared written statements. Keep to the point and be brief; however, a simple yes-or-no answer does not suffice. Do not be afraid to say, "I don't know" or "I will have to check the facts and report back," but do not say, "no comment." The reporter has an ethical responsibility to write a story that accurately represents the information and the interviewee. Be mindful that once the interviewee has agreed to "answer a few questions" all questions are fair game.

Typically, members of the media reporting on an incident will ask if anyone has been injured or killed. If someone has been killed, the obvious next questions will be asked to learn the disposition of the deceased

West Des Moines EMS Press Release

May 11, 2007

Further information: John Smith, EMS Assistant Chief/PIO,
555-123-4567, achief@web.com

For Immediate Release

WEST DES MOINES EMS ANNOUNCES
EMS WEEK 2007 ACTIVITIES

In recognition of the services that emergency medical personnel provide, the American College of Emergency Physicians (ACEP) has declared May 20–26, 2007, as Emergency Medical Services Week.

Wes Des Moines EMS provides life-saving care that dramatically improves the survival and recovery rate of those who experience sudden illness or injury in our community. Emergency medical services teams consist of emergency dispatchers, EMTs, paramedics, firefighters, educators, and others.

Events are planned around this year's National EMS Week theme, "Extraordinary People, Extraordinary Service," which exemplifies the excellent services provided every day, under any circumstances, by the nation's 750,000 EMS providers who serve their communities.

An open house will be held at West Des Moines new public safety facility located at 8055 Mills Civic Parkway, Saturday, May 19th, from 10:00 A.M. until 2:00 P.M. Food, station tours, and a display of Iowa's EMS helicopters will be available. In conjunction with these open house activities, the dedication of the State of Iowa EMS Memorial, which is located on the facility grounds, will occur at 10:00 A.M. This memorial is a tribute to Iowa's EMS providers who have died in the line of duty. The facility dedication conducted by the Iowa Masonic Lodge will begin at 11:00 A.M.

On Wednesday, May 23rd, from 4:00 P.M. to 7:30 P.M., West Des Moines EMS will host their Annual Bike Helmet Give-Away. The event will be held on the Jordan Creek Bike Trail in the 4000 block of EP True Parkway. Residents in need of children's bike helmets are encouraged to attend.

The week's events will conclude on Saturday, May 26th, with an ambulance and equipment display on the south side of Jordan Creek Mall from 11:00 A.M. through 5:00 P.M.

***

FIGURE 5.9 ■ Example of an EMS Media Release. *(Reprinted with permission of West Des Moines EMS)*

and if it was someone of prominence. They will try to obtain other information, such as the nature of the injuries, condition of survivors, and important medical information. The Health Insurance Portability and Accountability Act (HIPAA) privacy regulations do not allow personal medical information to be released without patient consent. It is best to provide general condition and generic information about the patient(s), then refer the news agency to the nursing supervisor for further detail. Most reporters want a description of the event, specifics of the rescue, relief efforts, causes, and any other information relating to witnesses, usual happenings, and other agencies involved.

LIFE-SAFETY EDUCATION PROGRAM

Life-safety education programs should be implemented based on market research. EMS managers should attend programs to see what is actually being done and said. Presenters should be thoroughly trained, including practice time and feedback. All presentations and publications should be age- and target-group appropriate, should hold attention, and above all should be professional. Visual aids are useful, and all programs should distribute something, preferably something the audience can use and keep (make sure the agency's logo and message are included).

It should go without saying that the presentation must clearly and repeatedly communicate the message. Formally and informally solicit audience interaction and incorporate their feedback into the presentation. This information is also useful for program modification, for new-program development, and to justify benefits such as funding. Presentations and written contacts by and between the public influences an agency's image. This is advantageous when launching a new program. Alert the local media, because they may wish to do a story about the program and actively inform the public.

COMMUNITY RELATIONS AND CITIZENRY

Community-relations programs seek to increase public recognition of issues involving the delivery of emergency services, thereby decreasing patient access time to the system. These programs teach the public how to interact effectively with dispatchers, provide pre-arrival care, and promote general illness and injury prevention. Community-relations programs may include:

◆ *Make the right call:* Sponsored by the U.S. Fire Administration and NHTSA, this campaign kit contains a variety of public-education materials designed to raise awareness about the role of EMS and when and how to access the EMS system. To request a copy of the kit, write to EMS PIER Campaign, c/o 1901 L Street NW, Suite 300, Washington, DC 20036.

- *Explorer post/junior paramedic:* This program teaches elementary students when and how to access the EMS system. Printed materials include coloring books, instructions for calling 911, and phone stickers. Students take materials home to review and discuss with parents.
- *Pre-EMS and CPR training:* Classes are often taught at a community center, church, or service club to encourage visits by the public. Some services provide training to large businesses on a fee basis. Resources are available through the American Heart Association, the Red Cross, and the National Safety Council.
- *Babysitter training:* Designed for 12- to 14-year-old children, classes cover safety, first aid, CPR, and system access.
- *Health fairs:* Fairs are used primarily to provide citizen education and promote EMS awareness. Activities include free blood-pressure and glucose finger-stick tests. One particular type of health fair is a teddy-bear clinic. Frequently conducted by local children's hospitals, a teddy-bear clinic demonstrates EMS procedures on teddy bears to reduce the anxiety a child would have in a real emergency.
- *Community Emergency Response Team (CERT):* This is an initiative from the Federal Emergency Management Agency (FEMA) in conjunction with local fire and EMS agencies. This program trains citizens in the coordination of civilian activities in times of emergencies or disasters.
- *House numbering:* This is a program undertaken by fire and EMS to help ensure a rapid response. With assistance from private vendors, house numbers or numbering plates are placed on residences in the community for more rapid location identification.

IMAGE-BUILDING PROGRAMS

Every EMS and fire agency needs to implement image-building and branding activities for their agency. Image-building programs are designed to focus attention on the positive aspects of the EMS organization and how the organization provides an essential community service.

Branding

It is vital for EMS agencies to build their image or brand within the community. A **brand** is a mark made to associate a class of goods identified as the products of a specific organization. **Branding** is a process that establishes direction, leadership, clarity of purpose, inspiration, and energy for an agency's most important asset—its brand. Branding differs from agency **identity** in that identity refers to a logo tagline and is usually

FIGURE 5.10 ■ Example of an EMS Logo Used on an Arm Patch.
(Reprinted with permission of West Des Moines EMS)

paired with an image. Logos should be clean, simple icons that become everlasting over time and represent the total organization. A logo should be visible on all communications, including Web sites, press releases, business cards, newsletters, and signage (Figure 5.10). An image is the public's perception of the agency, whether that perception is intended or not.

Branding is a process that requires a planned strategic focus and becomes integrated throughout the organization. Brands are formed by years of consistent effort and message reiteration. Good examples of brands include FedEx, the Marines, and Lexus. Brands are also undone in an instant, as we have seen when the media gets hold of a story of a fatal mistake or illegal or unethical behavior by a fire or EMS agency.

The EMS manager is the EMS organization's brand. Forty-five percent of an organization's reputation is based on the image of the leadership. Brands represent the organization's values, public image, visual look, service, and its latest news story. Brands are built from four components:

- *Accuracy.* Involves ensuring that the image truly represents the organization.
- *Consistency.* Requires that we select a look and then maintain it. An agency has to ensure that each experience with its members is consistent with the quality standards of the previous experience.
- *Repetition.* The average individual does not remember something unless he has heard it five to seven times.
- *Reach.* Refers to the number of people who have seen, heard, or had an experience with the EMS organization.

A brand helps to unify the mission with the vision and builds bridges with the community.

Citizen Relations

Encourage member participation in community affairs, such as civic groups, school boards, and neighborhood associations. Many social organizations gain strength from having their members involved in another organization. Networking and one-on-one communication help build an agency's image. Participation in civic activities such as community cleanups, Habitat for Humanity efforts, and charity fundraisers are prime examples of networking.

Fire and EMS services should maintain and vigorously promote a speaker's bureau on EMS topics. A **speaker's bureau** consists of employees, generally paramedics and EMTs, who have expertise or are comfortable speaking on certain subjects. Lectures on a variety of topics, such as CPR, heart-healthy living, stroke awareness, and injury prevention, are all useful EMS topics that add value for the community. Develop and actively maintain routine contacts with political and media personnel. Often a good public-service campaign and frequent media stories will supplement a speaker's bureau. Stay abreast of technological advances affecting EMS, and ensure public awareness of the organization's investment in research and development. Provide the media with the names and contact information of these subject experts so that when a reporter needs a quote for a story, he knows whom to call.

Periodically survey public opinion to stay in touch with community needs and perceptions concerning the service. One way to sample public opinion is to conduct a town-hall meeting in each neighborhood. A variety of methods exists to sample public opinion. Town-hall meetings scheduled at varying times and days of the week will oftentimes uncover perceptions, expectations, and needs of the community with regard to emergency services. EMS facility tours for local groups such as Boy Scouts, Girl Scouts, and day-care centers and open houses for the general public are easy to fit into daily operations, yet give citizens an intimate care experience and high perceived value.

Celebrate with the Community

Plan, organize, and host special events, anniversaries, and other organizational observances in a manner that emphasizes the organization's commitment to the community and the people it serves. An **EMS Week** is a perfect opportunity for a community celebration focusing on the local EMS service. EMS Week is a federally proclaimed annual event that occurs each spring and is sponsored by EMS organizations. A wide variety of EMS events focuses attention on EMS providers and organizations. The American College of Emergency Physicians (ACEP) distributes a high-quality information packet full of ideas and camera-ready art relevant for EMS Week for interested parties.

Identify the particular social needs of the community and develop specific programs to meet those needs. For example, the Home for the Holidays project provides free transportation for certain nursing-home patients to enable them to visit local family or friends during Thanksgiving. This program typically only requires a few hours in the morning and afternoon and is easy to implement. Contact nursing directors 8 to 10 weeks before the holidays to identify participants. Be sure each participant's physician has granted permission, and limit participants to a realistic number for the service. To increase coverage, plan the event with media representatives. Encourage them to ride along with participants who are going home. In many of the old neighborhoods of East Coast cities, fire and EMS workers often help neighbors with minor carpentry needs, such as an elder shut-in who needs a handrail in the bathroom or a wheelchair-bound citizen who needs a ramp in the home. This type of community service has been a longstanding tradition in emergency services.

Periodically evaluate all standard form letters, including collection letters, for professional tone, spelling, and appropriateness. Plan, develop, and distribute news releases concerning the EMS organization and human-interest stories. Maintain a prompt and efficient 24-hour press-information service to answer inquiries from the press and broadcast media. Develop and widely distribute public-service announcements via television, radio, and print media, conveying health, safety, and educational issues and highlighting the service's operation and logo.

Newsletters are effective, low-budget marketing tools. Printed educational material developed and distributed on a variety of EMS subjects is an important aspect of successful community-education programs. Materials must be well written, graphically interesting, and distributed through a variety of avenues including schools, supermarkets, and businesses. An alternative to the headache of an in-house publication is contracting with an outside firm to produce a customized client newsletter.

Slide presentations or videos can greatly enhance community presentations. The best audiovisual presentations are brief, lasting from six to eight minutes, and are universal enough for use in a wide variety of settings. Ensure that members are donning appropriate attire and that equipment is polished and in presentation condition at all times.

EMS agencies should participate as often as possible in standbys, displays, and parades. Placing units in positive, public view is a very effective image-building strategy, and should be done at every opportunity.

The ambulance ride-along experience is an excellent way for the public, the media, and potential employees to understand what goes on in EMS. The ride-along experience will be affected by the requirements of the HIPAA Privacy Rule (Figure 5.11). Therefore, the EMS manager should take steps to ensure that safeguards are in place to prevent accidental disclosures of individually identifiable health information to students and personnel who are participating. Even the name of a patient who is being transported is considered protected health information. The rise of media on cellular phones and other portable devices that have small cameras and audio recording capability present special challenges to upholding the HIPAA privacy rules as well. EMS managers must address those challenges by writing specific policies for the use of those media that will ensure patient privacy.

Planning Special Events

Open houses, medic-of-the-year banquets, award ceremonies, and other special events are excellent opportunities to showcase the service's facilities and generate positive coverage for the organization. (The guidelines shown in Figure 5.12 may assist you with

---

Sample Letter

[PLACE ON DEPARTMENT LETTERHEAD]

[INSERT DATE]

Dear **[INSERT NAME OF COORDINATOR OR VISITOR]:**
We are pleased that you have accepted our invitation or are participating in a ride-along with one of our crews as they provide vital medical services to the community. We believe this experience will assist you in understanding the role of the emergency medical services system that serves the community and other citizens in the **[INSERT COMMUNITY/COMMUNITIES]**.

As you know, the new HIPAA privacy regulations now require health-care providers to be more vigilant than ever in protecting the confidentiality of patient information. While HIPAA does permit disclosures of information to federal agencies and authorities for purposes of performing their oversight functions with respect to the health-care system and federal programs, HIPAA obligates us to limit the amount of patient information disclosed to the minimum amount necessary for any particular purpose. Consequently, our crews have been trained to treat all patient information, including, but not limited to, patient names and information about patient conditions and treatment with the utmost discretion. We request that you not seek information about our patients beyond what you incidentally observe in the course of the ride-along.

Finally, prior to the ride-along, we respectfully ask that you sign and return the attached Confidentiality Acknowledgment and liability waiver to help us discharge our ethical obligations to our patients to respect and protect their privacy and our legal obligations under HIPAA by documenting our efforts to safeguard patient information. This form is similar to the confidentiality statement we require our own employees to sign.

Please let me know if you have any questions.

Sincerely,
[NAME]
[TITLE]
[PROVIDER NAME]

Confidentiality Acknowledgment

The undersigned, for the purpose of medical education or community education related to the prehospital care system, will be participating in a ride-along on emergency equipment responding to medical emergencies. In the course of performing that function while participating in a ride-along with **[AGENCY/DEPARTMENT/COMPANY NAME]**, the undersigned will unavoidably come into contact with protected health information regarding patients of **[DEPARTMENT/AGENCY/COMPANY NAME]**. "Protected health information" includes patient names and other demographic information and all information about a patient's condition, treatment, and payment for medical services.

The undersigned acknowledges that this information is private and confidential and agrees not to request from **[EMS AGENCY NAME]** crews or subsequently disclose any protected health information, including that which I may incidentally observe or hear, to any party, except as necessary in the performance of my legally authorized congressional health-oversight functions or as otherwise required by law.

_____

Signature of Ride-along Participant

Date_____

FIGURE 5.11 ■ Sample HIPAA Privacy Rule Letter.

1. *Establish objectives.* Determine what should be accomplished at the event. Identify stakeholders, who will be showcased, and to whom you intend the message to be reinforced.
2. *Determine who is in charge of what.* Establish clear lines of responsibility to avoid the "I thought *you* invited the mayor" problems.
3. *Compile the guest list.* Identify who needs to be included to accomplish the objectives; remember VIPs who simply wish to attend. Request a timely RSVP to the event.
4. *Save the date.* Be sure to plan early to secure and schedule key attendees and participants. Prominent people, dignitaries, and celebrities are usually in high demand and fill their calendars early. Avoid holidays and other generally busy periods.
5. *Check facilities.* Make a final walkthrough to be sure everything is in proper order and that the facility is fully equipped and clean. Ensure adequate space for the number of guests.
6. *Distribute printed take-home material.* Plan a professional agenda and a fact sheet or brochure that is professional and reflects the organization. Give them to guests to take with them after the event.

FIGURE 5.12 ■ Special Event Guidelines.

event planning.) Invite the media to these events. If possible, the agency should issue a press release and include photos to be used by the media.

Each of these ideas supports the EMS organization's primary public image-building activity, which is to maintain consistent, positive, day-to-day interactions with the citizens of its community.

■ INTRODUCTION TO MARKETING———

Marketing is closely linked with PIER strategies, networking, and customer service. Developing solid public support is essential for all EMS organizations. In order to develop this support, EMS leaders must devote time and planning to effectively market and promote their agencies. Even services operating in a protected, noncompetitive environment should spend some time in marketing activities to upgrade public support. For marketing campaigns to be successful, EMS managers must develop a thorough understanding of their organization's current situation. They must identify whom the organization serves and whom it hopes to serve, as well as organizational strengths, weaknesses, opportunities, and potential threats. A well-defined marketing plan can help EMS managers identify these issues and define organizational goals. A marketing plan can also serve as an important internal

communications tool that effectively integrates the efforts of every department.

Marketing plans should be founded upon customer needs and expectations with consideration to realistic budgets, operations, and resources. The most effective way to identify customer needs is through market research. Typically, market research consists of developing both primary and secondary data. Primary data are collected directly from patients, through observation, interviews, or surveys. Secondary research involves using information that is already compiled into reports, documents, or records such as census records, annual reports, and accreditation information.

Common market-research tools include focus groups, written and mailed surveys, and telephone interviews. The focus groups are often used to develop an initial understanding of the issues and to narrow the number of variables to be included in surveys and interviews. If survey instruments are used, careful consideration must be given to the construction of the instrument and the sampling methods employed in order to avoid bias in the data. For example, a common method of collecting EMS marketing data is the feedback card mailed out to customers following use of the agency's services.

However, respondents tend to overstate the positive and understate the negative. As a result EMS agencies often draw invalid conclusions from the data. Use random sampling techniques and large sample sizes to help reduce bias. If carefully undertaken, market research will help EMS organizations develop a better understanding of their customers and what motivates them to perceive the service positively. In addition, market research can help ensure that marketing messages are appropriate for the target audience.

A classic marketing strategy is to segment the market. EMS agencies serve a variety of customers, and each should be categorized for different types of marketing strategies. For example, the market may be segmented by age, health status, experience with the service, geographic area, and so on. Market segmentation is important because different groups (e.g., senior citizens versus young adults) have different needs and values regarding the EMS system. To be successful, marketing strategies must identify and appeal to those various needs and values.

Marketing success also identifies ways to more effectively communicate with the various market segments. For instance, older citizens may best be reached through health fairs at the local senior center or direct mail, while young adults can be contacted at high school or college locations. To produce tangible marketing results, EMS organizations must set clear goals and objectives that answer five basic questions:

- What is to be done?
- Who will do it?
- When will it be done?
- Where will it be done?
- How will it be done?

The goals and objectives are based upon a firm comprehension of the organization's mission and vision. Broad marketing goals might include providing high-quality clinical and customer service, improving profitability, and increasing market share. In contrast to these broad goals, objectives should be specific expressions of what must be done. For instance, one objective might be to begin a 24-hour health-care referral service by August 1. Objectives should be specific, measurable, attainable, and written in a clear and logical manner.

Once marketing goals and objectives have been identified, the next step is to develop specific strategies and tactics to market the service. In order to identify the most effective marketing strategies, it is necessary to consider customer market mix and service mix. Specifically, these terms refer to determining which strategies will be most effective for different segments of customers and which service products should be marketed to the different customer segments. Several questions may help in matching customers with market strategy and service:

- What characteristics do all potential customers have?
- What characteristics are shared by the customers who are most likely to use the service?
- What are the characteristics of each major segment of the actual and potential market?

Finally, for a marketing plan to be successful, it must be continually monitored and adjusted to keep pace with changing organizational goals and customer values. A detailed, written marketing plan can facilitate this process and stimulate ideas about the best uses of organizational resources. A well-written plan should identify specific responsibilities and schedules, facilitate an organized approach to marketing efforts, and identify challenges.

DEVELOPING A MARKETING PLAN

A formal marketing plan should be clearly derived from and based on the organization's mission/vision and goals. Before actually developing a marketing plan, take the time to analyze the situation thoroughly. Consider the facts and assumptions that form the basis of the plan, including information about the economy, politics, environment, technology, competition, organizational resources, market share, current and potential opportunities, and other factors influencing the service's market approach. In addition analyze current marketing strategies and organizational structure. Assess the purpose, current marketing capabilities, strengths, and weaknesses.

Once the EMS manager defines and understands who, where, and what the organization is, he should set marketing objectives using all available resources. Describe results and specific objectives by asking questions, such as: What are the expected results? Where do you want to be next year? In two years? In four years? Next, develop the marketing strategies, policies, and procedures needed to achieve those objectives. Marketing strategies need to match the needs and values of the marketing segments the organization is trying to reach. From there, identify the marketing programs. Now the EMS manager can detail the specific tactics and courses of action that will achieve the manager's objectives.

Prepare to put the plan into action. Delegate action items and set realistic time frames for completion. Specify everything required to complete the action, including who is responsible to ensure completion, as well as where, how, and when. Remember to confer the necessary authority with the responsibility so that members can achieve their goals. Also, determine personnel requirements and identify the financial resources, costs, and risks associated with the plan.

Set the procedure for monitoring the progress of marketing strategies. Be sure to include standards of performance and how marketing performance will be measured. Specify details for updating your marketing strategy. The marketing plan and those involved in its implementation should meet regularly in order to ensure that the established goals and objectives remain focused on the organization's mission and vision.

Developing an effective marketing plan requires time and effort. While it may seem to be too time consuming at first, a marketing plan and customer-care plan can provide a long-term strategic road map for and investment in the organization.

ASSESSING CUSTOMERS

When it comes to customer service, the EMS manager should remember that every patient needs to be treated as the EMS manager would want his own family member or relative to be treated. In most cases when a lawsuit is initiated, it is not because of some significant level of medical malpractice by paramedics and EMTs but rather because someone disrespected or compromised the dignity of another person or family member.

Customers can be categorized as external and internal customers. Service for the **internal customer** relies on employee satisfaction and is a key indicator of customer satisfaction. Internal customers are individuals, entities, and organizations who are involved in or with the operation of the EMS system. Internal customers also are those who receive services provided by the support structure for the EMS system, a component of the system, or an individual working within the system. They are the EMS employees, volunteers, and members of the leadership councils or committees who plan and coordinate the system. In essence, internal customers compose the agency and those who interact to form the ongoing system and other support staff within the organization. A number of internal customer-service surveys help predict employee attitudes about delivering good customer service. Some surveys, such as employee-satisfaction surveys, and

other information such as attrition rates of employees and training hours in customer service per employee, can help an EMS manager track the impact of customer service programs in an EMS organization.

The **external customer** has traditionally been known as "Mrs. Smith," since named so by Alan Brunnicinni within his fire department customer-service manual. Mrs. Smith represents anyone who calls 911 and experiences service from the fire, EMS, and communications-center staff. This is where, when, and how first impressions and customer service get established. External customers also include hospital staff, nursing-home staff, law-enforcement agencies, neighboring fire and EMS agencies, contracted services, and elected officials.

EMS agencies should routinely sample the external customers beyond Mrs. Smith. EMS managers can do this by periodically visiting nursing homes, urgent-care clinics, and other health-care facilities or by sampling the external professionals who experience EMS services.

The concept of 360-degree customer service begins with assessing the level of service at each point where the customer experiences agency service standards. Survey cards, as shown in Figure 5.13, can create a

FIGURE 5.13a ■ Henderson Fire Department Cardiac Customer Satisfaction Survey Card. *(Reprinted with permission of City of Henderson Fire Department)*

Fire Department
Service Feedback

A Place To Call Home

To help us continually improve our service, please complete this form and return it to a City counter, fax it to (702) 267-2223, or mail it using the reverse side of this sheet. Your opinion counts. Thank You!

Please enter the date of your last interaction with the Henderson Fire Dept. (month/day/year): _____

1. What was the primary reason (need) for your last interaction with the City of Henderson Fire Department?

☐ Fire   ☐ Youth Firesetter
☐ Assist from a fall/Public Assist   ☐ Station tours/School visits/Events requests
☐ Fire Investigation   ☐ Other, please specify: _____

2. Please rate the following aspects of service provided by City employees at the Henderson Fire Department.

Aspects of Service	Excellent	Good	Fair	Poor	Very Poor	NA
Timeliness of response	☐	☐	☐	☐	☐	☐
Courtesy	☐	☐	☐	☐	☐	☐
Competency in handling an issue	☐	☐	☐	☐	☐	☐
Professionalism	☐	☐	☐	☐	☐	☐
Timeliness of resolving problem/addressing need	☐	☐	☐	☐	☐	☐

3. Please rate the overall job the City of Henderson Fire Department does in providing services.

☐ Excellent   ☐ Good   ☐ Fair   ☐ Poor   ☐ Very Poor

4. What is your <u>preferred</u> method of communication with the City of Henderson Fire Department?

☐ In-person   ☐ Phone   ☐ Email   ☐ Mail   ☐ Website   ☐ other: _____

5. How satisfied are you that your questions were answered and you were provided with adequate information or documents during your interaction with City of Henderson Fire Department personnel?

☐ Very Satisfied   ☐ Somewhat Satisfied   ☐ Somewhat Dissatisfied   ☐ Very Dissatisfied

6. Do you agree that City of Henderson Fire Department personnel interacted in a concerned and caring manner?

☐ Strongly Agree   ☐ Somewhat Agree   ☐ Somewhat Disagree   ☐ Strongly Disagree

7. How would you rate your overall experience with the City of Henderson Fire Department?

☐ Excellent   ☐ Good   ☐ Fair   ☐ Poor   ☐ Very Poor

8. What would you like to see the City of Henderson Fire Department do better? _____
_____

9. How satisfied are you in the <u>overall</u> job the City of Henderson (not just this dept.) does in providing services?
☐ Very Satisfied   ☐ Somewhat Satisfied   ☐ Somewhat Dissatisfied   ☐ Very Dissatisfied

☐ Please check here if you would like to be contacted by the City of Henderson regarding your comments.

Optional Name: _____ Company Name: _____ Phone: _____

FIGURE 5.13b ■ Henderson Fire Department Service Feedback Sheet.
(Reprinted with permission of City of Henderson Fire Department)

snapshot of how the agency is performing from the public's or patient's perspective. A survey on chest pain, orthopedic care, and shortness of breath can be used to identify efficient use of medication such as morphine or albuterol. The use of a Web-based survey tool known as Survey Monkey is gaining ground in fire and EMS service for various aspects of sampling service from other agencies or internally.

RESOLVING CUSTOMER COMPLAINTS

Informed patients know how EMS is supposed to work. If things are not working as they should, these patients are the first to know. For example, the public expects a quick response to a life-threatening emergency. If a first-response unit does not arrive in four to six minutes, the public knows the standard was not met. Research shows that when people in the United States have a negative experience with an organization, they are likely to tell 8 to 10 other people about the experience. But only about one in 20 will actually complain directly to the organization. That leaves another 19 people out there who do not give the organization a chance to improve, yet spread the image of bad service to others.

A **customer complaint** (call it a challenge or an opportunity) can be defined as any indication that the service or product does not meet the customer's or public's expectations. You might already know that in most government organizations, people have found that their complaints often go unresolved. In the last 30 years, the public's trust in the government has declined from a high of 70% to a low of 17%. When a complaint is not resolved, patients seek resolution at higher levels in the organization, thus further duplicating costs and eroding trust in government services. One study indicates that a written response to a complaint by a public official in the highest administrative role costs as much as $1000 to process.

An **EMS complaint** is defined as any indication that patient care or EMS service did not meet the customer's need or standard of care. EMS must actively listen to its patients and customers, so that their complaints can be resolved quickly and at the lowest level possible. For this to happen, EMS agencies must have a system in place that makes it easy for a customer or a patient to register a complaint (Figure 5.14). It also is important that everyone know the process for handling those complaints. One way to accomplish this is by having the complaint process displayed in a simple flow chart. Steps for building a complaint and compliment system are offered in Figure 5.15.

1. Make it easy for your customers to complain, and they will make it easy for you to improve.
2. Respond to complaints promptly and courteously with common sense, and you will improve customer service.
3. Respond to complaints on first contact, and save money by eliminating unnecessary additional contacts that escalate cost and destroy customer confidence.
4. Technology utilization is critical in complaint-handling systems.
5. Recruit and hire the best for jobs that require interaction with customers.

FIGURE 5.14 ■ Principles of Customer Service.

An effective strategy for preventing customer dissatisfaction is known as "call avoidance." It can be accomplished by providing written materials, posting information on the agency's Web site, and promoting standards and guidelines that encourage exceptional customer care.

The public and employees are great resources for designing a complaint/compliment system. The use of 800 numbers, automated call distributors, and online Web sites allow for public input. You will find that employees want to help, too. EMS agencies should use feedback forms, phone lines, letters, and surveys to identify root causes of poor service. Then, if a revealed expectation or need is in line with the agency's service standards, the agency should modify customer-care practices to meet that expectation or need.

Responding to customer complaints quickly and courteously with common sense will improve public trust toward government services. When a complaint is received, a frontline EMS manager should make every

1. Issue a policy statement that says your organization embraces complaints and views complaints as opportunities.
2. Establish an implementation team with representation from each stakeholder in the complaint-handling process, and identify each step in the process.
3. Establish a tracking system to record and classify complaints that allows for analysis, data collection, and reporting to top management. The difference between how you measure your process and the best possible process is known as "the gap." A gap analysis will show you what and where to improve.
4. Develop recommendations to improve your core processes and empower frontline employees to resolve complaints on first contact.
5. Create an action plan for implementing the approved recommendations.

FIGURE 5.15 ■ Steps to Build a Complaint and Compliment System.

attempt to resolve the complaint on first contact; this will save money by eliminating unnecessary additional contacts and subsequently builds customer confidence. Government research shows that resolving a customer complaint on first contact reduces the cost by 50%, so it is cost effective and worth the effort to equip and empower frontline EMS managers with clear guidelines that will help them resolve and investigate customer complaints.

The EMS manager should know that the investigation of a complaint requires that the organization provide due process and follow the labor agreement if it has a specific procedure for investigating complaints. Recommendations and training should be provided to improve the critical EMS processes and instruct EMS managers on how to resolve and investigate complaints. A suggested approach for EMS managers to investigate a complainant is shown in Figure 5.16.

Fire and EMS agencies should employ technology to a complaint-handling system. A database should be used to identify trends, times, and types of complaints. If electronic databases identify a trend, then fix it. A tracking system should classify complaints and generate gap reports. A gap is the difference between the organization's service and what is being done by the best in the industry. A gap report or analysis will show EMS leaders what and where to improve service. Electronic database information should be shared with everyone in the agency, so the organization and individuals can align services and products to meet agency standards. State-of-the-art imaging software can be used to produce graphs and charts indicating complaint sources, times, and locations.

An annual survey is another type of performance tracking. Database records should track a complaint or compliment longitudinally, including the demographics, actions taken, progress, disposition, and information gained to improve service. The successful and unsuccessful resolution should be tracked, with credit given to the managers who resolve these complaints.

Complaint Databases

Effective analysis of EMS customer service uses integrated and nonduplicated databases. These databases should include complaints from all sources, including telephone calls, surveys, focus groups, correspondences, complaint/concern cards, and personal visits. The database shown in Figure 5.17 is an example of a complaint form.

The key to customer service is to apply continuous quality improvement. Refine, design, and improve the delivery of EMS service, placing patient expectations first. Complaint databases can be used to analyze problem crews and overworked stations and may also help gauge the community relationships between EMS and ethnic groups. For example, complaints from a Hispanic neighborhood may indicate the need for Spanish-language cards or training for EMS providers. From management's perspective it also helps engineer changes in service delivery and identifies skills that managers may need in complaint resolutions. Many of the accreditation standards for fire and EMS require this data to be tied to education and training programs.

Internal Customer-Service Programs

So often we focus on internal and external customer service simultaneously. Frequently, taking care of the public eclipses our internal customer-care efforts; that is, we forget to "take care of our own." Make every attempt as EMS managers to recognize and reward

---

- ◆ Take a report and separate crew for individual responses.
- ◆ Interview witnesses.
- ◆ Gather physical evidence.
- ◆ Run background checks and criminal-history checks.
- ◆ Obtain intelligence information, and share information.
- ◆ Interview suspect.

FIGURE 5.16 ■ Complaint Information Process.

Minimum Database Information
1. Name:
2. Address or run number:
3. Telephone:
4. Manager assigned:
5. Action taken:
6. Due dates for follow-up:
7. Progress or disposition of the customer concern:
8. Other descriptive information about the event:

FIGURE 5.17 ■ Complaint Database Entry Form.

exceptional service. This empowers employees and increases chances that they will, in turn, provide a similar service to the public.

An internal customer or internal service provider can be anyone in the organization, including a coworker, another department, or a support person who depends upon us to provide products or services that in turn are used to create good public service for the external customer. In general, internal customers do not have a choice in many government agencies because they are limited in what they can improve regarding practices and policies. For example, if the EMS division does not like the fleet manager's policies, it cannot discontinue that service, replace the fleet manager, and alter the staff. An EMS organization should try to celebrate its successes and reward its people for providing customer service above and beyond the normal expectations of the agency. Great internal customer service creates employee satisfaction, employee loyalty, and employee retention.

Internal customers of an EMS organization are a rare breed described tongue in cheek as 25% adult, 25% child, 50% adrenaline junkie, 75% health-care technician, 10% whiner, and 100% schizophrenic. The youthful demographics of our workforce can create challenges. Younger employees, such as those who compose the majority of our workforce, generally place less value on pensions, retirement, health benefits, and job stability than on immediate gratification and economic advancement. To them work is all about feeling good, making money (now), and worrying about tomorrow when it arrives. It is not unusual for employees to change employers for a 50-cent- or dollar-an-hour pay raise. These are the same employees who ask if they can ditch their health insurance and take the premiums we pay on their behalf as cash in their paycheck. They will challenge everything but have incredible passion for what they do—at work and at play.

EMS managers can motivate field crews by providing a variety of formalized rewards. A strong positive impression can be made when managers make a personal visit to the station. Often the reinforcement of excellent internal customer service is completed around the dinner table. A lunch or meal provided by the EMS chief or other chief officers makes an undeniable gesture of appreciation to the employees. There are several commercial companies that provide customized rewards. Figure 5.18 shows a comprehensive award system using medals for various EMS service accomplishments in the state of Oregon.

A variety of awards that range from pins to acrylic plaques can be purchased to reward employees for good service, teamwork, and a variety of desired EMS behaviors.

Some ideas for internal customer service or employee appreciation are listed in Figure 5.19. A schedule or protocol for rewards is important to have in the standard operating procedures (SOPs); overuse of awards and praise can devalue any system of employee recognition. It is also important to ensure that an employee receiving recognition or rewards has not been disciplined or should not have a record of discipline that disputes the praise. The mixed messages of discipline and then reward can send the wrong message; however, meeting with the employee to recognize improvement where he once struggled is a great motivator to keep an employee focused on the intended path for growth.

A sincere and effective way to reward employees for outstanding job performance or service is to send a letter to the employee's family. As EMS workers face the daily responsibilities and often-tragic circumstances that are the nature of EMS, letting loved ones know how much the organization appreciates their service to the community provides reinforcement and support.

Another key to providing excellent internal customer service is weakening the tendency to build defensive walls. One strategy is to create forums for sharing information. Do this to the extent that your position in your organization will permit. Inform the paramedics and EMTs about agency goals and how each division contributes to accomplishing them. This will lessen the likelihood that they feel a need to "protect" their jobs by building walls around their "turf." Remember, a football quarterback enables his team to execute successful plays by ensuring every player understands what his teammates are doing in the play. Members of a football team do not advance the ball by keeping their plans secret from one another.

Similarly, colleagues in a company do not advance their plans by withholding information or assistance from one another. The quality-improvement division and training division cannot execute operations independent of one another. Training depends on quality improvement to help create a training plan for the entire agency's operations and service. Quality improvement depends on training so the paramedics and EMTs know how to collect the data that will help determine patient outcomes. Forums for sharing information can be as grand as a companywide assembly or as modest

Director's Medal Commitment to Quality   Meritorious Service   Community Service   EMS Cross

Medal of Valor   Life Saving   Unit Citation   Civilian Service

Excellence in   Years of
Education   Service

FIGURE 5.18 ■ Oregon State EMS Recognition Medals.
(Reprinted with permission of V. H. Blackinton & Company, Inc.)

- Pins, plaques, merchandise.
- Ribbons, medallions, or uniform additions.
- News releases, employee newsletters.
- Letters in personnel files.
- Paid time off above contract stipulations.
- Coupons for events, movies, or sporting events.
- Direct praise immediately following a job well done.

FIGURE 5.19 ■ Internal Recognition Programs.

as a chat in the hall. A shared lunch between two departments qualifies, as does an e-mail or a memo outlining what a particular department is doing and why.

Outstanding internal customer service is simply good business. Internal customer service flourishes in a high-communication environment. EMS managers should be proactive to provide colleagues with the information they need to do their jobs effectively. Offer it before they need it; in fact, offer it before they *know* they need it. EMS managers should think of the ways that the department's information, statistics, and other data can help others in the organization and share it with them. Most will appreciate the manager's interest and openness, will recognize the manager's keen insight, and eventually will repay the manager by knocking down their own walls.

Create or contribute to an environment in which status is provided to those who share freely and who do not build walls. Most people who construct defensive walls do it to protect their turf from encroachment by others in the agency. They fear that if others have what they have, including information, those others will make them obsolete. Make that fear groundless by rewarding employees and colleagues who do not protect their turf but who instead work to fulfill the goals of the organization. Reward behaviors via compliments, pats on the back, commendations at meetings, lunches, bonuses, letters of congratulation, and so on. All of this leads to open information sharing. Make it clear that territorial behavior sabotages the efforts of the EMS organization, while treating colleagues like valued internal customers contributes to the clinical and customer-service success in the organization.

Creating an Environment of Sharing and Helping

EMS managers can begin creating an environment of sharing and helping by modeling treatment of their fellow employees and other departments. EMS managers should understand that helping their colleagues do their jobs more successfully helps their organization and hence helps them. Therefore, these colleagues are the EMS manager's customers, so the manager should treat them like VIPs. Interruptions are not nuisances but opportunities to serve the internal customer. If EMS managers tend to view every interruption as an obstacle on their road to success, they should reexamine those interruptions. Being interrupted for sharing gossip is another story, otherwise known as a "pothole." An interruption by a colleague requesting call-volume figures for analyzing response-time performance is an opportunity.

Learn to identify every real need from a colleague as a "necessary lane change," and think of every necessary lane change as an opportunity to move the organization closer to its goals. EMS managers should take pride in helping their colleagues. They should enjoy their role in sharing information and providing services that help others get their jobs done. In most cases the manager's willingness to help others will lead others to readily assist the manager when he needs it. Similarly, internal customers should seek input on such things as facility design, vehicle specifications, dispatch procedures, and post locations. EMS managers who improve their internal customer service will go a long way in reducing personnel turnover.

When someone exceeds their expectations, most people feel delighted, excited, upbeat, and very positive about that person and his organization. EMS managers should think about what they can accomplish in their organization by exceeding the expectations of fellow employees. If payroll asks for time sheets by 3 P.M., provide them by 1 P.M. so payroll can relax, knowing they have the time sheets in hand. If human resources asks for a list of important points to cover in an employee orientation, take time to think about it and provide a thorough list.

Say "thank you." A simple and genuine thank-you goes a long way toward creating an atmosphere of sharing and helping. A focus on developing and practicing effective internal customer service helps organizations cut costs, increase productivity, improve communication and cooperation, boost employee morale, align goals, and harmonize processes. Replace interdepartmental competition with interdepartmental cooperation, and deliver better service to the external customer. Excellent service to the external customer is dependent upon healthy internal customer-service practices.

CHAPTER REVIEW

Summary

EMS organizations are part of the community, and the public observes their actions. The building of public trust requires careful cultivation of a relationship with the media, image-building programs, and customer-service assessments. How an agency handles its customers, both internal and external, requires comprehensive assessment and careful monitoring. EMS agencies should develop a marketing plan and have contingencies for media-worthy events. Internal rewards programs help build relationships with employees. Customer service and public relations are required skills for EMS leaders and managers. It is important to understand and establish how complaints and compliments are handled at the organization. Fire and EMS agencies should consider putting significant human resources into the development of an effective marketing plan and establishing relationships with local and regional media services.

Research and Review Questions

1. What does your agency do to ensure that EMS personnel actively listen to the customer?
2. How do leaders, supervisors, and officers in your organization view complaints?
3. How does your agency make it easy for the public or a customer to complain?
4. What does your organization do to make it easy for employees to resolve problems?
5. How does your agency track and analyze complaints?
6. How well are resource decisions aligned with customer needs?
7. How does your organization monitor, measure, and recognize employees who handle customer complaints?
8. How do you make sure the public understands fire and EMS services?

References

Birr, T. (1999). *Public and media relations for the fire service.* New York: Pennwell Publishing.

Brown, L. M. (2004). *Media relations for the public safety professional.* Sudbury, MA: Jones and Bartlett Publishers.

Bruegman, R. R. (2003). *Exceeding customer expectations: Quality concepts in the fire service.* Upper Saddle River, NJ: Pearson Education Inc.

Compton, D., & Granito, J. A., (2002). *Managing fire and rescue services.* Los Angeles, CA: ICMA.

Dick, T. H., & Taigman, M. (2003). Customer service in EMS. *Journal of Emergency Medical Services, 28*(3), 76–81.

Huff, R. (2006). Successful PR for volunteer squads. *Journal of Emergency Medical Services, 31*(5).

International Fire Service Training Association. (1999). *Public information officer.* Stillwater, OK: International Fire Protection Publications.

National Fire Protection Association. (1999). *Standard for professional qualifications for public fire and life safety educator.* Quincy, MA: Author.

National Highway Traffic Safety Administration. (2000). *Emergency medical services (EMS) public information, education, and relations (PIER) national standard curriculum.* Washington, DC: U.S. Department of Transportation.

Taigman, M., & Dick, T. (1991, September). StarCare. *Journal of Emergency Medical Services, 16*(9), 14–16.

CHAPTER 6

Risk Management and Safety

Accreditation Criteria

CAAS

202: Safe Operations and Managing Risk: Comprehensive safety standards are required to assure that the patients, employees, and the agency are protected from unnecessary risk (pp. 135–147).

Objectives

After reading this chapter the student should be able to:

6.1. Identify principles of risk management (pp. 127–130).
6.2. Identify how to calculate the risk in EMS operations (pp. 130–132).
6.3. Conduct an analysis of a system failure (pp. 132–134).
6.4. Create an infection control program for EMS agencies (pp. 135–145).
6.5. Recognize and identify the safety issues surrounding EMS operations and how to mitigate those events (pp. 148–152).
6.6. Diagram the progression and response to litigation against an EMS organization (pp. 132–134).

Key Terms

avoidance (p. 130)
communicable disease (p. 135)
designated infection-control officer (p. 138)
direct transmission (p. 141)
emotional intelligence (EI) (p. 135)

failure mode and effect analysis (FMEA) (p. 131)
indirect transmission (p. 141)
infectious (p. 135)
near miss (p. 128)
Needle Stick Prevention Act (p. 139)
pathogen (p. 135)

presumptive legislation (p. 138)
recognition-primed decision making (RPDM) (p. 134)
risk management (p. 127)
sentinel event (p. 132)
situational awareness (p. 128)
tactical alert (p. 152)

Point of View

Emergency-service organizations should provide the tools and training to perform any given assignment but only the user (the emergency responder) can perform the assignment safely. Safety is personal, and the choice is yours. The concept of risk management is a wide-spanning umbrella that encompasses several components. The primary goal is risk elimination followed by risk avoidance. These are best accomplished through risk-control activities. Even when the unfortunate happens, the risk-management umbrella still contains components known as loss control and loss management. These measures are initiated to reduce the amount of loss and return the situation to its preexisting state.

Safety is yet another risk-management component. In basic terms, safety is the state of keeping oneself from harm. Just like the general public, the emergency-services industry is faced with everyday safety and risk issues. These issues range from the potential of trips and falls, paper cuts, and metacarpal syndrome to being involved in motor-vehicle collisions. However, unlike the everyday citizen, emergency services are riddled with unique situations that place responders in environments the general public only reads about.

As you read this chapter, remember two important principles: *predictable is preventable*—most incidents that result in injury are predictable; and *frequency breeds severity*—the more frequent an occurrence, the more severe the consequences become over time. Think about your personal-safety beliefs and the risk-management and safety culture of your organization. The culture is established at the highest levels, but only you can make it work. Safety is personal!

Richard W. Patrick, Director
Medical First Responder Coordination Division
Office of Health Affairs—Medical Readiness
U. S. Department of Homeland Security

■ RISK MANAGEMENT

Risk management is any activity that involves the evaluation or comparison of risks and the development, selection, and implementation of control measures that change, reduce, or eliminate the probability or the consequences of harmful actions. In EMS, risk management is not just about preventing monetary losses. It is about preventing disability, loss of life, and irreparable business damage as a result of the provision of patient care. Risk management involves direct hands-on patient care, as well as various indirect aspects of patient care, including the development of effective training programs and the selection of qualified personnel. Regardless of its specific focus, the overall goal of risk management is to reduce the frequency and severity of preventable adverse events that create losses.

There are five possible risk-management strategies:

- Eliminate the risk by removing the hazard. For example, use a gurney that has a mechanical lifting motor to reduce the risk of back injury when moving overweight patients.
- Avoid the risk by avoiding the hazard. For example, do not take the route to an emergency that causes the ambulance to enter a known dangerous intersection.
- Acquire insurance. Often EMS agencies manage risk with insurance, which is considered risk sharing.
- Pool an amount of money to help an EMS organization recover from a loss associated with a risk.
- Transfer the risk. For example, some EMS systems only use specialized rescue teams to extract patients from technical-rescue scenarios.

Risk management has two primary focuses: organizational and operational. Organizational risk management focuses on the organization's efforts to protect itself and manage risk. To do so requires several systems working together, such as hiring practices, preventive maintenance, and safety training. The second focus of risk management is operational. It focuses on managing the risk of a specific event, incident, or patient. EMS leadership is responsible for the organizational risk management, and EMS managers are responsible for operational risk management. Both levels of EMS administrators need to design systems that are kept up to date and fully implemented. Loss or exposure to risk is predictable. If risk is predictable, then it can be eliminated, avoided, managed, or transferred.

Every risk-management program should be viewed as a process consisting of a set of components that

- ◆ Identification of potential risks so that uncertainties can be controlled.
- ◆ Measurement of risks to determine the probability of potential losses.
- ◆ Development of strategies to lessen risks.
- ◆ Implementation of those strategies.
- ◆ Monitoring of risk-management strategies and activities to ensure their effectiveness.

FIGURE 6.1 ■ Components of a Risk-Management System.

encourages a systematic approach. Those components are listed in Figure 6.1.

IDENTIFYING AND MEASURING RISKS

The proactive approach to risk management identifies the root cause of loss first and then a proximate cause. Risk managers look for causal factors. The amount of attention given to risk management is a choice by the leadership, organization, and the individual. Figure 6.2 identifies the balance between risk management and litigation.

There are no new ways to get in trouble, only new ways to get out of trouble. Things that go wrong are predictable and preventable. According to Gordon Graham, there are five causes of loss in the field of EMS. Gordon identifies them as a lack of quality people, policy, training, supervision, and organizational discipline. For example, hiring the right people is based on performance, conducting background investigations, and screening. When a workplace or on-scene injury is found to be the result of policy problems or of personnel arrogance, ignorance, or complacency, then risk-management activities need to respond.

NEAR-MISS REPORTING SYSTEMS

A near miss is an opportunity for an organization to implement management activities to prevent a loss. Analyzing near misses has become a national movement. A **near miss** is defined as an unintentional, unsafe act that could have resulted in an injury, a fatality, or property damage. Identifying a pattern or a set of contributing factors or human practices that led to an accident or near miss is an important function in any EMS organization. Typically, there are contributing factors to near misses, such as **situational awareness,** human error, decision making, and individual action. Identifying early patterns can help lead to strategies that can prevent injuries and loss within an organization.

FIGURE 6.2 ■ The Scales of Justice: Risk Management versus Litigation.

Because very few accepted standards in prehospital care have been validated through research, the risk manager must proactively seek out and analyze data and reports that suggest methodologies for aiding the detection and reduction of unsafe activities. Areas of EMS operations that should be considered in a risk-management program are listed in Figure 6.3.

Some aspects of prehospital care involve more risks than others. One area of increased risk for many EMS organizations is the no-transport call. These calls often represent a significant percentage of an agency's response. The legal issue regarding the paramedic's duty

- ◆ Vehicle collisions, breakdowns, and driving skills.
- ◆ Physical fitness of personnel and job-related injuries.
- ◆ Station inspections and building safety.
- ◆ Skills retention.
- ◆ Hazardous materials exposures.
- ◆ Stress management.
- ◆ Debriefing effectiveness.
- ◆ Documentation.
- ◆ Hiring selection criteria and employee attrition.
- ◆ Continuing education.
- ◆ Dispatching procedures.
- ◆ Protocols testing and protocol deviations.
- ◆ Response times.
- ◆ Nontransports.
- ◆ Vehicle down times.
- ◆ Mutual-aid interactions.
- ◆ Controlled-substances losses.
- ◆ Probationary supervision.

FIGURE 6.3 ■ EMS Risk Management Program Factors.

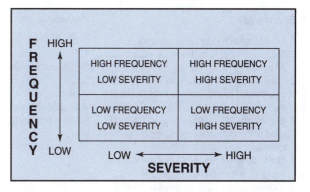

FIGURE 6.4 ■ Risk-Management Matrix.

to obtain the patient's "informed refusal" has not been addressed in appellate-court decisions. Therefore, risk management on these types of calls might first concentrate on improving the providers' physical-assessment skills and then secondarily deal with the legal complexities of consent issues.

Endotracheal intubations are another frequent source of liability. In these cases the issue again is assessment because of apparent failure to confirm the position of the tube. Because the result of improper tube placement is often death or severe brain damage, claims based on this medical error can be very detrimental. A possible plan to manage this risk might include (1) researching the "success rate" of intubations with identification of the criteria of "success," (2) providing re-education and skills labs or practice on a more frequent basis, and (3) having supervisors respond on all cardiac arrests to both observe techniques and provide assistance. Consideration should also be given to various technological methods that could be used to monitor the effectiveness of performance.

While examining the clinical and operational areas posing the greatest risk, it is also important to monitor underlying causes of deviations from protocols as well as patient complaints that have not materialized into high-risk areas. Risk managers must also be on the lookout for nontraditional risk; for example, the proficiency of endotracheal intubation when too many providers are in the system. A good tool for risk classification is the time-honored risk matrix in Figure 6.4. Each risk is placed in one of four areas. For

example, a severe and frequent risk would appear in the box at the upper right. A low-frequency event that is not serious is placed in the lower left. In general high-frequency and high-severity risks would be addressed first, and low-frequency and low-severity risks would be addressed last.

In addition risk-management activities should examine characteristics or factors within an EMS organization that may hinder or enhance risk-management efforts. For example, asking the following questions may reveal valuable information:

- ◆ Are new-member orientation and preceptorship programs providing inexperienced EMTs with sufficient information and training?
- ◆ Are the organization's selection techniques effectively identifying the most qualified personnel for employment?
- ◆ Are members from a particular training institution consistently lacking in skills?
- ◆ Are continuing-education programs consistently well attended?

In addition to studying factors within the organization, risk-management activities should also monitor extraorganizational risks created by the system within which the agency functions. Sample questions for evaluating risks in this arena include:

- ◆ What scene situations are most dangerous for EMS providers?
- ◆ What types of responses can benefit from more staffing on scene by reducing time or risk to the EMS crew, such as a paramedic supervisor to assist on scene rather than an EMT-staffed fire department pumper?

Of course, numerous additional questions for consideration could be listed. The EMS risk manager

Severity (S)	Probability (P)	Exposure (E)
0 No potential for damage or injury 1 Slight <5% or no lost time on equipment or by staff 2 Minimal 5%–24% lost time with no hospitalization or mechanical fix disrupting service 3 Significant 25%–50% required medical care with hospitalization or repair required 4 Major 50%–74% results in a permanent injury or major equipment repair 5 Catastrophic >75% fatalities or death, permanent loss of equipment	0 Impossible 1 Remote or unlikely in any condition 2 Unlikely in normal conditions 3 Chance 50% or less under normal conditions 4 Probability greater than 50% 5 Very likely to happen	0 No exposure 1 Below average exposure 2 Average exposure 3 Above average exposure 4 Great exposure

Total Risk = (S) × (P) × (E)

FIGURE 6.5 ■ Risk Value Analysis.
(Reprinted with permission of Volunteer Fire Insurance Services [VFIS])

must determine what questions should be asked and then obtain answers to those questions. For risk-management activities to be successful, they must include questions, answers, planning efforts, and the introduction and monitoring of programs to manage risks.

Risk management seeks to identify risk and measure the potential loss to an organization. Volunteer Firemen's Insurance Services (VFIS), Inc., has developed a risk-management formula based on three components: severity, probability, and exposure. The three components are converted to numerical values and then combined to create a total risk of that event, showing the potential loss to the organization (Figure 6.5). The value for severity is measured in lost time or equipment, injury, death, and adverse publicity if exposure to an accident or failure in the system occurs. Probability is calculated by estimating how many accidents or failures will occur if exposed to the hazard. Finally, exposure is calculated on the amount of time, number of events, and people or equipment involved in the amount of activity in which the exposure to a failure or accident may happen. This is usually expressed in the number of people exposed to certain hazards.

Once total risk is evaluated, a figure between 0 and 100 will be generated. As an EMS manager, you must on an annual basis assess the risk in your organization and create a plan to manage risk. Not all the risk can be managed; therefore, the high-risk/low-frequency and high-risk/high-frequency events should be the focus. Another approach may be to take the top 10% or 20% of the activities assessed. Based on

the calculated risk-management methods, an EMS manager can decide what techniques to put into place to counter the effects of the hazards. See Table 6.1 for a list of appropriate actions for the assessed risk levels.

METHODS OF MANAGING RISK

There are several methods of managing risk that every EMS organization should consider. One method is **avoidance** of risk activity. Policies such as refusing to allow relatives to ride in ambulances during patient transport or refusing to allow EMS providers to perform medical procedures without express authorization by the medical director are examples of this risk-management technique. This method also includes

TABLE 6.1 ■ **Reduction Strategy for Given Risk**

Calculated Risk	Threat Level	Actions
80–100	Very high	Discontinue, stop
60–79	High	Immediate correction
40–59	Substantial	Correction required
20–39	Possible	Attention needed
1–19	Slight	Possible acceptable
0	None	None

(Reprinted with permission of Volunteer Fire Insurance Services [VFIS])

avoiding lawsuits, which can be done by promoting professionalism, accurate report writing, and training.

Another method of risk management is the reduction or transfer of the degree or severity of risk. Routine physical-fitness training, clinical-skills testing, and scheduled vehicle maintenance are examples of risk-management activities that help reduce the risk of loss but do not eliminate risks completely.

A third way of managing risk is to accept it and then to insure or pool resources. This method is common in EMS operations, often reflected in the day-to-day running of calls, such as driving with the use of red lights and sirens, entering an apartment complex without police cover, or carrying a patient on a stretcher down a stairwell. These activities entail some amount of known risk to the EMT. Money is either put aside to self-insure to cover an estimated cost of a loss or a private insurance carrier may be employed to cover the cost of the risk. In some cases an insurance broker may employ multiple insurance companies to cover different levels of cost within an EMS organization.

A fourth way to manage risk is by way of insurance. Insurance transfers your monetary risk to someone else. However, because of the uncertainty of how juries will make their decisions and the lack of legal precedents in the field of EMS, the insurance industry may be hesitant to insure some EMS activities. Nevertheless, most EMS operations must have some kinds of insurance—at the very least motor-vehicle liability insurance—as a prerequisite to providing patient care.

EMS managers must decide which of these risk-management methods is the most feasible, affordable, and effective for their individual situations. Unfortunately, risk management is often dictated by the limited resources of the EMS organization. Expensive insurance may cause organizations to either underinsure their equipment or delay replacement. If wages are below standards, only poorly qualified personnel might be hired. Organizations might decide to save money by scaling back on the number of units on the street, which can lead to problems with the quality of care rendered. Clearly, the balancing and weighing of risks must be done constantly.

Analysis of System Failure

It would be great if EMS managers knew exactly what failures will occur in an agency and where. In industry under the ISO 9000 Standard, a process is used called the **failure mode and effect analysis (FMEA)**. The FMEA technique was used by the aerospace industry to find problems in aircraft before they ever leave the ground. In short, FMEA is a way of looking into the future and determining where the potential for failure is. This is one of the key tasks that would be conducted to implement risk-management procedures. In most agencies it takes a tremendous amount of time and energy from management to maintain this activity.

A slight modification can be made in FMEA by looking backward instead of forward. Looking back at past failures in the EMS system can give the EMS manager a better reference on those activities or systems that are at high risk for failure. This makes the process practical, requires less time to conduct, and reduces the number of possibilities. This modified approach to the FMEA process is used to determine what failures are occurring in an EMS organization and what their impact and frequency will be. An example is illustrated in Figure 6.6.

This simplified version demonstrates the power of this technique. Imagine performing this calculation for every failure in your EMS operation. While not every failure needs to be evaluated, in fact in most organizations 20% of the failures account for 80% of the losses. EMS managers should focus on the events that are most important.

Figure 6.7 lists several key indicators of EMS system failures. Some common definitions of failure include the following: any loss that interrupts the continuity of EMS delivery; loss of asset availability; failure of or unavailability of equipment; deviation from protocol or standard operating procedures and guidelines; falling short of target expectations such as ALS on scene in eight minutes or less 90% of the time.

The American Society for Testing and Materials (ASTM) has attempted to identify a series of sentinel events that may indicate a failure in the EMS system.

Failure Event	Failed Component	Frequency	Impact	Total Loss
Defibrillator failure	Battery failed	One failure/two years	$50,000*	$25,000/year*
*Based on a municipal organization with a statutory liability cap.				

FIGURE 6.6 ■ Analysis Chart.

- Interruption in EMS system delivery.
- Interruption in continuity of care.
- Unavailability of equipment.
- Deviation from SOPs or protocols.
- Not meeting target performance levels.

FIGURE 6.7 ■ EMS System Failures.

A **sentinel event** is a monitored event whose occurrence signifies the potential for a significant system or provider deficiency that should be examined or investigated by the appropriate regulatory or EMS managers of the appropriate regulatory or contracting entity. Sentinel events are listed in Figure 6.8.

Once you have defined what a failure in the system can be or what may be included, you must analyze it. The steps an EMS manager should take to analyze a system failure are defined in Table 6.2.

Several steps need to be put in place by managers before the team investigates. First, you need to narrow the focus on what system will be analyzed. For example, is an unmet response time a function of the dispatch system, driver training, or aspects of employee performance? Ensure that the definition of failure is agreed upon or has some consensus standard applied to it. In some cases it is helpful to draw or create a visual diagram of the system and the failing piece. Calculate the gap between what is the standard and what is the upper limit of the failure. For example, if the standard is an eight-minute ALS response and the actual measured response was 20 minutes, the gap was 12 minutes. Develop a worksheet or data-collection tool to query computer information or pull together interviews. Figure 6.9 lists the steps that should be in

- Any biomedical equipment failure.
- BLS and ALS response >12 minutes.
- Any arrest en route from the scene.
- Any defibrillation or pacing.
- Any vehicle failure with patient embarked.
- Procedural errors.
- All patient complaints against EMS personnel, organization, or system.
- Multiple responses to the same location within 24 hours.
- All utilization of physical restraints on patients.
- All medical air evacuations from the scene.
- Any unexpected diversion from an intended medical-receiving facility.
- Others as defined by your management team, regulatory agency, or labor group.

FIGURE 6.8 ■ Sentinel EMS Events.

place prior to having to conduct an investigation of an EMS system failure.

Conducting an Investigation

When conducting an investigation, it is important to identify the root cause of an accident or failure. When investigating an accident, it is important to investigate the past or historical events related to or surrounding the circumstances of the current incident. Accident investigation includes determining the who, what, where, and significance of the incident. One particularly important program to help establish data and identify the root cause of an accident involves training the crews on what to do when involved in an accident.

A supervisor's first response when notified of an accident should be to ascertain if the crew is safe and then how many civilians are injured. It is helpful to place in the vehicle new disposable cameras for the

TABLE 6.2 ■ How to Process a System Failure

Steps	Description
1. Perform preparatory work.	Develop a failure definition, list processes, gap analysis, and sentinel events.
2. Collect data.	Interview personnel, determine frequency and impacts, determine what the components of the failures are.
3. Summarize results.	Spreadsheets graphically display and identify redundancy.
4. Calculate loss.	Multiply frequency × impact for every failure in the list.
5. Determine "significant few."	Determine 20% or less of the failures that result in 80% of the losses.
6. Validate results.	Verify that the results are valid.
7. Issue a report.	Communicate results with field crews.

1. Define the system affected to analyze.
2. Define failure.
3. Conceptualize or diagram contacts.
4. Calculate the gap.
5. Develop data-collection tools.
6. Develop schedule and conduct interviews.

FIGURE 6.9 ■ Preparatory Steps for Investigating EMS System Failures.

crew to immediately photograph the scene and pre-addressed postcards to be given to witnesses who are inclined to leave the scene but who may be contacted later to provide information on the crash. Gathering and analyzing facts by way of sketches, photographs, video, and audio recordings are a comprehensive way to write a report. It is important to identify a fixed point of reference when photographing or measuring accident scenes. That way a reconstruction expert who gets the case days or months later after equipment has been moved and the scene disturbed can reference the measurements.

The equipment needed by every EMS manager to be able to conduct a proper investigation is shown in Figure 6.10. In a hard carrying case there should be a compass, accident forms, graph paper, mechanical pencils, rulers, tape, and a digital camera with extra batteries. Different-colored cones and paint to mark each vehicle are also needed. At a minimum, skid length, road width, area of impact, gouges and scuffs, point of rest, and the debris field should be photographed. It is important to mark the wheel location using a T at the base of the wheel, lining up with the axial. Details such as tools used, lighting, wetness or slippery surfaces, and missing safety equipment are important in looking for root causes. Maintenance records are also important to review to ensure medical equipment and vehicles were up to date on preventive and routine maintenance.

An EMS manager may also want to obtain a phase study from the administrator of the traffic-control device to ensure the sign system is working correctly. Finally, weigh the vehicles to ensure they were not overweight.

DOCUMENTATION

Without a paper trail documenting the work of the risk manager, identifying and solving risk-management problems can be very difficult. Documentation serves multiple purposes. It can validate testimony, assist in testimony by refreshing recall of an event, enable retrospective review without unnecessary duplication of collection efforts, and suggest efforts to reduce risk. Although paperwork does not necessarily demonstrate

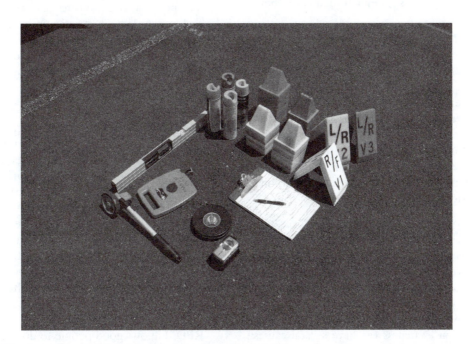

FIGURE 6.10 ■ Equipment for an EMS Manager's Accident Review Kit.

that an act has taken place, it can be convincing evidence. For example, vehicle maintenance logs and canceled checks do not prove a vehicle was repaired, but these documents can effectively persuade a jury.

Documentation that identifies a problem must also indicate what action was taken to correct the problem and that the task was successfully completed. Recording that a member must attend a continuing-education class is effective only if the documentation also shows that the member was aware of the situation and that the assignment was fulfilled. Organizations with well-defined recordkeeping policies will be better prepared to identify and implement effective interventions for reducing risks.

Successful Programs

For risk-management efforts to be successful, they require support from personnel at all levels of the organization. Problem identification often requires considerable fact-finding and information-gathering activities. Risk managers will need to be seen as trustworthy for personnel to be forthcoming regarding risk-management issues. Blindly imposing restrictions and recordkeeping requirements will only frustrate members and will likely defeat the very purposes of risk management. For the process to be successful, organizational members must believe that they are valuable contributors to the risk-management process—as opposed to feeling the process is being forced upon them.

Crew Resource Management

NHTSA, the International Association of Fire Chiefs (IAFC), and several other agencies have begun to look at ways to reduce medical and operational errors. Crew resource management (CRM) is a reduction strategy taken from the aircraft industry and the U.S. Coast Guard. It involves employing a system of layers that makes an attempt to trap human errors at various levels by employing a system to reduce risk. CRM is made up of five key components: communication, situational awareness, decision making, teamwork, and barriers.

- *Communication.* CRM teaches people to focus on a communication model, which involves speaking directly and communicating in a respectful or assertive way.
- *Situational awareness.* This refers to keeping people attentive to an event by teaching ways to perceive, observe, and recognize stress on personnel. Emphasis is placed on emergency calls, which require full attention. Loss of situation awareness can have several causes; they are listed in Figure 6.11.

Ambiguity	Open to more than one interpretation.
Distraction	Attention is drawn away from the original focus.
Fixation	Focusing attention on one item, excluding all others.
Overload	Too busy to stay on top of everything.
Complacency	A false sense of comfort that masks deficiencies and danger.
Improper procedure	Deviating from SOPs without justification.
Unresolved discrepancy	Failure to resolve conflicts or conflicting conditions.
"No one flying the plane"	Self-explanatory.

FIGURE 6.11 ■ Warning Signs of Loss of Situational Awareness.
(From International Association of Fire Chiefs Crew Resource Management Program. Reprinted with permission.)

- *Decision making.* CRM focuses on giving and receiving information through a closed feedback loop. It involves ensuring that messages are received and communications during EMS operations are repeated or acknowledged. This is similar to what occurs in the cockpit of an airplane between the pilot and copilot during a preflight check of the equipment.
- *Teamwork.* Teamwork is reinforced by a focus on leadership and followership that results in a mutual respect for the operation. This promotes teamwork for a common goal to provide the best patient care versus dwelling on personal performance or individual attitudes.
- *Barriers.* Finally, CRM recognizes and tries to eliminate barriers to any of the other four components of CRM.

Psychological Applications

The prevention of loss and risk management can be influenced greatly by using some principles of psychology, including how the brain responds to stresses and high-risk situations. **Recognition-primed decision making (RPDM)** is a retrieval process that draws on past experiences, recalling an incident and responding in a successful way based on what the outcomes were from previous experiences. For example, recognizing when a patient needs an elective intubation is an example of RPDM. Most new paramedics have not seen or intervened for a critical airway, and it often takes the experience of a patient deteriorating in front of them to learn aggressive airway control.

RPDM allows good people to do the right thing. If you have not experienced a task before you encounter it or you do not experience it on a regular basis, your brain cannot match the experience or process the information. High-frequency events do not create as many problems. It is the low-frequency events that lead to problems.

Emotional intelligence (EI) is an emerging science that looks at how to control emotions to recognize triggers and avoid a negative response. The basis of EI is facilitating thought by understanding changing emotions. Where higher forms of reasoning and learning occur, the brain is wired with much slower conduits that communicate with each other to access stored reasoning. In contrast, the midbrain (the area that processes the reflexes and emotions) is wired for high-speed communication. When confronted with a stress or event that produces a stimulus with a strong emotional response, the rapid communication channel causes a knee-jerk response often based on emotion and not logic, resulting in a failure or loss. If a person just takes a short pause in the decision-making process, the brain can engage its higher reasoning abilities and make a better decision. This is a powerful tool when engaging in high-risk activities or important communications.

Equipment Failure and Medication Reactions

EMS agencies have not been traditional contributors of reports to the Food and Drug Administration (FDA) on the failure of medical products. This is largely due to their unfamiliarity with the federal reporting system and the lack of surveillance systems to identify trends occurring in the EMS system. (Imagine the lives that might have been saved had EMS identified the cardiac effects of Vioxx or Celebrex.) Since EMS is often the common point of contact for many of the patients with cardiac events, bad outcomes could have been validated by information from EMS. In one system EMS identified consistent drops in blood pressure with the prehospital use of Procardia years before the FDA warned about the poor outcomes in cardiac patients. Biomedical equipment failure and adverse reactions to medications should be reported to the federal government at MedWatch by phone (FDA MedWatch, 1-800-332-1088) or on the Internet at http://www.fda.gov/medwatch/safety.htm.

■ INFECTION CONTROL

A **pathogen** is a biological agent that causes disease or illness in a person. Pathogens can enter the body through five primary routes (see Figure 6.12). EMS

- ◆ Inhalation.
- ◆ Contact with blood or other body fluids.
- ◆ Ingestion.
- ◆ Fecal-oral.
- ◆ An intermediate carrier or vector (such as a tick).

FIGURE 6.12 ■ Routes of Exposure.

personnel often provide care to patients who as well as being injured also have some kind of illness. Laws, standards, and regulations have been developed to protect providers from **infectious** exposure to those illnesses and when exposure has occurred to be sure providers get the medical attention they need.

LAWS, STANDARDS, AND REGULATIONS

NOTE: The legislation to reauthorize the Ryan White law (H.R. 6143) was signed in late December 2006 by President George Bush. However, the Ryan White CARE Act, which mandated that EMS personnel find out whether or not they were exposed to life-threatening diseases while providing care, was removed by congressional staffers. The reason for the removal is unclear. Some allege the staffers removed the clauses protecting emergency personnel because they did not understand why they were in the original Ryan White Act.

OSHA Standards

Occupational contagious diseases are infectious diseases that are contracted while a person is performing his work. An infectious disease is an illness resulting from the invasion of the body by bacteria, virus, fungi or parasites. A **communicable disease** is a disease that can be readily spread from one person to another. A disease can be infectious but not communicable. Occupationally contracted infectious diseases are considered compensable through the workers' compensation system, just like any other occupationally caused disease.

Facilities or agencies with 11 or more employees must maintain records of occupational injuries and illnesses as they occur. An occupational injury is an injury such as a cut, fracture, sprain, or amputation that results from a work-related accident or from exposure involving a single incident in the workplace. An occupational illness is any abnormal condition or disorder other than one resulting from an injury and caused by exposure to environmental factors associated with employment. Included are acute and chronic illnesses or diseases that may be caused by inhalation, absorption, or ingestion of or direct contact with toxic substances or harmful agents.

Most important to EMS providers are the classifications of bloodborne pathogens commonly experienced with EMS patients. Figure 6.13 lists the bloodborne pathogens as defined by OSHA and NIOSH that EMS personnel should be routinely protected from and educated on the hazards of.

All employees are required to comply with a health-and-safety program. Most facilities choose to extend this program to contractors, subcontractors, visitors, regulatory-agency personnel, and site owners or their representatives. Most of the OSHA standards have compliance directives. The rights and responsibilities of

Infection/Condition	Type of Precaution	Duration
Abscess, draining, major	Contact	Duration of illness
Adenovirus in infants and young children	Droplet and Contact	Duration of illness
Cellulitis, uncontrolled drainage	Contact	Duration of illness
Chickenpox (varicella)	Airborne and Contact	Until all lesions are crusted. Exposed susceptible health care workers (HCWs) should be excluded from patient contact from the 10th to the 21st day (28 days if VZIG given) after exposure
Clostridium difficile, enterocolitis	Contact	Duration of illness
Congenital rubella	Contact	N/A
Conjunctivitis, acute viral (acute hemorrhagic)	Contact	Duration of illness
Decubitus ulcer, infected major with no dressing or dressing does not contain drainage adequately	Contact	Duration of illness
Diphtheria, cutaneous	Contact	Until off antibiotics and two cultures taken at least 24 hours apart are negative
Diphtheria, pharyngeal	Droplet	Until off antibiotics and two cultures taken at least 24 hours apart are negative
Ebola viral hemorrhagic fever	Contact	Duration of illness
Enterocolitis, *Clostridium difficile*	Contact	Duration of illness
Epiglottitis, due to *Haemophilus influenzae*	Droplet	Until 24 hours after initiation of effective therapy
Furunculosis-staphylococcal of infants and young children	Contact	Duration of illness
Gastroenteritis, *Vibrio cholerae* Gastroenteritis, *Clostridium difficile* Gastroenteritis, *Rotavirus*, diapered or incontinent Gastroenteritis, *Shigella*, diapered or incontinent	Contact	Duration of illness
Hemorrhagic fevers (i.e., Lassa and Ebola)	Contact	Duration of illness
Herpes simplex, neonatal or mucocutaneous, disseminated or primary, severe	Contact	Duration of illness
Herpes zoster (varicella-zoster), localized in immunocompromised patient, or disseminated	Airborne and Contact	Duration of illness. Persons susceptible to varicella are also at risk for developing varicella when exposed to persons with herpes zoster lesions
Impetigo	Contact	Until 24 hours after initiation of effective therapy
Influenza	Droplet	Duration of illness
Lassa fever	Contact	Duration of illness
Lice (pediculosis)	Contact	Until 24 hours after initiation of effective therapy
Marburg virus disease	Contact	Duration of illness
Measles (rubeola), all presentations	Airborne	Duration of illness
Meningitis, *Haemophilus influenzae*, known or suspected	Droplet	Until 24 hours after initiation of effective therapy
Meningitis, *Neisseria meningitidis* (meningococcal) known or suspected	Droplet	Until 24 hours after initiation of effective therapy
Menigococcal pneumonia or menigococcemia (meningococcal sepsis)	Droplet	Until 24 hours after initiation of effective therapy

FIGURE 6.13 ■ Bloodborne Pathogens Exposure Standards Under 29 CFR 1910.1030.

Infection/Condition	Type of Precaution	Duration
Multidrug-resistant organisms, infection or colonization		
Gastrointestinal	Contact	Until off antibiotics and culture negative
Respiratory		
Skin, wound, or burn		
Mumps	Droplet	For 9 days after onset of swelling
Mycoplasma pneumonia	Droplet	Duration of illness
Para influenza virus infection, respiratory in infants and young children	Contact	Duration of illness
Adenovirus B19	Droplet	
Pediculosis (lice)	Contact	Until 24 hours after initiation of effective therapy
Pertussis (whooping cough)	Droplet	Until 5 days after patient is placed on effective therapy
Plague, pneumonic	Droplet	Until 72 hours after initiation of effective therapy
Pneumonia, adenovirus	Droplet and Contact	Duration of illness
Pneumonia, *Haemophilus influenzae* in infants and children (any age)	Droplet	Until 2 hours after initiation of effective therapy
Pneumonia, meningococcal	Droplet	Until 24 hours after initiation of effective therapy
Pneumonia, *Mycoplasma*, primary atypical pneumonia	Droplet	Duration of illness
Pneumonia, *Streptococcus*, group A in infants and young children	Droplet	Until 24 hours after initiation of effective therapy
Respiratory infectious disease, acute (if not exposed elsewhere) in infants and young children	Contact	Duration of illness
Respiratory syncytial virus infection, in infants, young children, and immunocompromised adults	Contact	Duration of illness
Rubella (German measles)	Droplet	Until 7 days after onset of rash
Shingles	Contact	Until 24 hours after initiation of effective therapy
Staphylococcal (*S. aureus*) disease, skin, wound, or burn; major	Contact	Duration of illness
Streptococcal (group A streptococcus) disease, skin, wound, or burn; major	Contact	Until 24 hours after initiation of effective therapy
Streptococcal (group A streptococcus) disease, pharyngitis in infants and young children	Droplet	Until 24 hours after initiation of effective therapy
Streptococcal (group A streptococcus) disease, pneumonia in infants and young children	Droplet	Until 24 hours after initiation of effective therapy
Streptococcal (group A streptococcus) disease, onset fever in infants and young children	Droplet	Until 24 hours after initiation of effective therapy
Tuberculosis, pulmonary, confirmed or suspected pharyngeal disease	Airborne	Discontinue precautions only after patient is on effective therapy, is improving clinically, and has three consecutive negative sputum smears collected on different days, or TB is ruled out
Varicella (chickenpox)	Airborne and Contact	Until all lesions are crusted. Exposed susceptible HCWs should be excluded from patient contact from the 10th to the 21st day (28 days if VZIG given) after exposure
Whooping cough (pertussis)	Droplet	Until 5 days after patient is placed on effective therapy
Wound infections, major	Contact	Duration of illness
Zoster (varicella-zoster), localized in immunocompromised patient, disseminated	Airborne and Contact	Duration of illness. Persons susceptible to varicella are also at risk for developing varicella when exposed to persons with herpes zoster lesions

Source: *Adapted from the Public Health Service, U.S. Department of Health and Human Services, Centers for Disease Control and Prevention, Atlanta, Georgia; Hospital Infection Control Practices Advisory Committee. "Guidelines for Isolation Precautions in Hospitals." Infect. Control Hosp. Epidemiol, Vol. 17 (1990). 51–80; Am. J. Infect. Control, Vol. 24 (1996): 24–52; http://www.cdc.gov/ncidod/hip/ isolat/isolat.htm (updated February 18,1997).*

FIGURE 6.13 ■ Continued

---

* 29 CFR 1910.1030 Bloodborne Pathogens (including needle sticks and other sharps injuries).
* 29 CFR 1910.20 Access to Employee Exposure and Medical Records.
* 29 CFR 1910.134 Respiratory Cleaning Procedures.
* CPL 2-2.69D Enforcement Procedures for the Occupational Exposure to Bloodborne Pathogens.

FIGURE 6.14 ■ OSHA Regulations Applied to EMS.

employees and employer are described by OSHA based on CDC guidelines. They include work restrictions, immunizations and vaccinations, and postexposure follow-up. OSHA has several standards that apply to the health and safety of workers (Figure 6.14).

The components within 29 CFR 1910 have some specific requirements for EMS managers to maintain certain documents and records. (Figure 6.15 contains excerpts from the OSHA regulations.) A **designated infection-control officer** is required to maintain and keep records for an exposure-control program. Often in smaller EMS organizations that responsibility falls on the EMS chief or officer.

NFPA 1581 Standard

The NFPA 1581 Standard on Fire Department Infection Control Program (2000) addresses the provision of minimum requirements for infection-control practices within emergency services. NFPA is a consensus standard designed to provide minimum criteria for infection control in the fire station, in the fire apparatus, during procedures at an incident scene, and at any other area where fire service members are involved in routine or emergency operations.

The EMS and fire service infection-control program must have a written policy statement. Such a statement should clearly define the department's mission in limiting the exposure of members to infectious diseases during the performance of their assigned duties and while in the fire-station living environment. Examples of generic policy statements are found in the appendix of NFPA 1581. The standard details the protection of the firefighter and other emergency responders while performing emergency medical operations. Personnel physical condition, protective clothing and equipment, and operational techniques are provided with minimum standards for infection control.

State or Local Legislation

Often state legislation protects workers who have been exposed to infectious diseases. For example, some states

---

Excerpts from 29 CFR 1910.1030 revised January 18, 2001; effective date April 18, 2001:

(c)(1)(iv) The Exposure Control Plan shall be reviewed and updated at least annually and whenever necessary to reflect new or modified tasks and procedures which affect occupational exposure and to reflect new or revised employee positions with occupational exposure. The review and update of such plans shall also:

(c)(1)(iv)(A) reflect changes in technology that eliminate or reduce exposure to bloodborne pathogens; and

(c)(1)(iv)(B) document annually consideration and implementation of appropriate commercially available and effective safer medical devices designed to eliminate or minimize occupational exposure.

(c)(l)(v) An employer, who is required to establish an Exposure Control Plan shall solicit input from non-managerial employees responsible for direct patient care who are potentiallly exposed to injuries from contaminated sharps in the identification, evaluation, and selection of effective engineering and work practice controls and shall document the solicitation in the Exposure Control Plan. . .

(h) (5) **Sharps Injury Log.**

(h)(5)(i) The employer shall establish and maintain a sharps injury log for the recording of percutaneous injuries from contaminated sharps. The information in the sharps injury log shall be recorded and maintained in such manner as to protect the confidentiality of the injured employee. The sharps injury log shall contain, at a minimum:

(h)(5)(i)(A) the type and brand of device involved in the incident,

(h)(5)(i)(B) the department or work area where the exposure incident occurred, and

(h)(5)(i)(C) an explanation of how the incident occurred.

(h)(5)(ii) The requirement to establish and maintain a sharps injury log shall apply to any employer who is required to maintain a log of occupational injuries and illnesses under 29 CFR 1904.

(h)(5)(iii) The sharps injury log shall be maintained for the period required by 29 CFR 1904.6. (9)

FIGURE 6.15 ■ Excerpts from OSHA Regulations on Bloodborne Pathogens.

have passed **presumptive legislation,** which puts the burden of proof on the employer, not the employee. The legislation covers public-safety officials under workers' compensation if they contract an illness (such as hepatitis C) on the job. Other states have source-patient testing legislation on the books. This type of legislation dictates that once a patient's blood is drawn for medical purposes, medical personnel do not need to receive further patient agreement to test the drawn blood for applicable bloodborne pathogens such as HIV, hepatitis B, or hepatitis C. This type of legislation is important because it assists exposed personnel and referred medical professionals with postexposure evaluations and prophylaxis treatment that acts to defend against or prevent a disease.

BUILDING AN INFECTION-CONTROL PROGRAM

By law and consensus standard, every EMS and fire agency is required to have an infection-control plan. Such a plan must have certain components (Figure 6.16). Infection-control plans can be contracted out to experts in the infection-control field and written by professionals.

Designated Infection-Control Officer

EMS agencies must assign a designated infection-control officer, who is to be responsible for maintaining a liaison with the physician, health-and-safety officer, and infection-control representative at health-care facilities and other health-care regulatory agencies. An infection-control program should have an experienced EMS manager or leader designated as the infection-control officer.

The designated infection-control officer is appointed by each ambulance service, EMS first-response organization, or fire-service agency for the purpose of receiving notifications of exposures to certain infectious diseases from health-care facilities. The officer also notifies the indicated care provider(s) of an exposure to an infectious disease dangerous to the public or the employee's health.

Under OSHA and NFPA guidelines, it is good practice to establish a designated infection-control officer. The officer has the responsibility of maintaining communication between the EMS or fire agency and all community health-care professionals. He also has the responsibility of examining compliance procedures and engineering controls, investigating exposure incidents, notifying members of exposure, properly documenting the exposure, and ensuring that members receive medical follow-up care.

- ◆ Designation of an infection-control officer.
- ◆ Definition of an exposure and key terms.
- ◆ Training and education.
- ◆ Vaccination and testing for hepatitis B and TB.
- ◆ Personnel protective equipment.
- ◆ Postexposure management program.
- ◆ Compliance monitoring.
- ◆ Record keeping.

FIGURE 6.16 ■ Components of an Infection-Control Plan.

The cost of maintaining an infection-control officer should compare well to the reduction in costs resulting from the elimination of exposures and management of risk.

Infection-Control Training

Training and education of emergency personnel is an important component of any infection-control program. It should include proper use of personal protective equipment, standard operating procedures for safe work practices in infection control, proper methods of disposal of contaminated articles and medical waste, cleaning and decontamination, exposure management, and medical follow-up. It also should provide information on epidemiology, modes of transmission, and prevention of diseases. Emergency responders should be educated on the diseases that have the potential for occupational exposure.

A two-day course entitled Designated/Infection Control Officer is designed to meet the requirements established by NFPA 1581 and OSHA. Understanding this role and the many aspects of the role of the infection-control officer is important to ensure that an EMS manager can oversee a risk-management program and help to be an advocate for the agency. The training program that certifies managers as infection-control officers reviews the various laws and the procedures to establish an infection-control program, including postexposure management programs. Once certified as an infection-control officer and active in that role for six months, an EMS manager can take the advanced infection-control officer course that covers updates in diseases, statistical data collection, program evaluation, and cost-benefit analysis.

Equipment Needs

Emergency-service agencies must provide certain protective clothing and equipment for each of their members during medical emergencies. Public Law 106–430, or the **Needle Stick Prevention Act,** was passed in 2000 and requires OSHA to update blood-borne pathogen standards. It also requires that employees participate in the evaluation and selection of devices. While the law focuses on equipment to prevent needle sticks, field personnel should be included in evaluating protective equipment. Some examples of equipment needed for infection control are listed in Figure 6.17.

Studies show needleless systems can reduce injuries by greater than 50%, which is a substantial

- ♦ Single-use medical gloves.
- ♦ Fluid-resistant clothing.
- ♦ Pocket masks and/or NIOSH-approved respirators.
- ♦ Splash-resistant eyewear and face-protection devices.
- ♦ Respiratory-assist devices.
- ♦ Approved sharps containers.
- ♦ Leakproof bags.

FIGURE 6.17 ■ Infection Control Equipment Needed for EMS.

savings to the organization, considering a needle-stick injury can cost an organization's health-care or workers' compensation system $1200 to $5000 per event and substantially more if the individual becomes infected.

Latex gloves now present a risk to responders, and NIOSH released a warning to EMS providers in 1997 indicating that proteins in latex have caused health-care workers to develop latex allergies. It is estimated that approximately 10% of all health-care workers have latex sensitivities. Several medical centers provide desensitization programs to help reduce reactions and protect employees. The Americans with Disabilities Act (ADA) recognized this, and latex allergies are considered a workers' compensation issue.

Decontamination Areas

The EMS agency should set aside an area in each station for the storage, cleaning, and disinfecting of emergency medical equipment and uniforms. Every agency should have a place for medical waste. The standard, NFPA 1581, outlines the recommended facilities for infection control within the department. These recommendations also comply with CDC and OSHA regulations. The station should be equipped with facilities for disinfection, cleaning, and storage.

The appendix of NFPA 1581 provides recommendations for the decontamination of firefighting apparatus and for including decontamination facilities when building new fire stations.

The Bloodborne Pathogen Standard under OHSA is found in 29 CFR 1910.1030, which lists the components and the major provisions for employees exposed to bloodborne pathogens.

There should be a room within the EMS or fire agency dedicated to decontamination of medical equipment. The room should be physically separated by four walls from living areas and well vented to the outside environment. The floor drains should be connected to a sanitary sewer system or to some sort of containment tank that can carry runoff to a disinfection tank.

A contract or agreement must be put in place with a qualified contractor to remove stored biohazards from the agency's facilities. It is important to make sure the contractor is licensed and bonded. Biohazard waste is regulated under hazardous-materials laws; should the EMS agency's wastes show up in an inappropriate area, the agency could be responsible for an expensive cleanup.

Cleaning and disinfection of equipment and clothing should be performed in the proper area and on a regular basis and immediately following an exposure incident. In 2000 OSHA ruled that if equipment is left at the hospital, it is the hospital's responsibility to clean it or make it safe for handling.

Consideration of infection-control measures should be applied to bathrooms, kitchens, sleeping areas, laundry facilities, equipment-storage areas, cleaning areas, disinfection facilities, and disposal areas. Under no circumstances should contaminated equipment or clothing be taken home for cleaning. The infection-control program outlined within OSHA's standard also addresses skin-washing practices, disinfectant handling and use, cleaning of contaminated emergency medical equipment, disposal of infectious materials, and the laundering of linens.

As an important factor in infection control, the OSHA standard addresses hand washing as follows:

> "Hands shall be washed as follows: (1) after each emergency medical incident; (2) immediately or as soon as possible after removal of gloves or other personal protective equipment; (3) after cleaning and disinfecting emergency medical equipment; (4) after cleaning personal protective equipment; (5) after any cleaning function; (6) after using the bathroom; (7) before and after handling food or cooking and food utensils."

As another important factor in infection control for field disinfection, the CDC recommends only alcohol-based foams or gels. Figure 6.18 is an example of an infection-control policy statement used by fire-based EMS organizations.

OCCUPATIONAL EXPOSURES

Pre-exposure

Pre-exposure activities need to be undertaken by EMS managers and the infection-control officer. Often these activities revolve around vaccinations and ensuring proper equipment is available when needed. The CDC

Purpose: To provide a comprehensive infection control system that maximizes protection against communicable diseases for all members of the Upper Pine Fire Protection district and for the public that they serve.

Scope: This policy applies to all members, career and volunteer, providing fire, rescue, or EMS in the Upper Pine Fire Protection District.

Upper Pine Fire Protection District recognizes that communicable disease exposure is an occupational health hazard. Communicable disease transmission is possible during any aspect of emergency response, including in-station operations. The health and welfare of each member is a joint concern of the member, the chain of command, and the department. Although each member is ultimately responsible for his or her own health, the department recognizes a responsibility to provide as safe a workplace as possible. The goal of this program is to provide all members with the best available protection from occupationally acquired communicable disease.

It is the policy of this department to:

- Provide fire, rescue, and EMS to the public without regard to known or suspected diagnoses of communicable disease in any patient.
- Regard all patient contacts as potentially infectious. Universal precautions will be observed at all times and will be expanded to include all body fluids and other potentially infectious material (body substance isolation).
- Provide all members with the necessary training, immunizations, and personal protective equipment (PPE) needed for protection from communicable diseases.
- Recognize the need for work restrictions based on infection control concerns.
- Encourage participation in employee assistance and critical incident stress debriefing (CISD) programs for the family and the employee.
- Prohibit discrimination of any member for health reasons, including infection or development of a disease after exposure, such as HIV, HBV, or HCV virus.
- All medical information is strictly confidential under HIPAA. No member's health information will be released without the signed written consent of the member.

FIGURE 6.18 ■ Sample Infection Control Policy Statement. *(Reprinted with permission of Upper Pine Fire River Protection District.)*

recommends vaccinations for hepatitis B, chicken pox, measles, mumps, and rubella and an annual flu vaccine. Some agencies vaccinate for hepatitis A and pneumonia. If your EMS agency is capable of deploying people into a disaster zone, EMS managers may choose to follow a moreaggressive vaccination schedule for the FEMA teams. Those vaccinations are included in Table 6.3.

In any vaccination process many agencies have sent employees to test for titers or blood markers that indicate the vaccination-produced antibodies for the disease. In 1999 OSHA required employers to pay for titer testing. For immune-competent EMS personnel, periodic testing or periodic boosting is not needed.

Post-vaccination testing (anti-HBs) should be done one to two months after the last dose of hepatitis B vaccine. If adequate anti-HBs (at least 10 mIU/mL) are present, nothing more needs to be done. If post-vaccination testing shows less than 10 mIU/mL, the vaccine series should be repeated and anti-HBs testing done one to two months after the last dose of the second series. This information should be recorded in the employee's health record.

In many states the exposure laws include presumptive legislation that assumes the employee would have only contracted the disease in the work environment.

The Chain of Infection

For an infection to occur, a certain sequence of events must take place. This sequence is often referred to as the chain of infection, which is made up of sections that link together to form the infectious process and include a pathogen, a mode of transmission, a route of exposure, and a susceptible host, such as the human body. EMS providers come into contact with pathogens every day. Most of the time the body's immune system destroys them before they can cause harm.

An EMS provider is considered exposed when he has been in contact with a pathogen. He is considered infected when a pathogen has entered the body and resulted in disease. Whether or not an exposure results in infection depends on the dose (the amount of organisms that enter the body), virulence (the strength of the organism), and host resistance (the ability of the body's immune system to fight infection) (Figure 6.19).

Modes of Transmission

Exposure occurs through either direct or indirect contact. A **direct transmission** occurs when an agent that causes disease, especially a living microorganism such as a bacterium or fungus, is transmitted directly from an infected individual to one who had been uninfected. For example, you could become infected with the hepatitis B virus if you had an open wound that came into contact with a patient's hepatitis B–infected blood.

An **indirect transmission** occurs when an inanimate object serves as a temporary reservoir for the infectious agent. For example, you could become infected with HBV if you come into contact with equipment that has dried infectious blood on it.

It is important to note that many diseases do not manifest themselves immediately. Therefore, it can often be difficult to track the source of an exposure.

TABLE 6.3 ■ FEMA Vaccination Schedule

Immunization	Nonmedical Personnel in Disaster Medical-Assistance Teams and Urban Search-and-Rescue Teams	Medical Providers and Disaster Medical-Assistance Teams and Urban Search-and-Rescue Teams	All Rescue Personnel and Designated Medical-Support Teams Who Deploy Overseas
Influenza ANNUAL	P	P	P
MMR (measles, mumps, rubella) 1X ONLY	P	P	P
Polio (IPV) 1X ONLY	P	P	P
Tetanus-diphtheria (Td) EVERY 10 YRS	P	P	P
Varicella (chicken pox) 1X ONLY	P	P	P
Hepatitis AV 2 DOSES	P	P	P
Hepatitis B 3 DOSES	P*	P	P
Rabies 3 DOSES			
Meningococcal (tetravalent) EVERY 3–5 YRS		P	
Typhoid EVERY 2 YRS		P	
Yellow fever EVERY 10 YRS		P	
TB skin test (PPD) ANNUAL	P	P	P

*Includes all administrative, communications, logistical, safety, and security personnel who would not normally be involved in direct patient contact. These personnel have until 1 July 2005 to complete at least the first two doses of the Hepatitis B series and must complete three doses by 31 December 2005 to meet deployability standards.

P = required for deployment.

Many of the symptoms of some diseases can be quite similar to the flu. If flulike symptoms do not subside in a normal amount of time with normal treatment methods, you may need to have blood tests performed to rule out other possible causes.

Defining an Exposure

A bloodborne-pathogen exposure can be a contaminated needle stick or blood or other potentially infectious material (OPIM) splashed in the eyes, mouth, nose, or open wound in the skin. OPIM may include amniotic fluid, pleural fluid, cerebral spinal fluid, and any body fluid with gross visible blood. The CDC does not consider tears, saliva, urine, sweat, or vomit to be risky body fluids. Cuts or scrapes caused by bloody or contaminated objects, as well as human bites, are also considered exposures. Note that when an EMS provider is bitten, both the provider and the

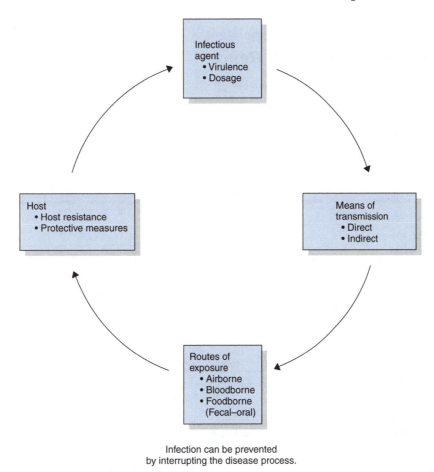

Infection can be prevented
by interrupting the disease process.

FIGURE 6.19 ■ Factors Involved in Infectious Disease Exposure.

Reporting Exposures

Your EMS agency must have standard procedures for the reporting and managing of exposures. To start, each member of the agency should know the name of and contact information for the designated infection-control officer. The official exposure-reporting form should include at least the requirements listed in Figure 6.20.

Exposure to an infectious or contagious disease requires prompt action, particularly if the individual does not have adequate immunity to the disease. Instructions for care of an exposure should include the following: The exposed area should be washed immediately. The incident should be reported to the infection-control officer within two hours of the exposure. The provider should then be treated by the

biter need to be followed, the latter because he may have gotten the provider's blood in his mouth.

agency physician as soon as practical but at least within 24 hours.

Whether occurring while on or off the job, all exposures to an infectious or contagious disease should become a part of the provider's confidential health file. In addition the information from a duty-related exposure should be made anonymous and added to the agency's health database. (See the exposure reporting form in Figure 6.21.)

♦ Name, date, time, and location of incident.
♦ Identification of possible pathogen involved in exposure.
♦ Description of the tasks being performed when the exposure incident occurred.
♦ Source of transmission.
♦ Portal of entry.
♦ Personal protective equipment used.

FIGURE 6.20 ■ Minimum Information Needed to Officially Report an Exposure.

FIGURE 6.21 ■ Sample Exposure Reporting Form.
(Courtesy of the U.S. Department of Labor, Occupational Safety and Health Administration.)

Documenting Exposures

Exposures should be documented and recorded in the responder's confidential medical record. The HIPAA requirements for privacy protection require any medical exposure to be kept with the strictest confidence within the organization. In most cases the management should only know that an employee accesses the health system for an exposure. No details should be released to anyone other than the employee. Relevant information for a medical record is included in Figure 6.22 with a sample log.

POSTEXPOSURE

The Centers for Disease Control (CDC) has published postexposure management procedures for exposures to the hepatitis B virus, the hepatitis C virus, HIV, and other infectious diseases. These procedures are widely accepted and are updated regularly to reflect advances in medical knowledge. A rapid HIV test now can be done using a blood draw, cheek swab, or finger stick and can produce a result within 10 to 20 minutes, providing immediate information to start prophylaxis with

---

Information Required for Exposure Incident:

* Date and time of exposure.
* Activity being performed by the worker when the exposure occurred.
* Details of exposure, including amount of fluid or material, type of fluid or material, and severity of exposure, including the depth of injury and whether or not the fluid was injected.

Contamination into broken or unprotected skin or mucous membranes like eyes or the mouth. The extent and duration of contact and the condition of the skin such as chapped, abraded, or intact.

* Description of source of exposure, including, if known, whether the source material contained the hepatitis B virus, hepatitis C virus, HIV, or other infectious diseases.
* Photo of sharps log.

FIGURE 6.22 ■ Required OSHA Record for EMS Personnel.
(Courtesy of the U.S. Department of Labor, Occupational Safety and Health Administration.)

OSHA's Form 300 (Rev. 01/2004)

Log of Work-Related Injuries and Illnesses

Attention: This form contains information relating to employee health and must be used in a manner that protects the confidentiality of employees to the extent possible while the information is being used for occupational safety and health purposes.

Year 20___ ___

U.S. Department of Labor
Occupational Safety and Health Administration

Form approved OMB no. 1218-0176

You must record information about every work-related death and about every work-related injury or illness that involves loss of consciousness, restricted work activity or job transfer, days away from work, or medical treatment beyond first aid. You must also record significant work-related injuries and illnesses that are diagnosed by a physician or licensed health care professional. You must also record work-related injuries and illnesses that meet any of the specific recording criteria listed in 29 CFR Part 1904.8 through 1904.12. Feel free to use two lines for a single case if you need to. You must complete an Injury and Illness Incident Report (OSHA Form 301) or equivalent form for each injury or illness recorded on this form. If you're not sure whether a case is recordable, call your local OSHA office for help.

Establishment name _____

City _____ State _____

Identify the person			Describe the case			Classify the case							Check the "Injury" column or choose one type of illness

Identify the person

(A) Case no.
(B) Employee's name
(C) Job title (*e.g., Welder*)

Describe the case

(D) Date of injury or onset of illness
(E) Where the event occurred (*e.g., Loading dock north end*)
(F) Describe injury or illness, parts of body affected, and object/substance that directly injured or made person ill (*e.g., Second degree burns on right forearm from acetylene torch*)

Classify the case

CHECK ONLY ONE box for each case based on the most serious outcome for that case:

Death (G)
Remained at Work: Days away from work (H); Job transfer or restriction (I); Other recordable cases (J)

Enter the number of days the injured or ill worker was: Away from work (K); On job transfer or restriction (L)

Check the "Injury" column or choose one type of illness (M): Injury (1); Skin disorder (2); Respiratory condition (3); Poisoning (4); Hearing loss (5); All other illnesses (6)

Page totals ▶

Be sure to transfer these totals to the Summary page (Form 300A) before you post it.

Public reporting burden for this collection of information is estimated to average 14 minutes per response, including time to review the instructions, search and gather the data needed, and complete and review the collection of information. Persons are not required to respond to the collection of information unless it displays a currently valid OMB control number. If you have any comments about these estimates or any other aspects of this data collection, contact: US Department of Labor, OSHA Office of Statistical Analysis, Room N-3644, 200 Constitution Avenue, NW, Washington, DC 20210. Do not send the completed forms to this office.

Page ___ of ___

(1) Injury (2) Skin disorder (3) Respiratory conditions (4) Poisoning (5) Hearing loss (6) All other illnesses

---

OSHA's Form 300A (Rev. 01/2004)

Summary of Work-Related Injuries and Illnesses

Year 20___ ___

U.S. Department of Labor
Occupational Safety and Health Administration

Form approved OMB no. 1218-0176

All establishments covered by Part 1904 must complete this Summary page, even if no work-related injuries or illnesses occurred during the year. Remember to review the Log to verify that the entries are complete and accurate before completing this summary.

Using the Log, count the individual entries you made for each category. Then write the totals below, making sure you've added the entries from every page of the Log. If you had no cases, write "0."

Employees, former employees, and their representatives have the right to review the OSHA Form 300 in its entirety. They also have limited access to the OSHA Form 301 or its equivalent. See 29 CFR Part 1904.35, in OSHA's recordkeeping rule, for further details on the access provisions for these forms.

Number of Cases

Total number of deaths _____ (G)
Total number of cases with days away from work _____ (H)
Total number of cases with job transfer or restriction _____ (I)
Total number of other recordable cases _____ (J)

Number of Days

Total number of days away from work _____ (K)
Total number of days of job transfer or restriction _____ (L)

Injury and Illness Types

Total number of . . . (M)
(1) Injuries _____
(2) Skin disorders _____
(3) Respiratory conditions _____
(4) Poisonings _____
(5) Hearing loss _____
(6) All other illnesses _____

Establishment information

Your establishment name _____
Street _____
City _____ State _____ ZIP _____

Industry description (*e.g., Manufacture of motor truck trailers*) _____

Standard Industrial Classification (SIC), if known (*e.g., 3715*)
___ ___ ___ ___

OR

North American Industrial Classification (NAICS), if known (*e.g., 336212*)
___ ___ ___ ___ ___ ___

Employment information (*If you don't have these figures, see the Worksheet on the back of this page to estimate.*)

Annual average number of employees _____
Total hours worked by all employees last year _____

Sign here

Knowingly falsifying this document may result in a fine.

I certify that I have examined this document and that to the best of my knowledge the entries are true, accurate, and complete.

Company executive _____ Title _____
( ) _____ - _____ Phone / / Date

Post this Summary page from February 1 to April 30 of the year following the year covered by the form.

Public reporting burden for this collection of information is estimated to average 58 minutes per response, including time to review the instructions, search and gather the data needed, and complete and review the collection of information. Persons are not required to respond to the collection of information unless it displays a currently valid OMB control number. If you have any comments about these estimates or any other aspects of this data collection, contact: US Department of Labor, OSHA Office of Statistical Analysis, Room N-3644, 200 Constitution Avenue, NW, Washington, DC 20210. Do not send the completed forms to this office.

FIGURE 6.22 ■ Continued

antiviral therapy immediately. Rapid HIV testing is the standard of care, should be immediately available, and can be used as a screening tool to help prevent EMS personnel from taking harsher drugs. Once the patient has tested positive, there is no need to test the EMS provider, and treatment should begin immediately.

Antiviral or pharmaceutical use is highly recommended for exposures to other diseases (such as tuberculosis) and can be found updated on the CDC Web site. The local health department or an infection-control committee at a local hospital also can provide valuable assistance regarding postexposure management procedures.

Postexposure evaluations must be made available to all personnel after an exposure occurs in accordance with OSHA guidelines. Figure 6.23 displays an algorithm for needle sticks. There is no significant argument against these postexposure tests, and the International Association of Firefighters (IAFF) certainly encourages such testing. If an individual's on-the-job exposure to blood or body fluids causes an infection, the agency has the responsibility to provide early diag-

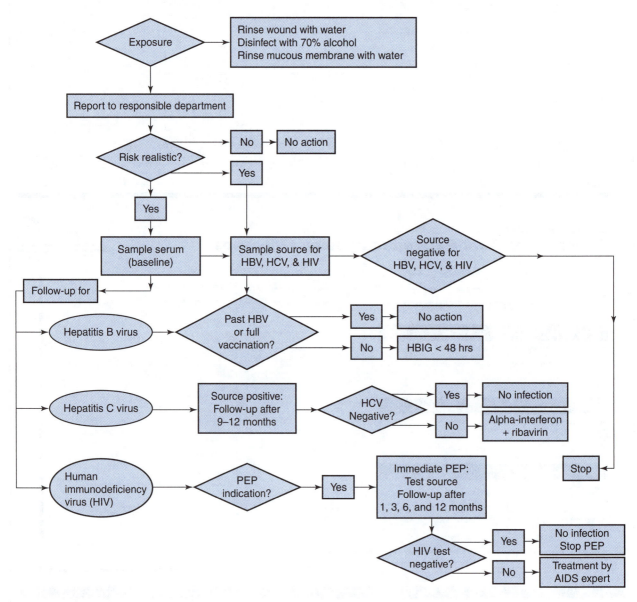

FIGURE 6.23 ■ Exposure Algorithm.

nosis and treatment to that individual. Postexposure management should include counseling for EMS personnel and their families. This can be supported by an employee assistance program.

Reporting Requirements

OSHA has requirements for recording workplace illnesses and injuries. These include the OSHA Forms 300, 300A, and 301. (See the OSHA Web site to view these forms.) Under OSHA requirements, employers must:

- Record needle-stick and sharps injuries involving contamination by another person's blood or other potentially infectious materials.
- Establish a procedure for employees to report injuries and illnesses and for telling the employees how to report.
- Guarantee employee access to individual OSHA 301 forms.
- Protect employee privacy (such as by not entering individual names for injuries resulting from sexual assault, HIV infection).
- Routinely update the forms, and save them for five years.

Privacy Issues

Careful attention must be paid to the means by which employers seek to create an infection-free workplace. Principles of public safety and efficient performance must be balanced against an individual's reasonable expectations of privacy. In particular the confidentiality of any employee medical records produced from department-required testing should be treated the same as any other medical record. Legal rights of individual employees should not be sacrificed as employers haphazardly rush to implement a testing program.

An important consideration to proper medical testing is confidentiality, something that is mandated by OSHA regulations. It is difficult to overemphasize the crucial importance of maintaining confidentiality. If personnel do not believe the records are confidential, then the records will be inaccurate and hence incomplete and much less useful. The confidentiality issue also gives rise to the following questions:

- Who will own the records?
- Who is authorized to see them?
- Where and how will they be stored?
- If computerized, are they really secured?

Confidentiality also has legal implications, since failure to maintain confidentiality can result in a lawsuit. Apart from the physical security of records, it is important to consider just how much information is needed for personnel functions. Management needs to know only that an individual can do his job or, if not, what restrictions apply. Specific medical diagnoses must not be revealed to management. If medical surveillance is to be meaningful, records must be as complete as possible. Only with confidentiality of records is this possible.

■ EMS SAFETY

SAFETY EDUCATION

Certain elements of safety education should be part of any EMS organization's risk-management program. EMS managers in charge of risk management should coordinate with training officers on a yearly or two-year cycle to conduct state and federal OSHA-required courses. Training only targets one piece of the risk-management puzzle and does not take into consideration motivation and attitudes. A safety culture is an umbrella that covers the organization, including training programs. Figure 6.24 is a list of important topics to be delivered as part of any safety program.

Emergency-service providers are required to make multiple decisions on each call. Embedded in this decision matrix is the concept of safety. Safety aspects range from immediate threats from the scene and environment and the people in it to long-term threats related to the lifestyle commonly associated with emergency providers.

The number of decisions EMS personnel make on every response is impressive. These include route selection, negotiating traffic, choosing the best scene approach with strangers in crisis, determining needed medical care, and determining appropriate transportation of the patient. Safety is the main consideration when making these decisions. There was a time when safety was an underlying theme, but it is not so any longer. The challenge is getting EMS responders to slow down and focus on the dangers that may be present, rather than the excitement and hype of the call itself.

The nature of emergency medical services involves hazardous work. Consider these common scene scenarios: traffic fatalities involving EMS units, downed power lines, responders held at gunpoint during a narcotics robbery attempt, infectious-disease exposures,

- *Hazard Communication*—OSHA 1910.101 requires training on any of the hazardous chemicals to which employees are continuously exposed.
- *Emergency Plan*—OSHA 1910.155 requires that twice a year an employee should practice how to activate an alarm system for various emergencies and understand evacuation plans for every facility in the organization.
- *Lock Out Tag Out*—OSHA 1910.211 requires that any employees using energized equipment need to understand the lock-out tag policies and procedures.
- *Respiratory*—Every employee who needs respiratory protection should be fit tested on an annual basis under OSHA 1019.134.
- *Access to Medical Records*—Employees must be trained on how to access and be informed about their rights to access their medical records according to OSHA 1910.1020.
- *Portable Fire Extinguishers*—EMS employees who are not firefighters need to be trained annually on the use of fire extinguishers under the OSHA 1910.155 standard.
- *Ergonomics and Equipment*—Each employee should be properly trained on lifting and operating equipment. Employees must also be educated on other work-related activities that could result in a physical injury. Employees are required to be trained on the main equipment used by the organization, including any equipment that could be hazardous to operate. Emergency vehicles operation is one area covered under OSHA 1910.1.
- *Hazardous Waste and Spills*—Under OSHA 1910.101 EMS employees need to be trained on the disposal and clean up of hazardous materials, including medical waste.
- *Hearing Protection*—New employees need hearing protection when hired and annually under OSHA 1910.151.

FIGURE 6.24 ■ Required OSHA Training for EMS Personnel.

hit-and-run incidents, and dog or other animal attacks. Hazards can be categorized as physical or environmental hazards, traffic, strenuous physical labor, and emotionally charged interactions. Safety issues cross all levels of certification, territorial boundaries, and population demographics. During the last decade many improvements have occurred in workplace safety for EMS personnel.

With the advent of modern EMS in some cities in the early 1970s, sophisticated equipment and education met an important social need. Three decades later this medical specialty has earned wide recognition and legitimacy within the medical community. An industry-wide trend promoting research forums and foundations demonstrates a sincere effort toward safety improvements. Organizations that highlight EMS safety are listed in Figure 6.25. There also are programs and initiatives from the National Association of

EMTs (NAEMT), the American Ambulance Association, the International Association of Fire Chiefs (IAFC), the International Association of Firefighters (IAFF), the National Fire Protection Association (NFPA), the National Volunteer Fire Council (NVFC), the National Institute of Occupational Safety and Health (NIOSH), the Centers for Disease Control and Prevention (CDC), the U.S. Fire Administration (USFA), the National Highway Traffic Safety Administration (NHTSA), and others.

WHY SAFETY?

A commitment to "all emergency responders returning home safely" may seem to be a simple and obvious statement. On the contrary it signifies a major commitment to safety by the EMS providers and their supporting organizations. Sometimes agreeing on such a simple statement (and daunting task) results in synergistic efforts and partnerships with huge impacts and outcomes. The fire service accepted losses and celebrated posthumous awards for line-of-duty deaths for over two centuries before adopting safety as the primary concern of responders. Basic tenets of workplace-safety programs are staying out of hazardous situations and taking little or no risk in nearly totally controlled environments and situations. The obvious conclusion for emergency responders is that they have little control over the circumstances under which they are summoned or must operate. In the words of a seasoned responder, "We go to situations that are out of control, where death or injury have already occurred, people are unable to reason, with the simple expectation of 'just fix it.'"

The effect of adrenaline has a great impact on people unaccustomed to emergencies, and they often overreact. The body's natural response to any perceived crisis is to release this "fight-or-flight" hormone. It can generate increased heart rate, faster breathing, sweaty palms, and explosive energy. The effects of being in EMS and gaining experience result in EMS personnel becoming desensitized or callused. By controlling the initial adrenaline response, they are able to think more clearly and assess the scene and medical aspect more effectively. The down side is that they tend to get complacent about safety. In addition there is some evidence that the constant up-and-down adrenaline rush has serious negative health effects on responders over time. Some of these effects are cardiac, digestive, sleep, and anger disorders.

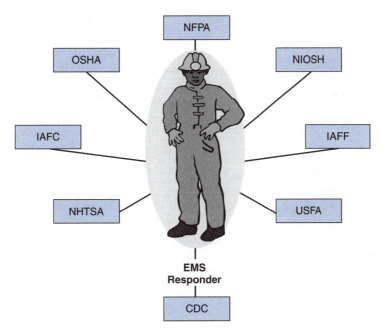

FIGURE 6.25 ■ Agencies Involved with EMS Safety.

Another basic yet complex aspect of EMS safety is related to responder health. Becoming a victim of the stresses of the occupation is an all-too-common outcome in the industry. A case in point is the relatively constant statistic of line-of-duty deaths, indicating that more than 40% of annual responder deaths are due to heart attacks. This statistic has remained constant for the last 20 years and will probably continue.

The benefits of adopting a commitment to healthy living—including exercise, good diet, and stress management—are the key. Response organizations have realized that by adopting a policy that supports wellness and fitness programs, they can reduce injuries and disabilities significantly. It is therefore incumbent on EMS organizations to have SOPs and policies concerning safety.

TOP FIVE INJURIES FOR EMS WORKERS

Back Injuries

Back injuries account for the majority of workplace injuries, time off work, and early retirement and disability in EMS personnel. A recent study administered fitness assessments related to the occupational activities of the prehospital provider, with the purpose of describing the incidence of occupational back injury and percentage of providers with known back-injury risk factors. When surveyed 47.8% of subjects re-

ported a back injury in the previous six months, but only 39.1% of these injuries were sustained while performing EMS duties. Though only 13% of these injuries resulted in missed work, 52.2% reported their injury interfered with daily activities. In addition in spite of the physical nature of the profession, EMS providers in the study were significantly overweight according to their body mass index (BMI) and may have lacked sufficient back strength and flexibility for safe execution of their duties.

The nature of patient care and emergency response will always place workers at risk for back injuries. An EMS provider may be required to carry equipment and patients over long distances or up or down multiple flights of stairs at any time of the day. At a minimum a prehospital provider must have sufficient lower-back strength and hamstring flexibility to prevent musculoskeletal injury while lifting.

Consider these occupational injury-prevention mechanisms to help reduce back injuries:

* Training and education on back injuries and lifting mechanics and evaluation of potential sources and situations that contribute to back injuries.
* Ergonomic-friendly stretchers, including motorized lifting stretchers.
* Vehicle chassis-lowering systems that adjust the height of the floor during loading and unloading of patients.

• Physical-fitness programs that target hamstring and lower-body strength and flexibility.

Exposure to Infectious Diseases

Exposure to blood and other body fluids occurs across a wide variety of occupations. Health-care workers, emergency-response and public-safety personnel, and others can be exposed to blood through needle sticks and other sharps injuries and by way of mucous membranes and skin exposures. The pathogens of primary concern are the human immunodeficiency virus (HIV), hepatitis B virus (HBV), and hepatitis C virus (HCV).

Responders and employers are urged to take advantage of available engineering controls and work practices to prevent exposure to blood and other body fluids. A well-designed infection-control program is the best offense and defense for infectious disease and includes training, reporting, and follow-up of potential exposures. The most important aspects of the infection-control program are training and coaching for EMS responders.

An emerging threat is the risk of contracting an antibiotic-resistant infection after a small break in the skin during an emergency response. Since many of these deaths occur weeks if not months later, often they are not seen or counted as line-of-duty deaths.

Injuries Sustained on Scene

There are on average about 110 firefighter line-of-duty deaths each year in the United States. It is not clear how many serious injuries and near misses also occur. One estimate from a safety expert says, "For every death that occurs, 30 serious injuries occur in the workplace," suggesting the problem of highway safety is much worse than we think. There are clear reporting standards for deaths in the workplace and almost no standards or mechanisms for reporting near misses. Of the over one hundred line-of-duty deaths each year, there are around 15 that come from being struck by vehicles during on-scene highway operations.

If vehicle design and warning systems could prevent the deaths of responders each year on highways, not to mention thousands of near misses, there would be a significant reduction in line-of-duty deaths. EMS managers need to create new policies and training to prevent injuries and death during roadway or freeway responses.

The Responder Safety Initiative is a program managed by the Emergency Responder Safety Institute with support from the National Traffic Incident Management Coalition (NTIMC), NIOSH, NHTSA, and the National Fire Academy. It was formed in 2002 to ensure the safe and efficient management of all incidents that occur on or substantially affect the nation's roadways. The institute supports a newsletter, education programs, and a Web site dedicated to reducing injuries and promoting safety on the roadway for responding police, fire, and EMS units.

Vehicle Crashes Responding Code 3, Lights and Siren

EMS personnel in the United States have an estimated fatality rate of 12.7 per 100,000 workers, more than twice the national average for all citizens. A 2003 NIOSH report documents 27 ambulance-crash-related fatalities among EMS workers over a 10-year period. Most crashes occur between 12:00 and 18:00, often in poor weather. Over the 10-year period there were 305 fatal ambulance crashes with over 36,000 nonfatal injuries, compared to 166 fatal firefighter crashes, with 29,000 nonfatal injuries. The data and case investigations identify unrestrained occupants as an important risk factor for injuries among EMS workers (Figure 6.26).

Two national sets of information that are being collected are the Fatality Analysis Reporting System (FARS) data on fatal accidents and the General Estimates System (GES) data on injuries sustained in ambulance crashes. Among all ambulance occupants, those riding in the patient compartment are associated with greater injury severity. As a safety and risk-management strategy, it is important to ensure that family members are restrained and in the front seat when transported along with the patient.

In one study of 27 incidents, fatal injuries were seen in 26% of the EMS workers who were drivers not wearing a restraint. Another 7.4% were fatalities who were unrestrained in the front right seat. Six EMS workers were killed when they were not wearing restraints while riding in the patient compartment. In the same 11-year retrospective study, it was found that the majority of fatal ambulance crashes occurred during emergency response, and rear-compartment occupants were more likely to be injured than those in front. Twenty percent of the fatalities in the ambulance module were actually restrained.

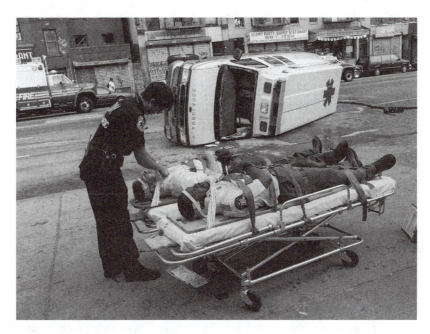

FIGURE 6.26 ◼ Seat-belt emergency response policies must be enforced. *(David Handschuh)*

Lap-belt restraint systems usually provided in patient compartments do not allow full access to the patient. The squad-bench lap belts position the EMS providers against the side wall, making it impossible for them to access the patient. If the EMS responders need to access the cabinets along the driver's-side wall, they must unbuckle the belt to stand up. If CPR or other procedures such as intubation or insertion of IVs must be performed, EMS personnel might need to stand over or kneel near the cot. For these reasons EMS workers often ride unrestrained, seated on the edge of the squad bench.

Unrestrained or improperly restrained patients who become airborne in a crash might pose an additional injury risk to EMS personnel and to themselves. The NIOSH findings are subject to at least three limitations: (1) NIOSH data records only crashes that involve motor vehicles traveling on a road customarily open to the public and that result in the death within 30 days of a person who was either a vehicle occupant or a nonmotorist. As a result fatal crashes on private property (such as driveways, parking lots, and private roads) are excluded. (2) NIOSH data do not determine which ambulance occupants were EMS responders, because it cannot be determined precisely by examining injuries by occupation, since

there are so many different configurations of EMS crews. (3) Data about nonfatal injuries to volunteer firefighters and EMS personnel are not included routinely in occupational-injury databases. CDC recommends that EMS employers ensure that EMS personnel use patient-compartment vehicle-occupant restraints whenever possible, ensure that drivers and front-seat passengers of EMS vehicles use the occupant restraints provided, consider equipping ambulances with patient cots that include upper-body restraints, and ensure that EMS workers who operate ambulances are qualified and trained appropriately.

Ambulance manufacturers should evaluate and develop occupant-protection systems designed to increase the crash survivability of EMS personnel and patients in ambulance patient compartments and ensure that such systems allow EMS providers mobility to access patients and equipment. The EMS manager should have a no-tolerance policy on seat-belt usage when crews are responding to emergencies.

Assaults on EMS

More and more incidents of assault on EMS personnel are being reported. High-profile cases of EMS providers being shot in Kansas City and Memphis have reiterated the risk involved in responding to calls where violence is reported. Numerous articles have been published reflecting anecdotal cases, and a study done in the Chicago Fire Department showed that over 90% of its EMS staff reports being shot at. While anecdotal information persists among EMS providers, a widespread scientific account of the seriousness of the problem of violence against EMS personnel has not been completed.

As an EMS manager, you should manage the problem of protecting your crews against acts of violence with equipment, training, policies, and procedures. EMS agencies need to have a radio designation that indicates there is an immediate threat to EMS personnel. Some agencies have established different levels of communication for increasing levels of threats. Emergency traffic communications for a "threat of violence" need to be understood by both the communications center and the field staff.

Policies need to be in effect that EMS crews are not to approach potential violent scenes without a police presence. Additional procedures should be in place for EMS responses to a **tactical alert** (a state of preparedness necessary to cope with civil disturbances). The use of body armor has been recommended for EMS crews serving in high-crime areas. Most agencies prohibit employees from carrying

◆ Assault with a deadly weapon.
◆ Domestic violence/family disputes.
◆ Shooting.
◆ Stabbing/cutting.
◆ Sniper incidents/police standby.
◆ Other violent crimes.
◆ Tactical alerts and urban unrest.

FIGURE 6.27 ■ Incidents with the Potential for Violence.

firearms, and a policy needs to be in place to state that fact. EMS managers should ensure that field crews are trained in response to calls with a high potential for violence. Figure 6.27 indicates the most common of such calls. EMS managers need to ensure that crews respond to calls with appropriate precautions taken against the possibility that they might encounter violent individuals or groups.

Any risk-management and safety program must have a labor-management team that analyzes workplace violence and determines how to protect EMS personnel in that environment. Episodes of violence perpetrated against EMS personnel need to be included in the national near-miss reporting system.

CHAPTER REVIEW

Summary

Risk management and safety operations go hand in hand within an EMS organization. The nature of emergency services places EMS personnel in hazardous environments with infectious patients, dangerous scenes, and physical and emotional stresses. Loss, injury, and things that go wrong are predictable and preventable, and EMS managers must look at causational factors. It is the responsibility of every EMS manager and leader to conduct comprehensive reviews and ongoing assessment of the risk-management and safety principles within the organization. Risk management and safety require quality policies, training, management, and organizational systems put in place to protect and educate employees.

Research and Review Questions

1. Use the risk-analysis formula to calculate the risk of a back injury to crew members in your agency for responding on a daily basis to a bariatric (over 300 pounds) patient.
2. Identify and evaluate your agency's response to an EMS crew threatened on scene with eminent danger from an armed assailant. How does your communications center handle such a scenario?
3. Identify the high-risk, low-frequency events in your organization. Using the ISO 9000 system, calculate the expected loss for the top two items on the list.
4. List the components required for OSHA training for EMS personnel.
5. Compare the procedures for bloodborne pathogen exposures in your organization, and build or write a procedure that meets the accepted standards for exposures.

6. List the top five hazards to EMS personnel.
7. Identify the infection-control equipment required in an ambulance.
8. Explain how disease is transmitted and the factors contributing to the transmission.

9. List the signs of a loss of situational awareness in an event you have experienced. Then identify applications that could benefit crew resource management.
10. Identify and explain the psychological factors involved with risk management and safety.

References

Ambulance crash-related injuries among Emergency Medical Services workers—United States, 1991–2002. (2003). *JAMA, 289(13)*, 1628–1629.

Crill, M. T., & Hostler, D. (2005). Back strength and flexibility of EMS providers in practicing prehospital providers. (June 2005). *Journal of Occupational Rehabilitation, Vol. 15, No. 2.*

Goldstein, J. (1990). Emergency: A vested interest in safety (*JEMS* communication). Carlsbad, CA: *JEMS.*

Grange, J. T., & Corbett, S. W. (2002). Violence against Emergency Medical Services Personnel. *Prehospital Emergency Care, 6(2)*, 186–190.

Kipp, J. D., & Loflin, M. E. (1996). *Emergency incident risk management.* New York: Von Nostrand Reinhold.

Lubnau, T. II, & Okray, R. (2002). *Crew resource management for the fire service.* Saddlebrook, NJ: Penwell Publishing.

Maguire, B. J., Hunting, K. L., Smith, G. S., & Levick, N. R. (2002). Occupational fatalities in emergency medical services: a hidden crisis. *Ann Emerg Med, 40*, 625–632.

Sachs, G. (2001). *Fire and EMS safety officer* (1st ed.). Upper Saddle River, NJ: Brady Prentice Hall.

Studnek, Jonathan R. (2007). On the job illness and injury resulting in lost work time among a national cohort of emergency medical services professionals. *American Journal of Industrial Medicine, 50(12)*, 921–931.

Tam, G. Y. T. (2006). Perceived effort and low back pain in non-emergency ambulance workers: Implications for rehabilitation. *Journal of Occupational Rehabilitation, 16*, 231–240.

United States Fire Administration. (July 1992). *Emergency incident rehabilitation* (FA-114). Washington, DC: Federal Emergency Management Agency.

United States Fire Administration. (November 1992). *Fire and emergency service hearing conservation programs manual* (FA-118). Washington, DC: Federal Emergency Management Agency.

Volunteer Fire Insurance Services. (2006). *Risk management for EMS manual.* York, PA: Glatfelter Insurance Group.

West, K. (1994). *Infectious disease handbook for emergency care personnel* (2nd ed.). Cincinnati, OH: American Conference of Governmental Industrial Hygienists.

Williams, W. W., MD, MPH. (1983). *Guideline for infection control in hospital personnel,* National Center for Infectious Diseases, Centers for Disease Control and Prevention, Atlanta, GA. National Technical Information Service (NTIS). http://wonder.cdc.gov/wonder/prevguid/p0000446/P0000446.asp

CHAPTER **7**

EMS Human Resources Management

Accreditation Criteria

CAAS

Objectives

After reading this chapter the student should be able to:

7.5   Identify the components of a positive discipline program and implement a due-process procedure involving a disciplinary action (pp. 173–175).

7.6   Create a list of and identify the warning signs of workplace violence and employ management activities to contain or prevent workplace violence (pp. 184–186).

7.7   Build an employee-screening and -hiring process (pp. 161–170).

7.8   Understand the application of labor laws that influence EMS operations (pp. 170–172).

Key Terms

assessment center process (p. 163)

baby boomers (p. 157)

collective bargaining (p. 169)

Delphi technique (p. 156)

Develop a Curriculum (DACUM) (p. 159)

Employee Assistance Program (EAP) (p. 169)

employee forecasting (p. 156)

Fair Labor Standards Act (FLSA) (p. 172)

functional analysis (p. 158)

Garrity Rights (p. 174)

generation X (p. 157)

indexing (p. 156)

job description (p. 158)

job specification (p. 158)

job-task analysis (p. 158)

Longitudinal Emergency Medical Technician Attributes & Demographics Study (LEADS) (p. 156)

Myers-Briggs Type Indicator (MBTI) (p. 168)

observation (p. 158)

performance (p. 158)

position-analysis questionnaire (p. 159)

rating errors (p. 178)

reliability (p. 161)

silent generation (p. 157)

simulation (p. 156)

trend analysis (p. 156)

validity (p. 161)

Weingarten Rights (p. 174)

workplace violence (p. 184)

work profiling (p. 159)

Point of View

John, a supervisor for a private EMS provider in a suburban community, is in a quandary. The problem? It is time once again for the annual performance evaluations of his EMS providers. John, himself a former paramedic, normally has no difficulties performing this job. However, the evaluation of one particular EMS provider is giving him fits. It is not that Sheila is doing a poor job. On the contrary, she is a good employee. Sheila is always on time for work, cares for her patients, and has above-average skills. She always seems to do everything "by the book." Yet something seems to be missing. Although Sheila performs the job description, she does not seem excited about her job or the company. And, she is not alone. Turnover is above average for John's service. Although John senses there is something wrong, he just cannot seem to pinpoint exactly what the problem is.

Within any EMS organization, is there anything more important than its people? After all, without skilled, dedicated, compassionate, and motivated EMS providers, no EMS agency can hope to fulfill the requirements of its mission. Human-resource management, a term popular within management circles, may be a misnomer. Perhaps *human-resource leadership* is a more accurate descriptor of this role. Why? Management's role is helping employees see themselves as they are within the context of their workplace environment. Human-resource leadership, on the other hand, allows people to see themselves as better than they believe themselves to be. And every human-resource function should focus as its primary goal on how to develop the capacity of each person who has chosen that organization as the place to provide care to the sick and injured people within any given community. Given the opportunity to excel, most employees will. The critical function of human resource leaders, therefore, is to determine how the organization can facilitate that process.

Mike Grill
Fire Captain/Paramedic
Sierra Vista Fire Department, Arizona

■ THE EMS WORKFORCE

The recruiting, selecting, and retention of employees will be one of the most challenging aspects of managing or leading an EMS agency. Shifting demographics in the United States will see a significant rise in the elderly. The rise in retirement of health-care workers will be placing a significant strain on the health-care system, especially since the number of people choosing EMS as a career or profession is declining.

The *EMS Workforce for the 21st Century: A National Assessment* states that "the percentage of the U.S. population 65 or older is currently about 12.5% and is expected to reach 16% by 2020 and 21% by 2050. The Bureau of Labor Statistics (BLS) projects that an additional 69,000 EMS workers will be needed by the year 2014, taking separation and replacement of workers into account." In contrast the National Registry of EMTs reports that 19 states saw a decrease in the number of paramedics and EMTs taking the National Registry exams. Even fewer minorities and women are choosing EMS as a career. And approximately 37% of the states report a current shortage of EMS workers. This mismatch between the increasing number of patients and the decreasing number of providers will result in an emergency-care crisis if it is not managed and led correctly.

The demographics of EMS workers are being thoroughly documented, using the **Longitudinal Emergency Medical Technician Attributes & Demographics Study (LEADS)**, which is conducted by the National Registry of EMTs. This project began in 1998, led by a team of researchers looking at longitudinal data for work activities, working conditions, and job satisfaction. An Institute of Medicine (IOM) study, Future of Emergency Medical Care in the United States, is under way to shape and define the status of health in the United States for the next decade. The EMS component of the IOM report, Emergency Medical Services at the Crossroads, affords the leaders of fire and EMS the opportunity to make the most of many of the recommendations made by the IOM.

The Health Resources and Services Administration (HRSA) and NHTSA have funded a project known as the EMS Workforce for the 21st Century. This is a three-phase project that is designed to examine the future of the EMS prehospital workforce in the Untied States. The project brings 17 national organizations and a number of federal agencies together to collect data and formulate a plan to ensure the viability of the future EMS workforce.

HUMAN RESOURCES ACTIVITIES

In response to the changing demographics of the EMS workforce, an EMS organization needs to do **employee forecasting**, an EMS management activity that estimates the number and type of personnel needed to meet the organizational objectives. Forecasting is often more of an art than a science and demands a certain amount of vision from the EMS leadership of the organization. It is tied to service demand, promotions, economics, and retirements and is conducted by looking at both supply and demand of employees.

The demand analysis can be conducted by either a top-down approach or bottom-up approach. The top-down approach uses trend analysis, indexing, or simulations. **Trend analysis** focuses on past hiring. For example, if the EMS agency had hired 12 paramedics every two years for the past 10 years, the trend would indicate a similar number for the next 10 years. **Indexing** uses the ratio of workload to employee, based on increases in call volume or other EMS activities such as injury prevention and public education. **Simulation** uses computer modeling and predicts employee needs with data collected in the computer-aided dispatching software or run-volume statistics. The bottom-up approach takes an inventory of current employee abilities, career plans, and interest and then matches that information in a nonstatistical way to the staffing needs of the organization. The **Delphi technique** can also be used in predicting staffing patterns. It is a method for obtaining forecasts from a panel of independent experts over two or more rounds.

The supply side of the forecasting process evaluates the internal supply of employees by conducting a skills inventory. This consists of each person's education, work experience, occupational interest, and skills. EMS managers should be constantly assessing the workforce for skills and abilities among employees who could be promoted and moved up or over in the organization. External supply is also factored into the supply of employees. EMS managers need to identify the suppliers of employees, and use other employees as a source to recruit new people to the organization.

GENERATIONAL BRIDGES

EMS managers and leaders will face an ever-changing workforce dynamic. In many areas EMS workers are in very short supply, and it may be necessary to reach across generations for help in filling key positions. It is important for the EMS manager to know how generations have been identified by human-resource professionals:

- *Silent generation.* The first is the **silent generation**, or as Tom Brokaw described it, the "greatest generation." Those are people born before 1946. They are uncomfortable turning over the hard work and sacrifice of a lifetime to younger workers.
- *Baby boomers.* The **baby boomers** were born between 1946 and 1964 and perceive today's generation as lazy and egocentric. If a boomer becomes disgruntled or unhappy in the workplace, he will poison the environment.
- *Generation X.* The **generation X** babies were born between 1965 and 1980 and see the boomers as demanding, not creative, and having no fun in the workplace. If a generation-X employee becomes unhappy in the work environment, he usually quits.
- *Generation Y.* These people are the most recent college graduates and were born after 1980.
- *Nintendo generation.* This is the latest group entering into EMS training. They are the most recent high school graduates, taking technical training and foregoing college. They are potential employees who have not had the interaction that workers of the past have had. They are accustomed to technology, but they have poor communication skills and a need for personal gratification in anything they do.

Several trends are occurring in the workplace. First, workers are increasingly staying in or reentering the workforce after they reach retirement age. While this is not always true in EMS due to the physical demands of the job, the average age of municipal EMS workers continues to climb. Younger workers are more quickly moving up the ranks, assuming more important roles than in the past. It is not uncommon to find a paramedic preceptor in his late twenties training someone in his late forties. Younger EMS officers are supervising older workers and are responsible for more complex management functions, due largely to the technology skills required for operating in today's EMS environment.

Many traditional organizational hierarchies have given way to more open communications and team-based structures that include all generations. Seniority counts for less and less today. EMS managers must figure out how to blend incoming generations X and Y with the baby boomers. One way is to remember that common experiences do not equal common

TABLE 7.1 ■ Summary of Generation-Gap Differences

Old (Traditional)	New (Modern)
loyalty	empowered
dedication	participative
respect	change oriented
trust	knowledge based
sacrifice	legitimate curiosity
discipline	electronically literate
service	need information
commitment	question authority
character	self-actualized
civility	very smart
honor	quick gratification
"obliged"	"spoiled"
conservative	body ornamentation
appreciative	contemporary music
grateful	contemporary clothing

attitudes; a bridge is built between the generations by merely recognizing and allowing differences. A summary of common generation-gap issues is presented in Table 7.1.

MOTIVATING THE GENERATIONS

The silent or greatest generation chooses formality over informality, often calling EMS managers and leaders by titles and surnames. They communicate face to face or on a regular telephone, rather than by electronic forms of communication, such as e-mail or text messaging. They often ask for things to be explained logically to them and respond to traditional forms of recognition, such as plaques, awards, or recognition by having a picture taken with their boss or supervisor.

Boomers are goal oriented and take steps to achieve those goals. They state objectives in people-centered terms, and actions reflect effort. They respond to inspirational speeches and function well in most team-centered activities. Boomers see widespread recognition from newsletters or press releases as an incentive.

Generation Xers need a leadership style that does not micromanage. Xers can be told what needs to get done and when, and they will operate with very little instruction. They are great at multitasking. Provide frequent feedback, keep the work environment fun, and ask for their opinions and reactions. Free time or additional time off is the best motivator for generation X.

Generation Yers must be constantly challenged with continuous learning and building skills. Know their personal goals, and try to match up tasks with those goals. They respond best to coaching and upbeat messages through informal routes of communication.

To sum it up an EMS manager's approach to building bridges across generational gaps has to involve an ongoing conversation with his people in two ways: first, by asking what can I do for you and, second, by explaining this is what I need from you. EMS managers will always have a bias, and it is important not to let those biases turn into prejudices. The rewards are enormous, and keeping biases under control starts with analysis of the job and creating a sound job description.

■ JOB ANALYSIS AND DESIGN

A job analysis is the systematic process of determining the skills, abilities, and knowledge required for performing each job in an organization. Creation of **job descriptions** and **job specifications** is required and is the foundation for compensation, selection, and promotion. A job analysis is required by law under the 1978 Uniform Guidelines on Employee Selection Procedures and should be conducted by EMS management and human resources professionals.

METHODS OF ASSESSING JOB STANDARDS

There are many methods of producing standards and designing jobs or positions within the EMS organization. Eight commonly accepted methods are used for establishing standards on which selection, **performance** measurement, and promotion can be based.

The first is to evaluate job-performance requirements, which is an approach that focuses on specific jobs. The procedure begins with a **job-task analysis**, then organizes tasks into duties and areas of responsibility, and finally converts this information into job-performance requirements. Tasks must be observable, discrete, and commonly carried out and must lead to specific output such as a product, service, or decision. Then information is collected about what tools, equipment, and materials are needed for the tasks to be performed effectively; how well the tasks need to be

performed; and what prerequisite skills and knowledge are needed by the job holder.

The second approach is called **functional analysis**. It begins with the organization's mission and the identification of the functions that enable the organization to achieve its mission. Each function is then progressively broken down into occupational areas or subfunctions. This method identifies the key purpose of each occupational area or subfunction, what needs to happen for each key purpose to be achieved, and the tasks performed by individuals. Thus functional analysis establishes standards, which reflect work activities grouped by purpose. Such standards can be common to more than one industry or sector if the roles performed by individuals and groups are the same.

Third is **observation**, a common-sense approach. To observe the jobholders actually performing their work must be one of the simplest and most valid ways of arriving at a set of standards. The method involves activity and time sampling, the recording of observations, and questioning of jobholders. The observations are made of jobholders' attributes and behaviors as well as content of tasks. The data gathered are then analyzed and summarized either in the form of narrative accounts or as tables of counts and frequencies.

The fourth approach is self-description, a method that uses job descriptions provided by the jobholders themselves. Diaries, logs, and narrative accounts are used. These may be unstructured or structured in various ways, such as time or by activity. Jobholders also may be asked to provide information about typical work routines and experiences. This method is particularly useful when the output of jobholders is difficult or impossible to observe, such as decision making in managerial jobs. The data gathered from jobholders are then analyzed singly, collectively, and comparatively, using content-analysis techniques until the structure of the jobs has been determined.

The fifth technique involves gathering detailed information about jobs from interviews. The interviews may be structured or unstructured. In the case of structured interviews, the interviewer uses a predetermined format or checklist of questions. In the case of the unstructured interview, supplementary questions are used to probe for detailed information. Unstructured interviews may be used to produce a list of questions that will be asked in a structured interview. Interviewing is mainly a descriptive technique, but there is no reason why jobholders cannot be asked to rank or rate a job criterion elicited from interviews.

Work profiling, the sixth job-analysis technique, is based on questionnaires. Different questionnaires have been developed for managerial/professional, service/administrative, and manual/technical jobs. The first part of each questionnaire focuses on job content, such as the tasks to be performed. The second part of the questionnaire is about job context. The questions may address skills, knowledge, qualifications, and training required. The completed questionnaires are rated and/or ranked, and the data is then reconfigured to produce reports for various purposes, such as task analysis, attribute analysis, job descriptions, person specifications, job/person matching, or assessment instruments for use in selection. The questionnaire technique and methods of analysis involved have much in common with those used in personality testing.

The seventh technique involves using **position-analysis questionnaires**, which contain a large number of job elements organized into six main areas. The elements are information input to the job, mental processes required to perform the job, job output, relationships, job content, and other job characteristics. Each job element is rated on scales relating to different aspects of the job, such as the importance of proportion of time spent. The data are then computer analyzed through an item analysis, dimension analysis, attribute analysis, and comparative analysis among jobs drawn from a large database of jobs.

The final method of job analysis is the checklist or inventories method, which is composed of a list of tasks. Ratings of one kind or another may be applied to each task. The list of tasks and ratings are then combined in the form of a questionnaire. The questionnaires may be standardized so that a large number of jobholders may be surveyed and the resulting data used to provide a consensus description of the content of a particular job as well as job attitudes. Job-task inventories provide useful models of particular jobs, which can be either adopted by organizations as standards or used as benchmarks.

JOB DESCRIPTIONS AND SPECIFICATIONS

After a job analysis has been completed, a description of the work can be created to develop performance appraisals and training programs. A job description is a document that provides information regarding the task, duties, and responsibilities of the job. Job descrip-

tions must have three key elements: a job title, job identification, and a job-duties section. The job title should provide some psychological importance and status, as well as describe the duties and relative level of the worker in the organization. The job-identification section indicates the location of the job, the person whom the employee will report to, the payroll code, and the date the job was last reviewed. In the job-duties section the tasks are usually arranged in order of importance, the responsibilities are listed, and the tools and equipment that will be used by the employee in the job are identified. Often a job description will include a statement that says, "such other duties as may be assigned."

A job analysis also allows for the creation of a job specification or a list of knowledge, skills, and abilities—plus the physical requirements—that an individual must have in order to succeed in the position. Knowledge refers to factual or technical information necessary to perform a task. A skill is an individual's level of proficiency at performing a particular task. Ability refers to the general capability that the person possesses. Job specifications are used for promotion, compensation, and selection.

THE DACUM PROCESS

One of the most popular techniques for conducting a job-task analysis is called **Develop a Curriculum (DACUM)** (Figure 7.1). The DACUM process for occupational analysis is one that involves the men and women in the organization who have the reputation of being the top performers at their jobs. For the process they are assigned to a short-term committee with a qualified DACUM facilitator. They become the panel of experts who collectively and cooperatively describe their occupation in the language of the occupation.

The panel works under the guidance of a trained facilitator for two days to develop a DACUM research

---

- ◆ An abbreviation for "Develop a Curriculum."
- ◆ An occupational analysis performed by expert workers in the occupation.
- ◆ An occupational skill profile that can be used for instructional program planning, curriculum development, training-materials development, organizational restructuring, employee recruitment, training-needs assessment, meeting ISO 9000 standards, career counseling, job descriptions, competency-test development, and other purposes.

FIGURE 7.1 ■ DACUM (day-kum).

chart. The chart contains a list of general areas of competence called "duties" and several tasks for each duty. Brainstorming techniques are used to obtain the collective expertise and consensus of the panel in identifying and describing their duties and tasks. Their findings for each are written on a card, and each card is then attached to a flat surface until all completed cards form a chart, which finally acts as a graphic profile of the duties and tasks performed by successful workers in the occupation.

The DACUM process has multiple uses, such as job analysis, occupational analysis, process analysis, functional analysis, and conceptual analysis. Many companies, community colleges, and government agencies have found the process to be very effective, quickly accomplished, and low cost. The DACUM process works because expert workers can describe and define their jobs more accurately than anyone else. EMS workers typically know all the operational tasks in the order they are to be performed. This demands a certain level of knowledge, skills, tools, and worker behaviors.

■ RECRUITING

The shrinkage in the EMS workforce has made it more and more difficult to recruit workers, especially in areas with poor working conditions and low pay. The demands of finding paramedics will only become more difficult as baby boomers place a higher demand on the medical system. So EMS must start to think differently about how to find and secure providers.

EMS could take a lesson from Fortune 500 companies and start sending personnel or recruiters to schools that are graduating paramedic students. Several years ago private ambulance firms identified such community colleges and paramedic training programs. As a result, the big West Coast ambulance services sent recruiters to the Midwest, where schools were turning out more students than there were jobs. As relationships developed among education coordinators, paramedic training institutions, and specific EMS agencies, the schools began to encourage their best and brightest to look at those agencies.

Traditional methods of advertising in trade journals, on Web sites, and on EMS billboards frequently meet with limited success. Paramedics and EMTs are often suspicious of organizations that repeatedly advertise. But there is another technique an agency can use: send current employees to regional and national conferences. While attending the conference, employees can rotate staffing in the exhibit area, and the additional expense of placing a booth at that conference and using it for recruiting is worth the benefit of finding new people. Attendees at most conferences are the kind of employees that embrace lifelong learning and bring a fresh perspective to the organization.

DISPARATE IMPACT

The terms "adverse impact" and "disparate impact" in the context of employment refer to hiring practices that appear to be discriminating on the basis of race, sex, or ethnic group. The Equal Employment Opportunity Commission (EEOC) Uniform Guidelines on Employee Selection Procedures require evidence of both statistical and practical significance in order to identify adverse impact, which is the initial burden in any disparate-impact case. The EEOC procedures require that if there are any changes to job requirements, protected classes and the workforce must be given access and time to achieve those standards. An employment decision or hiring requirement that causes substantial underrepresentation to a protected class would be considered disparate impact.

A rule of thumb for measurement of adverse impact is known as the "80% Rule," which says, "A selection rate for any race, sex, or ethnic group which is less than four-fifths (4/5) (or eighty percent) of the rate for the group with the highest rate will generally be regarded . . . as evidence of adverse impact. . . ." In layperson terms, an example would be this: if a majority of your employees are white males and less than 20% are from a protected class, you would need to defend your selection processes as being a BFOQ (bona fide occupational qualification).

THE GRIGGS DECISION

In 1971 the Supreme Court issued a unanimous opinion in the case of Griggs *v.* Duke Power Co. It held that for purposes of hiring and assignment to a laborer position, an employer's use of a high school diploma requirement and two standardized written tests, each of which disqualified a higher percentage of blacks than whites, violated Title VII of the Civil Rights Act of 1964. The Court stated that it was the intent of Congress to prohibit ". . . artificial, arbitrary, and unnecessary barriers to employment when the barriers operate invidiously to discriminate on the basis of racial or other impermissible classification."

- The degree of disparity is created by use of an acceptable standard.
- A demonstrated factual relationship between achieving the employment standard and successful performance of the job in question.
- Whether or not achievement of the employment standard is determined by a "neutral" entity external to the employer.
- Whether or not the employment standard focuses on innate and unalterable characteristics of candidates.
- Whether or not the job in question has a direct impact on public safety.
- Availability of effective alternative standards that create a lesser disparity.

FIGURE 7.2 ■ Key Factors in Assessing the Legality of Employment Factors That Create Disparity.

When testing for new EMTs or paramedics, the assessment tools you use must reflect a "business necessity," which is the legal yardstick for assessing the legality of such standards. The Court held that if an employment practice that operated to exclude blacks could not be shown to be related to job performance, the practice was prohibited. The Court did not provide additional guidance regarding the meaning of the phrase "business necessity."

In the 24 years since the Griggs decision, employers and courts have attempted to define "business necessity" and use it to evaluate a wide range of employment practices. Although the body of law that resulted is complex and difficult for EMS managers to apply, certain factors have emerged as keys to assessing the legality of employment practices that create disparity. The factors included appear in Figure 7.2. This legal requirement mandates that EMS organizations actively recruit and ensure that the selection criteria are based on necessary job tasks.

Agencies hire people with a variety of character traits, skills, and experience to be public servants. Some key skills that relate to EMS are the ability to do problem solving; skill in handling tense, stressful, and multitask situations; people skills; a strong desire to help; a strong sense of responsibility; and a compassionate customer-oriented attitude. These skills should be built into a career ladder for promotional advancement.

■ SELECTION

The process of selection is one of the most important tasks an EMS leader or manager can perform. Hiring the right people reduces risk to the organization and allows management to focus on the positive aspects of supporting and helping employees improve the organization and enhance their skills.

Each method of selection involves two very important tests—reliability and validity. **Reliability** is the degree to which interviews, tests, and selection procedures yield comparable data over the period of time. EMS managers need to be able to consistently rate the interviewees exactly the way they have done so previously. Likewise, a written test should yield the same results every time that it is administered. Reliability also refers to what extent a different method of selection yields the same results; for example, a medical-skills station (one method) and a medical written exam (a different method) both identify the same top candidate. EMS agencies often use oral board reviews to select employees. Managers must make sure there is interpreter reliability when more than one rater is used in the same selection method. Unless the selection methods are reliable, they can be challenged in court.

It is also essential that information be valid. **Validity** refers to what a test measures and how well it measures it. It is the extent to which data from the selection process predict the job performance. There are several types of validity. Criterion-related validity is when the selection method is predictive of the important elements of the work. Concurrent-validity techniques help establish criterion-related validity by assessing new employees and current employees. Predictive validity helps establish criteria by testing applicants and then retesting after they have been on the job for a certain period of time. Validity is required by the EEOC. Content validity exists when the selection method adequately samples the knowledge, skills, and abilities needed to perform in EMS. The closer the content is to the type of work, the higher the content validity. Content validity is the easiest type of validity to apply to a testing process. Construct validity measures theoretical outcomes in broader terms like intelligence, comprehension, and the ability to function under stress. Construct validity requires that the aptitude or human trait be linked to a necessary component of the job task or performance.

APPLICATIONS

The first phase of any selection process is the application. Applications in most organizations are on paper,

- Private organizations to which the interviewee belongs.
- Religious affiliations.
- Date of birth (except when that information is required for satisfying minimum-age requirements).
- Lineage, ancestry, national origin, descent, parentage, or nationality.
- Names and addresses of relatives other than a spouse and dependent children.
- Sexual preference or marital status.
- Height or weight, unless you can show that information is justified by business necessity.
- Existence of physical or mental disabilities.

FIGURE 7.3 ■ Topics Prohibited in Interview Questions.

though some agencies have gone to electronic submission forms. Many human resources firms and large employers use optical scanners to screen resume's and applications. This process seeks out key words in the document and provides a score based on criteria required for the job or desired by the employer.

Applications need to be constructed so they meet legal requirements. A list of questions you cannot ask potential employees on a job application—or in an oral board or interview—is included in Figure 7.3. Note that none of them addresses the skills needed to perform a job. It is permissible to ask if the applicant has any disabilities that would prevent him from satisfactorily performing the job.

WRITTEN EXAMS

Almost all agencies use a written exam to screen people or to bring the number of applicants down to a manageable number for the more expensive oral interviews and background screenings. The objective format uses true/false, multiple-choice, matching, or fill-in-the-blank questions, the traditional format for a written exam. It offers some variety, and most applicants are familiar with the format.

Testbanks are commercially available. It may be helpful to purchase the testbank software that is used in the local education program. Most commercial testbanks are validated and save time and effort by allowing you to choose from a large number of questions as well as to change questions as necessary.

Identification is another written-test process in which an applicant is asked to identify something, such as the cardiovascular system and its parts or travel routes on a map from one given point to another.

Short answers, paragraphs, and essay questions also give an applicant an opportunity to explain himself. Be sure to make the question(s) very clear. Short answers also give the EMS agency a chance to evaluate writing style, legibility, and grammar skills to see if the applicant can complete a prehospital report.

PHYSICAL-ABILITIES TEST

A nationally validated physical-abilities test with a Department of Justice evaluation does not exist for EMS workers. A popular physical-abilities test is one for fire-based personnel: the IAFF and ICHIEFS Candidate Physical Ability Test (CPAT). A labor-management task force successfully developed the test as part of the Labor-Management Wellness-Fitness Initiative in 1997. The CPAT was created to address the need for a holistic and nonpunitive approach to wellness and fitness in the fire service. In 1997 a joint task force of union and chief officers formed a task force that recognized that a number of municipalities were hiring people who would not be physically capable of a successful career in the fire service. The task force unanimously agreed to develop a physical-ability test for pre-employment testing of candidates. To do so the committee members reviewed six fire- and EMS-jurisdictions job-analysis and job-task surveys. The surveys then were validated with the local workforce. The physical-ability test was developed based on the resulting correlation.

The overall consensus of the committee members was that the test would be a good predictor of an applicant's ability to perform basic firefighting tasks. The task force was confident that the ability test would provide the fire service with physically competent recruits. The test also may be administered on a pass/fail basis. But municipalities may not rank candidates based upon CPAT completion times.

INTERVIEWS

Oral-board and interview questions should be prepared from job descriptions before you start interviewing. It pays to write down a set of questions focusing on the job duties and the applicant's skills and experience. Examples of interview questions include the following: Tell me about your experience in EMS or public safety. How much experience did you have in public safety or EMS? Explain how you typically go about organizing your workday. Have any of your jobs required strong leadership skills? By writing down the questions and sticking to the same format at all interviews for a position, you reduce the risk that a rejected applicant will later claim unequal treatment.

Tell the applicant about the job—the duties, hours, pay range, benefits, and career opportunities. Then get into the applicant's work history and relevant experience. Think twice before venturing beyond these fairly narrow topics. The rules of etiquette once dictated that you avoid discussing sex, religion, and politics in a social setting. Although that standard has been relaxed, it still applies to job interviews. You also must avoid focusing on an applicant's age, ethnicity, birthplace, and marital and family status.

The spontaneous and unpredictable nature of the job interview can create hidden traps for the unwary employer. Things that you say with the most innocent intent can be misconstrued as prejudicial and used later as grounds for a lawsuit. For example, a casual discussion about a female applicant's upcoming marriage could lead you to ask whether or not she plans to have children, which could lead the applicant to believe that you discriminated against her based on gender if she does not get the job.

PSYCHOLOGICAL RESOURCES

Liability issues related to an organization's personnel making decisions or taking inappropriate actions has created a new focus in human resources management on using psychological screening tools. More and more organizations are using such tools to help them identify successful matches for positions in the organization.

ACADEMIC PREDICTORS

A health occupations aptitude examination can predict an individual's readiness and capability for successful completion of an educational program designed to prepare qualified health-care personnel. The Psychological Services Bureau, Inc., (PSB) Health Occupations Aptitude Exam is composed of five separate tests that measure abilities, skills, knowledge, and attitudes important for success. The examination is the result of requirements and needs expressed by education professionals responsible for preparing qualified, competent health-care professionals. The PSB Health Occupations Aptitude Examination has been successful in helping to predict an individual's readiness and capability for successful completion of an educational program designed to prepare qualified health-care personnel. The examination addresses required prerequisite and acquired

educational achievements commensurate with the objectives of the preparation program.

Another test, called the Health Occupations Basic Entrance Test (HOBET), is published by Educational Resources, Inc. First, the HOBET provides an objective measurement of a person's critical-reading ability and compares that ability against the level of mastery required for success in college health programs. Second, it evaluates basic mathematics skills necessary to function not only in academic courses but also in clinical practice, for such tasks as calculating medication and ECG intervals. Third, it determines an effective speed for reading college-level material. Last, the HOBET identifies how a person approaches studying and which learning approach is most effective for that person.

ASSESSMENT CENTER PROCESS

The **assessment center process** (created by Development Dimensions International, Inc.) is becoming the preferred method of selecting managers and upper-level administrators within an organization. An assessment center is a series of exercises with which an employee is given the opportunity to demonstrate skills to a group of trained observers. The observers are called *assessors* and are usually above the level of the person testing for the position.

Assessors need to be trained in the assessment center method; in the dimensions, behavior, and skills required for the position being tested for; and in observing and recording those behaviors. Assessor training takes about three days. Assessment centers and their design are governed by the International Congress on the Assessment Center Method, and the methods and guidelines for design can be challenged if they are not conducted properly. The guidelines for conducting an assessment center are offered in Figure 7.4.

- Multiple assessors.
- Judgment resulting in an outcome that is based on pooling information from the assessors.
- The evaluation of the behavior observed and elicitation of reliable, objective, and relevant behavioral information.
- Use of simulated exercises that are job-related and observable and can elicit reliable, objective, and relevant behavioral information.
- The dimensions, knowledge, skills, and abilities that are determined by relevant job behaviors.
- The evaluating techniques for collecting data used to evaluate dimensions, knowledge, skills, and abilities.

FIGURE 7.4 ■ Assessment Center Guidelines.

- In basket.
- Oral presentation.
  - Personal biography.
  - Presentation to the media.
  - Presentation to city counsel.
- Written exercise.
- Group discussion/leaderless group discussion.
- Interview simulation.
  - Interview with the pubic.
  - Employee counseling.
  - Role play.
- EMS tactical scenario/MCI simulation.

FIGURE 7.5 ■ Assessment Center Exercises.

Assessment centers are commonly held over two or more days. There are basic exercises that are common among most assessment centers, and many of the same exercises are found in other assessments for promotion and occasionally for entry-level posi-

tions (Figure 7.5). The exercises are based on dimensions that are from observed behaviors. The dimensions need to be measurable, specific to the position, and based on or representative of the knowledge, skills, and abilities developed in a job-task analysis. These exercises should be made as real as possible.

Participants must demonstrate the appropriate behaviors. Generally, a rating is based on a Likert scale or a scoring range between one and four; one would be a rating for a behavior not demonstrated, and four would be a rating for a dimension that was clearly demonstrated. Most of the exercises seek to measure the skills of planning, organizing, scheduling, communicating, research/reports, and budgeting. Figure 7.6 is an example of a rating system for a tactical EMS scenario. Consistent instructions to the candidate and objective evaluation criteria are important to validity in assessment center testing.

EMS Captain Assessment Center: Multiple-Casualty Incident

Welcome. This is the assessment of a captain's skills in the management of an incident involving multiple patients. On the table you will find the department's EMS worksheet, a generic ICS/IMS worksheet, or tablet of plain paper. You may have your choice of one of the three tools to assist you in the exercise. This exercise will have three events. First, you will be played an audio tape of the dispatch. You will be given 30 seconds to process and respond to that information. After 30 seconds you will be escorted into the room and seated in the command seat and given a map and printout of the incident. You will have two minutes to process and respond to that information. Then you will be shown a series of pictures representing the incident two minutes after it was called in. You will have another eight minutes to complete the scenario. Other units coming on scene will radio you when they arrive. The same person will simulate those units.

PLAY DISPATCH RECORDING. START STOPWATCH.
AT 30 SECONDS ESCORT CANDIDATE TO COMMAND SEAT. HAND CANDIDATE DISPATCH SCREEN PRINTOUT AND MAP. START STOPWATCH.

Dispatch Information: The date is June 30. It is approximately 5:00 P.M. on a Friday. You are the EMS captain. An engine has passed command to you over the radio, opting to start extrica-

tion. You are dispatched to a rollover on U.S. 560 northbound just north of Molas Pass. This is reported as a van off the roadway. Other units en route are transporting paramedic rescue, the battalion chief, and state police.

AT TWO MINUTES, STOP AND REST STOPWATCH. START STOPWATCH FOR ANOTHER EIGHT MINUTES.

AFTER TWO MINUTES FROM THE TIME THAT THE CANDIDATE REQUESTS A TRIAGE OR MAKES A TRIAGE ASSIGNMENT, THE EXERCISE PROCTOR WILL RELAY THE FOLLOWING INFORMATION TO THE CANDIDATE.

Triage Information: There are 12 patients total. Five (5) are immediate—one is trapped, and one is a child that is not trapped. Three (3) are delayed—one is a child with significant road rash over the back and legs. Four (4) are minor injuries walking around the scene.

··

ASSESS BRIEF INITIAL REPORT: IT SHOULD CONSIST OF WHO AM I? WHERE AM I? WHAT DO I HAVE? WHAT AM I DOING? WHAT DO I NEED? WHO IS IN COMMAND?

PREDEFINED RESOURCES TO SEND TO MCI IN ADDITION TO THE INITIAL RESPONSE THAT A CALL COMES IN TO UPGRADE:

2 and 1 medical:	2 ALS engines (4 people)	1st alarm medical:	4 engines (4 people)
	2 ALS rescues (2 people)		3 ALS rescues (2 people)
	2 BLS trucks (5 people)		1 BLS truck (5 people)
	1 battalion chief		2 battalion chiefs

FIGURE 7.6 ■ Sample Instructions and Rating Form for EMS Tactical Scenario.

Candidate Name: _____

Evaluation Criteria	Unacceptable (6)	Acceptable (7)	Above Average (8)	Outstanding (9)	Excellent (10)	Score
Dispatch Phase						
Preincident information.	Fails to recognize 29 Delta is a high-mechanism accident.	Recognizes 29 Delta is a high-mechanism accident. Asks about or indicates probable extrication need.	Recognizes 29 Delta is a high-mechanism accident. Identifies site as high-speed roadway. Asks about or recognizes probable extrication need.	Recognizes 29 Delta is a high-mechanism accident. Recognizes reported vehicle has large numbers of passengers. Identifies site as high-speed roadway. Asks or recognizes probable extrication need.	Recognizes 29 Delta is a high-mechanism accident. Recognizes reported vehicle has large numbers of passengers. Identifies site as high-speed roadway. Asks for or recognizes probable extrication need. Asks dispatch for more information. Upgrades response on initial information.	
Response Phase						
Initial size-up.	Does not upgrade the response.	Upgrades response with specific units.	Upgrades response to 2 plus 1 medical response.	Upgrades response to 1st-alarm medical.	Upgrades to a 1st-alarm medical with helicopter or mutual-aid units.	
On-Scene Phase						
Brief initial report.	Achieved fewer than four of the components of initial report.	Achieved four of the brief initial report.	Achieved five of the brief initial report.	Achieved six of the brief initial report.	Achieved six of the brief initial report with great detail. Declares level-one MCI per city or county MCI plan.	
Assignments.	Does not establish rescue/extrication, treatment unit or group, transport group.	Establishes rescue/extrication group, triage unit, treatment unit or group, transport group. Does not establish helicopter landing assignment.	Establishes rescue/extrication group, medical group, triage unit, treatment unit or group, transport group, and/or landing-zone group.	Establishes rescue/extrication group, medical group, triage unit, treatment unit or group, transport group, and/or landing-zone group.	Establishes rescue/extrication group, triage unit, treatment unit or group, transport group, and/or landing-zone group. Establishes staging officer and calls for command staff (PIO liaison, safety).	

Continued

Evaluation Criteria	Unacceptable (6)	Acceptable (7)	Above Average (8)	Outstanding (9)	Excellent (10)	Score
Transport resources.	Requested fewer than six total transports.	Requested six total transports.	Requested seven total transports.	Requested eight total transports.	Requested eight total transports. Allotted for extra resources in case of rescuer injury. Geographically identifies closer resources (i.e., mutual-aid units). Calls for bus or other non-EMS unit to house minor injuries.	
Manpower resources.	Loses units and personnel on scene from the response. Does not assign adequate or appropriate personnel to IMS/ICS units (e.g., truck to extrication, ALS to treatment, rescues in transportation, treatment teams).	Tracks most units and personnel on scene. Does not assign adequate or appropriate personnel to IMS/ICS units (e.g., truck to extrication, ALS to treatment, rescues in transportation, treatment teams).	Tracks all units and personnel on scene. Assigns adequate personnel to IMS/ICS units (e.g., truck to extrication, ALS to treatment, rescues in transportation, treatment teams).	Tracks all units and personnel on scene. Assigns adequate personnel to IMS/ICS units (e.g., truck to extrication, ALS to treatment, rescues in transportation, treatment teams). Records officer in charge of IMS/ICS assignments.	Tracks all units and personnel on scene. Assigns adequate personnel to IMS/ICS units (e.g., truck to extrication, ALS to treatment, rescues in transportation, treatment teams). Records officer in charge of IMS/ICS assignments. Tracks private ambulance and helicopter resources.	
Safety	Does not address any safety issues. Fails to control traffic hazards or request law-enforcement support.	Generically mentions scene and roadway.	Prioritizes safety. Requests closure of roadway. Recognizes off-road and heat challenges, fuel leaks, and extrication safety.	Prioritizes safety. Requests closure of roadway. Recognizes off-road and heat challenges, fuel leaks, and extrication safety. Designates incident safety officer.	Prioritizes safety. Requests closure of roadway. Recognizes off-road and heat challenges, fuel leaks, and extrication safety. Designates incident safety officer. Addresses helicopter safety, rehab needs. Stages added crews for safety.	
Communications	Does not use the communication model.	Uses the communication model on most transmissions.	Uses the communication model. Repeats information for clarity.	Uses the communication model. Repeats information for clarity.	Uses the communication model. Repeats information for clarity.	

Continued

Evaluation Criteria	Unacceptable (6)	Acceptable (7)	Above Average (8)	Outstanding (9)	Excellent (10)	Score
Communications (cont'd)		Repeats information for clarity.	Requests progress reports.	Requests progress reports. Calls out patient care benchmarks (e.g., patient extricated, helicopter on and off scene, patient en route to hospital).	Requests progress reports. Calls out patient care benchmarks (e.g., patient extricated, helicopter on and off scene, patient en route to hospital). Notifies hospitals. Delegates or assigns information officer.	
Progress reports.	Fails to obtain triage report, make contact with law enforcement, or acknowledge and confirm control of hazards.	Confirms triage report, chief notifications, helicopter operations, hazards controlled, requested additional resources, extrication efforts.	Confirms triage report, chief notifications, helicopter operations, NHP contact, hazards controlled, staging, requested additional resources, extrication efforts.	Confirms triage report, chief notifications, helicopter operations, NHP contact, hazards controlled, staging, requested additional resources early, extrication efforts.	Confirms triage report, secondary triage report, chief notifications, helicopter operations, NHP contact, hazards controlled, staging, requested additional resources early, extrication efforts, hospital status, ETAs.	
Logical progression of incident action plan.	Does not build ICS/IMS structure. Does not designate any geographic issues (e.g., traffic approaches, treatment area, loading areas).	Builds ICS/IMS structure. Designates some geographic issues (e.g., traffic approaches, treatment area, loading areas). Asks for a progress report from one group or unit.	Builds ICS/IMS structure as resources become available. Reassigns triage team to treatment or other function. Designates some geographic issues (e.g., traffic approaches, treatment area, loading areas). Asks for more than one progress report.	Builds ICS/IMS structure as resources become available. Reassigns triage team to treatment or other function. Designates some geographic issues (e.g., traffic approaches, treatment area, loading areas). Establishes resources from left to right on IMS/ICS chart after triage function completed or assigned. Asks for timely progress reports from most of the group or units (more than two groups or two from the same group).	Builds ICS/IMS structure as resources become available. Reassigns triage team to treatment or other function. Designates some geographic issues (e.g., traffic approaches, treatment area, loading areas). Establishes resources from left to right on IMS/ICS chart after triage function completed or assigned. Asks for timely progress reports from each group or unit.	

FIGURE 7.6 ■ Continued

MYERS-BRIGGS

In the early 1900s Katherine Cook Briggs and her daughter Isabel Briggs Myers, expanding on the works of psychiatrist Carl Jung, created the **Myers-Briggs Type Indicator (MBTI)**, which identified valuable differences among people. After more than 50 years of research and development, the MBTI is the most widely used instrument for understanding normal personality differences. MBTI explains basic patterns in human functioning and is used for a variety of purposes in public-safety organizations. Figure 7.7 lists the applications of the MBTI.

Myers-Briggs indicates a specific personality type that is more successful in the public safety professions. The Myers-Briggs type ESTP (extrovert, sensing, thinking, perceiving) has a gift for reacting and solving immediate problems and persuading other people. The Myers-Briggs type ISTP (introvert, sensing, thinking, perceiving) has the ability to stay calm under pressure and excel in any job that requires immediate action.

The MBTI preferences indicate the differences in people that result from the four different assessments:

- *Extrovert or introvert.* Extroverted people focus on the external world. They direct their energy outward and receive energy by interacting with people or taking action. Introverted people focus on the inner world of ideas. They draw their energy and attention inward and receive energy from reflecting on thoughts, memories, and feelings.
- *Sensing or intuiting.* Sensing people like to take in information that is real and tangible. Intuitive people see the big picture, focusing on relationships and connections between facts.
- *Thinking or feeling.* Thinking people mentally remove themselves from the situation to examine the facts objectively. They get energy from finding what is wrong and fixing it. Feeling types are energized by appreciating and supporting others. They look for qualities to praise.

- *Judging or perceiving.* Judging types like to be in a planned and orderly situation to manage their lives. Perceiving types like flexible, spontaneous situations and seek to understand rather than control events and people.

There are no right or wrong answers or right or wrong types. Each of these assessments identifies normal and valuable human behaviors.

BACKGROUND CHECKS

More and more cases of falsified or embellished résumés are making their way into EMS services. EMS places employees in a person's home during his most vulnerable time, and it is important that a law-enforcement agency checks applicants for criminal backgrounds and character. It also is important to confirm that the applicant holds a valid certificate or license from the regulatory authority or the originating jurisdiction. Simply requiring photocopies of certifications opens an agency up to falsified credentials.

The number of falsified college degrees and degrees awarded from nonaccredited institutions is also on the rise. The Internet has become a frequent source of degrees that are purchased or awarded from organizations that do not meet educational standards. Federal officials have reported an increase in people who are buying phony credentials from "diploma mills" that require the buyer to do nothing more than pay their fee.

The Federal Trade Commission (FTC) emphasizes that while it is not always easy to know if academic credentials are from an accredited institution, there are clues to help spot questionable credentials on a résumé or application. According to the FTC, some of the telltale signs of a bogus degree are:

- *Out-of-sequence degrees.* When reviewing an applicant's education claims, be sure that the degrees earned are in the traditional order. For example, if an applicant claims a college degree but shows no signs of a high school diploma or GED, consider it a red flag.
- *Quickie degrees.* Generally, it takes time to earn a college or advanced degree. A degree earned in a very short time or several degrees listed for the same year are warning signs.
- *Degrees from distant institutions.* If an applicant worked full-time while attending school, the schools that are in locations different from the applicant's job or home may be a warning sign. Check their locations.
- *Sound-alike names.* If the institution has a name similar to a well-known school but is located in a different

---

- Self-understanding and development.
- Career development.
- Organizational development.
- Team building.
- Management and leadership training.
- Problem solving.
- Education and curriculum development.
- Diversity and multi-cultural training.

FIGURE 7.7 ■ Application of Myers-Briggs Type Indicator.

state, check it out. It should be considered a warning sign if an applicant claims a degree from a state or county where he never lived.

Federal officials recommend that agencies always check academic credentials, even when the school is well known. College degrees should be verified by official transcripts and phone interviews of college advisors or personal references in an academic setting. Because employers are less likely to check with schools for verification or academic transcripts, some applicants may falsify information about their academic background. To help an EMS agency verify academic credentials, the FTC suggests:

- Contact the school. Most college registrars will confirm dates of attendance and graduation, as well as degrees awarded.
- Research the school on the Internet. Colleges and universities accredited by legitimate agencies generally have a rigorous review of the quality of their educational programs. If a school has been accredited by a nationally recognized accrediting agency, it is probably legitimate.

MEDICAL EXAMS

The medical examination should consist of a physical exam, including a hands-on physical by a physician, with evaluation of respiratory, cardiac, ENT, neuro, musculoskeletal, genitourinary tract, and lymphatic systems. It should include critical assessment for chemical agents, physical agents, and carcinogenic exposures, with all vital signs such as height, weight, color vision, and blood pressure. There also should be an in-depth patient history (review past occupational exposures, personal/family health, and vaccinations), an extensive blood-work review for exposures, spirometry/pulmonary function test, electrocardiogram, chest X-ray, and annual TB test. If available, offer the hepatitis B vaccine series.

DRUG TESTING

Occupations involved with public safety have special status when it comes to the legalities of testing employees for drugs and alcohol. Drug testing for public-safety personnel may be performed in the following circumstances: pre-employment, post-accident, return to duty, when there is reasonable suspicion, and randomly. If your EMS agency requires a commercial

driver's license for your equipment, drug testing is required. The common drugs tested for are alcohol, methamphetamine, marijuana, cocaine, PCP, and opiates.

For pre-employment drug testing, the courts have ruled that an applicant is free to choose to seek a job that requires a drug test and therefore drug testing is not a violation of his privacy rights. The pre-employment standard also influences return-to-duty testing. If an employee is returning to duty after a drug counseling or **Employee Assistance Program (EAP)** intervention initiated by management, the employee does not have a right to privacy in the case of drug testing. The courts also have ruled that an employee is not entitled to privacy rights with regard to drug testing if a previous offense was identified and documented.

The U.S. Supreme Court, based on the Fourth Amendment of the U.S. Constitution, has made several rulings supporting drug testing of public-safety workers when there is reasonable suspicion. For example, in National Treasury Union *v.* Von Raab, the Supreme Court ruled that a government employee handling narcotics could be compelled to submit to random drug testing. This would allow for testing of paramedics who have the responsibility for drugs, such as Valium or morphine, which are regulated under the Narcotics Act.

In another Supreme Court case, Chandler *v.* Miller, the court required that government drug testing must employ a special-need test. The key test of any legal case that revolves around drug testing of public-safety workers requires a special classification or a showing that employees are in "safety-sensitive" employment. Paramedics and EMTs are assumed to be safety-sensitive employees, because they protect the public and are involved in dangerous and skilled activities. It is important to realize that those employees not involved in safety-sensitive positions cannot be randomly tested unless it has been agreed upon in a labor contract.

The National Labor Relations Board also says that if an employee is in a safety-sensitive position, he may be subject to random drug testing. Often drug-testing rules fall under a strong management-rights clause and become part of the work rules. Drug and alcohol policies should be part of every **collective bargaining** agreement and should be negotiated in a collaborative work environment. Drug-testing policies that are not well designed can cause poor morale, grievances, and expensive litigation.

In the future more cases will be argued. Appeals-court rulings on the East Coast vary greatly from rulings on the West Coast of the United States. It is important to employ a written policy and review this with the labor unit and legal counsel.

■ EMPLOYEE SATISFACTION AND RETENTION

It is expensive to recruit, train, and consistently employ an EMS provider. In many areas paramedics are at a premium and often in short supply. It is important to retain those employees once they are inside the organization. Employee turnover and satisfaction is a key performance indicator in an EMS organization. Retention is best accomplished by continued participation of employees in the operation of the EMS service. Some common strategies to increase employee satisfaction are listed in Figure 7.8.

Length-of-service award programs have been a long-standing way of rewarding employees and an effective strategy for retention. Five-year increments are used as a common denominator for service awards. Cumulative totals hitting specific benchmarks should be rewarded with time off, money, or other recognition awards. Longevity pay is a monetary incentive that is commonly tied to length of service. In a state supreme court case, Murray *v.* Rhode Island Department of Elementary and Secondary Education et al., the court ruled that longevity pay is a distinct "additive" to an employee salary, rather than a component thereof. Most collective-bargaining agreements or legislation includes longevity pay as a negotiable item; however, this does not mandate its inclusion into a salary structure. Longevity pay is a common tool for retaining employees.

Incentive plans vary and are often creative according to what the community is capable of offering. Some agencies provide incentive pay to employees based on transports, completion of prehospital-care reports, or total runs completed. Incentive plans are often written into union contracts. An educational incentive plan is another way of rewarding and retaining employees: offering financial incentives for associate, bachelor, and master degrees has been supported more than longevity-pay plans by human resources.

Job enrichment, a planned program for enhancing job characteristics, is another way of retaining employees, as well as increasing the effectiveness of an organization. It adds skills from the job classification above the employee's current one and leads to more training and autonomy. Job enrichment involves redesigning jobs to increase the level of responsibility, authority, control, and challenge. When it is successfully implemented, all core job dimensions are increased. Job enrichment has the potential of increasing the motivation of EMS workers.

Job enlargement is often confused with job enrichment but differs in that job enlargement involves adding more tasks at the employee's current level of operation. Job enlargement often requires some sort of monetary reward and may have financial concerns defined in a union contract. Job enlargement involves increasing the number of tasks a worker performs but keeping all the tasks at the same level of difficulty and responsibility. As with job rotation, this increases skill variety and to some degree increases the likelihood that an individual's core competencies will be included in required tasks. However, since responsibility and challenge are not increased, the motivational potential is limited.

■ LEGAL ISSUES IN HUMAN RESOURCES

DISCRIMINATION

Discrimination of various kinds continues to generate litigation. For an EMS manager, it is important to understand the federal and state laws that impact employment practices. The Title VII Civil Rights Act of 1964 prohibits discrimination against a person seeking employment based on the applicant's race, color, religion, sex, or national origin. In total the Civil Rights Act of 1964 prohibits discrimination in hiring, promotion, salaries, benefits, training, treatment of pregnancy, and other conditions of employment on the basis of race, color, religion, national origin, or sex. These protections are offered regardless of the citizenship status of the applicant or employee. Today most employment discrimination charges are filed under Title VII.

- ◆ Length-of-service award programs.
- ◆ Incentive plans.
- ◆ Recognition programs.
- ◆ Education and training opportunities.
- ◆ Job advancement and enrichment.

FIGURE 7.8 ■ Employee Retention Strategies.

The Civil Rights Act of 1991 (CRA 91) reversed several previous Supreme Court decisions and also raised the ceiling for damages for discrimination claims. The legislation countered a series of Supreme Court rulings, five of them from 1989, that had been considered adverse to bias victims. In addition, the act provided for awards of punitive damages in cases of intentional discrimination, set up a Glass Ceiling Commission to recommend ways to remove barriers to women and minorities seeking advancement, and banned the adjustment of test scores by race, a practice known as "race norming."

A new standard for disparate impact was reversed in Wards Cove Packing *v.* Antonio, a 1989 case that made it more difficult for protected groups to show disparate impact. In effect it reinstated disparate-impact criteria from Griggs *v.* Duke Power Co., the famous 1971 ruling that said hiring practices must be related to the job in question and consistent with business necessity. CRA 91 also makes clear that intentional discrimination occurs when race, color, sex, religion, or national origin is a motivating factor in an employment decision, even if there are other, nondiscriminatory reasons for the decision.

In addition to reversing these Supreme Court rulings, the new law in the Civil Rights Act of 1991 for the first time puts a limit on the monetary damages for victims of intentional discrimination under Title VII, the Americans with Disabilities Act (ADA), and the Rehabilitation Act.

The ADA prohibits discrimination of job seekers who have a physical or mental disability and is for those with a record of disability. In recent years employment actions have occurred that forced a municipal ambulance to reinstate a paramedic who had a hearing loss, and in a Michigan case an ADA application required a city to hire a person blind in one eye. Unlike civil rights cases, the ADA does apply to volunteers: in a 1994 court case, a fire-based EMS agency was forced to reinstate a volunteer firefighter after he was discharged upon disclosing he was HIV positive. A case in Pennsylvania resulted in a fire department being fined $10,000 under ADA guidelines after refusing to take care of an AIDS patient.

Back injuries have typically been the exception to EMS-related ADA complaints. Several court cases have been lost by back-injured workers who sued under ADA. Most physicians who operate in or review back-injury cases will not put their professional liability on the line, and constraints placed on workers, such

as not to lift over 100 pounds, have helped in court when the employer has rejected an applicant for field EMS positions.

The Pregnancy Discrimination Act is an amendment to Title VII that prohibits sex discrimination on the basis of pregnancy, childbirth, or related medical condition.

The Uniform Services Employment and Reemployment Rights Act (USERRA) prohibits the discrimination by employers against past or present job applicants who may be called to or volunteer for military service. The USERRA protects the right of an employee to be reemployed in the civilian job he left. Conditions for reemployment include the person being called to service gives advance notice to the employer, he must have had five years or fewer within the uniformed services, and the employee must return to work in a timely manner after the conclusion of service. If the employee is discharged with a less-than-honorable discharge, the benefits of USERRA can be denied. The employee must be restored to a job and the benefits he would have attained in the time he was absent due to military service. The employer cannot deny the employee initial employment, reemployment, retention in employment, promotion, or any benefit of employment. This includes health insurance, and employers are obligated to continue the employee's and his dependents' health-care coverage for up to 24 months. Once the employee returns, the employer must reinstate the health benefits exclusive of pre-existing conditions, except for service-connected illness or injuries. The U.S. Department of Veterans Affairs handles such complaints.

The Age Discrimination Act of 1967 protects those over the age of 40. Often agencies see older EMS employees as high risk for back injuries and more expensive to insure. It is important that EMS managers hold employees to job standards, as well as treat the more experienced workers with dignity.

The Immigration Reform and Control Act of 1986 protects against discrimination based on citizenship. This does not cover those who are in the country illegally, but it does cover the person who has a valid green card and is legally registered to work in the United States. With a shortage of health-care professionals in the United States, allied-health professionals have been coming from other countries. EMS professionals with valid work permits and green cards have protection under the Immigration and Reform Control Act.

The Family Medical Leave Act (FMLA) entitles a person to up to 12 weeks of unpaid leave for the care of a new child or family member in a 12-month period. This coverage includes both adopted and biological children and any immediate family member (spouse, child, or parent). FMLA states that the employee may not lose rank or pay level while on leave.

Recently, as a result of Hurricane Katrina, Congress passed the Volunteer Firefighter and EMS Personnel Job Protection Act (H.R. 3949). This act prevents an employer from terminating, demoting, or discriminating against an employee absent from a job for up to 14 days a year while serving as a volunteer emergency responder.

FAIR LABOR STANDARDS

In 1936 Congress enacted the **Fair Labor Standards Act (FLSA).** FLSA controls the working hours and payment for comp time or overtime of nonexempt employees. The FLSA established a minimum wage and created a 40-hour workweek standard, as well as record-keeping requirements. Several changes to the FLSA legislation have occurred since its inception. A change in 1963 mandated equal pay for equal work, and in 1966 the FLSA was amended to cover quasi-public employees in hospitals, schools, nursing homes, and transit firms.

In 1985 there was a congressional amendment that allowed local government the option of using compensatory time or additional time off, instead of cash payment for overtime. However, the FLSA allows for a maximum accrual of 480 hours. Compensatory time must be agreed upon between the agency and the labor unit, and an employee must be able to use any compensatory time awarded to him.

In 1986 a Department of Labor regulation established rules and policies for local government. This included compensatory time, rules for public-safety employees, recordkeeping, dual employment, and volunteers.

There are employees that are exempt from FLSA requirements. Any employee that holds an executive, administrative, or technical position can be exempt from regulations under the FLSA. An executive position as defined under the FLSA is one in which the person's primary duties consist of management of the organization, directing the work of two or more, and having the authority to hire and fire employees. An administrative position is for those employees who perform work directly related to management policies or general operations. An administrative position performs general supervision and does not devote more than 20% of its time to other duties. A professional-position exemption may be assigned under the FLSA for employees whose work requires knowledge of an advanced type in a field of science in which the work is intellectual and creative in nature. Professional exemptions are also applied to teaching positions.

Public-safety workers are regulated under section 7K of 29 CFR Section 553.201. A public-safety worker should have a regulated work period. A work period cannot be less than seven days or more than 28 days. A work period is not the same as a pay period, and the FLSA also defines a tour of duty as the period of time an employee is considered to be on duty for determining compensable hours. Tours of duty include regular shifts, special assignments, and any activities beyond scheduled time to complete assignments, yet shift swaps are exempted.

The FLSA stipulates that EMS employers keep payroll records, certificates, agreements, collective-bargaining agreements, and employee contracts for three years. Compensable hours include all hours fewer than 24 and any hours over 24; sleep and mealtime may be excluded if expressed in an agreement between the agency and the labor unit or employee. The exemption under the 7K rule depends on the 80/20 rule, which says that no more than 20% of an employee's work otherwise covered under a 7K exemption may be spent on nonexempt work. In 1992 a trend began in which single-tiered EMS agencies or civilian employees working for fire-service EMS divisions were exempt from a 7K exemption because more than 20% of their work was not related to fire activities.

Case law was established in West *v.* Anne Arundel County, in which a group of firefighters brought suit against the county for compensation for EMS duties. The Anne Arundel County fire services operated as a dual–role, cross-trained fire department providing EMTs on engines and paramedics on EMS transport units. A $2.9 million verdict went in favor of the firefighters and paramedics. A judge based this decision on function and the percentage of call volume that was EMS. This decision was immediately appealed to the U.S. Supreme Court in 1998, which refused to hear the case. A second case, Glynn County, Georgia, *v.* Falken et al., was ruled in favor of the employees when Glynn County failed to sustain an argument that nonemergency activity or time spent waiting for an EMS call should not be counted as regular job duties. Clearly these rulings would substantially increase the cost of delivering EMS and potentially force EMS into the private sector. The Bureau of Labor Statistics and the Department of Labor clarified the definition of a firefighter and extended the 7K exemption to ambulance and rescue activity.

Several legal cases in the 1990s led to HR 3985, which was sponsored by Maryland's representative Robert Leroy Ehrlich, Jr. HR 3985 incorporated the Department of Labor ruling and the occupational classification changes into federal law, modifying FLSA. Volunteering at an organization is also covered under the FLSA standard; an employee may not engage in volunteer work for the same employer if the volunteer work is substantially related to or incidental to his primary work. Cases have gone in favor of the applicant when employers have failed to hire a person because he has a history of exercising his rights under the FLSA law.

■ COLLECTIVE BARGAINING

In most fire-based EMS agencies and in some third-service and private ambulance providers, managers and leaders must contend with legally mandated procedures in a labor-versus-management environment. The union or labor association is a bridge between management and human resources. The International City/County Management Association (ICMA), International Association of Firefighters (IAFF), and International Association of Fire Chiefs (IAFC or ICHIEFS) have advocated and promoted a collaborative environment in which labor-management initiatives are opportunities for partnerships and cooperation. Collective bargaining is a process often mandated that outlines what can or will be negotiated between an employer and a labor group. Collective bargaining is governed by federal and state statutory law, administrative agency regulations, and judicial decisions. It is an accepted practice to meet and confer about the requirements of the bargaining process. The Bureau of Labor Statistics is authorized by the Labor-Management Relations Act of 1947 (Taft-Hartley Act) to mandate collective bargaining in labor agreements.

MANDATORY SUBJECTS OF BARGAINING

Both the employer and the employee representative are required to bargain over wages, hours, and other terms and conditions of employment. This has been defined over the years to include wages and fringe benefits, grievance procedures, arbitration, health and safety, nondiscrimination clauses, no-strike clauses, length of contract, management rights, discipline, seniority, and union security. Because contracts cannot address every problem that may arise at work, most collective-bargaining agreements include the establishment of a mutually agreeable procedure to settle differences in contract interpretation.

Furthermore, the grievance procedure is usually the means a worker has of enforcing the contract. This means that mediation and arbitration must be detailed in any contract negotiation. If the parties cannot agree on contract interpretation or proper enforcement, they may wish to call in an impartial outsider, or "arbitrator," to settle the question. This is only possible when an "arbitration clause" is negotiated into the contract.

An arbitration clause provides an alternative to time-consuming lawsuits. This category of contract language will often define how the arbitrator will be selected, who will pay what share, and what the arbitrator's scope of authority will be. Often both sides will agree that the arbitrator's decision is final or binding. This means the courts cannot review the arbitrator's decision unless it is clearly contrary to law.

GRIEVANCE PROCEDURES AND DISCIPLINE

Progressive discipline is a process for dealing with job-related behavior that does not meet expected and communicated performance standards. The primary purpose for progressive discipline is to assist the employee in understanding that a performance problem or opportunity for improvement exists. The process features increasingly formal efforts to provide feedback to the employee so he can correct the problem.

The goal of progressive discipline is to improve employee performance. The process of progressive discipline is not intended as a punishment for an employee but to assist the employee in overcoming performance problems and satisfying job expectations.

Progressive discipline is most successful when it assists an individual in becoming an effectively performing member of the organization. Failing that, progressive discipline enables the organization to fairly, and with substantial documentation, terminate employees who are ineffective and unwilling to improve.

Typical steps in a progressive-discipline system are outlined in Table 7.2. It should be noted that quality-improvement activities should become part of progressive discipline only after education and coaching have not been effective. QI must remain separate from discipline or there will not be buy-in from the field.

It is important that EMS managers understand the concept of due process in discipline. In all disciplinary proceedings, employees have the right to be informed of the accusation, to receive promptly a copy of the complaint, and to have access to relevant material to be introduced in order to guarantee the ability to prepare a defense. Employees have the right to be assisted without prejudice by an advisor who may be an attorney or union official. These are known as **Weingarten Rights**, which guarantee an employee's rights to assistance from union representatives during an investigatory interview and to have access to procedures for securing the appearance of reluctant as well as friendly witnesses, to examine all witnesses in disciplinary hearings, and to receive a timely and impartial decision.

No employee may be compelled to testify against himself, which is known as his **Garrity Rights** and is a

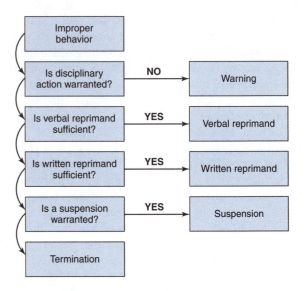

FIGURE 7.9 ■ Example of a Progressive Discipline System.

frequent source of litigation. If the employee chooses to testify, a negative inference may be drawn from any person's failure to respond to relevant questions in a judicial proceeding. A violation of these rights can result in litigation, constitute a gross procedural error in many union contracts, and establish grounds for appeal. Often in arbitration or district court, employees are reinstated for legitimate infractions due to a manager's not following the proper due-process procedures. It is helpful for an organization to develop a form to guide managers when approaching disciplinary issues (Figure 7.9).

CONFLICT RESOLUTION

EMS managers in their day-to-day operations will occasionally generate conflict. Conflict is a natural part of any workplace, and given the level of emotion in the EMS profession, a lot of conflict is generated and will need to be managed just like any other work process. Conflict is best managed in municipal EMS organizations by using a collaborative approach. When conflict is being addressed, an EMS manager should first start by asserting common interests and concerns. This sets the tone of further actions, develops a cooperative spirit, and promotes a positive attitude.

A collaborative approach involves listening and thinking carefully about how the manager can meet personnel needs. Develop criteria for what is a good solution by defining ideal solutions that consider the interests of both parties, with a focus on interests and

TABLE 7.2 ■ Progressive Discipline System	
Step 1	Counsel the employee about performance and ascertain his understanding of requirements. Ascertain whether or not there are any issues contributing to poor performance that are not immediately obvious to the supervisor. Solve these issues, if possible.
Step 2	Verbally reprimand the employee for poor performance.
Step 3	Provide a written warning in the employee's file in an effort to improve employee performance.
Step 4	Provide an escalating number of days in which the employee is suspended from work. Start with one day and escalate to five.
Step 5	End the employment of an individual who refuses to improve. In some programs, one additional paid day away from work before termination is considered a "decision-making leave," a day during which the employee is to consider the final consequences of his actions before he is separated from the organization.

not positions. Generate clear actions that define the problem, including symptoms, causes, and barriers. Generate broad solutions to the problem, and suggest specific actions to address problems. Develop ideas that are detailed, and make sure the other side agrees with the concepts. Implement the best suggestions, and communicate the realization that modifications will take place. Then as a last step evaluate results by checking to see if the results agree with the solution and verifying that both sides agree with the progress and the end results. A specific time frame for reevaluation should be set by management and labor to ensure progress is being made.

FACT FINDING AND NONBINDING ARBITRATION

Fact finding is typically a step taken to avoid a labor impasse that occurs before mediation and arbitration. In fact finding, a neutral third party is hired to review the relevant labor dispute and to render a finding of fact. The finding of fact is the independent arbiter's view of what has occurred in the labor dispute and what decision he would have made had he been empowered to make it. The difference between fact finders and nonbinding arbitration is that fact finders are usually those who are hired independently by either party.

Nonbinding arbitration occurs when both parties agree to present their dispute together. The purpose of fact finding and nonbinding arbitration is to push along negotiations by having an unbiased third-party opinion. Typically one side then uses the findings as a crutch in negotiations.

MEDIATION

Mediation is a process for resolving disputes with the aid of a neutral third party. The mediator's role involves assisting parties, privately and collectively, to identify the issues in dispute and to develop proposals to resolve the disputes. The mediator is not empowered to decide any disputes. Accordingly, the mediator may meet privately and hold confidential and separate discussions with both parties.

Mediation may be mandatory under the terms of certain laws or court rules, or it may be voluntary by agreement of EMS agencies and their labor units. Some laws and court rulings have rules requiring mediation of disputes at some point in the litigation process. Voluntary arbitration may be undertaken

when two parties agree in advance to submit any disputes to mediation. Such mediation clauses are common in agreements in which the parties seek to resolve their disputes in a manner that avoids hostility and preserves an ongoing relationship. Mediation agreements also may be made at the time a dispute arises.

A typical mediation might involve allowing each party to submit premediation briefs that succinctly set forth the essence of the dispute and each party's position. At mediation the mediator will typically conduct introductions, explain the mediation process, provide assurances of confidentiality, and give each party an opportunity to explain the dispute and the reasons behind each position. Many mediators will then meet privately with each party and provide an evaluation of the dispute, pointing out the strengths and weaknesses of each party's position. The mediator may then, again in private, assist each party to determine both parties' genuine interests and encourage each party to identify settlement proposals intended to address those interests. Typically, the mediator communicates settlement proposals to each party and helps each party determine how best to respond to a settlement proposal. In most cases records or proceedings from mediation cannot be used in judicial or administrative processes.

Arbitration may be voluntary. Voluntary arbitration refers to the arbitration of a dispute submitted to an arbitrator by agreement of the parties. Typically, parties to a dispute submit their dispute to arbitration to minimize the expense, delay, or publicity that they perceive will accompany litigation. Voluntary arbitration is consensual. Parties enter into an agreement to arbitrate or a submission agreement. The agreement to arbitrate may be entered into in advance of any dispute and may, for example, be included in a dispute resolution clause of a contract. Parties may agree to arbitrate a dispute at the time a dispute arises or at any time before a final judgment is entered in a court proceeding.

■ PERFORMANCE APPRAISAL AND INCENTIVES

Performance evaluations make very specific contributions to the organization. Performance refers to an employee's accomplishments of assigned task. In human resources the annual or periodic assessment of an employee's or work group's job-related strengths and weaknesses is known as a performance appraisal.

A performance appraisal often lays the groundwork for a promotion or salary increase for an employee. It should attempt to motivate a less-than-optimally performing employee who has the potential to achieve greater things in the organization. It also can lay the groundwork for progressive discipline or dismissal of an otherwise unacceptable worker. But the most important function of a performance appraisal is to take an inventory of a person's skills and abilities that can be used in new assignments for the organization.

A performance appraisal can have significant impacts on the organization's training and employee development. This makes the delivery of a performance appraisal the EMS manager's most important tool to help employees develop and maximize their potential. The key to a good employee evaluation is to evaluate the performance and not the person. The goal of a performance appraisal is to improve the quality of the work and the individual employees involved with that work.

BUILDING A PERFORMANCE APPRAISAL

When building a performance appraisal, an EMS organization needs to start with the job-task analysis. Performance appraisals are built on well-crafted job descriptions or a function assessment that describes work and the personnel requirements of a particular job. This includes assessing the knowledge, skills, and abilities (KSA) that an employee needs to be successful in the job. There are basic steps to an effective performance-appraisal process. The first is to review the legal requirements. Qualitative and quantitative attributes should be used to evaluate performance, but unless an attribute affects performance on the job, it is not appropriate or legal to pass your judgment on personal characteristics.

It is helpful to seek input from the various groups who currently use, or could benefit by using, performance-appraisal data. Some of these groups would never see individual member results but could make use of compiled data. They include EMS leadership, training officers, human-resources professionals, recruiters, and union or labor representatives.

Performance appraisals are based on a thorough job analysis. Identify performance-based, job-relevant EMS criteria jointly with medical direction, labor, and management. Criteria are, simply, those aspects of member performance that are measured during the process. They are the performance standards used to

evaluate member knowledge, skills, abilities, motivations, and behaviors. Only those aspects critical for effective performance should be identified and subsequently measured, evaluated, and developed. The list of criteria must be a thorough list of aspects of the job that are important for success, and thus the criteria also become goals toward which members should strive.

Continually review performance standards and revise as necessary to improve the quality and validity of the measures. Remember, this is an ongoing process and should be included in the strategic-planning cycles. No matter what your position, you should always provide input regarding the accuracy and effectiveness of the standards you use.

A labor-management partnership should develop the appraisal instrument. Both the EMS manager and EMS personnel must be clear on what constitutes average, below-average, and above-average performance. Review criteria must be clearly understood by members. If not, rewrite the criteria. Performance-review criteria must be clearly communicated to members before the beginning of the appraisal period. Members must be convinced that the measurement criteria are valid. If not, they will not respect the results and, most importantly, will not adopt the standards as personal goals.

EMS organizations need to decide how to select observers/evaluators for performance appraisals. The direct manager most often serves in this role. However, it is critical that the evaluator be perceived by members as qualified to review performance in order for the appraisal process to be successful. If not, the process is bound to fail. Just such a situation often arises in fire-based EMS systems that are dual-responsibility organizations. For instance, when a company officer has come up through the ranks as a firefighter with no advanced EMS/medical experience, that manager's credibility may be limited when assessing clinical skills. One solution might be to have the medical director or EMS manager with paramedic experience assist with, or assume responsibility for, the observation, evaluation, and even feedback and goal-setting discussions of those skills (Figure 7.10).

It is important that one EMS manager not be overloaded with personnel to evaluate. Courts have found that managers who evaluate performance must have had many opportunities to observe performance; their subjective evaluations must be only one performance measurement, and additional raters should be

1. MEDICAL KNOWLEDGE
    a. Adherence to protocols (pharmacology, procedure, administration). 1, 2, 3, 4
    b. Knowledge of pathophysiology (does assessment equal plan). 1, 2, 3, 4
    c. Demonstrates cardiology (correct ECG interpretation). 1, 2, 3, 4
    d. Knowledge of pharmacology (indications, administration, dose, and so on). 1, 2, 3, 4

2. SKILLS PERFORMANCE
    a. Performs intubation and airway control to 80% of standards on first attempt. 1, 2, 3, 4
    b. Completes vascular access on the patient at appropriate times. 1, 2, 3, 4
    c. Safe and appropriate drug administration (date check, appropriate route, and correct dose). 1, 2, 3, 4
    d. Provides c-collar, backboard, and KED when indicated (correct size, application, and sequence of application). 1, 2, 3, 4

3. CUSTOMER SATISFACTION
    a. Interface/communication with families. 1, 2, 3, 4
    b. Interface/communication with public. 1, 2, 3, 4
    c. Interface/communication with other agencies. 1, 2, 3, 4
    d. Accommodations/complaint submitted. 1, 2, 3, 4

4. PUBLIC EDUCATION
    a. Establishes and maintains a professional image in relationship with public. 1, 2, 3, 4
    b. Instructs CPR, injury prevention, and other classes to the public. 1, 2, 3, 4

5. CONTINUING EDUCATION
    a. Attends continuing education classes. 1, 2, 3, 4
    b. Submits or attends outside seminars and conferences. 1, 2, 3, 4

6. MAINTENANCE OF EQUIPMENT AND SUPPLIES
    a. Maintains medical inventory. 1, 2, 3, 4
    b. Keeps vehicle cleaned and disinfected. 1, 2, 3, 4
    c. Maintains vehicle preventive-maintenance schedule. 1, 2, 3, 4

1 = Unacceptable   *2 = Needs Improvement*
3 = Above Average   *4 = Outstanding*

Measurement Tools

◆ QA/QI process, ACLS and PALS test, database, infield observations by quarterly ride-along.
◆ Chart review, agency or regulatory-body QA audits, annual skills lab, preceptor.
◆ Letter of feedback, customer-service surveys, infield observations.
◆ Department records of payroll for classes taught.
◆ Department training records and computer printout.
◆ Department checkout sheets, battalion-chief or shift-commander inspection.

FIGURE 7.10 ■ Sample Paramedic Performance Appraisal.

used if additional, relevant information can be obtained. This has given rise to performance appraisals known as 360-degree evaluations, which use information from peers, subordinates, and supervisors.

It is important to train the observers in the organization. Managers must be sufficiently trained to measure job performance and conduct appraisals in an objective fashion. Training should be more than a class on "filling out the form." Observers and evaluators should receive sufficient training in developing a common frame of reference with other raters; knowing when and how to observe and document relevant behaviors; evaluating behaviors against the standards; avoiding rater errors; providing consistent, ongoing feedback; conducting the performance review and development-planning meeting; and monitoring and promoting member development. Raters should receive the opportunity to practice their skills and receive feedback until they are comfortable and their results are reliable. It is important to measure performance. Observations should be done using objective criteria and observed performance, documented, and the behavior evaluated against the performance criteria.

CONDUCTING APPRAISALS

A performance appraisal should be conducted annually or when an employee finishes a probationary period after a promotion or job change. The time period during which an employee's job performance is observed is called the *appraisal period*. When the observations have been collected, the results need to be presented to the member. Raters should be honest, open, objective, nonthreatening, and should focus on the specific and observable behavior. Raters should ask for input and feedback from the member being evaluated, should remain open to his ideas or suggestions, and should change the appraisal in response, if appropriate. Treat the session as a problem-solving session, and emphasize performance and expectations, not personality characteristics, which members are more likely to perceive as threatening and difficult to change. Avoid focusing only on negatives or positives.

EMS managers conducting performance appraisals should establish future performance goals. Develop specific behavioral performance goals and objectives jointly with members. Make certain the goals are within

1. Be prepared.
2. Arrange an uninterrupted meeting.
3. Identify performance gaps.
4. Find the cause of performance gaps.
5. Make a plan to improve performance.
6. Set performance goals.
7. Record it.
8. Do tracking and follow-up.

FIGURE 7.11 ■ Steps for Conducting a Performance Appraisal.

the member's ability and control and are realistic. Establish follow-up procedures for measuring progress toward goals. A step-by-step approach to performance appraisals is listed in Figure 7.11.

EMS managers and leaders should praise and reward performance on an ongoing basis following appraisal. Provide frequent, specific feedback on job performance. Coach and counsel as necessary to provide encouragement, support, assistance, and adjustments as the member works toward goals. Feedback is most effective when it is timely and noticeable. One of the basic rules from the U.S. Supreme Court is that there cannot be any conflict of interest, and appearance of conflict must be avoided as well. Evaluations should be applied consistently, and the same evaluation criteria should be applied to all. The evaluator has to be a manager or a supervisor with first-hand knowledge of the quality of work performed and of the personal behavior of the employees.

Maintain control by establishing ground rules. Stay focused on the topics at hand and keep the employee from going off in a different direction. Present improvements from a positive perspective as much as possible. If there is bad news, it should not be news to the employee. He should be involved in developing a plan for improvement, so that in every performance appraisal he has the opportunity to add his comments.

PROBLEMS AFFECTING THE PROCESS

While performance-appraisal processes differ, a number of common problems can diminish a system's effectiveness. Some of the most typical involve the way organizations apply performance appraisals, **rating errors** in the way raters conduct appraisals, and procedural errors.

Some EMS organizations misunderstand the purpose of the appraisal. The difficulties involved in accurately measuring EMS member performance often are not addressed by linking quality-improvement data and direct field observations with annual performance appraisals. That is, supervisors often do not go out on calls with members, and they should. An effective ride-along program at regular intervals is an important part of any employee-evaluation program. Otherwise, supervisors must rely solely on feedback from peers or patients, which can differ significantly from the member's perceptions of his performance. These are valuable sources of information if employed as a part of a 360-degree evaluation. If accurate, those differences could be a valuable learning experience for the member.

A lack of specific goals and expectations can create an ineffective performance appraisal. A good performance appraisal must be based on clear goals that are known and understood by both manager and member. But this can create differing perceptions of performance criteria. So both the EMS manager and the employee must be clear as to the duties and responsibilities of the job. Members should help the manager develop the performance criteria.

Rating Errors

EMS managers often make errors when conducting performance appraisals. These are errors in judgment that occur when one individual evaluates another. Because the rater usually is unaware that he is making these errors, it is important for the EMS manager to be aware of the common errors (Figure 7.12).

One common error, the contrast effect, is the tendency for a rater to evaluate a member relative to other members and not relative to the job requirements. This is a common occurrence when performance-appraisal systems have vague evaluation criteria. Ratings should reflect job-related criteria and not other employees.

A first-impression error is the inclination for a rater to make an initial favorable or unfavorable judgment about a member and then ignore or distort subsequent information to support the initial impression. An error of this type may come from a bias that the rater has for an employee. This compromises the integrity of the performance-appraisal system, rating someone on what is popular versus what is factual. A manager who has a personal bias should not conduct an employee appraisal without having concrete and objective criteria to evaluate. Not doing so can open managers up to litigation and discrimination charges.

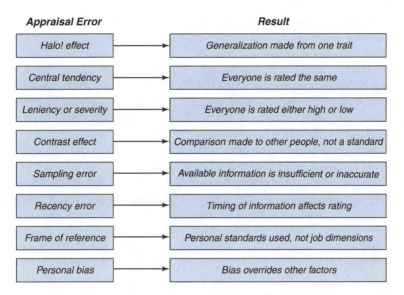

FIGURE 7.12 ■ Common Appraisal Errors.

The halo effect happens when an EMS manager makes inappropriate generalizations from one aspect of a member's performance to all aspects of job performance. The rater does not make any attempt to do an objective analysis of the specific job-related performance. For example, an EMS manager gives the employee an "excellent" rating on patient assessment and assumes that the employee is also excellent in ECG and pharmacology, when in fact the employee does not make that rating. This may provide a missed opportunity for education or career development.

The similar-to-me effect or frame-of-reference error is the propensity for a rater to judge more favorably those members most like him. Though it is a natural tendency to hire and favor people with similar values, beliefs, and actions, it can be prevented if the evaluation criteria are objective and reviewed by EMS leadership.

A central-tendency error is often called the "playing it safe" routine. It happens when a manager consistently rates a member on or close to the midpoint of the scale when the member's performance clearly warrants higher or lower ratings. This is often done by managers who do not believe in the performance-appraisal system. It is very common to see this error when an employee is at the top of the pay scale, and the EMS manager can avoid the extra work required by writing comments for any rating above or below the average.

The contrast-effect error occurs when a manager compares another employee to the person being evaluated and not to a specific job task or objective standard.

Positive and negative leniency errors occur when a manager is too hard or too easy in rating members. Leniency occurs with a weak EMS manager who is afraid to confront real problems with employee performance. Severity occurs when an EMS manager rates all employees below average due to unreal expectations or selective perception.

The recency error occurs when an EMS manager conducts a performance appraisal based on recent events and performance, instead of taking into consideration the performance of the entire period of evaluation. This can be avoided by keeping a tracking sheet or a file on the employee, which allows the EMS manager to recall events and performance over the entire time frame.

Finally, there are sampling errors, which occur when an EMS manager does not collect enough examples of the employee's work and does not observe or evaluate sufficiently. This is a common problem in EMS operations due to the number of EMS providers and lack of a manageable span of control of field supervisors.

Process Errors

Process errors occur when too much emphasis is placed on administrative aspects of the job—that is, rating on productivity (how many calls) versus the quality of the performance behaviors (clinical proficiency, decision making, and customer service). Often inadequate performance documentation produces a false picture of an employee's performance. The data used in making performance decisions must be specific and carefully documented (with dates, times, details, and so on). Documentation not only helps ensure the quality of the rating but is invaluable during performance discussions between manager and member and in justifying demotion or termination decisions at a later time, should that become necessary.

Ideally, the review should be an open, two-way discussion. The manager must give the member the opportunity to provide feedback. The member may need clarification, may object to some observations, and may even need to clear up misperceptions or misunderstandings. The manager should have the option to—and be willing to—change or adjust the ratings as a result of this discussion.

Failure to develop and implement an improvement plan makes a performance appraisal ineffective. Goals and specific plans for improvement and future development must be set by the end of the review. These goals and progress toward them should be documented (possibly on the appraisal form or instrument) and discussed during regular meetings between the manager and the member. They should also be discussed and considered when evaluating the member's progress during future reviews.

360-DEGREE EVALUATIONS

A current trend in EMS is to conduct a 360-degree performance appraisal, which is an appraisal that allows peers, subordinates, and superiors to rate the employee on performance. The Franklin Covey organization has a complete set of 360-degree evaluations that help an employee develop both professionally and personally. A very simple 360-degree evaluation is shown in Figure 7.13.

Rating	*Dimension*
	Oral Communication. Ability to accurately and clearly communicate information and directives, conditions, needs to groups or individuals with or without time for preparation (includes nonverbal gestures and use of aids where appropriate).
	Written Communication. Ability to communicate in writing, using proper grammar and syntax in an organized, accurate, and concise manner.
	Problem Analysis. Ability to identify problems, secure relevant information from both oral and written sources, identify possible causes of problems, and analyze and interpret data in complex situations involving conflicting demands, needs, or priorities.
	Judgment. Ability to evaluate courses of action, develop alternative courses of action, and reach logical decisions based on available information.
	Organization Sensitivity. Ability to perceive the impact of a decision on the rest of the organization, awareness of the impact of outside pressures on the organization, and awareness of changing societal conditions.
	Organization. Ability to efficiently establish an appropriate course of action for self and/or others to accomplish a specific goal and ability to make proper assignments of personnel and appropriate use of resources. This dimension also addresses time-management skills.
	Other Dimensions
	Initiative. Desire to actively influence events rather than passively accept them; to be self-starting, taking action beyond what is necessarily called for.
	Interpersonal Relations. Ability to perceive and react to the needs of others, paying attention to others' feelings and ideas, accepting what others have to say, and perceiving the impact of self on others.
	Independence. Ability to act based on one's own convictions rather than a desire to please others.
	Development of Subordinates. Ability to maximize human potential of subordinates through training and developmental activities.
	Persuasiveness. Ability to organize and present material in a convincing manner to gain agreement and acceptance.
	Listening. Ability to extract important information being presented in an oral communication and to convey an impression of interest in what others have to say.
	Decisiveness. Readiness to make decisions, render judgments, take action, and commit oneself to a course of action and to understand at what level a decision can best be made.

FIGURE 7.13 ■ Examples of 360-Degree Evaluations.

360-Degree Performance Feedback Form

What Are Dimensions?

Simply stated, a dimension is some characteristic—whether it be a knowledge, skill, ability, personal characteristic, or other hallmark of performance—that has been established through a job analysis as being critical to a successful execution of duties.

Elements of a Dimension

All dimensions have two essential elements. The first of these may be referred to as its title. Examples of dimension titles used in the assessment center include Oral Communication, Leadership, Written Communication, and so on. Titles are frequently used as a shorthand method of referring to a dimension, but that is only minimally important. Indeed, it is not necessary for a dimension to have a title, although for it not to have one would be highly unusual. Every dimension, however, must have a definition. The dimension's definition, the second essential element, specifically describes the important elements of that dimension. Many persons may look at a particular dimension, especially one like Leadership, and believe that they thoroughly understand the concept of leadership and can easily identify the characteristics of a good leader. However, what leadership means to one person—whether that is an assessor, a participant, a person performing a job, or someone supervising a person performing that job—may mean something significantly different to another. What one person sees as a good example of leadership, another may see as a bad example, and still another may not even see it as being related to leadership at all. For this reason, assessment center dimensions must have a clear definition, and all assessors must use that specific definition.

Please check one relationship box below:

☐ Boss
☐ Peer
☐ Subordinate

Ratings

The proper process will be to evaluate dimensions in an attempt to determine if I am a suitable candidate for promotion. You can help me by completing this survey. Below I have listed the dimensions, and I ask that you rate me based on your feelings about how competent I am in each dimension. Just put a numerical rating in the box to the right of the description of the dimension. The rating scale is as follows: 1, not competent; 2, somewhat competent; 3, competent; 4, more than competent; 5, highly competent.

360-Degree Performance Feedback Worksheet

(B1–B4 Bosses; P1–P9 Peers; S1–S7 Subordinates)

Raters → → → → → → Common Dimensions	B1	B2	B3	B4	TOT	P1	P2	P3	P4	P5	P6	P7	P8	P9	TOT	S1	S2	S3	S4	S5	S6	S7	TOT
Oral communication																							
Written commuciation																							
Problem analysis																							
Judgment																							
Organizational sensitivity																							
Organization																							
Other dimensions																							
Initiative																							
Interpersonal relations																							
Independence																							
Development of subordinates																							
Persuasiveness																							
Listening																							
Decisiveness																							

1. Dimension with lowest boss score: _____
2. Dimension with lowest peer score: _____
3. Dimension with lowest subordinate score: _____

(Continued)

Action Plan Development

I'm going to do the following about score 1:_____

_____

_____

I'm going to do the following about score 2:_____

_____

_____

I'm going to do the following about score 3:_____

_____

_____

FIGURE 7.13 ■ Continued

Often performance information can and should be gathered from a variety of sources. Research shows that appraisals completed by peers (other members of equal position) are the most reliable source of performance data for two reasons: Peers, who typically possess the technical know-how, have daily interaction with the member being evaluated and thus see how the member interacts with subordinates (if applicable), the supervisor, customers/patients, ED personnel, and so on. Further, their ratings tend to be fair and unbiased. Obtaining ratings from a number of peers allows the ratings to be averaged, which creates a more reliable result than a rating from a single evaluator.

Performance appraisal is a process that requires accurate measurement and evaluation of member behavior, with performance enhancement as the ultimate goal. The performance-review model should be a cyclical process that the EMS manager views as an ongoing cycle and a means to achieving goals—those of the organization, the manager, and the member.

■ SEXUAL HARASSMENT

Sexual harassment is defined as unwelcome sexual advances, requests for sexual favors, or other verbal or physical conduct of a sexual nature. A majority of sexual harassment complaints are brought by one coworker against another. To be considered unlawful sexual harassment, the harassing coworker's conduct must be unwelcome and interfere with the employee's work performance. Unlawful sexual harassment may include:

- Visual harassment, such as posters, magazines, calendars, and so on.

- Verbal harassment or abuse, which includes repeated requests for dates, lewd comments, sexually explicit jokes, whistling, and so on.
- Written harassment, such as love poems, letters, and graffiti.
- Offensive gestures.
- Subtle pressure for sexual activities.
- Unnecessary touching, patting, pinching, or kissing.
- Leering or ogling.
- Brushing up against another's body.
- Promise of promotions, favorable performance evaluations and grades, and so on in return for sexual favors.
- Demanding sexual favors accompanied by implied or overt threats to a person's job, promotion, performance evaluation, grade, and so on.
- Physical assault, rape.

All EMS and public-safety provider agencies are encouraged to develop a policy to address sexual harassment in their locations. These policies should include a notification to all employees and students that sexual harassment is a violation of law and is intolerable in either the educational or the employment setting. The policy statement should state that sexual harassment is considered a form of employee or student misconduct and that sanctions will be enforced against individuals engaging in sexual harassment and against supervisory, administrative, or managerial personnel who knowingly allow such behavior to continue. Policies should also include a procedure for the following:

- Making a complaint of sexual harassment.
- Identifying to whom complaints are to be made.
- Describing the form in which the complaint should be filed.

- Outlining the procedure the sponsor or employer will follow in investigating the complaint.
- Providing for a subsequent review to determine if sexual harassment has been effectively stopped.

In addition to filing a complaint, even in the case of agencies that do not have policies, individuals are also entitled to seek relief by filing a complaint with the:

- State Division of Human Rights.
- Federal Equal Employment Opportunity Commission.
- U.S. Labor Department's Office of Civil Rights.

Once developed, the policy should be widely distributed by providing a copy of it to all employees. It should be included in all new employee and student orientations and publicized within the workplace or educational setting. All employers developing policies should conduct appropriate training to instruct and sensitize all employees to the policy.

ETHICS AND CONDUCT

Public-safety leaders have been faced with more and more incidents of unethical behavior. In 2005 a series of scandals was reported in the Sacramento, California, press regarding the behavior of several fire companies in and out of the station. Stories of embezzlement and financial impropriety appear more frequently in the trade journals. EMS managers and leaders must start to incorporate the use of ethical systems to protect the organization and enhance patient care.

The word *moral* refers to the ability to tell right from wrong. In contrast, the word *ethics*, derived from the Greek word *ethos,* refers to customary or conventional rules of society, religion, and organizations. Those rules help us define behavior as right, good, and proper. Both morals and ethics are taught and developed from birth through adulthood. One model for ethical development is CORE (Figure 7.14).

Personal ethics is the ability to determine a correct train of thought, behavior, or attitude. A person's ethics are based on values, and it is important as an EMS manager or leader to inventory those values.

```
C—Character
O—Obligation
R—Results
E—Equity
```

FIGURE 7.14 ■ CORE Model.

Employees will observe a manager's or leader's behavior and make choices about their actions based on the manager's ethical perspectives. A strong ethical leader or manager develops a competitive edge, credibility, and efficiency due to trust.

Personal ethics are being replaced by the EMS organization's pressure to comply with the organization's culture, clouding the decision-making process that leads to responsible ethical behavior. Rarely in continuing education or initial EMS is training in ethics covered or explored.

Ethical considerations can be divided into three areas—personal, organization, and community. Ethical decisions are most often made from one of five different perspectives:

- *Utilitarian.* This approach makes an ethical decision based on what is most efficient.
- *Rights.* This approach seeks to maximize the employee's rights and sees a decision's purpose as a benefit to the employee.
- *Fairness or justice.* This approach seeks to make a decision that distributes benefits equally, based on the law.
- *Common good.* This approach seeks to do the most for the most people, and in making this decision an individual may receive a less-than-favorable personal outcome.
- *Virtue.* This approach to the decision process is inspired by supreme respect for the individual and a moral message.

The Williams Institute for Ethics and Management is a non-profit organization that specializes in research, education, and leadership development. It has developed several survey instruments that assess the ethical perspectives people use to make decisions. As an EMS manager, you can assess the ethical perspectives you use to make decisions by taking an ethics-awareness inventory developed by the Williams Institute.

When an EMS manager or leader is faced with an ethical dilemma, several key questions may be asked to guide the manager's decision-making process. It is important first to ask if the decision is legal and if the actions of the manager will influence or encourage others to break the law. Second, ask if the decision is within the parameters of department or organizational policy. Third, ask if the decision is fair to all parties involved. A final test is to ask how you would respond if the consequences of the decision were broadcast on local or national news and how your family or loved ones would respond to the information.

■ EMPLOYEE SAFETY

WORKPLACE VIOLENCE

Workplace violence is becoming more and more prevalent in EMS and public safety. In the last decade scene violence has escalated, injuring or killing EMS workers. On April 15, 2000, a Memphis shooting killed four, including an off-duty firefighter, a sheriff's deputy, and two firefighters. An engine company had responded to a call at a fire-department employee's home, and when the first fire engine arrived, the off-duty employee opened fire on members from his own department. Another example: On April 3, 2004, at Metropolitan Ambulance Service Trust in Kansas City, a paramedic and an EMT were killed in their quarters around 12:30 A.M. When the crew did not respond to an emergency call, attempts were made to contact the crew by pager, radio, and station phone. A call was placed to the local fire department to ask if they could check on the crew. When the fire department arrived, they found the crew dead from gunshot wounds.

Employers have a legal and ethical obligation to promote a work environment free from threats and violence. The economic loss to the EMS agency as a result of violence is in the form of lost work time, damaged employee morale and productivity, increased worker-compensation payments, medical expenses, and possible lawsuits and liability costs.

Defining workplace violence has generated extensive discussion. Some would include in the definition any language or actions that make one person uncomfortable in the workplace. Others would include threats and harassment. All would include any bodily injury inflicted by one person on another. Workplace violence is any violent act, including physical assaults and threats of assault, directed toward persons at work or on duty.

Social-science research has focused primarily on physical injuries. These studies specify that an average of 20 workers are murdered and 18,000 are assaulted each week while at work or on duty. A study conducted by the Loma Linda EMS faculty reported that violence associated with patient care is the primary source of nonfatal injury in EMS organizations today. The study reported that 61% of EMS workers were assaulted and 25% reported an injury as a result of an assault on duty.

Federal law puts differing pressures on employers and others concerned with preventing or mitigating workplace violence. On one hand EMS agencies are under a variety of legal obligations to safeguard employee well-being and security. Occupational safety laws impose a general requirement to maintain a safe workplace, which embraces safety from violence. For example, the General Duty Clause of the Occupational Safety and Health Act requires employers to have a workplace that is free from recognized hazards.

Most state and federal worker-compensation laws make employers responsible for job-related injuries. Civil rights laws require employers to protect employees against various forms of harassment, including threats or violence. In addition, employers may face civil liability after a workplace-violence incident. If there was negligence in hiring or retaining a dangerous person, for example, or a failure to provide proper supervision, training, or physical-safety measures, it can result in litigation and a monetary loss to the organization.

Workplace violence also is an increasing consideration for EMS leadership (Figure 7.15). EMS employers have an important role in violence prevention. Specific steps can be taken to reduce violence in the workplace and are included in Figure 7.16. An employer can be held liable for an employee who commits violence that results in another employee's injury.

Employees that commit workplace violence can be identified. Research shows that 85% of workplace-violence perpetrators exhibit clear warning signs before acting on their threats. EMS managers and

TYPE I:	The agent has no legitimate relationship to the workplace and usually enters the workplace to commit a robbery or other criminal act.
TYPE II:	The agent is either the recipient or the object of a service provided by the affected workplace or the victim, such as a current or former client, patient, customer, passenger, criminal suspect, or prisoner.
TYPE III:	The agent has an employment-related involvement with the workplace. Usually this involves an assault by a current or former employee, supervisor, or manager; a current or former spouse or lover; a relative or friend; or some other person who has a dispute involving an employee of the workplace. Violence can be committed in the workplace by someone who doesn't work there but has a personal relationship with an employee—the agent may be an abusive spouse or a domestic partner.

FIGURE 7.15 ■ OSHA identifies types of workplace violence.

- Adopt a workplace-violence policy and prevention program, and communicate the policy and program to employees.
- Provide regular training in preventive measures for all new and current employees, supervisors, and managers.
- Support—do not punish—victims of workplace or domestic violence.
- Adopt and practice fair and consistent disciplinary procedures.
- Foster a climate of trust and respect among workers and between employees and management.
- When necessary, seek advice and assistance from outside resources, including threat-assessment psychologists, psychiatrists, and other professionals; social-service agencies; and law enforcement.

FIGURE 7.16 ■ Management Activities to Reduce Workplace Violence.

leaders can identify a profile of a potentially violent person. Typically, people who commit workplace violence have a previous history of violence toward the vulnerable (such as women, children, animals). They are often loners, are withdrawn, and view change with fear. Often they have emotional problems, a substance-abuse history, depression, and low self-esteem. When their employment history is examined, managers may find that they have career frustration or anger over seniority issues in the work environment. EMS managers and frontline supervisors should look for outward warning signs (Figure 7.17).

Often there is a triggering event for workplace violence, such as being fired, laid off, or passed over for promotion. Disciplinary action, a poor performance review, or criticism from their supervisor or coworkers

- Violent and threatening behavior, hostility, approval of the use of violence.
- "Strange" behavior, such as becoming reclusive, having a deteriorating appearance and hygiene, and other erratic behavior.
- Emotional problems, such as drug or alcohol abuse, being under unusual stress, depression, inappropriate emotional displays.
- Performance problems, including problems with attendance or tardiness.
- Interpersonal problems, such as numerous conflicts, hypersensitivity, resentment.
- Employee appearing to be "at the end of his rope," such as exhibiting indicators of impending suicide or having an unspecified plan to "solve all problems."

FIGURE 7.17 ■ Behavior Warning Signs of Potentially Violent Employees.

can also be a triggering event. Bank or court action, such as a foreclosure, restraining order, or custody hearing can require immediate intervention. Often violence is the result of a failed or spurned romance or a personal crisis, such as a divorce or a death in the family.

EMS managers and leaders should create a workplace-violence program, with components that include:

- A statement of the employer's no threats-or-violence policy and complementary policies, such as those regulating harassment and drug and alcohol use.
- A physical-security survey and assessment of the premises.
- Procedures for addressing threats and threatening behavior.
- Designating and training an incident-response team.
- Access to outside resources, such as threat-assessment professionals.
- Training of different management and employee groups.
- Crisis-response measures.
- Consistent enforcement of behavioral standards, including effective disciplinary procedures.

PREVENTIVE PRE-EMPLOYMENT PRACTICES

Preventive measures can include pre-employment screening, identifying problem situations and risk factors, and security preparation and training of EMS managers and leaders. Pre-employment screening that identifies and screens out potentially violent people before hiring is an obvious means of preventing workplace violence. Pre-employment screening practices must be consistent with privacy protections and antidiscrimination laws.

A thorough background check can be expensive and time consuming but is a necessity today. The depth of pre-employment scrutiny will vary according to the level and sensitivity of the job being filled, the policies and resources of the prospective employer, and possibly differing legal requirements in different states. As an applicant is examined, certain red flags, such as a history of drug or alcohol abuse, past conflicts (especially if violence was involved) with coworkers, or past convictions for violent crimes, should exclude a candidate from employment. Other red flags can include a defensive, hostile attitude; a history of frequent job changes; and a tendency to blame others for problems.

> ◆ Increasing belligerence.
> ◆ Ominous, specific threats.
> ◆ Hypersensitivity to criticism.
> ◆ Recent acquisition of or fascination with weapons.
> ◆ Apparent obsession with a supervisor or coworker or an employee grievance.
> ◆ Preoccupation with violent themes.
> ◆ Interest in recently publicized violent events.
> ◆ Outbursts of anger.
> ◆ Extreme disorganization.
> ◆ Noticeable changes in behavior.
> ◆ Homicidal or suicidal comments or threats.

FIGURE 7.18 ■ Incipient Behaviors Leading to Violence.

IDENTIFYING CURRENT PROBLEMS AND RISKS

Problem situations or circumstances that may heighten the risk of violence can involve a particular event or employee or the workplace as a whole (Figure 7.18). No profile or litmus test exists to indicate whether an employee might become violent. Instead it is important for employers and employees alike to remain alert to problematic behavior that could point to possible violence.

EMS managers and human resources specialists must be alert to events linked with potential violence triggers, such as personality conflicts (between coworkers or between worker and supervisor); a mishandled termination or other disciplinary action; bringing weapons onto a work site; drug or alcohol use on the job; or a grudge over a real or imagined grievance. Risks can also stem from an employee's personal circumstances such as a breakup of a marriage or romantic relationship, family conflicts, financial or legal problems, or other emotional disturbance. Other problematic behavior may require EMS management intervention or employee discipline to curb a path to violent behavior.

Though a suicide threat may not be heard as threatening to others, it is nonetheless a serious danger sign. Some extreme violent acts are in fact suicidal—wounding or killing someone else in the expectation of being killed, a phenomenon known in law enforcement as "suicide by cop." In addition, many workplace shootings often end in suicide by the offender.

With regard to workplace violence, employers should make clear that zero tolerance in the original sense of the phrase applies; that is, no threatening or violent behavior is acceptable, and no violent incident will be ignored. Company violence-prevention policies

> ◆ Why has the offender threatened, made comments perceived by others as threatening, or taken this action at this particular time? What is happening in his own life that has prompted this?
> ◆ What has he said to others (friends, colleagues, coworkers, and so on) about what is troubling him?
> ◆ How does he view himself in relation to everyone else?
> ◆ Does he feel he has been wronged in some way?
> ◆ Does he accept responsibility for his own actions?
> ◆ How does he cope with disappointment, loss, or failure?
> ◆ Does he blame others for his failures?
> ◆ How does he interact with coworkers?
> ◆ Does he feel he is being treated fairly by the company?
> ◆ Does he have problems with supervisors or management?
> ◆ Is he concerned with job practices and responsibilities?
> ◆ Has he received unfavorable performance reviews or been reprimanded by management?
> ◆ Is there evidence of personal problems, such as divorce, death in the family, health problems, or other personal losses or issues?
> ◆ Is he experiencing financial problems, high personal debt, or bankruptcy?
> ◆ Is there evidence of substance abuse, mental illness, or depression?
> ◆ Is he preoccupied with violent themes, interested in publicized violent events, or fascinated with weapons, or has he recently acquired weapons?
> ◆ Has he identified a specific target and communicated with others his thoughts or plans for violence?
> ◆ Is he obsessed with others or engaged in any stalking or surveillance activity?
> ◆ Has he spoken of homicide or suicide?
> ◆ Does he have a past criminal history or history of past violent behavior?
> ◆ Does he have a plan for what he would do?
> ◆ Does the plan make sense, is it reasonable, and is it specific?
> ◆ Does he have the means, knowledge, and wherewithal to carry out his plan?

FIGURE 7.19 ■ Workplace violence assessment questions for managers to ask individuals familiar with the offender's behavior, both prior to and after any alleged threat or action.

should require action on all reports of violence, without exception.

That does not mean, however, that a rigid, one-size-fits-all policy of automatic penalties is appropriate, effective, or desirable. It may even be counterproductive, since employees may be more reluctant to report a fellow worker if he is subject to automatic termination, regardless of the circumstances or seriousness of his offense. In order to assess this risk, the questions suggested in Figure 7.19 should be asked of individuals familiar with the offender's behavior, both prior to and after any alleged threat or action. And it is important to remember that perpetrators of workplace violence can be both men and women.

One area in which managers express considerable concern is the restrictive effect of potential civil liability on disseminating information about past employees who have records of violence or other troubling behavior on the job. Those restrictions can significantly limit the potential employer's ability either to screen out dangerous people before hiring or to obtain information that would be highly relevant in a threat assessment when an incident has occurred. For example, though rules vary somewhat from one jurisdiction to another, law-enforcement agencies are ordinarily not allowed to disclose criminal records or inform employers if a worker or job applicant has been convicted of a violent crime, even though a conviction is a matter of public record. Similarly, strict confidentiality rules shield medical and mental-health records that can have direct relevance to assessing the risk of violent behavior.

EMPLOYEE ASSISTANCE PROGRAMS

Every emergency services organization should have an Employee Assistance Program (EAP) and men-

tal-health coverage. As a member of EMS, you must realize that the nature of this profession and what EMS personnel see and experience has a profound effect on mental health. The EAP provider or contractor refers people to a counselor, mental-health professional, or doctor. Most EAPs are corporate providers and have clients all over the United States.

When employees encounter a stress-related problem that affects job performance, they can call the EAP. The EAP will then analyze the situation over the phone and in many cases set up a counseling session with an approved or contracted EAP therapist. The therapist then can recommend whether or not the patient needs further counseling.

EAP can help with martial conflicts, stress, gambling, emotional conflicts, and drug or alcohol abuse. With most EAP contracts the first three to seven counseling sessions are paid by the EAP. The patient, employer, or insurance will cover the cost of continued counseling sessions. This varies by circumstance and company or insurance policy.

CHAPTER REVIEW

Summary

EMS managers and leadership need to constantly update and monitor the human resources activities in the organization. Recruitment, selection, maintenance, promotion, development, and succession planning create a timeline that requires management activities that may have legal and ethical consequences if not done properly. Measuring performance

is one of the most important functions in management and is often a controversial topic for EMS organizations. A healthy working relationship with human resources professionals is needed to ensure federal, state, civil service, and contractual requirements are matched appropriately with human resources and EMS leadership activities.

Research and Review Questions

1. As the EMS manager of your organization, obtain the job descriptions for the employee class over which you have responsibility. What changes would you recommend to the job descriptions and the related job tasks?

2. Design an assessment-center promotional exam for an EMS manager. The position will involve supervising 12 paramedics in a fire-based EMS system.

3. NFPA 1582 describes the medical-examination criteria for emergency responders. Design a process to incorporate it into an EMS organization.

4. Describe some uses of the Myers-Briggs Type Indicator (MBTI) in a conflict-management situation for paramedics.

5. Discuss the legal precedents that influence workplace-violence policies.

6. The General Duty Clause of the Occupational Safety and Health Act requires employers to provide a workplace that is free from recognized hazards. Identify several aspects of an EMS organization that are affected by this.

7. Review and analyze the common errors in performance appraisals.

References

Brown, W. E., Jr., Dickison, P. D., Misselbeck, W. J., & Levine, R. (2002). Longitudinal Emergency Medical Technician Attribute and Demographic Study (LEADS): An interim report. *Prehospital Emergency Care, 6*(4), 433–439.

Callahan, J. (2008, March 4). Six found dead in house on Lester. *Memphis Commercial Appeal.*

Cascio, W. F. (1998). *Managing human resources* (5th ed.). New York: McGraw-Hill.

Corbett, S. W., Grange, J. T., & Thomas, T. L. (1998). Exposure of prehospital care providers to violence. *Prehospital Emergency Care, 2*(2), 127–131.

Drabek, T. E., & Hoetmer, G. J. (1991). *Emergency management: Principles and practice for local government.* Washington, DC: International City/County Management Association.

Dubin, W. R., & Lion, J. R. (1996). Violence against the medical profession. In J. R. Lion, W. R. Dubin, & D. E. Futrell (Eds.), *Creating a secure workplace: Effective policies and practices in health care.* Chicago: American Hospital Publishing.

Edwards, S. T. (2005). *Fire service personnel management.* Upper Saddle River, NJ: Pearson Education Inc.

Erich, J. (2005, March). Wanted: Warm and willing bodies to fill vacant seats. *The Journal of Emergency Care, Rescue, and Transportation, 34*(3).

Hollingsworth, H. (AP). (2004, April 30). Two emergency workers slain in ambush. LJWorld.com. Retrieved from http://www2.ljworld.com/news/2004/apr/04/2_emergency_workers/

Institute of Medicine. (2003). *The future of emergency care in the United States health system.* Retrieved Aug 26, 2004, from http://www.iom.edu/project.asp?id = 16107

Karp, H., Fuller, C., & Sirias, D. (2002). *Bridging the boomer Xer gap: Creating authentic teams for high performance work.* Mountain View, CA: Davies-Black Publishing.

Maggiore, A. (July 1998). Courts apply ADA to EMS. *EMS Insider, 25*(7), 6–8.

McIntosh, A. (2007, April 19). Paramedic loses license in drug case. State revokes certification after he admits using stolen narcotics. *Sacramento Bee,* B1.

Michelson, R., & Maher, P. (2001). *Preparing for promotion: A guide for public safety assessment centers* (2nd ed.). San Clemente, CA: LawTech Publishing Co., Ltd.

Myers, I. B. (Revised by L. K. Kirby & K. D. Myers.) (1998). *Introduction to type (A)* (6th ed.). Palo Alto, CA: CCP, Inc.

National Highway Traffic Safety Administration. (2008). *EMS workforce for the 21st century: A national assessment.* Washington, DC: Author.

National Institute for Occupational Safety and Health. (1996). *Violence in the workplace.* Retrieved December 2006 from http://www.jrrobertssecurity.com/articles/violcont.html.

Occupational Safety and Health Administration. (1998). *Guidelines for preventing workplace violence for health care and social service workers* (OSHA 3148). Washington, DC: U.S. Department of Labor.

Schuler, R. S., & Huber, V. L. (1993). *Personnel and human resources management* (5th ed.). St. Paul, MN: West Publishing Company.

Sherman, A. W., Bohlander, G. W., & Churden, H. J. (1995). *Managing human resources* (10th ed.). Cincinnati, OH: South-Western Publishing.

University of Iowa Injury Prevention Research Center. (2001, February). *Workplace violence: A report to the nation.* Iowa City: Author.

Zemke, R., Raines, C., & Filipczak, B. (1999). *Generations at work: Managing the clash of veterans, boomers, Xers, and nexters in your workplace.* New York: AMACOM/American Management Association.

Management of EMS Education

8 CHAPTER

Accreditation Criteria

CAAS

106.06   Orientation Training: The agency shall have an orientation program for all new employees. At a minimum, the orientation program shall include a review of all policies/procedures and standards relating to the employee and the position for which they were hired; a mechanism for evaluation and feedback on the employee's progress in orientation; and a mechanism to verify and document successful completion of all required orientation standards (pp. 215–219).

201.04.01   Staffing: With input and approval from the Medical Director, the agency shall have established staffing certification and qualification standards for each level of service provided. These standards shall be reviewed, at minimum, once per year. The minimum acceptable staffing standard for patient care is two Emergency Medical Technicians (pp. 200–204).

106.06.03   With Medical Director approval and input, the agency shall have a Continuing Medical Education program that meets or exceeds local and state requirements. The Continuing Medical Education program shall be clearly linked to the agency's CQI Program and shall address each level of service provided. Individual CME requirements may differ among different levels of providers (pp. 200–202).

Objectives

After reading this chapter the student should be able to:

8.1   Understand the EMS Education Agenda for the Future (pp. 190–193).
8.2   Identify national resources to conduct EMS training (pp. 193–195).
8.3   Understand the national curriculums and application to each provider level (pp. 195–198).
8.4   Build an EMS refresher course for any level of EMS provider (pp. 198–202).
8.5   Identify EMS management-training programs and opportunities (pp. 220–222).
8.6   Conduct a training analysis of EMS needs (pp. 207–210).
8.7   Identify how to conduct training encompassing psychomotor skills, affective domain, and didactic knowledge (pp. 214–218).
8.8   Apply standard procedures to evaluate training in accordance with accreditation standards (pp. 194–196).

Key Terms

active learning (p. 203)
affective domain (p. 215)
asynchronous learning (p. 205)
Commission on Accreditation of Allied Health Education Programs (CAAHEP) (p. 194)
Continuing Education Coordinating Board for Emergency Medical Services (CECBEMS) (p. 199)
core elements (p. 192)

Crash Injury Management (p. 191)
distributive learning (p. 202)
EMS education portfolio (p. 197)
EMT–Basic (p. 192)
EMT–Intermediate (p. 192)
EMT–Paramedic (p. 192)
First Responder (p. 192)
intern (p. 214)
internship (p. 195)
National Standard Curriculums (NSC) (p. 195)

objective structured clinical examination (OSCE) (p. 204)
preceptor (p. 211)
primary EMS instructor (p. 198)
role modeling (p. 214)
seat time (p. 195)
secondary EMS instructor (p. 198)
standardized patient (SP) (p. 204)
synchronous learning (p. 204)

Point of View

My high expectations for student performance and achievement in EMS education are driven by several core beliefs. I sincerely believe that the most important "tool" you bring with you each day to your particular work environment is your humanity. If we lose sight of that, then all of the magic in the EMS "bag of tricks" may be rendered useless. I've seen patients receive enough IV morphine to stop a charging elephant, yet their pain was not relieved because no rapport had been created between the patient and the EMS provider, and the patient did not believe that the drug would work.

I believe that the role of the EMS provider is a unique one in the continuum of care and carries with it a significant responsibility not only to provide care but also to serve as a patient advocate. Unlike other patient-provider relationships, our patients do not get to choose who responds when a 911 call is made. We care for our patients at their most vulnerable, many times on the worst day of their lives. We are often the first "expert" eyes and hands on scene, and the judgments made in the prehospital arena may directly or indirectly influence what happens when that patient reaches the emergency department. To my mind it is essential that each patient feel that he has been cared for, not merely that care was received. We cannot disqualify a patient's perception of what constitutes an emergency simply because it does not match our own.

The charm and challenge of EMS is that no day "at the office" is like the previous day, and at any given point in time all of our education and experience is potentially on the line. As a result, each EMS provider must hold himself to the highest standards of service delivery and be committed to lifelong learning. It is not sufficient to know "just enough to pass the test" (a variant on the theory that "anything over 80% is wasted"). I believe that it's imperative to have a strong foundation in physiology, pathophysiology, and pharmacology to understand the "why," not just the "when" and "what," of our protocols.

These tenets have been the foundation of my approach to EMS education for the 20-plus years that education was my primary job. I know that at the time some of my students felt I was unrealistic and unreasonable, but I also know that many of those same students came back to say, "Thank you." The legacy of being an EMS educator is that you will influence patient care in countless ways and for far longer than you may believe through the actions, or inactions, of your students and the lives they touch.

Beth Adams, MA, RN, NREMT–P
Quality Manager
Fairfax County (Virginia) Fire and Rescue Department

∎ EMS EDUCATION

Learning something new and completely different liberates the mind. Facing a challenge, meeting it, and mastering it helps build confidence. As a frontline EMS manager, you will interface with various education activities. From initial paramedic or EMT training to refresher training, the manager will need to understand the requirements and frameworks of the national EMS system.

The technological advances and changes in the workforce of the future will focus on cognitive strategies, problem solving, and the ability to multitask. Employers will demand technically proficient, socially cognizant, and culturally aware EMS workers. Narrowly focused technical training is not sufficient for the individual or the EMS organization to be successful. As an EMS leader, you will need to map your own educational plan to develop your skills and abilities as a central figure in the organization.

Education is at the heart of every profession and helps distinguish a set of skills and abilities from a trade. Professional education involves developing a combination of professional competency and the attitudes that must accompany it (Figure 8.1). Professional education is distinguished from training in that it requires the learner to take information and put it together to create something new. That can include the validation of processes.

A BRIEF HISTORY

The first formalized EMS educational program was developed in the 1950s by the American College of Surgeons. Roughly at the same time, the American Red Cross published four advanced first-aid manuals, which were shortly followed by the American College of Orthopedic Surgeons' first textbook, known as the

---

- ◆ Conceptual competence—understanding the theoretical foundations.
- ◆ Technical competence—the ability to perform a task.
- ◆ Interpersonal competence—the ability to use written and oral communication.
- ◆ Contextual competence—understanding the environment in which EMS practices.
- ◆ Integrative competence—the ability to merge theory and practice.
- ◆ Adaptive competencies—the ability to anticipate change.

FIGURE 8.1 ∎ Educational Competencies for Professionals.

"Orange Book." It was entitled *Emergency Care and Transportation of the Sick and Injured*.

The National Academy of Sciences was one of the first agencies that made an attempt to standardize EMS training. It produced a white paper in 1966 called *Accidental Death and Disability: The Neglected Disease of Modern Society*, which called for the development of a prehospital care system that could drastically change the outcome of accident victims in a society that was taking to the highways and moving to suburbs in record numbers. The Highway Traffic Safety Bureau, which would later become the National Highway Traffic Safety Administration (NHTSA), offered a contract for the first EMT-ambulance-training program to a third-party vendor. That set the precedent for NHTSA's collaborative relationships with educational institutions.

The American Academy of Orthopedic Surgeons created the first curriculum in 1967 and published the textbook *Emergency Care and Transportation of the Sick and Injured*. This was the first well-designed and authoritative textbook for EMS personnel and is affectionately known as the "Omaha Orange" book.

In 1970 NHTSA created the First Responder program for law-enforcement officers, called **Crash Injury Management**, which was 40 hours of advanced first aid. In 1973 Congress passed the Emergency Medical Services Act, which included training and manpower among the key components of the EMS system. In 1975 the National Registry of EMTs (NREMT) was formed to credential a national testing process, and in 1977 the first nationally standardized curriculum for the EMT–Paramedic was adopted by NHTSA. In that same year the National Association of State EMS Directors developed the EMT–Intermediate certification, using a limited number of modules from the national paramedic curriculum.

In the 1980s the EMT curriculum was revised upward to 110 hours, and the paramedic curriculum was reorganized, with the EMT–Intermediate curriculum becoming its own stand-alone curriculum.

In the 1990s the *EMS Agenda for the Future* was formed, and the *National EMS Education and Practice Blueprint* was developed to map further clarifications in the EMS system.

In 1994 the current EMT–Basic curriculum, formerly called the EMT–Ambulance, was revised and focused on an assessment-based approach to prehospital care. The current First Responder curriculum was revised in 1995 and again in 1998.

THE PRESENT

The content of all levels of EMS training is currently based on NHTSA's national EMS standards for First Responders, EMTs, EMT–Intermediates, and EMT–Paramedics. It is important to understand that the NHTSA standards are minimum levels of education. Course work and curriculums are flexible enough to be modified by adding material that meets local needs. The rigor and educational methods should meet the standards required to get academic credit. Most education delivered in an academic setting meets an accreditation-system standard. That system evaluates the program policies, test security, academic standards, and course design. The accreditation standards are developed by a national consensus-building process and best-evidence medicine. Quality education is ensured by institutions adhering to curriculum standards, national accreditation, and national standardized testing.

In 1990 NHTSA held a consensus workshop on EMS training programs. A group of experienced EMS providers from all areas of EMS concluded that there was a national need for the development of a training blueprint for EMS. This included evaluation of the then current levels of prehospital-provider training and certification. The document became known as the *National EMS Training Blueprint Project*. The Blueprint Project was financially supported by the NREMT.

The goal of the Blueprint Project was to develop a national consensus on a framework for development of EMS training, education, and policies. The lack of outcome data and field practice on patient care forced the task force to use a peer-review and consensus process to validate the EMS training blueprint. In May 1993 national EMS organizations completed a consensus-building workshop and after some modification adopted the Blueprint document for presentation to the national EMS community.

The Blueprint divided the major areas of prehospital instruction and performance into **core elements**. Special populations (pediatrics, geriatrics, handicapped, and others) were not listed as separate core elements. The 1990 EMS Training Consensus Workshop recommended that EMS education on special populations be addressed throughout each curriculum. Pediatric issues, which are substantially different from those of adults, were included as separate key marks in the various core elements and in the OB/GYN and pediatric core elements. The final core elements are listed in Figure 8.2.

CORE ELEMENTS
PATIENT ASSESSMENT.
AIRWAY.
BREATHING.
CIRCULATION.
MUSCULOSKELETAL.
CHILDREN AND OB/GYN.
BEHAVIORAL.
MEDICATION ADMINISTRATION.
NEUROLOGICAL.
ENVIRONMENTAL.
EMS SYSTEMS.
ETHICAL/LEGAL.
COMMUNICATIONS.
DOCUMENTATION.
SAFETY.
TRIAGE AND TRANSPORTATION.

FIGURE 8.2 ■ Core Elements in EMS Education. *(Source: National Highway Traffic Safety Administration EMS Blueprint.)*

For each core element there are progressively increasing levels of knowledge and skill objectives, representing a continuum of education and practice. Each increasing level of skill and knowledge is inclusive of those at the prior level. The degrees of knowledge are described as essential, advanced, or comprehensive. Within the Blueprint, a few components of the core elements are not specified because either the technology is changing too rapidly or the EMS community has not yet had the opportunity to discuss the specifics of those issues.

The Blueprint recommends four levels of prehospital EMS providers, each corresponding to various levels of knowledge and skills in each of the core elements. The EMS **First Responder** uses a limited amount of equipment to perform initial assessment and intervention. The **EMT–Basic** has the knowledge and skill of the First Responder but is also qualified to function as the minimum staff for an ambulance. The **EMT–Intermediate** has the knowledge and skills of the preceding levels but also can perform essential advanced techniques and administer a limited number of medications. An **EMT–Paramedic** has the competencies of an EMT–Intermediate but also has enhanced assessment skills and can administer additional interventions and medications.

The *National EMS Education and Practice Blueprint*, the 1993 document created by the National EMS Training Blueprint Project, now acts as a guide to develop and modify the prehospital EMS-provider training curriculums. It identifies responsibilities for certain aspects of the EMS. The NREMT has been tasked with facilitating national certification and will conduct national testing and registration. The state EMS offices will ensure that any new levels of training programs and certification or licensing are compatible with the continuum of training. A national certification system formally linked with NHTSA will improve legal recognition and reciprocity between jurisdictions.

State EMS offices have the responsibility to evaluate education and practice for reciprocity and legal recognition. Curriculum developers and publishers ensure that various levels of prehospital EMS education curriculums reflect a continuum of education taken from the declarative material put out by the Blueprint. Curriculum and materials produced by publishers represent an increasing and inclusive knowledge and skill level, contain compatible educational knowledge and skill objectives, and allow effective bridging and linkages between sequential core curriculums.

The state EMS offices and others build effective bridging between various curriculums and certification levels. State EMS offices, NHTSA, and the NREMT have the responsibility to determine the appropriate level to incorporate new knowledge, equipment, and procedures into the testing processes. EMS managers who oversee EMS education may look for help from other medical and allied health professionals to provide EMS education.

Use of the Blueprint is designed to provide states the flexibility to build education and training systems that exceed national certification levels in some areas but still remain consistent with the continuum of education and practice provided by the Blueprint. The students and EMS workers certified at the state level would know where their certifications fall on the continuum and what they might expect if they moved to another state.

The EMT–Intermediate and EMT–Paramedic curriculums that are the basis for current instruction will continue to undergo routine revision by EMS educators, providers, physicians, and administrators. There is some indication that the EMT–Intermediate level of training is going to be disbanded and critical skills such as intravenous administration and esophageal airways will be added to the EMT–Basic level of practice. This change will affect the Blueprint documents and operational considerations in the future.

PLANNING FOR THE FUTURE

In 1996 the Department of Transportation published the *EMS Agenda of the Future*. One of the areas it emphasized was the EMS education system. The document focuses on where the federal government and EMS stakeholders see EMS education in the year 2010. The document states that EMS education will need to train providers to integrate into the healthcare system. In addition to emergency medicine, EMS education programs also will need to teach injury prevention, home health, treatment of chronic conditions, and community and public health.

Modern instruction using NHTSA curriculums requires that programs employ creative solutions to deliver the material. The *EMS Education Agenda for the Future: A Systems Approach*, created in June 2000, was developed by EMS administrators, physicians, educators, and EMS providers. It was created to answer strategic goals for EMS education and lists five integral components: core content, scope of practice, educational standards, accreditation, and certification (Figure 8.3).

National EMS Core Content
One of the five integral components of EMS identified in the *EMS Education Agenda* is core content. The National EMS Core Content is a comprehensive list of skills and knowledge needed for prehospital emergency care. The core content is specified by the medical community, with consensus building through regulators, educators, and EMS providers. The core content will be based on science and analysis of skills and abilities.

National EMS Scope of Practice
The second of the five integral components of EMS identified in the *EMS Education Agenda* is scope of practice. The National EMS Scope of Practice model

---

- National EMS Core Content.
- National EMS Scope of Practice.
- National EMS Education Standards.
- National EMS Education Program Accreditation.
- National EMS Certification.

FIGURE 8.3 ■ Five Components of the *EMS Education Agenda for the Future.*

places specific skills and knowledge in the core content into the four levels of EMS, defining the minimum knowledge and skills for each level. The scope of practice will be national in application, yet flexible enough for state and local add-ons. Since the scope-of-practice model will affect state and administrative law, the document will be overseen by EMS regulators. The national scope of practice will be the basis for each state's scope of practice and reciprocity.

National EMS Education Standards

The third of the five integral components of EMS identified in the *EMS Education Agenda* is standards. The National EMS Education Standards will take the national curriculums and spell out the minimum terminal learning objectives for each level of practices. The national standard curriculums will have been approved at every level of certification with educational objectives and content material. While a national standard has decreased flexibility and standardized content, there are still large differences in the length and quality of programs. The national EMS standard holds educational institutions accountable for the quality and content of their instruction. This requires education programs and institutions to evaluate educational methodologies and outcomes.

National EMS Education Program Accreditation

The fourth of the five integral components of EMS identified in the *EMS Education Agenda* is accreditation. The National EMS Education Program Accreditation is a mechanism for verifying educational programs for quality that is universal to all educational providers. This component applies to all levels of instruction and providers, and it will reduce the inconsistencies in the delivery of national curriculums.

Accreditation is a nongovernmental, independent, collegial process of self- and peer assessment. Accreditation ensures that an independent review of an institution's or program's quality measures up to a national standard. The process looks at three key areas:

- Faculty and administration self-assessment based on national standards.
- Peer visits that interview students, faculty, and staff.
- A commission that reviews the evidence provided by peers and self-study, which can result in accreditation of structure, process, and outcomes.

Currently this is being done by the **Commission on Accreditation of Allied Health Education**

Programs (CAAHEP) through the newly named subcommittee called the Committee on Accreditation of Educational Programs for the Emergency Medical Services Professions (CoAEMSP). Five states currently require paramedics to attend accredited paramedic programs before they can sit for state certification. A single national accrediting agency will be accepted by all state regulatory agencies.

National EMS Certification

The last of the five integral components of EMS identified in the *EMS Education Agenda* is certification. A national certification program will create a standardized testing process for all levels of providers that will be recognized nationally to ensure that entry-level personnel meet minimum competencies. National certification will be based on up-to-date practices and a psychometric methodology for entry-level providers. This will precipitate more licensure based on compelling education and passing an examination to ensure minimum competency.

Currently 44 states use some form of the NREMT exam to certify providers. A key component is that the students must graduate from an accredited paramedic or EMS training program. All aspects of the national EMS model will be guided by science and educational research. The future will embrace a variety of education methodologies including online and distance-education platforms.

Paramedic Education Program Accreditation

Educational institutions as well as EMS agencies can attain accreditation through CAAHEP and CoAEMSP. The goal of CoAEMSP is to make accreditation of paramedic programs mandatory and linked to certification and licensure. CoAEMSP is committed to making the process credible, timely, and valid. In the future the CoAEMSP will accredit EMT-Basic training programs and EMS programs that exceed one year. Annual fees are required to remain accredited.

The accreditation process is divided into three sections. The first section is the general requirements and is common to all educational programs. General requirements require sponsorship, resources, students, operations policies, and program evaluation. The second section is program specific to EMS and paramedic training and requires curriculum and a description of the paramedic roles, profession, and responsibilities. The curriculum must meet several standards. Incorporated in the curriculum section is the need for a description of the program, a plan of instruction, disclosure, structure, and content. Within the curriculum, clear guidelines and processes must be defined for didactic instruction, skills laboratories, clinical instruction, and field **internships**. The third section of the accreditation process requires the agency to maintain and administer the accreditation, and it outlines the program, sponsoring-institution responsibilities, and committee responsibilities.

If you manage a large EMS operation that conducts its own in-house paramedic training center, accreditation of the training program should be a strategic goal. Accreditation helps to ensure that the program meets national standards set forth by educational experts. For EMS managers without an educational background, navigating the accreditation process helps to ensure that the educational systems that will support quality learning are applied to the organization.

◼ CURRENT CURRICULUMS

PRE-EMS

Pre-EMS is an education level that describes a person trained in emergency care who functions as an integral element of EMS, providing life support and care while awaiting the arrival of licensed EMS personnel. The pre-EMS responder is a critical link in the chain of care for many ill or injured patients, because the care given and the information obtained from them can be vital to the patient's immediate care and long-term outcome. Pre-EMS responders by definition are laypersons including bystanders or passersby who are occupationally required to be trained in first aid, even though they may not be specifically required by law to perform first aid. Pre-EMS people receive basic life support (BLS) training to respond to victims with critical injuries.

The term "pre-EMS responder" can refer to a large group that may include community emergency-response teams; laypersons; public-safety employees; firefighters; police officers; highway patrol personnel; security guards; merchant marine sailors; and airline, railroad, and other public transportation vehicle crews. The early identification of a medical emergency combined with the delivery of prompt, effectively administered pre-EMS care can make a difference between life and death, rapid versus pro-

longed recovery, and temporary versus permanent disability. For pre-EMS responders to be optimally prepared, they must be ready physically, emotionally, and logistically. Knowing what to do and how to do it are of little value if people are not willing. Similarly to EMS systems, quality training from qualified educators and support from the local EMS agency are the foundation of and key to the effectiveness of a pre-EMS system. The American Red Cross and the National Safety Council provide the bulk of pre-EMS training. Publishers and trade organizations are beginning to enter this market, and several curriculums are available.

NATIONAL STANDARD CURRICULUMS

NHTSA oversees the creation and implementation of national course work and publishes national curriculums for EMS. Most of these curriculums undergo periodic revisions and are contracted to researchers or organizations to create, revise, and update the content. The states—or in some cases the local counties—determine the length of seat time required for the student to meet the certification standards. **Seat time** is a term used for an instructional method in which the student is actually in the classroom; it is usually measured in hours under a system called the Carnegie unit. It is important to understand that national curriculums do not have specific hours. The minimum number of hours for each curriculum is determined at a local or state level. The **National Standard Curriculums (NSC)** are provided by NHTSA on disk or by download from its publications Web site.

The entry-level EMS provider in this country is the EMS First Responder. The First Responder uses a limited amount of equipment to perform initial assessment and interventions and is trained to assist other EMS providers. For example, at the scene of a cardiac arrest, the First Responder would be expected to notify EMS (if not already notified) and initiate CPR with an oral airway and a barrier device. In most areas First Responder training is around 40 hours, yet states have various minimum hours. The current curriculum was released in 1995 and replaced the original First Responder curriculum, which was produced in 1978. A national standard refresher course, released in 1996, is also available.

The EMT–Basic makes up the majority of EMS providers in this country. The EMT–Basic provider has the knowledge and skills of the First Responder but is also qualified to function as minimum staff for an ambulance. The current EMT–Basic curriculum was released in 1994 and replaced the 1985 curriculum that used to be called EMT–Ambulance. At the scene of a cardiac arrest, the EMT–Basic would be expected to defibrillate and ventilate the patient with a manually operated device and supplemental oxygen. Refresher training posted by NHTSA in 1996 was designed to transition an EMT–Ambulance to an EMT–Basic.

The EMT–Intermediate has the knowledge and skills of the preceding levels but in addition can perform essential advanced techniques and administer a limited number of medications. At the scene of a cardiac arrest, the EMT–Intermediate would be expected to intubate and administer first-line ACLS medications. The current EMT–Intermediate course was released in 1999 by NHTSA and has a refresher program that was published in 2001. In many states the 1999 EMT–Intermediate curriculum is being used to accommodate the requirements of a paramedic, while the current 1998 paramedic curriculum is called paramedic specialist or advanced paramedic.

NHTSA has published a document called the *EMT-Paramedic and EMT-Intermediate Comparison*. It compares the objectives in the revised 1998 EMT–Paramedic National Standard Curriculum with those in the 1985 EMT–Paramedic curriculum and the objectives in the 1999 EMT–Intermediate National Standard Curriculum with those in the 1985 EMT–Intermediate Curriculum. State EMS offices, EMS education programs, and EMS providers can use the document to revise their current EMT–Paramedic and EMT–Intermediate programs to meet the revised standards. The National Council of State EMS Training Coordinators (NCSEMSTC) developed these comparisons, and there is some indication that the need for EMT–Intermediates will decline, as the skills that are included in the Intermediate curriculum could be moved into the Basic curriculum or be handled by paramedics.

The EMT–Paramedic is the highest level of nationally recognized prehospital training. The EMT–Paramedic has demonstrated the competencies expected for EMT–Basic and EMT–Intermediate providers but also can administer additional interventions and medications. At the scene of a cardiac arrest, the EMT–Paramedic might administer second-line

ACLS medication and use an external pacemaker. The current paramedic curriculum, produced by the National Association of EMS Educators under the direction of NHTSA, was published in 1998. Interest in an Advanced-Practice Paramedic level is being reviewed by a consensus team and NHTSA.

1998 Paramedic Training

The current paramedic curriculum was released in 1998 and preceded the 1985 paramedic course work. Many EMS managers received their paramedic training under the older 1985 curriculum, but it is important as an EMS manager to understand the changes in the 1998 course work and what to expect from new paramedic graduates being educated under the 1998 paramedic curriculum. The 1998 curriculum now calls for college prerequisites in algebra, reading, writing, and anatomy and physiology.

Some significant additions to content have been added to reflect wellness issues in the paramedic program. Content in the national curriculum now covers nutrition, physical fitness, disease prevention, injury prevention, and mental health in EMS providers. A new direction emerged in the 1998 curriculum, bringing the public-health model into the paramedic scope of practice. More involvement of EMS in injury prevention and community education is designed into the new course work, as well as more pathophysiology, encompassing cellular injury, genetics, inflammation, stress, and disease, which builds on the anatomy and physiology. An emphasis on better communication skills, patients with special challenges, and home-care interventions was added. Only a few new skills have been added, focusing on airway skills with digital intubation and transillumination. The scope of knowledge has not expanded; however, the depth of understanding of current topics in the paramedic curriculum has changed with regard to the physical exam, pharmacology, medical conditions, pediatrics, and geriatrics.

In many EMS education programs, the clinical experience and internships are patient-contact based rather than time based (for example, 5 live intubations, 30 geriatric assessments, or 30 chest-pain calls). While no exact number of patients is specified, a range and national average is in the curriculum. The 1985 paramedic curriculum was completed in 600 to 1000 hours; the new changes require 1000 to 1200 hours. A greater emphasis is being placed on competencies, rather than on a specific number of hours. Again the designation of hours is set by the local or state agencies and not the federal government or NHTSA. Figure 8.4 is a model of the current EMT–Paramedic National Standard Curriculum.

EMS Instructor Education

Through a partnership with the National Association of EMS Educators, NHTSA established a core curriculum for educating EMS instructors. The 2002 National Guidelines for Educating EMS Instructors provides the core content, knowledge, and skills necessary to be a successful EMS instructor. Taking a design from primary education, the new curriculum calls for the EMS instructor to build an **EMS education portfolio**. A typical portfolio contains the examples of educational products (lesson plans, tests, slide presentations, games, and so on) generated by an instructor that allows the instructor to present a representative body of work for review and comment.

COMPETENCIES
Mathematics, Reading, Writing

PRE- OR CO-REQUISITE
EMT or EMT–Basic Human Anatomy and Physiology

PREPARATORY		
Clinical/Field	EMS Systems/Role and Responsibilities of the Paramedic The Well-Being of the Paramedic Illness and Injury Prevention Medical/Legal Issues Ethics General Principles of Pathophysiology Pharmacology Venous Access and Medication Administration Therapeutic Communications Life-Span Development	Clinical/Field
	AIRWAY MANAGEMENT AND VENTILATION	
MEDICAL	**PATIENT ASSESSMENT**	**TRAUMA**
Pulmonary Cardiology Neurology Endocrinology Allergies and Anaphylaxis Gastroenterology Renal/Urology Toxicology Hematology Environmental Conditions Infectious and Communicable Diseases Behavioral and Psychiatric Disorders Gynecology Obstetrics	History Taking Techniques of Physical Examination Patient Assessment Clinical Decision Making Communications Documentation	Trauma Systems/Mechanism of Injury Hemorrhage and Shock Soft-Tissue Trauma Burns Head and Facial Trauma Spinal Trauma Thoracic Trauma Abdominal Trauma Musculoskeletal Trauma

FIGURE 8.4 ■ EMT–Paramedic National Standard Curriculum's Education Model.

SPECIAL CONSIDERATIONS		
Clinical/Field	Neonatology Pediatrics Geriatrics Abuse and Assault Patients with Special Challenges Acute Interventions for the Chronic-Care Patient	Clinical/Field
	ASSESSMENT-BASED MANAGEMENT	
	OPERATIONS	
	Ambulance Operations Medical Incident Command Rescue Awareness and Operations Hazardous-Materials Incidents Crime-Scene Awareness	

LIFE-LONG LEARNING
Continuing Education

FIGURE 8.4 ■ Continued

When a person completes an EMS instructor course, it is suggested that he be certified to a **secondary EMS instructor** level. This is a person who can assist with instruction, provide lab instruction, and help with clinical activities. The idea is to ease a person into instruction under the guidance of a primary EMS instructor. A **primary EMS instructor** is a member of the educational team and is the main educator in charge of a group of students who are attending a course, often referred to as the lead instructor or instructor of record at an agency or institution. In addition to providing and coordinating classroom instruction, the primary instructor also coordinates other aspects of the course or works closely with a program director in the coordination of a course. A national EMS instructor test is being offered by the National Association of EMS Educators.

Emergency Vehicle Operations Course

In 1995 NHTSA released the Emergency Vehicle Operators Course (EVOC) for Ambulance, and as the name suggests, the curriculum only addresses emergency-vehicle operations as they relate to the safe operation of ambulances. The U.S. Fire Administration distributes additional EVOC curriculums for other emergency vehicles. They consist of an instructor guide and participant manual, which allow for the inclusion of local and state laws and organizational operation procedures. This course is set up in three modules: classroom education, driving-range management, and on-the-job performance recommendations.

■ INSERVICE TRAINING

In a well-designed continuing-education program, most recertification cycles are two years for EMTs and paramedics, with the Canadian EMS services operating on a three-year certification cycle. During the first year, paramedics attend a continuing-education program that integrates video, printed programs, and interactive educational television. In the second year they receive a self-directed package of basic competencies and prescription training based on their QI assessment. In the third year the paramedics attend an academy for a week of skills and knowledge testing.

REFRESHER TRAINING

After initial certification or licensure, an EMS provider may experience a deterioration of skills and knowledge due to his failure to keep up with the constant changes in the field of medicine. If skills are not practiced or refreshed on a regular basis, a paramedic or EMT is at risk of outdating his skills and knowledge. It is important to

Preparatory	1 hour
Trauma	4 hours
Airway	2 hours
OB, Infants, Children	2 hours
Patient Assessment	3 hours
Medical/Behavioral	4 hours
Elective Topics	8 hours

FIGURE 8.5 ■ National Registry Requirements for EMT Recertification.
(Reprinted with permission of National Registry of EMTs.)

establish a continuing-education and refresher-training program that reflects the changes and updates to modern medicine and emerging science.

Most states operate under the NREMT recertification requirements. EMTs must complete a national standard state-approved EMT–Basic refresher program consisting of a minimum of 24 hours. Ideally, these can be completed through a state or formal EMT–Basic refresher course approved by the **Continuing Education Coordinating Board for Emergency Medical Services (CECBEMS)**.

The NREMT accepts up to 10 hours of the refresher course through distributive education, provided the program is approved by your state EMS office. The topics of a nationally accepted EMT–Basic refresher course are included in Figure 8.5. The NREMT refresher is based on the NHTSA guidelines for refresher training.

Most EMS regulatory agencies require a verification or copy of a certificate of attendance that indicates successful completion and that is issued by a recognized training agency with signatures from an approved course coordinator or EMS provider. Documentation may be in the form of an official letter signed by the training officer or physician medical director. EMS training records for EMT refresher classes must include topical content, dates of completion, and how the education was delivered (for example, classroom, inservice training, video training, and so on).

EMT Intermediate Recertification

NHTSA contracted with the NAEMSE to create the 2001 EMT–Intermediate: National Standard Curriculum Refresher Course. This course is intended for the EMS community to use to refresh their current Intermediate providers who have been educated to the 1999 EMT–Intermediate level. This course is not intended to transition current 1985 Intermediates to the revised 1999 level. For EMT–Intermediates to recertify on a national level with the NREMT, they must complete a state-approved National Standard EMT–Intermediate/99 Refresher course, adhering to the DOT EMT–Intermediate/99 Curriculum. In addition to the EMT–Intermediate refresher, an EMT–Intermediate must complete 36 hours of continuing EMS-related education as outlined in that document. All continuing-education hours, including the refresher, must be completed within the current reregistration cycle.

Paramedic Recertification

NHTSA published the 2001 EMT–Paramedic: National Standard Curriculum Refresher Course, which is an outline of the requirements for paramedic recertification. Traditional refresher programs refresh material already known by the students. The intent of such programs is to maintain a student's competence in knowledge and skill performance. The 2001 EMT–Paramedic: National Standard Curriculum Refresher Course embraces the same concept, but it also encourages the inclusion of new and expanded information. New and expanded information may be added to the course but not at the expense of content that is core material for the program.

The 2001 EMT–Paramedic: National Standard Curriculum Refresher Course is not designed to be continuing education for the participants and for the EMS community to use to refresh their current paramedic providers who have been educated to the 1998 EMT–Intermediate level. If a system wishes to incorporate additional information or a new intervention that requires a substantial amount of time to teach, the information must be offered in addition to the content of the refresher program. Moreover, this course is not a transition or bridge course for current paramedics to become certified at the revised 1998 EMT–Paramedic level. The course was developed by the National Association of EMS Educators (NAEMSE).

Most EMS regulatory agencies follow closely the NREMT recertification for paramedics. Most states require paramedics to complete a paramedic refresher that reflects the content recommended by NHTSA and the 1998 Paramedic Curriculum. The NREMT has established a refresher course requiring paramedics to receive 48 hours of education. The NREMT splits the content of refresher requirements into mandatory-core and flexible-core content areas. The subjects to be covered specific to the objectives are outlined in each division within the core-content areas. These topics can be delivered through a variety of formats, including lecture, simulated skill presentations, case- or run-review discussions, conference attendance, or a formal state-approved course. Distributive or online education may be used as a part of the mandatory or flexible core content for up to 10 hours. The NREMT Web site at www.nremt.org has more information pertaining to distributive education.

The NREMT's guidelines for paramedics and EMTs require a provider to obtain CPR certification at the health-care provider level based on the AHA 2005 Guidelines. Paramedics must obtain an ACLS card current every two years for recertification. Some systems also may require PALS.

CONTINUING EDUCATION

The additional education required to maintain certification that does not encompass the mandatory topics in refresher training is often called "continuing education." Most EMS regulatory agencies impose continuing-education requirements on people who hold paramedic licenses or EMT certificates. Continuing education requirements are intended to encourage EMS professionals to maintain their training and stay up to date on new developments. Depending on the field, some of these courses may be offered by regular colleges, but they are often provided in-house by the organization, conference corporations, or hospitals. Conferences and seminars are common in EMS and may be designed to satisfy professional continuing-education requirements. In 1999 NHTSA released the EMT–Paramedic and EMT–Intermediate Continuing Education Guidelines, which contain recommendations and guidance for the development of EMS continuing-education programs. The contents of these files are identical to the printed version now available on CD-ROM through NHTSA.

EMS providers under National Registry requirements must complete 24 hours of additional continuing EMS-related education. Most regulatory agencies follow the NREMT's lead and require some level of continuing education. A variety of subjects covered in the National Standard EMT Curricula or Basic/Advanced/Rescue topics will be accepted as continuing education. These topics may be presented through workshops, seminars, didactic sessions, or other state

PHTLS or BTLS	up to	16 hours
Emergency Driving	up to	12 hours
PALS	up to	16 hours
Dispatch Training	up to	12 hours
Auto Extrication	up to	16 hours
Teaching CPR	up to	12 hours
ACLS	up to	16 hours
AMLS	up to	16 hours

FIGURE 8.6 ■ Common Continuing Education Program Credits for NREMT.

approved inservice training. The NREMT accepts no more than 12 hours toward this section from distributive education that is approved by CECBEMS. A listing of approved programs is included on the CECBEMS Web site. A maximum of 12 hours may be applied toward this section in any one subject, with the exception of those listed in Figure 8.6. For example, a hazmat course that was 40 hours would only be awarded as 12 hours of continuing education toward this section.

An EMS manager is responsible for ensuring records are kept, including certificates of completion or records of education indicating the education completed, dates of completion, and validating signatures. Documentation may be in the form of an official letter signed by a training officer, physician medical director, or national conference provider. EMS documentation—whether electronic or handwritten—must include topical content, dates of completion, and how the education was delivered (classroom, inservice training, video training, or other methods). The NREMT and most authorities that oversee recertification will accept, on an hour-per-hour basis, EMT or paramedic course instruction. Instruction of an entire DOT EMT paramedic course will fulfill the current ACLS credentials.

Pediatric Education

The pediatric advanced life support (PALS) course from the American Heart Association has been one of the most popular pediatric-education platforms for paramedics. PALS provides paramedics and advanced EMTs with the information needed to recognize infants and children at risk for cardiopulmonary arrest and strategies needed to prevent cardiopulmonary arrest. Participants practice psychomotor skills needed to resuscitate and stabilize infants and children in respiratory failure, shock, and cardiopulmonary arrest. The course length is 16 hours and must be taught by PALS certified instructors. The certification is recognized for a two-year period. PALS now includes modules on trauma, counseling family members, and immobilization. AHA has included optional modules in PALS that can be interchanged by the course coordinator.

The American Academy of Pediatrics and AHA support advanced pediatric life support (APLS), which was implemented in 1984. It is a two-day course with the option for a PALS renewal as well as continuing-education credits, which are awarded on an hour-for-hour basis. A one-day course that has a distributive education self-study format is designed to help reduce the time and expense of conducting a multi-day course in the classroom. APLS covers more advanced topics than PALS, with more in-depth pathophysiology.

APLS covers a broad range of pediatric conditions and is case-based, which makes for more interesting, real-life learning. Some of the topics covered are the following: acute episodic medical illnesses, such as meningitis, sepsis, dehydration, pneumonia, diarrhea, renal failure, seizure, coma, hypotension, hypertension, and respiratory illness; problems associated with chronic disease, such as diabetic ketoacidosis, status asthmaticus, status epilepticus, congenital heart disease, cystic fibrosis, and gastrointestinal, metabolic, and neurological disorders; and step-by-step procedures, such as endotracheal intubation, placement of I/O and IV lines, umbilical artery and vein catheter placement, thoracic procedures, conscious sedation, and pain management. APLS can be modularized as smaller units in a 48-hour refresher course.

The National Association of EMTs developed and sponsors pediatric prehospital care (PPC), a problem-focused or assessment-based approach for common pediatric emergencies. This course is mainly hands-on practice. PPC integrates pathophysiology with assessment, using proper diagnostic skills. PPC is approved by CECBEMS for 16 hours for paramedic and advanced providers and recognized for three years. An 8-hour BLS version is available; the instructor course can be completed in 4 hours after initial certification.

The pediatric education for prehospital professionals (PEPP) course was built by many dedicated physicians, nurses, paramedics, EMTs, and EMS educators interested in improving the quality of prehospital care for children. The PEPP course is the culmination of the best and most recent educational efforts in prehospital pediatrics, including the pediatric components of the National Standard Curricula for EMT–Basics, EMT–Intermediates, and EMT–Paramedics. The PEPP course began in 1990 as a distant vision of the California Pediatric Emergency and Critical Care Coalition and the California EMSC Project, funded by the California EMS Authority. In 1992 the coalition's Pediatric Education for Paramedics (PEP) Task Force joined with the American College of Emergency Physician's Pediatric Emergency Medicine Committee and published the *Pediatric Education Guidelines for Paramedics*. The PEPP course emphasizes careful assessment and BLS care and supports prudent use of ALS interventions.

The American Academy of Pediatrics (AAP) and the American Heart Association have agreed that the ALS PEPP course may meet the requirements for a PALS course renewal as a part of the two-day ALS PEPP course. The modular format of the PEPP course allows the presentation of materials in smaller blocks of time, as long as course content and length do not change. A PALS course director must be responsible for supervising instruction and testing, and the course must be registered with a training center. Participants should have a current PALS provider card and must have access to the current PALS provider manual before, during, and after the course. The current PALS pretest should be distributed to all students at least one week prior to the course, and the completed pretest should be brought to the class. At least 84% is required to pass the PALS post-test and complete PALS practical evaluation stations. The AHA also allows a PALS course renewal through other pediatric courses.

The American Heart Association offers a two-day certification as a neonatal resuscitation provider (NRP). This course has been designed to teach an evidence-based approach to the resuscitation of the newborn. The causes, prevention, and management of mild to severe neonatal asphyxia are carefully explained so that health professionals develop best possible understanding and skill in resuscitation. Leadership for development of the neonatal resuscitation provider course came from both the American Academy of Pediatrics and the American Heart Association.

Trauma Continuing Education

The NAEMT in conjunction with the American College of Surgeons Committee on Trauma has led the country on trauma continuing education with a 16-hour course called Prehospital Trauma Life Support. PHTLS focuses on the management and transport of adult and pediatric trauma. Course content starts with baseline trauma scenarios, measuring performance based on national standards for trauma care. Short lectures on trauma on specific human systems are then supported by skills stations that cover rapid extrication techniques, chest decompression, and advanced airway techniques. PHTLS instructors are required to take an eight-hour instructor course, and certification is recognized for four years. An eight-hour PHTLS refresher course is available that focuses on skills and trauma in special populations.

If the majority of the people in need of trauma continuing education are at the EMT–Basic level, basic trauma life support (BTLS) may be more appropriate. Basic BTLS gives EMS First Responders and EMT–Basics complete training in the skills they need for rapid assessment, resuscitation, stabilization, and transportation of trauma patients. This is a 16-hour course designed for providers who are first on scene.

An advanced version of BTLS—Pediatric BTLS—builds on the basic BTLS courses with an emphasis on understanding and responding to trauma in children. This is an eight-hour course that teaches assessment, stabilization, and packaging of pediatric trauma patients. It also highlights techniques for communicating with young patients and their parents. BTLS and PHTLS are CECBEMS approved and are often funded by state offices of highway traffic safety.

The traumatic brain injury (TBI) course from the Brain Trauma Foundation offers continuously updated continuing education activities for all levels of health-care professionals (physicians, nurses, and EMS personnel) who care for TBI patients. The online activities include recorded presentations, live Web-based lectures, and interactive learning modules based on the latest scientific evidence. A four-hour course is approved by the CECBEMS.

MEDICAL CONTINUING EDUCATION

In almost every EMS system, the American Heart Association's advanced cardiac life support (ACLS) course is required. ACLS provides education on cardiac emergencies and respiratory emergencies. Recent updates to the curriculums have expanded the topics to include advanced airway control, death-and-dying issues, and special resuscitation scenarios. ACLS now can be recertified using a CD-ROM program that presents several scenarios of critical patients. The user must complete the cases, and after successfully finishing the distributive-learning portion, the participant must then bring the certification of completion that prints from

the CD to an AHA training center so he can be signed off on the skills stations. For those who have recertified with ACLS and do not feel challenged by the material, the experienced-provider course presents students with more advanced resuscitation scenarios.

The NAEMT created an advanced medical life support (AMLS) course for experienced EMS providers who focus on adult medical emergencies. The course is 16 hours, CECBEMS approved, and recognized for three years. AMLS provides in-depth, scenario-based adult medical emergencies and includes cases in respiratory failure, chest pain, altered mental status, and shock from a problem-solving approach. The curriculum is supported by the National Association of EMS Physicians.

Geriatric Education for Emergency Medical Services (GEMS), the American Geriatrics Society (AGS), and the National Council of State Emergency Medical Services Training Coordinators (NCSEMSTC) partnered to develop a program that will train prehospital professionals (first responders, EMTs, and paramedics) to deliver state-of-the-art care to older adults. The GEMS course content also addresses the geriatric objectives as identified in NHTSA's National Standard Curriculums. Each chapter contains multiple case studies, which prompt providers to think about what they might do if they encountered a similar situation in the field. A summary of each case study appears at the end of the chapter. Tips to help providers appropriately handle calls involving older adults also address myths or stereotypes. The GEMS course is innovative, comprehensive, and highly visual with case-based lectures, live-action video, hands-on skill stations, and small-group scenarios.

ACCREDITED EDUCATION

When an agency conducts EMS education, the EMS manager has three choices on the crediting of that course: keep it in house for credit within the agency, apply for regional or state credit, or apply for credit on a national level through the Continuing Education Coordinating Board for Emergency Medical Services. CECBEMS serves as the leader for continuing education in EMS. It promotes continuing education standards with a review process of education programs. CECBEMS supports **distributive learning** (when a portion of a course occurs outside the physical presence of the teacher) and online course work, supporting continuous-learning opportunities.

- American College of Emergency Physicians (acep.org)
- National Registry of Emergency Medical Technicians (nremt.org)
- National Association of Emergency Medical Services Physicians (naemsp.org)
- National Association of State EMS Officials (nasemso.org)
- National Council of State Emergency Medical Services Training Coordinators (ncsemstc.org)
- National Association of Emergency Medical Technicians (naemt.org)
- National Association of EMS Educators (naemse.org)
- American College of Osteopathic Emergency Physicians (acoep.org)

FIGURE 8.7 ■ Member Organizations in CECBEMS.

CECBEMS was chartered in 1992 by the sponsoring organizations listed in Figure 8.7. It has a fee structure to accredit courses that are one-time events, multiple events, and distributive-learning activities. Several national conferences and most Web-based EMS education programs apply for CECBEMS credit and NREMT awards credit for re-registration for CECBEMS-approved programs.

DESIGN OF EMS TRAINING

A study done on several major public-safety systems on the East Coast indicated that around 70% of EMS providers demonstrate attention-deficit behaviors. Research on adult learning suggests that a learning environment needs to be more creative than ever before, especially given today's up-and-coming multimedia generation. This includes attempting to make training as realistic as possible and using a variety of teaching and learning strategies. The gamer generation coming into the EMS profession is accustomed to programming that requires multitasking and increasing levels of difficulty.

Active Learning

EMS managers and leaders face many challenges in educating the workforce, and they need to design or incorporate active learning in EMS programs. **Active learning** is a process whereby learners are actively engaged in the learning, rather than passively absorbing lectures. Active learning involves reading, writing, discussion, and engagement in solving problems, analysis, synthesis, and evaluation.

For the EMS manager a few specific strategies can be used to ensure that active learning is employed in the educational design of any EMS education program. Demonstration is one form of active learning applicable to EMS. This technique has students working hands on, seeing, hearing, and touching equipment. Guided teaching is performed by asking questions, having groups debate content, and presenting problem-solving exercises. It is an excellent way to get EMS personnel involved in their education.

In Figure 8.8 Dale's Cone represents common retention rates for instructional techniques. Group inquiry is activity rich, and the instructional material

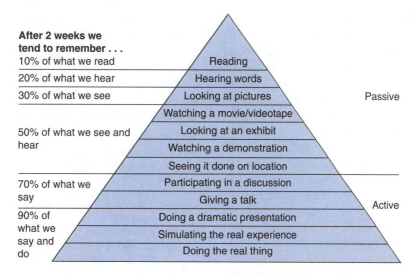

FIGURE 8.8 ■ Dale's Cone on Teaching.

should provide for the audience to ask questions that the instructor can answer or expand upon. One technique involves the instructor or the student reading and discussing content and then having small groups or individuals review handouts and articles and run reviews or audits, then discuss findings and attitudes in a larger group discussion. Much of the learning today involves a student or employee—individually, with partners, or in groups—finding his own information on the Web or from other sources. Such an activity, when based on the interests of the students, is an active learning technique.

Active learning is a multi-directional learning experience in which learning occurs teacher to student, student to teacher, and student to student. Active learning involves activity-based learning experiences: input, process, and output. These activity-based experiences take many shapes, including whole class involvement, teams, small groups, trios, pairs, and individuals. Activity-based experiences take many forms as well, including talking, writing, reading, discussing, debating, acting, role playing, journaling, conferring, interviewing, building, creating, and so on. Active learning involves input from multiple sources through multiple senses (hearing, seeing, feeling). It involves process, interacting with other people and materials, accessing related schemata in the brain, and stimulating multiple areas of the brain to act. Active learning involves output, requiring students to produce a response or a solution or some evidence of the interactive learning experience that is taking place.

Creative Ideas for EMS Training

The use of film is a common technique for providing education. Movie and television producers understand how to change the picture frame and create drama that keeps the viewer's attention. The use of short video clips is allowed under copyright regulations for educational purposes. Films that reflect the humanities can be quite relevant; for example, the movie *The Doctor* demonstrates aspects of compassion and patient-care issues that can be the topic of a facilitated discussion. Several episodes of the television dramas *ER*, *Rescue 911, Emergency*, and others highlight specific procedures that can be reinforced with lectures and skills labs. The animation in the television series *CSI* also may be helpful. (See Figure 8.9.)

The use of an **objective structured clinical examination (OSCE)** program is based on planned clinical encounters in which an EMT or Paramedic interviews,

Television	*CSI*	Animations of mechanism of injury
	ER	Procedures and assessments
Movies	*The Doctor*	Patient care and ethics
	Three Wise Men	GSW animation

FIGURE 8.9 ■ Television and Movie Media to Teach EMS.

examines, informs, or otherwise interacts with a **standardized patient (SP)**. SPs are individuals who are scripted and rehearsed to portray an actual patient who has a specific set of symptoms or clinical findings. SPs may be either able-bodied individuals or actual patients with stable findings. A great resource for SPs is local AARP members, senior-citizen groups, or other retired individuals. Such volunteers allow the EMS provider to assess a live patient and get honest feedback on skills from patients. Bringing in children or teens from within the EMS organization can help to expose EMS providers who have no children to pediatric emergency situations. Getting out of the classroom is another creative technique; consider taking employees to different clinical environments, such as a park with activity stations. Then create a scenario-based learning event at each station and have the participant rotate between each station. Other training ideas are listed in Figure 8.10.

Distributive Learning

In the EMS manager's toolbox there are more and more choices for delivering education on distributive-learning platforms. The term *distributive learning* refers to the portion of the contact hours of a course or an organized teaching/learning event that occurs outside the physical presence of the teacher. It is technology based and can be either synchronous or asynchronous. **Synchronous learning** occurs at a specific time when

---

- ◆ Teaching why versus memory.
- ◆ Scenario-based training.
- ◆ IMS for EMS.
- ◆ Mass-casualty simulations.
- ◆ Using real people as patients (seniors and children).
- ◆ Team training.
- ◆ Simulation software.
- ◆ Using human-patient simulators.
- ◆ Forming reading and journal clubs.

FIGURE 8.10 ■ Creative Trends in EMS Education.

the teacher communicates remotely, using chat rooms or online bulletin boards. **Asynchronous learning** occurs at various times without the teacher being present, often by Web-based programs, self-study, or CD-ROM learning programs. The foundation of distributive learning is the matching of instructional strategies, delivery systems, and materials to learner characteristics and course content.

Some principles about distributive learning are important to EMS managers. They include:

- Computers should be in an environment that is free from distractions.
- When budgets tighten or individuals miss planned training, distributive learning can fill the gap.

- The cost per hour for a training officer to remediate one EMT or paramedic makes it cost effective to employ some distributive-learning content in EMS programs.
- Distributive-learning programs must have verification or be followed up with testing to ensure the content was received. It is important to identify state, county, or city policies on EMS class attendance and distributive education and to develop a department policy on distance- and distributive-education programs (Figure 8.11).

Common programs used by colleges for distance education are Blackboard, WebCT, Centrelearn, and ANGEL. Several other Web-based commercial applications are available that are CECBEMS approved and can be applied to national and state recertification.

Attendance Policy
Definitions:

- "On campus" refers to classes held at the offices of the Education Center.
- "In house" refers to educational sessions held at a fire station.
- "DL" is short for distributive-learning educational modules posted on an Internet Web site.

Overview
In the event of an emergency or illness, a paramedic may be granted an "excused absence" and be allowed to request a rescheduling of his refresher class.

- Definition of "emergency" is a family emergency needing medical attention, injury to self that prohibits paramedic from attending class, and family emergency requiring paramedic's immediate attention.
- Definition of "illness" is personal illness needing the attention of a physician and personal illness of contagious nature (such as whooping cough).

If a paramedic is granted permission to reschedule, he must be rescheduled for the next mutually available refresher class. Paramedics are expected to arrive on time. It is the responsibility of any paramedic who will be late to a refresher class or CE conference to call EMS Education Center to inform the center staff of a late arrival. Any paramedic leaving a refresher class or CE conference will be required to make up the missed time.

ACLS and PALS Recertification
ACLS and PALS recertification will be done in house in the month of December each year. One-half of a fire department's paramedic roster will be done each year. All paramedics will be recertified within a two-year period. Dates for ACLS and PALS recertification will be done on dates that are mutually agreed upon between the EMS Education Center and each fire department. Fire-department administration will schedule its para-

medics to attend agreed-upon class dates, ensuring that class size meets minimum established by EMS Education Center. It is the responsibility of each paramedic to make sure he has current ACLS and PALS certifications as established by the American Heart Association.

Refresher Classes
Refresher classes will be offered each fall and spring semester. Attendance at one refresher class per semester is mandatory. EMS Education Center will publish the class dates six months prior to the dates offered. It is the responsibility of each paramedic to register for one refresher class for each of the fall and spring semesters during a two-year licensing period (a total of four on-campus classes in a two-year licensing period). At the end of each refresher class, the employing EMS agencies will be notified of a paramedic's attendance, the length of the class, and hours each paramedic attended. Those paramedics who have not attended either a regularly scheduled refresher class or have been granted an excused absence will be required to obtain six hours of refresher-class content. Arrangements must be made through the education manager at EMS Education Center. The required hours must address the same topic areas as the missed refresher class offered by the EMS Education Center.

Conference Attendance
EMS Education Center will offer three continuing-education (CE) conferences each academic year, September through June. Attendance at each of the conferences is mandatory. Paramedics who do not attend a CE conference must notify their fire department EMS administrator. Paramedics who do not attend a CD conference must present proof of obtaining an equivalent number of hours of CE in an EMS-related topic. Proof of attendance may be either a certificate of CEU or a conference agenda. Paramedics must sign in upon arrival at the CE conference and must sign out if leaving before the conclusion of the conference. The EMS Education Center will develop an "EMS System Update" presentation and post it on the DL Web site following each CE confer-

---

FIGURE 8.11 ■ CECBEMS Example of EMS Education Program.

ence. This presentation will cover updates to system policies, an orientation to new supplies, updates regarding health-information (patient-care record) issues, as well as other system elements. Each paramedic, whether he attended the CE conference or not, is required to review the EMS System Update within one month of the presentation being posted on the DL Web site. Since the system-update presentations deal with current EMS events, it is critical that this information be reviewed in a timely manner. If a paramedic is not able to review the update presentation within the one-month period, he must inform the department EMS officer of the delay and when he anticipates completing the presentation.

DL Requirements

A list of scheduled modules will be made available to the paramedics at least one month prior to the start of the semester. Five to six modules will be scheduled per semester.

Requirements to Maintain Full-Practice or Limited-Practice Status

In order for a paramedic to maintain full-practice or limited-practice status and be granted permission to practice under the medical control of the EMS Medical Director, a paramedic must attend one on-campus refresher class per semester, attend all CE conferences that fall within a given semester (or have made up any missed CE conference time), and complete all the required DL modules scheduled for a given semester.

Failure to Meet Requirements

Failure to complete the requirements to maintain practice status by the established due dates will result in a paramedic's losing his practice status and medical control. Both practice status and medical control will be suspended until such time as the paramedic completes the missed educational content and informs the education manager that he is up to date.

FIGURE 8.11 ■ Continued

Item Analysis (Questions 1-100)

Prints the number of incorrect responses per test question. A tally of responses will be printed when used with ballots/surveys.

Insert Pass 1 followed by Pass 2. Carefully remove the strip along the perforation prior to inserting pass 3.

Pass 1

Advantage

Test or Ballot Name

Class

Date

To display results as a percentage, mark this bubble.

21												1		4	2	3		2	2	7
Forms Scanned	0% to 4.9%	5% to 9.9%	10% to 14.9%	15% to 19.9%	20% to 24.9%	25% to 29.9%	30% to 34.9%	35% to 39.9%	40% to 44.9%	45% to 49.9%	50% to 54.9%	55% to 59.9%	60% to 64.9%	65% to 69.9%	70% to 74.9%	75% to 79.9%	80% to 84.9%	85% to 89.9%	90% to 94.9%	95% to 99.9%	100%

26.1	0	2	4	2	7	4	4	4	4	5	4	1	4	1	2	5	0	3	2	2	0	3	6	1	2
Class Avg	1	2	3	4	5	6	7	8	9	10	11	12	13	14	15	16	17	18	19	20	21	22	23	24	25
87.0%	0	4	2	4	0	0	0	0	0	0	0	0	0	0	0	0	0	0	0	0	0	0	0	0	0
Class Avg %	26	27	28	29	30	31	32	33	34	35	36	37	38	39	40	41	42	43	44	45	46	47	48	49	50
	0	0	0	0	0	0	0	0	0	0	0	0	0	0	0	0	0	0	0	0	0	0	0	0	0
	51	52	53	54	55	56	57	58	59	60	61	62	63	64	65	66	67	68	69	70	71	72	73	74	75
	0	0	0	0	0	0	0	0	0	0	0	0	0	0	0	0	0	0	0	0	0	0	0	0	0
	76	77	78	79	80	81	82	83	84	85	86	87	88	89	90	91	92	93	94	95	96	97	98	99	100

ADVANTAGE 1200™ #23570-RR 09/02 98825 US Patent No. 6,079,624 APPERSON PRINT MANAGEMENT SERVICES (800) 827-0210

Form 23570 Side 1
(for use with the Advantage 1200)

Pass 3

FIGURE 8.12 ■ Sample Apperson forms that can validate content.
(Reprinted with permission of Apperson Datalink.)

■ MANAGEMENT EFFORTS TOWARD EMS EDUCATION

EMS education systems should operate from information provided by the agency's quality-improvement process after it assesses its own performance. For EMS managers, there are administrative duties that are required to efficiently operate an education department or division. It is important that EMS managers meet with their entire EMS instructor staff quarterly and annually to solicit feedback on the EMS education-delivery system and to keep consistent information flowing among EMS instructors. It also is important to develop a budget. Budgets should include staff time to deliver classes, AV materials, and equipment. Rental space for classrooms, fees to organizations that have copyrighted material, and books also are important considerations for a budget.

EVALUATION OF TRAINING

EMS managers should evaluate their training programs. EMS leaders need to monitor the training division and programs to ensure a quality product and to determine if the agency is getting the correct outcomes. Education programs can be evaluated on different levels.

The first level is an evaluation based on reaction. Reaction evaluations are customer oriented and used to assess faculty, facilities, curriculum, and the process in which CEU or certification is processed. This includes the ease of registration and the convenience of course scheduling. Reaction evaluations should encourage feedback with written comments, measure the participant's reaction to the course, and assess how well content relates to medical practice or operational standards.

There is a variety of automated tools to use to assess educational delivery. An item analysis or a tally shift is a way to tabulate a participant's response. Each participant is given an analysis form, which is collected and run through a scanner. An item-analysis sheet can be used to summarize the responses, allowing the manager to see what questions and content were missed most often. Figure 8.12 is an example of forms from Apperson that can be used for test and item analysis of content items. Once the items are identified, remedial

Exam Item Analysis Report   Exams Graded: 11

Instructor: Mrs. Jenkins   Total Possible: 30   Average: 26.0 - 86.67 %
Exam Name: Biology - Chapter 4   Highest Score: 30 - 100.00 % Median: 26 - 86.67 %
Exam Date: Monday, September 19, 2005   Lowest Score: 22 - 73.33 %   KR20: 0.252

Correct answers are shown in italics

					Blanks	Multiples	Point Biserial	Correct	Percent Incorrect
Q 1	A (0, 0%)	B (10, 91%)	C (0, 0%)	D (0, 0%)	E (1, 9%)			.01	10, 90.9%	9.1%
Q 2	A (0, 0%)	B (0, 0%)	C (10, 91%)	D (1, 9%)	E (0, 0%)			.18	10, 90.9%	9.1%
Q 3	A (10, 91%)	B (1, 9%)	C (0, 0%)	D (0, 0%)	E (0, 0%)			-.29	10, 90.9%	9.1%
Q 4	A (0, 0%)	B (1, 9%)	C (0, 0%)	D (9, 82%)	E (1, 9%)			.30	9, 81.8%	18.2%
Q 5	A (0, 0%)	B (0, 0%)	C (0, 0%)	D (0, 0%)	E (11, 100%)			.00	11, 100.0%	0.0%
Q 6	A (0, 0%)	B (1, 9%)	C (0, 0%)	D (10, 91%)	E (0, 0%)			-.29	10, 90.9%	9.1%
Q 7	A (0, 0%)	B (0, 0%)	C (10, 91%)	D (0, 0%)	E (2, 18%)	☑		.30	9, 81.8%	18.2%
Q 8	A (9, 82%)	B (1, 9%)	C (1, 9%)	D (1, 9%)	E (0, 0%)	☑		-.02	8, 72.7%	27.3%
Q 9	A (0, 0%)	B (10, 91%)	C (1, 9%)	D (0, 0%)	E (0, 0%)			.01	10, 90.9%	9.1%
Q 10	A (0, 0%)	B (4, 36%)	C (1, 9%)	D (0, 0%)	E (6, 55%)			.15	4, 36.4%	63.6%
Q 11	A (0, 0%)	B (10, 91%)	C (1, 9%)	D (0, 0%)	E (0, 0%)			-.29	10, 90.9%	9.1%
Q 12	A (0, 0%)	B (0, 0%)	C (0, 0%)	D (10, 91%)	E (1, 9%)			.01	10, 90.9%	9.1%
Q 13	A (0, 0%)	B (3, 27%)	C (1, 9%)	D (0, 0%)	E (7, 64%)			.37	7, 63.6%	36.4%
Q 14	A (0, 0%)	B (0, 0%)	C (11, 100%)	D (0, 0%)	E (0, 0%)			.00	11, 100.0%	0.0%
Q 15	A (10, 91%)	B (1, 9%)	C (0, 0%)	D (0, 0%)	E (0, 0%)			.01	10, 90.9%	9.1%
Q 16	A (10, 91%)	B (0, 0%)	C (0, 0%)	D (1, 9%)	E (0, 0%)			-.14	10, 90.9%	9.1%
Q 17	A (0, 0%)	B (0, 0%)	C (10, 91%)	D (1, 9%)	E (0, 0%)			.01	10, 90.9%	9.1%
Q 18	A (0, 0%)	B (0, 0%)	C (0, 0%)	D (10, 91%)	E (1, 9%)			-.29	10, 90.9%	9.1%
Q 19	A (0, 0%)	B (0, 0%)	C (10, 91%)	D (0, 0%)	E (1, 9%)			.18	10, 90.9%	9.1%
Q 20	A (1, 9%)	B (0, 0%)	C (0, 0%)	D (10, 91%)	E (0, 0%)			.01	10, 90.9%	9.1%
Q 21	A (1, 9%)	B (0, 0%)	C (0, 0%)	D (0, 0%)	E (10, 91%)			-.14	10, 90.9%	9.1%
Q 22	A (0, 0%)	B (10, 91%)	C (1, 9%)	D (0, 0%)	E (0, 0%)			-.14	10, 90.9%	9.1%
Q 23	A (0, 0%)	B (10, 91%)	C (1, 9%)	D (0, 0%)	E (0, 0%)			.53	10, 90.9%	9.1%

FIGURE 8.12 ■ Continued

education, changes to training, and feedback to the instructors can be made to address the content, delivery, or evaluation of the educational program or class.

Using a computerized scoring and item analysis form can give an EMS manager a snapshot of the performance, and computer software then can be used to generate a variety of reports. Apperson's Datalink is an example of a software program that can generate reports on students, classes, and items on a test. Figure 8.13 shows computer screens generating reports based on item analysis and using that information to further analyze student and class proficiency.

Another way for an EMS manager to evaluate the organization's EMS training and operational training is to assess learning. An EMS manager evaluating learning should ask for what knowledge was learned, what skills were developed or improved, and what attitudes were changed. When collecting this information, use a control group if practical. Evaluate knowledge, skills, and attitudes both before and after with a pretest and post-test.

Behavioral assessment is employee and customer oriented and used to evaluate instructors, the facility, the curriculum, and the registration process. The best way to survey behavior changes after a course is complete is to conduct a 360-degree survey. When an EMS manager assesses behavior changes, he should give adequate time for the change. An assessment should be taken from the training participant and his company officers, field supervisors, and subordinates. If this is applied to paramedic training, the paramedic's EMT partner, fire captain or lieutenant, and other paramedics would be sampled to identify whether or not the desired change was achieved.

Results or an outcome analysis is the last type of assessment of education programs. This assessment is done by looking at operation statistics and the organization's data; for example, decreased employee injuries, increased employee retention, improved patient outcomes, and increased operational savings. When measuring for results, measurements must be taken both before and after the training. This works best with quality-improvement data obtained and used to identify the problem, then reassessed after the training event. Measurements should be made in intervals and repeated several times.

Be prepared to be challenged on the cost versus benefit of offering training. Often staff time and money can be saved by outsourcing training to a college or an agency that employs outside consultants. To evaluate the cost, the EMS manager must determine how much time in hours it would take to develop one training course. The development cost will have the number of times a course must be taught added to the hours that figure into the cost for the training officer's time. The total number of hours to deliver specific educational content then must be multiplied times the training officer's hours. Any additional costs for supplies are

Class Proficiency Report

Instructor:	Mrs. Jenkins	Total Possible:	30	Class Average:	26.0 - 86.67 %
Exam Name:	Biology	Highest Score:	30 - 100.00 %	Class Median:	26 - 86.67 %
Exam Date:	Monday, September 19, 2005	Lowest Score:	22 - 73.33 %	Class Size:	11
				Proficiency:	>=75%

Standard	Description	Students Proficient	%	
Biology				
3.004.1	Genetics	10	91%	
3.004.3	Biochemical Reactions	10	91%	
3.002.4	Scientific Observation	8	73%	
3.004.4	Cell Identification	8	73%	
	Standards Rating		82%	
	Overall Rating		82%	

FIGURE 8.13 ■ Computerized Item Analysis, Student Proficiency, and Class Proficiency Reports. *(Reprinted with permission of Apperson Datalink.)*

Student Proficiency Report ID# 0000000

Instructor:	Mrs. Jenkins	Total Possible:	30	Student Score:	26 - 87%
Exam Name:	Biology	Highest Score:	30 - 100.00 %	Class Average:	26.0 - 86.67 %
Exam Date:	Monday, September 19, 2005	Lowest Score:	22 - 73.33 %	Proficiency:	>=75%

		Responses			
Standard	**Description**	**Correct**	**Total**	**%**	
Biology					
3.004.1	Genetics	10	10	100%	
3.004.3	Biochemical Reactions	4	5	80%	
3.002.4	Scientific Observation	5	6	83%	
3.004.4	Cell Identification	7	9	78%	
	Proficiency Level	26	30	87%	
	Overall Proficiency	26	30	87%	

Missed Questions

#	Correct	Your Answer	Prescriptive Information
8	A	B	Textbook - Chapter 4.1
10	B	E	
16	A	D	Textbook - Chapter 4.4, 4.5
25	A	E	

Pass 2

Tally Report (Questions 1-50)

Prints a tally of marks made for each response selection.

Insert Pass 1 first and carefully remove the strip along the perforation prior to inserting the next pass.

Advantage

To display results as a percentage, mark this bubble.

Test or Survey Name _____ Date _____

21	0	21	0	0	0	1	1	19	0	0	17	3	0	1	0	0	1	1	19	0	0	1	0	6	14
Forms Scanned	1A	1B	1C	1D	1E	2A	2B	2C	2D	2E	3A	3B	3C	3D	3E	4A	4B	4C	4D	4E	5A	5B	5C	5D	5E
0	0	1	17	3	0	1	17	3	0	17	2	2	0	0	0	18	4	0	0	1	16	3	0	1	
6A	6B	6C	6D	6E	7A	7B	7C	7D	7E	8A	8B	8C	9A	9B	9C	9D	9E	10A	10B	10C	10D	10E		
0	17	3	1	0	0	0	0	20	1	0	0	0	4	17	0	1	20	0	0	20	1	1	0	0
11A	11B	11C	11D	11E	12A	12B	12C	12D	12E	13A	13B	13C	13D	13E	14A	14B	14C	14D	14E	15A	15B	15C	15D	15E
16	5	0	0	0	0	0	21	0	0	3	18	0	0	0	19	2	0	0	0	19	2	0	0	0
16A	16B	16C	16D	16E	17A	17B	17C	17D	17E	18A	18B	18C	18D	18E	19A	19B	19C	19D	19E	20A	20B	20C	20D	20E
0	0	0	0	21	0	18	3	0	0	15	6	0	0	0	1	20	0	0	19	2	0	0	0	
21A	21B	21C	21D	21E	22A	22B	22C	22D	22E	23A	23B	23C	23D	23E	24A	24B	24C	24D	24E	25A	25B	25C	25D	25E
0	0	0	21	0	0	0	0	4	17	0	0	0	19	2	1	17	3	0	0	0	0	21	0	0
26A	26B	26C	26D	26E	27A	27B	27C	27D	27E	28A	28B	28C	28D	28E	29A	29B	29C	29D	29E	30A	30B	30C	30D	30E
0	0	0	0	0	0	0	0	0	0	0	0	0	0	0	0	0	0	0	0	0	0	0	0	0
31A	31B	31C	31D	31E	32A	32B	32C	32D	32E	33A	33B	33C	33D	33E	34A	34B	34C	34D	34E	35A	35B	35C	35D	35E
0	0	0	0	0	0	0	0	0	0	0	0	0	0	0	0	0	0	0	0	0	0	0	0	0
36A	36B	36C	36D	36E	37A	37B	37C	37D	37E	38A	38B	38C	38D	38E	39A	39B	39C	39D	39E	40A	40B	40C	40D	40E
0	0	0	0	0	0	0	0	0	0	0	0	0	0	0	0	0	0	0	0	0	0	0	0	0
41A	41B	41C	41D	41E	42A	42B	42C	42D	42E	43A	43B	43C	43D	43E	44A	44B	44C	44D	44E	45A	45B	45C	45D	45E
0	0	0	0	0	0	0	0	0	0	0	0	0	0	0	0	0	0	0	0	0	0	0	0	0
46A	46B	46C	46D	46E	47A	47B	47C	47D	47E	48A	48B	48C	48D	48E	49A	49B	49C	49D	49E	50A	50B	50C	50D	50E

Form 23580 Side 1
(for use with the Advantage 1200)

ADVANTAGE 1200™ X23580 RR 0902 9ESS US Patent No 6,076,024 APPERSON PRINT MANAGEMENT SERVICES (800) 827-9219   1406

Pass 5

FIGURE 8.13 Continued

then tabulated, and a total cost for that segment of training can be calculated.

A vendor of distance-education learning platform usually can deliver the same content at less than it would cost the agency to create and develop it in house. For example, hearing-conservation training for an EMS agency with 30 people on three shifts may be required to be given twice a day, three times, for an hour each time. Add in four hours of time creating the training, then add in documentation, which would be another 10 hours. When the training officer makes $20 per hour for 10 hours at a total cost of $200 for hearing training, and the same course can be conducted on an OSHA-approved Web-based site for $2.50 per person, it is more effective to do Web-based instruction on this topic.

MAINTAINING EMS EDUCATION RECORDS

While most EMS recertification cycles occur every two years, recertification paperwork and educational records must be kept for the time frame required to submit the records to the EMS authority. The statute of limitations for medical malpractice in some states can be up to seven years. Therefore, EMS managers should keep records on employee training and certification for at least seven years, because those records are frequently summoned during the discovery phase of legal action. Electronic storage of EMS education records can be easily accomplished with a simple spreadsheet program. Several companies offer Internet-based programs with databases for recording and providing EMS CEUs and refresher training.

Several commercial programs are available to help maintain EMS education records. One of the most popular and comprehensive tools with excellent research capability is FISDAP, which is a skills-tracking system that allows EMS workers or students to enter data about their experiences, such as what they observed and skills they have performed. Figure 8.14 shows the entry screen for patient contacts and skills tracking, showing

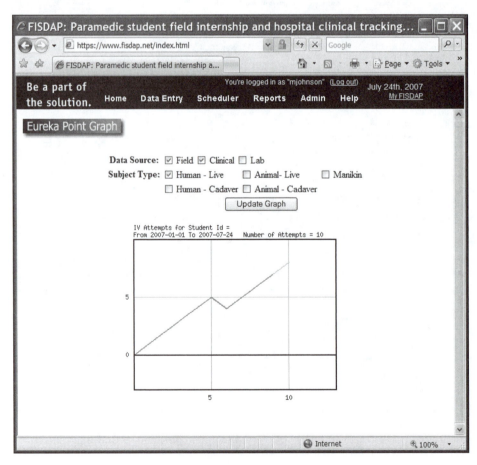

FIGURE 8.14 ■ Sample FISDAP Reports on Skills and Patient Contacts. *(Reprinted with permission of FISDAP.)*

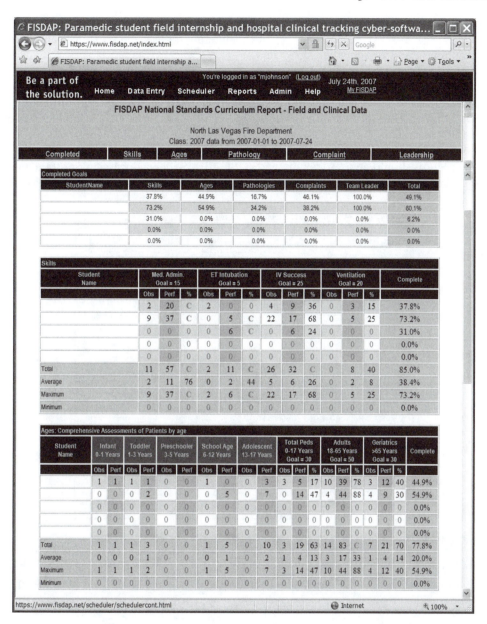

FIGURE 8.14 ■ Continued

when the student has reached competency with that particular skill or requirement under national recommendations. This includes the collection of working diagnoses, demographics, and scope or nature of the skills performed. Instructors have controlled access, and **preceptors**, classroom instructors, and program coordinators can have different levels of access. This creates an opportunity for critical-thinking exercises and facilitates accreditation and recordkeeping of EMS training and internship programs. Since several agencies participate in FISDAP, a concept called "eureka points" has

shown how often a skill or event must occur to achieve consistent performance, as shown in the graph in Figure 8.14 depicting EMS skills for IV starts.

Computer software created for handheld personal data assistants for access to personal data, continuing-education entries, and state recertification is available. Software solutions for EMS CEUs and education tracking should include the instructor, course location, hours, and type of EMS education. The software should be able to run reports and alert the user to upcoming recertification dates or employees who are low on CEU hours. One such system is illustrated in Figure 8.15.

Many of the online distance-education programs that are offering CEU courses on the Web also have the option to be used to tally or record EMS CEUs. JEMSprepare is a learning solution that provides over 100 courses and allows an EMS provider to create a personal journal to track CEUs from conferences, magazines, and online sources.

EMS training records should be user friendly and supported with good information-technology personnel.

Any software used for this purpose must be able to generate reports and create visuals from data tracking.

ATTENDING COMMITTEE MEETINGS

Often EMS regulatory authorities dictate the hours and type of EMS CEUs and refresher training required. The discussion of approved programs usually is undertaken at committee meetings. It is important that EMS managers assign or attain a position on regulatory committees that oversee EMS education. The

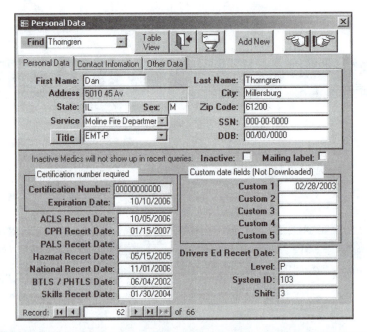

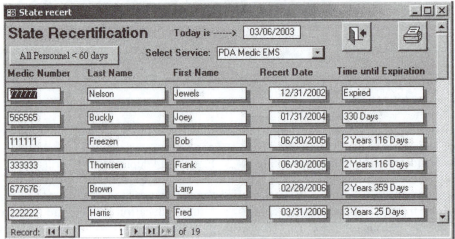

FIGURE 8.15 ■ Sample Reports Created on EMS CEU Software. *(Reprinted with permission of PDA Medic Software.)*

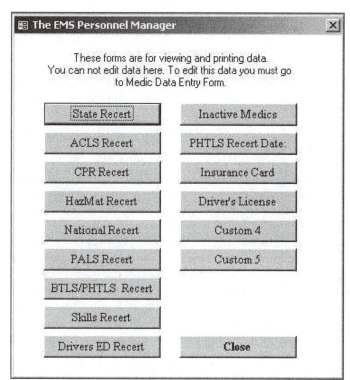

FIGURE 8.15 ■ Continued

input of provider agencies is important to ensure that mandated training is funded.

MAINTAINING CLINICAL AND TEACHING EXPERIENCE

It is important for EMS educators to maintain their clinical skills and to stay in touch with the current events in the field. One concept used in the college environment is taking a sabbatical, a process in which the educator disengages from actively teaching in order to enhance or refresh skills. Conceptual EMS education in an organization could be shut down for a month to allow the educator to be deployed onto an ambulance to refresh skills and assess the prehospital environment.

EMS managers also need to specify a line item in the operational budget to send EMS educators to conferences and to participate in national associations.

PSYCHOMOTOR EDUCATION

Skills are measured with a performance-based exam known as the objective structured clinical examination (OSCE), oral board, or skills challenge. Skills decline approximately 180 days after a course if they are not used or reinforced with another methodology. When teaching skills it is important to emphasize psychomotor skills. There has to be imitation and manipulation instruction.

A skill should be done with precision, meaning EMS personnel should perform the skill exactly as described and in the right time frame, independent of instruction. The next level of EMS skills instruction is called "articulation," when EMS providers can combine skills with harmony. The goal of skills training is naturalization, in which EMS personnel can complete more than one skill with ease or limited effort. To get people to the naturalization level, ask questions during the performance, include two or more skills during the process, and have them read something while performing the task. Skills sheets from the NREMT Web site are available free of charge. Textbook publishers routinely add checklists and step-by-step procedures to their textbooks as well.

PRECEPTOR PROGRAMS

The preceptor curriculum is based on the paramedic and EMT job analysis. The preceptor should be chosen as a person who demonstrates a high level of knowledge, clinical proficiency, and professionalism. A preceptor serves as a clinical instructor to new employees and students, assisting with the transition into the field environment. Most EMS agencies use some system to help promote the success of the new employee in the workplace; most have adopted the preceptor concept. The goal is to increase job satisfaction for the intern, preceptor, and all staff members and to promote retention.

An **intern**, who may also be called the "preceptee," is new to the EMS system and participates in an orientation and supervised-practice program. The preceptor has the responsibility of teaching the intern during a planned orientation program that helps introduce and integrate the intern into the EMS system or agency. Preceptors should teach the intern the customs, culture, and both formal and informal rules of the agency.

Four preceptor roles help make a new employee or student successful: role model, educator, facilitator, and evaluator. **Role modeling** is a process in which an individual identifies with and assumes the values and behaviors of another person, which ultimately results in behavior modification that is usually permanent.

EMS field-training programs around the country vary greatly as does preceptor training. A preceptor training program should teach preceptors what the job tasks are for an EMT or paramedic. Each of those job tasks needs to be broken down in a score sheet that identifies behaviors and actions that are unacceptable or need improvement and competency. Two basic procedures are required for preceptors to learn good teaching techniques. Knowing when to prompt a student to take action is a key preceptor skill: the preceptor allows students to perform until they come to a point when they no longer know what action is necessary, and then the preceptor provides prompts that allow them to continue their performance.

To build on the prompting process, the preceptor also must understand how to coach a new student or employee. A common coaching method is called the "seven-windows counseling process"; this puts the responsibility for performance back onto the student or employee after he and his preceptor arrive at a mutual understanding of the need for improvement (Figure 8.16).

TRAINING TO THE AFFECTIVE DOMAIN

EMS education programs often need to develop or deliver programs that reinforce components in the

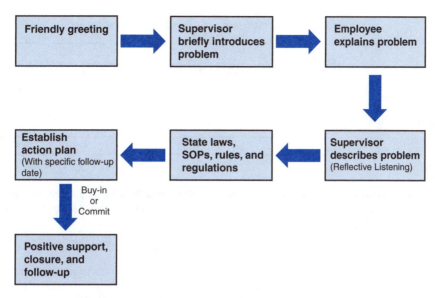

FIGURE 8.16 ■ Seven-Windows Counseling Process.

affective domain, which addresses interests, attitudes, opinions, appreciations, values, and emotions. Because entry-level employees many times do not have the same values as those in the managerial and leadership ranks, aligning these differences in beliefs in affective-domain categories is a real challenge for EMS managers.

The 1998 EMT–Paramedic Curriculum addresses the need for an affective-domain evaluation of paramedics. A sample of the affective-domain tool from Appendix V of the National Standard Curriculum for the EMT–Paramedic is shown in Figure 8.17. A hypothetical response to a counseling session dealing with a disruptive student is also included. The generational differences in the future will continue to create problems with attitudes and perspectives.

Ideally, one-third of an evaluation of a new paramedic should be an affective evaluation. Communications skills, customer-service skills, and patient advocacy are all reinforced by employing an affective domain. Communicating among crews, with other generations, and with diverse cultures are skills that need to be developed among incoming high school graduates.

Orientation

One of the more common places to start an education program with major components designed around the affective domain is in an employee-orientation program. Orientation training for new employees should be the joint responsibility of the organization's EMS and human-resources training staff and frontline EMS managers. There must be a clear understanding of the specific responsibilities of each, so that nothing is left to chance.

Generally, the EMS training staff should be responsible for providing information involving matters that are organization-wide in nature and relevant to all new employees. The EMS manager should concentrate on those items unique to the new member's workplace and convey the affective-domain component to the shift. For example, the EMS manager should tell the new employee the philosophy of work and how the management process works. The expectations of the manager for the employee are an important part of the process. Although the specific content of training will vary from organization to organization, the items shown in Table 8.1 are typically covered during new EMS member orientation training.

Orientation programs should be planned in advance. Several aspects regarding the orientation program should be considered in addition to the content:

- Hold the orientation as close as possible to the member's start date.
- Develop a schedule of activities and arrange for necessary individuals to be available to meet with the new member.

NATIONAL GUIDELINES FOR EDUCATING EMS INSTRUCTORS
AUGUST 2002

PROFESSIONAL BEHAVIOR EVALUATION

Student's Name:

Date of Evaluation:

---

1. INTEGRITY   Competent [ ]   Not yet competent [ ]

Examples of professional behavior include, but are not limited to: Consistent honesty; being able to be trusted with the property of others; can be trusted with confidential information; complete and accurate documentation of patient care and learning activities.

2. EMPATHY   Competent [ ]   Not yet competent [ ]

Examples of professional behavior include, but are not limited to: Showing compassion for others; responding appropriately to the emotional response of patients and family members; demonstrating respect for others; demonstrating a calm, compassionate, and helpful demeanor toward those in need; being supportive and reassuring to others.

3. SELF - MOTIVATION   Competent [ ]   Not yet competent [ ]

Examples of professional behavior include, but are not limited to: Taking initiative to complete assignments; taking initiative to improve and/or correct behavior; taking on and following through on tasks without constant supervision; showing enthusiasm for learning and improvement; consistently striving for excellence in all aspects of patient care and professional activities; accepting constructive feedback in a positive manner; taking advantage of learning opportunities

4. APPEARANCE AND PERSONAL HYGIENE   Competent [ ]   Not yet competent [ ]

Examples of professional behavior include, but are not limited to: Clothing and uniform is appropriate, neat, clean and well maintained; good personal hygiene and grooming.

5. SELF - CONFIDENCE   Competent [ ]   Not yet competent [ ]

Examples of professional behavior include, but are not limited to: Demonstrating the ability to trust personal judgment; demonstrating an awareness of strengths and limitations; exercises good personal judgment.

6. COMMUNICATIONS   Competent [ ]   Not yet competent [ ]

Examples of professional behavior include, but are not limited to: Speaking clearly; writing legibly; listening actively; adjusting communication strategies to various situations

7. TIME MANAGEMENT   Competent [ ]   Not yet competent [ ]

Examples of professional behavior include, but are not limited to: Consistent punctuality; completing tasks and assignments on time.

8. TEAMWORK AND DIPLOMACY   Competent [ ]   Not yet competent [ ]

Examples of professional behavior include, but are not limited to: Placing the success of the team above self interest; not undermining the team; helping and supporting other team members; showing respect for all team members; remaining flexible and open to change; communicating with others to resolve problems.

FIGURE 8.17 ■ 1998 Paramedic Curricula: Affective Domain Tool.

9. RESPECT	Competent [ ]	Not yet competent [ ]

Examples of professional behavior include, but are not limited to: Being polite to others; not using derogatory or demeaning terms; behaving in a manner that brings credit to the profession.

10. PATIENT ADVOCACY	Competent [ ]	Not yet competent [ ]

Examples of professional behavior include, but are not limited to: Not allowing personal bias or feelings to interfere with patient care; placing the needs of patients above self interest; protecting and respecting patient confidentiality and dignity.

11. CAREFUL DELIVERY OF SERVICE	Competent [ ]	Not yet competent [ ]

Examples of professional behavior include, but are not limited to: Mastering and refreshing skills; performing complete equipment checks; demonstrating careful and safe ambulance operations; following policies, procedures, and protocols; following orders.

Use the space below to explain <u>any</u> "not yet competent" ratings. When possible, use specific behaviors, and corrective actions.

_____

_____

_____

_____

_____

_____

_____

_____

_____

Continued

PROFESSIONAL BEHAVIOR COUNSELING RECORD

Student's Name: _____

Date of Counseling: _____

Date of Incident: _____

_	Reason for Counseling	Explanation (use back of form if more space is needed):
	Integrity	
	Empathy	
	Self-motivation	
	Appearance/personal hygiene	
	Self-confidence	
	Communications	
	Time management	
	Teamwork and diplomacy	
	Respect	
	Patient advocacy	
	Careful delivery of service	

Follow-up (include specific expectations, clearly defined positive behavior, actions that will be taken if behavior continues, dates of future counseling sessions, etc.):

_____

_____

_____

_____

_____

_____

FIGURE 8.17 ■ Continued

TABLE 8.1 ■ Types of New EMS Member Orientation

General Organizational Orientation

1. **Overview of the organization.** A brief history, what the organization does, where it does it, how it does it, the organizational structure. Key political and department heads. Regulatory official and other interfacing agencies.

2. **Policies and procedures.** Tardiness and absenteeism, vacations, holidays, grievances, grooming, trial period and evaluations, warnings, probation, suspension, promotion, transfers, training, and so on.

3. **Compensation.** Pay scale, overtime, holiday pay, shift differentials, when and how paid, time-clock procedures, etc.

4. **Benefits.** Insurance, retirement, tax-sheltered annuities, credit union, employee discounts, suggestion system, recreational activities, etc.

5. **Safety information.** Relevant policies and procedures, including exposure procedures and incident reporting procedures.

6. **Union.** Name, affiliation, officials, joining procedure, contract, and so on.

Specific Department/Unit Orientation

1. **Department/unit functions.** Explanation of the objectives, activities, and structure of the department/unit, along with a description of how the department's or unit's activities relate to those of other departments or units and the overall organization.

2. **Job duties.** A detailed explanation of the duties of the new EMS member's job, how the job relates to the activities of the department or unit, and expected standards of performance.

3. **Policies and procedures.** Those that are unique to the department or unit, such as work schedules, station duties, emergency vehicle operation, radio communications, patient-care protocols, time requirements (expected out-of-chute times, hospital-turnaround times, etc.), documentation procedures, on-scene command, infection control, proper lifting and moving techniques, etc.

4. **Department tour.** A complete familiarization with the department facilities, including layout, equipment, lockers, supply room, bulletin boards, and so on.

5. **Introduction to department members.**

6. **Narcotics refill procedures and policy.**

- Allow adequate time for the new member to meet with the EMS manager and others as required.
- Ask the new member to provide an evaluation on both the orientation program and the organization itself.

Core-Values Training

An example of an affective-domain program is the U.S. Marine Corps' core-values training, which is incorporated into the basic training of recruits. During the 12 weeks of boot camp, there are four to six core-value classes every week, each lasting at least one hour. Discussions of problems that are drawn from the experiences of the senior drill instructors, known as "senior circles," include topics that might deal with suicide or alcoholism. It is important to note that the discussion is "hats off," meaning that no one is penalized for any comments made or confessions offered.

The official core values of the Marine Corps are honor, courage, and commitment, which are intended to encompass integrity, discipline, teamwork, duty, esprit de corps, and other values. It is impressive that marines are taught that core values should function both externally and internally.

The Marine Corps is making the kind of effort necessary to change the behavior of its members for the better and has been effective in doing so. It has tracked the behavior of recruits, and about one-third will lapse back to their civilian values within a year; however, two-thirds of the recruits take on the Marine Corps values in civilian life. This program develops marines who take pride in being held to high moral standards.

I CARE Program

The I CARE program evolved following a lecture on ethics and values presented by Santa Rosa Junior College EMT instructor Chris Le Baudour. The lecture was presented on the second night of an 18-week, semester-long program. Halfway through his presentation,

Mr. Le Baudour had each of the students in the class complete a values activity to identify their personal core values. At the end of the class, the participants were given a homework assignment and challenged to identify five values that could be used to guide students on their journey to becoming EMTs throughout their career.

At the next class students brainstormed to identify the five values that would help guide them through the semester. Several words were offered up as suggestions and written on the board at the front of the room. Students became engaged in an active discussion on the relative merits of some of the values being offered. The brainstorming session lasted only 10 minutes and allowed the words offered by the class to remain on the board for the remainder of the class period. This process of a 10-minute brainstorming session continued for the next three or four classes until the class had identified and unanimously agreed upon the following five guiding values:

I —Integrity.
C—Compassion.
A—Accountability.
R—Respect.
E—Empathy.

After establishing the I CARE core values, several students suggested that the class come up with a way of recognizing fellow classmates for embracing the identified values. The idea was liked by all, and the discussion was tabled until the next class. At the next class meeting, one of the students brought in a catalog in which she had found a lapel pin displaying the words "I Care." A teaching assistant who saw the pin immediately offered to purchase the pins for the class in an effort to "get this great idea going." By the next class they had received the pins, and the I CARE program had officially begun. It was decided that they would take the first five minutes of each class to allow students to recognize one another for embracing the I CARE values.

LEADERSHIP TRAINING PROGRAMS

There are many prepared leadership training programs for management. Many programs are college-based. However, the bulk of EMS management training is conducted by consulting teams, industry trade groups, associations, and labor organizations.

Not-for-Profit Leadership Training

An educational package that emerged in early 2002 as a joint partnership between Phi Theta Kappa, a non-profit international leadership honor society, and the Community College System of California was originally called the California Public Safety Leadership and Ethics Training Program. The updated version is the International Public Safety Leadership and Ethics Institute (IPSLEI). IPSLEI and Phi Theta Kappa teach leadership using the humanities and has adapted its content to apply to public-safety issues. This leadership package targets presupervisory personnel in police, fire, EMS, and corrections. Figure 8.18 lists the titles of the four courses that make up the IPSLEI and Phi Theta Kappa program. These courses are designed to be delivered in partnership with a community college. Each course requires approximately 40 hours, and the participants are subjected to personal assessments and facilitated discussion and are required to complete a journal for credit.

Federal Leadership Training

The National Fire Academy (NFA) is open to all EMS managers, including those in emergency services other than fire. NFA offers a series of two-week courses designed to enhance EMS management and leadership skills. The first course is called Management of EMS (MEMS). This course focuses on management practices as they relate to EMS in the fire service. This interactive and fast-moving course enables participants to deal more effectively with day-to-day management issues that supervisory-level managers are likely to encounter. Personnel management, resource management, and quality-improvement techniques are some of the major components of this course.

Another course, Advanced Leadership Issues in EMS (ALEMS), is designed for upper-management people who have organizational responsibility for EMS operations in their agency or jurisdiction. Situational, scenario-based instruction is the foundation of this course, with an emphasis on problem-solving and decision-making techniques. Leadership techniques as

---

- Developing Your Personal Leadership Style
- Leading Others
- Organizational Leadership
- Ethics and the Organization

FIGURE 8.18 ■ The Four Courses of the International Public Safety Leadership Program.

they relate to establishing and directing EMS work teams are also an important part of this course.

The course EMS Management of Community Health Risk is a two-week course that targets EMS providers, supervisors, and program managers with the responsibility for development and implementation of community health and safety programs. During the class, students develop a community-specific Health Outcome Management Plan (HOMP), with the goal of a 25% reduction in preventable illness and injuries within a community. The course represents a major EMS prevention initiative at the National Fire Academy.

EMS Special Operations is a two-week course designed to enable EMS system managers to prepare their organizations to respond to special operations by identifying potential hazards, determining potential resource needs, determining how those resources may be acquired, and developing a plan that enables the effective control of these events. Events such as mass-casualty incidents, storms, earthquakes, technological emergencies, mass gatherings, dignitary visits, and terrorism can place an unusual demand upon the ability to provide continued EMS response to anticipated daily-call volume. It is only through effective planning and preparation for these unique events that EMS can continue to respond effectively. This program discusses many of the special operations and the burdens they place upon communities, EMS systems, and the responders within EMS.

The National Fire Academy's Executive Fire Officer (EFO) program offers opportunities for EMS leaders to build skills and networking opportunities. The EFO program requires that a student be a chief officer or in a critical or key leadership position. A bachelor's degree is required, and the student must commit to attending one class at the National Fire Academy every year over the next four years. Figure 8.19 lists the courses that are required at the National Fire Academy for Executive Fire Officer certification. While the EFO program remains focused on the executive fire officer, it is flexible enough for a participant to pick up skills and abilities necessary to manage an EMS organization. Many of the principles and skills covered are relevant for managers and leaders working in all hazard environments.

The opportunity to explore EMS topics comes with the applied-research paper. After every course students must complete an applied-research paper that is evaluated by some of the top researchers in the EMS management field. A variety of EMS topics can be explored and evaluated by the student and applied to the student's

---

- ✦ Executive Development
- ✦ Leading Community Risk Reduction
- ✦ Executive Analysis of Fire Operations in Emergency Management
- ✦ Executive Leadership

FIGURE 8.19 ■ National Fire Academy Courses Required for Executive Fire Officer Certification.

EMS agency; chiefs of departments or equivalent; chief officers who head major bureaus or divisions within a fire department, such as suppression, prevention, training, emergency medical services, and so on; or other individuals who are serving in key leadership positions.

Applicants for the EFO program must have attained a bachelor's degree or greater from a regionally accredited institution of higher learning. Requirements for entrance into the EFO program are defined in Figure 8.20.

A movement is underway by the National EMS Management Association to create a chief EMS officer certification similar to the chief fire officer designation offered by the Commission on Fire Accreditation. The chief EMS officer designation will require demonstration of certain competencies and involve a review of the applicant's credentials.

Other EMS Leadership Training Programs

For-profit and several trade and private organizations are beginning to offer leadership training. The National EMS Academy from the Texas Engineering and Extension Services (TEEX) and the California Fire Chiefs Association EMS Leadership Academy is one such organization that offers a course in EMS leadership. The EMS Academy course is an accelerated program based on immersion teaching and is offered

---

- ✦ Bachelor's degree. Advanced academic degrees will further strengthen the candidacy of the applicant.
- ✦ Unique perspectives that broaden the diversity of EFOP.
- ✦ Strength of the department chief's or sponsor's recommendation, commitment to supporting the applicant's participation, and description of the applicant's potential impact on the organization.
- ✦ Personal accomplishments and significant contributions to fire and emergency services and/or the community.
- ✦ Potential for future impact on the fire service.

FIGURE 8.20 ■ Requirements for Entrance into the Executive Fire Officer Program (EFOP).

several times a year. It is designed to transition employees to field supervision and upper-level EMS leadership positions.

Several private consultants offer EMS management and leadership training. Fitch and Associates offers a certification course entitled Ambulance Service Manager in conjunction with the American Ambulance Association that is a blended distance-education and on-site educational experience. Several trade organizations have offered EMS-leadership opportunities. The California Fire Chiefs Association EMS Leadership Academy has been put together for front-line EMS supervisors and support staff to fire-based EMS operations. The International Association of Firefighters (IAFF) conducts an EMS-management seminar periodically. One of the most attended EMS leadership and management conferences is the International Association of Fire Chiefs EMS Section Conference, Fire-Rescue Med. Held yearly, it brings together EMS managers and leaders from around the country. EMS leadership ideally should look to form or contribute to regional leadership opportunities.

The NAEMT EMS Expo and JEMS EMS Today offer a management track. Three conferences—Sand Key EMS conference, American Ambulance Association Trade Show, and Pinnacle EMS Conference—have an EMS-management focus.

- ◆ Drexel University, Philadelphia
- ◆ George Washington University, Washington, D.C.
- ◆ Johns Hopkins University
- ◆ Northwest Alabama Community College
- ◆ Shepard University, West Virginia
- ◆ Texas A & M
- ◆ University of Maryland, Baltimore County

FIGURE 8.21 ■ Schools Offering Academic EMS Management Degrees.

Academic EMS Management Programs

There are associate and bachelor degree programs that have some exposure to EMS management offered around the country. These programs target upper-level managerial and leadership positions. Several key institutions offer specific degrees in the EMS profession. The two universities that have maintained and been the flagship institutions offering EMS management degrees are George Washington University and the University of Maryland at Baltimore. UMB currently offers the only master's degree with an emphasis in EMS. Both institutions conduct outreach training and partner with many EMS organizations.

See Figure 8.21 for a list of the recognized education programs offering EMS-management degrees. The future will bring more academic offerings for EMS, as experienced providers become credentialed to teach in higher education.

CHAPTER REVIEW

Summary

An EMS leader needs to set a vision for EMS education in the organization, and it is paramount that those efforts are integrated with quality-improvement activities. EMS leaders must have a philosophy of lifelong learning and should see education as the most important aspect of the operation. EMS leaders must inspire and encourage education and sound training initiatives. Education demand on EMS systems will continue to increase as EMS services are challenged with bringing professional standards to the work. The EMS manager's responsibility is to ensure the EMS education system within the organization functions as it is envisioned by EMS leadership. EMS managers also must ensure that the system meets the regulatory and statutory requirements for EMS education and recertification. It is the EMS manager who will observe and respond to changes, needs, and feedback in the EMS education system.

Research and Review Questions

1. Taking an example from the Marine Corps core-values training program, create an EMS core-values training program with content for 12 sessions.

2. Design a two-year refresher-education schedule for your EMS agency that meets both the state and national standards for paramedic and EMT continuing education.

3. Explain and diagram the differences among the national standard curriculums. Draw a timeline, and place each curriculum on the timeline.
4. Develop an orientation program for a new employee at your agency.
5. Analyze your agency's quality-improvement data. Develop a six-month continuing-education schedule that will make a positive change in three of your agency's quality-improvement indicators related to patient care.

6. Explain the five components of the *EMS Education Agenda for the Future*.
7. Adapt an active-learning exercise for an EMS application.
8. Evaluate the 1998 paramedic affective-domain content, and discuss the pros and cons of this type of measurement system.

References

Alexander, M. E. (2005). *Foundation for practice of EMS education*. Upper Saddle River, NJ: Prentice Hall, Inc.

Cohen, H. C. (1997). *Is there a prevalence for attention deficit/hyperactivity disorder among EMS and fire service personnel?* (Executive Fire Officer Program Applied Research Project). Emmitsburg, MD: National Fire Academy.

Johnson, D. W., Johnson, R. T., & Smith, K. A. (1991). *Active learning: Cooperation in the college classroom*. Edina, MN: Interaction Book Co.

Kirkpatrick, D. (1998). *Evaluating training programs: The four levels* (2nd ed.). San Francisco: Berrett-Koehler Publishers.

Lightfoot, S. (1993). Getting your training programs on the right track. *Emergency, 25*(1), 39–41.

Lindsey, J. (2005). *Fire service instructor* (1st ed.). Upper Saddle River, NJ: Brady/Prentice Hall.

McClincy, W. D. (1995). *Instructional methods in emergency services*. Upper Saddle River, NJ: Prentice Hall, Inc.

National Association of EMS Educators. (1996). *Foundations of education: An EMS approach*. St. Louis, MO: Mosby, Inc.

National Highway Traffic Safety Administration. (2000). *EMS education agenda for the future*. Washington, DC: U.S. Department of Transportation.

Vahradian, S. (1995). Critical thinking strategies for the EMS classroom. *JEMS, 20*(3), 149.

Walz, B. J. (1993, October–December). Intermediate ALS providers' level of interest in paramedic education. *Prehospital Disaster Medicine, 8*(4), 340–344.

Financial Management

CHAPTER **9**

Accreditation Criteria

CAAS

104.01   Financial Policy: The service will make provisions and provide direction for the management of its fiscal affairs (pp. 225–251).
104.02   Budgeting and Financial Statements: The service shall develop and utilize a written budget and financial performance measurements (pp. 250–251).
104.03   If the patient billing/collection is a function carried out by or on behalf of the agency, the agency shall have written billing and collection policies that are in keeping with state and federal consumer protection and non-discrimination laws (pp. 226–233).

Objectives

After reading this chapter the student should be able to:

9.1   Discuss the Medicare Ambulance Fee Schedule Final Rule and identify strategies for optimizing reimbursement within its requirements and limitations (pp. 227–232).
9.2   Describe the historical development of and programs administered by the Centers for Medicare and Medicaid Services (pp. 226–227).
9.3   Identify the requirements of Medicare Part B as they apply to ambulance suppliers, including levels of service, medical necessity, physician certification, origins and destinations, vehicles and staffing (pp. 227–231).
9.4   Explain the alternative components used to fund ambulance service (pp. 246–251).
9.5   Describe the financial policies that are addressed in budgeting and in types of budgets (pp. 233–235).
9.6   Calculate the unit hour utilization and various benchmarks and seasonal fluctuations for various levels of service for their local areas (pp. 236–238).
9.7   Discuss various methods and considerations for costing out service (pp. 235–238).
9.8   Understand managed-care contracting strategies (p. 233).

Key Terms

Point of View

The snow fell steadily the night before the final meeting of Medicare's Negotiated Rulemaking Committee (NRC). The process of crafting a fee schedule for ambulance transports had taken more than 18 months, and now by morning more than 12 inches had accumulated. Concerns rose that the weather would preempt the final effort to adopt the first significant modernization of the federal reimbursement structure for the payment of ambulance transport services. Twelve hours later, after many tense negotiations to reach unanimity (one of the mandated rules of the process), the designated members of the NRC reached an agreement.

As a result of the efforts of the NRC mandated by the Balanced Budget Act of 1997, the federal government redefined payment policies for Medicare recipients to reflect a uniform system of reimbursement only differentiated by regional geographic cost differences. The changes developed by the NRC eliminated four different payment methodologies that had resulted in vast differences in reimbursement to providers. The new fee schedule reflected a uniform payment structure based on a relative-value system similar to the methodology by which physicians are paid for their services. While there are clearly positive aspects of the new fee schedule, many ambulance service providers witnessed significant decreases in their overall reimbursement. Other

practices, such as the balance billing of patients for the difference between the amount of Medicare payment and the total cost were eliminated, resulting in further reductions to total revenue.

The final chapter in the efforts of those engaged in redefining how the federal government reimburses providers of ambulance transport services has not been written. Efforts continue to implement elements that were not open for negotiation during the NRC process. At the top of that list is the implementation of a standardized coding system that utilizes a list of medical conditions developed by a subcommittee of the NRC as a substitute for the traditional ICD-9 coding system, which is based on diagnostic codes.

My impression, having participated as the principal negotiator representing the International Association of Fire Chiefs (IAFC), is that the ambulance service industry through its representation on the NRC was able to set aside individual differences and construct a reimbursement model that is fair and equitable to all participants. The participatory dialogue and spirit of cooperation reflected during the process embodies the dedication of all providers to ensure that the patient is ultimately the reason for our existence.

Chief Jack J. Krakeel
Director of Fire and Emergency Services
Fayette County, Georgia

■ FINANCIAL REQUIREMENTS AND BILLING

Revenue recovery is an important part of any EMS manager's or leader's responsibilities and is often the most misunderstood. The cost of service is rarely a concern among front-line EMS providers but becomes a significant challenge when EMS street-level providers are promoted to managerial positions. It is important for EMS managers to get a college-level education on public budgeting and health-care finance. Additional education should be provided by the parent agency in which the EMS agency or division derives its budget.

Ambulance rates and billing should be a balance between the cost of delivering EMS, collection of revenues, and nonservice-related monies coming into the organization. Rarely will ambulance rates reflect the actual cost of providing services, due to the cost of infrastructure required by municipal or fire-based EMS. The individual agency, states, and local jurisdictions set ambulance rates. Medicare and large insurance carriers, otherwise known as third-party payers, set reimbursement rates according to regional variations. Most insurance carriers must either pay 100% of the bill that is presented to them or allow the billing entity to balance bill the patient. The insurance company can limit what they pay, as long as it is in their contract with the beneficiary (such as Medicare's schedule of 80% or another set amount), but the patient is normally responsible for the remainder of the bill. The federal government determines rates of reimbursement for Medicare. Medicaid payments are normally the lowest reimbursement rates received by suppliers of ambulance service.

MARGINAL COSTING

As an EMS manager, it is important to calculate and know the **marginal cost** of an ambulance transport. The cost of providing the transport of a patient in a municipal EMS service divided by transport volume gives the agency a marginal cost per transport. In some areas revenue is collected and available to provide additional monies to enhance other EMS-related services. In some fire-based agencies, all costs need to be included in the delivery of ambulance transport. Some agencies erroneously do not figure the cost of firefighter/paramedics or firefighters/EMTs, explaining they are there as firefighters and their cost is absorbed

by the fire-suppression operations. Not only should firefighter salaries be included but a portion of the fire-suppression budget should be included according to the percentage of EMS calls to which nontransport or fire engines respond.

EMS agencies can anticipate that funding will come from several different sources. In most municipal EMS systems, funding comes from some form of public funds, which are most often from tax revenues. This could be from sales tax, property tax, or impact fee. Some government agencies create EMS operating levies to the public to specifically target tax monies to EMS (for example, the Medic One bond levy).

Another option for start-up cost and expansion of EMS services includes municipalities underwriting municipal bonds in order to finance EMS capital improvements or infrastructure cost. The ability for a government entity to underwrite a bond depends on statutory authority. Some jurisdictions require the voters to approve bond funding by a simple or a two-thirds majority. Other government agencies, the county or city executive, or governing body of elected officials can initiate a bond to fund certain activities EMS leaders and managers need to evaluate other potential funding services. Fee-for-service resources are brought into the organization from third-party payers, such as private insurance or privately paying customers.

Government reimbursements such as Medicare, Medicaid, and military or government dependent care are the primary revenue streams coming into an ambulance service. The EMS leadership in an organization needs to understand the information on the requirements of Medicare Part B as it relates to ambulance suppliers. EMS agencies are actually suppliers, according to the Centers for Medicare and Medicaid Services (CMS), unless the providers are hospital-based. (Note that CMS was previously called the Health Care Financing Administration.)

An EMS manager should understand how to develop and share strategies to optimize Medicare reimbursement. Other monies that come into the organization may be by way of grants, capitated agreements with health maintenance organizations (HMOs), and public or private philanthropist gifts or donations, which can help fund certain one-time purchases or specific programs. Because this type of revenue stream is sporadic and unreliable, it should not be counted on.

Civic-group funding and subscription programs offer another source of revenue and are popular in many smaller and rural jurisdictions.

MEDICARE

Medicare and Medicaid were established in 1965 by the Social Security Act. These programs covered only those over age 65 from 1965 to 1972, when they were expanded to cover those with disabilities. Medicaid provides coverage for low-income families; the aged, blind, and disabled; and those eligible for federally assisted income and is dispensed from state agencies. The State Children's Health Insurance Program (SCHIP) was established as part of the Balanced Budget Act of 1997 and allowed states to provide health insurance to more children. Administration of all three programs transferred from Social Security to the **Health Care Financing Administration (HCFA)** under Health and Human Services in 1997. HCFA was restructured and renamed the Centers for Medicare and Medicaid Services (CMS) in 2001 to oversee fee-for-service Medicare, Medicare Plus Choice (Medicare HMO) supplemental insurance, and the State Administered Medicaid and SCHIP programs.

In 1997 the Balanced Budget Act charged HCFA with reformulating the ambulance fee structure through negotiated rule making, which is conducted by bringing industry stakeholders together to negotiate the regulations specifying how ambulance services are reimbursed under Medicare. This is done to break deadlock in government and involves a facilitator to assist in reaching consensus. The Medicare negotiated rule-making process included developing definitions that link payment to the levels of service provided and identified regional cost for mileage and cost of living (Figure 9.1).

MEDICARE PART B

Under Medicare Part B, emergency medical services are classified according to the level of service pro-

---

- ◆ American Ambulance Association.
- ◆ American College of Emergency Physicians.
- ◆ American Health Care Association.
- ◆ American Hospital Association.
- ◆ Association of Air Medical Services.
- ◆ International Association of Fire Chiefs.
- ◆ International Association of Firefighters.
- ◆ National Association of Counties.
- ◆ National Association of EMS Physicians.
- ◆ National Association of State Emergency Medical Services Directors.
- ◆ National Volunteer Fire Council.

FIGURE 9.1 ■ Organizations Represented on the Negotiated Rulemaking Committee.

vided. Service providers must use ICD-9 coding formats; while the *International Classification of Diseases*, 9th Revision, Clinical Modification (ICD-9-CM) codes are not precluded from use on ambulance claims, they are currently not required by the Health Insurance Portability and Accountability Act (HIPAA) on most ambulance claims and generally do not trigger a payment or a denial of a claim.

Each patient must be assigned a primary and an alternative code. The primary code should reflect the condition on scene; a secondary code can reflect changes from the dispatch information versus what was found on scene. A narrative should have a transportation indicator in the prehospital-care report. There are seven levels of EMS service that can be provided, and under Section 414 of the Medicare/Medicaid Act, fees are adjusted based on one of nine regions throughout the county.

The first level of transport is the basic life support (BLS) level, which reflects those skills provided by EMTs. Medicare regulations very clearly mandate that any service provided by EMT basic-level personnel, including the establishment of an IV line or medication administration if the state allows EMT basic-level personnel to perform those skills, must be billed at the BLS level. Just because a paramedic is on the call does not mean it constitutes an advanced life support (ALS) call. Beginning in January 2006 billing charges under the Medicare program must be made primarily on the level of care provided. Agencies that once billed every transport as an ALS transport simply because a paramedic or ALS provider was on the scene are no longer allowed to do so under federal regulations. For Medicare and other federal programs, only calls during which ALS services are actually provided to the patient, when a "medically necessary" ALS assessment is provided, or when the information received at time of dispatch dictated an ALS response under a medically approved tiered-dispatch program are allowed to be billed at the ALS level.

Additionally, if an agency performs an ALS assessment on every patient in an attempt to gain ALS reimbursement, they are guilty of what is referred to as **upbilling**, a misrepresentation of provided services by billing for a more-expensive service. Upbilling can result in fines, criminal charges, or suspension of Medicare benefits.

In the second and third level of billing, two ALS levels are defined. ALS 1 includes at least one ALS intervention, including anything that could be considered a "medically necessary" ALS assessment—one that is required by the patient's condition and involves the use of ALS evaluation tools (such as an ECG, pulse oximetry, or blood-glucose readings). Simply having ALS-level personnel evaluate a patient without using any ALS-level assessment tools does not constitute an ALS assessment.

The second level of ALS transport is ALS 2. ALS 2 transports must include the administration of at least three intravenous medications or the performance of at least one of the following advanced skills:

- Manual defibrillation/cardioversion.
- Endotracheal intubation.
- Central venous line placement.
- Cardiac pacing.
- Chest decompression.
- Surgical airway.
- Intraosseous line.

The next level of service is related to **specialty-care transport (SCT)** units or interfacility transports requiring a level beyond the scope of paramedic care, such as a critical-care nurse or neonatal specialty nurse. SCT has a higher level of billing, and in some cases where the patient is an inpatient to a facility, a transport may be billed under Medicare Part A but must have a physician-documented medical reason for the trip.

A financial-reimbursement strategy is now available for ALS tiered response under the billing category for paramedic intercept (PI), which is classified as paramedic services provided by a nontransporting entity. There are limitations to this level of service. The billing agency must bill all recipients of service regardless of Medicare status, the area must be a rural service area, and the paramedic interceptor agency must be certified as an ALS service and qualified to bill Medicare. The service must be provided under contract with volunteer services that furnish only BLS-level care and are prohibited by state law from billing for any services. Currently, a paramedic intercept is only reimbursed in the state of New York.

Fixed-wing (FW) aircraft Medicare reimbursement is covered when the point of pickup is inaccessible by a land vehicle, great distances are involved, or there are other obstacles that are not defined by the rule-making committee. Rotary-wing (RW) aircraft reimbursement rules for helicopter transport have the same criteria as a fixed-wing transport.

Calculation of ambulance fees starts with a base rate adjusted for geographical cost differences. These geographical costs are factored in to adjust for regional differences and costs. Historically, Medicare reimbursement is raised every year based on the CPI–urban percentage increase. Medicare uses the physician's "practice expense" component of the Geographic Physician Cost Index (GPCI) and applies it to 70% of the base rate. GPCI cost is available from CMS for the regions in the United States. To calculate the ambulance bill for an ALS transport, start with the national base rate of $204.65. Then take 70% of that amount and apply the GPCI indicator to that amount. The remaining 30% of the $204.50 is then added to this adjusted figure.

The service then can charge for a loaded mile, which is the mileage during which the patient is actually on board the ambulance, at a rate of $5.47 a mile (the mileage reimbursement goes up each year). Each claim, other than a claim for a patient who died prior to transport, must have mileage associated with it. Mileage is rounded up to the next whole number (for example, 4.1 miles is billed as 5 miles). The rate on loaded miles is based on point of pickup, and the zip code for the point of pickup must be reported on each claim.

All ambulance services must participate in the Medicare program and are prohibited from billing the patient for any covered costs, other than the maximum 20% copay. Additionally, each agency must obtain a signed authorization form from the beneficiary (patient) giving the agency permission to bill for the transport service. If the patient is unable to sign, documentation as to why he could not sign must be provided (for example, he was unconscious).

Nonemergency transports need a **physician certification statement (PCS)** to bill Medicare (Figure 9.2). In January 1999 the HCFA released a final rule, and on July 8, 1999, it issued a program memorandum requiring a signed physician certification statement (PCS) for nonemergent transports. Medicare as well as many different health-insurance carriers (for example, Medicaid and HMOs) require a PCS. If the physician is not available, the PCS can be signed by a nurse, physician assistant, or nurse-practitioner. After 21 days of attempting to get a signature authorizing the transport, the EMS agency may bill for services and document the attempts to get a PCS.

Medicare and other insurance companies conduct prepayment and postpayment audits looking for appropriate documentation. There are serious fines and

PHYSICIAN CERTIFICATION STATEMENT (PCS)
FOR NON-EMERGENCY AMBULANCE TRANSPORT

Central Oregon Coast Fire & Rescue District
PO Box 505 • Waldport OR 97394 • (541) 563-3121 FAX: (541) 563-3190

Please print clearly and have physician sign where indicated below. Complete ALL sections of this form.
See reverse for important information on completing this form.

Section 1 – Beneficiary Information

Name: Last Name First Name Middle Initial	Age:	Date of Birth: ____/____/____	Sex: ☐ M ☐ F

SSN: _____-____-_____	Medicare No.: _____-____-_____	Part B? ☐ Yes ☐ No	Medicaid No.:

Diagnosis:

Date of Transport: _____/_____/_____

Note: The Physician Certification Statement (PCS) effective date must be no earlier than 60 days before the date services are furnished.

Section 2 – Transport Information

Transport From:	Discharge? ☐ Yes ☐ No
Transport To:	Admit? ☐ Yes ☐ No

Reason for Transport: (include name of service, treatment, or procedure the patient needs at the receiving facility)

Is the service, treatment, or procedure for which patient being transported available at originating facility? ☐ No ☐ Yes	If yes, why is transport necessary?

Section 3 – Medical Necessity Information – See Reverse for Definition of Medical Necessity

NOTE: LACK OF ALTERNATIVE TRANSPORTATION SERVICES **DOES NOT** CREATE A MEDICAL NECESSITY FOR **AMBULANCE** SERVICES.

Describe patient's condition (not diagnosis) at the time of pickup and/or discharge that necessitated utilization of an ambulance. (see reverse for HCFA definition of medical necessity)

Is the patient bed confined as defined by Medicare (HCFA) regulations? (see reverse for definition) .... ☐ Yes ☐ No

If the patient does not meet bed-confined criteria, can this patient be safely transported by wheelchair van?.............. ☐ Yes ☐ No

If No, why?

This Patient **(check all that apply):**

- ☐ Requires IV maintenance
- ☐ Requires care/monitoring by trained personnel during transport
- ☐ Required to be immobilized due to a fracture or possible fracture
- ☐ Is exhibiting signs of decreased level of consciousness
- ☐ Has decubitus ulcers & requires wound precautions

- ☐ Requires continuous oxygen
- ☐ Requires cardiac or other physiological monitoring
- ☐ Requires a Balloon Pump
- ☐ Is comatose & requires monitoring
- ☐ Requires isolation precautions (VRE, MRSA, etc.)

- ☐ Is ventilator dependant
- ☐ Requires airway maintenance or suctioning
- ☐ Requires restraints (Other than seat belts)
- ☐ Is seizure prone & requires monitoring
- ☐ Weight limit exceeds wheelchair or stretcher van safety limitations

☐ Requires other services or equipment (please list): _____

Section 4 – Ordering Physician Information and Signature

Print Name of Physician Ordering Ambulance Services:	UPIN:	Fax Number:

I certify that the above information is true and correct based on my evaluation of this patient. I understand that this information will be used by the Health Care Financing Administration to support the determination of medical necessity for ambulance service

➡

Physician's Signature	Date	Medical Support Staff Signature	Date

Please give this form to an ambulance crew member or fax to office listed on reverse.

FIGURE 9.2 ■ Physician Certification Statement for Non-Emergency Ambulance Transport.
(Reprinted with permission of Central Oregon Coast Fire & Rescue District)

TABLE 9.1 ■ Medicare Rates for EMS in 2005

BLS	$170.54
BLS Emergency	$272.86
ALS 1	$204.65
ALS 2	$468.99
SCT	$554.26
Mileage	
Base	$ 5.47
Rural Miles 1–17	$ 8.21
Rural Miles 18–50	$ 6.84

in some cases jail time for claims that do not match the documentation. Table 9.1 contains a sample of Medicare reimbursement rates.

An adjustment factor is allowed for emergency response that takes into account additional costs incurred with an immediate response. Emergency response can be assigned only to BLS and ALS 1 levels. ALS 2 transports are only reimbursed at the emegency level. Operational variations allow all ambulance companies, however organized, to be paid according to the fee schedule. An exception does exist for critical-access hospitals, allowing for a variance for higher charges when there is only one ambulance provider within a 35-mile drive. A 50% increase to both mileage and the base rate for air ambulance may be applied if the service is provided in a rural area. All ground ambulance providers receive a 25% bonus for all miles above 50.

Multiple patients also have a different billing formula and require an adjustment in the way they are

TABLE 9.2 ■ CMS Definitions of Service

BLS   *Basic Life Support.* Where medically necessary, the provision of BLS services as defined in the *National EMS Education and Practice Blueprint* for the EMT–Basic, including the establishment of a peripheral intravenous (IV) line.

ALS1   *Advanced Life Support, Level 1.* Where medically necessary, the provision of an assessment by an ALS1 provider and/or the provision of one or more ALS interventions. An ALS provider is defined as a provider trained to at least the level of the EMT–Intermediate as defined in the *National EMS Education and Practice Blueprint.* An ALS intervention is defined as a procedure beyond the scope of an EMT–Basic as defined in the *National EMS Education and Practice Blueprint.*

ALS2   *Advanced Life Support, Level 2.* Where medically necessary, the administration of at least three different medications and/or the provision of one or more of the following ALS procedures: manual defibrillation/cardioversion, endotracheal intubation, central venous line, cardiac pacing, chest decompression, surgical airway, intraosseous line.

SCT   *Specialty Care Transport.* Where medically necessary, in a critically injured or ill patient, a level of interfacility service provided beyond the scope of the paramedic as defined in the *National EMS Education and Practice Blueprint.* This is necessary when a patient's condition requires ongoing care that must be provided by one or more health professionals in an appropriate specialty area (nursing, medicine, respiratory care, cardiovascular care, or a paramedic with additional training).

PI   *Paramedic Intercept.* These services are defined in 42 CFR 410.40. They are ALS services provided by an entity that does not provide the ambulance transport. Under limited circumstances, these services can receive Medicare payment.

FW   *Fixed-Wing Air Ambulance.* Fixed-wing air ambulance is provided when the patient's medical condition is such that transportation by either basic or advanced life support ground ambulance is not appropriate. In addition, fixed-wing air ambulance may be necessary because the point of pickup is inaccessible by land vehicle or great distances or other obstacles (for example, heavy traffic) are involved in getting the patient to the nearest hospital with appropriate facilities.

RW   *Rotary-Wing Air Ambulance.* Rotary-wing air ambulance is provided when the patient's medical condition is such that transportation by either basic or advanced life support ground ambulance is not appropriate. In addition, rotary-wing air ambulance may be necessary because the point of pickup is inaccessible by land vehicle or great distances or other obstacles (for example, heavy traffic) are involved in getting the patient to the nearest hospital with appropriate facilities.

billed. If two patients are transported in the same vehicle, a special modifier must be used on the claim form after the base-rate billing code. Medicare will reimburse only 75% of the base rate for each patient according to the level of care received by the patient. If three or more patients are transported in the same vehicle, 60% of the base rate will be reimbursed by Medicare. Mileage must be prorated by number of patients transported.

EMS services also may bill for a response that is required for the pronouncement of death. Reimbursement is based on when death is pronounced. If the patient is pronounced before the ambulance is called, there can be no payment. If the patient is pronounced after the ambulance was called but prior to arrival at the scene, the deceased patient can be billed for a BLS nonemergency base rate without any mileage. If the patient is pronounced during transport, the normal reimbursement rates apply.

In the future, reimbursement issues will need to confront Medicare, Medicaid, and insurance reimbursement policies for treating and releasing patients, billing for nontransports, transporting to nontraditional destinations, first responder fees, and fees for nurses and physician assistants in the field. (See Table 9.2 for descriptions of the different levels of care.)

The Centers for Medicare & Medicaid Services (CMS) in 2005 added over $1 billion to help ambulance providers and other health-care providers recoup unpaid emergency health-care costs from undocumented workers. Section 1011 set aside this money until December 2008 (when the section is scheduled to expire) for emergency services of those undocumented workers who are uninsured or cannot afford emergency care. Each state will receive a specific amount of funding, and payment will be made directly to emergency services providers as long as they do not receive payment from any other organization. Along with those changes will be changes in the billing codes. Table 9.3 displays the differences between the old codes and the new codes. The Healthcare Common Procedure Coding System (HCPCS) is a set of health-care procedure codes based on the American Medical Association's Current Procedural Terminology (CPT).

Often unpaid claims by insurance companies, Medicare, and Medicaid occur from errors created on the provider's side of the billing process. The common sources of billing errors from providers are listed in Figure 9.3.

TABLE 9.3 ■ Excerpt from the Ambulance HCPCS Crosswalk and Definitions

New HCPCS Code	Description of HCPCS Code	Old HCPCS Code
A0430	Ambulance service, conventional air services, transport, one way, fixed wing (FW)	A0030
A0431	Ambulance service, conventional air services, transport, one way, rotary wing (RW)	A0040
A0429	Ambulance service, basic life support (BLS), emergency transport, water, special transportation services	A0050
A0428	Ambulance service, BLS, nonemergency transport, all inclusive (mileage and supplies)	A0300 (Method 1)
A0429	Ambulance service, BLS, emergency transport, all inclusive (mileage and supplies)	A0302 (Method 1)

Complete information is available on the CMMS Medicare Web site.

SCHEDULED TRANSFERS

EMS agencies are being called upon to conduct scheduled transfers from skilled-care facilities, nursing homes, or the patient's residence, which are not considered emergencies. The success in billing for these services lies in whether or not the transfer is medically necessary. A **medical necessity** is established when the patient's condition is such that transfer by any other means of transportation—for example, taxi, private car, or wheelchair coach—would endanger the health of the patient. A nonambulatory status, doctor's

- ◆ Inappropriate or inaccurate codes.
- ◆ Losing charge information.
- ◆ Billing carriers incorrectly or irregularly.
- ◆ No written billing policies and procedures.
- ◆ Inappropriate or outdated forms and documents.

FIGURE 9.3 ■ Causes of Claim Denial.

orders, or the patient being in a nursing home does not alone indicate medical necessity. The reason for the transfer by ambulance and exactly why the patient cannot ride in the car is required in the run report.

Appropriate documentation is required on all calls; however, the nonemergency calls are more closely looked at in a billing or audit process. A completed, signed prehospital care report with a narrative and services rendered and with a patient signature is needed. If the patient cannot sign, then the physician statement certifying the need for the transport must be attached (Figure 9.2). Nonemergency-transfer sheets from the facility and information from family or medical staff are required by most insurances and Medicare. A reason that the patient must go by stretcher is required by Medicare, HMOs, and insurance companies. Ideally, the agency should have a form that the provider may fill out prior to loading the patient onto the gurney that helps the provider identify the reason for the nonemergency ambulance transfer and to identify the required documentation.

The patient who is bed confined, bedfast, bedridden, or bed imprisoned is for Medicare purposes a patient who is unable to get up from bed without assistance, is unable to ambulate, and is unable to sit in a chair or wheelchair. The reason the patient must go by ambulance must be clearly defined. Most nursing homes do not understand what Medicare or insurance will pay for and how long the delays are in reimbursement. Some cities have enacted ordinances, and there are state laws holding nursing homes accountable for the bill if insurance or Medicare denies a claim because it is deemed not medically necessary. If an ambulance transport is scheduled for the family's convenience, it should be prepaid. This includes nursing home–to–nursing home transfers, dentist-office visits, or transfers to be closer to family. No insurance will pay for patient-convenience services or supplies.

FUTURE MEDICARE CHANGES

Expect that Medicare will become stricter on reimbursing for what is or is not considered an emergency. Under the CMS rules, the term *emergency* means service provided after a medical condition manifests itself by acute symptoms of such severity that the absence of immediate medical attention could reasonably be expected to result in any of the following: placing the patient's health in serious jeopardy, serious impairment to bodily functions, or serious dysfunction of any bod-

ily organ or part. Any ambulance transport that does not meet one of these three criteria is considered a nonemergency service.

In the future expect to see **performance-based reimbursement** being applied to ambulance reimbursement from Medicare, Medicaid, and large HMOs or insurance companies. This will force EMS providers to provide care and complete a transport based on medical necessity and **evidence-based medicine**. The latter will indicate which prehospital interventions and actions produce a positive outcome for the patient. In some systems this may mean a denial of reimbursement on medications, procedures, or transport cost with a specific medical condition that requires that intervention; for example, attempting to resuscitate a patient over 65 in cardiac arrest who presents in asystole and therefore statistically has no chance of survival. On the federal level the Medicare Payment Advisory Committee, or "Medpac," may decide not to pay for the cost of the EMS to work the arrest, knowing that evidence-based medicine indicates it will not have a favorable or positive outcome. Expect private insurance companies and large HMOs to follow this trend once it is established. Evidence-based medicine has already produced systems that ration health care based on outcomes seen in the Oregon Medicaid system, which only pays for health care that has shown proof of medical benefit.

BILLING THIRD-PARTY PAYEES

A **third-party payee** is generally referred to as anyone responsible for the ambulance other than Medicare, Medicaid, or the self-paying patient. Most often it is an HMO, a health-insurance company, or workers' compensation insurance. EMS agencies dealing with HMOs must plan on that source being slow to pay or not paying for a transport at all. Many EMS responders complain about the hassle and low reimbursement from HMOs; these problems can be reduced by making sure the charges are appropriate, there is quality patient and event documentation, and there are appropriate collections with up-to-date contracts.

EMS managers need to understand who their payer sources are and how their processes are conducted. Other payer sources are the U.S. military, Indian Health Services, Champus/Tricare, and prison systems. Reimbursement also may be obtained from commercial automobile insurance, homeowner insurance, victims of crime who have received restitution, and service groups that often contribute to financially challenged people. Ambulance reimbursement has been collected from bankruptcy proceedings, probate, victim restitution, small claims, special travel insurance, and embassies.

EMS managers need to make sure that providers, medical staff, peers, and billing clerks receive continuing education on reimbursement strategies. Billing for ambulance services may be itemized in contractual circumstances and when billing third parties or private payers. The advantages to itemized billing are that the supplies and true cost of the service and equipment used can be billed for. This maximizes cost recovery as compared to limits for service based on Medicare or private insurance contracts.

HMO CONTRACTING

When dealing with HMO contracts, EMS agencies can find themselves in a David-and-Goliath battle. A methodical process should be followed when contracting with HMO services. First, do not sign a standardized or template contract without a thorough review from a competent and experienced source. Negotiated fees for service should actually represent cost. In some contracts with HMOs, an agency may decide to give a discounted fee for service. The EMS agency can limit payment delays, denials, and down coding with good contract language.

A contract with an HMO should allow for reimbursement when a prudent layperson calls for an ambulance. Often special circumstances are placed in the contract for billing. This is known as **carve-out language** and should be seriously evaluated by EMS leadership. Ensure that all definitions are clear, since it is often a clerical person with limited medical background who is authorizing or denying a claim. It is important to establish prompt payment rules and enforce them. HMOs are notorious for delaying and denying payment to hold onto monies as long as possible.

HMOs can provide or supply several different payment options. In a **capitated contract** an HMO gives an annual lump-sum payment to the provider to cover an estimated number of patients under its plan. If an EMS provider can provide the service at a cost less than what is paid, an agency can generate a surplus. If the number of patients and the cost goes above the capitated amount, the service experiences a loss. Some HMOs receive discounted fees for service or set a fee schedule. Because of the volume, in many cases an HMO or large provider of health care can negoti-

ate a deeply discounted rate for EMS services. Certain cases can have a specific fee applied; for example, chest pain or cardiac arrest may have a specific cost that is reimbursed by the HMO at a higher rate. Group strategies and legislative limits can help with long-term HMO contracting.

HMOs offer a chance for a unique partnership with EMS. HMOs often have disease-management coordinators that oversee the process and practices for specific conditions, such as diabetes, asthma, and pregnancy. Through contracts and capitation an EMS agency can develop creative reimbursement strategies. Consider a person with diabetes who triggers a 911 response for insulin shock. Most of these patients without serious complications can receive D50 or glucagon and stay at home, saving the HMO several thousand dollars in expenses. The expense is incurred by the EMS agency because insurance usually does not pay unless the patient is transported. If a local fire department or EMS agency can contract specifically for capitated money to treat and release diabetics, the cost can be recovered by the EMS agency and money saved by the HMO through a comprehensive treat-and-release program.

■ FINANCIAL POLICY

BUDGETS

An EMS manager should be able to calculate direct cost for each element in the EMS system. Direct costs are those that can be assigned directly to a particular component of the operation and should include both start-up costs and ongoing costs. If a new program is to be implemented, the EMS system should recognize the nature and cost of each phase of that program's development, including long-term maintenance contracts. Figure 9.4 lists the basic or continuous costs or operating expenses each EMS agency will need to calculate.

- Emergency medical equipment and supplies.
- Vehicles and fuel.
- Facilities.
- Communications.
- Primary and support personnel, including salaries and associated benefits.
- Direct labor costs, including ongoing training and licensing.

FIGURE 9.4 ■ Ongoing or Continuous Cost.

- Insurance.
- Legal and consultant costs.
- Medical director salaries and insurance.
- Cost of contract services.
- Billing service costs.

FIGURE 9.5 ■ Indirect Cost for EMS Operations.

In addition to ongoing costs, most agencies also have **indirect costs**—for those resources that are necessary for logistics or infrastructure but cannot be traced directly to a specific product or service provided by the EMS agency. Figure 9.5 lists the common indirect costs associated with EMS operations. Some accounting systems identify variable and fixed costs. A **variable cost** is one that changes in total proportion to changes in the related level of total activity or volume. **Fixed costs** remain unchanged in total for a given time period despite wide changes in the related level of total activity or volume.

EMS BUDGETS

A budget is based on risk assessment and risk management. Many of those risks are based on demographics, call volume, and the loss versus benefit of certain programs and modes of service delivery. There are several types of budgets commonly seen in public safety. One is the **line-item budget**, which focuses on inputs. It lists specific items or services by division, department, unit, and occasionally in even smaller units. Line-item budgets are easy to balance and often require last-minute spending to empty accounts. This form of budget is often a problem for EMS agencies, when the money is not spent from an account or when a line item is perceived as not important and may be the object of budget cuts.

Line-item budgets typically do not focus as much on results as they do on the items being secured or paid for. Many government rules do not allow line-item monies to be floated from one line item to another without a review process or going back through the budgeting process. Line-item budgets work best for project work and important fixed costs such as personnel base salaries and benefits, vehicles, facilities, and other capital costs that do not require a degree of flexibility.

Line-item budgets provide a level of operational detail for EMS managers to time purchases, revise budgets, and control their day-to-day spending. The

integrative budgeting system is a modification of the line-item budget that uses three computerized categories: personnel, operations, and capital outlays. The line items are for operational use only, and the detailed line-item budgets are used to develop totals for the program budgets, which are appropriated without detailed control.

Another commonly used budget system is a service-based budget. It focuses on outputs or a specific service. Service budgets encompass all aspects of a specific type of service and all the processes that support that service. Processes in a service-based budget are often called cost centers, which include decision-making resources targeted to a goal or organizational objective authorized by the legislative body.

Service-based budgets are more appropriate for EMS agencies because they focus on services and not dollars. These budgets serve not as a control but as tools for decision making and delivering quality service. Service-based budgets require full costing or recording of both direct and indirect costs. Costs such as administrative support, overhead, managerial salaries, benefits, and other related costs are incorporated into full costing, the cost of resources used in a period that cannot be traced easily to either products or production processes.

An **activity-based cost system** is a method that assigns costs to activities and cost objects based on the consumption of resources, unlike the traditional costing approach in which costs are allocated to products based on some arbitrary bases such as labor. An activity-based cost system describes various activities (such as unit-level response, unit hours, EMS education, customer service, and facilities) that drive an EMS agency's costs. The integration of an activity-based cost system into a cost-value model recognizes the existence of multiple cost drivers and results in the production of better information for management.

An attribute-based cost system integrates cost with quality, function, and deployment. An attribute-based cost system contribution analysis provides a means for examining performance in the EMS system and customer levels. It thus improves management decision making and uses multiple internal as well as external cost centers.

Originally called the planning-programming-budgeting system, the **program-based budget** became very popular in the 1960s. Planning and itemization of the range of services being conducted by an EMS organization is done in a systematic fashion. The programming aspect is more specific in identifying cost of personnel, materials, and facilities. A program-based budget links the system cost with results and limits most line items for lump sums to major objects in the EMS operation; for example, personnel, supplies, capital outlays, and discretionary monies. Traditionally, government expenditures have been considered as inputs rather than outputs. The program budget is derived from this concept; it classifies expenditures in terms of the outputs to which they are categorized. Program budgets become more difficult as the program they are supporting becomes more detailed and closely monitored. Under a pure program budget, all costs are summarized for each program rather than placed in broad categories with a true cost of the item.

Performance-based budgeting also focuses on results by developing workload and unit costs, such as cost per response or cost per capita for EMS activities, and then funding those activities based on the number of responses. Generally, performance-based budgets have specified categories—personnel cost, maintenance cost, operations, and development cost. Often performance-based budgets are used for shared functions. For example, an EMS agency shares a percentage of the cost of operating a communications center based on the number of calls that communications center handled for the EMS agency. Performance-based budgeting gives an excellent historical view of EMS operations based on call volume or outputs from the operation.

Zero-based budgeting is another form of budgeting that requires EMS managers to justify the existence of a budget for the program or item. Zero-based budgeting takes a significant amount of administrative time and frequently revisits the same issues each year. Consequently, they have lost popularity in favor of other budgeting processes and are employed only when a city or department is facing budget shortfalls or there is an economic downturn.

ECONOMIES OF SCALE

When an EMS manager is preparing a budget, one option that can help share or reduce the cost of EMS operations is to employ economies of scale. An economy of scale is achieved when an increase in efficiency of EMS services is seen as the number of EMS units and response capability being produced or placed in service increases. Typically, an agency can achieve economies of scale by lowering the average cost per unit through increased production, since fixed costs are shared over an increased number of responses. For example, a large metropolitan EMS provider that furnishes ALS services to the smaller BLS-equipped suburbs decreases the overall cost of that service because the service is subsidized by the jurisdiction or a fee for the service is collected. There are two types of economies of scale:

- *External economies.* The cost per EMS unit depends on the size of the response area, not the agency.
- *Internal economies.* The cost per unit depends on the size of the individual agency.

Economies of scale give large EMS providers access to a larger market by allowing them to operate in larger geographical areas. But for the more traditional (small to medium) EMS agency, size does have its limits, so after a point an increase in size (output) actually causes an increase in production costs. This is called *diseconomies* of scale. It is seen in many metropolitan areas that have landlocked suburbs where the choice is adding additional EMS units versus asking a neighboring department for mutual aid.

Economy of scale is the basis for dual-role cross-trained fire service EMS providers. The future trend is to see economies of scale start to be employed by EMS providers and dispatch centers and in the consolidation of 911 communications centers. A growing number of smaller fire and EMS services are merging into one larger organization to take advantage of economies of scale. A contiguous community or operating area with a population from 250,000 to 500,000 provides the most efficient levels of service. If the population served is smaller than 200,000, the clinical and response-time reliability will suffer or substantial levels of subsidy for more units will be required to offset the economies of scale after a service reaches 200,000. With the continuing crisis in the health-care system, the future may see EMS agencies being used to supplement home health care, school health programs, and other municipal services.

■ COSTING OUT SERVICE

In many EMS departments budgets often depend on how many EMS units are placed in service or are needed to handle requests for service. A delay in the dispatch or a wait time for a transport unit develops

$$\frac{\text{Total Expenses}}{\text{Total Number of Responses}} = \text{Total Cost Per Response}$$

FIGURE 9.6 ■ Cost-Per-Response Formula.

when EMS units are busy and the requests for EMS from citizens exceed the units available. If there are not enough units, citizens will wait too long for EMS to make a difference. On the other hand, if an agency fields too many EMS units, then the cost will increase and the units will not be very busy. It is important to have information on the basic cost of operation and utilization to help plan and budget for services.

COST PER RESPONSE

EMS providers regardless of provider type need to calculate cost per response. This is determined by dividing each provider's total expenses by the total number of responses. To calculate cost per response, the complete cost of the service must be used, including overhead that is allocated in a one-month period, divided by the total responses in the same one-month period. This should also be done with an annual number of responses and yearly expenses using the formula in Figure 9.6.

The cost per response evaluates the systems cost from an external perspective and is used primarily for purposes of comparison to other service-delivery models in EMS. The number is frequently used by county and city managers to compare cost of privatization of government services.

CALCULATING UNIT HOURS

The concept of total cost per unit is a measure of the total cost of providing coverage during a given accounting period divided by the total number of unit hours of coverage provided during that same accounting period (Figure 9.7). This measurement can be used to identify whether or not unit-hour costs are excessive given the quality of care being delivered in comparison to other similar EMS systems. If unit-hour cost becomes excessive and response times are not being met, then EMS leadership should consider

moving vehicles or adding unit hours to ensure response times are met. Other factors, such as a top-heavy administration, high wages, poor economies of scale, and overtime pay, can drive up unit-hour cost. An example using very simple numbers for calculating a cost per unit for a fire-based EMS unit is shown in Figure 9.8.

Vehicle operating cost can be figured in a variety of ways. One of the more common ways is to create an inventory list of all equipment carried on the ambulance, as well as the vehicle itself. The inventory list should contain the cost of each item and its anticipated life span. Dividing the cost of the equipment by its anticipated life span gives you a cost per year that can be added into the unit-hour cost, as shown in Figure 9.9.

The yearly cost can be divided by the number of days of the year and then by a 24-hour shift, resulting in an overall cost of a little under two cents an hour or about 45 cents per day. When each piece of equipment is expensed and included, the small amounts for each piece add up. The vehicle-maintenance costs include mileage, maintenance, and price of the vehicle amortized over five years. These records are used to determine the average cost per year and, again using simple division, the average cost per hour. Disposable supplies billed to the patient are not included in the cost per hour. If your department or agency does not bill for soft supplies, then the line item for soft supplies in the budget should be divided to ascertain an hourly and daily cost of those supplies.

CALCULATING UNIT-HOUR UTILIZATION

Determining how busy a unit is helps calculate cost. A process called **queuing theory** helps determine how many ambulances are needed during each hour, day, and week for the EMS operation. Queuing theory is the study of waiting lines and the consequences of those lines. It has been applied in a variety of situations where waiting in lines occurs. It is an excellent approach for agencies in determining how many EMS units are needed to handle calls. Queuing theory can estimate how busy EMS units are, the probability of a

$$\text{Cost Per Unit Hour} = \frac{\text{Total Costs of Providing Coverage During Given Accounting Period}}{\text{Total Number of Unit Hours Covered During an Accounting Period}}$$

FIGURE 9.7 ■ Cost-Per-Unit Hour.

A fire department has five ambulances covering a 24-hour period. Each ambulance is staffed with a firefighter/paramedic at $23 per hour and firefighter EMT at $18 per hour including benefits. There is a shift supervisor who makes $28 per hour as an EMS supervisor. The cost to operate the vehicle, including fuel, maintenance, and hourly time cost for the fixed facility to house the unit, is $45 per hour.

a. Units hours − 24 hours × 5 units = 120 units hours
b. Cost Paramedic salary 24 hours × $23 × 5 = $2760
   Firefighter/EMT salary 24 hours × $18 × 5 = $2160
   Vehicle operating cost 24 hours × $45 × 5 = $5400
   Total cost for 24 hours = $10,320
c. Cost per unit hour − $10,320/120 = $86

If an EMS supervisor is deployed in the system at a salary of $28 × 24 hours = $672 with additional cost increasing the cost per unit to approximately $91 per unit hour. Additional cost for dispatching, billing, and personnel can be added onto this same formula; however, most agencies share these resources or fall under a separate budget item or agency budget.

FIGURE 9.8 ■ Example of Cost Per Unit for a Five-Unit EMS Organization.

$$\frac{\text{Oxygen Equipment Bag and Equipment (\$825)}}{\text{Five-Year Life Span (5)}} = \$165 \text{ per year}$$

FIGURE 9.9 ■ Costing Out EMS Equipment.

the number of EMS calls and the average amount of time a crew spends on an ambulance call. Unit-hour utilization can be used to determine how busy EMS units will be based on an anticipated workload.

The same formula also can be used to estimate the probability that a citizen will have to wait for an ambulance and the number of EMS units needed to meet the response-time objectives of an EMS system. Another term for tracking utilization is "in-service ratio." Definitions are often in conflict between the private-ambulance industry and the fire service. Fire services measure this number based on total calls versus private ambulances, measuring only actual calls that result in transport. This gives a skewed number because the dry-run rate or nontransported EMS calls in many systems can reach 50% or higher. The U:UH ratio is calculated by dividing the total number of calls during a given accounting period by the total number of unit hours of coverage provided during the same accounting phase. The higher the utilization is, the lower the cost per patient served.

Unit-hour utilization can be measured by using two different time frames (Figure 9.10). It is commonly calculated on a daily and a weekly basis and is an indication of appropriate deployment of resources. The first formula in Figure 9.10 is for a 24-hour ambulance running an average of 5.5 calls per day. A second formula is displayed using a weekly calculation. If an EMS manager wanted to monitor the entire system or battalion of EMS units, comparing unit hours and unit-hour utilization is a way of looking at vehicle deployment and workload and then budgeting for additional hours or personnel.

call, how much waiting time there is, and the number of EMS units needed to satisfy operations.

Just as important as cost per hour, the measurement of workload of an EMS vehicle and its crew has no set upper or lower limit. Measurements of productivity are common in the industry; the key term is **unit-hour utilization** (U:UH). Unit-hour utilization is the measurement of workload for EMS agencies that was developed from queuing theory. It is an excellent resource for EMS agencies to use to determine the line between having too few ambulances and putting an ambulance on every corner. Queuing theory provides a simple mathematical formula to determine the number of ambulances needed per hour, per day, and during seasonal fluctuations. To make calculations on unit-hour utilization, the EMS manager must know

FORMULA:

U:UH Ratio = $\dfrac{\text{Total Number of Calls in Budget Cycle (Day, Week, or Month)}}{\text{Total Number of Unit Hours of Coverage Provided During Budget Cycle}}$

EXAMPLE:
Daily U:UH rate = 5.5 calls per day/24 hours = 0.22 U:UH
Weekly U:UH rate = (840 hours = 5 units staffed 24 hours a day, 7 days a week)
    310 responses for the week/840
    = 0.37

FIGURE 9.10 ■ Formula for Unit Hour Utilization with Example.

BENCHMARKS OF UNIT-HOUR UTILIZATION

Unit-hour utilization has been challenged by the fire service as a measurement tool applicable only to private industry. City, county, and agency administrators need to strive to identify a way to measure workload in EMS. The International Association of Firefighters (IAFF) has established the term "in-service ratio" to define the total amount of time on task that an EMS unit has, including such activities as restocking, training hours, and other time intervals that traditionally have been outside the private-ambulance industry measurement. This gives EMS administrators a more accurate indication of the day's events and the true workload for an EMS crew.

A closer look at utilization of EMS vehicles can be expected in the future. Three additional measurements are being made in some high-performance or heavy-workload systems: UhU-T, UhU-R, and UhU-TD.

- *Unit-hour utilization-transport (UhU-T).* A measurement made on workload that excludes calls for which a patient is gone on arrival, assessed and not transported, or canceled en route.
- *Unit-hour utilization-responses (UhU-R).* Only measures responses and not transports. This again is a misleading measurement of workload, because each response that results in a transport may require much more time than just a response. In many EMS systems the no-haul or dry-run rate is 40% or higher, and in those systems, if UhU-R is used to receive funding, the cost may be inflated for the actual workload or too low if the transport ratio is higher, with long transport times.
- *Unit-hour utilization-total deployment (UhU-TD).* Measures the workload based on the total time deployed in the field of an EMS unit. However, current measurement standards under the total-deployment

measure do not include lost time for EMS units due to training or vehicle breakdowns.

SEASONAL VARIATIONS

Often calls fluctuate with the seasons. Consider a fire-based EMS system with five EMS units staffed on a 24-hour basis (Figure 9.11). In the first three months of the year, the EMS units transported 3684 patients. The average time was 38 minutes for each transport. The total number of unit hours for the three months has to be adjusted: 60 minutes multiplied by 24 hours times five units multiplied by the three-month time period. Plugging these figures into the formula produces a utilization of 22%. Later in the year during the fall season, the call volume jumps to 5829 and the ambulance wait time grows to 65 minutes. The corresponding increase in workload is then calculated to be 58%, indicating issues with excessive workload. Some fire departments have identified triggers to add more units. Consider adding two part-time units for 12 hours for the fall season only. The bottom of the equation then receives 129,600 hours, and that drops the average unit utilization to 48%, which is still high but manageable.

That produces a number that identifies the time necessary to treat and transport a patient, which should be an average for each crew or unit. A wide variation in times can be expected—some days will be busy, with longer times, compared to others that may be shorter on days that are slower. Fluctuations will occur, and the use of unit-hour utilization should be averaged over a given period (for example, monthly, quarterly for seasonal tracking, and by crew to identify individual-performance issues). Once this is calculated an analysis can be done on what areas can be adjusted or how many additional units need to be provided for a more efficient response. The utilization period covers the time period from when the unit is assigned to a

$$\text{Unit Utilization} = \frac{(3684, \text{ total number of calls}) (38 \text{ minutes call time})}{(60 \text{ minutes}) (5 \text{ units}) (90 \text{ days}) (24 \text{ hours})} = 0.22$$

$$\text{Unit Utilization} = \frac{(5829, \text{ total number of calls}) (65 \text{ minutes call time})}{(60 \text{ minutes}) (5 \text{ units}) (90 \text{ days}) (24 \text{ hours})} = 0.58$$

$$\text{Unit Utilization} = \frac{(5829, \text{ total number of calls}) (65 \text{ minutes call time})}{(60 \text{ minutes}) (5 \text{ units}) (90 \text{ days}) (24 \text{ hours}) + (60 \text{ minutes}) (2 \text{ units}) (90 \text{ days}) (12 \text{ hours})} = 0.48$$

FIGURE 9.11 ■ Seasonal Variations in Utilization.

- ◆ Out-of-shoot time.
- ◆ Response time.
- ◆ On-scene time.
- ◆ Transport-to-the-hospital time.
- ◆ Patient-drop or restock time.

FIGURE 9.12 ■ Time Increments.

call to the time that unit becomes available for the next call (Figure 9.12).

Assume that an EMS unit runs eight transports in a 24-hour shift. This produces a unit utilization of 33%, which means the transport unit uses 33% of the shift transporting a patient. Unit utilization measures the amount of time that each unit is busy in a given time period. Some agencies measure only transports versus all calls. The IAFF has advocated also measuring the other requirements of utilization, including the time that units spend in training.

How often will an ambulance get a call? This can be determined based on historical data. It requires that EMS managers track the number of calls on a daily, weekly, and quarterly basis. The other time increment that is required is the average amount of time it takes a unit to run a call or the "in-service" time (Figure 9.13). The in-service time is defined as the time it takes an EMS unit to complete a call—from the time it is dispatched until the time it reports and is available to respond to another call. This can be affected by long waits at the emergency department due to long hospital drop times, extended scene times, or long responses (the time it takes to get to the scene).

ESTIMATING AMBULANCE DELAYS

The formula for the probability of delay and the length of waiting for an ambulance are complicated and involve using statistical formulas called Poisson and exponential distribution. To simplify the mathematics, Tables 9.4 and 9.5 are included with calculations of up to seven EMS units for the probability of a

$$\text{Unit UH Ratio} = \frac{(C)(T)}{60(N)}$$

C = number of calls in a given period.
T = time in minutes from dispatch to completion.
N = number of units.

FIGURE 9.13 ■ In-service Ratio Calculations.

delay and the length of the delay, respectively. These tables require the following calculation with the average number of calls and average time per call:

c—Number of calls in a given period.
t—Time in minutes from dispatch to completion.
    Table Key = ct/60

In the Table Key, "c" is the number of calls per hour and "t" is the average time per call. The use of 60 minutes converts the average time to hours for the convenience of using Table 9.4. The Table Key is actually the amount of work expected each hour. As an example, if c is equal to four calls per hour and t is 50 minutes, then the Table Key is 3.33 ($4 \times 50/60$). If this is then applied to Table 9.4, matching the number of units (4) to the Table Key of 3.3 (rounded from 3.33), you will find that the probability of a delay is 64%.

Taking the same numbers and applying them to Table 9.5, you will see that on average there will be 3.03 people waiting for EMS units, and by adding just one EMS unit, the number of people waiting for an ambulance drops to fewer than one or 0.62. The length of time a person will wait to get an ambulance can be figured by taking the figure from Table 9.5 and multiplying it by 60 to achieve a number in minutes and then dividing by the average number of calls. In this scenario 3.03 × 60 divided by 4 average calls per hour = a 45-minute wait for an ambulance. While this formula can work for emergency and nonemergency calls, in most emergency situations a mutual-aid ambulance will be summoned or a unit will drop the low-priority call to respond to a life-threatening call.

MANAGING UNIT HOURS

A common question is: how many EMS transport units does a system need? Once the probability of a delay and the unit-hour utilization have been figured, you can calculate the number of units needed to meet the agency's objectives. There is no national standard to determine what unit-hour utilization should be employed. In fire-based operations, given the other duties required of EMS units (such as training, inspections, and so on), some invalidated literature suggested a unit-hour utilization of 0.28 to 0.32. Private-ambulance-industry EMS units routinely operate at 0.40 to 0.50. The unit-hour utilization should be carefully discussed with the labor-management team.

To determine the number of EMS transport vehicles to meet a given unit-hour utilization, assume that

TABLE 9.4 ■ Probability of Delay

Calls for Service	Number of Units					
	2	**3**	**4**	**5**	**6**	**7**
1.0	33.3	9.1	2.0	0.4	0.1	0.0
1.1	39.0	11.5	2.8	0.6	0.1	0.0
1.2	45.0	14.1	3.7	0.8	0.2	0.0
1.3	51.2	17.0	4.8	1.1	0.2	0.0
1.4	57.6	20.2	6.0	1.5	0.3	0.1
1.5	64.3	23.7	7.5	2.0	0.5	0.1
1.6	71.7	27.4	9.1	2.6	0.6	0.1
1.7	78.1	31.3	10.9	3.3	0.9	0.2
1.8	85.3	35.5	12.9	4.0	1.1	0.3
1.9	92.6	39.9	15.0	4.9	1.4	0.4
2.0		44.4	17.4	6.0	1.8	0.5
2.1		49.2	19.9	7.1	2.2	0.6
2.2		54.2	22.7	8.4	2.7	0.8
2.3		59.4	25.6	9.8	3.3	1.0
2.4		64.7	28.7	11.4	4.0	1.3
2.5		70.2	32.0	13.0	4.7	1.5
2.6		75.9	35.4	14.9	5.6	1.9
2.7		81.7	39.1	16.8	6.5	2.3
2.8		87.7	42.9	19.0	7.5	2.7
2.9		93.8	46.8	21.2	8.7	3.2
3.0			50.9	23.6	9.9	3.8
3.1			55.2	26.2	11.3	4.4
3.2			59.6	28.9	12.7	5.1
3.3			64.2	31.7	14.3	5.9
3.4			68.9	34.7	16.0	6.7
3.5			73.8	37.8	17.7	7.6
3.6			78.8	41.0	19.7	8.6
3.7			83.9	44.4	21.7	9.7
3.8			89.1	48.0	23.8	10.9
3.9			94.5	51.6	26.1	12.2
4.0				55.4	28.5	13.5
4.1				59.3	31.0	15.0
4.2				63.4	33.6	16.5
4.3				67.5	36.3	18.1
4.4				71.8	39.2	19.9
4.5				76.2	42.2	21.7
4.6				80.8	45.3	23.7
4.7				85.4	48.5	25.7
4.8				90.2	51.8	27.8
4.9				95.0	55.2	30.1
5.0					58.8	32.4
5.1					62.4	34.9
5.2					66.2	37.4
5.3					70.0	40.0
5.4					74.0	42.8
5.5					78.1	45.6
5.6					82.3	48.6
5.7					86.6	51.6
5.8					90.9	54.8
5.9					95.4	58.0
6.0						61.4
6.1						64.8
6.2						68.4
6.3						72.0
6.4						75.7
6.5						79.5

TABLE 9.5 ■ Queue Length (Lq)

Calls for Service	Number of Units					
	2	3	4	5	6	7
1.0	0.33	0.05	0.01	0.00	0.00	0.00
1.1	0.48	0.07	0.01	0.00	0.00	0.00
1.2	0.68	0.09	0.02	0.00	0.00	0.00
1.3	0.95	0.13	0.02	0.00	0.00	0.00
1.4	1.35	0.18	0.03	0.01	0.00	0.00
1.5	1.93	0.24	0.04	0.01	0.00	0.00
1.6	2.84	0.31	0.06	0.01	0.00	0.00
1.7	4.43	0.41	0.08	0.02	0.00	0.00
1.8	7.67	0.53	0.11	0.02	0.00	0.00
1.9	17.59	0.69	0.14	0.03	0.01	0.00
2.0		0.89	0.17	0.04	0.01	0.00
2.1		1.15	0.22	0.05	0.01	0.00
2.2		1.49	0.28	0.07	0.02	0.00
2.3		1.95	0.35	0.08	0.02	0.00
2.4		2.59	0.43	0.10	0.03	0.01
2.5		3.51	0.53	0.13	0.03	0.01
2.6		4.93	0.66	0.16	0.04	0.01
2.7		7.35	0.81	0.20	0.05	0.01
2.8		12.27	1.00	0.24	0.07	0.02
2.9		27.19	1.23	0.29	0.08	0.02
3.0			1.53	0.35	0.10	0.03
3.1			1.90	0.43	0.12	0.03
3.2			2.39	0.51	0.15	0.04
3.3			3.03	0.62	0.17	0.05
3.4			3.91	0.74	0.21	0.06
3.5			5.17	0.88	0.25	0.08
3.6			7.09	1.06	0.29	0.09
3.7			10.35	1.26	0.35	0.11
3.8			16.94	1.52	0.41	0.13
3.9			36.86	1.83	0.48	0.15
4.0				2.22	0.57	0.18
4.1				2.70	0.67	0.21
4.2				3.33	0.78	0.25
4.3				4.15	0.92	0.29
4.4				5.27	1.08	0.34
4.5				6.86	1.26	0.39
4.6				9.29	1.49	0.45
4.7				13.38	1.75	0.53
4.8				21.64	2.07	0.61
4.9				46.57	2.46	0.70
5.0					2.94	0.81
5.1					3.54	0.94
5.2					4.30	1.08
5.3					5.30	1.25
5.4					6.66	1.44
5.5					8.59	1.67
5.6					11.52	1.94
5.7					16.45	2.26
5.8					26.37	2.65
5.9					56.30	3.11
6.0						3.68
6.1						4.39
6.2						5.30
6.3						6.48
6.4						8.08
6.5						10.34

$$3.5 \times 45/60 \times 0.32 = \text{Requires 8 units}$$

FIGURE 9.14 ■ Calculating the Number of Units Required for a Given Call Volume.

there are 3.5 calls per hour, with an average time for each call at 45 minutes. The goal in this example is to maintain enough units to meet a unit-hour utilization of 0.32 (Figure 9.14).

If an agency wants to ensure there is enough excess capacity not to incur a lengthy delay, a chi-square mathematical algorithm chart with either a 5% or 10% confidence interval can be used to determine the number of units to ensure an extremely low probability of a delay in answering a call. Look for a delay of less than 5%, since most performance objectives look to meet 95% of life-threatening calls in eight minutes or less. Anything more would be considered a delay. Using the previous example, you need a table key ($3.5 \times 45/60 = 2.62$ translated to degrees of freedom). Round this to 3, and it indicates that eight units should be able to maintain a standard of less than 5% of the time a person will be waiting for a unit to be available.

EMS leaders need to understand the effects of manipulating unit hours and increasing utilization on EMS crews. Utilizations that are too low produce a deterioration of EMS skills or "rust out"; too high, and they burn out crews. When unit hours are low, the cost per patient is high, and when unit-hour utilization is high, the cost per patient is low. Unit-hour utilization also needs to take into account the population density, size and shape of the service area, traffic patterns, and most importantly the response-time standards that the agency is trying to meet or comply with as required by contractual mandates. Matching units to call volume is a common challenge for EMS managers; other factors, such as poorly trained dispatching, lack of electronic technologies to track vehicles, and inflexible scheduling of crews, can result in utilizations that do not even out the workload.

DETERMINING NUMBER OF PERSONNEL

Once the total number of vehicles needed is known, the number of personnel to staff those vehicles must be determined. It is not as simple as placing two personnel in each vehicle for each shift; some systems operate with three personnel on an ambulance. Many fire-based EMS crews on the eastern coast of Florida operate with three personnel—two medical technicians and an officer or EMS lieutenant—on each vehicle. Some agencies figure in full-time equivalents in a traditional staffing plan that uses a formula to identify the number of personnel to staff for coverage of sick time and vacation days used by full-time employees. The worksheet in Figure 9.15 can be used to figure the number of personnel to fill one position. This worksheet gathers information to calculate the number of personnel to staff a field position 24 hours a day, 7 days a week. For agencies that do not staff on a 24-hour basis, enter the actual days and hours of work. Enter the number of shifts/platoons and the annual hours worked per group with the hours on and hours off for the shift rotation. The workweek hours then need to be entered. The agency must know the average leave used by employees, including sick leave, on-duty injury, vacation, training leave, and other leave. Subtract the average leave used from the hours worked per group, and enter as hours actually worked. The total annual hours of work divided by the hours actually worked calculate a staffing factor.

The staffing factor is then moved to the marginal personnel requirements. If the system is bringing on more people to increase from BLS to ALS services, new positions are required. The worksheet shown in Figure 9.16 can be used. The equalizing per shift requires a new position to be multiplied, using the new field positions per shift, by a staffing factor. Then divide the results by the number of shifts, and enter this number in the "Equalized per Shift" column. If the resulting number is not a whole number, then round up. The worksheet uses the staffing factor of number of shifts. Multiply the "required per shift" by number of shifts/platoons, and enter the results as "total personnel required." This is the number of people required to fill positions 24 hours a day, 7 days a week, including employee leave coverage.

To estimate the marginal-personnel cost of an EMS system, wages for each position need to be calculated. Record the number of new or existing employees to be reassigned, their current and step increases in wages, and any overtime or Fair Labor Standards Act (FLSA) costs in Figure 9.17. The number of new employees as calculated in the marginal-personnel worksheet is carried over to the summary sheet. Some systems have opted for a pattern called **constant staffing.** A constant-staffing plan uses one person for each budgeted position on each unit and fills vacations

Hours of work to be covered in 1 year	
Days of work	365
Hours of work	24
Total annual hours of work	8760
Number of shifts/platoons	
Hours worked per group	(8760 divided by # of shifts)
Shift rotation	Hours on
	Hours off
Work week (hours)	
Average leave used per person (hours)	
Average sick leave	
Average on-duty-injury leave	
Average vacation leave	
Average training leave	
Average holiday leave	
Average bereavement leave	
Average other leave	
Average leave per employee	
Hours actually worked by average employee	
Staffing factor	(Number of employees required to fill one position within the department)

FIGURE 9.15 ■ Staffing Factor Calculation Worksheet.
(Source: International Association of Firefighters EMS Implementation Guidebook. *Reprinted with permission.)*

Apparatus	Training Level	New Field Positions	Staffing Factor	Number of Shifts	Equalized per Shift	Required per Shift	Total Personnel Required
EMS Engine							
	CFR						
	EMT–B						
	EMT–I						
	EMT–P						
EMS Transport							
	EMT–B						
	EMT–I						
	EMT–P						
Batt. Chief/Supervisor							
	Total						

FIGURE 9.16 ■ Marginal Personnel Worksheet.
(Source: International Association of Firefighters EMS Implementation Guidebook. *Reprinted with permission.)*

Year/Phase	1	2	3
Estimated Transport Volume			
Estimated Actual Increase in Transport Volume			
Personnel			
Number of new personnel			
Wages			
Employer-paid payroll taxes			
Hiring costs			
Cost of benefits			
Health			
Dental			
Vision			
Disability			
Miscellaneous			
Pension contributions			
Workers' compensation costs			
Longevity pay			
EMS incentive/premium pay			
Training and Certification			
New-hire recruit-program attendees			
Recruit-program costs			
New-hire EMS-training attendees			
EMS-training costs			
EMS continuing education attendees			
Continuing education costs			
Personnel requiring exams and certification			
Examination and certification costs			

FIGURE 9.17 ■ Wage Calculation Worksheet.
(Source: International Association of Firefighters EMS Implementation Guidebook. *Reprinted with permission.)*

and sick time with current staff on overtime. This saves expenses on benefits, training, and protective clothing, which are commonly one-third of the cost of a new employee.

SUBSIDY-PRICE TRADE-OFF

To calculate a system's subsidy-price trade-off, collect all the estimated and total payments for all emergency and nonemergency services from all sources. This includes payments received from local tax support, private payers, third-party payers, HMO contracts, Medicare, Medicaid, donated funds, subscription fees, and so on. That number is then divided by the population of the primary service area and plotted on the horizontal axis of the subsidy-price trade-off chart in Figure 9.18.

The second point on the chart is calculated by estimating the current average total bill, which includes the base rate, mileage, and all add-on charges. That average is then graphed on the vertical axis and a line drawn between the two points. The diagonal line can then be compared with proposed changes in the service or budgetary changes. For example, if fire department A is operating with a subsidy of $24 and has an average bill of $550 for a transport, any point on the diagonal line combining any change in subsidy and ambulance bill, as long as the changes plot on the diagonal line, the system will operate the same.

Most EMS agencies require a costing-out summary. The worksheet in Figure 9.19 uses all of the relevant cost to start up, transition, or maintain an ambulance service. A summary should be based on estimated transport volume and the annual increases. The number of new personnel is calculated in the marginal-personnel requirements. Total wages, hiring cost, and the cost of benefits need to be entered into the summary worksheet. Pension, workers' compensation, and longevity pay or incentive

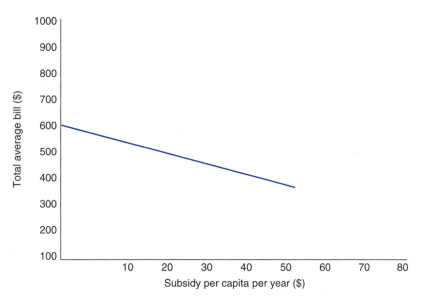

FIGURE 9.18 ◼ Subsidy-Price Trade-Off.

Year/Phase	1	2	3
Apparatus			
New transport units			
Purchase/lease cost of transport units			
Equipment			
BLS equipment package for transports			
ALS equipment package for transports			
Monitor/defibrillator for transports			
Transport vehicle radios (dispatch/medical control)			
Transport vehicle cellular phone			
BLS equipment package for engines			
ALS equipment package for engines			
Monitor/defibrillator for engines			
Medical control radio for ALS engines			
Disposable supplies			
ALS disposables			
BLS disposables			
Equipment repair/replacement			
Turnout gear for cross-trained/dual-role employees			
Operational			
Transport fuel costs			
Transport maintenance costs			
Medical director costs			
Medical liability insurance			
EMT–B cost			
EMT–I cost			
EMT–P cost			
Billing service charges			

FIGURE 9.19 ◼ Summary of EMS Operational Cost.
(Source: International Association of Firefighters EMS Implementation Guidebook. *Reprinted with permission.)*

pay need to be recorded and may also include education incentives. Certification cost, continuing education, and the cost of vehicles and equipment are figured in the summary (refer to Figures 9.17 and 9.18). Additional cost for firefighting gear, fuel cost, maintenance cost, medical liability insurance, and the cost of the Medical Director's salary must be added to the summary sheet in Figure 9.19. If an EMS billing company is used, its cost or the cost of its service must be added.

■ FUNDING MECHANISMS

The majority of local governments are funded by taxes. Taxes typically come from property taxes, local income taxes, and a general sales tax. In many cases there is a special tax sometimes called a millage rate that is collected for fire and EMS.

To properly determine the amount of property tax, several values are needed. These values are the fair market value (FMV), assessed value (ASV), and millage rate. The assessed value is typically 40% of the fair market value of the property. The millage rate is determined by the proposed budget for EMS services divided by the tax digest, which is the total number of assessments in that district. The millage rate is defined as dollars per $1,000 of assessed value. The 2006 millage rate for a county EMS and fire system set at $30.08 per $1,000 would be the decimal equivalent of 0.03008. As an example, the county fire and EMS millage rate was set at 0.03008. If your property has a FMV of $100,000, the ASV would be 40% or $40,000. Deduct any exemptions and multiply the ASV of $40,000 by the milage rate of 0.03008 to get a close estimate of person's property taxes. Assuming this example is for a resident with no exemptions, the total tax bill would be approximately $1,203.

Sales tax, sometimes called **gross-receipts tax,** is taken from the sales of local businesses. Often municipalities will annex geographic areas with commercial activities to collect sales tax revenue. A gross-receipts tax is a tax on the total amount of the sale price of all sales or the total amount charged or received for the performance of any act, service, or employment as done as part of or in connection with the sale of goods, wares, or merchandise.

In 2004 the state legislature authorized Pennsylvania municipalities to enact a new "Emergency and Municipal Services" (EMS) tax. The EMS tax can be enacted by any municipality in an amount not to exceed $52 per year. Employers are required to deduct and remit this tax from employees whose headquarters location levies such a tax. The City of Pittsburgh assesses a flat tax of $10 on all property within the city's EMS service area. In the city of San Leandro, California, a tax per "benefit unit" (BU) is $9.90 per year (Figure 9.20). The number of benefit units is determined by the use to which the owner or occupant has put the property. The state of Utah assesses a surcharge from all criminal fines, penalties, and forfeitures imposed by the courts, and 14% of the amount collected is allocated to the EMS grant program.

Other taxes may include **impact fees**. These are often assessments on new construction, new business, or activities as a one-time fee to establish infrastructure in the district or community. Each state has different legislation that allows municipalities to assess fees for various government and school expenses. Some communities impose a fee per square foot of new construction; others charge a fee only to commercial, not residential, square footage.

The state of Maryland applies an EMS tax to fund EMS and trauma systems, including state police EMS helicopters. Maryland created a dedicated EMS fund in the late 1980s by assessing a fee on all vehicle registrations. Support for the program is so strong that the fee has been raised from $8 to $11.

Idaho passed a law requiring an annual $1 fee for each driver's license issued, to go into a dedicated EMS fund. The fund receives $1.4 million annually and has rescued Idaho's embattled EMS programs by helping to pay for extrication equipment, ambulances, and communication tools for rural programs. Hotel, lodging, and gambling taxes also can help provide funding for EMS services in some communities for the impact that transient populations have on emergency services.

AMBULANCE MEMBERSHIP PROGRAMS

Membership or subscription programs have been implemented by several local fire and EMS organizations. An ambulance membership program is a fixed-price prepayment of all or part of fee-for-service ambulance-service bills for a certain period of time (Table 9.6). Federal regulations qualify two types of EMS subscriptions programs. Under the first type of program, the provider may not bill third-party payers

Section 2-13-140. Amount of Tax

The tax per "benefit unit" (BU) is Nine Dollars and Ninety Cents ($9.90) per year. The number of benefit units shall be determined by the use to which the owner or occupant has put the property, as follows:

One living unit	1 BU
Two to five living units	3 BU
Six or more living units	1 BU per unit
One-story store	2 BU
Store first floor w/office/apts. above	4 BU
Miscellaneous commercial	2 BU
Department store	5 BU
Discount house	5 BU
Restaurant	4 BU
Shopping center	7 BU
Supermarket	4 BU
Comm./indus. condominium	4 BU
Warehouse	2 BU
Light industry	4 BU
Heavy industry	6 BU
Miscellaneous industrial	2 BU
Nurseries	2 BU
Quarries	2 BU
Wrecking yards	2 BU
Terminals, trucking	2 BU
Improved government-owned property	2 BU
Golf courses	2 BU
Schools	5 BU
Churches	4 BU

Other institutions	3 BU
Lodge halls	7 BU
Clubhouses	4 BU
Car washes	2 BU
Commercial garage/auto repair	2 BU
Service station	2 BU
Funeral homes	2 BU
Nursing/boarding homes	6 BU
Hospitals	5 BU
Hotel	5 BU
Motel	5 BU
Mobile home parks	5 BU
Banks	4 BU
Medical-dental offices	4 BU
One- to three-story office	4 BU
Over three-story office	7 BU
Bowling alleys	4 BU
Theaters, walk-in	4 BU
Theaters, drive-in	4 BU
Other recreational	4 BU

In order to accommodate the increased costs of providing emergency medical services, the annual tax per "benefit unit" shall be adjusted annually by an amount not to exceed the Consumer Price Index (All-Urban Consumers, San Francisco Bay Area) unless the City Council finds and determines that said adjustment would result in the tax exceeding the cost of providing the services. Under no circumstances shall the total adjustment provided for herein exceed Five Dollars and Ten Cents ($5.10) per benefit unit.

FIGURE 9.20 ■ Excerpt from a Municipal Ordinance to Offset EMS Cost.

for services rendered and all the costs are supplied by the membership program. The second type of program is the preferred method and more common. With it, the provider can bill and collect from third-party payers or insurance companies and the deductibles and copays are taken care of by the membership program. It generally offers less exposure to the EMS provider and lower cost to the user.

An example of a subscription program, the FireMed program in Eugene, Oregon, was one of the first programs brought online by fire-based EMS.

TABLE 9.6 ■ Membership Billing Example

Actual Ambulance Services	$475.00
Medicare Allows	$308.75
Medicare Reimbursement	$247.00
Out of Pocket Expenses Covered by Membership	$228.00

FireMed offers ambulance and wheelchair transportation, emergency services, nonemergency services, and mutual-aid services.

Ambulance membership programs should be linked to the goals of the organization. Membership programs can provide additional personnel, equipment, enhanced training, a mechanism to improve marketing and customer service. Ambulance memberships work best when the public is sensitive to ambulance rates, the service is perceived as high quality, the EMS agency has name recognition, and the area has the demographics that can support the program. Consider a statistically valid public-opinion poll to determine membership-program success. If EMS providers wish to start an ambulance membership program, they must receive a letter from Medicare granting permission to do so; Medicaid patients already are exempt from unreimbursed covered services.

Some successful features of a subscription program are listed in Figure 9.21. It is important to realize that a subscription program may bring an increase in

- The program covers nonemergency and emergency transport as a full-service program.
- Renewals will be dependent on the reputation of the service.
- The EMS and subscription programs must provide excellent customer service.
- The program is marketed through a concentrated advertising effort, and members can only join during a specific time of the year coordinated with annual budgets.
- High transport rates and aggressive collection efforts are necessary to provide citizens with an incentive to purchase a subscription.
- The subscription is managed as a stream of income for multiple years with renewals.

FIGURE 9.21 ■ Components of a Successful Subscription Program.

utilization and may burden 911 resources. It is important to track revenue estimation by monitoring membership fees and increased member utilization. Expense estimations need to be monitored; for example, administration, marketing, increased call volume, and start-up cost. Each community is different in its experience with subscription services, and the revenue brought in needs to be monitored and adjusted with any increase in utilization. Policies and procedures should address misuse, and field crews should be educated on the program's mechanics.

It is important to ensure that any membership agreements include language that complies with Medicare, Medicaid, insurance, and state EMS regulations. Mandatory Medicare assignment has reduced the benefits of ambulance membership programs. The current Medicare regulations require a provider to receive what Medicare will pay and not bill additional charges to a Medicare patient. Any contract needs to establish the location, membership definitions, and when membership starts and stops. Provisions should be in the contract for revoking membership, member requirements, exclusions, reciprocal coverage, and location of services. EMS agencies need to designate a specific time of year and location to sign up for the program. It may be more cost-effective to place the form on a Web-based application.

NONTRANSPORT MVA REIMBURSEMENT

A common EMS response is for motor-vehicle collisions, yet in most systems, due to new vehicle safety technology, a significant portion of these calls do not require an ambulance transport. Most medical insurances only pay when a patient is transported. Motor-vehicle accident in-

surance allows for emergency services to be reimbursed, usually up to $350, so a cost-recovery program can recover the cost of the EMS First Responder services to MVAs. In an accident the auto-insurance deductible will already be met by the damage to the vehicle.

Policies that allow ambulance billing for auto insurance per incident on the average will have already met their deductibles. The fee can be collected from the vehicular insurance without a resident's ever receiving a bill. The area's insurance rates will not be impacted by the collecting of these fees, because only 0.7% of the residents are involved in accidents and the state insurance commission must approve any rate hikes.

Currently, your community's taxpayers are paying for EMS resources to respond to accidents by nonresident drivers. Studies show that nonresidents account for 40% to 85% of the accidents, depending on the nonresident traffic. In most communities this revenue source is not collected due to limitations on documentation. While most EMS agencies collect a medical release and demographics on noninjury crashes, few collect insurance information. This type of program requires EMS field crews to capture the automobile insurance from a nontransport motor-vehicle accident (Figure 9.22).

FIRST RESPONDER FEES

In many states, municipalities, and service areas, the private ambulance service is the primary EMS provider and transporting agency. In locations that utilize the fire service to support the private-ambulance business model, the fire service is reimbursed for its first response or assistance. EMS managers can request or place in a local ambulance ordinance the reim-

FIGURE 9.22 ■ Nontransported patients may be billed for service and reimbursed from automobile insurance.

bursement cost for ambulance services, often called marginal cost.

Marginal cost can include the cost of training, supplies, equipment, fuel, and accelerated maintenance and vehicle-replacement schedules. A good place to start calculating marginal cost is by computing the cost per response. Somewhere between the fire agency's cost per response and the ambulance service's unit-hour cost is a figure that will add to fire service resources, yet not significantly impact the ambulance company's profit margins.

Medicare and the Centers for Medicare & Medicaid Services (CMS) allow for an ALS-intercept charge. Paramedic-intercept services are those provided by paramedics who operate separately from the ambulance transport service, such as an ALS squad responding to assist a rural volunteer ambulance service transporting a patient. Often it involves a paramedic jumping on board an existing BLS transporting ambulance. This also could be an ALS ambulance from a private agency that meets a rural unit and transfers the patient into a different ambulance to allow the rural unit to return to its home town.

The intercept must be provided in rural areas, be provided under contract with one or more volunteer services, and be medically necessary based on the condition of the beneficiary receiving the service. In addition, the volunteer service must be prohibited from billing as a volunteer service and only provide BLS service. Payment for the ALS intercept is the difference between the ALS rate and the BLS rate. Since the full ALS rate in a non-intercept scenario is based on two parts—the ALS and the transport—the intercept rate is less than the full ALS charge. The ALS rate minus the BLS charge is the difference that will be provided to the ALS responder.

FUTURE REIMBURSEMENT ISSUES

As the health-care system becomes more taxed, EMS systems especially in urban areas will have to be creative in the delivery of their services. In many systems fewer than 50% of the calls for service result in no transport or treat and release. EMS agencies have no standardized mechanism to collect a fee for this service. In other systems nontraditional transport is used for certain patients.

Some systems budget bus-token or taxicab vouchers for patients to find their own ways to their health-care providers. The Phoenix Fire Department operates a program called an alternative-response (AR) van. These vans are staffed with a social worker and an EMT to handle low-level assault calls, domestic violence calls, and EMS responses that are more social problems than medical emergencies. There is no mechanism to reimburse for this service, and the justification is solely to reduce the need for more expensive ambulances or disrupting response times for emergencies.

With hospital overcrowding becoming more of a problem, the transportation of patients to nontraditional destinations is being evaluated. For example, an EMS provider transports a patient to a clinic and is reimbursed by insurance or Medicare. Under most insurances and current Medicare regulation, an ambulance service will be reimbursed only when the patient is delivered to a Joint Commission on the Accreditation of Healthcare Organizations (JCAHO)–certified hospital.

BILLING SERVICE

In some organizations the task of billing and collection is contracted to a vendor or firm that has expertise in health-care billing. Medicaid, Medicare, and private insurance companies generally require electronic billing and arrange for transfer of money by wire or electronic transfers of deposits. Many EMS organizations do not have the infrastructure or staff to keep up with this activity. Before contracting with a billing service, an EMS agency should develop a list of the services that are required. A billing-service contractor can provide training and may improve collection rates. Since many billing companies take a percentage of the billing revenue, their incentive is to maximize collections.

One option is to establish an **enterprise account,** which is often set up in government budgets to contain money within a specific operation; for example, appropriating money collected from ambulance transports only to EMS operations within the fire service instead of allowing that money to pay for other non-EMS functions or be returned to the general fund.

When billing a patient, it is customary to send no more than four statements. The first bill should go out within 48 to 72 hours as an initial invoice and as a non-itemized bill asking for insurance information and signature. After 15 days, send a second bill with an itemized statement and another insurance request, if needed. After 30 days, the agency or billing vendor should send an itemized statement with a past-due

- ◆ Step-by-step description of billing process with time frames.
- ◆ Training for staff both initial and ongoing.
- ◆ A clear explanation of financial terms.
- ◆ Responsibilities for computer software, support, or additional cost.
- ◆ References.
- ◆ Key personnel and their responsibilities.

FIGURE 9.23 ■ Contracts and Proposal Inclusion for Billing Services.

message. After 55 days, send an itemized statement as a final notice. After 60 days, a patient should be sent final notice with a message informing the patient that his account will be referred to collections. If there is no response after 90 days, the account should be sent to a collection agency. Some municipal agencies do not send ambulance bills to collections for fear of a public backlash when they ask for additional bonds or funding. Most agencies allow for a grace period in billing of five to seven days before action is taken.

When using a billing contractor, it is important that any payments to the EMS organization forwarded immediately. (Be sure that is detailed in the contract with the billing contractor.) Many billing companies float money by holding it and collecting interest or paying other expenses. A monthly statement should be provided from the billing agency. The billing contractor should provide the EMS agency with a monthly bill. Contractually, it is important to ensure that HIPAA and federal regulations that mandate how Medicare billing should be done are met. It is important to ask and verify that any ambulance billing contractor carries insurance for errors or omissions and negligent acts throughout the contract. A billing contract with an outside firm should include the information in Figure 9.23. EMS managers should require customized reports on billing activities on at least a quarterly basis.

Collections generally progress through a process designed to maximize revenue (Figure 9.24). A "soft collection" is simply a billing sent to the patient or insurance provider. Often several mailings or electronic attempts are sent. A more stern final-payment request, often sent by registered mail with a follow-up phone call, is considered a hard collection. Bills for service typically progress to small-claims court, superior court, or a collection agency that may decide to file for an asset seizure as a final attempt to secure payment.

[ ] Identify levels of service to be provided by the department or agency.
[ ] Calculate base rate for each level for each year of phase-in using the Ambulance GPCI (Physician Practical Expense factor).
[ ] Review current billing rate to determine estimated effect (increase or decrease in revenue). If current rates are lower than allowed fee, consider restructuring fee schedule.
[ ] Review collection rates/bad debts/claim denials. Contact intermediary to determine cause of claim denial, isolate primary causes, review existing billing cycles and secondary billing procedures.
[ ] Review existing billing policies and procedures. Assess advantages versus cost of electronic billing if not currently using this method. Are forms in use current? Are claims submitted regularly? What is the average claims-processing time?
[ ] Revise and update billing policies and procedures to address fee-schedule requirements.
[ ] Assess knowledge and experience of current billing personnel. Are personnel knowledgeable regarding whom to contact for questions? Is correct contact information for questions to insurance carriers available?
[ ] Consider advantages of billing service. If currently using a billing service, contact and discuss their actions regarding the fee schedule. What changes are required in the current submission practices of the organization? Will they provide training to organization personnel?
[ ] Assess current knowledge level of personnel regarding billing requirements.
[ ] Arrange for training officer to conduct training of personnel. *CMS Training Manual.* Intermediary-provided training.
[ ] Ensure resources are available to all personnel: HCPCS codes, zip codes, etc.

FIGURE 9.24 ■ Ambulance Billing Checklist.

FOUNDATIONS AND NONPROFITS

EMS leaders may opt to develop a not-for-profit foundation or association to help supplement special initiatives within a fire-based EMS agency. Research grants, estate gifts, and endowments from other nonprofit organizations often have restrictions on the money that require it to stay outside the normal government budgets. This allows for a more targeted approach and for other service groups or private individuals to advance a program or aspect of EMS service that may not be a priority or affordable by the EMS agency's governing or funding agency.

The Medic One Foundation in Seattle does *not* supplement the normal operating budgets of the King County EMS providers, but it does support key areas of the Medic One program. Medic One Foundation money supports paramedic training, continuing education, quality assurance, targeted research, and special research projects. Foundations and associations have a parallel fundraising mission that often finds the public leaving estate gifts. Several million dollars has passed through the Medic One Foundation through gift planning, research money, and contributions from corporations.

A foundation or nonprofit association must have a board of directors of at least three people and operate under state and federal law as a corporation. Directors and board members should reflect people in the community, with an emphasis on successful professionals. To qualify for federal tax-exempt status under 501(c)(3) of the Internal Revenue Code, the nonprofit must be organized and operate for some religious, educational, charitable, scientific, or literary purpose or for testing for public safety, fostering of national or international amateur sports, or prevention of cruelty to animals or children—purposes permitted under this section of the code. Nonprofits may also be formed for other purposes pursuant to different sections of the Internal Revenue Code.

To qualify for federal tax-exempt status as a nonprofit under a different section of the code, your corporation must comply with the requirements of that federal tax-code section. The business purpose of the nonprofit must be listed in the articles of incorporation, and for the organization to apply for tax-exempt

- ◆ Total gross billings.
- ◆ Collection rates.
- ◆ Billing mix.
- ◆ ALS versus BLS.
- ◆ Miles billed.
- ◆ Transfers versus primary 911 billings.
- ◆ Payee mix.
- ◆ Accounts receivable turnover rate.
- ◆ Bad-debt expenses.
- ◆ Contractual allowances.
- ◆ Write-offs.

FIGURE 9.25 ■ Components of an EMS Financial Report

status, it is very important that its purpose be well described in the articles.

FINANCIAL REPORTS

EMS managers will need to assist or prepare a financial report. Financial reports help to provide accountability to the parent agency and elected officials. In some EMS agencies it is required by law that the agency submit a financial report 501(c)(3) nonprofits as an example. Financial reports help identify the health of the organization and indicate the ability of EMS administrators to manage the organization's money and budgets. The elements of an EMS financial report are included in Figure 9.25.

CHAPTER REVIEW

Summary

For an EMS manager or leader, the cost of an agency's services must be calculated as accurately and with as much detail as possible. The cost should be tracked to aid the EMS manager in making key decisions on operations and the placement of resources. Billing is often a complex and ever-changing process that requires expert knowledge of Medicare regulations and billing techniques. EMS agencies can maximize their revenue streams by identifying alternative income or establishing funding mechanisms for other aspects of EMS service. Financial reports should be reported to the management team and used in strategic planning. Financial monitoring and participation in the budgeting process are key functions for EMS managers.

Research and Review Questions

1. Identify the cost of your service based on the models or formulas provided in the chapter.
2. Create an EMS financial report for your agency.
3. Review your agency's privacy policies and make recommendations for change in the system.
4. Prepare a budget for EMS standby activities at the closest college football facility. Use the current schedule of games at that location.
5. Cost out the difference between a two-paramedic ambulance crew versus paramedic-and-EMT teams.

6. Figure the cost of a three-person and four-person ALS engine versus a BLS engine calculating only the EMS equipment on the vehicle and not the fire equipment.

7. Obtain your agency's call statistics and calculate unit-hour utilization for each EMS unit. Identify the workloads for each unit in the organization.

References

Centers for Medicare & Medicaid Services. (2001). *Negotiated Rulemaking Committee minutes*. http://www.cms.hhs.gov/apps/media/press/release.asp?Counter=610

Centers for Medicare & Medicaid Services. (2002). Ambulance fee schedule final rule. Retrieved 2004 from http://www.hcfa.gov/regs/CMS1002FC.doc

Centers for Medicare & Medicaid Services. (2002). Medicare and Medicaid Provisions of Balanced Budget Act of 1997. Retrieved 2004 from http://www.hcfa.gov/regs/bbaupdat.htm

Centers for Medicare & Medicaid Services. (2007). *CMS manual on Medicare ambulance fees*. Retrieved November 20, 2007, from http://www.cms.hhs.gov/AmbulanceFeeSchedule/

Code of Federal Regulations. (2001). §42CFR410.40: Coverage of ambulance services. Washington, DC: U.S. Government Printing Office.

Code of Federal Regulations. (2001). §42CFR410.41: Requirements for ambulance suppliers. Washington, DC: U.S. Government Printing Office.

Health and Human Services. (2002). HHS fact sheet. http://www.hhs.gov/news

Lee, R. D., and Johnson, R.W. (1994). *Public budgeting systems* (5th ed.). Gaithersburg, MD: Aspen Publications.

Medicare Program: Revisions to Payment Policies under the Physician Fee Schedule for Calendar Year 2001. (2001). *Federal Register, 65*(212). Washington, DC: U.S. Government Printing Office.

Newkirk, W. L. (1984). *Managing emergency medical services: Principles and practices*. Reston, VA: Reston Publishing Company.

Oberlander, J., Marmor, T., & Jacobs, L. (2001). Rationing medical care: rhetoric and reality in the Oregon Health Plan. *CMAJ, 164*, 1583–1587.

Wirth, S., Wolfberg, D., & Page, J. O. (2003). *Better billing: The ambulance service model compliance plan*. Mechanicsburg, PA: Page, Wolfberg & Wirth, LLC.

Wirth, S., Wolfberg, D. & Page, J. O. (2004). *The ambulance service guide to HIPAA compliance* (3rd ed.). Mechanicsburg, PA: Page, Wolfberg & Wirth, LLC.

Worth, D. (2002, February). Legal consult: Get hopping on HIPAA. *EMS Insider, 29(2)*.

Medical
Practice

Accreditation Criteria

CAAS

201.01   Medical Oversight: Strong leadership from the agency's medical director is key to establishing current, appropriate clinical standards (pp. 255–272).

Objectives

After reading this chapter the student should be able to:

10.1 Understand and define the role of the physician medical director for an EMS service (pp. 255–258).
10.2 Understand the selection process and qualifications desired in a medical director (pp. 267–269).
10.3 Differentiate between on-line and off-line medical control (pp. 265–267).
10.4 Identify the areas of responsibility that need physician involvement in an EMS organization (pp. 259–261).
10.5 Understand and create a system that involves the physician in due process to apply discipline to an EMS worker (pp. 260–261).
10.6 Identify the training and opportunities to promote professional development for a physician involved with or entering the field of EMS (pp. 267–269).
10.7 Understand the role of the EMS medical director in developing on-line and off-line medical control (pp. 264–267).
10.8 Describe the network opportunities for a physician medical director (pp. 258–259).

Key Terms

algorithmic protocol (p. 262)
American College of Emergency Physicians (ACEP) (p. 258)

communications protocols (p. 262)
delegated practice (p. 255)

descriptive protocols (p. 262)
destination protocols (p. 263)
due process (p. 261)

Point of View

A common feature of any quality EMS system is an engaged and active medical director. From a legal standpoint every EMS system must have a medical director who assumes ultimate responsibility for patient care. However, the level of medical-director involvement can range from a physician's simply signing a few legal documents and performing a periodic review of problem records to a full-time medical director's passionate involvement in every aspect of system design and operation. In my experience the more involved the system medical director is, the better the system. Most physicians are not familiar with the practice of prehospital medicine. Many believe that medical care begins when the patient arrives at the hospital door. Of course, nothing could be further from the truth—there is a growing body of bona fide empiric research that demonstrates that prehospital care saves lives and the quality of prehospital care provided can impact the patient's subsequent medical care.

The traits of an EMS-system medical director are those that we look for in any medical consultant: ability, availability, and affability. First and foremost, an EMS medical director must be competent in the practice of emergency medicine because prehospital care is in the domain of emergency medicine. He should be board certified in emergency medicine or in a related specialty with demonstrated competence in emergency medicine. Fellowship training in EMS is an added benefit. You cannot expect a physician to serve as a system medical director unless he is competent in all of the procedures expected of field providers. Medical directors must intimately understand the unique environment in which EMTs and paramedics practice. Many of us learned this by working as EMTs and paramedics prior to medical school. Others have learned this through the devotion

of time on ambulances and in the other aspects of the EMS system. Some have actually taken EMT and paramedic education in order to understand how to function in the relatively austere prehospital environment. Regardless, the first trait of a successful medical director is ability; if he lacks ability, field personnel will never develop the needed respect for the medical director that is required for the system to function effectively.

The second quality an EMS system medical director must have is availability. That is, he must be available to deal with any unusual problems that occur in the system. Many EMS systems have become so busy that they now employ full-time medical directors or contract for the provision of medical direction through a hospital or medical school. Availability goes beyond providing on-line medical direction. The medical director must be engaged in all aspects of EMS system operation. A good medical director will feel equally comfortable in the boardroom, the communications center, the classroom, and the field. He will get out on the streets on a regular basis and interact with field providers. The staff should become comfortable with the medical director and not be intimidated when working side by side with him. The medical director should also be available for educational sessions, for policy and procedure development, and to any field provider on a regular basis.

Finally, a common feature of successful EMS medical directors is affability. The EMS medical director must respect the EMTs and paramedics and treat the relationship as collegial instead of totally supervisory. However, because medical directors will sometimes have to make disciplinary decisions with due process, there must always remain a supervisory and management relationship. In turn the medical director must be respected by the field providers and seen as an advocate not only for the patient but also for the provider. Often when problems arise related to patient care, blame is placed on prehospital providers. The medical director must be willing and able to represent field providers as an advocate in hospitals and the medical community as a whole. A good medical director will remain open to suggestions and comments from field providers. While the development of off-line protocols must be evidence based, a good medical director will involve various system members (EMTs, paramedics, supervisors, managers, dispatchers, supply staff) in the development of protocols and procedures. These should never be written in a vacuum.

In summary an EMS system is only as good as the system medical director. Systems devoted to quality care should seek out a medical director who shares the system's vision and passion for patient care. When designing an EMS system, it is important to budget for system medical direction. Like many things in life, with medical direction, you get what you pay for. If you simply want a physician who will sign the requisite legal forms and review a few charts every couple of months, your system will be doomed to mediocrity. If you want a quality, cutting-edge EMS system, you need a quality medical director who is able, available, and affable.

> Bryan E. Bledsoe, DO, FACEP
> *Co-chair,*
> *Curriculum and Education Board for U.S. Special*
> *Operations Command (USSOCOM),*
> *MacDill AFB, Florida*

■ MEDICAL LEADERSHIP

In successful EMS systems one of the most common elements is strong and visible physician leadership. EMS members must understand that in the course of performing their professional health-care duties they serve as an extension of their system's medical director. The medical director is the key to quality patient care and the competency of EMS members. Paramedics are empowered by the medical authority or bureaucracy within the city, state, or local government to practice under the license of a physician. By his presence in the organization, the medical director extends medical accountability and clinical supervision and provides clinical training and continuing education of prehospital providers.

The medical director also provides his medical license to the EMS members. By law the arrangement by which a paramedic practices under a medical director is known as **delegated practice**. With such a major investment, the medical director needs to be involved in the management of the EMS system, must understand its operation, will direct its clinical practice, and should know and trust the personnel involved (Figure 10.1).

MEDICAL AUTHORITY

The medical director must collaborate with EMS leaders on the EMS system or organization's mission. An EMS system must have a single medical authority that oversees patient care and that holds the providers responsible for delivery of patient care to industry standards. Some systems have a medical advisory board that may also serve as the area authority. In those cases it is often still one chief medical officer who has the statutory authority for certifications and uses the board in an advisory capacity.

The medical director is the link between the government medical authority and the provider agency. The medical director ensures accountability that prehospital procedures and the action of providers are in accordance with acceptable medical practice. The physician as medical director is the final say when it comes to medical issues, so the medical director's role needs to be clearly defined in department SOGs, SOPs, or contracts. He is responsible for recommending certification, recertification, decertification, and suspension of prehospital personnel and should oversee the development and management of a medical oversight board to review medical issues.

ROLE OF THE AGENCY MEDICAL DIRECTOR

The medical director is a key figure in the leadership of an EMS organization (Figure 10.2). He provides leadership that sets the standards and the tone for the agency or system as it relates to patient care. An EMS medical director can function in several settings: He can be assigned or contracted to a single agency to supervise

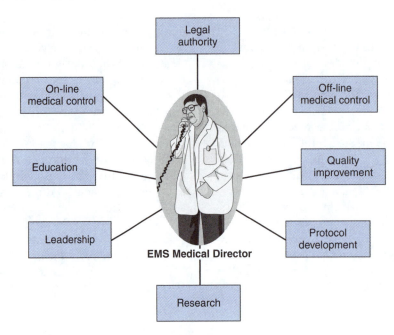

FIGURE 10.1 ■ The Medical Director's Role in the EMS System.

patient care. An entire EMS system that encompasses local, state, regional, or national programs with multiple services will use a medical director to supervise or administer operations. Educational programs that provide varying levels of EMS education are required to have a medical director; for example, in advanced training

FIGURE 10.2 ■ Communication between the EMS crews and the physician medical director must be open and frequent. (© *Edward T. Dickinson, MD*)

EMS medical direction is required for both coursework and student evaluation.

The medical director must ensure compliance with patient-care standards. This role is commonly referred to as clinical supervision, and often it is conducted in conjunction with other health-care providers, such as nurses, physician assistants, educators, and EMS supervisors. Clinical supervision of EMS members may be prospective, immediate, or retrospective. **Prospective supervision** (off line/indirect) refers to written standards, policies, procedures, and treatment protocols. **Immediate supervision** (on line/direct) involves real-time medical direction by a physician to a field provider. On-line direction requires the medical director to know the capabilities and limitations of providers, provider treatment protocols, and provider standing orders. **Retrospective supervision** (off line/indirect) includes medical audit of a specific response or overall system quality management.

The medical director is responsible for ensuring that prehospital personnel under his jurisdiction and supervision meet certification or licensure standards; this includes some oversight to ensure that the 911 communications specialist and dispatch-center personnel meet initial qualifications. The medical director should provide routine review of education programs, book adoptions, and training scenarios, and at some point during the recertification process, the medical director needs to implement a checks-and-balances review of the paramedics under his license. This can be accomplished with standardized testing through the NREMT, the state, or the local regulatory authority. A physician signature and verification of skills are required for NREMT recertification.

To assist EMS members with clinical direction, the medical director must provide treatment and triage protocols. **Treatment protocols** are standards of practice for an EMS system and apply to all parties of the system. **Standing orders** are used to guide members when direct medical control is not established or available. Triage protocols provide patient-destination policy (point-of-entry plan), designate a facility for transport, and prevent medical-control facilities from usurping patient-destination decisions.

The interface between EMS medical control and EMS operations is a team effort. EMS members must know the medical director and recognize that every

clinical action they take, or in some cases fail to take, reflects upon that medical director. Such a reflection may shine only on the director's reputation but if things go wrong may extend as far as medical regulatory action or civil or criminal action against the director—and the EMS provider—in the court system.

A medical director must be an advocate for the EMS system and the organization's mission. This may include participating in lobbying efforts for the system, engaging in political and social events to increase the visibility of EMS, and recruiting members for the EMS oversight or advisory board. He must promote public education and information on the prevention of emergencies. It is important that the medical director maintain a liaison with the medical community, including emergency departments, physicians, prehospital providers, and nurses.

A medical director needs to develop and maintain a network of other physicians with specialties to keep an EMS system functioning effectively (Figure 10.3).

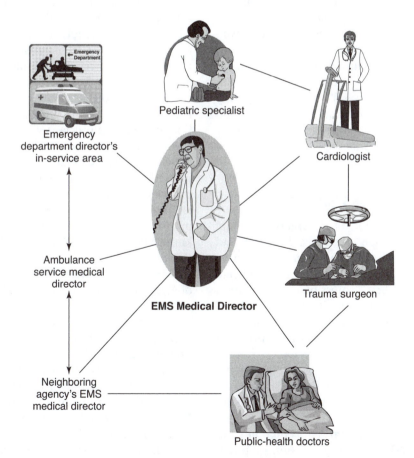

Emergency department director's in-service area

Pediatric specialist

Cardiologist

Ambulance service medical director

EMS Medical Director

Trauma surgeon

Neighboring agency's EMS medical director

Public-health doctors

FIGURE 10.3 ■ Other physicians network with the EMS medical director.

The medical director serves as the interface and liaison between the EMS system and the medical community.

The medical director should be a member of one or more of the associations that represent EMS. Often he also is affiliated with a particular hospital. To ensure that the EMS system and its patients are put first, the EMS manager should apply an ethical test to the physician who has managerial responsibilities, financial interest, or contractual agreements with hospitals. The medical director should have no potential for unethical conduct or conflict of interest.

The Rural Fire-Based EMS Medical Directors' curriculum and training project was developed through funding from the Critical Illness and Trauma Foundation and the Office of Rural Health Policy. In three years this project converted the *Guide for Preparing Medical Directors* into an electronic format that will reach the rural audience. The guide is a basic EMS medical director's training curriculum that targets physicians who have assumed the EMS medical director's role but lack an EMS background. Although this occurs more frequently in rural and frontier environments, this training can be beneficial to anyone who participates in the duties of the EMS medical director, regardless of position or environment. The curriculum became available in 2001; it was the product of a cooperative agreement between the National Highway Traffic Safety Administration and the Maternal and Child Health Bureau with the National Association of EMS Physicians and the American College of Emergency Physicians.

PHYSICIAN ORGANIZATIONS FOR EMS

EMS-based medical direction has several professional organizations that provide resources, direction, and continuing education. Early in a physician's training the **Society for Academic Emergency Medicine (SAEM)** may be the first professional organization encountered. SAEM's mission is to promote improved patient care by advancing research and education in emergency medicine. SAEM provides direction for emergency-medicine residency training programs and conducts research on educational methodologies that are used in the instruction of emergency medicine residency and fellowship programs.

Formed in 1984, the **National Association of EMS Physicians (NAEMSP)** is an organization of physicians and paraprofessionals that provides leadership and fosters excellence in hospital emergency medical services. NAEMSP promotes publications and products that connect, serve, and educate medical professionals in the prehospital arena and hospital emergency departments. NAEMSP fosters research and training and holds an annual conference that showcases EMS research and provides education on current topics related to emergency medicine. NAEMSP provides grant monies for projects, publications, and fellowship opportunities in EMS.

The **American College of Emergency Physicians (ACEP)** exists to support quality emergency medical care and promote the interests of emergency-medicine physicians. ACEP focuses on increasing knowledge of emergency medicine by promoting emergency-medicine residencies, identifying best practices through quality improvement, and ensuring unobstructed access to emergency medicine. ACEP sees the physician as the person having responsibility to provide the leadership necessary to define, manage, and improve the quality of care in the prehospital setting. ACEP produces several computer-assisted training programs, books, CD-ROMs, and videotapes on a variety of emergency-medicine topics. ACEP is actively involved in lobbying for government regulations and reimbursement for emergency medicine.

Key positions and committees in the American Heart Association and the American Stroke Association are led by physicians with an interest or roots in EMS. The scientific assembly of the AHA focuses on the clinical research that supports the practice of EMS in cardiovascular emergencies. Some of the AHA's programs are taught only by physicians, and the medical oversight that is required in the EMS quality-improvement process will require that a physician know and understand the ACLS, PALS, and CPR guidelines for patient care. Periodically, a physician may be asked to review or sign off on the competency of a paramedic to complete the algorithms for certain cardiac-arrest scenarios.

Several other key EMS organizations have limited participation by physicians. The International Association of Fire Chiefs EMS Section maintains a close relationship with NAEMSP. Several physicians and representatives are involved with the Advocates for EMS, a lobbying organization located in Washington, D.C., designed to advance the political agenda of EMS on a national level. The National Association of EMTs has physician involvement in many of its programs and committees. Several of the key training programs have physician advisors, and the NAEMT maintains a liaison with several of the major physician groups.

PHYSICIAN INVOLVEMENT

In large EMS systems most EMS managers will have little direct contact with the system medical director. While that trend is being reversed in communities like Austin, Texas, and Boston, Massachusetts, a physician actively involved and personally accountable, with knowledge of each paramedic, is still the exception more than the rule. In small systems the contact between system managers and the medical director may be more frequent and less structured. Generally, any question or formal recommendation from EMS managers regarding specific medical policy, procedures, or clinical changes should go to the medical director via the system chain-of-command. Routine questions regarding clinical procedures, standing orders, protocols, or patient care may be handled on a more informal basis directly with the medical director's office or during any regularly scheduled meeting with the medical director.

Unless otherwise defined in state or local statutes or requirements, a medical director must have authority over all clinical and patient-care aspects of an EMS system. These aspects include certification, recertification, and decertification of nonphysician prehospital providers (including EMS dispatchers); establishing, implementing, revising, and authorizing system medical protocols, policies, and standard operating procedures or standing orders; establishing criteria for the level of initial emergency response; establishing patient-destination protocols; establishing guidance for any concurrent medical direction of EMS providers; and establishing procedures or protocols under which nondispatch or nontransport of patients may occur. Disaster planning and management of a hazardous-materials response are other areas that the medical director should be involved in developing.

The medical director's involvement in the certification process includes responsibility for establishing certification requirements and criteria, providing initial training, doing testing, and providing continuing medical education. The medical director's input into the certification-recertification of EMS personnel in some states is a gatekeeping function for the medical-practice act. The physician medical director or county medical director approves individuals to be able to work in a specific area. The medical director must ensure the appropriateness of initial qualifications of personnel involved in patient care and emergency medical dispatch.

Support of Medical Directors

For its part an EMS system must provide a medical director with the necessary resources (personnel, vehicles, and equipment) and authority to accomplish his assigned medical responsibilities. To accomplish the clinical supervision, the agency or organization must provide effective quality-improvement processes for system improvement and patient care. EMS agencies must provide or ensure that the medical director has access to and subscribes to EMS research. A mechanism and a budget should be in place to ensure that the EMS medical director maintains up-to-date knowledge.

Issues regarding liability and protection of the medical director's activities are often solved by making the medical director an employee of the EMS service. This is especially beneficial when the agency has government immunity under state statutes. If such protection is not provided under state law for municipal agencies, the EMS agencies should budget for and provide malpractice insurance for the physician. Smaller items, such as a helmet, coat, and radio, should be provided to identify the physician on scene or at fire and rescue events. A training budget must be in place to ensure that the physician is refreshed and updated on incident-command training, triage, prehospital education, and remediation and disciplinary practices.

The 911 Communications Center

The 911 communications center that services an EMS agency or jurisdiction should have medical direction over the clinical aspects of triage protocols, pre-arrival instructions, and quality improvement. The local EMS agency medical director ideally should also be the medical director for the 911 communications center. However, the medical director would not be involved with personnel actions or discipline in the communications center.

Most communications-center employees are covered by public-safety unions, and the physician is not prepared to function in the due-process or procedural events required under collective bargaining or union contracts. The medical director should have the authority to suspend an emergency medical dispatcher (EMD) if the dispatcher's medical actions do not meet acceptable performance. The medical director should have a written agreement dictating his responsibilities to an emergency medical dispatch program. This should include what the medical director's role is and his responsibility in the due process of an investigation or disciplinary action. The medical director also should have the time to dedicate to the assistance and improvement of the communications center.

Most physicians are not exposed to the complexities of the communications center. It is the responsibility of EMS leadership to provide the physician medical director with an appropriate orientation to the communications center. This orientation should include information that is specifically related to the medical director's role within the communications center and the emergency medical dispatcher program. The medical director should be involved in all aspects of the emergency medical dispatcher program planning and implementation. He should understand the human resources and technology that are required to operate a communications center. This includes an overview of the legal issues, training, and emergency medical dispatcher protocol development.

RELATIONS WITH THE MEDICAL DIRECTOR

Most of the time medical directors and EMS managers work effectively together. To some extent this is surprising, since they have a number of things working against them. Both are accustomed to being in control, and both are definitely not accustomed to having their decisions and directions questioned. Further, the lines of authority between them often become blurred.

There are ways to avoid or make the most of differences when they arise. Although it may sound trite, communication and mutual respect are the keys to building and maintaining a useful working relationship with a medical director. Planning and outlining responsibilities clearly is critical, but accept that there will always be gray areas. These gray areas often center on personnel and the use of resources. Discuss these situations as they arise, negotiating and compromising as necessary.

EMS managers and medical directors must not just develop common goals but must agree on objectives and action plans as well. For example, this does not mean simply agreeing to provide good customer service; it also means agreeing on how to achieve that goal within system and situational constraints. In the past there was a trend toward medical directors who simply "signed off" on the necessary paperwork so that the system could legally exist. Considering that providers function under the medical director's medical license, this is a risky practice indeed. Further, it clearly is not in the best interests of patient care.

The current trend is toward medical directors who monitor and control all aspects of medical practice, including policy decisions. The EMS leader bears the responsibility of ensuring that the medical director is significantly involved in all medical aspects of the system. Ideally, this is addressed during the recruitment and selection process, but sometimes this requires influencing an existing situation and relationship. In extreme cases leaders may need to alter the existing agreement or contract in order to ensure they are providing adequate levels of care, using their resources effectively, and avoiding legal liabilities.

Medical directors should be involved in all aspects of prehospital medical care. They bear the ultimate responsibility for medical control and must approve (and often develop) all guidelines before they are instituted. Medical directors also have supreme authority over providers as they deliver care and must monitor their competency (Table 10.1). Field participation by the medical director is an invaluable tool in medical control. Medical directors should be deeply involved in quality control, risk management, customer service, and education programs. They should be partners with the EMS manager, with complementing and even overlapping functions and responsibilities. Ideally, medical directors are active leaders and advocates for the EMS system and its patients.

DISCIPLINE

One of the most contested areas inside the scope of practice of the physician medical director is discipline.

TABLE 10.1 ■ The Medical Director's Area of Authority

Area of Authority	Medical Director	Chief
Medical issues	Has overall authority	Makes recommendations
EMS operations	Makes recommendations	Has overall authority

Medical directors are tasked to make certain that EMS providers are delivering quality patient care. Medical directors as employees of the jurisdiction are given that responsibility; however, they have a limited capacity to recommend or administer discipline to employees. Case law has established that a medical director has the ability to limit the practice of a paramedic or an EMT under the license of the medical director. In a Minnesota case—Hennepin County *v.* Hennepin County Association of Paramedics and Emergency Medical Technicians—the medical director was able to limit the practice of a paramedic, although it was unclear whether or not the medical director has the authority to terminate employment.

No organization or person does the right thing every time. By its nature EMS requires paramedics and EMTs with an incomplete knowledge of the situation to make complex life-or-death decisions very rapidly and during continuous distraction. It is a tribute to EMS professionals that through dedication and training they get things right the vast majority of the time. When mistakes happen, the challenge of the medical director is to decide a course of action. Several considerations need to be assessed prior to taking action. Prior to any decision, it is imperative that the investigative process be completed.

Generally, to help focus the investigative process, the physician must consider the following:

- Whether or not, given the situation faced by the medics, the mistake is understandable.
- Whether or not the mistake at hand indicates a need for further training.
- Whether or not the mistake was simply an anomaly.
- Whether a specific medic or the entire service needs to be addressed about the mistake.

All possible information should be gathered before action is taken to handle a potential problem. Feedback should be collected from everyone involved, including the medics, physicians, and nurses at the receiving hospital and, perhaps, the patients and their families. Obtain run records, ED reports, and, when relevant, inpatient-care records and autopsy records. Obtaining courtesy privileges at all hospitals to which a service transports often gives the medical director access to a patient's medical record. Often what initially appeared to be a dangerous mistake turns out to be a reasonable decision based on the information available to the medic at the time.

Many states give the medical director the right to limit the practice of EMS personnel under his direction. The medical director can discipline an EMS worker by suspending him, terminating him, or only in specific circumstances that may be set forth in a contract.

In fire-based EMS systems, the medical director's authority is influenced by the public employee's property right in the job. Public employees are entitled to the rights of **due process** when faced with disciplinary charges that may affect their continued employment. Due process includes notification of the employee regarding the pending decision, a hearing in which the employee has the opportunity to present his side, and representation by legal counsel at a hearing if the employee so chooses. In Brodie *v.* Connecticut 401 U.S. 371, these rights were guaranteed, and only when there is an immediate threat to safety or health can the hearing be postponed. Discipline must be suitable for the offense and not be subjective to the individual. Employers need to develop—in a collaborative fashion with the labor unit—policies and procedures concerning discipline and the boundaries within which the medical director will operate.

Many states have collective bargaining, and the union will be required to represent and protect the employee under the labor agreement. Terminations under such a system must be for cause. Under NLRB *v.* Weingarten, Inc. 420 U.S. 251, a grievant may also have Weingarten Rights (an employee's right to assistance from union representatives) during interviews when management is investigating the employee for disciplinary actions. Employees must make a clear request for union representation when being investigated and cannot have retribution leveled against them for requesting the representation. Any union officer may be called to represent employees and has the right to counsel them on the answers, engage in private conversation prior to the interview, and ask the employer to clarify a question. Feedback from medical directors can strengthen an EMS system and an employee's performance and ideally will focus on quality-improvement activities and not employee discipline.

■ MEDICAL CONTROL

The essence of off-line medical control is that, ideally, there is a single, identifiable physician who assumes responsibility for the overall quality of care in an EMS system. While many physicians from different disciplines

- Protocols for treatment, triage, communications, and transport as well as standing orders.
- Standards for participation by communications resource hospitals and physicians.
- Medical standards for the categorization and designation of emergency receiving facilities.
- Medical standards for ongoing system review and quality assurance.
- Standards for training and continuing education of all classes of EMTs.
- Standards for participation by training programs for all classes of EMTs.

FIGURE 10.4 ■ Responsibilities of EMS Medical Directors.

might be involved in the development of protocols, policies, and procedures for the system, there still must be a single physician who bears the ultimate responsibility for the quality of prehospital medical care.

Because of this scope of responsibility, it is essential that the medical director have authority commensurate with it. This does not necessarily imply administrative authority over nonmedical system components. However, the medical director must have ultimate authority in medical matters. This includes the authority to approve participation or suspend prehospital providers and resource hospitals. Other specific responsibilities of the medical director include establishing and implementing the regulations listed in Figure 10.4.

PROTOCOLS

The medical director must establish a set of protocols that defines the overall system design. These protocols must address and identify the scope and involvement of prehospital providers, the role of **on-line medical control**, and the mechanism for quality assurance.

Types of Protocols

Protocols are generally written in two types of formats, either descriptive or algorithmic. **Descriptive protocols** are written as a narrative, describe a condition or situation, and usually contain educational material. An **algorithmic protocol** is a series of logical steps based on recognized or objective facts or conditions and does not have educational components.

Treatment protocols are guidelines that define the scope of prehospital intervention that will be practiced by providers. They are a means of establishing uniform quality of care and become the basis for the design of

training, continuing education, and refresher training programs for prehospital providers. Additionally, they form the basis for ongoing monitoring of the system and for quality assurance. Treatment protocols are therefore the first protocols that new or refining systems should develop.

Because they define the scope of health-care delivery within an EMS system, treatment protocols are the responsibility of the off-line medical director and are ideally developed with input from medical advisory committees. Because of the dynamic nature of the practice of medicine, treatment protocols must be reviewed on an annual or biannual basis and revised accordingly to reflect the current state of the art in emergency medical practice. Whenever major changes are made in treatment protocols, there needs to be appropriate inservice or continuing-education training regarding the changes. No changes in protocols should be instituted until providers have been reeducated and uniform implementation of the changes can be expected.

Standing orders are more specific than treatment protocols and are normally included in a protocol when delay in treatment might have a harmful effect or outcome. Standing orders are quite specific and define those prehospital interventions that are authorized by an EMS system without on-line, real-time medical control. Because standing orders authorize nonphysicians to perform invasive procedures in the absence of direct orders, they must be developed and signed by the off-line medical director. Standing orders should be standardized throughout the system to ensure uniformity and prevent confusion.

An important source of quality is the agency's **communication protocols**. These define the method and timing of communications with on-line medical control. They are closely related to standing orders in that they might define those situations where on-line voice communication is not necessary. As this is a medical decision, it must be established by the medical director, ideally in consultation with a medical advisory committee.

Another important aspect of the communication protocol is the designation of a specific on-line medical control source for each identified field unit. The designation will determine whether the same resource will be used for all communications regardless of the patient's destination or if direct contact is to be made with the receiving hospital. When the designated on-line medical-control resource is not the facility that will receive the patient, there may be concern if the physicians at the receiving facility wish to have input into field treatment of patients for which they subsequently will be responsible. This is an issue that must be addressed and resolved by the off-line medical director in the best interests of patient care. For purposes of system standardization and quality control, the primary on-line medical-control authority should always be the same for a specific unit.

An important aspect of patient care that the physician director often overlooks is the triage protocol. The term *triage protocol* refers to the sorting of patients at multiple-casualty incidents according to the seriousness of injuries. Several types of triage systems are available, and the medical director should monitor to ensure that all aspects of the EMS system understand and use the chosen one efficiently.

Destination protocols are protocols that define the appropriate receiving facility for a patient with any given medical condition. For example, protocols may call for transport of trauma patients to designated trauma centers. Often this involves bypassing closer, nondesignated facilities. Thus, these protocols are probably the most controversial, attracting the most attention in the course of their development. Determining the appropriate receiving facility clearly involves medical decisions and must be established through off-line medical direction. Again, this is ideally accomplished through an advisory committee of physicians from various specialties. Hospital representatives should be included in the development of these protocols.

Transport protocols define the proper mode of transport for particular cases. They are based not only on the nature of the medical emergency and the patient's condition but also on transport times to facilities. More specifically, they address the use of air versus ground transport in given situations. Transport protocols also address the level of expertise or certification of a transport provider. For instance, should a paramedic accompany every ALS transport, or are there times when an EMT–Intermediate will suffice?

Special protocols may need to be developed to address special circumstances that might be encountered within the EMS system. These would include when a run is called off or the patient cannot be found, when the patient refuses treatment, when the patient is obviously dead, when the patient has "do not resuscitate" orders or a living will, when to terminate cardiopulmonary resuscitation, whether or not to transport a patient against his will, and when a physician is on the scene.

Protocol Development

The development of protocols follows a process similar to developing a standard operating procedure (SOP) or guideline (SOG). The physician medical director, in collaboration with management, conducts a well–planned, standardized, and comprehensive process to develop protocols. A nine-step process to build a protocol is outlined in Figure 10.5.

Medicine continues to change and requires that protocols and procedures be reviewed or new ones developed. Assembling a team to build protocols reflects the principles of total quality management by encouraging participation and feedback by those EMS providers who will be using or who will be regulated by the protocols. Participation helps build compliance and buy-in on the protocol once it is implemented.

When selecting a team to develop and review protocols, a variety of viewpoints should be solicited for the process. Some examples of participants include operational or line EMS personnel who have the greatest stake in the implementation, a member with a perspective on the safety of the process or procedure, managers who handle the budget and political implications, a mutual-aid or regional-response representative, someone representing local government or the regulatory agency, legal counsel or others who are familiar with the content and how it will affect current regulation or laws, a member with technical expertise, and a training representative.

The team should have management span of control and should be periodically supplemented with people who have expert knowledge in certain areas being reviewed or created. Members should be rotated on and off the protocol team at staggered intervals to ensure continuity of the process and to help bring in new ideas and perspectives. Top-level management should also be involved to help ensure that the organization supports the process.

Organizational support is key, and EMS managers should make sure that computer and clerical support is available. EMS leadership needs to ensure an adequate meeting space, an agenda, and accountability for the time spent in meetings and retreats. EMS administration should support attendance of team members through monetary support, by supplying time off to participate in meetings, or another type of reward. It is important to establish procedures and to agree on how

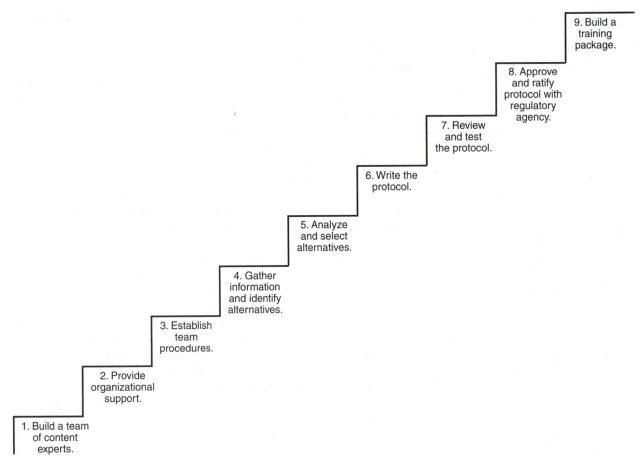

9. Build a
training
package.

8. Approve
and ratify
protocol with
regulatory
agency.

7. Review
and test
the protocol.

6. Write the
protocol.

5. Analyze
and select
alternatives.

4. Gather
information
and identify
alternatives.

3. Establish
team
procedures.

2. Provide
organizational
support.

1. Build a team
of content
experts.

FIGURE 10.5 ■ Steps for Building Protocols.

the business of the team will be conducted. This in-cludes establishing rules covering leadership, voting, consensus building, and how to keep the work on track. Flow charts, checklists, and computer software should be used to ensure that the work stays on track and is finished in a timely fashion.

Information gathering and research is important when establishing protocols. Alternatives and linkages to the needs of the organization and the community need to be considered by the team. Research from the Web, medical-archive services, journals, and feedback from the quality-improvement processes are excellent sources of input for the development of protocols.

Once a protocol is formed, the team needs to ana-lyze and evaluate any possible alternatives. This in-cludes determining feasibility by asking, Can this work in the street, and how will responders react? Imple-mentation factors in, as does whether or not the agency has the financial and operational resources to

commit to supporting the protocol with training re-sources, equipment, and human resources. The proto-col must be tested for compliance with regulations, current standards of practice, ethical constraints, and safety. Preparation of a protocol requires considera-tion of whether or not it can survive politically when exposed to the public, the medical community, and the EMS workers who will use or be constrained by the protocol.

When writing a protocol, the level of detail needs to be determined. Some agencies prefer brief proto-cols, and others see the need to place educational ma-terial into the protocol. Generally, protocols should not be training manuals. A protocol needs to be writ-ten clearly and concisely in plain English, using bullets or outline formats with a target audience in mind. Since not all patients or EMS situations are the same and because often they do not fit a protocol, they should be designed with some flexibility. Ideally, they will be developed with a standardized format with title and numbering system, effective date, and review or expiration date. The medical director should sign or authorize each protocol, identify the scope, and provide references.

A peer review of the protocol by paramedics and EMTs should be conducted prior to full implementation. This period should allow for the collection of written or verbal feedback. A protocol should go out as a draft document to the field and then have management implement the protocol after ensuring adequate training. A comprehensive process will produce a good product. It is incumbent on EMS managers and leaders to *not* make last-minute changes, or the changes could be unsupported, and participation will fall off. The finished product is an example of good quality management.

Some areas that could have contributions made by the physician medical director include protocols or SOPs for risk management. Protocols that need physician input are those that could encompass procedures for infection control, protective clothing, lifting and moving patients, and hostile situations. Consider the use of a protocol for chemical sedation. This procedure clearly is about the safety of EMS workers, yet key elements need review by the physician medical director due to the numerous procedures that could be conducted.

Operational issues also need physician involvement in protocol development. All patient-care, medical-device-use, and dispatch protocols need physician review. Through a comprehensive network in the medical community, the physician can help develop destination guidelines, methods of transportation, and helicopter and ambulance operations. From a managerial standpoint, physician medical directors may be asked to help make recommendations for protocols that specify data collection, standards of care, and quality-improvement processes.

Finally, special-operations protocols need the benefit of physician involvement or review. Situations involving mass gatherings, hazardous materials, technical rescues, and rehabilitation need to have physician input. A physician should also oversee the medical procedures needed for patients buried in rubble and the treatment necessary to deal with compartment syndrome prior to their removal. Many of these special-operations protocols are defined by FEMA Urban Search and Rescue Teams and have national standards defined for specialized rescue.

FIGURE 10.6 ■ Physicians or their designees provide on-line medical control.

ON-LINE AND OFF-LINE MEDICAL CONTROL

The underlying goal of on-line medical control is real-time (via radio or telephone) direction of prehospital providers in the delivery of emergency medical care (Figure 10.6). The medical director, or more often a designee—such as a mobile intensive-care nurse who has taken special certification on the protocols and had an orientation to the prehospital environment—provides direct voice communication to the field practitioner in an on-line medical-control model.

The difference between on-line medical control and off-line medical control is that off-line control refers to administrative and protocol control exercised by a medical director or committee of physicians and EMS providers. Quality on-line medical control requires the presence of on-line physicians who are involved in the design and implementation of the system, participate routinely in on-line direction of care, are well acquainted with EMS personnel, and are involved in EMS medical audits and quality-assurance programs. In some systems, in order to receive an order for scheduled narcotics, a conversation with a physician, not a nurse, is often required. Some situations, usually of a more serious nature or those not covered in the protocols, would require speaking to a physician.

Medical-Control Design

EMS systems vary in configuration as well as in the amount of control given to physicians at the state or local level. In most states, legislation defines the duties and responsibilities of physicians involved in the system, with most specifying a regional design with a hierarchy of medical control. The design of the local medical-control system is dependent on the regional design as well as on the responsibilities assigned to practicing physicians in that area. Often the agency responsible for providing EMS personnel has considerable influence over local medical control system design. In addition local political considerations often dictate certain aspects of the on-line system.

Regardless of its actual configuration, however, every EMS system should have a medical director (off-line medical control) whose responsibilities include overseeing the system, monitoring its performance, and ensuring quality patient care. This physician should be a member of the on-line team that provides ongoing direction of patient care and EMS team performance.

Three general models of on-line physician direction have been developed:

- *Hospital-based direction.* Provided by radio or telephone communication from a hospital-based physician who is usually located in an ED; this is the most common model in EMS systems.
- *Mobile medical direction.* This model is less well known but has several advantages. Emergency physicians provide medical direction on a rotating basis to field teams through portable radios and telephones, allowing for the medical management of the EMS system through a consortium of hospitals or physician groups.
- *In-field direction.* Direction to field teams in this model can occur when a physician is present at the scene of an incident.

On-line medical control involves physician direction of out-of-hospital providers in the delivery of emergency medical care. Such direction implies the ability to effectively monitor the performance of the field team, ensuring quality medical care and prompt patient transport. An essential element in the design of any system is the on-line physician's ability to alert the hospital team of the patient's impending arrival and to suggest appropriate preparations.

Off-Line Medical Direction

Off-line medical direction is composed of prospective medical direction, protocol compliance, and retrospective review of the performance of the prehospital-care practitioners. Off-line medical direction activities include the implementation of standing orders, protocol development, training, and certification.

Physician medical directors should look to create standing-order systems. In this arrangement, EMS field teams apply prescribed protocols according to the presenting signs and symptoms of the patient. Physician consultation is not required, and a report is given to the hospital team on arrival. Protocol systems provide field teams with a set of patient protocols designed as guidelines to patient care in a field setting. In such a system, orders are either presumed or direct.

With presumed orders the protocol may be initiated, but the physician is contacted at a point designated in the protocol. With direct orders, patient-care protocols may be applied only after direct orders from the physician. Some allowance is usually made for extenuating circumstances, such as technical breakdown or interference in communications. Research comparing standing-order systems to systems that require on-line medical control has demonstrated that paramedics have a high degree of compliance with protocol selection and that there is no change in the application of care between the two systems.

Medical audits are part of the off-line responsibilities of medical direction. When conducting medical audits the physician should review procedures, provide direction for remedial education, and define limitations of patient care. A good practice is to ensure that for every case requiring remedial education a case is presented that highlights exceptional or best practices by a crew. Too many cases pointing out deficiencies will only produce a climate of fear and handicap the quality-improvement process.

Every ALS run should be reviewed for correctness by medical direction. Periodic reviews of BLS calls and targeted review of specific runs should be done by the medical director. More important is that the EMS medical director get out into the field to actually observe the care of the medical providers. Field audits or observations should be random.

Two approaches to field observation can be beneficial. One approach consists of the physician being in a vehicle, circulating around the city and randomly responding to calls, which provides a more objective view of the care. The other approach is to ride with the crew as part of the shift work. This provides the physician with an informal communication channel at the dinner table and allows the physician to see a call from start to finish. The longitudinal look—seeing the call from start to finish—allows the physician to see all aspects of the system, including dispatch and how the crews are received at the hospital. When EMS agencies have the physician riding along with the EMS crews, crews may be on their best behavior and often may not give the physician an accurate view of care in the field. This is often called the **Hawthorne Effect;** EMS providers improve their care or adhere to protocols when they are being observed. A certain amount of time can be spent with crews responding from the station to build camaraderie, but real progress in patient care can come only when the crews perform to the standard without supervision or without knowing that they are being watched.

Training and testing is a prospective method of providing medical direction. When the medical director implements training on the standards of practice and core skills and abilities of prehospital providers, he is providing the framework for direction of patient care. Testing is another form of medical control in that the content and selection of medical references to be tested is reflected in the standards and practices in the organization's medical direction.

The physician, in conjunction with the training staff, should evaluate prospective continuing medical education (CME) programs for content. The physician must understand that most accreditation processes require his involvement to ensure a CME program is driven by the QI process to address system issues. EMS management and the medical director should develop a process to evaluate provider skills and knowledge on a routine basis.

The physician medical director may be used to support or lobby for operational procedures and legislative activities. This may include testifying for EMS legislation, making presentations to chief officers, and participating in the regulatory process at the local government authority. The pros and cons of such configurations relate to several elements inherent in each local system. When providers are given more responsibility, there is a greater need for specific and detailed protocols. In addition, there is greater need for on-line physicians to become well acquainted with providers through daily involvement in the system, presence in the streets, and frequent exposure to EMS personnel through training and on-line supervision of care.

The advantage of "presumed orders" is more expeditious patient management; however, presumed-order protocols must be subject to a strict quality-improvement program. A "direct order" system can be cumbersome and inefficient, and without the benefit of the same level of audit and quality assurance as other designs, it cannot, of itself, guarantee good quality. Regardless of how a system is specifically designed, prehospital management must be based on written protocols developed by clinicians involved in the system on a daily basis. These guidelines should set the standard of care, not only for providers but also for on-line physicians. A physician should act as a role model and demonstrate knowledge of the EMS protocols as this will help field personnel in learning their protocols.

Interrelations

The EMS system is unique in that it seeks to deliver quality emergency medical care via a team approach. As such, it includes a wide range of players by both design and implementation. In the implementation phase it requires interactions between basic- and advanced-level EMTs, emergency and intensive-care unit nurses, emergency physicians and other medical specialists, on-line medical directors, and the off-line medical director.

Because the off-line medical director is responsible for EMS system design as well as its daily implementation, he must have expertise and be comfortable in dealing with each of those subgroups. The off-line medical director will need to form advisory committees consisting of representatives from various interested groups. In addition, he will need to sit as an advisor on other committees that do not directly address medical care but have an impact on the system in such a way that the input of the off-line medical director is important.

SELECTING AND HIRING A MEDICAL DIRECTOR

Depending on the nature of the given system, there will be various individuals or groups of individuals who are charged by law or tradition with the selection of the medical director. This authority should be familiar with the qualifications for medical directors as specified by the American College of Emergency Physicians (ACEP) (Figure 10.7). In addition, the selection committee should seek out the advice, if not the consent, of key medical leaders within the community.

1. License to practice medicine or osteopathy.
2. Familiarity with the design and operation of prehospital EMS systems.
3. Experience or training in the prehospital emergency care of the acutely ill or injured.
4. Experience or training in medical direction of prehospital emergency units.
5. Active participation in hospital emergency-department management of acutely ill or injured patients.
6. Experience or training in the instruction of prehospital emergency units.
7. Experience or training in EMS quality-improvement processes.
8. Knowledge of EMS laws and regulations.
9. Knowledge of EMS dispatch and communications.
10. Knowledge of local mass-casualty and disaster plans.
11. If possible, should be certified in emergency medicine (ACEP).

FIGURE 10.7 ■ ACEP Recommendations for EMS Medical Directors.

The physician medical director should go through the normal hiring process. He should be actively involved with the EMS system, be familiar with EMS, and have an unrestricted license to practice locally. The medical director should have the support of local emergency physicians, local and state medical societies, and hospital associations and their representatives. It is an option to contract with a physician for his services or to employ him within the agency. In many states if a fire-based medical director becomes an employee of the fire department that he represents, certain state laws limit his liability as an employee of a public-safety organization.

Ideally, a medical director should be board certified in emergency medicine and emergency medical services by the American College of Diplomatic Medicine. To achieve board certification in emergency medicine, a physician would most likely have to complete an emergency medicine residency; a list of the top 10 emergency medicine residencies is published annually to aid the selection committee in evaluating potential medical directors. It is preferable to have a physician that has actually been in an ambulance and has the motivation to seek out field experiences. It is unique to EMS medical direction that a physician medical director needs an understanding of labor relations, incident command, and disaster preparedness.

The selection of an EMS medical director should include evaluating the residency training program and fellowship programs where the physician trained. An **EMS fellowship** program is an additional one to two years of training after a physician has completed residency training and medical school; most fellowships are linked closely with teaching institutions or agencies supporting research. The EMS fellow has typically worked one or two shifts per week in the sponsoring institution's emergency department. In some systems the local EMS agency or a private corporation will provide some financial support for the EMS fellowship.

Emergency physicians seeking to gain additional expertise in EMS began to enter fellowship training in the late 1970s. This trend has grown over the years, and currently over 70 EMS fellows have completed training. The fellows have successfully secured employment as EMS medical directors for large urban EMS systems, have become faculty members at academic centers, and include respected EMS researchers. They are truly EMS leaders, holding key committee positions in the National Association of EMS Physicians (NAEMSP), Society for Academic Emergency Medicine (SAEM), and American College of Emergency Physicians (ACEP).

One of the premier residencies that may be of benefit to fire-based EMS services is provided by the International Association of Firefighters (IAFF) in conjunction with Johns Hopkins Bloomberg School of Public Health. The IAFF provides a two- to three-month internship for physicians. After completion they leave with an understanding of the demands placed on firefighters and many of the health and safety issues that surround fire and emergency services. The University of Pittsburgh's Department of Emergency Medicine offers a two-year fellowship in emergency medical services, providing superior training and experience in all aspects of contemporary prehospital care.

The fellowship should include an overview of the EMS system and EMS system designs, covering such subjects as the history of EMS systems, components of an EMS system, legislation, and unique services. An EMS-trained fellow should understand the certification levels of EMS personnel. The training levels, skills sets, and health-safety issues should be understood by the EMS-trained physician. The role and scope of the medical director and how physicians interact in the EMS system need to be defined in a fellowship program. Physicians will need to have functional knowledge of communications systems, radio protocol, EMS vehicles, and equipment. The EMS fellow will know the types of EMS agencies, providers, and air-ambulance capabilities. Education regarding legal issues involving EMS is important for every EMS physician.

A physician completing an EMS fellowship will need experience in planning for and delivering EMS during mass gatherings and disasters. Today this also would include some knowledge of weapons of mass destruction and terrorism. Funding for EMS systems, community efforts, education, and research agendas should all be included in an EMS fellowship. Finally, operations management is another important core skill for which an EMS fellowship–trained physician will need some knowledge.

Ideally, a set of EMS-experience activities would also be documented in a medical director's training. Many emergency medicine residency or fellowship programs offer field experiences with extensive EMS ride-a-longs. The medical director should have participated in in-field activities as an observer, educator, and care provider. There should be training in the use of the system's EMS equipment, emergency-driving techniques, and rescue and extrication techniques. Experience as a base-station physician, in the communications center, and as a flight physician are important additional characteristics that may be necessary for an EMS medical director. Some experience in hazardous-materials training, statistics, and disaster exercises is necessary to most EMS systems.

AFFILIATION AND ETHICAL CONSTRAINTS

Even though the medical director is usually a member of one or more of the representative EMS associations and is often affiliated with a local hospital, he also must be an advocate for patient care. EMS managers need to apply an ethical test to physicians who have managerial responsibilities, a financial interest, or a contractual agreement with hospitals to ensure that the EMS system and its patients are put first. The medical director should avoid potential unethical conduct or a conflict of interest. In many cases legal representation for the organization should draft a contract and place ethical guidelines in that contract.

A physician contract should be in place between the agency and the physician medical director. The items that are in the contract should cover the job task of physician oversight of the EMS system, including medications administered by providers, by defining the roles and responsibilities. Clearly defined

compensation and the mechanism for how the work will be measured (e.g., hours, per chart, meeting times, and so on) need to be included in the contract. Often a physician is contracted, and neither EMS administration nor the field crews ever see the physician or receive any contact with the medical director. A specific amount of contact time should be incorporated into a medical director's contract, and any contact with the medical director should include a specified amount of administrative and secretarial support.

Often the medical director will need to become part of the EMS service to take advantage of liability caps that are provided to public-safety personnel through state law and Good Samaritan coverage. If the contract does not include liability protection in that fashion, a separate-section provision should be made to provide liability protection for the physician medical director. In addition, if the physician is to be provided equipment (such as a car, turnout coat, radios, and so on), that equipment should be spelled out in the contract.

Finally, if the physician medical director will serve as a resource for occupational health, then the terms of the contract need to ensure that the physician is familiar with NFPA 1500 and is knowledgeable about infectious and communicable diseases and the OSHA requirements for health-care workers.

Recent EMS literature reflects a growing concern about the importance of intense physician supervision of EMS systems. Although there is some evidence to suggest that physician observation increases outcome, most physician supervision remains primarily a matter of philosophy and theory. A physician can be a powerful ally for an EMS agency and a solid resource for an EMS administrator. But it is incumbent that the EMS administrator manage the medical director and provide a framework for the integration of a physician into daily operations.

While the roles a physician has as an EMS medical director are many (Figure 10.8) above all the physician has to be the leader in quality patient care. Each role the physician medical director fills ultimately is distilled down to quality improvement.

- ◆ Discipline.
- ◆ Education.
- ◆ Monitoring.
- ◆ Advocacy.
- ◆ Research.

FIGURE 10.8 ■ EMS Physician Roles.

QUALITY LEADERSHIP

Quality management is probably the most important task a medical director performs. Medical directors have the responsibility and often the legal liability to ensure that all care provided under their authority is of the best quality possible. Quality management is the tool that makes this possible. Several actions are required to accomplish that goal. They are typically done in a routine review of prehospital care reports conducted by the EMS agency's leadership.

Someone within the EMS organization routinely should review all EMS patient-care records for completeness. The EMS chief, field supervisor, or a designated quality-improvement (QI) coordinator should maintain statistics on medical practice (including scene times, number of attempted IVs and intubations, number of completed IVs and intubations, number of cardiac arrests managed) for each paramedic in the system. The data collection should be analyzed by the physician medical director to determine necessary areas of improvement for the entire service. This requires the physician to develop the skills and abilities needed to screen for problems that might otherwise go undetected. Random and focused run review and in-field observation by the medical director are necessary functions.

The medical director should conduct a medical review of a reasonable percentage of EMS runs. In a small service, that may be 100% of the calls, while in a larger system there may need to be a specific focus on high-risk and advanced life support calls. If the medical director can only review a small percentage (such as 10% or 20%) of the calls, management within the agency should be designated to review a large portion of the rest. The medical director should review at least the types of calls listed in Figure 10.9.

- ◆ All calls in which emergency transport to the hospital occurred.
- ◆ All cardiac arrests.
- ◆ All calls in which airway support was performed.
- ◆ All refusals of care.
- ◆ All prehospital terminations of resuscitations and DOAs.
- ◆ Ten percent of the 911 center or public-safety answering center calls.
- ◆ A random sampling of the remaining calls.

FIGURE 10.9 ■ EMS Responses Requiring Physician Attention.

OTHER SOURCES OF INFORMATION

What is written on a prehospital care report may not adequately detail what happened on a call or whether or not there were issues in the quality of medical care. A physician medical director should actively solicit input from other sources, including providers within the system, first responders, ambulance companies, hospital staff, receiving-facility physicians, and the patients. Physician medical directors also should monitor customer-service data collected from other aspects of the EMS operation.

The EMS medical director should meet with EMS operations personnel and medical directors of all the emergency departments and agencies that transport at least once a year to ensure that they have up-to-date contract information and can express any concerns they may have. An anonymous reporting system should be in place for agency providers to submit comments or concerns. Patient-satisfaction surveys should be reviewed for evidence of poor customer service.

MEDICAL-INCIDENT REVIEW

In conjunction with the labor unit, a physician medical director will be asked to assist in developing a process to identify potential issues with patient care. The EMS medical director must have a clear, consistent, formal process for investigating incidents and providing discipline or retraining as needed. Such a process should have the following characteristics:

- ◆ The procedure must be written in the agency's standard operating procedures as a policy.
- ◆ The procedure must include the EMS operations and administrative staff.
- ◆ All providers must sign a statement on day of hire that they have read and understand the policy, including potential disciplinary actions up to and including decertification.
- ◆ The procedure must differentiate between minor and major infractions.
- ◆ The procedure must clearly describe progressive disciplinary and due-process actions and their triggers.
- ◆ The course of action must include grounds for and the process of immediately suspending a provider's privilege to practice.
- ◆ The medical director or his designee must be involved in all incident reviews; and the medical director must be involved in all reviews that may involve decertification.

- Records, protected from discovery if possible, must be maintained of all medical-incident reviews regardless of the outcome.
- The process must provide written feedback for individual medics and may include retraining.
- The process must identify patterns of behavior that need individual or group retraining.

■ SITUATIONS REQUIRING IMMEDIATE INVOLVEMENT

UNRECOGNIZED ESOPHAGEAL INTUBATION

A missed intubation is like a police shooting. It is a high-risk event that almost always has a bad outcome and forces a serious look at the actions of the person, the system, and procedures. Although an endotracheal tube (ETT) can become dislodged when the patient becomes agitated, it should not become displaced with simple movement of the patient.

Missed or failed intubation occurs more often than many would imagine, and a response to this problem should be consistent. The EMS medical director should meet with the paramedic involved to discuss the situation with EMS administrators and union representation. If a pattern develops, refresher training should be required for the individual paramedic or, perhaps, all paramedics in the service.

Colorimetric carbon dioxide detectors, waveform carbon dioxide detectors, and esophageal intubation detectors are commercially available. With all of the tools available to EMS to ensure a reliable airway, a missed intubation is a catastrophic failure. If a specific paramedic has recurrent problems with unrecognized esophageal intubations, the paramedic should be placed on probation and assigned only with a senior paramedic or preceptor or should be suspended or de-certified until the problem has been corrected.

INAPPROPRIATE MEDICAL CARE

In a case involving inappropriate medical care, the medical director should conduct a thorough investigation. Policies should be revised, refresher training should be provided, or discipline of medics should be carried out, as appropriate. The indication of inappropriate medical care is the occurrence of a patient injury, often because of the following situations: (1) patients are dropped from the ambulance stretcher, (2) patients are injured during the application of restraints, or (3) EMS crews fail to recognize a serious medical problem or refuse to transport a sick or injured patient. There are a number of cases in which paramedics responded to a call and provided less than the needed standard of care to a patient. These situations can typically be corrected with refresher training and written policies.

Ambulance gurneys and bariatric patients continue to be a common cause of injuries during care and to the EMS providers. Many ambulance stretchers are not designed to be in the up position when patients are rolled long distances and are not meant to carry the load of extremely large patients. The EMS service may need to invest in new stretchers or require that stretchers are in a down position when rolling patients. A bariatric stretcher or equipment can be implemented to help deter injuries. This policy eventually will reduce back injuries among medics; in this way it is possible for the physician medical director to become an advocate for preventing injuries to patients.

SUBSTANCE ABUSE AMONG MEDICS

This problem requires action by the EMS-system chain of command and the medical director. The EMS agency must have personnel policies and procedures to address this situation. The EMS medical director is responsible for appropriate use of all narcotic medications purchased by the EMS service under the Drug Enforcement Agency (DEA) number of the medical director. The medical director's responsibilities in these situations are outlined in Figure 10.10.

- Act as a patient advocate by protecting patients from the risk of a medic's acting while intoxicated or not providing therapeutic interventions.
- Act in the best interest of the patients and medic involved.
- Comply with legal requirements to report impaired medics to the state EMS and the NREMT licensing or certification authority.
- Work with the EMS agency's attorney during every step to ensure that the medic is treated fairly and to protect the director and EMS agency from a lawsuit.
- Realize that the potentially impaired health-care professional is handled best by consulting with an expert in the areas of addiction to medicine and substance abuse.
- Carefully handle the process of confrontation, ensuring due process and compliance with union or agency policies.

FIGURE 10.10 ■ EMS Physician Medical Director Procedures for Monitoring of Narcotics.

The physician medical director should understand how to report through the organization's chain of command and how not to compromise any potential law-enforcement investigation if a narcotic irregularity suggests issues with a particular medic or EMS crew. The medical director should be aware that most state medical societies are a valuable resource in these matters. The EMS medical director should routinely inquire about, randomly check, and verify narcotics logs.

CHAPTER REVIEW

Summary

EMS systems are served well by EMS medical directors, who understand the EMS system as well as they understand clinical medicine. EMS agencies should strive to educate physicians, providing medical direction in field operations, incident-command training, and other critical elements of the EMS agency response. Communication between the EMS medical directors and EMS administration, supervision, and personnel must be direct and ongoing to ensure effective patient care. Feedback from EMS personnel to the physician medical director can strengthen the performance of EMS personnel if limited to quality assurance rather than employee discipline.

Research and Review Questions

1. Contrast the concept of delegated practice of a paramedic with the nursing profession and other allied-health professions.
2. You are tasked with orienting a new medical director to your service. Develop an outline of pertinent information and training opportunities to assist in this process.
3. Research the application process, and identify three benefits for a medical director of being a member of each organization: NAEMSP, ACEP, NAEMT, IAFC (EMS Section).
4. How did the case of Hennepin County *v.* Hennepin County Association of Paramedics and EMTs influence

discipline issues on a national scale? Create a system of discipline for your organization. Then describe how it would be used for a paramedic that consistently violates protocol.

5. Contrast the pros and cons of descriptive versus algorithmic protocols.
6. Is a medical director that operates in a volunteer (non-paid) position liable in the same way as if he were in a compensated position? What are the implications of the Good Samaritan laws as they relate to the duties of a medical director?

References

Brame, K., & Davis, P. (1992). Fire service medical control: Who do those doctors think they are? *Fire Chief, 36*(3), 53–55.

Federal Emergency Management Agency U.S. Fire Administration. (1999). *Guide to developing effective standard operating procedures for fire and EMS departments* (FA-197-508). Washington, DC: Author.

Fitch, J. J., Keller, R. A., Raynor, D., & Zalar, C. (1993). *EMS management: Beyond the street* (2nd ed.). Carlsbad, CA: JEMS Communications.

Foley, S. (1998). *Handbook on fire service occupational safety and health standards*. Quincy, MA: National Fire Protection Association.

International Association of Fire Fighters. (1999). *Implementation guide to EMS*. Washington, DC: Author.

Keseg, D. P. (2001). *Medical direction of emergency medical services* (3rd ed.). Irving, TX: American College of Emergency Physicians.

Kipp, J., & Loflin, M. (1996). *Emergency incident risk management*. New York: Van Nostrand Reinhold.

Krentz, M. J. (1989). Off-line medical control. In W. R. Roush (Ed.), *Principles of EMS systems: A comprehensive text for physicians* (pp. 95–100). Dallas, TX: American College of Emergency Physicians.

Krohmer, J. R., Swor, R. A., Benson, N., Meador, S. A., & Davidson, S. J. (1994, Jan–Mar). Prototype curriculum for a fellowship in emergency medical services. *Prehospital Disaster Medicine,* 9(1), 73–77.

Polsky, S., Krohmer, J., Maningas, P., McDowell, R., Benson, N., & Pons, P. (1993). Guidelines for medical direction of prehospital EMS. *Annals of Emergency Medicine,* 22, 742–744.

Shanaberger, C. J. (1991, July). Determining domain. *Journal of Emergency Medical Services 16*(7), 107–108.

Stewart, R. D. (1989). On-line medical control. In W. R. Roush (Ed.), *Principles of EMS systems: A comprehensive text for physicians* (pp. 101–108). Dallas, TX: American College of Emergency Physicians.

Accreditation Criteria

CAAS

203.01.01 Vehicle Specifications: All vehicles operated and used by the agency in the delivery of patient care shall be designed and maintained in good working order, in accordance with applicable federal, state and local specifications. All vehicles must adhere to manufacturer's gross vehicle weight recommendations. The agency shall have documented guidelines for total weight restrictions for each vehicle (pp. 277–278).

203.02.01 Preventive Maintenance: The agency shall have a program to document all vehicle maintenance, both scheduled and unscheduled. Summary reports of all vehicle maintenance records shall be provided to management at least quarterly (pp. 285–288).

203.03 Medical Equipment: Medical equipment shall be sufficiently stocked and maintained to allow for delivery of quality patient care (p. 276).

Objectives

After reading this chapter the student should be able to:

11.1 Discuss various considerations for calculating the cost of an EMS service (pp. 281–283).
11.2 Distinguish between functional and direct services for operating budgets (p. 275).
11.3 Describe the inspection processes for equipment (pp. 275–278).
11.4 Compare and contrast the concepts of unit-hour utilization and inservice ratios when determining workload (p. 283).
11.5 Calculate cost per capita and cost per response for EMS runs (pp. 281–285).
11.6 Apply cost-out strategies for a variety of EMS system components (pp. 284–286).
11.7 Track and apply costing mechanisms for soft supplies (pp. 284–285).
11.8 Understand the types of inventory systems and replacement plans (pp. 288–289).
11.9 Track fleet maintenance and vehicle cost including failure rates (pp. 285–287).
11.10 Determine the equipment needed in the system and the specifications of that equipment (pp. 277–279).
11.11 Understand and apply federal, state, and local specifications and procurement processes for ambulances, biomedical equipment, and durable equipment (pp. 279–281).

Key Terms

Point of View

Within the realm of possibility of the fleet manager, especially the EMS fleet manager, lies tremendous opportunity. Aligning the professional goals of fleet personnel to those of the organization and serving as a catalyst to see them through is significant. As we in fleet management impact, in a positive way, the person and the talent of those in our charge, we are able to contribute to the vision of the organization. In that way we weld the moment of contribution to the mission of EMS.

It has been my pleasure to work with so many dedicated and talented technicians during my 33 years of EMS experience. Many have unselfishly contributed and moved on, some to other industries and a few dear ones on to retirement. Countless others continue the delivery of professional service to their organizations and are as committed to EMS as their medic-team members. The future of EMS has at its core the delivery of the product to those in need. Dedicated fleet-maintenance employees will undoubtedly continue to be an integral part of that delivery. The best is yet to come!

Bill Vidacovich
Vice President, Acadian Ambulance Service, Inc.
Lafayette, Louisiana

■ FLEET MANAGEMENT AND EQUIPMENT

The delivery of EMS services cannot be accomplished without technologically sophisticated medical equipment and state-of-the-art emergency vehicles. Many EMS managers never see or understand the mechanics of equipment purchase and specification and are often left to compensate for poorly designed equipment and surges in the cost of EMS operations caused by vehicle expenses and equipment failures. When considering equipment and vehicle management in EMS, several models and best practices can be applied to fleet management and equipment procurement. EMS managers should be directly responsible for the specification development, bid process, purchase, and maintenance of vehicles and of equipment. Fleet and equipment management are vital functions of EMS service delivery and are major cost components.

A number of factors must be considered in establishing the cost of operating an EMS system. First, determine the types of services provided by the organization. **Direct services** are those provided to the public; they include patient-care activities, public-education programs, **preventive maintenance (PM)** programs, and special events. **Functional services** are those provided within the agency; they include yearly physicals, employee assistance programs (EAP), training, uniforms, office support, and utilities. An EMS manager must itemize all the services provided by the organization as the first step in identifying and costing out each key aspect of the operation.

FLEET MANAGEMENT

Fleet administration requires EMS managers to supply vehicles and services to support EMS activities. These services cover a wide range of job tasks (Figure 11.1).

- ◆ Procurement and specification of vehicles.
- ◆ Sale or disposal of obsolete equipment and vehicles.
- ◆ Preventive maintenance.
- ◆ Mechanical upkeep.
- ◆ Registration and licensing of vehicles.
- ◆ Fueling.
- ◆ Roadside assistance and towing.
- ◆ Computerized fleet management.
- ◆ Specialty-response equipment design and construction.
- ◆ Establishing a bid process and approval criteria, including vendor evaluation and qualification.

FIGURE 11.1 ■ Job Tasks for EMS Manager Responsible for Fleet Management.

Modern EMS agencies position this function under a logistics chief officer or frontline supervisor, which gives proximity to the day-to-day operations of an EMS organization and allows access to the National Incident Management System (NIMS).

How EMS services are provided varies greatly, depending on the size, complexity, and budget of the fleet. Some fleet requirements are mandated by the local health department or by the performance contract of the municipality. Some fleet-management operations provide all services listed in Figure 11.1 and more, while others outsource many or all of those tasks. An EMS manager must consider replacement policies, contracting and government bidding standards, whether to purchase or lease apparatus and equipment, and replacement criteria. A contingency plan should be put in place to provide reserve equipment or equipment on loan during planned and preventive maintenance. In the event that there is equipment failure, a procedure should be put in place for using reserve equipment, such as borrowing an **ambulance** or cardiac monitor from a neighboring department or perhaps a local vendor who will loan you a demo unit in the event your reserve unit is unavailable or you need additional resources.

Verifying and Maintaining Checkout

The cornerstone of any EMS system's equipment reliability and maintenance program is in the vehicle and equipment checkout program. Accountability for the inspection and inventory of equipment and vehicles should be imbedded as part of an organization's culture. It is important for EMS leadership to instill in the company officers, paramedics, and EMTs that regular inspections need to be conducted every shift, cycle, and month by on-duty EMS personnel.

Equipment checks should be established as the first action at the start of a shift. Ensuring that equipment is ready for service should be reinforced during employee orientation, taught during inservice training, and included in management objectives as the very first thing to be done at the start of a shift. Equipment not used often should be inspected monthly and tagged with a breakaway seal.

Vehicle checkouts should be completed during every shift on mechanics, as well as fluids, belts, and tires. Vehicles should be inspected on the outside for damage and leaks and to ensure working lights and warning devices. Mileage and engine hours should be recorded daily. Critical medical equipment should be on the vehicle–checkout list, including oxygen, backboards, and soft supplies. Any equipment that requires batteries, such as suction devices, laryngoscope handles, and cardiac monitors, must be checked before any vehicle is made available for a call. Responding to a call without working equipment has resulted in significant litigation and financial losses for EMS organizations. Two of the first items usually requested by attorneys in medical malpractice cases are the prehospital care reports and the vehicle checkout report.

Regulatory inspections are those conducted by the government authority that licenses ambulances. In most locations these inspections are mandated annually. The regulatory agency inspects the vehicles for items mandated by law or regulation, and frequently expired drugs or equipment that is on a state inventory—yet rarely used within the agency—are found to be missing. Regulatory inspections also verify insurance coverage, proper licensing, and personnel certification. The driver's licenses of EMS employees should be inspected regularly to identify those whose licenses have been revoked or suspended. In conjunction with annual inspections, regulatory agencies often check for current and valid driver's licenses and conduct routine criminal-background checks. Any of the inspections can occur on a random basis. As a management tool EMS leadership should conduct random audits or inspections to ensure compliance. Most accreditation agencies will conduct spot checks when doing a site visit. Management must establish the importance of checkout procedures to avoid any errors or interruptions in EMS service.

Technology has improved vehicle checkout. Bar codes placed in certain areas of the vehicle allow use of handheld or optical scanners to screen inventories. In some fleets a bar code is placed on the dipstick of

FIGURE 11.2 ■ Example of an Inventory Control Device for EMS.

the vehicle, and the crew must scan the bar code, indicating an oil check. Similar tabs can be placed in the wheel wells and on key medical equipment. Scanned bar codes on batteries can indicate life spans and shelf-life benchmarks for taking equipment out of service. Locking tabs on bags or equipment, such as pediatric equipment, that are seldom used saves time at checkout and provides security (Figure 11.2).

Ambulance Specifications

EMS managers and the organization should create specifications for EMS vehicles within the system. Local health departments often have standards for EMS vehicles and mandated equipment, usually listed in the local board of health regulations or in an operating agreement. State EMS authorities usually have specifications for EMS vehicles that are reflected in state law or state statute. For example, Montana has specifications calling for back-up alarms, a specific tread depth, and onboard CO detectors to protect crews and patients during long idling times in deep snow. Virginia's laws simply refer to the federal standard plus locks for narcotics.

Nationally, one set of standards exists: the Federal Specification for the Star-of-Life Ambulance (KKK-A-1822E), prepared by the General Services Administration in June 2002 (Figure 11.3). Federal specifications that require the copyrighted Star of Life to be displayed on the side of an ambulance must be followed. Laws regarding bloodborne pathogens, public-safety radios, and federal motor-carry safety regulations apply to federal ambulance specifications. Additional information from the Society of Automotive Engineers, the National Truck Association, the American College of Emergency Physicians (ACEP), and the American Society for Testing and Materials (ASTM) was used in the revision of the federal specification for ambulances.

FIGURE 11.3 ■ Federal Standard for Ambulances.

The **Ambulance Manufacturers Division (AMD)** of the National Truck Equipment Association (NTEA) has for the last 25 years represented a majority of ambulance manufacturers in the United States. The NTEA is the only trade association representing the nation's manufacturers and distributors of commercial trucks, truck bodies, truck-equipment trailers, and accessories. NTEA members include companies that produce every conceivable configuration of commercial trucks, including highly specialized vehicles, such as ambulances and other emergency-response vehicles. The NTEA provides its nearly 1600 member companies with resource materials, technical assistance, and business improvement programs. Headquartered in suburban Detroit, the NTEA interacts directly with the major truck chassis manufacturers on product compatibility issues. From its government-relations office in the nation's capital, the NTEA keeps its members advised of changing regulations affecting commercial trucks and lobbies on the industry's behalf.

The AMD was founded in 1976 and became a division of the NTEA to further enhance its depth of professionalism and its position in the industry. Currently composed of approximately 60 companies, the AMD has consistently maintained representation of over 90% of the ambulance production in the United States. Since its founding, the AMD

- ◆ Coordination with the General Services Administration (GSA) in the development and revision of the Federal Specification for the Star-of-Life Ambulance (KKK-A-1822E).
- ◆ Development of performance test procedures for ambulances.
- ◆ Developed current AMD standards for:

AMD001	Static Load Test for Ambulance Body Structure.
AMD002	Body Door Retention Components Test.
AMD003	Oxygen Tank Retention System.
AMD004	Litter Retention Test.
AMD005	Ambulance 12-Volt DC Electrical System.
AMD006	Sound Level Test Code for Ambulance Compartment Interiors.
AMD007	Carbon Monoxide Levels for Ambulance Compartment Interiors.
AMD008	Load Test for Ambulance Patient Compartment Grab Rail.
AMD009	120 VAC Electrical Systems.
AMD010	Water Spray Test for Ambulances.
AMD011	Ambulance Equipment Temperature Test.
AMD012	Ambient Temperature Test.
AMD013	Weight Distribution.
AMD014	Cooling System Test.
AMD015	Ambulance Main Oxygen System Test.

FIGURE 11.4 ■ AMD and NTEA Activities.

has worked closely with all state and federal regulatory agencies. Some of the more significant projects with which it has been directly involved are listed in Figure 11.4.

AMD standards are those to which all newly manufactured ambulances must conform. Before a standard is adopted, it is scrutinized by review panels from the AMD, industry experts, the U.S. General Services Administration (GSA), and all interested parties. All comments are considered, and appropriate revisions are made before a standard is voted on for adoption. All of the AMD tests listed in Figure 11.4 have been incorporated into the latest revision of KKK-A-1822.

Most AMD members maintain staff engineers to keep their companies abreast of technological advances applying to the manufacture of ambulance bodies, electrical systems, environmental systems, and other ambulance components. These advances are incorporated into new ambulance models, thereby continuously improving the industry through competition. No governmental agency dictates that AMD members make these improvements; they are

done voluntarily to upgrade the product, make it more reliable, and provide even more dependable life-support capabilities.

Ambulance Types, Classes, Configurations

Vehicle types are divided into either class I or class II, representing two-wheel and four-wheel drive, respectively. The federal standard also uses three ambulance types, with further divisions into classes and floor configurations. The type I conventional truck, with cab chassis and a modular ambulance body, is shown in Figure 11.5. The type I ambulance can

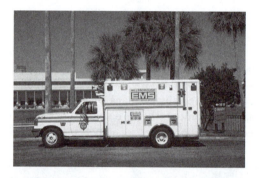

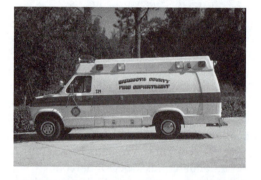

FIGURE 11.5 ■ Types of Ambulances: (a) Type I, (b) Type II, (c) Type III.

have an additional-duty (AD) unit when modified for neonatal, critical-care transports, or rescue or fire-suppression package. This is often referred to as a "medium-duty rescue." The type II is a standard van with an integral cab-body ambulance. A type III ambulance is a cutaway van, cab-chassis with integrated modular ambulance body. The type III ambulance also can have a classification of additional duty (AD).

ALS ambulances may be either type I or type II, as recommended in the federal standard. These ambulances have two standard configurations in the federal specification: "A" for ALS and "B" for BLS.

Ambulance specifications are best put together by those that use them the most—line EMS personnel. An ambulance-specification committee led by an EMS manager in the organization should be used for building and equipping ambulances. The committee also can be used to specify equipment. To make the process work, the members of the committee must understand their roles in research, writing, and investigating information.

Specification Processes

The first step in purchasing or leasing equipment is to create a procurement process for your agency, whether for a new ambulance or a cardiac monitor. Defining the specifications for the vehicle or equipment is really one small part of the process. While some EMS agencies are content to order equipment off the shelf, modern management principles related to empowering the workforce suggest that the employee will feel there is more ownership and pride in the equipment when employees have input into the design and construction.

Vendors usually provide a set of specifications and place in those specifications conditions or statements designed to exclude other vendors. Rarely is that set of specifications designed to a performance standard. In many cases the vendors reserve the right to substitute or change the specifications. For this reason it is important to use the request for proposal (RFP) process and include key performance criteria and penalties if the equipment fails to meet those standards. Similar to construction contracts, bid specifications may include monetary penalties for late delivery or for not meeting performance criteria. When purchasing equipment, the EMS manager should ask for several different requests for proposals or contracts used by other agencies for purchases and employ a committee

to review the content of these RFPs and use the best of the content to create a custom request for proposal.

Procurement Process

A simple approach to procurement is to apply the incident management system to the entire process under the title of a procurement or specification committee. The committee will have an EMS leader that manages the process, and the operations person will supervise the research and evaluation of the products. A planning section will create the RFP or bid request. The logistics person will arrange for the testing and demonstration and will make sure that coordination of facilities is achieved. Finally, a finance person will qualify bidders, account for cost, monitor budgeting, and ensure payment and transfer of money. Each of these sections needs to be populated with field personnel and end users of the equipment.

A timeline should be established and a budget arranged with a 10% to 15% emergency allocation or reserve fund for unforeseen issues. Notes and records need to be kept on the decision-making process. Often there is a spending benchmark that requires certain rules to be followed that are defined by ordinance or state law. Many agencies are handicapped by federal bids, which do not allow for an RFP process, or when the purchasing is done by another department within the government or corporate structure.

Ethical Considerations with Bidders

Large contracts often provide tempting opportunities to supply **graft** to EMS managers or committee personnel. Graft is the acquisition of money, position, etc., by dishonest or unjust means, as by actual theft or by taking advantage of a public office or any position of trust or employment to obtain fees, perquisites, profits on contracts, legislation, pay for work not done or service not performed, etc.; or illegal or unfair practice for profit or personal advantage. Establishing ground rules is important in any procurement process to ensure vendors are dealt with fairly. Figure 11.6 is a list of ethical ground rules to apply to any RFP process.

While travel and meal expenses are not acceptable prior to the award of a contract, they may be included as part of the RFP. After an award is made, provision will be made in the inspection process for staff to travel to a site for inspection of the equipment during construction, training at the manufacturing site, or to take delivery of the equipment. It is common for ambulance builders to have staff flown to the site for

- ◆ Be fair to all vendors.
- ◆ Do not give preferential treatment to any vendors.
- ◆ Do not let vendors know how much money you have to spend.
- ◆ Accept no gratuities, food, travel, or other gifts.
- ◆ Do not disclose the results until the vendors have been notified.
- ◆ Do not let vendors influence department personnel.
- ◆ Do not let vendors slander competitors.
- ◆ Include the ground rules in the information that goes out to the vendors.
- ◆ Ask vendors to confirm receipt of the ground rules.

FIGURE 11.6 ■ Ethical Ground Rules for Vendor Relations.

delivery of the vehicle to the agency. In most cases, if you specify that you will make your own travel arrangements and cover expenses for construction review or vehicle delivery, the manufacturer may be able to provide a better bid for the vehicle.

RFP Construction

The RFP process is the key to matching the product to your needs. This process requires a time-consuming amount of research and collection of information. Often bids, specifications, and technical reviews need to be collected from other organizations or from vendors prior to developing the organization-specific RFP. An RFP typically has several components and should allow for the vendors to ask questions (Figure 11.7).

Vendors should be qualified by the committee or the EMS manager overseeing the process. A stable operational history, solid customer support, FDA approval, and up-to-date technology are important in a

- ◆ Scope or type of equipment and services.
- ◆ Procedures.
- ◆ Timeline with key dates.
- ◆ Instructions for submission.
- ◆ Quantities, pricing, discount terms, warranties, trade-in discounts.
- ◆ Presentation requirements and evaluation criteria.
- ◆ Options for collaborative or pooled bids.
- ◆ Proposal formats and deadlines.
- ◆ Rebuttal or clarification-question procedures and deadlines.
- ◆ Insurance requirements and minimum qualifications for bidders.
- ◆ Points of contact and communication processes.
- ◆ Award dates and notification processes.

FIGURE 11.7 ■ RFP Components.

profile of a vendor. Most companies should support your product for its life cycle, which is often seven years for vehicles and eight years for medical equipment. An EMS manager should contact the vendor's most recent customers and at least three other customers who have used their product for several years.

Scoring and Evaluating Bids

A simple and easily understood scoring process is beneficial to EMS personnel who are participating in the bid and the RFP process but who may have limited analytical skills. A simple 100-point scoring system that weighs the specifications from most important to least important helps secure the best vendor. A point system needs to reflect each area, including finance, operations, background with customer service, vendor presentation, and functional ability. Field evaluations of equipment must be part of the process; when this method is used to test equipment, a scoring system and summary of all the comments of the users will be collected and given to the members scoring the field test.

Field testing may require the product to be consumed or destroyed. For example, protective equipment or material that is supposed to be resistant to breakage or certain types of damage may need to be tested. This requires the material's claimed resistance to be supported by testing documents or put through an actual demonstration.

Vehicle Inspections

When an EMS manager takes delivery of an ambulance, a series of procedures is recommended. A source inspection should be completed by EMS management or leadership prior to shipment from the manufacturer and should include workmanship, quality conformance, and a first-production inspection. First-production inspections ensure that the manufacturer is conforming to the standards. They should be done at the manufacturing plant, and the cost of getting EMS managers to the factory should be borne by the manufacturer.

A **destination examination** also should include a check of all ambulance controls, electrical systems and devices, doors, windows, cabinets, and accessories, as well as a road test at highway speeds, a brake test, and a test for rattles and squeaks.

The performance test is listed in the federal specification and encompasses 18 key points that in total look at all the vehicle systems. A **road test** of new vehicles involves driving a total of 150 miles, with 75 miles of that on highways at a speed of 70 mph; 30 miles on city streets at 30 mph, 15 miles on gravel or dirt roads at 35 mph, and 5 miles on cross-county operations that are in muddy or open-field areas. A class 2 vehicle that has a four-wheel-drive rating must demonstrate an additional 20 miles in a cross-country operation. A **water-spray test** subjects the vehicle to a water spray at 25 psi for 15 minutes in order to look for any evidence of a leak. Finally, the **oxygen-system test** must be tested, and the system should be pressurized to 150 psi with dry air or nitrogen and be able to hold that pressure for four hours.

A great tool for anyone going to pick up a new ambulance is a checklist prepared in advance that details the tests and inspection points to be completed before acceptance of the vehicle. Today's modern electrical systems are much better, but a detailed check of the lighting systems and running batteries and alternator under a load—with all the emergency and non-emergency lights on for a period of 30 minutes—can determine if there are any potential problems.

If a vehicle fails any of these tests and the vendor cannot fix them in a reasonable number of attempts, the EMS leadership should have the option to terminate the contract.

The life of a vehicle or equipment should be estimated, and warranties should be identified, including shipping or transport charges, parts, and labor. The liability of the vendor if the equipment fails or injures a patient should be identified. Copyright infringement, third-party claims from the public, and the cost of arbitrating or negotiating a settlement in an alternative dispute-resolution system should be identified, as well as with whom expenses will be incurred or shared.

Typically in a contract for vehicles or medical equipment, the minimum number of hours of labor per unit should be specified and the response time after notification of a maintenance problem or equipment failure should be arranged according to the level of severity of the trouble. A forward-thinking contract delineates the vendor's responsibilities to provide enhancements, updates, or upgrades. Contracts with maintenance should cover the life span of the equipment.

TRAINING AND INSERVICE

Crashes are a common cause of litigation and risk for an EMS agency. Emergency-vehicle collisions are often serious, and some court cases have resulted in criminal charges being levied against EMS workers involved in them. Consequently, when a new vehicle arrives EMS personnel should receive appropriate education and training to ensure they can operate the new emergency equipment well. Training also should include how the vehicle is to be checked out and maintained according to manufacturer's specifications.

If your service is changing vehicle types—from a Type III to a medium-duty rescue, for example—then a driving course and time spent in the operation of the vehicle are necessary, because turning radius, backing up, and mirrors will be somewhat different in these kinds of vehicles. Any vehicle training must include an explanation of what privileges an emergency-vehicle operator has in your specific state.

The Emergency Vehicle Operator Course (EVOC) and Coaching the Emergency Vehicle Operator (CEVO) are two programs designed to help train EMS personnel to operate emergency vehicles.

■ MONITORING COST IN FLEETS

A key function of EMS managers is to oversee processes and equipment. This often means that improving productivity and reducing administrative times related to equipment can reduce overall operational costs. It is therefore important to calculate and estimate cost for fleets and equipment. It is important to have the cost of one crew, in one unit, operating each hour. Much like the airline industry, when equipment is out of service, there is a cost associated with not having the equipment available as well as resources assigned to fix the equipment.

VEHICLE COSTING

Vehicle costing starts with the initial cost of the vehicle. The purchase price, taxes, warranty, and delivery charges are the initial cost of the vehicle. The cost of maintenance, repairs, and nonwarranted parts are recorded and fed into the cost for EMS vehicles, as shown in Figure 11.8. It is important that a **job card**, similar to that shown in Figure 11.9, be filled out for any repair or maintenance on the vehicle; it should include the vehicle number, odometer reading, hours and cost of labor, and any miscellaneous expenses around the event, such as towing. Fuel cost for each vehicle also needs to be tracked. Overhead cost such as support personnel, office, cell phone, and filing or computerized applications to operate the system should be included as a portion of the expenses.

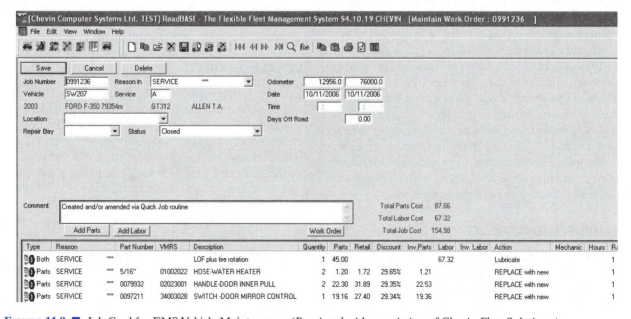

FIGURE 11.8 ■ Example of Fleet Tracking Software. *(Reprinted with permission of Chevin Fleet Solutions.)*

RECORDS MANAGEMENT

Cost-Tracking Software

A computer-based system that tracks all information about vehicles, such as the Fleet Accounting Computer Tracking System (FACTS), should be in place. Like FACTS, the software should be able to track vehicle identification numbers, license-plate numbers, manufacturers, vendors, purchase prices, vehicle ages, and work orders for repairs. The system

FIGURE 11.9 ■ Job Card for EMS Vehicle Maintenance. *(Reprinted with permission of Chevin Fleet Solutions.)*

$$\text{Cost per Unit Hour} = \frac{\text{Total Cost of Providing Coverage During Given Accounting Period}}{\text{Total Number of Unit Hours Covered During an Accounting Period}}$$

FIGURE 11.10 ■ Cost Per Unit Hour.

links specific information on each vehicle with operational data, such as labor hours, costs of parts, and frequency of parts usage, so that the office can assess its performance against established targets. Most fleet-management software enables automated parts ordering and allows users to determine which parts are in stock.

A fleet-management software system should be able to use the information to calculate costs for vehicles by type and age, which helps planning for future purchases and maintenance. It should be easy to use by technicians and EMS managers who have no computer experience, and it should allow EMS managers and frontline supervisors to collect and analyze performance data and establish benchmarks to continue improving competitiveness.

Such systems can be used to track vehicles from the time they are acquired until the time they are disposed of. EMS managers can use them to track the initial cost of the vehicle, set up a preventive-maintenance schedule, issue work orders, and do fuel and warranty tracking. The systems also can be used to help determine the best time to replace a unit.

Calculating Unit-Hour Cost

The concept of total cost per unit is a measure of the total cost of providing coverage during a given accounting period divided by the total number of unit hours of coverage provided during that same accounting period. This measurement can be used to identify whether or not unit-hour costs (Figure 11.10) can help identify excessive cost compared to the quality of care being delivered. Unit-hour costs are used to compare similar EMS systems, the type of delivery vehicle (such as a fire engine or ambulance), or how crews are configured (EMT/paramedic or paramedic/paramedic). If unit-hour costs are excessive and yet response times are not being met, the EMS leadership should consider moving vehicles or adding unit hours. Other factors, such as top-heavy administration, high wages, poor economies of scale, overtime pay, and many other causes, can drive up unit-hour cost.

Here is an example that uses very simple numbers for calculating a cost per unit: A department has five

ambulances covering a 24-hour period. Each ambulance is staffed with a paramedic at $23 per hour and a firefighter EMT at $18 per hour, including benefits. There is a shift supervisor who makes $30 per hour as an EMS supervisor. The cost to operate the vehicle, including fuel, maintenance, and hourly time cost for the fixed facility to house the unit, is $45 per hour. Figure 11.11 explains costing out a unit hour.

Cost per Mile

A common analysis of vehicle costs is called **cost per mile.** It assists in helping managers create comprehensive plans for fleet management, including replacement plans, selections of vehicles, and evaluations of effectiveness of maintenance programs. Cost per mile is calculated by taking the specific vehicle cost, including the fuel, insurance, maintenance, and indirect cost to support that vehicle in a budget cycle or calendar year, and dividing it by the annual mileage accumulated on that vehicle. In some systems, managers group certain vehicles, such as nontransport, transport, and administrative vehicles.

Another option in cost-per-mile analysis is to calculate the cost in the rolling year versus the life to date. A rolling year is the specific year the vehicle is on the road; life to date calculates the entire time the vehicle is in the inventory. However, the cost of the vehicle must be depreciated over the life span of the vehicle.

There is no industry-specific standard for replacing ambulances. Some contractual agreements and accreditation standards require frontline EMS response vehicles to be replaced every five years. In Figure 11.12 the cost per mile is figured using industry-accepted standards for other fleet operations. Cost per capita is another way of comparing annual cost of a vehicle to the public or the service population.

---

a. Unit hours: 24 hours × 5 units = 120 unit hours
b. Cost: Paramedic salary, 24 hours × $23 × 5 = $2760
   Firefighter/EMT salary, 24 hours × $18 × 5 = $2160
   Vehicle operating cost, 24 hours × $45 × 5 = $5400
   Total cost for 24 hours = $10,320
c. Cost per unit hour: $10,320/120 = $86

FIGURE 11.11 ■ Cost Per Unit Hour.

```
Cost of vehicle: initial price, $155,000 / 5 years     $31,000
Annual maintenance cost                                $15,000
    Overhead                                           $25,000
    Fuel                                               $ 8,000
                                                       _____
                                                       $79,000

    Annual mileage                                      32,000
Annual cost/mile: 79,000/32,000 = $2.46 / per mile to operate

Annual Cost per Capita
30,000 population / 32,000 = 0.93 cents / per person to operate
that unit without staff.
```

FIGURE 11.12 ■ Figuring Cost Per Mile and Cost Per Capita.

$$\frac{\text{Oxygen equipment bag and equipment \$825}}{\text{Five-year life span (5)}} = \$165/\text{year}$$

FIGURE 11.13 ■ Life Span and Cost Per Year for Equipment.

Most managers who oversee fleet operations use computer software that produces reports and queries for any aspect of the fleet, including fleet running costs and utilization, as well as accident and risk management. Many of these systems also monitor fuel cards. A common and more variable expense is fuel cost. Some fleets have opted to move to alternative fuel sources, and in the future the cost breaks will require that EMS leadership consider changing to those fuels.

Most fleet managers track fuel cost for each vehicle to help identify maintenance needs and times to replace vehicles. Tracking fuel cost on a spreadsheet or by computer program combats fuel fraud, rogue drivers, and vehicle abuse. Some computer programs provide detailed audits or track vehicles and assignments to comply with personal-mileage-reporting requirements. This is important, because some EMS staff members have taken vehicles home for personal use.

EMS fleet managers must make sure all vehicles are licensed and insured. Another feature of using fleet-management software is that it helps to manage any type of regulatory compliance, such as permits, certifications, registration renewals, physicals, and inventory inspections from government agencies.

■ MEDICAL EQUIPMENT

EMS managers will be responsible for the maintenance, specification, and purchase of biomedical equipment—any medical device, medical supply, or tool used to provide care to a patient. The cost-effectiveness of such equipment is rarely tracked or rationalized before budgeting. EMS agency defibrillators,

ancillary equipment, biomedical services for preventive maintenance on defibrillators, routine nonwarranty work, training equipment, trainer and provider certification—all become part of the expense of implementing new equipment.

It is common to measure the cost per life saved when budgeting for biomedical equipment. For example, some Canadian studies have determined that the implementation of defibrillators on emergency apparatus results in a cost benefit in an urban area of $6776 and $49,274 in a rural area for each life saved. Typically, that does not include the cost effectiveness of other equipment carried on the vehicle. Often the EMS agency's inservice training, ambulance, radio equipment, medical equipment (such as stretcher, oxygen, and first-aid supplies), defibrillator, telemetry radio, and drug box are not stated.

Figure 11.13 shows the typical costing-out service for medical equipment carried on most EMS vehicles. The cost of biomedical equipment should be added to the overall vehicle operating cost and can be figured in a variety of ways. One common way is to create an inventory list of all equipment carried on the ambulance, including the vehicle itself. The inventory list should contain the cost of each item as well as its anticipated life span. Dividing the cost of the equipment by its anticipated life span gives you a cost per year that can be added into the unit-hour cost.

The year cost can be divided by the number of days of the year and then by a 24-hour shift, resulting in an overall cost of a little under two cents an hour or about 0.45 cents per day. When each piece of equipment is expensed and included, the small amounts for each piece add up. The vehicle-maintenance cost includes mileage, maintenance records to determine the average cost per year, and—again using simple division—the average cost per hour. Disposable supplies billed to the patient are not included in the cost per hour.

If your department or agency does not bill for soft supplies, then the line item for soft supplies in the budget should be divided to ascertain an hourly and daily cost of soft supplies.

■ VEHICLE MAINTENANCE

It is important to establish a schedule for preventive maintenance—a schedule of mechanical checks or changes designed to prevent critical failures in vital equipment. Preventive-maintenance schedules should be based on manufacturer recommendations, or schedules should be determined within the organization. While no specific standard is in place for ambulance fleets, many EMS agencies avoid placing an ambulance in front-line service for more than five years.

The NFPA 1911 Standard for Inspection, Maintenance, Testing, and Retirement of Inservice Automotive Fire Apparatus 2006 covers industry-accepted practice for emergency-vehicle inspections. The standard has several checklists and forms in the appendix applicable to EMS fleet operations. The quarterly or annual fire-inspection form covers the ambulance chassis, engine, and vehicle components outside the patient-care compartments. EMS fleet managers should determine the factor(s) that most seriously affect the particular reliability of an EMS vehicle and the changes required to improve the performance of that vehicle or device. This is done by identifying common events or common causes of failures.

VEHICLE AND EQUIPMENT FAILURES

The annual **critical vehicle-failure (CVF) rate**, or the average frequency at which something fails, should be calculated. This is done by identifying a specific mileage or operating-hour benchmarks for the entire fleet. A common benchmark is the number of incidents for the fleet every 100,000 miles, 1000 engine hours, or 100,000 engine hours. Mileage and hour benchmarks work well, since wear and tear on equipment is based on use and not calendar days.

Common areas to plan for in preventive maintenance are replacement of items such as tires, batteries, belts, fluids, brakes, and other parts determined to need replacement on a scheduled or regular basis. A common rate of failure in public-safety and ambulance fleets suggests that out-of-service incidents occur at a rate of 0.87 vehicle failures per 100,000 fleet operating miles. A vehicle failure above this rate in public safety may indicate that a vehicle should be taken out of service. Failure rates can be reduced when problems are identified in a vehicle checkout.

Failure Rate Calculation Example		
Ambulance Number	**Hours**	**Failure**
Ambulance 1	1000	No failure
Ambulance 2	1000	No failure
Ambulance 3	467	Mechanical failure
Ambulance 4	1000	No failure
Ambulance 5	630	Mechanical failure
Ambulance 6	590	Mechanical failure
Ambulance 7	1000	No failure
Ambulance 8	285	Mechanical failure
Ambulance 9	648	Mechanical failure
Ambulance 10	882	Mechanical failure
Totals	7502	6

Failures/total hours x 1,000,000 = Estimated failure rate is 799.8 failures for every million hours of operation; divided by 1000 hours of operation, which converts to a 0.79 vehicle failure for every 1000 hours of operation.

FIGURE 11.14 ■ Calculating Vehicle Failure Rates.

As an EMS manager you will want to establish a preventive-maintenance program. To do this, you must estimate the vehicle-failure rate of ambulances in the fleet; if you take units out of service prematurely, service interruptions and undue cost may result. Ten ambulances or EMS vehicles should be tracked until they have a breakdown, a field crew identifies the need to take the vehicle out of service, or the vehicle reaches 1000 hours of operation, at which time the measurement period is terminated for that fleet. A sample of how to calculate vehicle-failure rates appears in Figure 11.14.

These numbers then can be used to estimate the number of interruptions of service anticipated, given a specific number of vehicle hours or unit hours of vehicles on the street. Once the frequency is calculated, you can track the average cost in personnel and parts to repair the vehicle. These calculations help when budgeting for the number of repairs and estimating when a unit should be placed into a preventive-maintenance program. For example, from the information in Figure 11.14, when a daily checkout identifies operating hours that surpass 1200, you can estimate that one failure will occur in a vehicle between 1200 and 1300 hours.

Vehicle-failure rates, fleet operating costs per mile, and the other measures employed in the above examples are key performance indicators or measures of results. Comparisons of key indicators are a common benchmarking process. When the same key indicator is compared over time for the same organization, the benchmarking method is a sequential-trend analysis. But when we compare the same key indicator for several organizations, this is considered a lateral benchmarking method.

A critical vehicle-failure rate can be applied at many different levels in the hierarchy of a vehicle. Figure 11.15 is one way of examining a vehicle failure. It shows that there are a number of different levels at which it may be sensible to undertake a critical vehicle-failure rate. Key is the level at which the end effects produce a hazard, such as when the agency has a vehicle failure that results in the inability to respond or deliver a patient to the hospital. When performing a critical failure rate, the automotive industry tends to use three parameters for each failure—severity, occurrence, and detection—and gives each one a score between 1 (not critical) and 10 (extremely critical). The Insurance Services Organization (ISO) has guidelines that allow for application-specific scoring of each of these categories. Safety-analysis guidelines propose a scoring scheme along the following lines:

♦ *Severity.* The hazards associated with each failure should be categorized into 10 different levels, ranging from 1 (least severe) to 10 (most severe) and based on safety and impact on the operation.

♦ *Occurrence.* This is the number of events and the probability of random failures in a given mileage benchmark or a specific number of hours of engine use. The occurrence of random faults ideally should be scored objectively, using a percentage per certain hours or mileage.

♦ *Detection.* A meaningful interpretation of this number is a measure of the degree to which preventive systems and vehicle checkouts can be detected. It should be made up of the probability expressed in a percentage.

Historically, the automotive industry has multiplied these three scores to give a **risk priority number (RPN)** and then concentrated on dealing with those faults, or failures, that have a high RPN. The RPN should be used with care, since it is a one-dimensional assessment of three dimensions of information. Once a failure has been found, it is important to document the path through the systems that failed. Mapping the components and systems that failed as in Figure 11.15 can bring a visual perspective to the systems an EMS manager oversees. Determining a critical-path method of tracking vehicle failures helps in the design of better

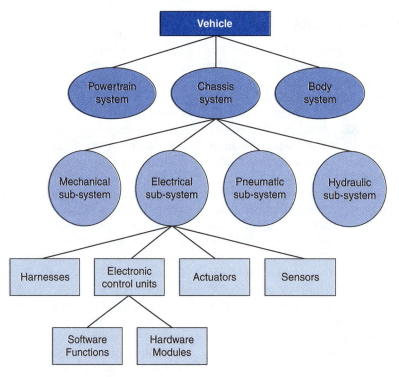

FIGURE 11.15 ■ Documented Vehicle Failure of Process for Electrical System.

vehicles and prepares the maintenance people for the types of repairs most likely to occur.

MAINTENANCE

Planned Maintenance

In today's climate of budget constraints and limited resources, it is absolutely imperative that EMS system managers take positive steps to keep their vehicles and equipment on line. This is especially true for small- and medium-size services. Faulty or poor equipment and vehicle maintenance can be an EMS system manager's major exposure to liability through contributory negligence.

A quality maintenance program supported by user involvement and awareness is the key to success, and money is the bottom line. Although planned maintenance programs can save thousands of dollars per year, they have not been universally implemented. However, there is a downside to a planned maintenance program when system and maintenance managers fail to take into account the useful service life of vehicles and equipment. Extending the useful service life of equipment or vehicles beyond established standards may lead to failure of a piece of equipment during critical patient care or an inability to respond to a situation. Legal liability in either situation may be much more costly than the prudent replacement of the equipment or vehicle.

Someone needs to have responsibility for both preventive and **corrective maintenance** of a system's equipment and vehicles. To make the system work, each member must be involved as part of a team maintenance effort to prevent or uncover and report problems. Unless problems are reported, they cannot be corrected before failures occur. EMS managers must insist that EMS members be accountable for vehicle and equipment maintenance documentation, and managers must be accountable to review maintenance records for compliance with requirements and schedules.

Computer tracking is vital to ensure that maintenance is performed proficiently. Computer systems should be used to track the time it takes to complete each maintenance task. This allows the EMS manager to determine how repair times will affect budgets and to predict and evaluate maintenance operations. Tracking the frequency and types of maintenance services also allows EMS managers to more accurately determine when a vehicle needs to be serviced or replaced.

Preventive Maintenance

The objective of a preventive-maintenance (PM) program is to prevent a malfunction or failure, not just correct it after it occurs. Preventive maintenance should be a key component of managing vehicle recalls. Vehicle warranties also help to prevent unnecessary costs to the EMS agency.

Equipment and vehicle manufacturers normally provide routine maintenance and inspection recommendations for their products. These recommendations should be used by an EMS maintenance manager as the basis for developing the system's routine maintenance requirements and PM schedule. Individual team members should be accountable for performing and documenting routine inspections, cleaning, and maintenance, using standardized checklists as part of regular shift changes. Using a standardized checklist makes the task easier and helps to avoid overlooking any requirements. Relying on individuals to take responsibility is essential to maintaining equipment and vehicles.

An extensive PM schedule permits EMS managers to plan for equipment and vehicle downtime, ensures that equipment and vehicles receive more detailed inspections and maintenance procedures, and ensures that equipment and vehicle warranties remain in force. An extensive PM schedule also facilitates coordination between operational managers and fleet-maintenance managers.

In the case of vehicles, a manager must be aware of the number of hours the vehicle has been in service as well as the number of miles traveled since the last service period. Allowing a vehicle to idle for any length of time is extremely hard on its engine, especially if it is a diesel engine. Excessive idle time may be detected when reviewing vehicle records if recorded engine hours increase disproportionately to the vehicle time/distance usage data.

It is especially important to inspect and rotate vehicle tires. On August 23, 2004, a 22-year-old male career firefighter/EMT died after the ambulance in which he was riding reportedly hydroplaned, left the road, overturned, and crashed into a tree. NFPA 1915 states that "as part of the maintenance program to ensure safe travel, tires should be inspected for advanced and unusual wear, which can reduce the ability of tread to grip the road in adverse conditions." Visually check tires for uneven wear, looking for high and low areas or unusually smooth areas, and check for signs of damage. Tires should be replaced if there are deficiencies, such as cuts in the sidewall that penetrate to the cord, defects, or a tread depth of 4/32 of an inch or less on any steering axle or 2/32 of an inch or less on any nonsteering axle at any two adjacent major tread grooves anywhere on the tires.

The NFPA 1915 Standard for Apparatus Preventive Maintenance can be applied to EMS vehicles. It is important for EMS managers to set up schedules to ensure that apparatus have appropriate licenses according to state and local registrations; this includes current Department of Motor Vehicle registrations, insurance certificates, and smog checks. Key components should be brake inspections and replacements, fluid-level checks and replacement, tire rotations and checks for sidewall damage, tread depth, and belt separations. Lighting and siren warning systems should be inspected and replaced as soon as possible when the system is identified as defective.

Corrective Maintenance

A corrective-maintenance program depends upon each user identifying specific problems and reporting them to the maintenance manager. A simple standardized format should be used to identify the equipment or vehicle and the specific problem, including as many details as possible that might aid in accomplishing the corrective action. Managers need to review vehicle and equipment incident reports to detect trends that might indicate serious maintenance deficiencies or poor operator practices.

After corrective action is accomplished, the results must be documented in order to compile the equipment and vehicle maintenance repair history and to record the costs incurred. A well-documented record of repairs and costs is an invaluable indicator when it is time to make cost-effective decisions regarding replacement of the equipment and vehicles. It is important that all maintenance routines carefully articulate the plans and schedules to provide reserve equipment or loaner equipment from vendors during preventive and scheduled maintenance.

Replacement Considerations

EMS managers with fleet responsibility should consider many factors when deciding whether or not to replace a vehicle. A written replacement policy that is calculated using economic factors such as depreciation cost, increasing maintenance cost, service life, and warranty limits is important. The goal is to reduce overall equipment costs (procurement, maintenance, and disposal).

- ◆ Initial cost (purchase price, leasing fees, transport).
- ◆ Depreciation/resale value (cost per capita, market factors, trade-in value).
- ◆ Fuel economy (miles per gallon).
- ◆ Maintenance cost.
- ◆ Cost per mile to operate.

FIGURE 11.16 ■ Replacement Cost Factors.

Accurate, factual, and computerized data are vital for making certain that a vehicle is replaced at the best possible time. If the annual maintenance cost plus trade-in value exceeds the cost of a new vehicle, then it is time for replacement. However, the actual decision about whether or not to replace a vehicle is much more complex, involving historical and projected cost and usage data. One benchmark used is an increase in the cost per mile or cost per capita to operate the equipment. Financial factors to consider are included in Figure 11.16.

One way to graphically represent safety and repair issues such as accident records, breakdowns, and cost per mile to operate is to use a **bathtub curve** as shown in Figure 11.17. The bathtub curve describes a particular form of the hazard-analysis function that comprises three parts:

- ◆ Decreasing failure rate, or early "infant mortality" failures.
- ◆ Constant failure rate, known as random failures.
- ◆ Increasing failure rate, known as wear-out failures.

The bathtub curve is often employed to represent the failure rate of a product during its life cycle; namely, the product experiences early "infant mortality"

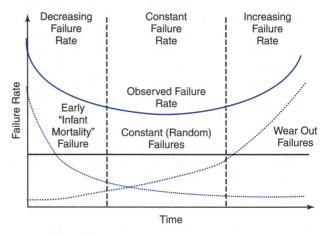

FIGURE 11.17 ■ The "Bathtub" Curve Hazard Function.

failures when first introduced, then exhibits random failures with a constant failure rate during its useful life, and finally experiences "wear-out" failures as the product exceeds its design lifetime. The EMS manager should attempt to plot vehicles and equipment on this graph.

Feedback

Feedback should be provided through the system to indicate to all concerned what the actual problem was, the corrective action taken, any indicators that might point out future problems, and any recommendations for incorporating into the system PM program. In addition, maintenance should be a regular topic for discussion at staff meetings, training sessions, or safety meetings. Involving everyone in regular discussions extends the team concept and makes it easier to keep everyone informed and involved. Maintenance programs on emergency vehicles, biomedical equipment, and other essential EMS equipment are conducted by specially trained personnel. Apparatus should be maintained by certified emergency-vehicle technicians, and EMS providers using biomedical equipment should be trained and certified by manufacturers.

TRAINING AND CERTIFICATION

EMS managers overseeing fleet operations ideally need an associate's degree in management, automotive diesel technology, or a related field. Most such positions require several years of hands-on experience in automotive maintenance and repair. Typical requirements for the mechanics who maintain city and county fleets are that they be high school graduates with course work in automotive mechanics and have experience as automotive mechanics or have an equivalent combination of training and experience. A well-documented training program for your fleet-service people is a must. Remember, if a vehicle is involved in an accident, law enforcement, lawyers, or regulatory agencies may examine your mechanics training records.

The Emergency Vehicle Technician Certification Commission (EVTCC) certification is the minimum standard set in many states and is one of the most common ways for fire mechanics and emergency-vehicle technicians to demonstrate their qualifications to work on emergency vehicles. Deciding which tests to take depends on the kind of vehicles worked on. Those who work on ambulances should take the **E-series** of tests. Additional certification tracks and fleet manage-

ment are available for other disciplines. Other technicians and mechanics who service county and municipal fleets go on to achieve the highest emergency-vehicle technician (EVT) certification by taking the EVTCC tests and seeking certification by the **National Institute for Automotive Service Excellence (ASE)**. In order to receive these certifications, a technician must pass all tests required for a specific level and then submit an application to the EVTCC office. Continuous training is critical and should include certifications, such as those offered by ASE. Smaller fleet operations that cannot offer their own training often find appropriate classes available at local colleges or from original equipment manufacturers.

Certificates demonstrate the skills and experience needed to perform high-quality maintenance and repair services; further, they instill confidence in the fleet's drivers and customers that the fleet operation is professionally run. Regardless of which path your agency chooses, all technicians must recertify their skills every five years to ensure that those skills are still current.

To receive ASE certification, mechanics must have at least two years of experience and pass a test that covers on-the-job diagnostic and repair procedures. Many jurisdictions have earned the ASE Blue Seal of Excellence Award because at least 75% of their mechanics are ASE certified and they have at least one mechanic who is ASE certified in every area of service work performed.

ACCIDENT REPORTS

Accident history and repair costs need to be recorded and a number of analysis reports produced; for example, accident details by driver and location and data on near misses. The accident file itself should have two main elements: the key information about the incident (who was driving, which vehicle, date, time, location, and so on) and which law-enforcement agency and risk manager or EMS supervisor was involved in the investigation. It also will contain a comments section and space for recording damage details as well as body-shop estimates. If an ambulance is involved in an accident with a civilian vehicle, there may be a large number of third parties involved in the incident (including witnesses). EMS managers need to maintain these reports in a separate but related file. This allows you to record an unlimited number of details rather than restricting you to one or two third parties.

CHAPTER REVIEW

Summary

The task of an EMS manager who oversees fleet operations is to serve field personnel by making available to them an ambulance of a type and design favored by those who use it. For the sake of the patient and the family, that ambulance should rarely if ever malfunction in the field and should be provided at a competitive operating cost. It is important for EMS managers to determine whether or not a particular system reliability measure meets the organization's requirements. The factors that most seriously affect a particular reliability of the fleet and equipment need to be reviewed and presented to the EMS agency leadership, as well as the changes required to improve on and identify common events or common causes of failures. A poorly managed vehicle or fleet does not protect the public and the agency. An agency's fleet should reflect the organization; it should provide professional service and present an image to employees and to the public that reflects pride and competence.

Research and Review Questions

1. What are your agency's common methods for costing out vehicles? How does this compare to industry standards?
2. Review your department or agency vehicle-checkout sheet. Determine if there is a method to reduce the time and increase reliability of the checkout process.
3. Calculate your agency's vehicle failure rate for the last year.
4. Design a training program for fleet mechanics for your agency. Evaluate the training that is currently conducted for vehicle-maintenance technicians.
5. What factors are used to calculate or identify when a vehicle needs to be replaced?
6. Explain the differences between planned, preventive, and scheduled maintenance.

References

Acai, S. (1996). VDA: Quality management in the automotive industry, quality assurance before series production: Vol. 4 Part 2: *System FMEA, Failure Mode and Effects Analysis.* Verband der Automobilindustrie e.V.

Drabek, T. E., & Hoetmer, G. J. (1991). *Emergency management: Principles and practice for local government.* Washington, DC: International City/County Management Association.

General Services Administration. (2002). *Federal specification for the Star-of-Life ambulance* (KKK-A-1822E). Washington, DC: Author.

International City/County Management Association. (2002, July). Fleet management. *IQ Report, 34*(7).

Interpreting the new ambulance specifications. (1980, October). *JEMS, 5*(8), 22–7, 30.

Key, C. B. (2002, November). Operational issues in EMS. *Emergency Medicine Clinics of North America, 20*(4), 913–927.

NFPA 1915: Standard for Fire Apparatus Preventive Maintenance Program. (2000). Quincy, MA: National Fire Protection Association.

Novak, J. L., & Bosheers, K. (1996). *The QS-9000 documentation tool kit* (2nd ed.). Upper Saddle River, NJ: Prentice Hall.

O'Brien, R. J. (2005). Buying stuff. *Fire EMS.* Penwell Publications, NJ.

QS 9000: Quality system requirements. Chrysler Corporation, Ford Motor Company and General Motors Corporation. (1995). Essex, UK: Carwin Continuous Ltd.

Career Development and Staff Focus

Accreditation Criteria

CAAS

103.03   Management Development: The agency shall demonstrate its commitment to the ongoing development of its leadership (pp. 292–295).

103.03.01 Management Training: The agency shall have a process to provide managers with initial and ongoing management training (pp. 299–304).

Objectives

After reading this chapter, the student should be able to:

12.1 Create a map of a career plan and personal growth path to an EMS leadership position (pp. 306–309).

12.2 Understand how to develop and participate in a mentoring program (pp. 306–307).

12.3 Identify activities that enhance professional development (pp. 304–306).

12.4 Recognize, select, and participate in staff development opportunities (pp. 293–296).

12.5 Identify education pathways for career and staff development (pp. 297–300).

12.6 Identify common experiences needed to be an effective EMS manager or leader (pp. 300–301).

12.7 List possible outside activities that enhance professional growth within the organization (pp. 304–306).

Key Terms

administrative skills (p. 301)

articulation (p. 299)

certified ambulance manager (p. 294)

chief fire officer designation (CFOD) (p. 294)

chief medical officer designation (CMOD) (p. 294)

Degrees at a Distance (DDP) Program (p. 299)

executive fire officer (EFO) (p. 294)

Point of View

Career development is a critical component of the working EMS professional. All too often, we enter the field as young providers, love the job, and forget that the future will be here all too soon. Many don't plan ahead and simply take classes in a haphazard fashion or experiment with possibly new careers with no clear goal in mind. Don't get me wrong. There is nothing incorrect about working in the field as an EMS professional for an entire career, but one may want to move into other aspects of EMS or something altogether different.

Many in the emergency services want to move up through the organizational chart. Unfortunately, there are only a few top slots at any organization, and many simply do not reach that goal. Instead, one should try to discover what he truly likes and then relate it to a future career. Many careers have great commonalities, so one may need to simply gain further education that would be applicable to many career areas.

A lot of us in the emergency services got there through a happenstance or on a whim. It worked for many and me as well. I did not prepare goals. I simply entered a paramedic program. When I gained employment, I soon saw my future at that organization. I wanted to teach and ultimately be the training director. Upon gaining the training position, I found that education was more my calling and started working at Cowley College as an EMS program director. At that point, I developed a personal career development plan to help me and the college. I achieved that goal and started a new academic department that has been very successful. I now plan to finish a doctoral degree and work as an academic vice president at a college. I will do that.

Having a plan and the guts to do it is all it takes. There are a plethora of methods to formulate the goals. Essentially, it takes a considerable amount of thought and the ability to achieve. I would recommend that everyone do this, at any age or stage of a career, because it is extremely beneficial. Emergency service providers are a great group of folks and can accomplish anything they want. Plan ahead and do it!

Slade Griffiths, M.Ed.
Department Chair
Cowley College, Kansas

■ NEED FOR PROFESSIONAL DEVELOPMENT

There is an interesting phenomenon with regard to planning and goal setting: Perhaps 3% of people, organizations, and governments that develop a plan actually set goals and pursue them with conviction. It has been measured that these three-percenters outperform and outproduce more than the other 97% of people, organizations, and governments combined. The secret to their success is quality management and professional development. If you develop people, you develop quality.

Many people debate the concept of "professional" development. A profession is a group of people who have an identified body of knowledge and whose training is universally recognized. The NREMT contributes to the profession by helping EMS providers meet the common professional standard of independent assessment and reciprocity. Professions also have a code of ethics, a system of accountability, and discipline.

EMS is facing a significant loss of leadership and experience in the next decade as aging baby boomers retire. What will be lost are leadership skills and the prehospital medical professional's knowledge and experience gained in the days when there were more emergencies. The benefits from new safety systems, engineering, and injury-prevention programs are reducing the seriousness of vehicle crashes. Therefore, the experience level from which EMS agencies will have to draw for managers and leaders will continue to decline.

There are several reasons why individuals in emergency-service organizations fail in planning for professional development: peer jealousy, lack of motivation, unclear career paths, lack of support from leadership, and lack of knowledge of the process. The term **professional development** seems overwhelming to new EMS managers, and it is more easily identified as staff or career development. Ask the question, "What do you want to be when you grow up?" and the response from an adult may be very similar to an adolescent's. Most people cannot answer that question with any detail—compared to the three-percenters, who can give a lot of detail.

Professional development is a necessity when you consider the skills differences between a field paramedic or clinician and a manager. The emergency-response environment is fast paced, always different, often with high stakes, versus the management environment that is nontechnical, has a political element, and involves being accountable to the public. The best field paramedics do not necessarily make the best managers.

As an EMS manager or leader, you will need to focus on not only developing yourself but also your staff. Studies often show that managers who do not train others to fill their positions are less likely to be promoted. For example, incident-management and FEMA U.S. Army Reserves task forces often establish personnel equally qualified in key positions at least three deep. The process of ensuring that there are three people in the organization who can fill a position in the event of a vacancy or absence, whether planned or in an emergency situation, can help maintain continuity in the organization's operations. If you do not develop your people, often they will seek out other opportunities to enrich their skills and abilities, leaving less-trained or poor-quality workers to do the work.

Professional development can be divided into four basic categories: technical skills, managerial skills, self-development, and eclectic experience. **Technical skills** are gained from training and certification related to a trade or profession. **Managerial skills** include analysis, planning, problem solving, and communications (especially listening!). On an individual level self-development in the areas of ethics, vision, and leadership must be ongoing. A manager must have experience in different roles, cultures, individuals, and groups.

PROFESSIONAL DEVELOPMENT

The International City/County Management Association (ICMA) created the Next Generation Initiative to inspire young and midcareer professionals and to help the senior public manager prepare, develop, and motivate the next generation. A part of the Next Generation program is the ICMA coaching program. The program goals are:

- Supporting the professional and personal development of new and aspiring managers.
- Encouraging an ethos of mentoring and coaching at all levels in the local government management profession.
- Providing opportunities for successful local government managers to share their expertise in ways that are effective, efficient, and personally rewarding.

The International Association of Fire Chiefs (IAFC) *Officer Development Handbook* defines professional development as the planned, progressive lifelong process of education, training, self-development, and experience. Perhaps the most important word in that description is "planned." Remember that only about 3% of people and organizations have a plan. So an individual commitment to plan for professional development will put you in the top 3% immediately. Several EMS-oriented organizations have or are creating professional- and career-development models.

The International Association of Fire Chiefs (also known by the acronym ICHIEFS) has a standing committee and ongoing activities focused around five core activities: mapping, measuring, mentoring, motivating, and multiplying. This program parallels similar activities for fire- and EMS-chief development through the International Association of Chiefs of Police. The activities by ICHIEFS can be applied to the career development for EMS officers and leadership. ICHIEFS has collaborated in its professional-development model with the National Fire Academy's Fire and Emergency Services Higher Education Initiative on the chief fire officer designation (CFOD). The CFOD is a designation for officers at the administrative level or higher. A combination of education and experience are submitted as part of a portfolio to the Commission on Fire Accreditation International, now known also as the Center for Public Safety Excellence, which performs credential review and peer assessment. The CFOD process evaluates the entire professional development of an individual. The components of the

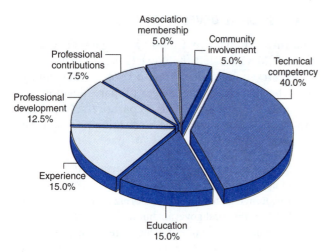

Association membership 5.0%
Community involvement 5.0%
Professional contributions 7.5%
Technical competency 40.0%
Professional development 12.5%
Experience 15.0%
Education 15.0%

FIGURE 12.1 ■ Components of the CFOD Process. *(Reprinted with permission.)*

CFOD process are shown in Figure 12.1. An EMS manager or leader can easily apply EMS-related experiences to this model.

The National Highway Traffic Safety Administration (NHTSA) through the implementation of a quality improvement guide identifies a career-development plan for EMS leadership as part of the key to enhancing quality in EMS activities. NHTSA builds on the Department of Commerce Malcolm Baldrige National Quality Award. The EMS activities directed by NHTSA have only limited focus on career development and instead have focused on clinical issues and system development. Relatively few academic programs exist in EMS management.

The American Ambulance Association (AAA) in partnership with private consulting firms has a program that leads to a **certified ambulance manager**. An internal management-training program through the AAA has an ongoing offering for ambulance industry managers to develop management and leadership skills. The material is not tied to college credit and is recognized internally through the ambulance association when selecting ambulance company operations, executive officer, and field supervisors. While the focus is on maximizing ambulance reimbursement and improving efficiency in ambulance deployment, participants receive an intense focus on business administration skills.

An initiative of the U.S. Fire Administration (USFA) and the National Fire Academy, the **executive fire officer (EFO)** program can be used as a career-development model, which is open to any government-sponsored or nonprofit EMS agency. This four-year program, consisting of four core courses, each two weeks long, also requires applied research projects for each year. The EFO has become a standard in the industry, and it is a common requirement for promotional opportunities or hiring of chief officers. Application procedures for the EFO are available at the National Fire Academy Web site. This program is open to any government-funded EMS organization. The executive fire officer program can enhance an EMS officer's career development if he uses the applied research project required in each of the four classes to focus on an EMS problem.

The National EMS Management Association (NEMSMA) has an open-source initiative to develop knowledge, skills, and abilities required for a chief EMS officer designation. In an open forum information is being solicited as to what is required to become a professional EMS manager. This information is being developed as part of the movement toward a national EMS administration. The **chief medical officer designation (CMOD)**, designed to apply career-development activities and education similar to the EFO, is available from the National Fire Academy to EMS providers who may or may not be affiliated with fire-based EMS organizations. The Center for Public Safety Excellence (CPSE) has taken up the credentialing of the chief EMS officer designation.

CPSE has developed a standardized format for evaluating credentials for EMS chiefs. A map for professional development ideally contains four elements: education, training, experience, and self-development. The elements that enable someone to do a job are **knowledge, skills, and abilities (KSAs)**, and throughout a career KSAs become enhanced by the four elements. Recognition for improving in these areas is now accomplished by attaining the **chief fire officer designation (CFOD)**. There are quantifiable measures recognized in the form of certifications, academic degrees, diplomas, licenses, certificates, transcripts, and continuing-education units or credits (CEUs). Remember, the goal of these activities is to strengthen a person's KSAs, not to get more merit badges.

When considering promotions and new hires, a savvy leader will always measure a candidate's ability to perform as well as the credentials he has. The higher one goes in the management structure of the agency, the better rounded one must become. Entry-level EMS providers can effectively do their jobs with mostly technical training; chief officers and organizational

- Planned by staff groups that include representatives of all stakeholders in the EMS system.
- Task specific, focusing on actual uses of technology that will serve EMS administrative goals, including those identified by a staff-needs assessment.
- Led by staff members, who use the technology in EMS operations and instruction technology and who are proficient in teaching adults and in modeling effective patient-care strategies.
- Adequate in the time allowed for EMS managers to learn, practice, and apply new concepts and techniques.
- Offered to staff members who have access to hardware and software they can practice with in their offices during inservice training or downtime in the station.
- Sensitive to EMS members' personal needs and schedules, offering credit, stipends, and/or release time.
- Flexible in allowing EMS personnel to use what they learn in a variety of ways appropriate to their individual situations.
- Supported at the district and organization levels with adequate, ongoing staff and time for person-to-person and small-group instruction.
- Accepted throughout the EMS community as an ongoing activity, not a single event.
- Continuously evaluated and improved.

FIGURE 12.2 ■ Characteristics of a Successful Organizational Staff Development System.

executives must have much more academic education and training.

So how can you plan your professional development and become one of the three-percenters? Successful professional development needs an organizational effort. As the EMS leader in your organization, you need a multidisciplinary approach and system in place. Some general activities and principles in an organization's professional development planning are included in Figure 12.2. EMS as a profession has had little in the way of professional development, with only a handful of EMS degree programs and very few programs that offer professional development focused on fire-based EMS providers. Internally, consider a comprehensive system that uses the organization's resources.

IAFC *OFFICER DEVELOPMENT HANDBOOK*

Through several years of hard work and consensus building, the Professional Development Committee of the IAFC has produced the *Officer Development Handbook*. Perhaps one statement in the handbook says it all: "...professional development is a journey, not a destination." Career and professional development as described in that text focuses on the four elements: education, training, experience, and self-development. Also, four levels of expertise are defined, with the four elements as a guide on the path: supervising fire officer, managing fire officer, administrative fire officer, and executive fire officer. The National Fire Protection Association (NFPA) standard for fire officer professional development is the backbone of the handbook.

One of the key principles of the IAFC development plan approaches professional development with a focus on the general-education recommendations (Table 12.1). The IAFC and the **Fire and Emergency Services Higher Education (FESHE)** model promote general education. The approach is to focus on professional development that is competency based by taking general-education courses.

The National Fire Academy (NFA) and the Fire and Emergency Services Higher Education (FESHE) conferences have crosswalked training and education with the National Fire Protection Association (NFPA) competencies, and a fire service professional-development model was built to the NFPA 1021 standard. With a little imagination, this model can be applied to the hierarchy of an EMS division or agency (Figure 12.3).

QUALITY IMPROVEMENT AND PROFESSIONAL DEVELOPMENT

NHTSA and the Baldrige Criteria

Since the enactment of the National Highway Safety Act of 1966, and the formal beginning of EMS, the common goal of EMS systems has been to reduce unnecessary death and disability. While this goal remains constant, the industry is confronted more than ever before by the public with the demand that EMS provide the highest quality service at the lowest possible cost. There are clear expectations for improved health, improved quality, and improved efficiency.

A Leadership Guide to Quality Improvement for Emergency Medical Services (EMS) Systems is a NHTSA publication that EMS leaders can use to improve quality within their organizations. The manual provides a guide for integrating continuous quality-improvement (QI) practices into EMS operations to the extent that those practices become an essential and seamless part of normal EMS routines. Specific activities are suggested within developmental stages. While these activities may differ depending on the jurisdiction of the organization, the developmental stages of QI integration will be the same for local, regional, or statewide

TABLE 12.1 ■ Example of IAFC Development Handbook Competencies for Supervising Fire Officer

Component	Content
General	Firefighter II or Technical Skills
General Knowledge	Organizational structure; procedures; operations; budget; records; codes and ordinances; IMS; social, political and cultural factors; supervisory methods; labor agreements.
General Skills	Verbal and written communications; report writing; incident management.
Human Resource Management	Using human resources to accomplish assignments safely during emergency, nonemergency, and training work periods; recommending action for member problems; applying policies and procedures; coordinating the completion of tasks and projects.
Community and Government Relations	Dealing with public inquiries and concerns according to policy and procedure.
Administration	Implementing departmental policy and procedure at the unit level; completing assigned reports, logs, and files.
Inspection and Investigation	Determining preliminary fire cause; securing the scene; preserving evidence.
Emergency Service Delivery	Conducting preincident planning; developing incident action plans; implementing resource deployment; implementing emergency incident scene supervision.
Health and Safety	Integrating health and safety plans; policy and procedures into daily work unit work activities; conducting initial accident investigations.

Source: IAFC Officer Development Handbook, page 5. Reprinted with permission.

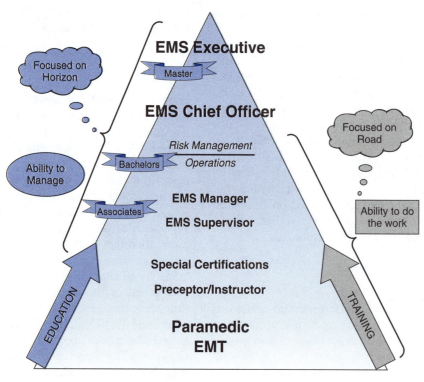

FIGURE 12.3 ■ Professional Development Model for EMS Leadership.

EMS organizations. These developmental stages start with building potential for success by developing an awareness and appreciation that QI is a worthwhile endeavor. The second stage requires expanding the workforce's knowledge of and capability in QI practices and techniques by providing results and feedback. Several texts and courses are offered by the National Association of EMS Physicians and the American College of Emergency Physicians. Finally, EMS leadership will need to fully integrate the strategic quality planning process and related quality improvement actions into the daily EMS operation.

The Baldrige program identifies seven key action areas or categories, and the QI information in the manual is organized according to those categories. One chapter outlines the leadership principles and development efforts needed by senior leadership and management to integrate quality improvement into the strategic planning process and throughout the entire organization and to promote quality values and QI techniques in work practices.

Health-care organizations that follow the Baldrige program have the option of asking for an external review of their progress. They report benefits gained by simply applying the Baldrige guidelines and recommendations, including improvements in service and patient-care delivery; economic efficiency and profitability; patient and community satisfaction and loyalty; and health outcomes.

A complete copy of the 2008 Baldrige Health Care Criteria is available at the National Institute of Standards and Technology. The Baldrige Criteria offers a complete package for the organization. Since most people start in a management position rather than a senior leadership position, the most influential place to begin on the quality-improvement trek is with self-improvement, or professional development.

Malcolm Baldrige and Professional Development

Malcolm Baldrige was Secretary of Commerce from 1981 until his death in a rodeo incident in July 1987. Baldrige was a proponent of quality management as a key to this country's prosperity and long-term strength. He took a personal interest in the quality-improvement act that was eventually named after him and helped draft one of the early versions. In recognition of his contributions, Congress named the award in his honor.

The Malcolm Baldrige National Quality Award is given by the president of the United States to businesses—manufacturing and service, small and large—

and to education and health-care organizations that apply and are judged to be outstanding in seven areas: leadership; strategic planning; customer and market focus; measurement, analysis, and knowledge management; human-resources focus; process management; and results.

The Baldrige Criteria for Performance Excellence provides a systems perspective for understanding performance management. It reflects validated, leading-edge management practices against which an organization can measure itself. With their acceptance nationally and internationally as the model for performance excellence, the criteria represent a common language among organizations for sharing best practices. The criteria are also the basis of the Malcolm Baldrige National Quality Award process.

What is significant about the Baldrige criteria is the template they provide for individuals and organizations to pursue excellence and organizational performance. What is the significance of this for EMS managers? Organizations are capable of nothing without a road map (a plan) and people (human resources) to drive the organization in a direction (goals and objectives). An organization can have the plan and the goals and the objectives, but it is dead in the water without trained and committed people. Conversely, an organization can have well-trained and motivated staff, but without a plan they will wander aimlessly in pursuit of their mission. The Baldrige criteria offer a comprehensive view of the many aspects an organization needs to have excellent performance.

EDUCATION OPPORTUNITIES

It is important that EMS managers and leaders understand that professional development distinguishes between training and education. Training involves understanding what to do and is anchored in the past. It focuses on job skills, or skills that have practical application. Training has a specific outcome; for example, computer-program instruction or training on a specific piece of equipment. Education is geared toward the future. It is often about life skills and cognitive knowledge based on theory. Education seeks to meet general outcomes.

COLLEGIATE EDUCATION

To receive paramedic certification in the future, you will have had to attend an accredited program, which usually requires the resources of a college or university and results in a two-year degree. In most colleges degree programs are divided into general education, specialized degree requirements, and electives. General education courses (English, math, social science) are required by educational institutions for a degree, but often chasing down college credit in general education puts an adult learner back in class with recently graduated students and sidetracks a person's professional development.

General-education courses often seem to impose limits on achieving a degree when adult learners return to school. A few strategies can help mitigate those barriers. Testing, credit by examination, or nontraditional education credits that allow students to demon-strate knowledge and skills are offered by most colleges and universities as a way for students to earn credit for general education without attending a class. Online or distance education programs can offer general education over the Internet or by correspondence from home, where adult learners can attend to other responsibilities and make efficient use of their time.

Core courses have been established for a model EMS management degree similar to those offered by the National Fire Academy, which has established a model core curriculum and degree structure for associate's and bachelor's degrees. Figure 12.4 identifies the model course curriculums established by FESHE

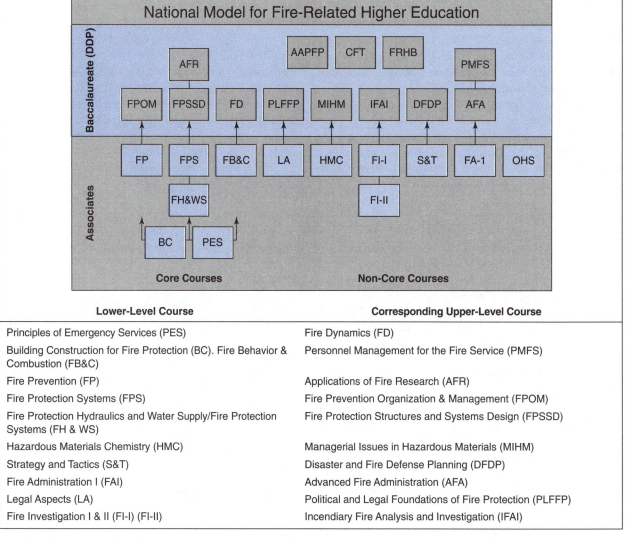

Lower-Level Course	Corresponding Upper-Level Course
Principles of Emergency Services (PES)	Fire Dynamics (FD)
Building Construction for Fire Protection (BC). Fire Behavior & Combustion (FB&C)	Personnel Management for the Fire Service (PMFS)
Fire Prevention (FP)	Applications of Fire Research (AFR)
Fire Protection Systems (FPS)	Fire Prevention Organization & Management (FPOM)
Fire Protection Hydraulics and Water Supply/Fire Protection Systems (FH & WS)	Fire Protection Structures and Systems Design (FPSSD)
Hazardous Materials Chemistry (HMC)	Managerial Issues in Hazardous Materials (MIHM)
Strategy and Tactics (S&T)	Disaster and Fire Defense Planning (DFDP)
Fire Administration I (FAI)	Advanced Fire Administration (AFA)
Legal Aspects (LA)	Political and Legal Foundations of Fire Protection (PLFFP)
Fire Investigation I & II (FI-I) (FI-II)	Incendiary Fire Analysis and Investigation (IFAI)

FIGURE 12.4 ■ DDP Model Course Curriculum.

initiatives. FESHE established a core group of courses for every emergency services degree, and noncore classes were added to help customize degree programs for various EMS disciplines. FESHE also has developed 13 courses for a bachelor's degree.

Identify potential bachelor's-degree and advanced-degree programs that may be required for promotion to a managerial or leadership position as early as you can. Choose an institution and a program that will accept all the credits you have earned toward your associate's degree. Some colleges and universities do not transfer all the courses. When one degree transfers in its entirety to another college or university as part of a higher degree, it is called academic **articulation**.

The **Degrees at a Distance Program (DDP)** is sponsored by the National Fire Academy. It includes a significant Web-based and online program for emergency-service personnel that awards a bachelor's degree. DDP provides an alternative means for fire-service personnel to earn a bachelor's degree or to pursue college-level learning in a fire- and EMS-related course concentration without the requirement of having to attend on-campus classes. The program is supported with a curriculum designed by the U.S. Fire Administration, which subsidizes these courses for emergency personnel. DDP is especially suited for working adults whose time is challenged by shift work and family obligations, making attendance in traditional campus classrooms difficult. Information on the colleges and universities that offer DDP degree programs is available on the U.S. Fire Administration Web site.

EMS MODEL CURRICULUM

A national EMS-management curriculum committee was formed as part of the National Fire Academy's Fire and Emergency Services Higher Education conference. A significant aspect of this effort is the partnership with the National Association of EMS Educators, the national organization of leaders in the development and delivery of EMS education programs. The committee developed working management-level definitions, using public-safety and EMS industry practices. Four levels were established:

- *EMS supervisor.* A crew chief or leader of a single unit or crew.
- *EMS manager.* Manages more than one crew or supervisor.
- *EMS chief officer.* Oversees more than one manager, is responsible for a major component of an EMS or-

---

- ◆ Foundations of EMS Systems.
- ◆ EMS Operations.
- ◆ Human Resource Management.
- ◆ Management of EMS.
- ◆ Quality Management and Research.
- ◆ Injury Prevention.
- ◆ EMS Educator/Instruction.
- ◆ Management of Ambulance Services.
- ◆ Safety/Risk Management.

FIGURE 12.5 ■ FESHE EMS Core Courses.

ganization, is a middle manager or frontline supervisor.
- *EMS executive.* Head of the organization or senior staff.

To help develop these four levels, a series of core courses was identified. Similarly to many other technical professions, a core group of courses is usually required. For example, a management graduate at most accredited institutions would require a course on personnel management, operations management, and organizational behavior. The courses in Figure 12.5 have been identified as covering baseline knowledge that would be required of an EMS manager or leader.

These courses do not have numbers attached to them, which allows the courses to be offered in a variety of different locations. In an area heavy with certified paramedics, for example, these courses may be offered as part of an associate's degree program. In other areas they may be offered at accredited institutions in bachelor's degree programs.

OPTIONS AND OPPORTUNITIES

University Programs

Universities and some state and private colleges offer advanced degrees such as master's or doctoral degrees. At this level it is thought that EMS leaders and managers need to study the same degree material that a city or county manager studies. Upper-level managers need to understand budget, advanced quality improvement, and political economies and be able to write and speak well in public. Advanced degrees require that a candidate pass a standardized test, such as the graduate record exam (GRE) or the graduate management admission test (GMAT) to enter study of the management sciences. Common advanced degree majors for EMS managers or leaders are public administration, business administration, and health-care administration. One of a handful of degrees in EMS management is offered by George Washington University (Figure 12.6). Advanced

FRESHMAN YEAR

EHS 200	Concepts of EHS	3	EHS 302	Clinical Concepts (EMT)	4+
ENGL 100	Composition	3	ECON 101	Principles of Microeconomics	3
<GFR>	<Science Elective>	3	<GFR>	<Science Elective>	3
MATH 100/115	Mathematics	3	<GFR>	<Language/Culture Elective>	4
<GFR>	<Language/Culture Elective>	4	PHED	Physical Education	1.5
EHS 100	Freshman Seminar	1	EHS 101#	Living Learning Experience	1
TOTAL		17	TOTAL		16.5

SOPHOMORE YEAR

ECON 121	Principles of Accounting I	3	ECON 122	Principles of Accounting II	3
SOCY 101	Basic Concepts of Sociology	3	PSYC 100	Introduction to Psychology	4
POLI 100	American Government	3	POLI 250	Public Administration	3
<GFR>	<Arts/Humanities Elective>	3	<GFR>	<Arts/Humanities Elective>	3
STAT 121	Introduction to Statistics	4	EHS 360	Instructional Issues in EHS	3
<PHED>	<Physical Education>	1.5			
TOTAL		16.5	TOTAL		16

JUNIOR YEAR

EHS 300	EHS Theory & Practice I	3	EHS 400	Theory & Practice II	3
EHS 301	EHS Planning	3	EHS 351	Financial Mgt. & Budgeting	3
EHS 352*	Microcomputer Applications	3	SOCY 351	Medical Sociology	3
POLI 353	Gov't Budgeting & Fin. Adm.	3(M)	POLI 350	The Policy-Making Process	3(M)
POLI 354	Public Mgt. & Personnel Sys.	3	POLI 352	Administrative Law	3(M)
EHS 350^	Supervision & Operations	3			
TOTAL		15	TOTAL		18

SENIOR YEAR

EHS 430	Research Topics in EHS	3*	EHS 450	EHS Practicum and Seminar	15
EHS 467	Health Economics	3			
POLI 450	Seminar in Public Admin.	3(M)			
Elective EHS Elective		3			
<GFR>	<Arts/Humanities Elective>	3			
TOTAL		15	TOTAL		15

SUGGESTED ELECTIVES

Fall/Spring	EHS 310	Seminar in EHS Management
Spring	EHS 311	Stress & Burnout
Fall	EHS 320	Disaster Management
Spring	EHS 330	Search & Rescue
Fall	EHS 345	Death & Dying
Spring	EHS 350	Supervision & Operations

FIGURE 12.6 ■ Sample EMS Management Degree Sheet.
(Reprinted with permission of The George Washington University Medical Center.)

degrees commonly require research and a thesis or professional paper that will be evaluated and challenged by a committee of professors. The concept of having your ideas or research challenged is commonly called the defense of a thesis or research.

As part of career development as an EMS manager, it is important to identify an education plan. Attending college and paying for it is a balancing act between academic studies, work, and family. Financial and time commitments need to be planned out ahead of time to reduce stress and coordinate with the demands of other aspects of life.

Trade School and Sponsored Certifications

The National EMS Academy, based in the Texas Engineering and Extension Service (TEEX), operates a leadership institute to develop current and future EMS leaders. One course offered covers supervisory effectiveness, focusing on mapping managerial skills in preparation for participating in hands-on learning scenarios. The supervisory course includes four instructional blocks: managing your job, relating to others, developing teams, and critical thinking; it is operated in conjunction with Southern Louisiana Community College. Several partnerships or for-profit companies offer accelerated or abbreviated leadership and management training programs.

Skills Acquisition

A skill is the ability to do something in an effective manner. Individuals who want to improve skills needed for career development should be adding to their basic skills. Focus on the skills that are related to technology and to the EMS practices that affect your current position. Four types of skills are important: conceptual, administrative, interpersonal, and technical.

Conceptual skills are those that require general analytical ability, logical thinking, and conceptualization of complex and ambiguous relationships. **Administrative skills** usually involve planning, negotiating, and coaching. **Interpersonal skills** are those skills necessary to make greater contributions to the organization by interpersonal processes and the ability to understand feelings, attitudes, and motives of others from what they say and do. This often includes the ability to establish effective and cooperative relationships. Figure 12.7 displays many of the common skills necessary for EMS managers and leaders. Making an assessment of the skills used for business, human resources, financial resources, labor relations, and technology or computer skills can provide a priority list for career development.

Technology is overtaking the EMS services at a pace faster than we can imagine. Understanding wireless, geographic information system (GIS), and computer software programs that use e-mail and the Internet search features are a necessity for today's managers and leaders. Courses on these technologies are typically offered through workforce development or continuing education divisions at colleges and universities. Each year one of these skills should be targeted for improvement; for example, learning how to make charts or spreadsheets in Excel® spreadsheet software or advancing your reading level by taking a speed-reading course. Stephen Covey calls this sharpening the saw, or taking time to reenergize yourself with new skills and abilities.

Skills and knowledge attained for emergency response are important for skill development of people attempting to advance their EMS careers. The size and frequency of emergencies generating patients and presenting as complex challenges will force EMS leaders to perform under pressure with limited resources or for extended periods of time. Hurricane Katrina, the events of September 11, and the Loma Priata earthquake are examples of the challenges for EMS personnel who serve in the command position.

Any career path for EMS professionals needs to require incident command system (ICS) 100, 200, and 700 training. ICS training is now available online from the U.S. Fire Administration and FEMA. Additional training and practice, using simulation and serving in various roles, is an invaluable process. Attending the National Fire Academy command and control or EMS special operations course provides several simulation opportunities.

Real-world experience can be obtained for EMS providers by securing a wildland red card from the National Wildfire Coordinating Group (NWCG) and receiving a credentialing to serve as a medical unit or multi-casualty branch director. Wildland incidents rarely produce large amounts of casualties, yet the experience earned by interfacing with a large-scale emergency operation is good preparation for operating in more serious circumstances. Some departments send observers to real incidents to obtain experience.

Competencies

Acquisition of skills should transition into mastery of those skills, also known as competencies. Competencies are important because agencies write job descriptions and promotional tests to match competencies. Each position requires a different set or level of competencies. Similarly to those required for a fire chief, competencies have been established for EMS managers and executives through the FESHE process at the National Emergency Training Center. Table 12.2 shows the competencies for the various levels of EMS management.

Job Experience and Rotations

When considering a long-term plan to secure a position in EMS, it is often important to map out a progression of job experiences and rotations to gain knowledge of the operations. Most management studies cite a five- to seven-year window before a manager

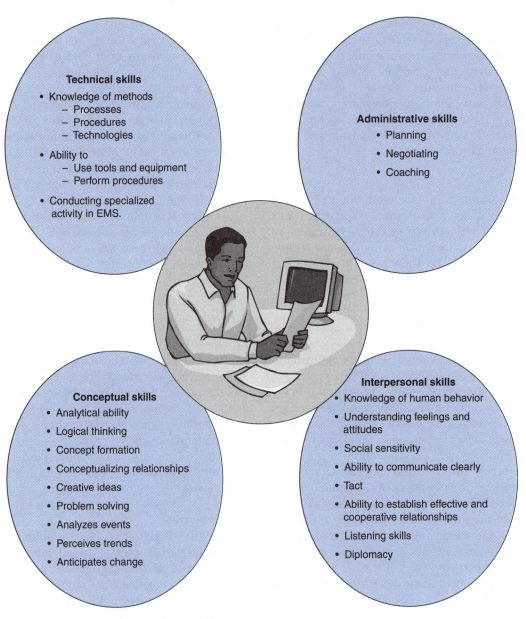

Technical skills
- Knowledge of methods
    - Processes
    - Procedures
    - Technologies
- Ability to
    - Use tools and equipment
    - Perform procedures
- Conducting specialized activity in EMS.

Administrative skills
- Planning
- Negotiating
- Coaching

Conceptual skills
- Analytical ability
- Logical thinking
- Concept formation
- Conceptualizing relationships
- Creative ideas
- Problem solving
- Analyzes events
- Perceives trends
- Anticipates change

Interpersonal skills
- Knowledge of human behavior
- Understanding feelings and attitudes
- Social sensitivity
- Ability to communicate clearly
- Tact
- Ability to establish effective and cooperative relationships
- Listening skills
- Diplomacy

FIGURE 12.7 ■ List of Skills Needed for EMS Management Positions.

becomes obsolete in his position. Job enrichment is the concept of rotating to different assignments. These assignments may include time assigned to a hazardous-materials unit, specialized rescue teams, and fire-based resources that do not have a majority EMS function.

Consider rotational assignments as a frontline supervisor on an engine company to gain perspective of the approach and needs of a company for EMS. Secure the same management credentials and certifications that are required of fire officers, such as coaching, counseling, time management, and leadership.

An EMS manager or leader needs to serve as an instructor of some aspect of EMS for career development. At some point an EMS leader or manager will have to speak in a public forum, and there is no better preparation for public speaking than working at it as a teacher. The more programs an EMS manager is

TABLE 12.2 ■ Excerpt from Competencies for EMS Management Levels Certifications.

Education		Competencies/Standards	Training
NFA Courses-ACE Recommendations	*College Courses	EMS Officer Competencies	State or Local Courses
R-822 or R-154 Advanced Safety Operations ISO F729 F730 HSO	Personal Health and Wellness (A)	SEMSO-10 Understanding and implementing the basic principles of health, fitness and wellness.	
	American Government (A) Political and Legal Foundations of Fire Protection (FBC)	SEMSO-11 Understanding the basic concepts of government at the federal, state, and local levels.	
ED (R123), EL (R125),TPM (R342), CLTO (R815)	Human Resource Management (A) Personnel Management for the Fire Service (FBC)	SEMSO-12 Understanding functional areas of human resource management and laws; job analysis, testing; performing interviews, selection, training and performance evaluation.	
R-152 R-247 Special Operations		SEMSO-13 Must have knowledge of special operations: hazmat, technical rescue, mass care and interoperability	
NAEMSIS, NAEMSE Instructor Course		SEMSO-14 Coaching, counseling, mentoring and team dyamics	
R-150	Fire Administration I (FAC) Introduction to Fire and Emergency Services Administration (A) Fire and Emergency Services Administration (FBC)	SEMSO-15 Understanding and performing basic responsibilities of supervisors including patient care, supervision, delegation, problem solving, decision-making, communications, billing and leadership.	
		SEMSO-16 How to conduct discipline with due process	
		SEMSO-17 Understand the concepts of research, statistics used in research/clinical performance in order to implement a quality process	
ICS 100-400, 700, 800		SEMSO-18 Demonstrate an ability to function within the National Incident Management System and appreciate the issues with interoperability	
Benchmark: Associates in EMS Management or EMS Administration			

Source: FESHE

*College Courses: A-Associate's B-Bachelor's, G-Graduate, NAEMSE is the National Association of EMS Educators. IPSLEP is International Public Safety Leadership and Ethics Program in conjunction with Phi Theta Kappa.

- CPR instructor
- PHTLS or BTLS instructor
- ACLS instructor
- PALS instructor
- PEPP, GEMS, ATLS
- EMS instructor

FIGURE 12.8 ■ Instructor Opportunities for EMS Development.

credentialed to teach, the greater the opportunity to refine presentation skills and build credibility. Figure 12.8 identifies some common instructor opportunities that should be sought by paramedics and EMTs who aspire to be managers or leaders.

■ PERSONAL SKILLS

ATTITUDE AND PERSPECTIVE

A great deal is said about attitude and how it affects the progress of a person's career in EMS. A successful attitude requires a person to develop the ability to use power and position to accomplish a purpose, create value, and control the destiny of yourself and others. As part of your career plan, teamwork will need to be fostered, and opportunities to serve on a team should be sought and taken. Teamwork involves working effectively and productively with others. What should be learned is the ability to respect others, work toward a consensus, meet deadlines, and support decisions.

Management perspectives require looking at the bottom line of the processes in the organization, while leadership requires looking at the orientation of people. Leadership skills require experience in inspiring, influencing, guiding, and motivating people. Management skills require opportunities to direct, monitor, and control systems, processes, and outcomes.

Interpersonal skills require effective communication, both written and verbal. The successful progression of a career in EMS requires building rapport and relating well to all kinds of people. A person must be sincere in the communications process and treat people with dignity and respect. Managers and leaders must have effective listening skills and good writing and presentation skills.

Problem-solving and decision-making skills are important to any managerial or leadership position.

With the changing environment of EMS, creativity and innovation are required to solve more-complex problems. When confronting problem issues that require management activities, it is important that the manager has good goal-setting and action-planning techniques. One of the key skills for goal setting is writing down the steps to get to a goal and the necessary actions.

Making effective decisions is also a necessary skill for managers and leaders. Placing yourself in situations that require decision making provides the repetitive sequence that helps build logic processes. The logic process or the problem-solving process often involves teaching yourself to employ reasoning and the scientific method when confronted with a decision.

INTERPERSONAL EFFECTIVENESS

The job of manager entails getting the best performance from people through teamwork. Contrary to some old-school approaches, the manager exists to support the team with resources, training, and coaching that leads to a good outcome. Teamwork skills include providing the tools and training that fit the jobs individuals perform. A new approach to teambuilding requires EMS managers to teach the team members good followership techniques and expectations. Managers are in the relationship-management "business" and must strive to keep people challenged and appreciated at all times. Communication with individuals includes active listening, good conservation, and effective interpersonal feedback.

Being an effective EMS manager or leader also means being efficient. For a manager, it may often seem as though there is not enough time to get everything done. There will always be competing interests and priorities that demand attention, beyond the available hours of the job. Managing multiple priorities and being productive is the challenge. Using project management tools, delegating, and being able to track what is going on at any given moment is a good place to start. Boeing, the aerospace company, tells its managers they should not be doing anything that a team member can do and that, in effect, they are making and managing clones of themselves. Other manager and leader skills include becoming an empowered leader, member coaching/teaching coaching for improved performance, creating a motivational workplace through effective delegation, resolving conflict between others, giving recognition, and leading change.

Working on personal effectiveness with assertive communication, such as conducting successful meetings, also is important, as well as working with internal and external groups to achieve consensus. Robert Greenleaf is responsible for a concept known as Servant Leadership. The essence of servant leadership is that leaders must serve the followers. To put this in context, servant leaders will constantly strive to meet the needs of the followers in order to empower and support their ability to serve the organization. Servant leadership perhaps is the ultimate form of selflessness a leader can demonstrate, because the focus is on the follower, not the leader.

Entering a New Position

Many times when a person assumes a new role in an organization, errors are made in the first month that take years from which to recover. Changing a career position often brings a new set of people to manage, interact with, and follow. It is important to meet with everyone within the first week and tell people your priorities and how best to communicate with you. It is important on these first visits not to tackle problems but instead to put out a personal vision and then listen. Do an assessment of the strengths and weaknesses of the new position, the organization's use of this position, and your skills for the position.

Your office or work space also reflects how you are perceived. Rearrange or change the furniture, if possible. Personalize the space by adding something that shows who you are and your credentials. Favorite books, awards, and pictures of your crew, family, or partners make you a real person to the people in your office space, not just management. A conversation piece is often a nice way to make people feel comfortable when business may be unpleasant.

Stabilizing Personal Issues

It helps to establish credibility by letting people know your background. It also helps to do what you say you will; even the smallest failures in follow-through can damage integrity, which often means doing what is right versus what is popular. Realize that even when you are not on the job, you are on the job, and this will be reflected by the people in the community who become your champions—and your critics. Navigating the career path from the line to manager and on to a leadership role requires establishing relationships. Be careful to never establish relationships that will embarrass you or your department. Many leaders and managers do not live in their agency's community, and this is seen by many as not supporting the position or the organization, because most would argue that both are attached to the community.

LIAISON ACTIVITIES

Internal Customers

When interacting with internal customers, remember to smile. A smile is contagious and sends a positive message. It is important to remember that your employees are your customers. Be friendly toward firefighters, EMTs, and paramedics. Develop shared goals with your employees. It is your job to build their skills and help them improve themselves. This includes making the health and fitness of your staff your concern. Everyone should be treated with respect, and it should be known they are responsible for their own actions. The union and union officials also are your customers, and it is important to develop a collaborative working relationship with labor leaders.

Government Relations

Remember that most EMS organizations belong to a government entity and are not isolated agencies. Along the career path, it is important to support the city manager, mayor, commissioners, or council member programs. These often include initiatives upon which politicians campaigned. Once the decisions have been made to move forward, it is important to be supportive and establish loyalty. Include the city or county manager in EMS department activities.

When it is within your power, look to provide EMS department resources to other departments. This could include other EMS, fire, and public-safety departments or departments within your governmental structure. For example, if the wastewater treatment plant needs a CPR class, that resource should be offered, as it might be their vacuum truck that makes the difference between life and death in your trench rescue.

Community

At every point in a career path in EMS you will have an interaction with community politics. As you ascend the career ladder, part of your professional development should be to know your policy makers and the top political issues. Stay away from political controversies that do not involve EMS. There may be other issues in the community that are not political; for example, charitable, leisure, religious, and social activities. The presence of EMS representatives at these events fosters the idea of community and makes the city or county manager look good.

An excellent opportunity to build community skills and networking opportunities is to become involved with a charity event or organization. The activities of a nonprofit charity give a manager a chance to build skills in leadership. It also provides community liaisons and experience in handling finances and building consensus.

The demands of modern life often do not allow for as much leisure time as most people would like. One aspect desired in an EMS manager and mandatory in an EMS leader is an attachment to the community that reaches beyond employment in the EMS service. Making this connection requires attendance and participation in local community events. In preparing to move up the career path in EMS, public-speaking skills in front of community groups are required. It is important as part of professional development to create and deliver public-education presentations and help coordinate public information.

An EMS manager should look at affiliating or assisting with a community-service organization such as a church, a nonprofit, or a local chapter of a national charity. Often involvement with a service organization such as the Kiwanis, Rotary, or Chamber of Commerce brings not only a service perspective but also excellent networking opportunities. These interactions can provide opportunities for EMS agencies to partner with businesses and private industry for customer-service and community programs on such subjects as helmet safety and car seats.

National Outreach

As part of staff development, your organization should support involvement in one or more trade organizations that involve EMS. Participation in a national, state, or regional trade group such as NAEMT, NAEMSP, or the National EMS Management Association provides opportunities for leadership in those organizations before being placed in a position of leadership in the local or home agency. EMS managers should consider sitting on boards or committees to gain experience in working with others and achieving collective goals.

■ A PERSONAL CAREER PATH PLAN

One predictable path for career planning has five steps. The value of such a plan is its use of the previously mentioned information to build a personal career path. The five steps are:

1. *Pre-employment preparation.* Prepare for a position by identifying the type of work; the setting; and the knowledge, skills, and abilities necessary to qualify for the position.
2. *Competing for and capturing a position.* Find a place and a position to apply for and complete the written, physical, and problem-solving portions of the entrance test. Complete the interview portion of the hiring process, and accept an offer of employment.
3. *Establishing a personal career ladder.* Identify the areas of expertise and the positions that result in promotion and opportunity.
4. *Defining a personal career path.*
   - Career positions.
   - Education and training.
   - Life and career experience.
5. *Succession planning.* Preparing to exit the position and ensuring that a pool of trained people is available to fill your position.

It is important that on an annual or biannual basis, you sit down and write out your personal career-path plan. A variety of tools is available to assist in putting your career path in writing; goals and plans do not exist until they are written down. The Franklin Covey program employs one aspect of the Seven Habits of Highly Effective People known as "sharpening the saw." This looks at what a person would do to improve upon himself. A timeline approach similar to what is used in the strategic planning process is helpful in setting up personal career goals to sequence goals by dates and time.

■ MENTORING

The Merriam-Webster dictionary defines a **mentor** as a trusted counselor or guide. A mentor also is an individual, usually older and always more experienced, who helps and guides the development of another individual who is called a **protégé**. This guidance is not done for personal gain. A successful mentor assists someone through personal and professional growth and development by coaching and guiding. Mentors do not do the work for you; they provide you with options and experience, allowing an aspiring manager or leader to make better decisions.

Effective mentoring is a lifelong experience, with several people assisting an EMS manager or leader at various stages in his career and life. Often this process creates an end result that those who have been well mentored become good mentors themselves, so the process perpetuates itself.

A mentor has traveled roads similar to the ones you will encounter and provides shortcuts, more efficient strategies, and sometimes just a sounding board for your thoughts and ideas. When reflecting on life experiences, people often say, "If I had only known." That is what a mentor can help with, since he has already "been there and done that." One of the most valuable assets your career can have is a good mentor, and it is not uncommon to have more than one mentor helping you navigate different paths of professional development.

FINDING A MENTOR

A good place to look for a mentor is right in front of you. Look around at work. Is there an individual whom you admire and respect or someone who has always impressed you with their insight and perceptiveness? Maybe it is your boss or your boss's boss, or maybe it is a leader in a different division. It could even be the older individual who is not currently a top executive of your organization but who has lots of experience. Approach that individual and ask if he would consider being your mentor. Depending on the individual, and your current relationship, your proposal will vary in the amount of detail provided and how it is delivered. At the very least let the person know what and why you selected him and what you hope to learn from the association. It is important to select a community leader or a person of national recognition that you respect as a potential mentor. If appropriate for the specific individual, you also can discuss amounts of time to be committed and what you will contribute. Do not put it off. What can you lose? Even if the person declines to be your mentor, and few will, he will be flattered that you asked.

MENTORSHIP EXPECTATIONS

Often people enter a mentoring relationship without knowing exactly what to expect from the experience. A critical assessment of the protégé's skills and abilities should be the first step for a mentor. Ideally, this is an honest and comprehensive assessment of the protégé and involves asking the right questions about the protégé's goals. The mentor is expected to help take a person from theory to application. Mentors are expected to help protégés avoid the political errors and navigate the political processes. Mentors help people become involved in the community and in networking. In many instances the mentor serves to reassure or provide positive reinforcement during the struggles involved with career development. Ideally, mentoring provides coaching, goal setting, counseling, career mapping, tutoring, and general advice. Often such assistance includes specific advice on political issues, financial concerns, and relationship issues.

Most EMS chiefs support mentoring and cite barriers such as a lack of support internally and externally in the department, favoritism, not having a pool of qualified mentors, cost, time, and training requirements as issues that prevent strong mentoring programs.

■ PROFESSIONALISM

Academics will argue that EMS has not received "professional" status, but the term *profession* has many definitions and criteria. Most people identify scientific research and peer review as the foundations for EMS professionalism. Professionalism involves investigating the current writings in the field, which means reading all the books that are involved with promotions or that are required reading for the people you manage. EMS personnel also should consider contributing to trade magazines, professional organizations, and EMS-related research.

Part of being a professional is assessing your skills and other areas for improvement. The use of a 360-degree evaluation can assist the EMS manager or leader with that assessment. There are a large variety of 360-degree devices offered by several management institutes and commercial customized training programs.

A 360-degree feedback assessment is a powerful tool for helping individuals improve, grow, and develop their interpersonal skills. Traditional performance evaluations are subjective and one-sided, providing only the supervisor's impression. Consequently, most traditional evaluation systems are ineffective in providing true and objective feedback. No matter what position you hold in the department, it is important to understand how others (boss, crew, team, peers, and the public) perceive your effectiveness as a leader, as well as your strengths and weaknesses. Input from a

360-degree assessment comes from supervisors, peers, and subordinates and is valuable in improving organizational and individual growth.

■ SUCCESSION PLANS

The day you reach the next level in your organization, you should be planning how to backfill and supply someone to replace you. One-fifth of the country's largest corporations will be losing 40% of their top-level talent, a statistic from the corporate world that could easily be applied to EMS services. Over the next 15 years a decline in the number of people ages 35 to 44 will create a large void in the middle-manager and leadership populations. Approximately 34% of the federal civilian workforce is over 50 years old. If one-third of your workforce retires, who will replace your most experienced EMS employees? **Succession planning** ensures that there are highly qualified people in all positions, not just today but tomorrow, next year, and five years from now.

A good strategic plan looks at building human resources that can be three deep in a position. Often managers think that preparing people to fill their positions makes those people a threat, when in actuality not doing so often limits a person's upward opportunities.

HOW SUCCESSION PLANNING HELPS

Succession planning establishes a process that recruits employees, develops their skills and abilities, and prepares them for advancement, all while retaining them to ensure a return on the organization's training investment. Succession planning involves:

- Understanding the organization's long-term goals and objectives.
- Identifying the workforce's developmental needs.
- Determining workforce trends and predictions.

In the past, succession planning typically targeted only key leadership positions. In today's EMS organizations, it is important to include key positions in a variety of job categories, including quality-improvement coordinators, EMS preceptors, and EMS training officers. With good succession planning, employees are ready for new leadership roles as the need arises, and when someone leaves, a current employee is ready to step up to the plate. In addition, succession planning can help develop a diverse workforce by enabling decision makers to look at the future makeup of the organization as a whole.

Personal succession planning requires taking an inventory of what it took to get you to your current position, and then creating a procedure to ensure others get the same development and education. One of the common complaints is that people are placed into positions and then given some education after they have been in the role for a period of time. Consider as part of your professional development building an EMS management-training program within your organization to help develop your subordinates.

Succession planning also requires that you take an inventory of people within and even outside the organization whom you can use to fill positions within the EMS division. This includes doing a skills inventory and assessing the interest level. It helps match personnel with those who could be assigned tasks that are progressively increasing in levels of complexity and responsibility. Motivation to succeed is an attitude to look for. Ideally, you hope to identify people with a passion for EMS.

SUCCESSION PLANNING

A process or plan can help an agency focus on the succession-planning process. A succession plan is like any human-resources function and requires a complex process. The components of a succession plan are listed in Figure 12.9.

When exiting an organization, whether for retirement or promotion, your role is to let the people who worked for you know where they stand in the organization. Your departure gives you the opportunity to let people know how the chief, peers, and subordinates

---

- Develop a communication strategy for the success-planning process.
- Identify expected vacancies.
- Determine critical positions.
- Identify current and future competencies for positions.
- Develop a recruitment strategy.
- Create assessment and selection tools.
- Supplement human resources functions to include active recruiting, training, and staffing.
- Identify gaps in current employee and candidate competency levels.
- Develop individual development plans for employees.
- Develop and implement coaching and mentoring programs.
- Assist with leadership transition and development.
- Develop an evaluation plan for the succession management plan.

FIGURE 12.9 ■ Steps to Conduct Succession Plans.

perceive those around you. It is important to leave them with an idea about where the organization is going. For example, upcoming changes, expansion plans, reorganizations, or a new technical area into which the department may want to move. Help build technical skills for those who are in your group. When leaving, or if you are witness to another EMS manager or leader leaving your organization, find out or announce what the plans are for the future. Offer or receive assistance in making the transition. The person leaving may be a resource in the future or a source of unbiased reflection on difficult issues.

PUTTING OFF SUCCESSION PLANNING

Not many people want to think about their own deaths, and a serious illness or disability is not much easier to think about. Retirement often seems a long way off, or you do not think you even want to retire. And for many leaders who have spent most of their lives building up a company, whether it is a family-owned business or a corporation you have helped mold and grow, it is difficult to disassociate yourself from the company. The thought of anyone else running the company is often painful.

For these reasons many people put off succession planning. Quite simply, their emotions get in the way. But it is those feelings of pride and ownership in a company that really should push you to make a succession plan, not ignore it. Succession planning is the act of creating a plan for the inevitable—your departure—because you care so much about the company. If you did not care, you would not be where you are today.

Procrastination also occurs because leaders think they are too busy to plan for this life-changing event. But the truth is that it is another important part of a thriving business that should not be ignored, just as boosting the bottom line and keeping customers happy is a big part of business. The small amount of time it takes to create a succession plan will save the hundreds of hours of time spent cleaning up the mess made by a departure with no plan.

Finally, leaders may opt not to choose a successor, because they think no one can handle the company as well as they do. While that might be true in some cases, isn't it better to find someone now and start giving them the skills needed so they are prepared, instead of dropping them into the position at the last minute with no preparations?

CHAPTER REVIEW

Summary

Only 3% of individuals in an organization produce a professional development plan. Those who do not will be subject to the will of their organizations, personal circumstance, and blind luck to determine the outcome of their careers. By developing a plan, following the plan, and asking a mentor to assist you, it is possible to attain 97% more success than the person you work next to on a daily basis. Ask a new recruit what their professional goals are, and they will likely give you a blank look. Ask a 10-year veteran and he will acknowledge the importance of a professional plan. A 20-year veteran will most likely lament, "I wish I would have done more." Traditionally, emergency service organizations have not had the tools or leadership to pursue or embrace professional development. This is an area of much needed attention in modern emergency service organizations, with a very large positive payback when professional development is a priority.

Research and Review Questions

1. Create an EMS professional-development algorithm for a senior paramedic who has asked to be considered for management.
2. Choose one competency from the International Association of Fire Chiefs professional-development program and apply it to your EMS experience, education, or seminar.
3. Define a succession plan and write a policy, SOG, or SOP for your agency.

4. Identify the components of the chief medical officer designation (CMOD) program.
5. Identify the closest post-secondary educational institution providing EMS-management education.
6. List the additional humanities skills that are needed for an EMS manager or leader and identify how to obtain the training or experience necessary to attain those skills and abilities.

References

Argyris, C., Bennis, W. G., Thomas, R. J., & Harvard Business School Press. Harvard Business Review *on developing leaders.* (2004). Boston: Harvard Business School Publishing Corporation.

Bruegman, R. R. (2005). *The chief officer: A symbol is a promise.* Upper Saddle River, NJ: Pearson Education/Brady Publishing.

Edwards, Steven T. (2005). Fire service personnel management. Upper Saddle River, NJ: Pearson Education Inc.

Hesselbein, F.; Goldsmith, M.; & Beckhard, R. (1996). New York: Peter Drucker Foundation for NonProfit Management.

International Association of Fire Chiefs. (2003). *Officer development handbook.* (2003). Fairfax, VA: Author.

International City/County Management Association. (2005, January 13). Overview program announcement, Cal-ICMA. Retrieved January 20, 2005, from https://www.icma.org/main/bc.asp?bcid = 351&hsid = 8&ssid1 = 2190&ssid2 = 2203.

Mink, O. G; Owen, K. Q.; & Mink, B. P. (1993). *Developing high-performance people: The art of coaching.* Reading, MA: Addison-Wesley Publishing.

EMS Quality Management

Accreditation Criteria

CAAS

201.06.01   CQI Program: The agency's CQI Program shall include prospective, concurrent, and retrospective initiatives designed to improve the care delivered by the agency's providers. All aspects of the Clinical CQI Program shall be developed in conjunction with the Medical Director (pp. 319–321).

201.06.02   Clinical Indicators: The agency's CQI Program shall have measurable clinical indicators that are regularly assessed for compliance with established thresholds. These indicators shall include, at a minimum, the following: timely, accurate patient assessment; timely medical interventions delivered in accordance with established protocols; success of skills; clinical documentation quality; and outcome data (p. 322).

101.06.03   With Medical Director input and approval, the agency shall have a Continuing Medical Education Program that meets or exceeds local and state requirements. The Continuing Medical Education Program shall be clearly linked to the agency's CQI Program and shall address each level of service provided. Individual CME requirements may differ among different levels of providers (p. 334).

Objectives

After reading this chapter, the student should be able to:

13.1   Define the activities involved with quality assurance (pp. 318, 328).
13.2   Define the activities involved with quality improvement (pp. 319–329).
13.3   Apply QI techniques to various aspects of EMS operations (pp. 320–322).
13.4   Identify the techniques to measure quality indicators in EMS operations (p. 322).
13.5   Locate and identify other sources of quality data information that can improve EMS operations (pp. 321–328).
13.6   Create and implement a customer service assessment as part of a quality-improvement program (p. 319).

13.7 Understand and create a process that helps document trends that require increased education or modification of the EMS system (pp. 330–334).

13.8 Evaluate and apply the historical aspects of quality improvement to modern EMS efforts (pp. 318–321).

13.9 Chronical the history of CQI activities (pp. 313–318).

Key Terms

benchmarking (p. 327)

continuous quality improvement (CQI) (p. 313)

critical indicator (p. 332)

critical to quality (CTQs) (p. 316)

customer (p. 313)

effectiveness (p. 319)

Kaizen (p. 316)

key driver (p. 322)

objective (p. 322)

PDCA cycle (p. 313)

performance measure (p. 315)

qualitative (p. 319)

quality (p. 315)

quality assurance (QA) (p. 318)

quality control (QC) (p. 315)

quality of care (p. 319)

quality improvement (p. 312)

quality indicators (p. 319)

quantitative (p. 319)

scientific management (p. 313)

sentinel events (p. 327)

service audit (p. 321)

Six Sigma (p. 316)

stakeholder (p. 330)

strategic quality planning (p. 330)

total quality management (TQM) (p. 315)

Point of View

In the early 1990s the Long Beach (CA) Fire Department was developing a quality-improvement process based on the NHSTA guidelines. They were one of the first fire departments to institute the concept of trend analysis and link the quality measurements to their educational and operational plan. If a trend analysis identified a weakness in the care of the EMS operation, a series of questions were asked. First, is this a system problem in which the design or delivery mechanisms are the problem? If so, some system that is used to provide care is not adequate to achieve the goals of the organization. Second, if not a system problem, then it is important to look for or analyze the procedures involved with what is identified as a problem. A new procedure or process may be needed; a lack of educational support for an existing procedure, or a revision of an existing procedure, might be the solution. Lastly, if the first two questions do not find a solution, then the EMS or quality-improvement manager must look to the individual involved with the problem and ask, is there a personal or performance-related issue that triggered the problem identification? Often the personal problems need referral to remedial education, disciplinary action, or the employee assistance program.

Richard Resurreccion, EoD
Human Resource Development Consultant
Long Beach (CA) Fire Department

■ QUALITY IN EMS: A NECESSITY

Most people in EMS come to work with the philosophy to do the best job possible for the public and provide prehospital care that will make a difference. In most cases they never see results or are never given feedback on their performance and how they impacted the system, unless it involves a problem or perceived problem. Over time a lack of feedback on performance creates apathy. **Quality improvement** activities are essential to keep the workforce motivated, provide excellence in patient care, and reduce the risk to the organization. There are two key themes to keep in perspective when managing EMS quality-improvement activities: do the right thing, and do the right thing well.

Doing the right thing involves ensuring efficacy—performing the procedures and treatments that are rooted in evidence-based medicine and appropriate for the patient. Doing the right thing well also requires EMS services to be available for the patient who needs them. Service delivery must be timely, with care conveyed promptly. EMS quality-improvement activities need to be effective with regard to the tests, procedures, treatments, and services that are provided. It is also important to accurately measure the continuity of care with respect to other health-care providers and services, because a bad outcome for the patient may have a root cause different from the patient's final destination; that is, in your agency, in another agency, or from a different provider in the health-care system. Safety is another quality indicator. Emergency care should be safe for the patient and those who provide the service. Finally, quality care will need to be efficient, respectful, and compassionate to those to whom service is provided.

These necessities are useful for outlining the performance indicators that monitor the processes and outcomes of the important aspects of emergency medical service delivery.

QUALITY IMPROVEMENT IN EMS

A Short History

Quality-improvement (QI) activities are conducted to guarantee that quality patient care is provided and that there is continuous improvement in service delivery. QI activities are conducted by EMS managers and leaders, emergency department physicians, firefighters, nurses, and the public. It is important that everyone touched by the EMS system feel that he is part of the QI system. QI is the sum of all activities undertaken to continuously examine and improve products and services. QI activities are prospective, concurrent, and retrospective, depending on when they are conducted relative to an event. Quality measurement tools include tests, databases, in-field observations, chart review, **customer** service surveys, and checkout sheets.

The application of quality-improvement techniques in EMS can be traced back to classical management and applications in the private sector. **Scientific management** was the start of a formal look at the economic dimensions and motivations of work that would result in improved efficiency. Fredrick Taylor, author of *The Principles of Scientific Management* (1911) and whose principles were applied to the steel industry, is considered the father of scientific management. Since

scientific management lacked any consideration for the psychological aspects of the work, Frank and Lillian Gilbreth, who closely followed Taylor's work, improved upon scientific management by studying human motivations and the psychological aspects of work.

French industrialist Henry Fayol began looking at management activities in the late 1800s by defining functions and processes conducted by managers. In the 1920s a Harvard business professor, Elton Mayo, initiated a five-year series of studies on the efficiency of work and its relationship to working conditions, which came to be known as the Hawthorne studies. The Hawthorne studies focused more attention on understanding individuals, attitudes, and groups and less on organizational structure and efficiency of the work. One of Mayo's conclusions, the Hawthorne Effect, was that the knowledge of being observed causes alterations in behavior. Since the 1920s new and improved analyses have been done on the management process and the improvement of quality.

In modern times W. Edwards Deming began applying quality processes to modern manufacturing. Deming began working in Japan in 1950 and was instrumental in building the Japanese industry into an economic world power. His strong humanistic philosophy is based on the idea that problems in a production process are due to flaws in the design of the system, as opposed to being rooted in the motivation or professional commitment of the workforce. Under Deming's approach, quality is maintained and improved when leaders, managers, and the workforce understand and commit to constant customer satisfaction through **continuous quality improvement**.

If Deming's principles were applied to the prehospital area, EMS would focus on a productive system for providing quality patient care and the appropriate use of resources, not revenue. Deming sees management turnover as affecting quality and so, ideally, EMS managers staying in their positions with little turnover would ensure quality and productivity. Outcome measurements including items considered multipliers from the hard data that may not be easily quantified, such as customer satisfaction, should be measured under Deming's principles (Figure 13.1).

In 1920 Walter Shewhart developed the **PDCA cycle**, a four-step model (plan, do, check, and act) to execute change (Figure 13.2). Deming put this

Deming's Seven Deadly Diseases	EMS System Parallel
1. Lack of constancy in purpose.	1. Focus other than on quality patient care and appropriate use of resources.
2. Emphasis on short-term profits.	2. Emphasis on revenue, which undermines system quality and productivity.
3. Evaluation of individual performance/merit rating.	3. Performance rating builds fear and creates rivalry and bitterness.
4. Mobility of management.	4. Workforce turnover; long-term EMS personnel are necessary for quality and productivity.
5. Running a company on visible figures alone.	5. Ignoring elements in EMS systems that may not be measurable, such as the multiplier effect of a happy patient or family member.
6. Excessive medical costs.	6. Costs related to worker compensation and on-the-job injuries.
7. Excessive costs of warranty.	7. Costs of litigation against system providers.

FIGURE 13.1 ■ Deming's Seven Deadly Diseases Applied to EMS.
(Reprinted with permission of International Association of Fire Fighters [IAFF].)

model into practice as a policy to improve quality in addition to decreasing the cost of providing services. This process uses the "plan" part of the PDCA cycle as a study or process to collect data and evaluate results. The "do" aspect of the cycle puts the plan into action and then must "check" to see if the plan has worked. Finally, managers "act" to stabilize, enhance, or abandon the plan. Managers must be ready to consider that some solutions may not work and to establish what went wrong if the anticipated gains did not happen. PDCA is a constant cycle, and any improvement realized by carrying out one PDCA cycle will become the baseline for the goal on the next PDCA cycle. The process of improvement is unending, although the dramatic improvements of initial PDCA efforts may be difficult to sustain. Deming's principles are applied to almost every aspect of EMS operations. Deming also developed his famous "14 Points" to transform management practices (Figure 13.3).

Deming advocated using simple statistical techniques as the primary approach to improving performance in EMS. When problems arise in EMS delivery, the causes should be verified by data. All human activity will contain variations in performance, and some simple statistical analysis can reduce the inaccuracies. Random variations must be established for every EMS procedure or service.

Two other modern QI methods, supported by the Joint Accreditation Committee on Hospital Organizations, provide a background for EMS-related QI or quality assurance. One is promoted by Philip Crosby and the other by Ernst and Young. Philip Crosby advocates that organizations should redesign operations to encourage doing the job right the first time. He challenged organizations to think of how processes could be redesigned to reduce errors and to reach a goal of zero defects. The basic principles in Crosby's philosophy of quality are listed

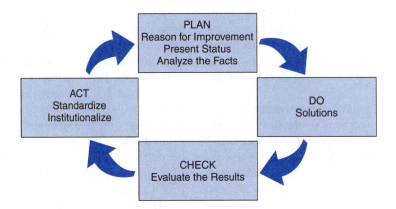

FIGURE 13.2 ■ The PDCA Cycle.
Source: Bruegman's *Exceeding Customer Expectations.*
(Reprinted with permission of Pearson Education.)

1. Create constancy of purpose for improvement of product and service.
2. Adopt the new philosophy of Total Quality.
3. Cease dependence on mass inspection.
4. End the practice of awarding business on price tag alone.
5. Improve constantly and forever the system of production and service.
6. Institute training.
7. Institute leadership.
8. Drive out fear.
9. Break down barriers between staff areas.
10. Eliminate slogans, exhortations, and targets for the workforce.
11. Eliminate numerical quotas.
12. Remove barriers to pride of workmanship.
13. Institute a vigorous program of education and retraining.
14. Take action to accomplish the transformation.

FIGURE 13.3 ■ Deming's Principles of Quality Management.

in Figure 13.4. Figure 13.5 lists the steps for the Ernst and Young process.

Dr. Joseph M. Juran made many contributions to the field of quality management. He revolutionized the Japanese philosophy on quality management and helped shape their economy into the industrial leader it is today. Juran's work incorporates the human aspect of quality management, which is referred to as **total quality management (TQM)**. Juran's concept of top management involvement is embodied in the Pareto principle (or 80–20 rule), the need for widespread training in **quality**, or fitness for use. This project-by-project approach to quality improvement helped to establish the NHTSA quality-improvement philosophy for EMS.

Juran's approach is based on the idea that a QI program must reflect the strong interdependency among all operations within an organization's production processes. Juran advocated a quality trilogy that included quality planning, **quality control**, and quality improvement. Quality planning is the process of understanding what the customer needs and designing all aspects of a system that is able to reliably meet those needs. Designing an EMS system to do less is wasteful because it does not meet patient needs or the organization's goals. When the quality of the EMS system is put into operation, quality control is

---

*   "Quality is free," meaning the absence or lack of quality is costly.
*   Quality is conformance to requirements.
*   Design or redesign processes to meet a goal of "zero defects."
*   Achieving a "zero defects" culture is dependent on "management maturity."

FIGURE 13.4 ■ Crosby's Quality Principles.

---

*   Identify and define the problem.
*   Measure the impact on customers.
*   Prioritize possible causes.
*   Research and analyze root causes.
*   Outline alternative solutions.
*   Validate that solutions will work.
*   Execute solutions and standardize.

FIGURE 13.5 ■ Ernst and Young QI Steps.

used to constantly monitor performance for compliance with the original design standards. If performance falls short of the standard, plans are implemented to deal quickly with the problem. Quality improvement (QI) occurs when new, previously unobtainable levels of performance are achieved. QI requires the continuous comparison of the system to itself using **performance measures**.

Juran also proposed the idea of "the vital few and the useful many," which helps prioritize QI projects. In any organization there will be a lengthy list of possible ideas for improvement. The assets to actually put into practice new ideas is limited, and EMS leaders must choose those few vital projects that will have the greatest impact on improving the ability to meet customer or patient needs—in other words, those that make the biggest impact on patient outcome, cut waste, or are able to gather the resources necessary for the project. The National EMS Performance Measurement Project and the National EMS Outcomes Project used Juran's concepts to focus the agency's efforts on a few vital projects. In the original pilot, EMS efforts focused on four outcomes: time of access to a public-safety answering center to time of dispatch, respiratory care, identification, and matching ALS resources to ALS calls.

Total Quality Management

Total quality management (TQM) is a management system aimed at continuously improving performance at every functional level, focusing on customer satisfaction. An EMS operation using TQM involves both leadership and employees. TQM assumes that most problems result from the inability of the system to perform, rather than the individual's inability to perform. TQM in an EMS system requires three elements: First, EMS managers and leaders must have an absolute commitment from top management. Second, it must be easy to identify measurable and accurate indicators of quality. Third, there must be involvement in the quality-improvement process by EMTs, paramedics, and support personnel in all quality-improvement methods.

* Focus on customers, both external and internal.
* Seek continuous improvement in products, services, and processes.
* Acknowledge and bring problems to the surface.
* Foster an environment of openness, honesty, and frankness.
* Use teams as the primary unit of organization.
* Manage projects cross-functionally.
* Nurture supportive relationships at all levels of the organization.
* Develop self-discipline, personal responsibility, and accountability.
* Foster organizational communications in all directions.
* Set up employees for success; enable their achievement.

FIGURE 13.6 ■ Kaizen Quality Principles.

Several quality programs build off the TQM process. **Kaizen** is a program that encourages a continuous and gradual improvement through evolution, rather than through revolution. Kaizen works well in EMS agencies because it uses a team-based approach to foster a gradual change in a team environment. Kaizen allows teams with specialized experience to come up with solutions to problems and advance the organization. Many teams are cross functional. For example, a dispatch problem may have a team that is composed of a communication operator, paramedic, EMS manager, computer technician, and EMT. This allows for a more coordinated response to problems. Kaizen has 10 basic principles, which are listed in Figure 13.6.

Six Sigma

Another offshoot from the early 1980s is **Six Sigma**. While the TQM concept thrives in some organizations and fails in others, Six Sigma's foundation of TQM principles helps it remain a popular QI program. Six Sigma's background stretches back over 80 years to the Japanese management style and more recently to changes in the top Fortune 500 companies, such as GE, Motorola, and American Express. The National EMS Management Association has adopted Six Sigma as one strategy to improve an organization. Six Sigma is a way to manage an EMS organization by putting the patient or customer first, using facts and data to find better solutions to service. Six Sigma targets three areas: improving customer satisfaction, reducing the cycle time, and reducing defects in service. Improvements in these areas usually provide an opportunity to build a reputation for top-performing service and patient care.

Six Sigma is a statistical measure of the performance of a process or product. It is a goal that reaches near perfection for performance improvement,

applying a system of management to achieve lasting EMS leadership. In the analysis of those systems, EMS leadership will need to identify protocol requirements, customer expectations, and the process for delivering EMS; these items are called **critical to quality (CTQ)**. EMS managers must understand how well the organization performs on all CTQs. In this context the term *sigma* represents standard deviation. Six Sigma uses standard deviation to describe how much of a variation exists in a set of data, group of items, or other processes in EMS that contribute to the CTQs. A major difference between Six Sigma and TQM activities of the past is that under Six Sigma, management plays a key role in monitoring outcomes and accomplishments. Another major difference is the use of statistic analysis.

To explain in detail the math involved, a calculator is necessary. And the example will be about a trauma call. But first, here are three basic data definitions:

* The "unit" or item being delivered will be an EMS response to the trauma call.
* The "requirements" that make the unit good or bad for the patient will include an eight-minute or less response time, a less than ten minute scene time, and correct treatment for each patient with no protocol deviations.
* The "defect opportunities" are the number of times per unit that the requirements were not met or service defects were delivered.

As an illustration, look at 500 trauma calls. Then figure there are four defect opportunities that are important to the critical criteria: response time, scene time, protocol compliance, and procedures applied to assessment. In this hypothetical example, 25 of the 500 trauma calls did not meet response time, 10 did not meet scene times, 7 had protocol violations, and 16 did not have the appropriate care applied or an injury on assessment was missed. To calculate Sigma, take the total defects and divide by the total calls multiplied by the number of defects opportunities:

$$\frac{(25 + 10 + 7 + 16)}{(500 \times 4)}$$

The result is 58 ÷ 2000, or 0.029; this is the defects per opportunity (DPO). Six Sigma operates with a baseline of one million opportunities for mathematical purposes, so this would be calculated as 29,000 defects per million trauma calls. This number is now applied to the Sigma calculation in a calculator with a standard deviation function, or match it to Table 13.1, and you

TABLE 13.1 ■ **Sigma Abridged Conversion Table**

Yield	Sigma Level	Defects per 1 Million	Defects per 100,000	Defects per 10,000	Defects per 1,000	Defects per 100
99.99966%	6.0	3.4	0.34	0.034	0.0034	0.00034
99.99946%	5.9	5.4	0.54	0.054	0.0054	0.00054
99.99915%	5.8	8.5	0.85	0.085	0.0085	0.00085
99.9987%	5.7	13	1.3	0.13	0.013	0.0013
99.9979%	5.6	21	2.1	0.21	0.021	0.0021
99.9968%	5.5	32	3.2	0.32	0.032	0.0032
99.9952%	5.4	48	4.8	0.48	0.048	0.0048
99.9928%	5.3	72	7.2	0.72	0.072	0.0072
99.989%	5.2	110	11	1.1	0.11	0.011
99.984%	5.1	160	16	1.6	0.16	0.016
99.977%	5.0	230	23	2.3	0.23	0.023
99.966%	4.9	340	34	3.4	0.34	0.034
99.952%	4.8	480	48	4.8	0.48	0.048
99.931%	4.7	690	69	6.9	0.69	0.069
99.903%	4.6	970	97	9.7	0.97	0.097
99.87%	4.5	1,300	130	13	1.3	0.13
99.81%	4.4	1,900	190	19	1.9	0.19
99.74%	4.3	2,600	260	26	2.6	0.26
99.65%	4.2	3,500	350	35	3.5	0.35
99.53%	4.1	4,700	470	47	4.7	0.47
99.38%	4.0	6,200	620	62	6.2	0.62
99.18%	3.9	8,200	820	82	8.2	0.82
98.9%	3.8	11,000	1,100	110	11	1.1
98.6%	3.7	14,000	' 1,400	140	14	1.4
98.2%	3.6	18,000	1,800	180	18	1.8
97.7%	3.5	23,000	2,300	230	23	2.3
97.1%	3.4	29,000	2,900	290	29	2.9
96.4%	3.3	36,000	3,600	360	36	3.6
95.5%	3.2	45,000	4,500	450	45	4.5
94.5%	3.1	55,000	5,500	550	55	5.5
93.3%	3.0	67,000	6,700	670	67	6.7
91.9%	2.9	81,000	8,100	810	81	8.1
90.3%	2.8	97,000	9,700	970	97	9.7
88%	2.7	120,000	12,000	1,200	120	12
86%	2.6	140,000	14,000	1,400	140	14
84%	2.5	160,000	16,000	1,600	160	16
82%	2.4	180,000	18,000	1,800	180	18
79%	2.3	210,000	21,000	2,100	210	21
76%	2.2	240,000	24,000	2,400	240	24
73%	2.1	270,000	27,000	2,700	270	27

(*Continued*)

TABLE 13.1 ■ Continued

Yield	Sigma Level	Defects per 1 Million	Defects per 100,000	Defects per 10,000	Defects per 1,000	Defects per 100
69%	2.0	310,000	31,000	3,100	310	31
66%	1.9	340,000	34,000	3,400	340	34
62%	1.8	380,000	38,000	3,800	380	38
58%	1.7	420,000	42,000	4,200	420	42
54%	1.6	460,000	46,000	4,600	460	46
50%	1.5	500,000	50,000	5,000	500	50
46%	1.4	540,000	54,000	5,400	540	54
42%	1.3	580,000	58,000	5,800	580	58
38%	1.2	620,000	62,000	6,200	620	62
34%	1.1	660,000	66,000	6,600	660	66
31%	2.0	690,000	69,000	6,900	690	69
27%	0.9	730,000	73,000	7,300	730	73
24%	0.8	760,000	76,000	7,600	760	76
21%	0.7	790,000	79,000	7,900	790	79
18%	0.6	820,000	82,000	8,200	820	82
16%	0.5	840,000	84,000	8,400	840	84
14%	0.4	860,000	86,000	8,600	860	86
12%	0.3	880,000	88,000	8,800	880	88
10%	0.2	900,000	90,000	9,000	900	90
8%	0.1	920,000	92,000	9,200	920	92
7%	0.0	930,000	93,000	9,300	930	93

will find 29,000 equates to a Sigma of 3.4. The range for Six Sigma is from one to six, with the number six indicating perfect service delivery and the number one indicating poor service delivery. An organization with the number 3.4 on the Sigma chart falls in the middle for performance on trauma patients. Quality-improvement data fits nicely into the Six Sigma model. The six themes of Six Sigma are presented in Figure 13.7.

Six Sigma uses the DMAIC process to solve problems: define, measure, analyze, improve, and control. To implement Six Sigma, EMS leadership must have the ability to see the "big picture" and gather data. Once a problem is identified, the ability to break

through old assumptions and work collaboratively will yield solutions. Six Sigma thrives on change, and EMS managers and leaders must embrace change as a natural part of the management process.

EMS Quality Improvement

EMS quality improvement officially began with an initiative from NHTSA. In 1995 NHSTA created a document called "A Leadership Guide to Quality Improvement in Emergency Medical Services" and conducted a series of national seminars on basic quality-improvement techniques. Prior to NHTSA's efforts, efforts of many individual agencies to improve quality in their organizations focused on quality assurance. **Quality assurance (QA)** has been defined as an essential component of medical control that employs the processes of audit, review, analysis, and evaluation of prehospital performance after the fact. QA consists of activities that measure EMS performance against a standard, using benchmarks. The goal of QA is to maintain a medically correct and consistent level of care by identifying deviations from the prehospital standard. It

---

- ◆ Genuine focus on the customer or patient.
- ◆ Data- and fact-driven management.
- ◆ Processes are where the action is.
- ◆ Proactive management.
- ◆ No boundaries on collaboration.
- ◆ Drive for perfection, tolerate failure.

FIGURE 13.7 ■ Six Themes of Six Sigma.

often falls short of effecting real change and is seen by field personnel as punitive. A more modern philosophy is to look more closely at the full system.

JUSTIFYING QUALITY IMPROVEMENT

Most EMS systems currently are operating under standards designed years ago with little or no scientific background to determine if the service was efficient and effective. Most EMS providers collect data on operational activities; however, they do not use the data effectively. In the next decade it is inevitable that EMS providers will be confronted with threats of reduction of service hours, staffing decreases, or budget reductions. Local decision makers will require justification of costs expended. It is important to evaluate the EMS system to determine the value of the service provided and to carefully examine the medicine for outcomes. Any analysis of the system needs to look at the individual parts of the system and make fine adjustments.

Analysis of the quality of EMS services is either a **qualitative** or **quantitative** measurement. Quality improvement requires a genuine drive to provide good patient care; to evaluate health-care costs; to analyze policy, patients, and the system; and to defend against litigation. **Quality of care** is defined by the Institutes of Medicine as the degree to which health services for individuals and populations increase the likelihood of desired health outcomes and are consistent with current professional knowledge. Quality is often seen by the provider as patients getting what they want, and often that is a quick response and ride to the hospital. In contrast, EMS managers look at quality as getting the job done right the first time.

Many EMS organizations have attempted to establish quality measurements and standards. NFPA 1710, CAAS, ASTM F-30, and the American Heart Association, along with local and state health authorities, have attempted to establish consensus standards. Accreditation is a process by which an agency evaluates and recognizes a department or organization as having met certain predetermined standards or qualifications. Quality improvement is a necessity for attaining accreditation from the Committee of Fire Accreditation International and from the Commission on Accreditation of Ambulance Services.

The QI program in EMS has the responsibility to evaluate utilization of staff, deployment of vehicles, and capabilities of the system. QI activities help reinforce the health and safety of EMS workers and patients. The program helps ensure quality of training and continuing education of personnel and that the EMS agency is in compliance with state and local regulations. QI programs can help resolve complaints and maintain conformity of clinical practice guidelines. QI activities give a correct picture of the overall performance of the system.

Most EMS agencies collect data; however, many do not know how to generate or produce valid reports. EMS has very little data on the **effectiveness** of the services it provides and few **quality indicators** have been defined. It is important through a labor-management partnership that EMS agencies identify, define, and develop EMS system indicators of quality. A performance indicator is a point of comparison used to answer the question "how are we doing" for a specific issue. Key components of performance indicators focus on providers, standards of care, hiring processes, the training process, supervision, system certification, appropriateness of vehicles and equipment, hospital facilities, continuous quality-improvement (CQI) programs/processes, clinical education, and anything else that either directly or indirectly affects patient care.

COMPREHENSIVE QUALITY MANAGEMENT

Modern quality management concepts take an approach to measuring EMS activities by looking at processes retrospectively, concurrently, or prospectively. An effective methodology for evaluating EMS system internal (organizational) quality and the management and evaluation of external patient care has become critical to the success of EMS programs. In the context of EMS, quality becomes a measure of how well customers are treated clinically and how well their expectations of care and service are met. Modern EMS techniques are being questioned today to ensure that they are making an impact in the outcomes of patients. In addition, comprehensive quality management programs need to provide EMS organizations with the information to efficiently manage their resources, identify their system's strengths and weaknesses, and monitor the effects of change on their system. EMS systems usually use one or more of the recognized techniques for quality-improvement activities.

Patient care and operations in most systems are conducted by a retrospective process. An older, more traditional approach to quality management, it tends to look back at system performance. Retrospective

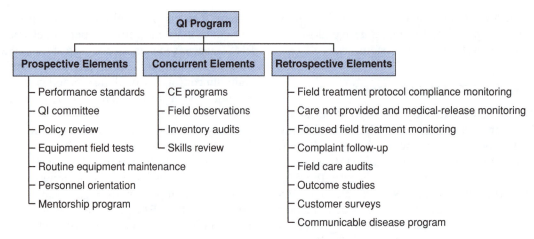

FIGURE 13.8 ■ Long Beach Fire Department's Quality Improvement Program Elements.

processes seek to correct a problem after it has occurred. Quality assurance (QA), the most commonly known retrospective process, involves looking at reports after they are completed and using audits to match paperwork with protocol compliance. Managers use "red flags" to note system failings, such as a lack of information on run reports, not meeting desired response times, failure to follow SOP, or failure to follow proper protective protocols. When questioned by quality managers, EMS providers are more apt to view retrospective quality activity as looking only for mistakes and view any feedback as punitive rather than constructive.

Chart reviews are often the most common type of QA activity conducted in an EMS agency. QA of run reports can be done by clerical personnel with the same efficiency as EMS managers when the activity is to ensure the proper paperwork has been filled out correctly. EMS managers and medical directors retrospectively reviewing procedures and care then become the hallmark of quality assurance activity.

Concurrent quality initiatives involve in-field evaluations by EMS managers, field training officers, and the physician medical director or staff. Observations should be random, and a summary or documentation should be provided to the QI committee. Physician and EMS managers should also conduct "ride-alongs" with the staff to observe crews from start to finish. Front-line EMS supervisors should also make observations of and report any opportunities for education to EMS managers.

Prospective activities include observing and conducting primary EMS and recertification training.

Case reviews, clinical rotations, and quarterly skills practices are required to ensure sustained proficiency. Attendance at outside continuing education conferences and ongoing modification in clinical protocols and policies are considered prospective. This also includes research and following medical advances and changes in the practice of EMS. Figure 13.8 displays a sample of QI programs that encompass retrospective, concurrent, and prospective approaches to EMS improvement.

EMS managers have begun to use a more proactive vision in their approach to quality management. EMS operations are now looking ahead to foresee problems before they happen, while still ensuring that their EMS system meets customer or patient expectations. This type of trend analysis has become the most popular QI technique. Computer databases or customized EMS reports can identify trends that need the application of an educational prescription. Everyone must be involved in open discussions to identify problems, develop solutions, and implement corrective procedures. However, the process must originate from top-to-bottom, and often it is the EMS manager who leads a QI committee that identifies and finds solutions to improve patient care and EMS operations.

Responsibility still belongs to management, but EMS providers are encouraged to submit input and assist in making organizational decisions. When EMS providers feel they are being heard and making an impact in system matters, overall system quality improves. This extends to involving fleet maintenance managers in discussions involving vehicle and equipment maintenance and safety measures. The more

EMS members "buy-in" to a quality program activity, the more effective it becomes. In a participative system, management provides the tools and guidance to permit responders to do their jobs and permits them to make quality decisions within established protocols.

■ IMPLEMENTING QI PROGRAMS

BUILDING A SUCCESSFUL PROGRAM

There are several stages involved in implementing a QI program in an EMS system. An EMS organization or system will begin by educating the EMS leaders and managers on the process, theory, techniques, and benefits of quality improvement. EMS leaders and members of the QI committee should be able to articulate these benefits, believe that these benefits can be achieved, and develop a plan of action for realization of the goals. Early stages of implementation of a QI process require strong leadership and commitment at the local, regional, and state level to do all of the following:

+ Learn and understand quality-improvement strategies.
+ Assess thoroughly the present quality levels of each EMS organization or system.
+ Establish action plans for training and orientation in quality improvement.

Service Audit

When EMS is compared to other private ambulance or other emergency service providers for any aspect of service, the focus is often only on emergency services and not on the service as a whole. The first task of an EMS leader or manager is to analyze the EMS system for strengths and weaknesses; this is known as a **service audit**. This is even more important when EMS providers are asked whether budget requests and efficiency questions make a difference to the outcome. It is incumbent on EMS leaders and managers to conduct a service audit on a periodic basis. A service audit is an assessment of all services being provided by the organization and is detailed in Figure 13.9. Some organizations have called this assessing a value-added service.

Establishing Goals

An important goal of every EMS system should be to continuously evaluate itself and constantly strive for performance improvement. In order to do this, it is

- ◆ ALS first response.
- ◆ BLS first response.
- ◆ ALS transport.
- ◆ BLS transport.
- ◆ Auto extrications.
- ◆ Rescue (water, below-grade, high-angle, hazardous-materials, confined-space).
- ◆ Special events response and standby.
- ◆ Wilderness EMS and searches.
- ◆ Tactical EMS call-outs.
- ◆ Hazardous-materials responses.
- ◆ Aero-medical evacuations.
- ◆ Disaster response.
- ◆ Shelter EMS.
- ◆ Multiple-casualty incident (MCI) responses.
- ◆ CPR training.
- ◆ Community emergency response team (CERT) training.
- ◆ EMS education.
- ◆ All personnel trained as EMTs and paramedics.
- ◆ Sports EMS and standbys.
- ◆ ALS engine responses.
- ◆ Critical incident stress debriefing (CISD) call-outs.
- ◆ Injury risk management and prevention.

FIGURE 13.9 ■ Components of an EMS Service Audit.

first necessary to establish achievable goals for every behavior inherent in the system. The role of quality management is then to measure the system's actual performance against those goals. If data analysis reveals that an EMS system is meeting performance goals, new goals should be set that encourage higher levels of performance. If, however, it appears a system is continuously falling below standards, the system should be analyzed to determine the causes of the problem. Once cause has been identified, a plan should be developed and implemented to correct the problem.

It is imperative that each EMS system look inward to establish goals of behavior. In some cases, it may be appropriate to adopt published goals of other systems. For example, a maximum five-minute response time to all patients would be ideal, based on current studies; however, it may not be reasonable for your system when you have only five ambulances to cover 400 square miles and meet the needs of 60,000 patients. When goals that are compatible with your system do not exist as a local, state, or national written standard, you are faced with either guessing or studying available data. Although studying data may not in itself set the goal for your system, it will at least tell you where you are in establishing new goals of service that others can use as benchmarks.

Identify and Develop Standards of Care

To measure performance and improve quality activities in an EMS system, EMS workers need to be measured against a standard. Measurements of EMS should encompass every aspect of patient care longitudinally across the health-care system, from injury prevention to rehabilitation. EMS managers should look for recognized standards on a national level. As a group, members of an EMS agency should look to identify local standards and review their own procedures for standards that are set by the agency's own interest through a labor management process; for example, customer service standards or enhancements to accepted medical practice, such as ensuring pain management.

EMS standards need to be designed based on the components that focus on measurements encompassing a structure, process, or outcome. These three components help create performance goals and performance measures as a percentage. Measuring EMS system results is necessary to assess how well the system is doing in one of those three areas, as well as to measure the impact of efforts to improve performance in each portion of the system.

Structure results focus on the necessary resource components of the system, such as EMS staffing, equipment, and the population being served. Other examples commonly include square miles, number of ALS units, number of AEDs, number of calls, number of call types, and equipment and drugs carried.

Process results examine the effectiveness of the design and delivery of work processes, productivity, and operational performance. Conducting a facilitated intubation without desaturating the patient using a three-person team is a process that can improve the patient's outcome. Time increments, procedure success rates, medications administered, and adherence to protocols are other common forms of process results measured by EMS agencies.

Outcome results encompass more than mortality and morbidity. Outcomes also look at the effectiveness of patient care, support services, and fulfillment of public responsibilities. Such public responsibilities might include, for example, improvements in disaster response and public health emergencies. The outcome that is most important in terms of EMS effectiveness and improvement is patient-care results. The most commonly accepted definition of quality care is an increased rate of survival from a life-threatening event. Other, broader patient health-care results are also important and include a variety of changes in the pa-

D—*Death:* Did the patient survive to hospital discharge?

D—*Disability:* Was there an improvement in the patient's functional status as a result of the patient care rendered?

D—*Discomfort:* Was there improvement in the patient's symptoms (such as alleviation of pain or improved breathing)?

D—*Dissatisfaction:* Was the patient (and/or family) satisfied with service rendered?

D—*Destitution:* Was the treatment provided at the lowest cost to the patient, the payer, and to society as a whole?

FIGURE 13.10 ■ Patient Health Status Measurement.

tient's health status. Figure 13.10 lists the "5 Ds" and can help EMS managers examine the results of EMS operations.

Objectives and Performance Indicators

Whether a standard is developed around a structure, process, or outcome, it must relate to the organization's objectives. Correct measurement of objectives requires performance indicators to determine if objectives are being met. **Objectives** are measurable statements that are consistent with the system's or agency's mission, vision, and key drivers. Clear operational definitions are needed for each objective. When well-defined, objectives serve as performance indicators against which system progress can be assessed or compared. Broad organizational objectives tend to come from the top down and, ideally, quality-improvement projects are often created from the bottom up and should be consistent with the broader organizational goals. Some objectives will be similar or even identical to state, regional, and local agency objectives. Other objectives will be unique to each type of EMS structure or will vary among similar structures. Structure and process have equal influence on outcomes.

Performance indicators sometimes are called key performance indicators or key drivers. **Key drivers** are quantifiable measurements that reflect the critical success factors of an EMS organization. If a key driver is going to be of any value, there must be a way to accurately define and measure it. Key drivers usually are long-term considerations and must include ways to collect data on the indicator (Figure 13.11). The definition of what they are and how they are measured should not change but should be referenced.

The goals for a particular key driver may change as the organization's goals change or as it gets closer to achieving a goal. Common sets of performance indicators

Peripheral Intravenous (IV) Skill Success Rate—Adult	
Type	**Process**
Definitions	
Success rate	Percentage (%) of successful placement of peripheral intravenous access device by EMS personnel per patient case.
Patient case	An individual patient for whom EMS personnel have performed one or more attempts to puncture the skin with a needle catheter device with intent to gain access to peripheral venous circulation.
Attempt	Needle tip penetrates patient's skin.
Success	Access to peripheral venous circulation as evidenced by ability to infuse intravenous fluids.
Adult	Patients who have reached the age of 15 years or older.
Reporting format	% success rate per (aggregate summary).
Reporting formula	Number of successes divided by total number of patient cases × 100 = %.
Data Points	
Inclusion criteria	All patients age 15 years or older treated by EMS personnel.
Numerator	Total number of patients for whom peripheral IV was successful.
Denominator	Total number of patients for whom IV was attempted.
Minimum points	n = 30
Reporting period	Monthly or annually (minimum 12 consecutive months).
Data source	Patient care documents (documented by EMS personnel).

FIGURE 13.11 ■ Sample Performance Indicator Definition Sheet.

Reporting example	Peripheral IV success rates.
Reporting period	Month of 7/07.
Numerator	Total number of successful peripheral IVs (N = 1769).
Formula	Numerator/denominator × 100 = % (1769 / 2021 × 100 = 87%).
	Summary indicator reported item = 87% success rate on peripheral IV on adult patients.
Analysis	Rate is within the researched benchmarks at like departments or systems.
Process	Variation-special causation.
Outcome	Benchmark comparison—best practices.
State benchmark	To be determined by baseline data collection.

Benchmark references:
1. *91% success rate. Los Angeles.* Jones, S.E., Nesper, T.P.,& Alcouloumre, E. (1989). Prehospital intravenous line placement: A prospective study. *Annals of Emergency Medicine, 18,* 244–246.
2. *71% success rate. Arizona.* Spaite, D.W., et al. (1994). A prospective comparison of intravenous line placement by urban and nonurban ALS personnel. *Annals of Emergency Medicine, 24,* 209–214.
3. *80% success rate. Pittsburgh.* Carducci, B. (1994). Intravenous maintenance with saline lock in prehospital environment. *Prehospital and Disaster Medicine,* 9(1), 67–70.

FIGURE 13.12 ■ Sample Reporting Definition Template.

are available to EMS agencies from NHTSA, IAFF, and NEMSMA. The IAFF has a validated set of indicators free to IAFF fire services, and the NHSTA minimum data or NEMSIS set are available through the federal government. NEMSMA has an ongoing project using an open source format for consensus building. The EMS performance project has completed a set of performance indicators and has established long-range goals to refine EMS systems to a national level.

Once the data collection points and processes are defined, there should be a reporting format similar to the key indicator definitions. This should include what,

when, how, and an analysis of the skill or process. A sample of a reporting definition template is offered in Figure 13.12. Reporting templates help identify areas of improvement or allow the quality-improvement activities to move to the next item if standards are being met.

Once the performance reporting template is completed, then it is time to create an action plan to meet the objectives and a compliance requirement. This plan includes the steps or a method to achieve success with the process or skill. It is helpful to include reference to current standards and scientific research to support the standard. Figure 13.13 is a sample of performance objectives or an action plan template to respond to a report from a measure of a key indicator.

EMS agencies should embrace national standards to collectively move the service delivery forward and advance the body of knowledge needed to justify EMS activities. EMS organizations also need to develop their own set of performance indicators unique to the organization. This process starts by asking a

Objectives: Appropriate/timely patient interventions.

Performance Indicator: Endotracheal Intubations.

Procedure-Oriented Objectives:

1. Finalize the computerized intubation data collection and reporting system by December 2007.

2. Implement a statewide skills requirement for endotracheal intubation.

3. Identify anesthesiologists and hospitals in each region willing to sponsor paramedic intubation experiences in the operating room.

Outcome-Oriented Objectives:

Improve the intubation success rate across the entire state to 90%.

Evaluation of Compliance:

1. Statewide skills requirement implemented in March (annual requirement of two live intubations).

2. Intubation success rate of 90% across entire state with exception of rural response areas.

3. 50% statewide compliance with skills requirement.

4. Four hospitals agreeing to sponsor paramedic intubation training; three others considering it. No hospitals in rural areas have agreed to participate.

Action Plan:

1. Assemble intubation CQI team with members from regional and local EMS agencies, rural, urban, and suburban hospitals, and state anesthesia association to work on plan to improve intubation success rates and increase intubation clinical experiences.

2. Continue to hold quarterly training sessions throughout the state for local EMS companies regarding data collection and reporting of intubation attempts and successes.

FIGURE 13.13 ■ Sample Performance Objectives.

question related to the service delivery and system performance. With a group of providers and management, all components of the question need to be defined and then an indicator built for the question. The indicator then needs to have buy-in from the field personnel and be evaluated as a beta test, meaning that a limited number of people test the indicator and collect data prior to the indicator being applied system- or agency-wide.

A focus group made up of customers is another effective method to use in evaluating the success of the relevance of a performance indicator. Focus groups can be used at the state and EMS organization level, as well as at the local provider level.

Similar to focus groups, customer surveys provide another way to evaluate the outcomes of the strategic plan. Surveys are generally less expensive than focus groups but are often limited by poor response rates and variable reliability of the data. Even so, surveys assessing the customer and patient are the foundation of a quality-improvement plan. Surveys are sent with questionnaires and often use a Likert Scale, where the customer can select the degree of satisfaction on a scale from one to five. Some areas to consider assessing are the patient's satisfaction, actions of the employees and overall performance. Some agencies add a few short questions for the customer to fill in or add an opinion. Using customer feedback to understand customer expectations and needs is a way to organize the system and educate patients.

IAFF Performance Indicators

EMS system performance indicators should be evaluated for relevance to system design or process and must be clearly defined. EMS system performance indicators must be measurable and should be adjusted for system demographics. There are few documented quality indicators by which to evaluate EMS systems. The International Association of Fire Fighters also categorizes performance indicators into three categories: structure, process, and outcomes.

Quality-improvement measurements that focus on the system structure look at the attributes of the care provided and measure patient outcomes; for example, the "resource deployment for each response area" and the "percentage of life-threatening calls that are met within eight minutes or less 90% of the time" are system measurements. Another system process that should be monitored is the interrelated components forming a prehospital system and what is actually done when care is provided.

One aspect of EMS is the call-processing time. Can a 911 operator process a call in less than 30 seconds and get the call to the first-due BLS unit so it can be at the patient's side within four minutes? The system should be measured in an integrated manner to determine if the components of the entire EMS system are functioning correctly to achieve a positive outcome for the patient. A patient outcome may be that a chest-pain patient reports pain relief and normal vital signs upon arrival at the hospital. The IAFF EMS Committee identified 15 key drivers with 27 different measures. The performance indicators are listed in Figure 13.14.

Response times have been defined by the NFPA and CAAS. Response time is made up of four compo-

- Response time.
- Call-processing time.
- Turnout time.
- Travel time.
- Staffing.
- Deployment.
- Road structure coverage capability.
- Patient care protocol compliance.
- Patient outcomes.
- Defibrillation availability.
- Extrication capability.
- Employee illness and injury.
- Employee turnover.
- Quality program.
- System-user opinion.
- Multiple-casualty event plan.

FIGURE 13.14 ■ IAFF Key Drivers.
Source: Bruegman's *Exceeding Customer Expectations.*
(Reprinted with permission of Pearson Education.)

nents: call processing time, turnout time, travel time, and an option to record the time from arrival to being at the patient's side. Call processing is the time from call intake by the unit dispatching agency until unit notification, including answering the phone or alarm, gathering information, and initiating a response by dispatching appropriate units. The measurement for this indicator is to assess the number of EMS calls that are processed in 90 seconds or less.

Turnout time is the time from response unit notification to vehicle wheels rolling toward the incident location. This includes personnel preparation for response, boarding the responding vehicle, placing the apparatus in gear for response, and wheels rolling toward the emergency scene. The percentage of all EMS calls in which turnout is 60 seconds or less should be measured.

Travel time is the time from responding vehicle wheels rolling toward the incident until the arrival of the vehicle at the incident location. The measure of performance for travel time is what percentage of all EMS calls achieves first responding unit travel time of 4 minutes and 0 seconds, or less. What percentages of all EMS calls achieve transport unit travel time of 8 minutes and 0 seconds or less? Finally, what percentages of all EMS calls achieve an ALS unit travel time of 8 minutes and 0 seconds or less?

Staffing as a performance indicator includes both the number and level of training of personnel deployed in the system. Ask: What percentage of ALS calls receives a response of two EMTs and two paramedics within 8 minutes or less? What percentage of BLS-level calls receives a response including two EMTs within 4 minutes? Deployment is the mobile response unit staffed and equipped to respond immediately to a request for emergency medical assistance. For what percentage of EMS calls is a unit immediately available to be dispatched?

Road capability is the capacity to provide equality of response to all features of the street network within the jurisdiction. Road structure is an identifiable data point in most communities as compared to a more unreliable proportion of the population served. This indicator is measured by what percentage of the jurisdiction's road miles has projected coverage by a first response unit within specified response time-frames. Percentages for first responders should be 5 minutes 0 seconds, and a transport-capable unit in 9 minutes 0 seconds.

Protocol compliance occurs when EMS personnel operated or performed according to the appropriate treatment for the specific injury or illness as defined by the medical authority. This is reported as the percentage of ALS and BLS calls in which EMTs and paramedics follow an appropriate recognized protocol. Patient outcomes relate to protocol compliance and patient status following prehospital treatment or care relative to the patient signs and symptoms. Outcomes should be measured based on the status of the patient at first contact following the EMS experience. Ask: What percentage of patients encountered improved, what percentage had no change, and what percentage got worse or died after EMS interfaced with them? The availability of defibrillation will have a substantial impact on patient outcomes. The availability of defibrillator-trained emergency response personnel and a defibrillator available for use within five minutes of a call are key indicators. This should be measured by asking: In what percentage of calls that need defibrillation was the first shock delivered within five minutes from the time of collapse?

Getting an extrication tool and trained rescuer to the scene is another key driver. A hydraulic extrication tool should be available for rescue of victims of illness or injury trapped or confined in an area from which they cannot readily be removed. You should ask: What percentage of calls that needed a hydraulic extrication tool for extrication of an ill or injured person had one available on scene within 8 minutes of initial responding unit notification?

Measuring employee illness and injury as a result of participating in an EMS encounter, including employee exposures requiring evaluation and medical follow-up, have been identified as a key driver. This is measured by recording what percentage of EMS calls resulted in reported employee injury and illness.

Employee turnover is defined as any employee leaving employment with the department for any reason and thus no longer participating in the EMS system. A key indicator for the health of the EMS system is annual turnover, which is calculated by the total number of EMTs or paramedics leaving the organization in one year divided by the total number of personnel in the agency.

Training and the amount of continuing medical education (CME), including opportunities for outside training, are key indicators in the system. Training indicators should be tied to a quality management program. Another indicator is whether or not the agency has a comprehensive quality management program. A quality program is one that includes total quality management, continuous quality improvement, and quality assessments. The programs also include direct field observation by a designated medical quality officer or medical director. Measurements for these programs include identifying whether or not it is an ongoing, well-defined program with trained observers. System-user opinion is another key indicator that identifies system and process performance. Data for measuring the percentage of users who are satisfied with the system should be collected from friends and family of the patient and from other agencies, such as law enforcement, that utilize EMS services. This includes patients from nontransport calls. The last performance indicator under the IAFF program is the presence of a specific disaster and multi-casualty plan for the agency.

At the request of the local IAFF president, the key drivers and software systems are available to EMS leaders. The IAFF has spent considerable expense to validate its version of performance indicators and has participated in many national forums to establish EMS performance indicators and correlate to the U.S. Department of Transportation's NEMSIS database. The IAFF system offers the likelihood of buy-in from labor. The software is Web-based and password protected. The power behind this system is the ability to compare the performance of fire-based EMS in one city to fire-based EMS in other cities on a national basis.

State EMS Directors Performance Indicators

In an attempt to incorporate a larger number of services in quality-improvement activities, the state EMS directors and EMS managers have begun an initiative to define performance indicators that complement the IAFF program. The state EMS directors look to incorporate an approach to quality from a traditional health-care model that may be measured and improved in the following six dimensions:

♦ Competency and credentialing on both a professional and/or organization level.
♦ Appropriateness and accessibility.
♦ Resource utilization and cost efficiency.
♦ Effectiveness by assessing desired outcome (quality assurance).
♦ Safety/risk management, medical and legal.
♦ Customer satisfaction.

The performance indicators were blended with the results from a survey of focus groups of paramedics that suggested that 15 EMS quality indicators are important to providers. They are listed in Figure 13.15.

Other performance indicators suggested by paramedics in various focus groups also indicated research activities, crew and equipment appearance, accuracy of communication, and public confidence. Most of those suggested key indicators have neither established measurements nor an identified way to measure them accurately. The use of valid performance measures

---

♦ Patient satisfaction.
♦ Managerial satisfaction.
♦ Patient outcomes.
♦ Dispatch accuracy.
♦ EMS crew satisfaction.
♦ Call quality.
♦ Partner performance.
♦ Response times.
♦ Complaints.
♦ Paramedic wellness/occupational illness.
♦ Equipment practicality.
♦ EMS cost-effectiveness.
♦ Public confidence.
♦ Crew and equipment appearance.
♦ Innovations/research.

FIGURE 13.15 ■ National Performance Indicators Defined by the State EMS Professionals.
(Reprinted with permission of National Association of State EMS Officials.)

cannot be substituted in improving quality in EMS systems. It is important to adopt validated criteria or conduct your own research to contribute validated new performance indicators.

Benchmarking

Once a set of key drivers has been selected or developed, an agency can start to compare itself to other EMS agencies or to national performance standards. This process of comparing performance standards or setting operating targets for a particular function by selecting the top performance levels within or outside an EMS agency is called **benchmarking**. Benchmarking is a means for determining how well a unit or organization is performing compared with similar units in the organization or externally. Highlights should be on **sentinel events**, which signal problems in the system.

Benchmarking is about systematically learning from the best in business or government and using that information to improve one's own performance. It is the process of identifying, understanding, and adapting outstanding practices from organizations anywhere in the world to help your organization improve its performance. For benchmarking to be worthwhile, an organization needs to understand the gap between its own performance and best practices and then take actions to close that gap.

Any aspect of an EMS operation can be benchmarked. In EMS benchmarking involves searching for and copying new ideas and best practices for the improvement of processes, products, and services that make an impact. Benchmarking can be done by conducting a literature review, consulting experts, identifying current practices, or replicating another agency's best practices. You can conduct a comprehensive benchmark study and obtain 80% to 90% of what you need by just using the telephone, e-mail, or an electronic survey to ask what others are doing as best practices.

One of the biggest mistakes an EMS organization makes when conducting a benchmarking plan is to look only at the benchmarks within EMS. It is important to look at other companies or organizations outside EMS to see what they are doing as best practices. Benchmarking will help you find out who performs the business process very well and who has practices that are adaptable to your own organization. Communication among individual members of the EMS workforce, self-directed work teams, organ-

TABLE 13.2 ■ ICMA Benchmarking Process

Step 1:	Identify what to benchmark (process, effectiveness, productivity measures).
Step 2:	Identify benchmark partners (state-of-the-art agencies, similar agencies).
Step 3:	Collect data.
Step 4:	Determine performance gaps (difference between your agency and others).
Step 5:	Communicate findings (present findings, gain acceptance from stakeholders).
Step 6:	Establish improvement goals (decide what you will improve).
Step 7:	Develop action plans (decide how you will improve identified areas).
Step 8:	Implement actions and monitor results (take action, make changes).
Step 9:	Recalibrate benchmarks (periodic review of benchmarks).

izations, and systems speeds adoption of successful innovations.

Benchmarking is increasing within EMS organizations, as well as with other organizations outside EMS. Regional and state EMS organizations support interorganizational benchmarking and communication. The International City/County Management Association (ICMA) has a nine-step process for benchmarking. Table 13.2 explains the steps of benchmarking in an agency.

The most important benchmark for accreditation is achieving results from QI efforts; for example, increased patient satisfaction and health status from improved EMS services and quality of care. Economic benefits also result, including cost savings, increased profitability or operating surplus, and more efficient use of resources.

Benchmarks should be easily understood, and they should be built with buy-in from the EMS providers. It is important that benchmarks are communicated to all EMS staff members. Changes that make a large impact might be implemented on a trial basis and then converted to policy. Benchmarks have to be realistic. Setting the bar for attainment of a goal requires proficiency. Results have to provide opportunities to educate providers and staff. Everyone is accountable in a QI system and evaluations need to be adjusted.

Sentinel Events

A sentinel event is an unexpected occurrence involving death or serious physical or psychological injury or the risk thereof. Serious injury specifically includes loss of limb or function. The phrase, "or the risk thereof" includes any process variation for which a recurrence would carry a significant chance of a serious adverse outcome. Such events are called "sentinel" because they signal the need for immediate investigation and response. When a sentinel event occurs, an analysis and action plan should be done to correct the cause(s) within 30 to 45 days.

Cardiac-Arrest Outcomes

The result most commonly applied to EMS is survival after a prehospital cardiac arrest. This result is particularly attractive as an outcome measure. It focuses on the clearly definable clinical condition of sudden cardiac arrest with a clearly definable outcome of return of spontaneous circulation or discharge neurologically intact. Treatment is standardized nationwide (i.e., advanced cardiac life support). EMS advanced life support (ALS) has been shown to impact patient outcome positively. A wealth of literature exists to provide benchmarks for comparative purposes, identifying that survival is time dependent. The data definitions are fairly standardized under the Utstein criteria to measure success. (Utstein is an internationally recognized standard for measuring cardiac arrest outcomes.)

As a result, an EMS system can use cardiac arrest as a measure of the system's structural components (e.g., response), process (e.g., consistency of ALS care rendered), and outcome (e.g., survival to hospital discharge). Reporting survival rates as a function of the population served per year aids in adjusting for varying survival rates. This helps the QI initiative adjust the factors that may impact survival, such as bystander CPR, system access, system response, and actual scene care rendered so that they can be examined and evaluated. Those results then may be used as input to the strategic quality plan for subsequent improvement action.

Focusing on cardiac arrest has its limitations because cardiac-arrest cases comprise only a small percentage of the care provided by an EMS system. In addition, access to hospital patient data to determine outcome is not consistently available. Despite these limitations, cardiac-arrest survival data is useful as one of the many benchmarks with which to assess system performance and is often a common source of media inquiry.

DEVELOPING CQI PROCESSES

Programs that support QI activities begin with the hiring process. Hiring the right people who embrace quality and are open to change or ideas to improve will bring positive results. When building a QI team, keep in mind that it should be multi-disciplinary. Training programs need to support the standards established by QI systems. Processes and programs need several layers with a team approach and should focus on the touch points, where the system interacts with people. Also, some may feel that the idea of making improvements is an admission that the current way of doing things is flawed or that those responsible are poor performers. Improved performance cannot occur unless personnel feel comfortable that they can speak truthfully and are confident that their suggestions will be taken seriously.

EMS managers should structure the CQI processes to focus on the above-average and excellent performing EMS providers. In most EMS systems 80% of management activities are often focused on the 10% to 20% of employees who perform poorly. These employees rarely show improvement and, in most union environments, it is often lengthy and costly to remove these people from the organization. A more modern CQI approach is to focus on the top performing 10% to 20% of EMS providers. If 80% of the activities are focused on reinforcing those behaviors, the idea is that the center or remaining average or below-average performers will improve to garner the attention of management and leadership. Everyone wants to be on the winning team. Figure 13.16 illustrates the shift in focus for modern quality improvement.

Quality Assurance

Quality assurance (QA) can be defined as a system for the maintenance of medically correct and consistent levels of prehospital care. Quality assurance activities often involve audit sheets and are often the only processes an EMS manager has time to complete. Ideally, QA processes include:

- Identification of errors or deficiency in patient care.
- Verification of proper completion of run reports.
- Verification of completion of prehospital personnel procedures and skills.
- Identification of educational opportunities, including opportunities to improve writing, grammar, and spelling.

Two types of quality assurance systems are used: manual and computer based. Manual processes in-

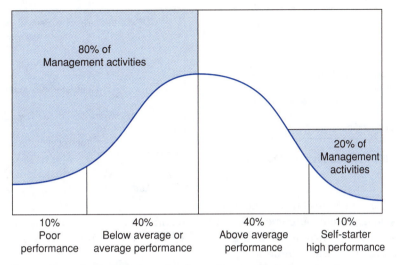

Comparison of current approaches to modern quality improvement

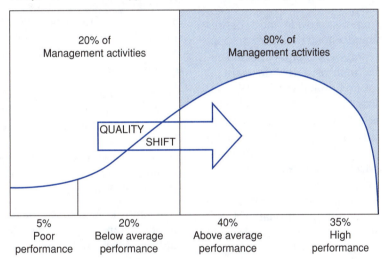

FIGURE 13.16 ■ EMS Quality-Improvement Curve with Left Shift from Modern Quality-Improvement Techniques.

volve the inspection of each chart by a person monitoring the quality process. Both computer-based and manual systems can be used for QA of prehospital ALS services. A computer-based system requires more personnel time and is more expensive but generates more reports than a manual one does. Whatever the method, a minimum of 30 consecutive charts should be collected from only the most recent calls of patient types that are being evaluated in the QA process. It is important to take the most recent charts to avoid the Hawthorne effect; that is, if EMS crews know that a specific type of chart is being audited, the measurement will be biased and not measure the true performance of the EMS system. The steps in conducting a QA audit are listed in Figure 13.17.

Many systems have no database from which to extract information. However, until an ideal quality management system is adopted, EMS system managers must compile data to evaluate the efficiency of their ongoing systems in terms of the most cost-effective means of delivering quality patient care to the emergency scene and the subsequent transport of the patient to a medical receiving facility. The extent and quality of medical care provided by a system must be examined periodically and the results documented.

EMS response extends from the initial call into and through to the receiving facility's emergency department. To measure this continuum requires standardized forms and data definitions. Data collected on each run should include dispatch and system response

- ◆ Scan run sheets.
- ◆ Match demographics.
- ◆ Prepare audit sheet.
- ◆ Tally and monitor ALS skills sheets.
- ◆ MD or QI/QA person review run sheet and audit sheet.
- ◆ Match with hospital outcome.
- ◆ Identify discrepancies and make copies to send to prehospital providers.
- ◆ Prepare a quarterly or periodic report.

FIGURE 13.17 ■ Procedures for Conducting Quality Assurance.

(QI Personnel salaries × Percentage of time spent on chart review)
+
(Audit sheets cost)
+
(Computer maintenance time)
+
(Software cost divided by number of charts)
=
Total cost of QA activities

FIGURE 13.18 ■ Cost Analysis of Quality Assurance and Chart Audit Initiatives.

(times, and so on), patient prehospital treatment, patient turnover status, and patient outcome (ED and post-ED). The system must also include minimum and measurable performance standards. Within legal limits, the data collected should conform to acceptable standards of confidentiality.

Current EMS run reports provide most of this information but are a misunderstood tool. EMS responders too often view run reports as a means of documenting their mistakes rather than as a tool to assist in improving the delivery of prehospital care. EMS managers need to review as many run reports as possible before they enter medical review, make sure they are complete, and critique them with responders based upon established run standards. Critiques should emphasize the positive as well as identify improper activities for corrective action before major problems develop.

The cost to review each run report should be calculated (Figure 13.18) when all operational and personnel costs are calculated. Some studies estimate that the cost per review per chart is between $2 and $4. This cost helps create a budget item for QA activities and can be applied to a cost benefit analysis when measured against the number of errors of prevented or controlled. If the agency has a person performing quality assurance, the hours devoted to chart review need to be added into the cost. For example, a person on a 40-hour work week who spends six hours each week reviewing charts would have those hours calculated on a quarterly or periodic basis. If a committee is used to assist or conduct the chart review, the hourly cost for each person in the meeting should be added into the cost of QA activities. Computer repair or programming services, if they are obtained from people outside the organization or from another interagency department, should be added to the cost of quality assurance activities. It is important to track support cost to determine future needs for an information

technology or computer specialist to support larger EMS quality-improvement operations.

QI Process Steps

While specific activities may differ depending on the jurisdiction of the organization, the developmental stages of QI integration will be the same for local, regional, or statewide EMS organizations. These developmental stages are:

- ◆ Building potential for success by developing an awareness and appreciation that QI is a worthwhile endeavor.
- ◆ Expanding workforce knowledge of and capability in QI practices and techniques.
- ◆ Integrating fully the **strategic quality planning** process and related quality-improvement actions into the daily EMS operation and education programs.

Conduct QI courses for front-line employees, including listening techniques, sensitivity training, and cultural diversity. Establish patient-to-provider networks to provide effective, ongoing communication for feedback and information gathering (civic groups, call-in phone line, surveys). Solicit feedback through newsletters, Internet home-page postings, local television spots, and articles in local newspapers. Characterize and chart the specific requirements for various groups of patients and **stakeholders**, using information gathered from market research, complaints, surveys, focus groups, and new customers. Compare customer satisfaction levels with similar EMS providers. Offer training through local training agencies, mass media, EMS symposia, satellite uplink programs, continuing medical education, or retreats.

EXPANDING KNOWLEDGE

After the plan is established and the QI structural foundation is fully integrated into the strategic planning process, an emphasis is placed on ensuring that the entire workforce of an EMS organization or system is informed about and participates in the development of the strategic QI plan. Paramedics and EMTs need a working knowledge of basic QI philosophy, tools, and techniques so they can be full partners in the strategic QI planning process. At the end of this phase, all EMS workforce members should be able to identify their internal and external customers, how to measure the quality of the services provided or received, and how to identify and resolve quality problems in their own work. There should also be a new sense of openness and partnership between staff and management within organizations and among local, regional, and statewide systems.

EMS leaders must involve the entire staff of the EMS system. Organizational leaders such as battalion chiefs, EMS field supervisors, and union officials are key figures in getting the system to work. It is important that EMS leaders adopt a philosophy to facilitate change in the processes. Senior leadership involvement is a must since QI activities are as important as other management tasks (including budgeting, human resource management, purchasing, and training), and leaders can integrate QI into every aspect of EMS operations. The idea of change and adjustment for quality needs to reinforce the vision and goals. In-

volvement and empowerment of employees at every level is important for a QI program to be successful. Leaders and managers need to provide the coaching and feedback necessary to make that happen.

The dispatch center is an important part of the quality-improvement program. The initial experience for a patient or family is the communication center, and often some of the actions taken may make a tremendous difference in the outcome of the patient. First responders, ALS transport units, and receiving facilities provide input to patient care; their activities affect the outcome of the patient and should be measured as key components of quality improvement. Providers need to understand that patient care is number one. They must be involved, and they must understand that feedback from the system is an opportunity to improve. Field providers must see the goal of a QI system as providing quality, cost effectiveness, and best-evidence medicine.

A quality analyst and clinical educators must be linked into the QI process to build a response to any identified needs. The federal, state, and local governing boards often mandate quality activities. Lastly, the medical director should oversee and provide input and credibility to the QI process (Figure 13.19).

Local, regional, and state EMS systems should have in place the structure and process that allows evaluation

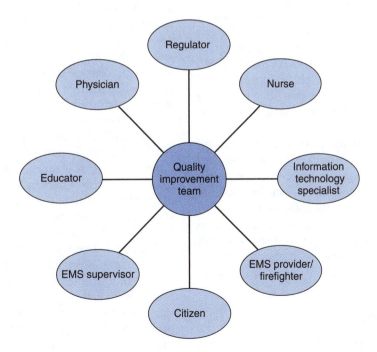

FIGURE 13.19 ■ People to Involve in Quality Improvement.

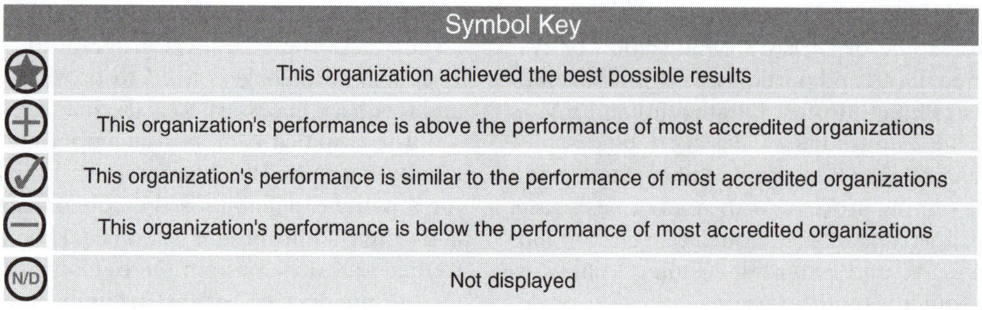

Symbol Key	
★	This organization achieved the best possible results
⊕	This organization's performance is above the performance of most accredited organizations
✓	This organization's performance is similar to the performance of most accredited organizations
⊖	This organization's performance is below the performance of most accredited organizations
N/D	Not displayed

FIGURE 13.20 ■ JCAHO Quality Rating System.

and comparison of the quality indicators identified in the strategic plan. EMS organizations should be able to take action to attain the quality targets identified in their plans, determine the success of their efforts, and when negative results occur, revise and restart their action plans for improvement. The efforts of initial QI committees should also begin to show benefits, typically in the areas of measuring and reporting **critical indicator** quality levels.

FULL INTEGRATION AND MONITORING

Once QI activities are fully implemented and a clear impact on the organization's work processes is underway, EMS personnel must feel empowered to take action in the work settings to identify, set up, and assess new patient-care methods and approaches. EMS crews can take self-correcting action by assessing timely information on performance levels for IAFF or NHTSA key quality indicators. At all levels, traditional EMS management gives way to leadership that helps the workforce maintain and improve the quality of its work.

At the state and regional level, there is less emphasis on regulatory inspection of local agencies and more emphasis on providing resources, comparative information, and coaching to local agencies that are accountable for developing and following their own strategic QI plans. A second set of eyes, such as a regulatory agency or hospital personnel involved in patient care, on a random and occasional basis provides an outside quality check to an EMS organization's quality processes.

JCAHO ACCREDITATION

Since 1951 the Joint Commission on Accreditation of Healthcare Organizations (JCAHO) has been the national leader in setting standards for health-care organizations. When a health-care organization seeks accreditation, it demonstrates a commitment to giving safe, high quality health care and to continually working to improve that care. Many of the JCAHO principles have been applied to EMS, and it is a natural migration for JCAHO standards to find their way from a physician medical director to EMS agency. JCAHO ratings are based on five categories. Figure 13.20 illustrates the JCAHO rating scale for health-care organizations.

A JCAHO standard is a generalized goal that is an achievable model of excellence and is used to define expectations. Expectations are based on key performance indicators that have been validated by evidence-based medicine. An indicator is an objective behavior or outcome that can be measured to determine compliance with a standard. A threshold is an established level or percentage of acceptable compliance that indicates when further evaluation should be initiated. When care drops below the threshold, it is considered a sentinel event.

JCAHO also requires a root-cause analysis on sentinel events as part of the accreditation process. Anytime there is a serious physical or psychological injury or medical intervention that results in a chance of an adverse outcome, it needs to be investigated. The root-cause analysis and an action plan must be submitted within 45 days of the event or of becoming aware of the event. This same concept is applicable to EMS systems, and often EMS regulatory agencies require similar time frames to report errors or serious compromises in quality.

JCAHO's 10-step monitoring and evaluation process serves as the model with which to manage quality of care provided in the EMS setting. That 10-step process is displayed in Figure 13.21.

1. Assign responsibility.
2. Delineate scope of care.
3. Identify important aspects of care.
4. Identify indicators.
5. Establish thresholds for evaluation.
6. Collect and organize data.
7. Evaluate care.
8. Take actions to solve problems.
9. Assess actions and document improvement.
10. Communicate relevant information to the organization-wide QA program.

FIGURE 13.21 ■ JCAHO Steps for Developing Medical Quality Improvement.

HUMAN-RESOURCES NEEDS

An ideal quality management system will first identify a system's strengths and develop the means to reinforce those strengths. An ideal system will go on to identify a system's weaknesses and develop means to correct those weaknesses as well as monitor the corrective action. Lastly, an ideal system will monitor how well established goals are met and assist in identifying new goals or areas requiring change.

To work internally, a CQI program requires trust between management and EMS system members. Collaboration should be incorporated into management of the quality process. A standing QI committee should be formed and managed by an EMS officer. The committee should be well rounded, because a comprehensive program will touch every aspect of the EMS operation (Figure 13.22). Basic-level providers, medical direction, EMS officers, training staff, and an outside representative can provide unique perspectives to any QI system. EMS members must be trained in QI activities and encouraged to participate.

More than half of the country's large EMS programs now have full-time quality activity managers and

- ◆ EMS officer (leadership role).
- ◆ Physician medical director (medical science support).
- ◆ Paramedics (representing each shift).
- ◆ EMTs.
- ◆ EMS educator.
- ◆ ER nurse representative.
- ◆ Information technologist or clerical.
- ◆ Union representative.
- ◆ Dispatch representative.

FIGURE 13.22 ■ Suggested Members of Quality Improvement Committee.

use a combination of in-house and outside monitoring and analysis. Any system must provide for feedback to all EMS members and an avenue for communication with system managers. Quality activity managers must facilitate feedback, make data available, and assist EMS first-line managers with training to make the system responsive. However, EMS members must remember that even the most responsive quality system is limited in what it can accomplish. EMS managers must review and prioritize all QI recommendations in order to implement those that will have the maximum impact on their EMS system.

Response time reliability and compliance with established performance criteria and standards are major issues for EMS managers and take significant resources. An EMS agency or division should have a dedicated QI specialist. If the agency does not have in-house people qualified to manage and conduct QI activities, one solution is to subcontract that operation to an outside consultant. A data system specialist also should be assigned to the QI division and be part of the committee efforts. The need for a computer specialist who understands relational databases and programming is critical to the mechanics of any EMS quality-improvement activity. Also clerical support for the QI process is a necessity. With HIPAA requirements and other related privacy issues, clerical support to maintain orderly and timely processing of the prehospital care report (PCR) and other related documents is important.

PROVIDER FEEDBACK AND ACCESS

The program needs to be communicated and every crew and employee needs to have everything they need to get started. A decision must be made as to when and how often reports are dispensed to the crews. EMS leaders must decide early in the program if the reports will be dispensed by area, crew, or department. A quality-improvement program needs management support and marketing, and the best way to achieve this is to communicate the program and its results.

Statistical analysis is the last of the most common activities conducted for QI techniques. This requires computer support and database management skills. Statistical analysis is more often a QI technique used to identify trends to improve on medication. Statistical analysis provides a direction for prospective and concurrent quality activities. Databases and data

- ◆ Number and type of EMS calls for service.
- ◆ Number of nonemergency and emergency transports.
- ◆ Number of patient refusals.
- ◆ Average scene time overall and per unit.
  - – Trauma patients.
  - – Medical patients.
- ◆ Average call time overall and per unit.
- ◆ Average response time overall and per unit.
  - – Emergency calls.
  - – Nonemergency calls.
- ◆ Unit hour utilization or in-service for EMS and first response vehicles.
- ◆ Reports on specialty or studied calls (trauma scene times, cardiac arrest, stroke, and so on).

FIGURE 13.23 ■ Monthly Quality Improvement Information.

collection vary greatly from agency to agency. Monthly and quarterly reports are needed to provide tools for EMS managers to improve patient care and EMS operations. Three layers of reports should be produced by EMS managers detailing the QI activities of the EMS agency. On a monthly basis, the items in Figure 13.23 should be presented to management and EMS crews.

On a quarterly basis, statistical information can be analyzed to produce trends in patient care and EMS operations. Results from quarterly audits will need to be provided to the QI committee and the EMS medical director. Evaluation of those results by the QI committee and the EMS medical director will be provided to EMS leadership and the EMS chief.

Twice a year the statistics should be collected and published for a semi-annual report to the EMS, fire, or government boards. These reports are comprehensive and involve some survey information. Figure 13.24 presents some ideas for biannual QI activities.

- ◆ A report on interfacility, interagency, and employee evaluation of services.
- ◆ The results of medical studies as defined by the quality-improvement committee.
- ◆ The results of any audit that comes from the government regulatory or medical authority.
- ◆ The results of any ambulance inspections by the regulatory agencies.
- ◆ Advanced airway control success rates and cardiac arrest survival rates.
- ◆ Paramedic and EMT continuing education reports.

FIGURE 13.24 ■ Biannual Reporting of EMS Quality Improvement Activities.

LINKING RESULTS TO EDUCATION

The management focus should be on improvement through education. One common technique in which education is employed in the QI process is called academic detailing, often known as spot teaching. This involves showing and not telling a team, crew, or group the issues surrounding the activities; for example, patient care charts, labs, discharge data, or proof of performance. This kind of hard data supports statements made by EMS managers and removes the possibility of personal or opinionated material. The organization should reward successes or provide incentives.

Educational programs need to be developed around the QI process. This education should begin with the new-hire process and then touch everyone in the system. Short education programs should be offered regularly. The basis for QI education should follow national, state, or local standards. Education programs or methods need to target QI hot topics and low-frequency high-risk activities. EMS workers can improve their lives through education and ever-broadening career and life opportunities. EMS needs not just good people; it also needs people who are growing through education and life experiences. Management, as well as members of the workforce, must continue to experience new learning and growth.

DISCIPLINE

When a medical quality issue results in discipline, it may indicate the QI process has failed at some point. This requires that clear policies and procedures are in place. Information should be factual, and a methodical approach and cooling off period should be used when discipline is necessary. All feedback should be documented and statistical analysis done on employees who do not comply with the QI programs. Progressive discipline with due process should be used, and a customized plan for individual improvement needs to be made with the physician medical director. Repeated failure to comply with QI programs or medical standards should ultimately result in decertification or reassignment.

It is important to ensure that an EMS provider's due-process rights are protected when he enters the discipline process. An employee must be presented with the information for which he is being investigated as well as advised of all rights, including union representation. When a situation is flagged by the QI system, in most cases it is an opportunity for coaching, education, and counseling by a supervisor or medical authority. Situations that involve repeated poor performance or gross negligence or that result in a potential loss to the organization are all possible scenarios that would place an EMS provider into the disciplinary process.

Attitude toward a patient situation or poor customer service fall under the responsibility of the chief's chain of command; whereas in a system that operates under an EMS medical director, the physician may revoke or suspend an EMS provider's credentials. The delegated practice acts in most EMS systems put the legal responsibility for paramedics under the license of a physician. Often employment requires certifications, and the loss of those certifications results in termination or demotion.

QI EFFORTS THAT FAIL

Many agencies attempt to provide QI activities but fail to achieve any results or make an impact on the system. Barriers to success can be divided into three levels: knowledge, attitudes, and behaviors. Often the failure is in the leadership, and the lack of management buy-in is reflected by leadership's attitude toward the program. A quality-improvement program can be seen as a cause for punitive actions to providers, which may cause EMS providers to develop an attitude that they need to "cover" themselves. This results in a lack of stakeholder involvement. In addition, many times a plan is not patient-driven and does not rely on evidence-based medicine. The lack of an established standard of care is a common problem in many areas of the country. Late feedback and not communicating the results devalues a QI program. Creating performance measures that are unattainable or that lack provider access to the results signals that the program is not important. Adult learners and employees in empowered organizations require feed-

---

> *Any data that may be directly linked to specific patients is confidential and absolutely must not be discussed with persons not associated with the Arlington EMS system. This is a confidential communication regarding medical staff peer review. As such, it is protected under Article 449b, Texas Revised Statues, Section 161.031, et Seq., Texas Health and Safety Code and 42 USC 11.111 et Seq., the Federal Health Care Quality Improvement Act.*

FIGURE 13.25 ■ Sample Privacy Statement for QI Material.

back. Quality-improvement programs need to be included in the strategic planning and should be reviewed periodically for effectiveness.

QI AND THE LAW

Some state courts protect the use of quality-improvement material and maintain its confidentiality. In many states QI material cannot be used in court, so protecting the process protects the participants. However, in other states EMS quality-improvement reports are not protected from legal discovery. States must follow their statues or laws protecting QI function unless the jurisdictions are involved in investigations in civil rights or other constitutional violations that require information from a QI process. It is important for EMS managers to educate themselves on their state and local laws (Figure 13.25).

All documents involved in EMS quality improvement need to be marked with legal terminology to keep information confidential. Patients should be identified with numbers and any information distributed should be collected and then shredded as per HIPAA requirements. Often HIPAA is used as an excuse to limit or prevent agencies from being able to conduct true peer-driven QI. The possibility of discovery during the QI process could result in the information entering the legal process.

CHAPTER REVIEW

Summary

All EMS organizations participate in quality management, some through well planned and monitored approaches and some through crisis management of sentinel events. It is the choice of the organization, in most cases, to define how it will handle quality management. As is the case with many of the nonemer-gency aspects of EMS, quality management is often relegated to a lower priority and is "tolerated" by an overworked management team. Quality management is an investment in preventive maintenance where a significant effort can be made up front before a potentially catastrophic failure occurs in the system. Quality management also offers a positive and nonpunitive way to focus the organization on continuous improvement, when done properly. It all starts with informing the leaders, managers, and members of the organization on the benefits, methods, and applications of quality management. If your organization is weak in quality management, seek out an experienced department to mentor your organization.

EMS has expanded from its narrowly focused beginning of impacting trauma deaths. Now EMS is involved in providing community health needs through injury prevention and screening programs. EMS is a part of a variety of special operations in an all-hazards environment, including hazardous materials, confined space rescue, and tactical EMS. Now EMS must be able to integrate services with other emergency agencies on a national level in incidents of national significance. The public expects—actually demands—that those services are of high quality. Webster defines quality as "a peculiar and essential character, an inherent feature, a degree of excellence, and superiority in kind." Providing quality EMS today requires far more equipment and a much broader education for both providers and managers than just a few years ago. Simply receiving grant money does not guarantee quality service.

Agencies must focus on programs and response procedures that provide quality services to both the community and the providers. EMS providers focus on providing quality help to others. Therefore, it is easy for them to place their individual needs second, particularly in significant operations. You need only look at the aftermath of the September 11 attacks on the World Trade Center to see the effects on many responders who worked in that environment for weeks. Quality management must be based on a needs assessment; ensure that the design, development, and implementation of both existing and new services are effective and efficient in terms of the system and its performance, and make the inherent safety and care of those who will deliver the service a priority.

Research and Review Questions

1. Contrast the QA and QI approaches, and discuss the motivations of each on the employees of an EMS organization.
2. Identify benchmarks from a Fortune 500 company and discuss their application to an EMS agency.
3. One of the methods defined by Edward Deming is called "quality circles." Research the concept and then describe its uses in EMS.
4. Identify an Internet source for each of the following: NHTSA performance indicators, JCAHO performance standards (EMS), and the IAFF EMS performance standards. Rank these groups in order of "most user friendly" for an EMS organization to access and gain information on quality management.
5. Identify three sources of EMS quality management software, and contrast the features of each.
6. Create a report of the statistics you would produce for your agency on a quarterly basis.
7. Design a targeted quality-improvement initiative for an issue identified in your organization.
8. Develop a key indicator for your agency that revolves around a common skill or procedure in EMS.

References

Andrews, S. L. (1991, January–February). QA vs. QI: The changing role of quality in health care. *Journal of Quality Assurance, 13*(1) 14–15, 38.

Balestracci, D., Jr. (1994). *Quality improvement: Practical applications for medical group practice.* Englewood, CO: Center for Research in Ambulatory Health Care Administration.

Bruegman, R. R. (2004). *Exceeding customer expectations: Quality concepts in the fire service.* Upper Saddle River, NJ: Pearson Education Inc.

Carey, R. C., & Lloyd, R. C. (1995). *Measuring quality improvement in healthcare: A guide to statistical process control applications.* New York: Quality Resources.

Davis, E. A. (1990, July–September). Methods of medical control and quality assurance used in EMS systems. *Prehospital and Disaster Medicine, 5*(3).

Emergency Medical Services Systems Act of 1973, Public Law 93–154, 93rd Cong. (1973).

International City/County Management Association. (1993, February). *Benchmarking: Achieving superior performance in fire and emergency medical services.* (MIS Report). Washington, DC: ICMA Press.

Jannig, J., & Sachs, G. (2003). *Achieving excellence in the fire service.* Upper Saddle River, NJ: Pearson Education.

Johnson, J. C. (1992). Introduction to quality improvement. In S. S. Polsky (Ed.), *Continuous quality improvement in EMS* (pp. 1–19). Dallas, TX: American College of Emergency Physicians.

McLaughlin, C. P., & Kaluzny, A. (2006). *Continuous quality improvement in health care* (3rd ed.). Sudbury, MA: Jones & Bartlett.

Narad, R. A. (1990). Emergency medical services system design. In W. R. Roush & P. B. Fontanarosa (Eds.), *Emergency medicine clinics of North America: EMS/prehospital care* (pp. 45–52). Philadelphia: W. B. Saunders, Harcourt Brace Jovanovich.

O'Leary, M. R., et al. (1994). *Lexikon: Dictionary of health care terms, organizations and acronyms for the era of reform.* Oakbrook Terrace, IL: Joint Commission on Accreditation of Healthcare Organizations.

Pande, P., & Holpp, L. (2002). *What is Six Sigma?* New York: McGraw-Hill.

Roush, W. R., & McDowell, R. M. (1989). Emergency medical services systems. In W. R. Roush (Ed.), *Principles of EMS systems: A comprehensive text for physicians.* Dallas: American College of Emergency Physicians.

Swor, R. A., & Bocka, J. J. (1990). A paramedic peer review quality assurance audit. *Prehospital and Disaster Medicine, 5*(3).

Swor, R. A., & Pirrallo, R. G. (Eds.). (2005). *Improving quality in EMS* (2nd ed.). Dubuque, IA: Kendall/Hunt Publishing Company.

Taylor, F. W. (1911). *The principles of scientific management.* New York: Harper & Brothers.

United States Fire Administration. (1993). *EMS safety techniques and applications.* Emmitsburg, PA: Author.

CHAPTER 14

Incident Management

Accreditation Criteria

CAAS

CAAS does not offer accreditation criteria for the topic of this chapter.

Objectives

After reading this chapter, the student should be able to:

14.1 Apply the National Incident Management System to a mass-casualty incident (pp. 342–344).
14.2 Describe the major components of an incident-management system (pp. 341, 344).
14.3 Describe the functions of the incident commander at all EMS incidents (pp. 341–342).
14.4 Describe the federal typing of the EMS resources (pp. 345–352).
14.5 Identify the component of an EMS strike team and EMS task force (pp. 349–352).
14.6 Map the federal requirement and resources for NIMS training and EMS (pp. 347–348).
14.7 Differentiate between types of incident management teams and their applications (pp. 340–341).
14.8 Develop an incident-management system for an EMS incident (pp. 353–354).

Key Terms

area command (p. 343)
branches (p. 344)
command staff (p. 342)
division (p. 343)
EMS task force
    (p. 349)
FIRESCOPE (p. 342)
general staff (p. 342)
group (p. 343)
incident action plan (IAP)
    (p. 341)

incident commander (IC)
    (p. 341)
incident management (p. 340)
incident-management team
    (IMT) (p. 340)
mass gathering (p. 352)
mass-casualty incidents
    (p. 342)
National Wildfire Coordi-
    nating Group (NWCG)
    (p. 340)

patient-transportation
    group supervisor (p. 344)
predetermined response
    matrix (p. 349)
rescue/extrication group
    supervisor (p. 344)
treatment unit leader
    (p. 343)
unified command (UC)
    (p. 343)

Point of View

In many communities, as EMS providers we spend time annually practicing for a bus accident or plane crash, when in reality the most likely major incident is a multi-patient auto collision involving severe entrapment. The typical response involves fire, police, EMS, and many rescuers. If your experience is like that of many others, these incidents can quickly turn into a three-ring circus. Frankly, there is no excuse for the type of chaos that often develops. The reality is if your day-to-day system for managing small multi-agency, multi-patient events results in chaos, it won't get any better with larger-scale ones. When the big one hits, you will only do what you do every day. The solution is to aggressively practice the principles and techniques of incident command at small rescues/MCIs, so when the big one occurs, it's only an extension of daily practice. It's a pretty sad commentary that as emergency responders we will practice for esoteric, once-in-a-career emergencies, when our performance at smaller, common emergencies leaves much to be desired.

In many respects the fire service is light-years ahead of EMS in its ability to function during tactical or rescue operations. Nearly every working fire requires a high degree of organization, working with multiple units and large numbers of people. Firefighters are often well versed in local operating procedures for fire-ground operation. They also have roles well defined by their position on the responding apparatus. They are used to following orders. Many EMS providers are more "free spirited." EMS usually responds to calls and deploys personnel to emergencies, with resourcefulness and free thinking encouraged. In addition the routine direct supervision of EMS personnel at calls is rare. Consequently, the majority of EMS personnel are not used to taking orders or working in highly directed tactical or rescue operations. This problem becomes readily apparent during multiple-casualty incidents, where individual patients are treated by clusters of EMTs while others go untreated.

The bottom line is this: when the big one hits, you will only do what you do every day. To become proficient in complicated tactical and rescue operations, EMS needs to have the basics of the incident-command system woven into daily practice. Command and control must be part of the routine for all types of emergencies; otherwise we will simply do what is comfortable or familiar. EMS personnel will freelance rather than take the disciplined, highly organized approach necessary for success. The agencies and personnel who are very good at ICS are the ones who use it at everyday incidents—because they use it every day!

The one factor most commonly associated with incident fiascoes is the lack of strong, visible command. Good command saves lives! The best example is the relatively small number of EMS fatalities on 9/11/01. Given the magnitude of the response, many more lives would have been lost if not for FDNY EMS command and control of their people. They established a perimeter, gave orders, and prevented the death or serious injury of many EMS responders.

Jon Politis, MPA, NREMT-P
Chief, Colonie (NY) EMS Department

■ NATIONAL INCIDENT MANAGEMENT SYSTEM

In Homeland Security Presidential Directive-5 (HSPD-5), President Bush called on the secretary of the Department of Homeland Security (DHS) to develop a national incident management system to provide a consistent nationwide approach for federal, state, tribal, and local governments to work together to prepare for, prevent, respond to, and recover from domestic incidents, regardless of cause, size, or complexity.

On March 1, 2004, after close collaboration with state and local government officials and representatives from a wide range of public-safety organizations, DHS issued the National Incident Management System (NIMS). It incorporates many existing best practices into a comprehensive national approach to domestic incident management, applicable at all jurisdictional levels and across all functional disciplines.

The NIMS represents a core set of doctrines, principles, terminology, and organizational processes to enable effective, efficient, and collaborative

incident management at all levels. To provide the framework for interoperability and compatibility, the NIMS is based on a balance between flexibility and standardization. The recommendations of the National Commission on Terrorist Attacks Upon the United States (the "9/11 Commission") further highlight the importance of using the NIMS. The commission's report recommends national adoption of the NIMS to enhance command, control, and communications capabilities.

BENEFITS OF THE NIMS

The NIMS provides a consistent, flexible, and adjustable national framework within which government and private entities at all levels can work together to manage domestic incidents, regardless of their cause, size, location, or complexity. This flexibility applies across all phases of incident management: prevention, preparedness, response, recovery, and mitigation.

The NIMS provides a set of standardized organizational structures, including an incident command system (ICS), multi-agency coordination systems, and public-information systems, as well as requirements for processes, procedures, and systems to improve interoperability among jurisdictions and disciplines in various areas. There are six components of the NIMS; command and management (including ICS) is only one of them.

DHS recognized that the overwhelming majority of emergency incidents is handled on a daily basis by a single jurisdiction at the local level. However, there will be instances in which successful domestic incident-management operations depend on the involvement of emergency responders from multiple jurisdictions, as well as personnel and equipment from other states and the federal government. These instances require effective and efficient coordination across a broad spectrum of organizations and activities.

The success of the operations will depend on the ability to mobilize and effectively utilize multiple outside resources. These resources must come together in an organizational framework that is understood by everyone and must utilize a common plan, as specified through a process of incident action planning. This will be possible only if all the response entities unite, plan, exercise, and respond using a common system, the NIMS.

Types of Command Teams

In August 2003 the U.S. Fire Administration (USFA) convened a focus group of stakeholders and experts from across the country to best determine the means to develop all-hazard **incident management teams (IMTs)** across the country. In the wildland fire community, the USFA and the **National Wildfire Coordinating Group (NWCG)** recognized five types, or levels, of IMTs. The focus group agreed to stay with this model for the all-hazard emergency-response community. The IMT types, including certifying level and basic makeup, as recommended by the focus group are:

* *Type 5: Local Village and Township Level.* Consists of emergency-response providers from a small- to medium-size municipality or a group of smaller jurisdictions who are part of a mutual-aid agreement. It is envisioned that Type 5 IMTs would be developed in, but not limited to, areas serviced by smaller volunteer or combination departments that individually may not have adequate resources but jointly could support an IMT. Type 5 would in most cases respond and operate within the jurisdictional boundaries of those communities that are signatories to the agreement. The responsibility for certifying the readiness of this IMT shall reside with the local authority having jurisdiction (AHJ) or its designee.
* *Type 4: City, County, or Fire-District Level.* Consists of emergency-response personnel from a larger and generally more-populated area, typically within a single jurisdiction. This level of IMT may be developed within a larger city, county departments, or fire districts. The membership will involve personnel from emergency-response and public-safety agencies or organizations within the jurisdiction. This team would primarily respond and operate within the city, county, or fire district having jurisdiction. The responsibility for certifying the readiness of this IMT rests with the county or regional authority having jurisdiction (AHJ) or their designee.
* *Type 3: State or Metropolitan-Area Level.* Consists of personnel from different departments, organizations, or agencies within a state or metropolitan region who have trained together to function as a team. The teams are intended to support incident management at incidents that extend beyond one operational period. Type 3 IMTs will respond and operate throughout the state or larger portions of the state, depending upon state-specified laws, policies, and regulations. The responsibility of certifying the readiness of this IMT rests with the state or with a regional council of government or its designee.

- *Type 2: National and State Level.* Consists of federally or state-certified personnel. This IMT has less staffing and experience than a Type 1 IMT, and is typically used on smaller-scale national or state incidents. Type 2 IMTs are currently in existence and operate through the U.S. Forest Service.
- *Type 1: National and State Level.* Consists of federally or state-certified personnel and is the most robust IMT with the most experience. These IMTs are fully equipped and self-contained. Type 1 IMTs are now in existence and operate through the U.S. Forest Service.

Elements of Effective Incident Management

To be effective, an incident management system must be suitable for use regardless of the type of jurisdiction or agency involvement. These may include single jurisdiction/single agency, single jurisdiction/multi-agency, and multi-jurisdiction/multi-agency involvement. The organizational structure must be adaptable to any incident or event, applicable and acceptable to users throughout a community or region, readily adaptable to new technology, and capable of logical expansion from the initial response to the complexities of a major emergency.

Common elements in organization, terminology, and procedures are necessary for maximum application of a system and use of existing qualifications and standards. In addition, they ensure the ability to move resources committed to the incident quickly and effectively with the least disruption to existing systems. Effective fulfillment of these requirements must be combined with simplicity to ensure low operational maintenance costs.

Application of Business Techniques

Tasks that business managers and leaders perform include planning, directing, organizing, coordinating, communicating, delegating, and evaluating. The responsibilities of the **incident commander (IC)** include gathering and evaluating information relative to planning and size-up, as well as development and communication of plans.

The IC must be involved with directing available resources to accomplish incident goals through operational and command responsibilities. To ensure proper incident management by coordination of overall operations of command, tactical operations, and support functions, a responsive organization must be developed. The IC must be able to communicate effectively within the organization and assess feedback from an entire incident. The use of terms that are understood by all resources is critical to the IC's ability to manage the incident.

Gathering and assigning resources functionally and geographically are also included in the IC's responsibilities. Overall effectiveness of the **incident action plan (IAP)** must be evaluated continually, based on the results of previous operational decisions. Using these data, the IC should modify the action plan as necessary. Although the IC may delegate functional authority, he should always retain ultimate responsibility for the incident. If the IC chooses not to delegate authority for one or more functions, he must perform the functions required by the incident.

Factors Affecting Incident Management

Although many similarities exist between business and incident management, several factors make incident management more difficult. One factor unique to the management of emergencies is danger. For example, while the number of deaths and injuries to EMS providers is not collected on a national basis, a total of state reports indicate that on average, 123 firefighter fatalities and 100,000 injuries (accounting for the loss of at least one workday) have occurred over each of the past 10 years. Since 1977 that statistic has held consistent at 120 per year, or one every three days. Danger to civilians is also a serious problem. In 1999, 3347 civilians died in fire incidents, and 95,000 were injured. Total property loss was $11.1 billion.

Untold numbers of responders have been exposed to toxic materials, resulting in immeasurable numbers of injuries and future pain and suffering. Among the health risks for responders are hepatitis B, AIDS, and other infectious diseases.

Incident management is carried out in a constantly changing environment—the situation may get better or worse, but it seldom stays the same. The dynamics of a constantly changing environment present additional challenges to the IC. Effectiveness of the incident action plan depends on factors that may be difficult to assess or confirm. Danger increases with the presence of hazardous contents within the involved buildings or vehicles. Dynamics of the incident may create difficulty in gathering accurate and current information, especially because of the limited time available at an incident scene. Additionally, emergency personnel reporting to the IC may not be able to judge the total picture.

If reports that follow an incident mention compromised responder safety, poor management of resources, or the inability to expand the command organization to meet the demands of the incident, these issues may have a negative effect on public perceptions about the department. The public has a right to expect that departments will be ready for any type of incident. Because there is no guarantee that adequate resources will be available for every incident, preparation to handle every incident with available resources, regardless of size or complexity, is needed.

Emergency-response personnel must consider the physical environment, command structure, and proper ICS procedures during planning. Incident outcomes may be forecast by thinking ahead about the situation while planning, as well as during an incident. At major incidents a planning section is instituted to conduct this forecasting.

The complexity of an incident complicates overall incident management. Command activities include strategic goal setting, developing and implementing action plans, controlling and coordinating incident operations, using all available resources, considering safety in decision making, providing logistical support, and evaluating the action plan. In addition to stabilizing an incident, the IC also is responsible for managing or delegating medical treatment, acting as a liaison with other agencies, ensuring the safety of personnel, and responding to media requests.

Multiple priorities of life safety, incident stabilization, and property conservation must be maintained, often with limited resources. Property-conservation considerations include impacts on structures and the environment. Some examples of incidents with complex problems include a **mass-casualty incident**, structure fires with serious rescue problems, hazardous-materials incidents requiring major evacuation, and wildland fires extending from one jurisdiction to another.

Complicating matters further, interagency cooperation may be required from mutual-aid departments, local utility companies, law-enforcement agencies, public-works departments, public-health departments, and state and federal agencies. Time constraints may cause confusion. Where business managers may have weeks or months to devise strategies, the IC has only seconds. Persons calling for assistance are sometimes unable to describe the scope of the incident fully. If public-safety communicators (dispatchers) do not understand the value of information received, they may unknowingly withhold critical facts and information. Command officers sometimes fail to gather information from all sides of the incident or to access plan information.

NIMS Incident-Command System

The development of new branches of government such as the Department of Homeland Security and the Office of Domestic Preparedness identified a need for a national direction on a specific command system to manage large-scale events. With the exception of the way the intelligence function is handled, the principles and concepts of NIMS are the same as similiar systems, including Firefighting Resources of California Organized for Potential Emergencies or **FIRESCOPE**, the incident-management system (IMS), and the National Fire Academy's ICS. The application of "national" implies a mandate that the NIMS will be necessary for all emergency and government responders.

ICS Management Characteristics

As part of NIMS, the incident command system (ICS) is based on proven management tools that contribute to the strength and efficiency of the overall system. The ICS management characteristics taught by the Department of Homeland Security (DHS) in its ICS training programs are displayed in Figure 14.1.

ICS Command Staff

Command comprises the incident commander (IC) and **command staff**. Command-staff positions are established to assign responsibility for key activities not specifically identified in the **general staff** functional

---

- ◆ Common terminology.
- ◆ Modular organization.
- ◆ Management by objectives.
- ◆ Reliance on an incident action plan.
- ◆ Manageable span of control.
- ◆ Predesignated incident mobilization center locations and facilities.
- ◆ Comprehensive resource management.
- ◆ Integrated communications.
- ◆ Establishment and transfer of command.
- ◆ Chain of command and unity of command.
- ◆ Unified command.
- ◆ Accountability of resources and personnel.
- ◆ Deployment.
- ◆ Information and intelligence management.

FIGURE 14.1 ■ Components of an EMS System.

elements. These positions may include the public-information officer (PIO), safety officer (SO), and the liaison officer (LNO), in additional to various others as required and assigned by the IC.

Unified Command

Unified command (UC) is an important element in multi-jurisdictional or multi-agency domestic incident management. It provides guidelines to enable agencies with different legal, geographic, and functional responsibilities to coordinate, plan, and interact effectively. Because it is team-based, unified command overcomes much of the inefficiency and duplication of effort that can occur when agencies from different functional and geographic jurisdictions or agencies at different levels of government operate without a common system or organizational framework. The primary difference between the single-command structure and the UC structure is that in a single-command structure, the IC is solely responsible for establishing incident management objectives and strategies. In a UC structure the individuals designated by their jurisdictional authorities jointly determine objectives, plans, and priorities and work together to execute them.

General Staff

The general staff includes incident management personnel who represent the major functional elements of the ICS, including the operations section chief, planning section chief, logistics section chief, and finance/administration section chief. Command staff and general staff must continually interact and share vital information and estimates of the current and future situation and develop recommended courses of action for consideration by the IC.

Area Command

Area command is activated only if necessary, depending on the complexity of the incident and span-of-control considerations. An area command is established either to oversee the management of multiple incidents that are being handled by separate ICS organizations or to oversee the management of a very large incident that involves multiple ICS organizations (Figure 14.2). It is important to note that area command does not have operational responsibilities.

- ✦ Sets overall agency incident-related priorities.
- ✦ Allocates critical resources according to established priorities.
- ✦ Ensures that incidents are managed properly.
- ✦ Ensures effective communications.
- ✦ Ensures that incident-management objectives are met and do not conflict with each other or with agency policies.
- ✦ Identifies critical resource needs and reports them to the emergency operations center(s).
- ✦ Ensures that short-term emergency recovery is coordinated to assist in the transition to full recovery operations.
- ✦ Provides for personnel accountability and a safe operating environment.

FIGURE 14.2 ■ Area Commander Responsibilities.

Divisions and Groups

The terms **division** and **group** are common designators used in the NIMS to define tactical-level management positions in the command organization. Divisions represent geographic responsibilities; for example, Branch C (the rear of the facility). Groups represent a functional (job) responsibility, such as the treatment group. The use of divisions and groups in the command organization provides a standard system to divide the incident into smaller, more manageable elements.

As initial assignments are ordered to incoming resources, the IC should begin assigning EMS officers to appropriate division or group responsibilities. By doing this at all small incidents, the department is preparing itself to manage effectively the resource-intense incidents that occur much more sporadically. Divisions and groups are under the control of a "supervisor" (e.g., medical group supervisor or transportation group supervisor). Units are controlled by a "leader" (such as a **treatment unit leader** or triage unit leader).

Complex emergency operations often exceed the capacity of one officer to manage the entire operation effectively. Divisions and groups reduce the span of control to more manageable, smaller-sized elements. Span of control is the number of people or units that report to a single supervisor. This allows the IC to communicate principally with persons in these organizational positions, rather than individual resources or members (hence, the IC has control of communications). The number of divisions or groups that can be managed effectively by the IC varies. Normal span of control is three to seven, but in fast-moving, complex operations, a span of control of no more than five divisions or groups is indicated. In slower, less-complex operations, the IC may be able to handle more.

Branches are an organizational level having functional or geographic responsibility for major aspects of incident operations. A branch is organizationally situated between the operations section chief and the divisions or groups in the operations section, and between the logistics section chief and the units in the logistics section. Branches are established to assist in maintaining the proper span of control during major incidents when there are numerous divisions or groups operating.

NIMS ICS AND EMS

Triage, Treatment, and Transportation

The ICS is used during all EMS incidents, even if it does not appear to be. Someone is always in charge, although he typically is not referred to as the incident commander. Remember, a main premise of the ICS is that there is an incident commander at all incidents who is responsible for evaluating priorities, establishing strategic goals and tactical objectives, identifying resource needs, developing an incident action plan, making sure the plan is carried out, and evaluating the results of the action. This holds true for all EMS incidents, whether the total response to the incident is two EMTs in an older-model ambulance or involves hundreds of responders from dozens of agencies, complete with state-of-the-art command vehicles.

At major EMS incidents, whether multiple-casualty incidents (MCIs) or complicated incidents with only a few patients, the IC normally establishes the rescue/extrication group position early in the incident. It often is assigned to the first resource in the area, and the **rescue/extrication group supervisor** is responsible for managing the rescue of entrapped victims. Personnel operating within the extrication area generally do initial triage on the patients and coordinate the transport of patients to the treatment areas. Often they work hand in hand with the triage unit. The rescue/extrication group is operating within the hazard zone with potential risks to personnel and patients, and appropriate action should be taken to provide safeguards.

The medical group supervisor reports to the incident commander and supervises the triage unit leader, treatment unit leader, and medical-supply coordinator. The medical group supervisor establishes command and controls the activities within a medical group, in order to ensure the best possible emergency medical care to patients during an MCI.

The triage unit leader position is often filled shortly after the rescue/extrication unit leader, before the medical group supervisor position. This position is responsible for the triage and tagging of all patients at major incidents. Triage and tagging may take place either in the extrication area or at the entry to the treatment area. In either case, close coordination must be maintained with the treatment unit and the rescue/extrication group. Personnel assigned to triage must have the basic medical skills needed to make appropriate triage decisions.

The treatment unit leader position typically is the next to be established. He will establish an area where patients can be collected and treated. Central treatment areas maximize the limited resources of rescuers in incidents that involve large numbers of patients. It is in the treatment area that extensive treatment and advanced life support care are conducted. The treatment unit leader is responsible for the overall management of patient-care delivery in the treatment area.

The **patient-transportation group supervisor** has a substantial challenge. He is responsible for the overall management of patient movement from the scene to the receiving hospitals. He must obtain all required transportation, and he must cause patients to be transported to the appropriate facilities. Hospitals will need to be notified, and there will be an almost continuous flow of radio communication between the group and the receiving hospitals (either by direct radio communications from the scene to the hospital or by relay through a dispatch center).

Special attention must be given to the needs of the patient and decisions made as to whether transport to a specialty center is required. In addition to patient condition, the receiving hospital's ability to handle additional patients and the overall impact on the EMS system must be part of the decision-making process on patient destinations.

With this basic command organization in place, additional arriving resources are assigned to existing divisions, groups, and units (Figure 14.3). The basic IMS for EMS can get an EMS manager or incident commander through 80% of the multiple-casualty incidents. These additional resources work for, report to, and communicate to the division/group supervisor or unit leader.

Air Operations at EMS Incidents

It may be helpful to designate landing zones (LZs) for any air evacuations that may be necessary. Good incident management preplanning, for both planned and spontaneous events, should always address the unexpected need for helicopter landing spots. For example,

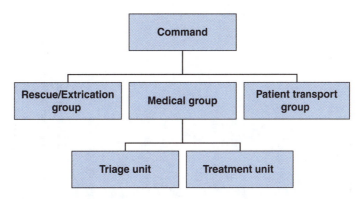

FIGURE 14.3 ◼ Basic Incident Management Structure for EMS Incidents.

most large-scale EMS incidents will require medical evacuations as the transportation system becomes choked with spectator traffic and as law enforcement deploys SWAT teams at events or for VIP access and egress by helicopter. Using a global positioning system (GPS) unit to predesignate landing zones, noting the latitude and longitude for each site on the plan, should be part of incident and target-hazard preplanning.

Helicopter landing zones are managed by helispot managers and are temporary landing and takeoff areas. At short-term incidents an incident commander may call this a landing-zone group, and the group leader has the same responsibilities. This group function is often an engine company that has the capability to secure the landing site and water down a landing zone to prevent dust and debris from the rotor wash as the helicopter takes off and lands. When dealing with air operations at an EMS incident, managers should ensure that personnel are familiar with the responsibilities of the helispot and helibase manager (Figure 14.4).

The most dangerous time for a medical air operation is the takeoff and landing of the helicopter. The primary LZ—the one closest to the site that has refueling capabilities—is considered the helibase and is managed by a helibase manager. This portion of the preplanning necessitates meeting with the agency or company providing the aero-medical services.

Helicopter-resource capabilities have been designated by the federal government and are displayed in Table 14.1. Each team or unit can work a maximum of one12-hour shift, depending upon individual policies and procedures. Fuel tankers or other supply points must be identified. Backup supplies and some equipment may be required, depending upon number of patients and type of event. Communication equipment may be programmable for interoperability but must

be verified. Communication frequencies to ground incident command must be provided.

EMS officers should plan for augmenting existing communication equipment. Landing zones (space, clearance, and weight restrictions) must be considered. The typical civilian air ambulance requires an LZ of 150′ × 150′. Ground safety assurance and traffic control are important support requirements for injury and crash prevention. This support may be significant depending upon the size and the location of the incident.

If the incident will have multiple helicopters and returning flights, the incident commander may elevate the air operations to a branch designation. Under an air-operation branch leader are the helicopter

---

Helispot Manager

- Report to helispot and coordinate with helibase.
- Inform helibase of helispot activity.
- Coordinate air traffic control.
- Ensure dust abatement and slope parameters are followed and that rotor clearance is sufficient.
- Manifest patient, personnel, and cargo loading.
- Call out benchmarks and record of helicopter activities on the radio.
- Coordinate with pilots for loading and unloading and safety problems.

Helibase Manager

- Brief pilots.
- Ensure helibase is posted and cordoned off.
- Coordinate helibase air traffic.
- Ensure fueling and repair services are identified.
- Ensure dust abatement.
- Supervise loading of personnel and cargo.
- Ensure security.
- Ensure aircraft firefighting services are provided.
- Display and organize helispots and radio frequencies.

FIGURE 14.4 ◼ Helispot Manager Responsibilities.

TABLE 14.1 ■ Federal Classification of Helicopter Ambulances

Resource: Air Ambulance (Rotary-Wing)

Category: Health & Medical (ESF #8)   **Kind:** Aircraft

Minimum Capabilities

Component	Metric	Type I	Type II	Type III	Type IV	Other
Team	Care provided	Advanced Life Support	Advanced Life Support	Advanced Life Support	Advanced Life Support	
Personnel	Minimum staff	Same as Type II	Same as Type III	3 pilot 2 paramedics or 1 paramedic and 1 nurse or physician	2 pilot 1 paramedic	
Team	Transport	Same as Type II	2 or more litter patients	Same as Type IV	1 litter patient	
Aircraft	Rotary-wing with these capabilities	Same as Type II, plus Full SAR including hoist capabilities	Night operations IFR	Same as Type IV	Night operations VFR	
Equipment		ALS ambulance equipment	Same as Type III	Ability to deploy a medical team; MICU equipment (i.e., ventilators & infusion pumps, medications, blood)	ALS ambulance equipment	
Comments:	Emergency medical services team with equipment, supplies, and aircraft for patient transport & emergency out-of-hospital medical care.					

- Each team/unit can work a maximum of 12-hour shifts, depending upon individual policies & procedures.
- Aircraft maintenance requirements may occur during deployment. Aviation maintenance must be planned. Hangar facilities should be planned for all extended operations. Fuel tankers or other supply points must be identified. Backup supplies and some equipment may be required depending upon number of patients and type of event.
- Communication equipment may be programmable for interoperability but must be verified. Provide communication frequencies of ground incident command. Plan for augmenting existing communication equipment.
- Landing zones (space, clearance, and weight restrictions) must be considered. The typical civilian air ambulance requires an LZ of 150' × 150'.
- Ground safety assurance and traffic control are important support requirements for injury and crash prevention. This support may be significant depending upon the size of the incident and the location of the incident.

Source: Federal Emergency Management Agency (May, 2005). FEMA 508-3 Typed Resource Definitions—Emergency Medical Services Resources. www.fema.gov

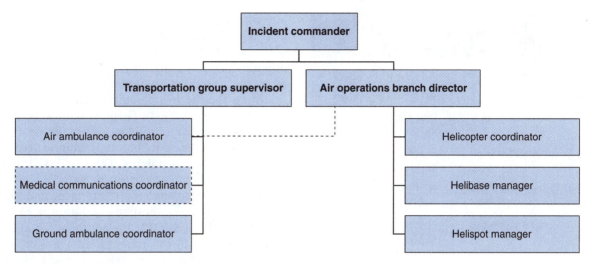

FIGURE 14.5 ■ Air Operations at an EMS Incident.

coordinator, helibase manager, and helispot managers. The air-operations branch will speak with the air-ambulance coordinator under the transportation group supervisor. The components of the air-operations branch are shown in Figure 14.5.

NIMS ICS versus FIRESCOPE ICS

The ICS organization has five major functions: command, operations, planning, logistics, and finance and administration. In the NIMS ICS a potential sixth functional area to cover the intelligence function can be established for gathering and sharing incident-related information and intelligence.

The information and intelligence function provides analysis and sharing of information and intelligence during an incident. Intelligence can include national security or classified information but also can include operational information such as risk assessments, medical intelligence, weather information, structural designs of buildings, and toxic-contaminant levels. Traditionally, information and intelligence functions are located in the planning section. In exceptional situations, however, the IC may need to assign this role to other parts of the ICS organization. Under the NIMS ICS, the intelligence and information function may be assigned in only one of the following ways:

* Within the command staff.
* As a unit within the planning section.
* As a branch within the operations section.
* As a separate general-staff section.

DHS and ICS

One of the first steps in becoming compliant with the NIMS requires states and local governments to institutionalize the use of ICS (as taught by the DHS) across the entire response system. This means that ICS training must be consistent with the concepts, principles, and characteristics of the ICS training offered by the various DHS training entities. ICS training courses need not be taught by a DHS employee or at a DHS facility, although they can be. Organizations that are developing ICS training courses should be sure to review their materials and revise them if they are not consistent with DHS concepts and principles.

Available NIMS ICS Training

According to the NIMS, ICS training should be developed around the National Wildfire Coordinating Group (NWCG) model. The minimum training for first responders in the NWCG model is detailed in Table 14.2.

DHS, through its many training bodies, makes ICS training available. The National Fire Academy (NFA) and the Emergency Management Institute (EMI), both part of DHS/FEMA's U.S. Fire Administration, offer training developed around the NWCG model. NFA works closely with NHTSA's EMS Division to ensure that the needs of prehospital EMS providers are met. Both NFA and EMI work closely with various offices within the U.S. Department of Health and Human Services, including the Centers for Disease Control and Prevention, to ensure that the training needs of hospital emergency department staff are met.

TABLE 14.2 ■ Incident Management Training	
ICS-100, Introduction to ICS	All First Responders
ICS-200, Basic ICS	All First Responders
ICS-300, Intermediate ICS	First Responder Supervisors
ICS-400, Advanced ICS	First Responder Supervisors
ICS-700, Orientation to the NIMS	All First Responders
ICS-800, Federal Response Plan	First Responder Supervisors

NFA and EMI offer ICS train-the-trainer classes at their National Emergency Training Center in Emmitsburg, Maryland, and the Noble Hospital Training Center in Anniston, Alabama; and through state fire, EMS, and emergency management training offices. A variety of other ICS training programs are also available. The NIMS Integration Center is working with federal and state training providers to ensure that their ICS course offerings are consistent with the NIMS.

Incidents involving multiple casualties produce a tremendous amount of stress on EMS crews; for that reason the skills and abilities of EMS leaders to manage an incident are important to reduce crew stress, maximize patient outcomes, and ensure the public trust. The National Fire Academy has a course available entitled Incident Command for Emergency Medical Services. It is a 16-hour curriculum designed to assist EMS providers in managing an incident. Web-based courses are available on the Virtual Campus through FEMA and provide various levels of incident-management training. Regional and national conferences now include an incident-management course routinely in their programs.

According to the NIMS Integration Center, responders who have already been trained in ICS do not need retraining if their previous training is consistent with DHS standards. Since NIMS ICS is based on FIRESCOPE and the National Interagency Incident Management System (NIIMS), any training developed or provided by FIRESCOPE and NIIMS is consistent with NIMS ICS.

The Future of NIMS ICS Training

Over time the NIMS Integration Center will continue to define the critical components of NIMS ICS, and training providers should update their courses accordingly. With so many training bodies and compa-

nies offering ICS training, it will be impossible in the near term for the center to certify each training program as "NIMS ICS compliant." But the center will provide NIMS ICS training and make the training materials available to others who offer ICS training.

More specific ICS modules, such as those developed by FIRESCOPE to facilitate the use of ICS in situations other than wildland fires, should be reviewed and updated to become additional components of NIMS ICS training. The FIRESCOPE ICS modules include multiple-casualty, hazardous materials, high-rise, wildland/urban interface, and urban search-and-rescue applications. As groups like the NWCG and FIRESCOPE update their ICS training modules, the NIMS Integration Center will be an active participant. Those ICS modules can form the basis for a suite of ICS training materials in which responders from all disciplines and at all levels of government can learn how to fit into the ICS structure and how to work with other responders.

Updates and revisions to existing ICS training modules should include the modifications necessary to allow for multiple methods of delivery. To ensure that all responders adopt and use ICS, ICS training must be provided in numerous ways. Classroom instruction, field training, independent study, and distance learning are all valuable training methods. The more materials and options that the center and its partners, the training providers, can provide, the more responders will be trained to use ICS.

ICS training also should encourage and support integrated training opportunities, where law-enforcement, fire, public-health, emergency medical, emergency-management, and public-works personnel from a jurisdiction are trained together on using ICS. While the response disciplines may need specific tools and training to understand how they fit into the ICS structure, everyone should learn the same incident command system.

Mass-Casualty Terminology

Under the NFPA 450 standard, a catastrophic mass casualty is a situation with 100 or more casualties that significantly overwhelms available emergency medical services, facilities, and local resources. An EMT's first exposure to the incident management system will likely occur at a multiple-patient incident. Several important definitions categorize events that involve various needs and demands on the EMS system.

Multiple-casualty incidents (MCIs) are events that generate several victims and that often make a traditional EMS response ineffective. A small medical incident is one involving a few patients but that can be managed within the agency. Research shows that most small incidents generate between 5 and 10 patients and account for 80% of MCIs. An expanded medical incident requires a regional response. This is broken into three categories: a major medical response requiring regional or multi-regional resources; a disaster, which requires state- and federal-level resources; and a catastrophe, which requires international resources.

A natural disaster is one that is weather related, such as a flood, earthquake, tropical storm, tornado, extreme heat or cold, and pandemic events. Man-made disasters include building collapses, fires, explosions, transportation accidents, terrorism, and civil disorder.

An MCI will need to be managed with a heightened response, including the requesting of additional resources from other agencies. When agencies are called upon by other jurisdictions, the request is for mutual aid. Multiple-casualty incidents typically do not overwhelm the hospital capabilities of a jurisdiction or region but may exceed the capabilities of one or more hospitals within a locality. There is usually a short, intense peak demand for health and medical services, unlike the sustained demand for these services typical of mass-casualty incidents.

An EMS agency should plan to designate in the dispatch center a **predetermined response matrix**, which sends a specific group of resources to an incident. For example, Phoenix Fire dispatches two ambulances, two engine companies, a truck company, and a chief officer to incidents involving multiple patients. Some agencies have developed a first-alarm medical assignment similar to a first-alarm fire assignment.

By definition mass-casualty incidents involve sufficient casualties to overtax current resources. The nature of incidents can vary widely, will require regional resources, and will likely receive federal resources (Figure 14.6). Often these events may require extended operations and require the incident commander to put in place an infrastructure to meet the needs of the incident. Some EMS systems have specified the number of patients that designates a mass-casualty event. Check with your local authority for those guidelines. Examples of multiple- or mass-casualty incidents include those listed in Figure 14.6.

■ RESOURCE ALLOCATION

When preparing for a response to an MCI, the federal government has now categorized EMS resources into single resources, ambulance strike teams, and **EMS task**

- ◆ Major vehicular crash with multiple victims.
- ◆ Fires with burn or smoke-inhalation victims.
- ◆ Environmental disasters.
- ◆ Public-transportation accidents (aircraft, train, bus, and so on).
- ◆ Mining or construction accidents.
- ◆ Industrial incidents.
- ◆ Building collapses.
- ◆ Hazardous-materials incidents.
- ◆ Chemical, biological, radiological, nuclear, or explosive (CBRNE) incidents.
- ◆ Planned events, such as celebrations, parades, concerts.
- ◆ Sporting events, such as college or high school football, basketball, or major PGA or LPGA golf tournaments.

FIGURE 14.6 ■ Types of Multiple- and Mass-Causality Incidents.

forces (Table 14.3). As a single resource, ambulances have been identified by type and capability. Each ambulance team will need to be able to work 12-hour shifts (Figure 14.7). Backup supplies and some equipment required according to the number of patients and the type of event may be necessary; for example, backboards, oxygen, or added medications. Communications equipment must be programmable for interoperability. Fuel supply and maintenance support must be available. Environmental considerations related to temperature control in the patient-care compartment and pharmaceutical storage may be necessary for locations with excessive ranges in temperature. Security for the vehicle may be required for periods of standby, when the crew cannot be in attendance. Decontamination supplies and support for crews may be required for responses to incidents with the potential for infectious disease.

An ambulance strike team is a group of five ambulances of the same type with common communications and a leader (Table 14.4). It provides an

FIGURE 14.7 ■ Successful Management of EMS Incidents Requires Organization of Resources.

TABLE 14.3 ■ Ambulance Resources

Resource: Ambulance (Ground)

Category: Health & Medical (ESF #8) **Kind:** Team

Minimum Capabilities: Component	Metric	Type I	Type II	Type III	Type IV	Other
Team	Care provided	Advanced Life Support	Advanced Life Support	Basic Life Support	Basic Life Support operations	Non-transporting emergency medical response
Personnel	Minimum staff	2 paramedic and EMT	2 paramedic and EMT	2 EMT and first responder	2 1 EMT and first responder	1
Vehicle	Transport	2-litter patients	2-litter patients	2 litter patients	2 litter patients	
Personnel	Training and equipment	Same as Type III	Non-HazMat response	Meets or exceeds standards as addressed by EPA, OSHA and NFPA 471, 472, 473 and 29 CFR 1910, 120 ETA 3-11 to work in HazMat Level B and specific threat conditions All immunized in accordance with CDC core adult immunizations and specific threat as appropriate		BLS or ALS equipment/supplies
Comments:		Emergency medical services team with equipment, supplies, and vehicle for patient transport (Type I-IV) and out-of-hospital emergency medical care.				

- ◆ Each team unit can work 12-hour shifts. Backup supply and some equipment required according to number of patients and type of event.
- ◆ Communication equipment may be programmable for interoperability but must be verified. Plan for augmenting existing communication equipment.
- ◆ Environmental considerations related to temperature control in patient care compartment and pharmaceutical storage may be necessary for locations with excessive ranges in temperature.
- ◆ Security of vehicle support required for periods of standby without crew in attendance. Fuel supply and maintenance support must be available.
- ◆ Decontamination supplies and support required for responses to incidents with potential threat to responding services or transport of infectious patients.

Source: Federal Emergency Management Agency (May, 2005). FEMA 508-3 Typed Resource Definitions—Emergency Medical Services Resources. www.fema.gov

TABLE 14.4 ■ Ambulance Strike Teams

Resource: Ambulance Strike Team

Category:	Health & Medical (ESF #8)			Kind: Team		

Minimum Capabilities:

Component	Metric	Type I	Type II	Type III	Type IV	Other
Team	Scope of Practice	Advanced Life Support	Advanced Life Support	Basic Life Support	Basic Life Support	
Personnel	Minimum number	2 staff (paramedic and EMT) transport per ambulance	2 staff (paramedic and EMT) per ambulance	2 staff (EMT and driver) per ambulance	2 personnel (1 EMT and 1 driver) per ambulance	
Personnel	See Note 1	Same as Type III	Non-HazMat response	Meets or exceeds standards as addressed by EPA, OSHA, and NFP471, 472, 473, and 29 CFR 1910, 120 ETA 3-11 to work in HazMat Level B and specific threat conditions All immunized in accordance with CDC core adult immunizations and specific threat as appropriate		
Equipment	See Note 2	5 Type I Ambulances; Capable of transporting minimum of 10 litter patients total (2 per ambulance)	5 Type II Ambulances; Minimum capability of 10 litter patients	5 Type III Ambulances; Minimum capability of 10 litter patients	5 Type IV Ambulances; Minimum of 10 litter patients	
Personnel	Training See Note 3 See Note 4	ICS 300 HazMat FRO Course WMD Awareness Course 3 years of EMS experience				
Supply	Go-Pack See Note 5	X	X	X	X	X

Comments: An Ambulance Strike Team is a group of five ambulances of the same type with common communications and a leader. It provides an operational grouping of ambulances complete with supervisory element for organization command and control. The strike teams may be all ALS or all BLS.

Support elements needed include fuel, security, resupply of medical supplies, and support for a minimum of 11 personnel (if 2 crew per ambulance) or 16 (if 3 crew per ambulance). Temperature control support may be required for medical supplies in some environments. Vehicle maintenance support required.

Note 1: Can be deployed to cover 12-hour periods or 24-hour ops depending on number of ambulances needed at one time. Should be self-sufficient for 72 hours.

Note 2: Emergency Medical Services team with equipment, supplies, and vehicle for patient transport (Type I-IV) and out-of-hospital emergency medical care.

Note 3: Required training, ICS 100 and 200, Basic MCI Field Operations (8 hours).

Note 4: Strike Team Leader - Ambulance Course (8 hours), 1 year leadership experience in a related field.

Note 5: Equipment and supplies to meet minimum scope of practice (ALS or BLS). Equipment and supplies to meet minimum requirements of State agency that provides regulation.

Source: Federal Emergency Management Agency (May, 2005). FEMA 508-3 Typed Resource Definitions—Emergency Medical Services Resources. www.fema.gov

TABLE 14.5 ■ Federal Classification of Emergency Medical Task Force

Resource:		Ambulance Task Force					
Category:	**Health & Medical (ESF #8)**			**Kind: Team**			
Minimum Capabilities:		*Type I*		*Type II*	*Type III*	*Type IV*	*Other*
Component	**Metric**						
Personnel	Supervisor/Leader See Note 1	1					
Vehicle	Ambulances See Note 2	Any combination of different types of ambulances assembled for an EMS mission, with common communications & a leader.					
Personnel	Training	ICS 100 and 200					
		Basic MCI Field Operations (8 hours)					
		Task Force Leader-Ambulance Course (8 hours)					
		One year Leadership experience in a related field					
Comments:	Any combination of ambulances, within span of control, with common communications and a leader. This resource typing is used to distinguish between a Task Force of Ambulances and an Emergency Medical Task Force (any combination of resources).						
	Note 1: Must have own vehicle with communications capabilities - both enroute and at scene - to all other units under the leader's supervision.						
	Note 2: Emergency Medical Services team with equipment, supplies, and vehicle for patient transport (Type I-IV) and out-of-hospital emergency medical care.						

Source: Federal Emergency Management Agency (May, 2005). FEMA 508-3 Typed Resource Definitions—Emergency Medical Services Resources. www.fema.gov

operational grouping of ambulances complete with a supervisory element for organization command and control. The strike teams may be all ALS or all BLS. The four types of ambulance strike teams are based on ALS versus BLS capabilities and the level of hazardous-materials training. The crew will have to be certified to support services that are needed, including fuel, security, cache of medical supplies, and support for a minimum of 11 personnel (if two crew per ambulance) or 16 (if three crew per ambulance). Temperature control support may be required for medical supplies in some environments.

An EMS task force is a team comprised of five resources in any combination (within the span of control) of resources (such as ambulances, rescues, engines, squads) assembled for a medical mission, with common communications and a leader or supervisor

(Table 14.5). Only one type of EMS task force exists and must be self-sufficient for 12-hour operational periods, although it may be deployed longer, depending on need. Support elements needed are fuel, security, restock medical supplies, and support for a minimum of 11 personnel, depending on staffing of individual units. Vehicle-maintenance support is required. For instance, an EMS task force might comprise two ALS teams, three BLS teams and a supervisor.

■ PREPLANNING SPECIAL EVENTS

Multiple casualties can happen at planned events. Frequently as an EMT you may be assigned as a standby crew at a large event that is classified as a **mass gathering**—any collection of greater than 1000 people at one site or location. This applies to all types of events, including concerts and sporting events or other venues in which a large number of people are crowded into a fairly limited area that is somewhat isolated from EMS.

One factor that may affect patient volumes at these events is whether the setting is an indoor or outdoor event. Consider the 2005 visit from Pope John Paul II in Denver that triggered a mass-casualty event when temperatures soared and ill-prepared parishioners crowded together and suffered from heat exhaustion. Extremes in temperature, both hot and cold, can generate more patients. The access routes in and out of an incident or event should be taken into consideration. An airplane crash on Long Island, New York, was complicated by the one road into the site's being clogged with the personal vehicles of volunteer rescue personnel and the cars of sightseers.

The size of the event crowd also can dictate what generates calls. The term "crush load" describes when a mass of humans presses forward at physical barriers or narrow exit, suffocating some by imposing weight on the thorax and preventing them from taking a breath. Crush loading at events has killed hundreds over the years. Recent events include a Rhode Island nightclub, a WHO concert, and a European soccer event when a sudden surge of the crowd created fatalities from asphyxiation and traumatic injuries.

EMS managers may want to develop and implement municipal ordinances that control a venue's seating arrangement. Festival seating involves large crowds with unassigned seats, which can affect the mobility of a crowd and make the event prone to crush loading. This presents the need for more EMS resources, greater security, and the ability to move EMS crews through large crowds.

Certain events by their very nature suggest the type of patients that may be seen. For example, rock-concert attendees usually have emergencies involving drugs, alcohol, and minor trauma. Sporting events can produce major or minor traumatic injuries to the athletes. Environmental or heat exhaustion from extremes in temperature can affect the spectators and generate patients. The amount and coordination of when and to whom alcohol is being served is another key factor that EMS must take into account during incident management of a special event. The Emergency Management Institute (EMI) recommends one man-

ager for every 250 attendees during special-event coverage.

INCIDENT ACTION PLANS

Large-scale events will typically have a written or oral incident action plan (IAP) worked out beforehand. The incident commander will make the decision on whether the IAP is written or verbal. For large-scale incidents the planning section creates and puts together the written action plan with the approval of the incident commander. The IAP includes the overall incident objectives and strategies established by incident command (IC) or unified command (UC). The planning section is responsible for developing and documenting the IAP. In the case of UC, the IAP must adequately address the overall incident objectives, mission, operational assignments, and policy needs of each jurisdictional agency. This planning process is accomplished with productive interaction between jurisdictions, functional agencies, and private organizations. The IAP also addresses tactical objectives and support activities for one operational period, generally 12 to 24 hours. The IAP contains provisions for continuous incorporation of "lessons learned," as identified by the incident safety officer or incident management personnel as activities progress. The essential elements of any written or oral incident action plan are shown in Figure 14.8.

The incident action plan must be made known to all incident supervisory personnel. This can be done through briefings, by distributing a written plan prior to the start of the operational period, or by both methods. Official federal forms are used in a large incident when a written incident action plan is required and are excellent tools for use in EMS preplanning of scheduled events. The forms that are needed to complete an incident action plan are listed in Figure 14.9.

- Statement of objectives appropriate to the overall incident.
- Organizational chart that describes what aspects of the ICS will be staffed.
- Operational period.
- Assignments to accomplish the objectives, which are normally prepared for each division or group and include the strategy, tactics, and resources to be used.
- Supporting material, such as a map of the incident, communications plan, medical plan, traffic plan, and so on.

FIGURE 14.8 ■ Components of an Incident Action Plan (IAP).

- ◆ Incident objectives: ICS Form 202.
- ◆ Organization assignment: ICS Form 203.
- ◆ Assignment list: ICS Form 204.
- ◆ Supporting plans: ICS Forms 205, 206, and so on.
- ◆ Sign-in form: ICS 211.
- ◆ Report activities of a specific assignment: ICS 214.
- ◆ Radio assignments: ICS 205.
- ◆ Medical plan: ICS 206.

FIGURE 14.9 ■ Federal Forms in an IAP. (Ask your instructor to download form examples from Pearson's IRC.)

MEETING WITH INVOLVED AGENCIES

Using as an example a diplomatic or dignitary event at the state capitol, it would be wise to meet with the commanding officer (or his designee) of the local law-enforcement agency to avoid the problems that would arise if it became necessary for a medical aircraft to land at an installation without prior notice—not to mention that landing a medical helicopter in the middle of a special event may present a few challenges. Included in an incident action plan should be the routes of travel and the travel time to each treatment area in the plan, to avoid potential conflicts with road or bridge closures and other infrastructure complications. Also, the responding scene units may be from a distant response district and might not be familiar with the area. If necessary, provide a map in the final document. Incident action plans work best when they are practiced with agencies in a drill prior to the incident.

CHAPTER REVIEW

Summary

Throughout the transition to NIMS, it is important to remember why both NIMS and its component ICS are critical to incident management. Most incidents are local, but when emergency responders are faced with the worst-case scenario, such as September 11, 2001, all responding agencies must be able to interface and work together well. The NIMS, and in particular the ICS component, allows that to happen but only if a good foundation has been laid at the local level. If local jurisdictions adopt a variation of ICS that cannot grow or is not applicable to other disciplines, the critical interface between responding agencies and jurisdictions cannot occur when the response expands.

It is important that everyone understand that with the establishment of the NIMS, there is only one ICS. As agencies adopt the principles and concepts of ICS as established in the NIMS, the incident-command system can expand to meet the needs of the response, regardless of the size or number of responders. The key to both the NIMS and ICS is a balance between standardization and flexibility.

Research and Review Questions

1. Research and discuss the origins of the incident command system.
2. Who is the sponsoring agency for the National Wildland Coordinating Group, what is their mission, and what is their area of expertise?
3. What is the intent of HSPD-5, who does it affect, and what are the requirements for the use of NIMS?
4. Contrast the staff command positions with the general command positions in an ICS situation.
5. Identify the areas in the IMS operation that hold the following titles: commander, chief, supervisor, and leader.
6. What are the three main functions of a medical group? Describe each of them.
7. Diagram a simple ICS organization chart with a medical group assigned under the operations section.
8. List all the ambulance and EMS resources in the local area. Place those resources on the list into groups that represent regional resources based on ambulance strike teams and EMS task forces. Assume that all units have agreed to a regional mutual-aid agreement to be placed into grouped resources.

References

Brunacini, A. V. (2002). *Fire command* (2nd ed.). Quincy, MA: National Fire Protection Publications.

Christian, H. T., & Maniscalco, P. M. (1998). *The EMS incident management system EMS operations for mass casualty and high impact incidents*. Upper Saddle River, NJ: Prentice Hall.

Epheron, H. (2002, January/February). Safety in numbers. *NFPA Journal,* p. 40.

International Fire Service Training Association. (2000). *Incident Management System Model Procedures Guide for Structural Firefighting* (2nd ed.) Stillwater, OK: Fire Protection Publications.

Nicholson, J. (2002, January/February). Looking back. *NFPA Journal,* p. 80. Retrieved 2008 from http://findarticles.com/p/articles/mi_qa3737/is_200201/ai_n9071502

U.S. Fire Administration. (2003). *Incident Command and Control Simulation Series: Self-Study 2.0* (CD-ROM). Emmitsburg, MD: Author.

CHAPTER 15

Interagency Relations and Operation

Accreditation Criteria

CAAS

102.01   Mutual Aid: The ambulance service shall develop and maintain mutual aid relationships with other ambulance/EMS organizations in its immediate and neighboring service areas (pp. 362–363).

102.02   Disaster Coordination: The service shall play an active role in the regional disaster plan and response (pp. 364–365).

102.03   Conflict Resolution: The agency shall develop and maintain a means to resolve conflicts among personnel of all organizations directly or indirectly involved in patient care (e.g., other ambulance service providers, police and fire departments, medical personnel etc.) (p. 358).

102.04   Inter-Agency Dialogue: The agency shall maintain on-going dialogue with area EMS agencies, public safety agencies, hospitals, and other healthcare and government officials to facilitate improved relationships and improved service coordination (pp. 360–361).

Objectives

After reading this chapter, the student should be able to:

15.1 Compare and contrast mutual aid and automatic aid (pp. 362–363).

15.2 Identify the hierarchical organization of resources that respond to disaster from a regional, state, and federal perspective (pp. 368–370).

15.3 Identify the intrastate and interstate mutual-aid components (pp. 366–368).

15.4 Understand the needs and organizational applications of area and unified command (p. 361).

15.5 Identify the components of the national response plans (p. 368).

15.6 Match and define the emergency-support functions within the national response plan (pp. 369–370).

15.7 Navigate the progress and communication chain for an agency to secure resources from local, state, and federal resources (pp. 370–376).

Key Terms

all-hazards approach (p. 362)

assisting agency (p. 360)

automated resource management system (ARMS) (p. 367)

automatic aid (p. 362)

disaster medical-assistance teams (DMATs) (p. 372)

disaster mortuary operational response teams (DMORTs) (p. 372)

Emergency Management Assistance Compact (EMAC) (p. 365)

emergency operations center (p. 361)

Emergency Planning and Community Right-to-Know Act (p. 363)

emergency support function (ESF) (p. 365)

hazard analysis (p. 361)

local emergency-planning committees (LEPCs) (p. 363)

mutual aid (p. 362)

Mutual Aid System Task Force (MASTF) (p. 366)

National Fire Service Intrastate Mutual Aid System (IMAS) (p. 366)

National Response Plan (NRP) (p. 362)

Patient Self-Determination Act (PSDA) (p. 360)

risk analysis (p. 361)

state emergency-response commission (SERC) (p. 363)

urban search and rescue (USAR) (p. 374)

veterinary medical assistance teams (VMATs) (p. 372)

vulnerability analysis (p. 361)

Point of View

The Mutual Aid Box Alarm System (MABAS) originated in Illinois in the 1970s. It is a formalized mutual-assistance system that now covers portions of four states. It is broken down into divisions (regions). It is used on a daily basis and covers emergencies ranging from a routine incident, such as a multiple-vehicle crash that requires several surrounding communities to respond, to something as large as the Hurricane Katrina response. This plan designates the equipment and personnel response in an organized, scalable manner. MABAS eliminates the need or desire to freelance. Since it is formalized, the personnel and equipment are covered in case of injury or worse. Under the plan regional communications centers can dispatch anywhere from 300 to 600 additional ambulances, as well as fire equipment in an organized manner. Two incidents illustrate this:

On the night before Thanksgiving 2005, a commuter train traveling 70 mph struck 15 vehicles sitting on a grade crossing. It was a long diagonal crossing and the vehicles became trapped when the gates came down. Within three minutes of the accident, with just two radio calls, units from 18 communities were responding with specified equipment. The incident commander knew how many ambulances, that engines and trucks would be on the scene, and that all critical command functions would be covered. Within minutes, unified command was established. Fire and EMS units communicated on a common frequency. Triage was completed, and the injured were transported from the scene within the first 30 minutes.

The second example was when a tornado struck a small, rural northern Illinois community, resulting in building collapses and entrapment. The local agency was quickly overwhelmed. MABAS was activated, and designated regional task forces were responding from as far away as 100 miles within 30 minutes. Without advanced preparation and interagency cooperation, this would not have been possible.

Rick Nosek
EMS Coordinator
Schiller Park Fire, Illinois

■ INTERAGENCY COMMUNICATION AND COOPERATION

Whenever one system or agency comes up against another, there is a potential for challenges. These problems can range from basic interpersonal communications to more difficult interprofessional issues, such as turf control, differences in procedures, differences in rank structure, or competition for funding. When interfacing with another system or agency, the goal of EMS members must be to avoid problems or adapt to the situation to achieve a cooperative working interface. Unfortunately, EMS managers are generally faced with reacting to a problem after the fact rather than having the opportunity to be proactive. For EMS providers the priority is ensuring delivery of the best patient care possible. And the patient is not concerned with interagency issues at all—the patient wants quality care and transportation.

EMS managers and system decision makers must remain cognizant of interagency issues and factor consideration of those issues into their decision-making process. In addition, managers and decision makers must solicit input from field providers in order to avoid serious "street impact" from their decisions. This is especially important since managers may be far removed from responders where and when the impact of their decisions is felt.

Interagency problems can escalate quickly whenever the EMS response involves mass casualties or the scene of a disaster. Because mass-casualty and disaster scenarios may involve multiple EMS agencies as well as police and fire agencies, preplanning for major incident command and control can significantly reduce interagency problems. Mutual-aid and cooperation agreements need to be in place and standard procedures written to guide implementation. Various incident command system (ICS) models are available to assist with the management of response resources at the scene of large-scale emergency operations.

Interestingly, both OSHA and EPA have written a requirement for establishment of an ICS into some of their regulations. FEMA has taken the process a step further, combining ICS management with tactical action plans (fire/ hazmat, and so on) in an integrated emergency management system (IEMS). IEMS is a detailed planning tool for use when multiple agencies are involved at a disaster scene that will require a prolonged period of operations to conclude. Intrastate and interstate mutual-aid initiatives

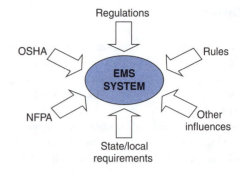

FIGURE 15.1 ■ Interagency Influences on an EMS Agency.

and federal inventory and credentialing systems have been created under FEMA.

■ INTERFACE WITH EXTERNAL AGENCIES

Every day EMS managers are involved with rules, regulations, and other influences generated by external agencies and impacting upon their organization and personnel (Figure 15.1). Managers must know the extent of involvement and whether or not the particular agency has regulatory power over their operations. The best practice is to comply with applicable rules and regulations. If a situation arises that makes compliance difficult, the manager must work to achieve compliance or a reasonable solution to the problem. During this period, the EMS manager must be careful to document the problem and the action taken to achieve resolution. Failure to comply with applicable rules and regulations can cause administrative nightmares, preclude effective delivery of certain health-care services, and in extreme cases result in litigation or shut down the entire EMS system.

Some of the interfaces and influences affecting EMS systems include OSHA, CDC, and FEMA. Each of those agencies has promulgated guidance, standards, and recommendations for health-care providers. In states covered by federal OSHA, the guidelines apply only to the private sector. In states that have their own OSHA programs, both the public and private sectors must comply. Of particular interest currently are provisions regarding bloodborne pathogens and infectious-disease control. OSHA regulations require health-care employers to have follow-up procedures for HIV/HBV exposure, but because many EMS systems are not, by legal definition, a health-care service, the National Fire

Protection Association (NFPA) has issued requirements that can apply to all EMS systems.

The impact of these regulations on an EMS manager is how to ensure total compliance and how to communicate up the chain. Protocols, SOPs, and standing orders must be developed for washing and decontamination, minimizing needle sticks, specimen handling, and so on. Immunization of all EMS members must be accomplished for HBV, or at least offered, if it cannot be mandated due to a labor agreement. The employee does have the option to refuse, which means that if he contracts hepatitis B a workers' compensation claim may be limited or denied. The employer should develop some form of waiver for employees who decline vaccinations.

Personal protective clothing and equipment must be purchased and made available to EMS members. Training must be developed and implemented; individual training must be documented and records retained for three years. Medical records for personnel at risk of exposure must be maintained for the duration of employment plus 30 years, and those records must be kept confidential. Obviously, compliance is necessary for the health and safety of the individual EMS responders, but there is a substantial cost, in dollars and time, to the EMS system managers. Failure to comply may be even costlier should workers' compensation claims result or lawsuits be filed. Failure to comply also may result in substantial fines. For example, failure to comply with 1993 OSHA regulations regarding working in confined spaces could result in fines of up to $70,000 per violation.

HEALTH-CARE SYSTEM INTERACTIONS

Regulatory agencies tend to have many hats to wear and are often given marching orders by the political leadership of the day, a situation that ensures continual changes as political agendas and administrations change. For emergency service organizations this can appear to be inconsistency and disorganization. There is no simple answer to this situation, only the awareness that change is a constant in this business. When dealing with regulatory agencies, try to understand what it is they want from your organization, and sometimes just asking in an approachable way will get you an answer. You always have a right to ask for clarification of requirements from such agencies, so do not be afraid or reluctant to ask. Remember the lesson on managing relationships: good relations with people will always produce better results than having no relations.

Competing Emergency Service Interests

Some communities experience competition between emergency service groups and agencies. A facilitator once described the process of unraveling turf issues between these competing groups as "herding cats, all with paramedic attitudes." The best way to get these groups to work with each other is to have a neutral facilitator working together with them on neutral ground, asking each person the following questions: What does your organization contribute to the needs of the community? What unique talents does your group bring to the citizens? How can the talents of all the groups be applied together to enhance the service provided to the community?

Keep in mind that turf issues can be decades old, and working through them will require patience, commitment on the part of the participating organizations, and professional-level facilitation.

Tiered or Dual-Response Agencies

Turf issues also exist within organizations at the interdepartmental, opposing-shift, and dual-response levels. A common place for disagreements is in combination career-and-volunteer agencies. Generally, an outsider will quickly recognize that each group is simply looking out for its own interests and not that of the combined total. A reasonable approach to improving this situation will involve having each group defining its contribution to the whole mission and a facilitation of the contributions each group makes to the community. A basic understanding of teamwork and the expectations of the agency that everyone will "play well together" help and must be delivered by the leaders of the organization. Of course, the leaders then need to walk the talk of getting along and using accepted behavior.

There is plenty of room for a wide diversity of opinions and vocations in emergency service organizations. The trick is getting all that diversity focused on the customer, the mission, and the community. New leaders in organizations have the opportunity to improve the organization's behavior and change some of the dysfunctional elements, such as rumor mongering and backbiting. One of the reasons negative elements and behaviors exist is that they are tolerated and enabled by leaders that refuse to deal with them. It takes courage and skill to handle entrenched behaviors, but the cost of not doing so is poor performance for the entire organization. When a leader is contemplating the difficult task of coaching and correcting behaviors, it is of some use to remember this saying: "Better a horrible end than horror without end."

EMS Interaction with Home Health

Many EMS managers are dealing with home-health agencies that present advance directives. In 1990 the **Patient Self-Determination Act (PSDA)** was passed as part of the Consolidated Omnibus Budget Reconciliation Act (COBRA). Although PSDA is directly applicable to various medical agencies that accept Medicare/Medicaid, many provisions of the act now affect EMS systems. PSDA recognizes a patient's right to self-determination, including the right to refuse lifesaving treatment. PSDA also requires institutions to ask patients if they have executed any advance directives, such as a living will, durable power of attorney for health care, or a do not resuscitate (DNR) order. Because of the law, more people have executed advance directives, and hence EMS members are much more likely to encounter them. Several states have enacted prehospital DNR legislation, but even with SOPs in hand, the EMS member is faced with making life-or-death decisions, while legal and ethical pressures add to the burden.

MULTIPLE-AGENCY INTERACTIONS

Single Command with Assisting Agencies

Moving from the day-to-day interfaces and routine regulatory oversight of an EMS agency to being an EMS manager in an operational setting requires knowing how to communicate, secure, and organize resources from other agencies or jurisdictions. In most command scenarios EMS will be an assisting or cooperating agency. Under the ICS system, a single command may request help from other agencies for specific issues without developing a larger command structure. The common interfaces with agencies that can be considered as assisting agencies are listed in Figure 15.2.

Under ICS an **assisting agency** is an agency directly contributing tactical or service resources to another agency. A supporting agency is an agency supplying assistance other than direct tactical or support functions or resources to the incident-control effort (such as the Red Cross or a telephone company).

Hazardous Materials Response

Not too long ago a hazardous materials (hazmat) response was left to trained fire service response teams. However, in 1989 OSHA published rules entitled "Worker Protection Standards for Hazardous Waste

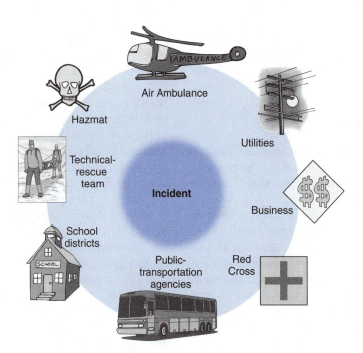

FIGURE 15.2 ■ Agencies Assisting and Supporting EMS Operations.

Operations and Emergency Response," or the HAZ-WOPER standard. HAZWOPER requires all emergency services to develop and implement a written incident plan to handle hazmat emergencies. With the crossover into hazmat response, EMS planners, managers, and responders were forced to modify their incident-response procedures to preclude EMS responders becoming victims themselves.

As part of planning, EMS managers must consider the action or requirements of various governmental agencies. The Department of Transportation has issued hazmat guidelines and recognition tools, OSHA has mandated training for emergency medical responders, FEMA is providing funds for training, and the EPA is providing hotline support. The crossover to hazmat response has also resulted in funding requirements for personal protective clothing and equipment.

By virtue of other agency involvement, EMS managers are forced to take action to ensure compliance with rules, regulations, and standards. In house they must ensure adequate training to protect the health and safety of EMS members, as well as provide the protocols and SOPs necessary to ensure the delivery of quality medical care.

Unified Command and Area Command

ICS incorporates a unified command (UC) and an optional area command when important elements need to be managed in multi-jurisdictional or multi-agency domestic incidents. UC is applied to ensure participation of all agencies with jurisdictional authority or functional responsibility for any or all aspects of an incident, as well as those able to provide specific resource support. The UC structure works best when it is used to:

- Contribute to the process of determining overall incident strategies and selecting objectives.
- Ensure that joint planning for tactical activities is accomplished in accordance with incident objectives.
- Ensure the integration of tactical operations.
- Approve, commit, and make optimum use of all assigned resources.

An area command is established either to oversee the management of multiple incidents that are each being handled by a separate ICS organization or to oversee the management of a very large incident that involves multiple ICS organizations—either situation is likely for incidents that are not site specific, geographically dispersed, or evolve over longer periods of time (such as a bioterrorism event). In this sense acts of biological, chemical, radiological, or nuclear terrorism represent particular challenges for the traditional ICS structure and will require extraordinary coordination between federal, state, local, tribal, private-sector, and nongovernmental organizations.

Area command avoids unnecessary competition for the same resources, such as when there are a number of incidents in the same area and of the same type (for example, two or more hazardous-material or oil spills and fires). The use of area command is an approach that can enhance a large operation by assisting a local or regional response by building on those procedures used for a single incident. These steps might include the activation of **emergency operations centers** or a unified command, which will grow into an area command. When incidents do not have similar resource demands, they are usually handled separately and are coordinated through an emergency operations center (EOC).

If the incidents under the authority of the area command are multi-jurisdictional, then a unified area command should be established. This allows each jurisdiction to have representation in the command structure. Area command should not be confused with the functions performed by an EOC. An area command oversees management of the incident, while an EOC coordinates support functions and provides resources support.

EMS "ALL-HAZARDS" APPROACH

An all-hazards approach is a process by which hazards in a community are identified (**hazard analysis**), persons or segments of the population that are vulnerable to the hazards are identified and quantified (**vulnerability analysis**), and then after mitigation actions have been identified, projected vulnerabilities are refined (**risk analysis**). Once it is determined who is at risk from which identified hazards, you must prioritize planning efforts based on the most urgent needs. The same techniques described in Chapter 6 can be applied to the community.

Having determined the at-risk populations, your next steps are to identify the resources that will be needed to respond effectively to a given hazard and to determine the assets and voids with regard to those needed resources. If there is a lack of a particular resource, the EMS agency must identify how those items can be acquired regionally or requested from higher levels of government. Going through the process of a hazards risk assessment to determine needs and available resources for each type of hazard is a step toward creating an inventory of needed and available resources. These resources then can be brought to bear in a modular fashion on any and all events that may occur in the community.

As greater resources are brought into play, a corresponding increase in the command-and-control structure managing these resources must be implemented. An **all-hazards approach** is defined as the advance development of plans, procedures, and resources that can be assembled in a modular fashion and brought to bear upon either a crisis or a scheduled event for the purpose of reducing medical risks to the EMS system's customers and response personnel.

Plan Components

All these plans have common characteristics that include:

- Hazards analysis.
- Vulnerability analysis.
- Basic plan consideration (for example, scope, activation, authorities, and concepts of operation).
- Resource identification.
- Support functions or mechanisms for resource deployment.
- Recovery activities.

Demands on organizational resources constantly increase due to heightened customer expectations. Because of these changing conditions, greater and greater resources must be assembled to manage special operations and unique events that may occur in the community. The resources that are available individually, as well as the capabilities of those resources, are limited. You must identify in advance additional resources that you can bring to bear on virtually any type of event to which you may be called to assist. In order to accomplish this, EMS leadership must:

- Identify potential hazards.
- Determine the at-risk populations for each of those hazards.
- Identify the resources that would be required to manage events caused by those hazards.
- Determine what assets and voids in resources you have.
- Find resources to fill those voids.
- Develop a methodology for deploying those resources to a particular event.

There are multitudes of statutory, regulatory, and planning resources that can assist in this process at a local, then regional, state, and federal level. These also can be tremendous sources for an incident commander or agency to fill those resource voids.

Resource Acquisition

The **National Response Plan (NRP)** provides information on how to obtain resources and assistance that may be required during a disaster or emergency. However, many special-operations events occur that never become federalized. The resources of the local community, the region, and sometimes the resources of the state are needed to control these incidents. Through the cooperation and assistance of neighboring communities or the state in which you live, the vast majority of incidents are handled effectively. That grassroots ability is possible through effective local, regional, and statewide planning. These plans can take many forms, but most commonly you will find them as state EOPs, mutual-aid agreements, and regional resource-response plans.

Mutual-Aid Agreements

Most organizations have mutual-aid or automatic-aid agreements with neighboring communities. These agreements specify that another agency will help, and in return the other agency can expect help when necessary. In some cases, there is statewide mutual aid that eliminates the need to have to renew signed agreements on an annual basis. These plans also identify specific authorities with regard to command, compensation, and liabilities. **Mutual aid** requires an agency's command officers to make a specific request. In contrast, **automatic aid** is sent by the dispatch center without a command officer's input and is done on a prearranged matrix.

Mutual-aid agreements have been a backbone of emergency services for decades. During incidents that overwhelm local resources, organizations have the ability to call upon neighboring resources rapidly, whether these resources are unique or similar. One limiting factor regarding such agreements is that the agreement does not relieve an agency's responsibility to provide a given level of service to its own community.

Many mutual-aid agreements are hampered by a belief by managers or command officers that they are to send help only when helping will not adversely affect the system. Many systems already are taxed by resource and financial constraints, and when a disaster, such as a hurricane or earthquake happens, it further strains the regional resources. In order to meet resource needs, mutual-aid agreements need to be complemented by regional or statewide resource response

In consideration of the mutual commitments given herein, each of the Signatories to this Mutual-Aid Agreement agrees to render aid to any of the other Signatories as follows:

1. *Request for aid.* The Requesting Signatory agrees to make its request in writing to the Aiding Signatory within a reasonable time after aid is needed and with reasonable specificity. The Requesting Signatory agrees to compensate the Aiding Signatory as specified in this Agreement and in other agreements that may be in effect between the Requesting and Aiding Signatories.
2. *Discretionary rendering of aid.* Rendering of aid is entirely at the discretion of the Aiding Signatory. The agreement to render aid is expressly not contingent upon a declaration of a disaster or emergency by the federal government or upon receiving federal funds.
3. *Invoice to the Requesting Signatory.* Within 90 days of the return to the home work station of all labor and equipment of the Aiding Signatory, the Aiding Signatory shall submit to the Requesting Signatory an invoice of all charges related to the aid provided pursuant to this Agreement. The invoice shall contain only charges related to the aid provided pursuant to this Agreement.
4. *Charges to the Requesting Signatory.* Charges to the Requesting Signatory from the Aiding Signatory shall be as follows:

   a. *Labor force.* Charges for labor force shall be in accordance with the Aiding Signatory's standard practices.

   b. *Equipment.* Charges for equipment, such as bucket trucks, digger derricks, and other special equipment used by the Aiding Signatory, shall be at the reasonable and customary rates for such equipment in the Aiding Signatory's location.

   c. *Transportation.* The Aiding Signatory shall transport needed personnel and equipment by reasonable and customary means and shall charge reasonable and customary rates for such transportation.

   d. *Meals, lodging, and other related expenses.* Charges for meals, lodging, and other expenses related to the provision of aid pursuant to this Agreement shall be the reasonable and actual costs incurred by the Aiding Signatory.

5. *Counterparts.* The Signatories may execute this Mutual-Aid Agreement in one or more counterparts, with each counterpart being deemed an original Agreement, but with all counterparts being considered one Agreement.
6. *Execution.* Each party hereto has read, agreed to, and executed this Mutual-Aid Agreement on the date indicated.

Date_____

Entity_____

By_____

Title_____

FIGURE 15.3 ■ Sample Mutual Aid Agreement.

plans. But even with the best of intentions, without detailed operational plans execution of such agreements can be destroyed by personal interpretations or local politics and can have disastrous effects. A sample mutual-aid agreement is included in Figure 15.3.

In large-scale, multi-jurisdictional incidents, where self-dispatching of personnel is common, mismanagement of resources and general disorganization could lead to injuries, fatalities, or to not being able to control the emergency in a timely fashion. Many parts of the country have only informal, "handshake" agreements. Some communities have signed agreements, but frequently there is little beyond the signatures that separate them from informal agreements, since many do not contain detailed plans, protocols, or training exercises to support their integration into an operational process.

■ LOCAL EMERGENCY PLANNING

The Superfund Amendments and Reauthorization Act of 1986 (SARA) Title III, or the **Emergency Planning and Community Right-to-Know Act**, required the governors of every state to establish a **state emergency response commission (SERC)** and to designate **local emergency-planning committees (LEPCs)**. The purpose of this law was to require emergency-planning efforts at the state and local levels and to increase the public's awareness of hazardous chemicals within their communities. Depending on the state, the LEPC can be designated as the entire state, a portion of the state, or any county or other geographical designation.

Facilities that store or use materials that have been identified by the EPA as an extremely hazardous substance (EHS) must be reported to each state's SERC and subsequently to each LEPC, as well as to the local fire department. Once these facilities have been identified, the LEPC and the local committee are required to plan for a potential release of the EHS if the quantity stored exceeds predetermined planning-level requirements called the threshold planning quantity (TPQ).

During the planning process, the facility must designate a facility emergency response coordinator to manage the planning process with the LEPC and to provide information at the time of an emergency. In addition, the facility itself must develop an ERP for its employees and assist with the development of the

LEPC and the local plan. The information found within these facility-specific plans is extensive. The plan itself is developed in accordance with several planning guidance documents, including *Chemical Emergency Preparedness Program* (U.S. EPA, 1985) and *Technical Guidance for Hazards Analysis: Emergency Planning for Extremely Hazardous Materials* (U.S. EPA, 1987).

The three major components of a facility-specific plan developed by the LEPC include:

- *Hazards analysis.* Identifies the chemical hazards that are present at the facility.
- *Vulnerability analysis.* Identifies the vulnerable population of the surrounding community should an accidental release of the EHS occur.
- *Risk analysis.* A determination of the potential for an actual release from the facility, based upon historical release data and onsite work practices.

Generally, the LEPC plans also identify the resources capable of responding to technological emergencies of this nature. However, these resources are not deployed based on the LEPC plan. Actual deployment of resources is determined by the ERP of the authority having first-response duties or the community's local ERP. However, the LEPC plan does provide extensive information to aid in the development of those response policies.

The chemicals that are stored at these facilities, the EHS, all have been identified as presenting an extreme risk to the local community, should an actual release occur. These materials are all volatile liquids or gases that if released from containers would migrate off-site and create harm. It behooves any EMS provider to assess the presence of these EHS facilities within the community and to ensure that the EMS function has input into the planning process with regard to the medical effects of a potential release.

REGIONAL RESOURCE RESPONSE PLANS

Many states and regions already have developed or are currently in the process of developing resource response plans. The concept is that communities from an entire region and even from throughout a state will adopt a single plan that allows for the sharing of like resources. These plans identify the method of mobilization, organization, and operation, including financial considerations, well in advance of the need.

Resource response plans are based heavily upon the concepts of incident management. An agency

---

- ◆ The state emergency response plan (ERP) may not need to be activated, because this is a freestanding plan. However, the state may assist by playing a coordination role.
- ◆ Like/in-kind resources are defined in advance.
- ◆ Financial and liability responsibilities are identified ahead of time.
- ◆ Tracking of resources by tasking number ensures proper reimbursement and limits self-dispatch.
- ◆ Accountability for all resources and personnel is ensured when they are dispatched through an ERP.

FIGURE 15.4 ■ Advantages of an All-Hazards EMS Regional Response Plan.

requiring a specific resource contacts a lead organization (many times the state EOC) and places a request for a specific resource, task force, or strike team. This lead agency then determines the closest available source of the needed items from the membership of the plan and contacts one or more organizations to assemble what is needed. The advantage to this system is that no one agency is overtaxed by the demand, yet the resource is assembled as one functional group. As the emergency grows and needs increase, the lead agency contacts member organizations in other areas to assemble yet more and more task forces or strike teams. Again, the advantage is that no one organization or even region becomes overburdened by the requests. There are many other advantages to this type of system, as listed in Figure 15.4.

As stated earlier, the incident management system (IMS) and the use of task forces and strike teams are heavily emphasized in these agreements. Examples of such organizations of resources are in Figure 15.5.

Type I Ambulance Strike Team
- ◆ Five ALS ambulances.
- ◆ Two crew members for each ambulance. At least one must have ALS-level certification/licensure with hazardous material technician training.
- ◆ One supervisor.
- ◆ Common communications among all.

Debris-Removal Task Force
- ◆ One front-end loader.
- ◆ One backhoe.
- ◆ One five- to six-cubic-yard dump truck.
- ◆ Two crews for each piece.
- ◆ Two forepersons.
- ◆ One supervisor.

FIGURE 15.5 ■ Predesignated Resources for Regional Responses.

Ideally, these resources are included in a predefined response matrix that dispatch or command officers can access immediately for a timely response. In many cases this type of organization can be staged ahead of an anticipated event or impending natural disaster.

In addition to predefined resources, the requesting member can specify a unique resource at the time of the request. These could include EMS task forces, a disaster medical-assistance team, or an urban search-and-rescue team. Responding units should have a means of common communication between units capable of working outside their home jurisdictions. This concept of having regional resources able to communicate with each other is known as *interoperability*. Its ensurance is a goal of the federal government. In the absence of interoperability, the agency requesting regional resources needs to provide a means of communication to the responding agencies.

STATE EMERGENCY OPERATION PLAN

Under the Stafford Act, states are encouraged to engage in emergency planning and are provided funding to augment the state's planning efforts. Therefore, each state has an emergency operation plan (EOP) for the deployment of state resources and to guide coordination activities with the national response plan (NRP). Many of these plans closely resemble the national response plan in organizational concept and through the use of an **emergency support function (ESF)** structure. Although these plans may have specific components for the most likely hazards to be faced by the state (such as earthquakes and hurricanes), a properly developed plan also will have sufficient flexibility to allow it to be applied modularly to any emergency or disaster for which the state is called upon for assistance.

During the state planning process, planners evaluate their state for the hazards that are likely to affect it. The planning process estimates vulnerabilities and then determines resource needs for mitigating the emergency. Planners identify where resources required to support the response to the emergency may be obtained. Many times these resources may be available from the state government, its contractors, or from other public jurisdictions or private organizations within the state. However, there are some resources to which the state does not generally have direct access. Ambulances and urban fire-suppression

vehicles are examples of resources that are in limited supply, and to meet the resource needs, the state plan must identify sources outside state government from which these resources may be obtained; namely, other local fire and EMS providers throughout the state as well as outside it.

The state accesses resources from the federal government through the request for activation of the NRP. In order to obtain federal activation, the state must demonstrate that the disaster or emergency exceeds its abilities to respond to the incident, even though the state EOP has been activated. As a general rule, this concept of demonstrating need also applies to the local jurisdiction's ability to access the state EOP. Once local fire and EMS officials have determined that the incident exceeds their capabilities to respond, those local agencies contact city or county officials to request activation of the local EOP by declaring a local state of emergency and, subsequently, by activation of the state's EOP.

INTERSTATE RESOURCE-RESPONSE PLANS

In the wake of some of the disasters of recent years, a concept of interstate resource agreements has developed and was invaluable in Louisiana during Hurricane Katrina in 2005. These agreements, which are placed into statute with the participating states, allow for the movement of resources from one state to another. Although the NRP may have been activated already, these plans provide for access to resources that may not be available from the federal government. Specifically, the ready acquisition of like or in-kind ambulances, fire apparatus, and law enforcement officers can be achieved through such plans.

The largest of such plans is the **Emergency Management Assistance Compact (EMAC)**. This interstate agreement began after Hurricane Andrew in 1992, when it was recognized that significant resources were needed that could not be provided through the NRP. Consequently, a majority of the member states of the Southern Governors' Association entered into a shared-resources agreement called the Southern Regional Emergency Management Compact (SREMAC). This plan enabled the states to share like or in-kind services and resources in a disaster, thus reducing some of the overall burdens on the NRP. The compact gained popularity and has grown into what is now EMAC, which was used extensively during the aftermath of Hurricane Katrina.

EMAC is enacted by the member states by adoption into statute provisions, allowing state emergency-management officials to share state resources with other member states, if the requesting state has declared a disaster EMAC. The Emergency Management Assistance Compact is a congressionally ratified organization that provides form and structure to interstate mutual aid. Through EMAC, a disaster-impacted state can request and receive assistance from other member states quickly and efficiently.

The EMAC resource pool is based upon each state's conducting an "assets and voids" analysis as part of comprehensive emergency planning. During this analysis, each state identifies the resource voids and needs it anticipates having during a particular disaster or major emergency, as well as the assets it can provide to another state. These assets and voids are cataloged by EMAC for use when the plan is activated.

EMAC is also unique in that each state agrees to have an EMAC advance team in its EOC during the time of the disaster. This creates a trio of resource coordinators: the state coordinating officer (SCO) from the affected state, a federal coordinating officer (FCO) if the NRP was activated by a presidential declaration, and an EMAC advance management team (A-team). These three coordinators jointly decide who can provide the needed resource the fastest and the most cost-effectively.

Financial responsibility for EMAC-provided resources rests with the state that requests the resources. Generally speaking, these charges are based on a three-year prorated cost of the equipment plus fuel and personnel expenses. Administrative overhead is not calculated into the cost unless that overhead can be attributed directly to the deployment of the resource. Currently, EMAC is developing like or in-kind standards for debris clearing; medical, fire, and law-enforcement resource groups in both task-force and strike-team configurations; and other such teams that can be developed as needed. Additionally, licensure and training requirements for a given resource (such as paramedic qualifications and capabilities) all can be assessed in advance, and states can specify the member states that best meet their current level of service.

FEDERAL MUTUAL-AID INITIATIVES

The federal government has developed two projects that will help EMS service's ability to support the national response plan and implement concepts presented in the National Incident Management System (NIMS). This initiative was authorized under a Homeland Security Presidential Directive (HSPD-5) to aid all federal, state, tribal, or local jurisdictions in locating, requesting, and ordering resources to assist neighboring jurisdictions when their local capability is overwhelmed. The initiative will provide the nation with the highest capability to rapidly and easily exchange disaster resources during times of need.

The first project is the **National Fire Service Intrastate Mutual Aid System (IMAS)**. IMAS was funded in July 2006 by the NIMS Integration Center within FEMA. The goal of the IMAS project is to support the creation of formalized, comprehensive, and exercised intrastate (that is, within states) mutual-aid plans. Assistance is being provided to states through FEMA partnerships to develop formal, comprehensive mutual-aid plans for efficiently mobilizing and deploying fire, police, and EMS assets to incidents within their states. The plans that are produced will provide a mutual-aid model that can be adopted and adapted to suit the needs of other emergency services and disciplines.

The second project focuses on interstate (between states) mutual aid and is being built by a multi-agency team known as the **Mutual Aid System Task Force (MASTF)**. In conjunction with IMAS, the Fire Service Mutual Aid System Task Force has developed plans for an interstate mutual-aid system. That IMAS project has strengthened the foundation of effective interstate mutual aid by ensuring the existence of a system of states experienced in providing mutual aid within their own group. The MASTF effort will help shape an improved interstate mutual-aid system.

The goal of the MASTF is to bring a fire-service perspective to recommendations to improve the sharing of resources across state lines. The task force established a system based on lessons learned from recent national disasters about how to most efficiently and effectively mobilize and deploy fire-service resources to large-scale incidents across state lines. The relationships are defined in Figure 15.6. IMAS and MASTF have designed a map of mutual-aid systems to support EMAC requests more efficiently. MASTF has finished its project, and the IAFC has now created an Emergency Management Committee to continue work on the coordination of national resource mutual aid.

FEMA recently released a glossary of terms, definitions, and resource typing for an initial 60 resources. This product provides a foundation for facilitating the use of common terminology across the nation.

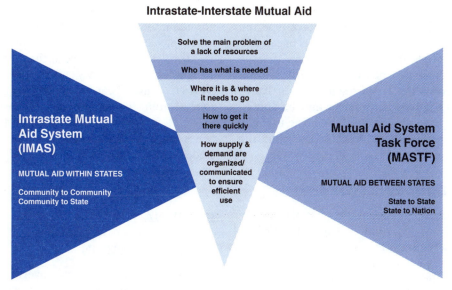

Intrastate-Interstate Mutual Aid

Solve the main problem of a lack of resources

Who has what is needed

Where it is & where it needs to go

How to get it there quickly

How supply & demand are organized/communicated to ensure efficient use

Intrastate Mutual Aid System (IMAS)

MUTUAL AID WITHIN STATES

Community to Community
Community to State

Mutual Aid System Task Force (MASTF)

MUTUAL AID BETWEEN STATES

State to State
State to Nation

FIGURE 15.6 ■ Intrastate-Interstate Mutual Aid.
(Source: www.ICHIEF.org. Reprinted with permission.)

Participation in the national mutual aid and resource management initiative is strictly voluntary.

The following discipline groups have been formed to "resource type" critical-response assets: animal health, emergency management, EMS, fire, hazmat, health and medical, law enforcement, public works, and search and rescue. The initiative will expand to include additional discipline groups if the need arises. FEMA used a working group made up of federal, state, and local representatives with the goal of addressing resource management issues. *Resource typing* was completed by the working group and is the categorization of resources that are commonly exchanged in disasters via mutual aid, by capacity or capability. Resource typing organizes resources into groupings of like resources (and in kinds) according to minimum standards for ease of ordering and mobilizing.

AUTOMATED RESOURCE MANAGEMENT SYSTEM

The **automated resource management system (ARMS)** is a secure system that maintains an inventory of typed federal, state, local, and privately owned resources. ARMS has four main functions: locating, requesting, ordering, and tracking resources. A long-term goal of ARMS is to expand to include GPS/GIS and mapping capabilities in the future. ARMS will not play a role in prioritizing resources requested during multiple events. Prioritization of resources requested and ordered through the ARMS will always be a function of the appropriate level of incident management, whether it be a local Incident Commander, a local or state EOC, an area command, or a multi-agency coordination system (MACS).

As part of the strategic plan, a state user or designee will be able to enter the ARMS at any incident command post, dispatch center, EOC, or other approved entry point, and search for a capability, function, or a specific resource by name or identifier. ARMS will provide a picture and verification of the information on each resource searched, including whether or not the resource is available for deployment. The user will be able to see the resource's location, the estimated travel time, and the requesting and approval process. Once the resource request is made and the supplying entity grants approval, information will be provided back to the user on the date and time the resource was dispatched, its estimated arrival date and time at its destination, and any support requirements. Resources ordered are expected to come with full operating capability, including crews and support equipment.

ARMS will allow a user to search the resource inventory by function, resource name, brand name, location, owner, and type of incident. This will allow the user the opportunity to determine the kind of resource that can best meet the need. By searching for a function, the user may discover a potential resource that can perform the necessary task. By reviewing the resource type, the user also will be able to determine what level of capability is most appropriate for the need. In some cases, a Type II or III resource may be closer than a Type I and able to fulfill a need until a Type I resource can arrive. In other cases, multiple Type II resources may be able to provide the same capability as one Type I. The user can use the kind and type information to determine the most appropriate resource to request.

NATIONAL CREDENTIALING SYSTEM

A nationwide system for credentialing has been established under the IMAS initiative. Currently, FEMA has established a nationwide system of credentialing by working with existing state or discipline-specific credentialing bodies toward national recognition for multi-jurisdictional response under mutual-aid agreements. Current credentialing bodies will continue to issue the credentials; however, those credentials will be entered into a national database and validated. The nationwide credentialing system will be used to request and credential people before they are placed into the role at an emergency scene or at the site of a disaster. The main components of the proposed credentialing system are shown in Figure 15.7

New training, certification, and qualification standards have been developed for the nationwide credentialing system. The National Wildfire Coordinating Group (NWCG) and the NIMS integration center have defined the prerequisites, which establish an ability to perform the specified task and include education, applicable licenses, training, and professional experience. Existing training courses should be evaluated for mission-essential tasks list (METL) relevance and a gap analysis conducted for the development of additional courses.

---

1. *Eligible participants.*
2. *Certifications and qualifications standards.* To include mission-essential task analysis and required training with associated course curriculum.
3. *Credentialing organization.* To include how the authorized credentialing organizations are reviewed and accredited.
4. *Credential.* To include identification information as well as certifications.
5. *Record-keeping system.* To include a centralized database with multiple access points.

FIGURE 15.7 ■ Credential for Regional Response System.

NIMS awareness training for all levels will be incorporated into course curricula for credentialing personnel for incident-management roles prescribed by NIMS. The core curriculum would include general-awareness courses for all emergency-services professionals, such as NIMS training, ICS, and an emergency-management 101 course from the Emergency Management Institute (EMI). The intention is to work with the various disciplines and jurisdictions in a way that will bring current credentialing efforts into a nationwide system without duplicating efforts.

■ THE NATIONAL RESPONSE PLAN—

DISASTER CONDITION

A disaster or emergency may overwhelm the capabilities of a state and its local governments to provide a timely and effective response to meet the needs of the situation. For example, the occurrence of a large, catastrophic earthquake in a high-risk, high-population area will cause casualties, property loss, and disruption of normal life-support systems and will affect the economic, physical, and social infrastructure of the whole region.

A disaster or emergency has the potential to cause substantial health and medical problems, with hundreds or thousands of deaths and injuries, depending on factors such as time of occurrence, severity of impact, existing weather conditions, area demographics, and the nature of building construction. Deaths and injuries will occur principally from the collapse of buildings and other structures and from collateral events, such as fires and mudslides.

A disaster or emergency may cause significant damage, particularly to the economic and physical infrastructure. An earthquake could trigger fires, floods, or other events that will multiply property losses and hinder the immediate emergency-response effort. An earthquake or hurricane could cause significant damage to or destroy highway, airport, railway, marine, communications, water, waste-disposal, electrical-power, natural-gas, and petroleum transmission systems.

NATIONAL RESPONSE PLAN ASSUMPTIONS

The national response plan (NRP) assumes that a disaster or emergency, such as an earthquake or act of terrorism, may occur with little or no warning at a time of day that produces maximum casualties. The plan also deals with other types of disasters, such as hurricanes, that could result in a large number of casualties and cause widespread damage and with the consequences of any event in which federal response assistance under the authorities of the Stafford Act is required. In all cases the plan assumes that the response capability of an affected state will be overwhelmed. The large number of casualties and the heavy damage to buildings, structures, and the basic infrastructure will necessitate direct federal government assistance to support state and local authorities in conducting life-saving and life-support efforts.

As the result of persons being injured and others being trapped in damaged or destroyed structures, the likelihood of a significant number of deaths within 72 hours will require the immediate response of federal search-and-rescue personnel—and medical personnel, supplies, and equipment—to minimize preventable deaths and disabilities. Federal departments and agencies may need to respond on short notice to provide effective and timely assistance to the state. Therefore, the NRP provides preassigned missions. Prior to 2004 the National Emergency Management Association (NEMA) was funded by the Department of Homeland Security (DHS) to create Model Intrastate Mutual Aid Legislation. Such enabling legislation is a necessary step in developing effective intrastate mutual-aid systems.

The declaration process under the plan will be carried out under PL 93-288, as amended, and as prescribed in Title 44 Code of Federal Regulations (44 CFR), Part 205. Based on the severity and magnitude of the situation, the governor will request the president to declare a disaster or an emergency for the state, and the president will issue a declaration, as warranted. The president also will appoint a principal federal official (PFO) to coordinate the overall activities under the declaration.

In certain situations the president may declare an emergency with or without a governor's request, as specified in Title V of PL 93-288, as amended. Under Title V, the president may direct that emergency assistance be provided, either at the request of a governor [Section 501(a)] or upon determination by the president that an "emergency exists for which the primary responsibility for response rests with the United States . . ." [Section 501(b)]. Examples of national emergencies declared by the president include the bombing of the Alfred P. Murrah Federal Building in Oklahoma City in 1995 and the terrorist attack on the World Trade Center in New York City in 2001.

THE PLANNING FUNCTION OF OPERATIONS

During the period immediately following a disaster or emergency requiring federal response, primary agencies, when directed by FEMA, will take actions to identify requirements and mobilize and deploy resources to the affected area to assist the state in its life-saving and life-protecting response efforts.

Agencies have been grouped together under 15 emergency support functions (ESFs) to facilitate response assistance to the state. If federal response assistance is required under the plan, it will be provided using some or all of the ESFs as necessary. Each ESF has been assigned a number of missions to provide response assistance to the state. The designated primary agency, acting as the federal executive agency and with the assistance of one or more support agencies, is responsible for managing the activities of the ESF and ensuring that the missions are accomplished. ESFs have the authority to execute response operations to support state needs directly. The primary and support agency assignments by each ESF are shown in Table 15.1.

TABLE 15.1 ■ National Response Plan Emergency Support Functions

ESF	Function
ESF-1	Transportation
ESF-2	Communications
ESF-3	Public Works and Engineering
ESF-4	Firefighting
ESF-5	Emergency Management
ESF-6	Mass Care, Housing, and Human Services
ESF-7	Resource Support
ESF-8	Public Health and Medical Services
ESF-9	Urban Search and Rescue
ESF-10	Oil and Hazardous-Materials Response
ESF-11	Agriculture and Natural Resources (Animal and Plant Disease and Pest Response)
ESF-11	Agriculture and Natural Resources (Natural and Cultural Resources and Historic Properties)
ESF-12	Energy
ESF-13	Public Safety and Security
ESF-14	Long-Term Community Recovery and Mitigation
ESF-15	External Affairs

Specific ESF functional missions, organizational structures, response action, and primary- and support-agency responsibilities are described in the Functional Annexes to the NRP plan. ESFs will coordinate directly with their functional-counterpart state agencies to provide the assistance required by the state. Requests for assistance will be channeled from local jurisdictions through the designated state agencies for action. Based on state-identified response requirements, appropriate federal response assistance will be provided by an ESF to the state or, at the state's request, directly to an affected local jurisdiction.

A principal federal official (PFO) will be appointed by the president to coordinate the federal activities in each declared state. The PFO will work with the state coordination officer (SCO) to identify overall requirements, including unmet needs and evolving support requirements, and coordinate these requirements with the ESFs. The PFO also will coordinate public information, congressional liaison, community liaison, outreach, and donations activities and will facilitate the provision of information and reports to appropriate users. The PFO may appoint a federal coordination officer (FCO) to head a regional interagency emergency response team (ERT), composed of ESF representatives and other support staff. The ERT provides initial response coordination with the affected state at the state EOC or other designated state facility and supports the FCO and ESF operations in the field. The PFO will coordinate response activities with the ESF representatives on the ERT to ensure that federal resources are made available to meet the requirements identified by the state.

A national interagency emergency support team (EST), composed of ESF representatives and other support staff will operate at the Department of Homeland Security, FEMA headquarters, to provide support for the PFO and the ERT. The catastrophic disaster response group (CDRG), composed of representatives from all departments and agencies under the plan, will operate at the national level to provide guidance and policy direction on response coordination and operational issues arising from PFO and ESF response activities. The CDRG also is supported by the EST and will operate from FEMA headquarters. Activities under the plan will be organized at various levels to provide partial response and recovery, using selected ESFs, or to provide full response and recovery, using all ESFs.

ORGANIZATION OF EFFORTS

The organization to implement the procedures under the plan comprises standard elements at the national and regional levels. The overall response structure is shown in Figure 15.8. It is designed to be flexible to accommodate the response and recovery requirements specific to the disaster. The response structure shows the composition of the elements providing response coordination and response operations activities at the headquarters and regional levels, but it does not necessarily represent lines of authority or reporting relationships. In general national-level elements provide support to the regional-level elements, which implement the on-scene response operations in the field.

National-Level Response Structure

The national-level response structure is composed of national interagency coordination and operations

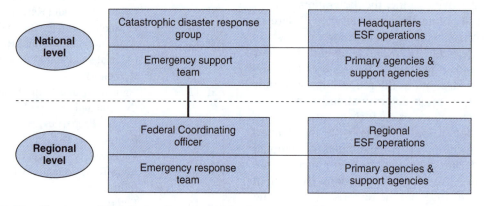

FIGURE 15.8 ■ Classification and Jurisdiction Responsibilities for Large Incidents.

support elements from the participating departments and agencies. Overall the CDRG and EST at FEMA headquarters support interagency coordination activities. These elements will be augmented by department and agency operations support elements at other locations.

The national-level response structure is composed of the following specific elements:

* Catastrophic disaster response group.
* Emergency support team.
* Agency operational centers.

Regional-Level Response Structure

The regional-level response structure is composed of interagency elements operating from various locations as shown in Figure 15.9. Initially, representatives from the ESFs and FEMA will assemble at the regional operations center (ROC) located at the FEMA regional office (or federal regional center). As needed, an advance element of the ERT (ERT-A) will deploy to the field to assess or begin response operations as required. When fully operational, the regional-level response structure will include the FCO and ERT in a disaster field office (DFO), with regional ESFs conducting response operations to provide assistance to each affected state.

Regional Operations Center

The regional director at a FEMA regional office activates the ROC. It is staffed by FEMA and representatives from the primary agencies and other agencies, as needed, to initiate and support federal response activity.

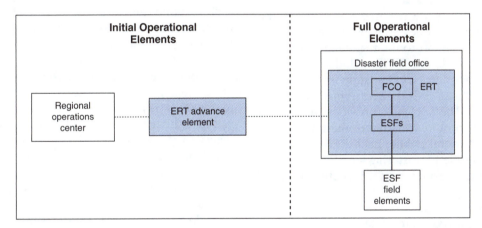

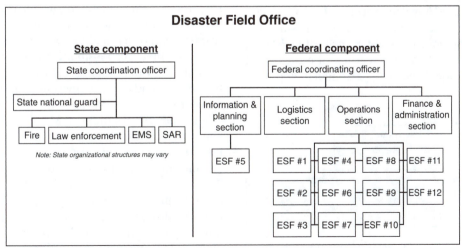

FIGURE 15.9 ■ Regional-level Disaster Responses.

- Gathers damage information regarding the affected area.
- Serves as a point of contact for the affected states, national EST, and federal agencies.
- Establishes communication links with the affected states, national EST, and federal agencies.
- Supports deployment of the ERTs to field locations.
- Implements information and planning activities (under ESF #5).
- Serves as an initial coordination office for federal activity until the ERT is established in the DFO in the field.
- Supports coordination of resources for multi-state and multi-regional disaster response-and-recovery activities, as needed. The organization of the ROC is shown.

FIGURE 15.10 ■ Regional Operations Center Functions.

The ROC functions are listed in Figure 15.10, and the organizational structure appears in Figure 15.11.

Emergency Response Team

The emergency response team (ERT) is the interagency group that provides administrative, logistical, and operational support to the regional response activities in the field. The ERT includes staff from FEMA and other agencies that support the federal coordinating officer (FCO) in carrying out interagency activities. The ERT also provides support for the dissemination of information to the media, Congress, and the general public. Each FEMA regional office is responsible for fostering an ERT and developing appropriate procedures for its notification and deployment.

Disaster Medical Assistance Team

The Department of Homeland Security (DHS), through the National Disaster Medical System (NDMS), fosters the development of **disaster medical-assistance teams (DMATs)**. A DMAT is a group of professional and paraprofessional medical personnel (supported

by a cadre of logistical and administrative staff) designed to provide medical care during a disaster or other event. Each team has a sponsoring organization, such as a major medical center; public-health or public-safety agency; nonprofit, public, or private organization that signs a memorandum of agreement (MOA) with the DHS. The DMAT sponsor organizes the team and recruits members, arranges training, and coordinates the dispatch of the team. To supplement the standard DMATs, there are highly specialized DMATs that deal with specific medical conditions, such as crushing injury, burn, and mental-health emergencies.

Other teams within the NDMS section are **disaster mortuary operational response teams (DMORTs)** that provide mortuary services, **veterinary medical assistance teams (VMATs)** that provide veterinary services, and national nursing response teams (NNRTs) that will be available for situations specifically requiring nurses but not full DMATs. For example, the NNRTs might assist with mass chemoprophylaxis (a mass-vaccination program) or could be called in when the nation's supply of nurses is overwhelmed in responding to a weapon of mass destruction (WMD). Other teams are the national pharmacy response teams (NPRTs) that will be used in situations such as those described for the NNRTs but where pharmacists, not nurses or DMATs, are needed, and the national medical response teams (NMRTs) that are equipped and trained to provide medical care for potentially contaminated victims of weapons of mass destruction.

DMATs deploy to disaster sites with sufficient supplies and equipment to sustain themselves for a period of 72 hours while providing medical care at a fixed or temporary medical-care site. In mass-casualty incidents, their responsibilities may include triaging patients, providing high-quality medical care despite the adverse and austere environment often found at a disaster site, and preparing patients for evacuation. In other types of situations, DMATs may provide primary medical care and may serve to augment overloaded local health-care staffs. Under the rare circumstance that disaster victims are evacuated to a different locale to receive definitive medical care, DMATs may be activated to support patient reception and disposition of patients to hospitals. DMATs are designed to be a rapid-response element to supplement local medical care until other federal or con-

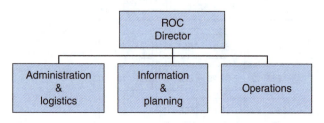

FIGURE 15.11 ■ Regional Operations Center Organization.

FIGURE 15.12 ■ DMAT Deployment at a Planned Event. *(Reprinted with permission of EMS Innovations, Inc.)*

tract resources can be mobilized or until the situation is resolved.

DMAT members are required to maintain appropriate certifications and licensure within their discipline. When members are activated as federal employees, their licensure and certification are recognized by all states. Additionally, DMAT members are paid while serving as part-time federal employees and have the protection of the Federal Tort Claims Act, in which the federal government becomes the defendant in the event of a malpractice claim.

DMATs are principally a community resource available to support local, regional, and state requirements (Figure 15.12). DMATs deploy frequently for scheduled large events. However, as a national resource they can be federalized in preparation and in coordination with local law-enforcement, fire, hazmat, EMS, hospital, public-health, and other first-response personnel plans to more effectively respond in the first 48 hours of a public-health crisis or scheduled event.

Metropolitan Medical Response System

The Metropolitan Medical Response System (MMRS) Program began in 1996 and currently is funded by DHS. The primary focus of the MMRS Program is to develop or enhance existing emergency-preparedness systems to effectively respond to a public-health crisis, especially a WMD event. Through preparation and coordination, local law-enforcement, fire, hazmat, EMS, hospital, public-health, and other first-response personnel plan to more effectively respond in the first 48 hours of a public-health crisis (Figure 15.13). MMRS officially became part of the

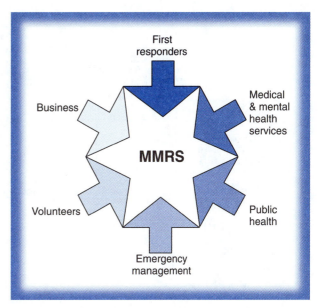

FIGURE 15.13 ■ MMRS Operational Model and Components. *Source: FEMA.*

new Department of Homeland Security on March 1, 2003, and as of 2005 there are 125 MMRS jurisdictions in the United States.

There are nine classifications of technical-rescue incidents according to the National Fire Protection Association (Figure 15.14). As an EMS manager, you need to be aware of the clinical, operational, and logistical needs of patients and EMS personnel involved with technical-rescue incidents. This includes supporting specialized medical training for these disciplines; for example, additional medical training to deal with incidents that may produce crush injuries at the site of a search-and-rescue operation.

- ◆ Water rescue.
- ◆ Rope rescue.
- ◆ Rescue from confined spaces.
- ◆ Wilderness search and rescue.
- ◆ Trench rescue.
- ◆ Vehicle and machinery rescue.
- ◆ Dive search and rescue.
- ◆ Collapse rescue.
- ◆ Any other rescue operations requiring specialized training.

FIGURE 15.14 ■ Categories of Technical Rescue Incidents.

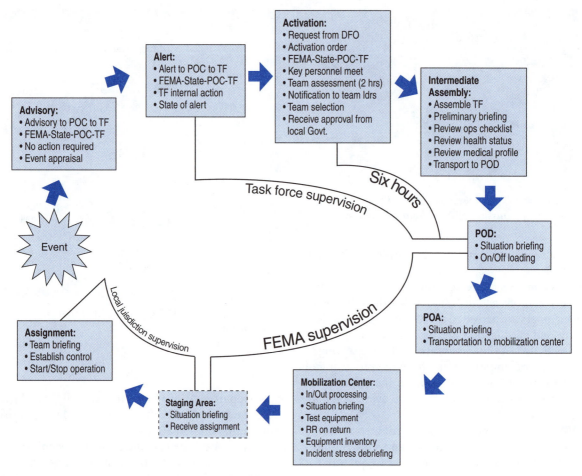

FIGURE 15.15 ■ Sequence in FMEA USAR Team Response.
Source: FEMA USAR Manual.

For an EMS manager it is important to establish interagency cooperation and train with technical rescue teams before an incident. When EMS is summoned to a technical-rescue incident, the minimum qualifications require rescuers to be certified at the BLS level and to have rescue awareness training. All personnel at the site need to understand that they must operate under an incident management system.

Awareness-level training represents the minimum capabilities of responders who in the course of their duties could be called upon to respond to or be first on scene of a technical-rescue incident. Awareness-level operations could involve search, rescue, and recovery of victims in technical-rescue scenarios. At this level the EMS manager and field crew are generally not considered rescuers but rather support personnel for the operations.

The EMS crew conducting rescue operations will need to have specialized training. Operational-level training is designed for responders who will have the capability of hazard recognition, equipment use, and techniques necessary to conduct a technical rescue. Rescue operations are usually supervised by a rescuer certified at the technician level. Technician-level certification has the capability of hazard recognition, equipment use, and techniques necessary to perform and supervise a technical-rescue incident.

The federal system for **urban search and rescue (USAR)** relies on local fire and EMS agencies to organize into 60- to 72-person FEMA USAR teams.

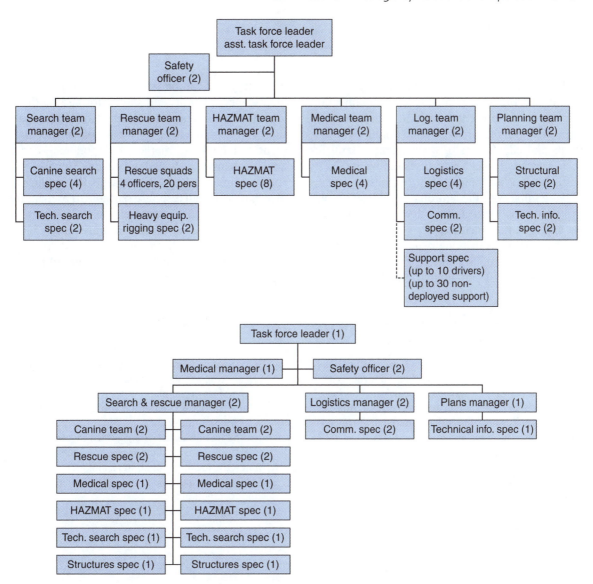

FIGURE 15.16 ■ Medical Group Organization Under USAR Task Force.
Source: FEMA USAR Manual.

There are 28 teams stationed throughout the United States. These teams require federally mandated training and a specialized equipment list. Requests for FEMA teams must go through a chain of command. Figure 15.15 shows the sequence to get a FEMA USAR team into a disaster in your area.

Within each FEMA USAR team is a medical group designed to care for the team and any potential victims. The composition of a USAR task force is listed in Figure 15.16. The medical-specialist and medical-team managers are EMS specific and have requirements for patient care and team care. Type I teams have six personnel, and Type II teams have three medical personnel. While it may not be economical to support a FEMA USAR with your organization, EMS agencies can contribute to regional, state, or other teams. The medical-specialist course is offered in several locations and is required as part of the mandated training for FEMA EMS operations at a USAR event.

CHAPTER REVIEW

Summary

Establishing good relationships with other agencies is paramount for the emergency service organization that expects to ensure continuity of operations and utilize mutual-aid resources. Understanding the expectations and capabilities of a state and federal response is crucial during large disaster events. Most local jurisdictions and emergency response agencies should be familiar with the National Response Framework (NRF) and other guiding documents used for disaster and terrorism incidents.

The bottom line is that emergency response is ultimately a local responsibility, backed up with local, state, and federal resources. Every organization is in charge of its destiny and can be effective when it practices good interagency relations and stays informed on the larger aspect of emergency management outside its jurisdiction. The rapidly changing nature of disaster management dictates that emergency responders keep up to date on changes to the federal system. The National Response Framework (NRF) presents the guiding principles that enable all response partners to prepare for and provide a unified national response to disasters and emergencies. It establishes a comprehensive, national, all-hazards approach to domestic incident response. The National Response Plan was replaced by the National Response Framework effective March 22, 2008.

Research and Review Questions

1. Build a timeline, and list the agencies and notifications that would arrive at your location if an F5 tornado devastated 15 miles of area starting in your jurisdiction.
2. List a set of federal resources that you can identify.
3. Identify three resources to aid an emergency-service organization in learning and instituting ICS.
4. What are the current requirements for emergency service organizations with regard to the National Incident Management System (NIMS)?
5. What is the intent of the National Response Plan, and how does a local agency request aid under the guidelines of the NRP?

6. Locate a mutual-aid agreement for your agency or neighboring agency, and discuss the procedures and components of the plan.
7. Identify the closest federal resources or specialty teams to your community.
8. List the resources that are grouped together in your community for specific disaster or large-scale events.
9. Match the local, regional, state, and federal resources that could be summoned for each of the emergency-support functions.

References

Metcalf, B. (2006, November). Mission: Mobilize, integrate, respond. *FireRescue* Magazine, *24 (11)*, 92–93.

Department of Homeland Security. (2006). *National Response Plan*. Washington, DC: Author.

Data Collection and EMS Research

Accreditation Criteria

CAAS

201.03.01 Patient Care Records: Medical records shall be kept on all patients contacted. At a minimum, the records shall contain incident location and location type; date and call times; patient name, gender and DOB; agency, vehicle and crew identification; assessment of the patient, including vital signs; impression of patient condition; treatment and response to treatment; and disposition of the patient (pp. 382–388).

202.05.01 Incident Reporting: The agency shall have a written policy/ procedure addressing the process for employees to report incidents or unusual occurrences. At a minimum, this shall include the process and documentation of: definition of incidents requiring reporting; investigation of incidents; resolution of incidents; feedback to involved individuals; and how incidents are tracked for any trends and what is to be done with this information (pp. 398–402).

Objectives

After reading this chapter, the student should be able to:

Key Terms

clinical research (p. 380)

data (p. 379)

experimental design (p. 381)

field (p. 380)

ICD-9 codes (p. 379)

implementation fidelity
   (p. 392)

injury severity score (ISS)
   (p. 387)

integrated information
   management system
   (IIMS) (p. 379)

LEADS project
   (p. 384)

mean (p. 390)

mixed method
   (p. 381)

morbidity (p. 396)

mortality (p. 395)

optically scanned report
   (p. 379)

prehospital care report
   (PCR) (p. 379)

qualitative (p. 381)

quantitative (p. 381)

range (p. 390)

raster (p. 391)

relational database
   (p. 380)

reliability (p. 381)

spatial data (p. 391)

trend (p. 380)

Utstein criteria
   (p. 388)

validity (p. 381)

vector (p. 391)

Point of View

Medical research has been around for thousands of years. Currently, EMS is being placed under the microscope to prove that what we do makes a difference. Some EMS providers see this as an obstacle. I see this as a challenge and an opportunity. We should all be open to scrutiny. If you are a true believer in EMS, then take this opportunity to unequivocally prove that *we do make a difference.*

In the December 1997 issue of *Annals of Emergency Medicine*, Dr. Michael Callaham said that the perceived benefit of EMS has largely been based on historical precedent or anecdotes rather than fact. He went on to say, "It is possible to document exactly how much scientific support there is for the efficacy of our present scope of EMS practice, and it is impressively deficient. There is virtually no aspect of EMS that could meet the current requirements of the Food and Drug Administration for approval as safe and effective new therapy." Again, we should not argue this point but take it as a challenge and an opportunity.

We must remember that what we do is medicine, and medicine is based on science, and science is based on research. Much of the practice of EMS medicine we are doing today is significantly different from the EMS medicine we practiced 10 years ago. EMS is not any different from other areas of medicine; much of the practice of medicine is unproven or unanswered.

Much of the EMS research being done is not by EMS professionals. Some of the research is incredibly flawed. We as EMS providers must be able to understand research so we can challenge poor methodology.

We can be in control of our destiny. Research can be a tool to pave our way to that destiny. Be in control of your future. Understand, participate, and be a principal investigator in EMS research.

Baxter Larmon, PhD, MICP
Professor of Medicine, UCLA School of Medicine
Director, Prehospital Care Research Forum at UCLA

■ THE OUTCOME AND RESEARCH CHALLENGE

EMS is often a patient's first access into the health-care system. In early 2000 the federal government began an effort to collect data and quantify the impact of EMS. Since EMS is designed to reduce morbidity and mortality, a system needs to be designed to quan-tify the impact EMS has. As the health-care system faces increasing demands, EMS will need to validate its existence by demonstrating how well the system performs and at what cost. The Institute of Medicine report *Emergency Medical Services at the Crossroads* challenges the very heart of EMS by asking the question, "What does EMS do that makes a difference?" EMS also needs to ask, "Is the performance worth the cost?" The cornerstone of those efforts will be EMS research.

EMS as a specialty and a profession has been in place now for more than 40 years. Despite years of dedicated service, EMS systems are treating and transporting patients with almost no evidence that the care provided is effective and efficient. Few agencies are actively engaged in applying the scientific method to EMS operations, and an even smaller number of services have allied themselves with research institutions. Very little scientific research on EMS management procedures has been produced, and efforts to sustain an EMS journal devoted to the management sciences have yet to be established. As health costs increase and every expense is scrutinized by private insurance, federal, state, and local government money will be put toward services that prove their worth in making an impact on patient outcomes.

EMS AND THE RESEARCH AGENDA

The *EMS Agenda for the Future* identified the need for advancing quality research in the field of EMS. Its Implementation Guide identified the need for a national EMS research agenda as one of the top 10 action items, and this action plan has given rise to the National EMS Research Agenda. The Research Agenda was a cooperative agreement between NHTSA, the National Association of EMS Physicians, the Maternal Child Health Bureau, and the Health Resources Administration.

The process of creating a research agenda has given rise to the EMS Outcomes Project (EMSOP), which has produced a prioritized list of conditions for adults and children that are variables for EMS study. The EMS for Children Project at the National Children's Medical Center in Washington, D.C., supported a similar study to identify research topics for children. It is important for research to be applied to EMS to ensure that the best possible patient care is provided in the prehospital setting.

INTEGRATED INFORMATION MANAGEMENT SYSTEMS

EMS reports are the foundation of any EMS research effort. Often an EMS report is called a **prehospital care report (PCR)**. EMS reports connect patient care with dispatch data and patient outcomes. They serve as a defense mechanism for litigation and help reduce patient-care errors by ensuring a smooth transition of the patient-care information from EMS provider to hospital.

There are three basic types of EMS reporting technology: the optically scanned report, the manually entered report, and the paperless system. The first type, the **optically scanned report**, is filled out in the field by EMS providers and sent to a clerical person, who in turn feeds the form into an optical scanner that tabulates the data and enters it into the computer. Optical scanners read ink or pencil and can scan for specific words.

The second and most common method of EMS reporting is the manual-entry method, in which clerical personnel manually enter specific data elements from PCRs generated in the field. This allows for better quality control. The King County Medic One program utilizes this system to enter cardiac-arrest and chest-pain data for patients in the Seattle system. It provides reliable data and is able to generate more detailed physiological conditions than those listed in the ICD-9 codes. (The *International Statistical Classification of Diseases and Related Health Problems* **ICD-9 codes** are the classifications that allow medical providers to bill government insurances including Medicare and Medicaid.) This process provides an additional quality check, and if a clerical person identifies an error or omission, the PCR can be returned to the provider to correct the information.

The third type of data collection is the paperless system, such as software that uses point-and-click data entry. Many of these programs will write the narrative for the report after sufficient data has been entered into the system. This ensures a consistent narrative format and allows the EMS provider to narrate the information. Automated processes help reduce the time on task for EMS crews, and drop-down menus in the computer-based programs help standardize the data. Some forms of pen-based computer systems allow for data collection by converting the EMS provider's handwriting on a screen. Once this data is collected, it can be sent quickly to the EMS or fire agency's database, which reduces the need for a data-input person to collect the information. The efficiency of the process allows for a quicker turnaround time on the billing, and once it is stored in a computer, the data can be sent by telephone modem, infrared, or wireless networks.

Collection of data from any EMS system includes large amounts of computerized data that originates from the prehospital care reports, communications centers, and other databases. **Integrated information management system (IIMS)** is a name commonly used to describe the system that collects data for an EMS system. An IIMS is used to manages databases that are a collection of records that must be organized. **Data** are

TABLE 16.1 ■ IIMS Planning

- What will the system do?
- Who should enter the data?
- What data should be gathered?
- Where should data acquisition occur; should it be centralized or decentralized?
- What software and hardware should be used?
- In what format should the data be reported?
- To whom should data be reported?
- How will the system function?
- What safeguards will be needed?
- Who will need or conduct training?
- What will the system cost to maintain and install?

the facts or figures; a single datum is related to other data and is displayed in a specific place in the database known as a **field**. In broader terms information professionals call the sharing of data between two different collected sources a **relational database**. For example, many electronic EMS PCRs relate or transfer their data to the National EMS Information Systems (NEMSIS) or to the National Fire Incident Reporting System (NFIRS) from the EMS agency's databases. Relational databases are sometimes confused with a network, which is a group of computers tied to printers, modems, data, programs, and other devices.

An effective system must establish validity by managing information that is related to the EMS agency's needs, and that information must be produced in a timely and accurate fashion to generate reliable data. When establishing an IIMS, it is important to answer several questions, which are listed in Table 16.1.

To assist in making decisions about how the system will be set up, how it will function, training needs, costs, and so on, a top-down and a bottom-up management analysis must be performed. A top-down analysis identifies the type of data management required, and a bottom-up analysis should identify how data currently is collected and may suggest means for acquiring data.

■ DOMAINS OF EMS RESEARCH

There are three areas of interest in the EMS research community that should be applied to an EMS organization: clinical, educational, and systems research. **Clinical research** involves the study of direct patient-care activities. Educational research examines the methods of preparing, remediating, and providing recertification activities to EMS providers. Systems research studies the effects of EMS system designs and operational methods for resource utilization.

Clinical research should be conducted by every EMS agency regardless of its size or level of care. Unfortunately, there are very few people trained to conduct clinical research, and this presents challenges to agencies with limited personnel. There is very little EMS research that proves that medical treatment and procedures performed in the field have any impact on the outcome of the patients. Typically, there are fewer than 30 clinical studies of EMS produced in the United States per year.

For an EMS manager a plan to partner with a research facility of a local university is an important collaboration to build. A major university can provide statisticians, faculty, and student workers to assist with the development and collection of scientific data. EMS managers should attempt to generate research on cost effectiveness of medications or procedures, resource utilization, efficacy of protocols, and injury prevention. A major effort is needed to implement research that will identify ways to protect patients and reduce medical errors.

Systems research has been conducted to some extent by private ambulance providers. Systems research is more complex and often requires support from the disciplines of engineering, economics, and epidemiology. Often mathematical and computer simulation packages are needed to capture the appropriate data. This type of undertaking requires a shift from traditional EMS quality-improvement activities to one based on overall system performance; an example is response-time performance and utilization of EMS units. Systems research has some barriers that make it difficult to capture the data accurately; for example, arrival on scene as a data point is measured differently by different organizations. Some agencies go as far as to identify the time at the patient's side versus identifying the time when the vehicle arrives on scene by a computer-aided dispatch system.

Research on education seeks to link the training that involves course work, educational methods, and practice to the delivery of EMS services. Research education is designed to identify **trends** in EMS response and performance issues and then help to establish whether or not an education package can improve them. One trend in educational research is to focus on distance-education platforms and the use of Internet training to meet continuing education requirements.

TYPES OF RESEARCH

Research is often misinterpreted or misunderstood by the reader. So it is important to understand that research can be qualitative or quantitative in nature (Figure 16.1). **Qualitative** research is often anecdotal, composed of soft data that does not have enough numerical or objective measurements to be statistically significant. Qualitative studies are often based on observation of a single phenomena or a case study composed from material from interviews and reviews of documents. Qualitative research often has problems with reliability.

Reliability refers to the consistency of the results or measurements that the research produces and is the degree to which a measure is free from random errors. **Validity** is the extent to which the research accurately measures what it is designed to measure. A lack of validity is a threat to quantitative research. Increases to validity and reliability can be solved by using a process called *triangulation*, in which a combination of historical documents, archived material, and interviews is used together to come to a conclusion. This is often done with a literature search.

The **mixed method** is becoming the most common type of research; it uses one type of research to set up the other. For example, the qualitative method can be used to identify the problem through interviews with EMS providers who describe difficulty with traumatized airways; then a quantitative research model is applied to the problem, and a statistical conclusion is made. The reverse to this sequence also is considered a mixed method. For example, a research project may identify statistical data on poor response times in a district and then conduct interviews on the problem to reach a conclusion.

Quantitative research, also known as experimental research, is based on objective and measurable factors. Many quantitative applications are one of three types: experiment, quasi experiment, or nonexperiment. An **experimental design** defines a population and applies a treatment, intervention, or change, then measures the effects. The term *treatment* is used to describe a change, an intervention, or a modification to the groups. A quasi experiment measures a population that has already been treated or has received an intervention, and then it measures the difference between groups. A nonexperiment measures the difference between two groups without identifying the intervention, change, or treatment applied to the groups. Experiments can be single or double blind. Subjects in the population are randomly assigned instead of specifically selected. Proving cause-and-effect relationships or a correlation between treatments and outcomes can be completed only by using an experiment. But experiments can produce good data and poor data.

In many cases organizations make changes based on one scientific study that has been poorly interpreted. The MAST garment is an excellent example of a scientific study that changed practice but should not have done so without further study, evidence, and interpretation. In one study involving 911 patients, military antishock trousers (MAST) were placed on patients when indicated on even days and withheld on odd days. All of the other factors, such as response, scene, and transport time were similar. In open-system or penetrating trauma, there was no statistical difference in the outcomes of the patients. However, the ones who received the MAST had significantly longer hospital stays at tremendous cost. Consequently, in many jurisdictions the MAST were removed from EMS vehicles. But no consideration was given to other studies that showed tremendous benefits for pelvic fractures. Even more important is the subset of the EMS patients in the original study with blood pressures lower than 50 mmHg systolic, who had a 300% increase in survival rates with the MAST. The number of patients was fewer than 20, and it was thought that it was not statistically significant.

RESEARCH METHODOLOGIES

Several different methods are used by drug companies, medical equipment manufacturers, epidemiologists,

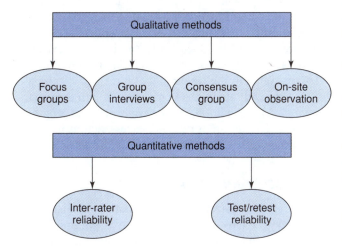

FIGURE 16.1 ■ Categories of Research.

and quality-improvement professionals. Most often EMS managers need to refer to publications on research. The validity of research progresses from the less reliable, starting with what is commonly referred to as an urban legend, and progressing to anecdotal information or simple stories about a personal experience. The next level is news articles, which are often researched and vetted, meaning the information is verified by another source. Industry journal articles such as those placed in *JEMS*, *EMS Magazine*, or other EMS publications comprise the next level of research information, often with several sources to substantiate the information. A case study is the beginning of what is considered scientific research and is usually a factual account of an event described in detail. Finally, a peer-reviewed journal is considered the most comprehensive review of research and is accepted in the eyes of most scientists and clinicians who want to establish evidence-based practice.

When reviewing information on the science of EMS, managers should look for the most expensive and complex research method, which is known as the *randomized control trial*. This allows the researcher to assign host, treatments, and environments to experimental and control groups. A prospective study has ethical constraints in that not all patients receive the same treatments, and potentially life-saving techniques or devices may be withheld from a population of patients. The amiodarone antiarrhythmic study was a double-blind, randomized control trial, with the EMS crews administering one of three unmarked vials containing amiodarone, lidocaine, or normal saline. In most communities and EMS organizations where this kind of objective analysis is being conducted, the public must be educated on the study. The public has the right to refuse to participate in prehospital randomized studies by obtaining a wristband. Since it can be impractical to get informed consent from a patient in an emergency, most ethics review boards approving this kind of research favor the wristband as a way to conduct informed consent.

Case-control studies are observational designs in which a group of individuals or events are chosen according to whether they do or do not have a characteristic or risk factor of which the cause is to be studied. Those individuals or events are compared with controls that share important characteristics with the group being studied. An example would be an injured person in a vehicle rollover; the control would be a noninjured patient in a vehicle that did not roll over on the same section of roadway. The research focuses on what individual, environmental, and vehicle factors were observed between the two events. This type of study requires the collection of retrospective data.

Another type of qualitative or observational study is the cohort study, which can be retrospective or prospective. A cohort study involves a group of individuals with common characteristics or exposures that are observed over time. The case group is exposed to the injury problem, and the comparison group is not. Cohort studies allow for incident rates to be determined. Prospective cohort studies follow groups for their exposures in the future to an event, and retrospective studies reconstruct records of exposures and records for outcomes.

A cross-sectional study, sometimes called a *snapshot*, is used by researchers and quality-improvement personnel to ensure protocols and standardized therapies are being followed and producing a designed outcome. A cross-sectional study makes an observation of a group during a certain period of time; for example, the intubation success rate of a group of paramedics in a three-month period.

An important consideration of any type of research is the size of the sample. The sample size is used to evaluate the power of the design—that is, its validity and reliability. Always ask: are there sufficient events or hosts to study in order to draw a conclusion on the outcomes of the treatments applied to the event?

Research in EMS is focused on **implementation fidelity**, which researches how closely the design of a program equals the actual implementation of the program. A program with low fidelity is one that is implemented but does not address the issues or perform to its original design or goal. Paramedic preceptor programs are often low fidelity—that is, instead of a true evaluation of their medical skills or education, few if any weaknesses are documented and often every student or employee passes. A high-fidelity program is demonstrated in the research that tracked the cardiac-arrest survival rate in Las Vegas casinos with the placement of AEDs in hotel casino properties. The program saw a 70% save rate for cardiac-arrest victims in those properties with automatic external defibrillators.

DATA ENTRY AND DATA QUALITY

Research and data on EMS incidents and casualties rely very heavily on the street-level paramedic or EMT entering information into the computer and the inputs being available for analysis. While manual analysis certainly is possible, it usually is avoided because the tedious calculations quickly overwhelm the ability to perform analysis in any significant manner. The advantage of a computer is that it processes data quickly and accurately.

Most EMS agencies have a computer system of some sort, ranging from personal computers (PCs) for small departments to local area networks (LANs) or wide area networks (WANs) for large metropolitan or regional settings, where multiple agencies agree to share a system. Whatever the case, the data is typically entered into a custom software program that is purchased by the state or local EMS agency. A custom vendor's software must be compatible with both the NEMSIS and NFIRS programs.

Any program you purchase should contain a good error-checking routine. Data quality is always a problem, and the old adage "garbage in, garbage out" certainly applies to EMS organization reports. The entry program should, for example, check each item to make sure a valid code has been entered. Whenever the program encounters an error, it should give an opportunity to correct the error before it becomes part of the data-base. Response times obviously cannot have hours greater than 23 and minutes greater than 59, so an entry program should check hours and minutes for valid numbers and allow corrections to be made immediately. When a service in an area begins to exceed a 10-minute response time on more than 10% of emergency responses annually, this data point can be used to add additional EMS coverage. GIS and data can be combined to identify station coverage for drive times that would meet a six-minutes-or-less response time.

Figure 16.2 is a graphic related to deployment analysis. This is an enlarged view of one small section of a fire district that provides EMS services. It spans 277 square miles and represents a response-performance map that is updated monthly by the organization. The boxes represent roughly one-quarter square mile. If a box is green (in the illustration it appears as gray), the agency met fractile performance (90%, six minutes). If a box is yellow (in the illustration it appears as light gray or white), the agency did not meet performance standards. The top number in the center of the box is the 90% fractile (90% of all calls in that quad had a response time of that time or less). If the top number is

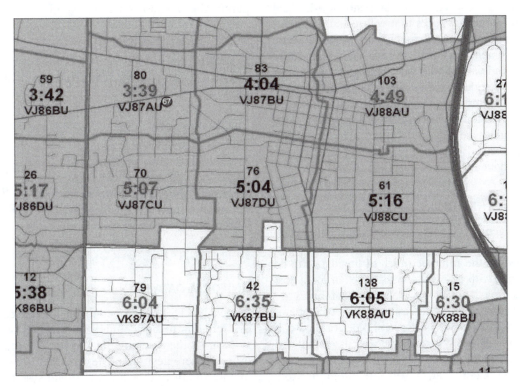

FIGURE 16.2 ■ Deployment Analysis Using GIS.
(Courtesy of Tualatin Valley Fire Rescue. Reprinted with permission.)

black, the performance in that quad is the same as or better than the previous month's performance. If the top number is red (in the illustration it appears as light gray), performance is worse than it was the previous month. The middle number is the number of emergency calls (code 3 for all; typically, Charlie, Delta, Echo) that occurred in that particular quadrant in the previous six months. The bottom number is the GIS map identifier. With a map like this, you can see at a glance where performance is slipping, and it becomes easy to zero in on an area to establish what the root causes are for poor performance.

The data collected to describe an incident is the foundation of any EMS system. Therefore, editing and correcting errors is a system-wide activity, involving local, state, and federal organizations. All errors resulting from the edit/update process need to be reported to EMS agencies, and the submission of corrections from EMS agencies is essential. This is especially important for "fatal errors," which prevent the data from being included in the NEMSIS and NFIRS databases.

Agencies collecting EMS data need to establish data-quality procedures if they intend to take full advantage of their data. There should be a system in place to double-check the collection and data-entry work. Field checks and relational checks that will reveal unacceptable and unreasonable data can be built into the system. Data-management personnel can use these techniques to improve and validate the data. Data-entry programs should include code-checking routines to identify errors in individual items in the report and errors reflected through inconsistencies between items. Because entry programs cannot be expected to find all errors, EMS agencies also need data-quality procedures to ensure that correct data are entered into their systems.

DATA-COLLECTION TOOLS AND TRENDS

Most EMS agencies already are collecting some data on their operations, and this data can be used for research. With the use of the Internet, much of this data can be cross-matched with other databases; for example, a hospital's or census data systems. Relational databases are commonly used to produce research on EMS operations and their effectiveness. Several types of data-collection templates and systems will be discussed in detail later in this chapter. One trend in cardiac emergencies is the use of Code Stat, a software program from Medtronics that merges with many of the computerized software applications designed for prehospital care reports. This data then can be cross-referenced with census data to identify the area in which the most cardiac emergencies are occurring. Many of these projects aim to describe the demographics and provide more of a description of the system than real science does.

One demographic project is the Longitudinal EMT Attributes and Demographic Study, or **LEADS project**, being conducted by the NREMT in cooperation with NHTSA. The LEADS project is a 10-year-long data-collection project started in 1999 and is designed to monitor changes in the EMS workforce. The survey collects information from 5000 EMTs on a wide range of topics: education, finances, management, patient care, and the impact of their careers on their personal lives. The survey also rotates a specific topic that focuses on a timely topic; data has been collected on ambulance safety, sleep, and common health issues. A committee at the NREMT oversees the data and provides material to publish results from the project in trade journals.

The NREMT also conducts a practice analysis every five years to determine if the material in NREMT tests reflect what EMTs and paramedics do in the field. The practice analysis looks at the kind of skills being used and the frequency of critical patients. This process ensures that important functions of the paramedic and EMTs are evaluated.

LONGITUDINAL DATA

A specific type of medical emergency, such as a hip fracture, cannot be tracked from onset through rehabilitation because the linkages among the data collected by various medical providers do not exist. This makes measuring outcomes and scientifically identifying causes of negative impacts on the system unattainable. This can result in errors in patient care due to incomplete medical histories or a lack of integration by the provider network. Much like a hazardous material, the substance should be able to be tracked from cradle to grave.

■ NATIONAL DATA-COLLECTION EFFORTS—

UNIFORM DATA SET

In 1993 NHTSA developed through a consensus group 81 elements to be included in an EMS information system. Of the 81 elements 49 were essential and another 32 were desirable. These elements were created in an attempt to develop a national system to establish a baseline for quality improvement and benchmarking activities. A first step in standardizing the national data is to create a set of national definitions for each element in the database. The uniform data set would become the basis for a national system after the 1996 *EMS Agenda for the Future* called for creation of a national database.

HOSPITAL COLLECTION OF EMS DATA

Hospital emergency departments, in conjunction with the CDC's National Center for Injury Prevention and Control, established a uniform data set. This collection of information is called the Data Elements for Emergency Department Systems (DEEDS) and is very similar to the NHTSA EMS data set. Until recently, though, there was very little hospital adoption of the DEEDS program.

NHSTA formed a collaborative process to establish a uniform data set that would identify the key elements to be collected. The uniform data set provides common definitions for the prehospital system to link to and use to analyze other data elements that are obtained from dispatch centers, hospitals, and other health facilities. The benefits of using the uniform prehospital EMS data set are improving patient care and providing feedback to EMS providers and system administrators. For frontline EMS managers and leaders, the uniform data set allows management to analyze the system performance.

Implementation of a uniform data set allows for comparison of performance from system to system and validation of processes through multisite practice and analysis. The uniform data set permits regions to make decisions on injury prevention and to accurately evaluate programs, and managers can track the patient from initial injury or illness and link that data with that of other health-care providers and facilities to get a real cost of the patient's event. It is imperative that EMS see the importance of linking the EMS data set with other related reports. Other common databases include the crash statistics from the Office of Highway Traffic Safety, hospital records, and outcome data.

FEDERAL COORDINATION OF RESEARCH

The Agency for Healthcare Research and Quality (AHRQ) is a government agency that has created 12 evidence-based centers throughout the United States. Those centers are commissioned to study evidence-based medicine guidelines on an assortment of topics. The AHRQ prioritizes research and is an excellent resource for performance-improvement programs.

Currently, the uniform EMS data set contains 81 data points that document the patient identification, status, treatment, medical direction, and disposition

and describe the emergency response. There are 49 essential elements that are critical to the basic operation of any EMS agency. Examples of the essential elements are date of birth, time to arrival at the scene, and cause of injury. There are 32 desirable elements that help justify services on a regional basis. Desirable elements include such items as chief complaint, procedure attempts, and revised trauma score.

It is imperative that EMS see the importance of linking the EMS data set with other data sources. The power behind the type of statistical data that the uniform data set achieves is that it can serve EMS managers by justifying equipment and personnel. Some barriers are to be expected in implementing the data set. First, a legal authority for collecting the data needs to be established, and with HIPAA, confidentiality needs to be ensured. Financially, there is a cost and ongoing expense for data input and querying data for reports and analysis. Many EMS agencies are employing data-analysis or computer-information specialists to manage the data collection and staffing programs. Politically, EMS managers need to obtain the cooperation of all the stakeholders in the system to get accurate data. The role of the government needs to be defined and a consensus built on the use and management of the data. Finally, the technology is a hurdle—most paramedics and EMTs see data input as a nonimportant function of their job. Incentive programs need to be tied to performance and accuracy of data input.

An eight-hour course offered by NHTSA is designed to provide guidance for those wanting to build EMS databases incorporating the Uniform Prehospital Data Set. The project is managed by NHTSA, which contracts organizations and researchers to investigate the development of the system.

NATIONAL EMS INFORMATION SYSTEM

The National EMS Information System (NEMSIS) is a fully implemented and useful EMS database that resides or is controlled at a local, state, and national level (Figure 16.3). To date there are 48 states and two territories that have interoperating agreements to promote and implement the system. The National Association of State EMS Directors, National Association of EMS Physicians, NHTSA, and the Health Resources and Services Administration (HRSA) developed the database with grant money from the federal government. The purpose was to evaluate patient and EMS system outcomes, facilitate research, and develop a nationwide EMS training curriculum.

THE PORTABILITY OF NEMSIS DATA

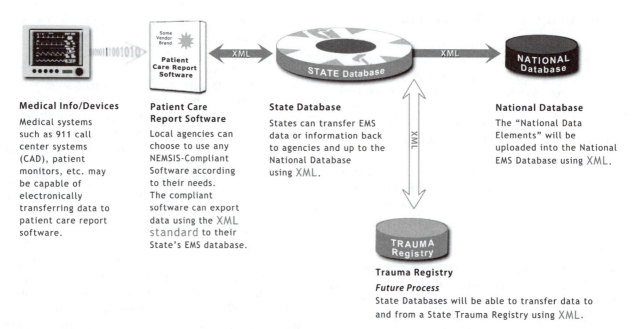

Medical Info/Devices

Medical systems such as 911 call center systems (CAD), patient monitors, etc. may be capable of electronically transferring data to patient care report software.

Patient Care Report Software

Local agencies can choose to use any NEMSIS-Compliant Software according to their needs. The compliant software can export data using the XML standard to their State's EMS database.

State Database

States can transfer EMS data or information back to agencies and up to the National Database using XML.

National Database

The "National Data Elements" will be uploaded into the National EMS Database using XML.

Trauma Registry
Future Process
State Databases will be able to transfer data to and from a State Trauma Registry using XML.

FIGURE 16.3 ■ National EMS Information System (NEMSIS). *(Source: NEMSIS, www.nemsis.org. Reprinted with permission.)*

This system is easily integrated with EMS patient run records and can be routinely used for performance measurement and EMS research. NEMSIS starts by collecting data from medical devices such as cardiac monitors, 911 computer systems, and patient monitors such as implantable defibrillators or insulin pumps. Information is then collected from prehospital care reports from the local agency in software that is in an XML standard to the state EMS database. The state database then can transfer the data either back to the agency or up to a national database in the same XML format. EMS services and hospitals can be linked with local, state, and national databases and the trauma registry to enable data to flow both ways. Even the smallest EMS agency can report in by phone or Internet access. This type of pooled data can be used to evaluate the clinical care provided to patients.

NEMSIS uses performance indicators defined in the NHTSA *Guide to Performance Measures*, which is based on the Uniform Prehospital Dataset definitions. Version 2.2.1 (October 2008) is the latest dataset and is available through various vendors; there are 52 partici-pating states and U.S. territories. NHTSA maintains a Web site and upgrades the dataset periodically. Presentation software to educate paramedics and EMS managers is available to download on the Web site in PowerPoint and PDF formats. The NEMSIS system data elements are categorized in Figure 16.4. The goal of the NEMSIS system is to have a national standard set of EMS data to make objective decisions about EMS based on evidence.

A small subset of the NHTSA Uniform Prehospital Dataset, consisting of 68 different elements, is the minimum dataset each state is recommended to collect and submit. Each state can choose to collect different information and require the EMS provider agencies to submit different material. The national EMS database is the central storage agency for the data. Most of the vendors provide that material in XML format to allow it to be imported and exported in and out of several types of programs.

The Department of Transportation has made progress regarding EMS software and NEMSIS compliance. The NEMSIS Technical Assistance Center

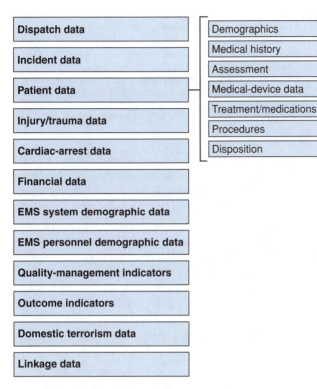

Dispatch data		Demographics
Incident data		Medical history
Patient data		Assessment
Injury/trauma data		Medical-device data
Cardiac-arrest data		Treatment/medications
Financial data		Procedures
EMS system demographic data		Disposition
EMS personnel demographic data		
Quality-management indicators		
Outcome indicators		
Domestic terrorism data		
Linkage data		

FIGURE 16.4 ■ Capture Data Elements Categories.

(NEMSIS TAC) provides a certification process for EMS software developers, with test cases and documentation, to become NEMSIS compliant. The initial developers represented 11 EMS software packages that work on Web-based, Windows, Palm OS, or Pocket PC platforms. There is no charge for NEMSIS compliance testing, and EMS managers should ensure that software has been tested and is in compliance with NEMSIS TAC standards.

EMS OUTCOMES PROJECT

A major research initiative funded by NHTSA is designed to measure patient outcomes from EMS interventions. The EMS Outcomes Project (EMSOP) will build on the NHTSA uniform data set to prioritize and identify EMS activities. The goal is to identify the frequency of EMS events, the impact of EMS treatments, and the clinical outcomes of the treatment provided. EMSOP is being managed by the National Association of State and EMS Administrators.

In the future EMSOP will establish clinical outcomes that can compare EMS systems performance. It will also give rise to a national standard of care for EMS providers that is created from evidence-based

medicine. A significant amount of current EMS practice has not been validated scientifically compared to other medical specialties, and very little EMS research is being undertaken by provider agencies. EMSOP is an attempt to put forth a national movement to look critically at the science behind the EMS activities and ensure it is valid, reliable, comprehensible, and available to the public and the provider agencies.

TRAUMA REGISTRY

An EMS service that transports to a level-one or level-two designated trauma center is included as part of a regional trauma system. A level-one trauma center designation requires that the trauma center conduct research in trauma care. As part of this system, each EMS provider is given a designation number, and outcome scores are correlated with each agency based on the injuries the patient receives. Prehospital factors such as response time, scene time, and transport times are all analyzed. This set of data collected on trauma-center patients is known as a *trauma registry*.

Trauma-center registries are a valuable descriptive and qualitative management tool for trauma centers and trauma systems. EMS providers should be linked electronically to the trauma center, as this helps in tracking the patient from the initial event to rehabilitation and discharge. Two national trauma databases collect data from participating trauma centers and state EMS offices: the National Trauma Data Bank (NTDB) maintained by the American College of Surgeons, and the National Pediatric Trauma Registry maintained by Tufts University. The NTDB is designed to provide regional and national benchmarking for trauma systems. The data will provide evidence-based medicine to build better guidelines for trauma care and options to measure performance. Over 1.1 million cases from 43 states and U.S. territories are in the NTDB, and an annual report is distributed every October. The information is HIPAA compliant, and security and anonymity of the data is guaranteed by the American College of Surgeons.

Research in trauma care focuses on two relational databases that assess the patient's outcomes. Patients that enter into a trauma system receive an **injury severity score (ISS)**. The ISS is based on anatomical injuries and is calculated by adding the sum of the squared scores on the abbreviated injury scale for the three most severely injured body regions and the age of the patient. Field data is collected and converted into a revised trauma score (RTS), which then assigns values from zero to four based on defined intervals of the Glasgow Coma Scale, systolic blood pressure, and respiratory rate. The two scoring systems are combined, and based on the patient's age, RTS, and ISS scores, the probability of survival may be computed and the patient outcome compared with similar patients. A scattered plot graph then can be used to show the patients who survived but should have died due to the injuries, and those patients who died but should have lived. This allows for the scientific analysis of therapies and skills. In some cases it allows for analysis of errors in the care of iatrogenic injuries. A "Z" score is calculated from an ISS score to identify outcomes as a mortality indicator. Trauma can also be assessed using the A Severity Characterization of Trauma (ASCOT) or the Trauma and Injury Severity Score (TRISS).

MOTOR-VEHICLE CRASH DATA

Motor-vehicle crash data is collected and maintained through state departments of transportation and law-enforcement agencies. Each of these sources should also attempt to be linked to EMS information systems. NHTSA uses a probabilistic linkage in the Crash Outcomes Data Evaluation System (CODES) to match data in 25 states from law enforcement, EMS, and emergency departments. CODES helps generate outcome data for collisions and aids in developing highway traffic-safety programs.

CARDIAC-ARREST DATA

Data on cardiac arrest is generally collected and reported as a registry for quality improvement and as a research report that looks at outcomes and the interventions related to those outcomes. The **Utstein criteria** have been the gold standard for measuring cardiac-arrest survival since 1991. They help EMS managers conduct quality-improvement activities and compare arrest statistics against other systems. Outcome data after a cardiac arrest is dependent on critical interventions such as defibrillation, CPR, and advanced life support. The Utstein criteria have a reporting template in an algorithm format that instructs a quality-improvement coordinator or EMS researchers on which cardiac-arrest patients to include in the database.

The Utstein algorithm looks at attempted and nonattempted resuscitation, the location of the cardiac arrest, and the first monitored rhythm. The outcome data is collected on the patient's neurological status at several points, usually upon discharge from the hospital, at 30 days, at six months, and at one year. To deter-

- ◆ Time of witnessed/monitored arrest.
- ◆ Time when call received.
- ◆ Time of first rhythm analysis/assessment of need for CPR.
- ◆ Time of first CPR attempts.
- ◆ Time of first defibrillation attempt if shockable rhythm.

FIGURE 16.5 ■ Time Elements for Cardiac Arrest Applied to Utstein Criteria.

mine a valid neurological score, one of the most commonly used assessments is the cerebral performance category (CPC). Also recorded are whether the arrest was witnessed or unwitnessed and the time it took to start CPR before EMS arrival, all of which data are linked with the etiology of the cardiac arrest. It is important for EMS units to synchronize their clocks with the dispatch computer to get an accurate time. The time elements that need to be recorded for the Utstein criteria are listed in Figure 16.5. The Utstein criterion that has been modified for reporting cardiac-arrest outcomes is displayed in Figure 16.6.

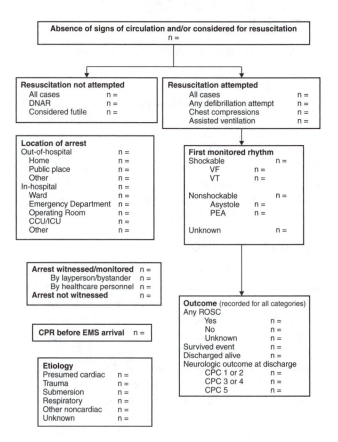

FIGURE 16.6 ■ Utstein Criteria for Measuring Cardiac-Arrest Outcomes.

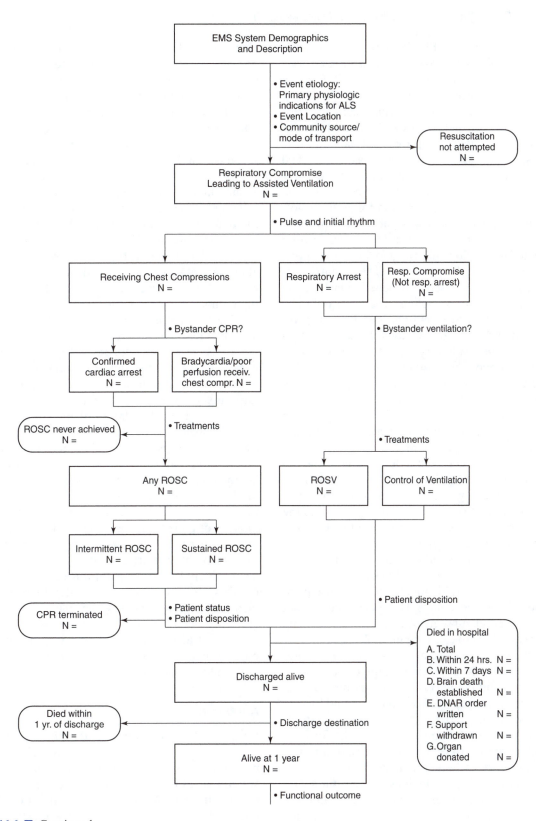

FIGURE 16.6 ■ Continued

389

In 1995, as part of the FDA's approval of the intravenous drug amiodarone, Wyeth-Ayerst Laboratories set up the Advanced Resuscitation of Refractory VT/VF IV Amiodarone Evaluation (ARRIVE) registry to collect data on out-of-hospital cardiac-arrest rates. This data can help EMS systems using amiodarone with their quality-improvement data and cardiac-arrest outcomes. The system has over 4000 patients in the registry, and it continues to grow.

In 1991 when the recommended guidelines for the uniform reporting of Utstein criteria for adult out-of-hospital cardiac-arrest data was published, it did not fit pediatric patients. The pediatric Utstein criteria were created to develop uniform guidelines for reporting clinical pediatric ALS research. The pediatric Utstein criteria were purposely broadened to include the very important group of children requiring only airway and ventilation interventions, because improvements in outcome are likely to come from prevention of progression of respiratory failure or shock to cardiac arrest. To make best use of these guidelines, the pediatric Utstein criteria emphasized simplicity while ensuring that the guidelines would be compatible with recommended data collection in adults. The pediatric Utstein criteria used many of the same definitions detailed in the Utstein style for adults.

The template begins with a description of the demographics of the EMS system of interest, a definition of the event etiology, a description of the patient's source and mode of transport, as appropriate, and the location of the event. The first numerical input is a determination of the number of children in the broad category Respiratory Compromise Leading to Assisted Ventilation. The number of patients in that section then provides the denominator for subsequent calculation of incidence rates. The specific number of patients entered at subsequent levels in the template permits researchers to calculate multiple rates.

Investigators can calculate a large variety of outcome rates from the reporting template because of the multiple combinations of denominators and numerators possible. Reported outcomes should be presented as rates or percentages; for example, rate of admissions per total number of resuscitations attempted. The appropriate outcome rate to report may differ among various systems and locations. As a minimum, the task force recommends that all studies report the denominator of cardiac arrests and the numerator of number discharged alive (the outcome would be the cardiac-arrest survival-to-hospital-dis-charge rate). However, it is recognized that the number of survivors will be a small percentage of total cardiac arrests in children. Therefore, collection of additional outcome rates is strongly encouraged. For example, the task force recommends enumerating the number of children with sustained return of spontaneous circulation (ROSC) and survival for more than seven days as additional core outcome data.

Because some template sections will contain only a small number of patients, calculation of all possible rates or percentages may fail to adequately express the potential clinical variability that exists. Reporting confidence intervals (CIs) can improve interpretation of study data by indicating the degree of uncertainty about an observation. Confidence intervals give a **range** of values based on observed data; the greater the number of patients, the narrower the CI. A range of clinically acceptable CIs may be calculated; usually a CI of at least 80% is used.

A recent analysis illustrates the advantage of reporting CIs. A study of out-of-hospital arrest outcome in New York City included over 3000 patients; the Utstein style was combined with CIs to express outcome data. If ventricular fibrillation onset was witnessed, **mean** survival was 5.3% (99% CI, 2.9% to 8.8%). Despite the large total number of patients, the threefold variation in possible mean survival should be noted in this subgroup analysis. The CI more accurately reflects the potential variability of the observation of interest, if the study is repeated with a similar number of patients. Depending on the research question asked, patients who subsequently experience one or more additional arrests during hospitalization may be considered as one person for analysis, whether or not they are successfully resuscitated following subsequent arrests. Alternatively, they may be analyzed separately to determine the impact of multiple arrests on outcome. EMS managers should implement a database for collecting the information in this format for both children and adults.

AIRWAY DATA

Cardiac-arrest data is frequently a high-profile number and is requested by government officials and news media when budget or political situations arise in the system; for example, tax increases or requests for additional funding. Although cardiac-arrest data is only one small portion of the call volume, it is the securing and maintenance of the airway that presents more liability issues to EMS. The collection of airway data is a vital part of any research or QI program being conducted by an EMS service. Many of the studies conducted on prehospital use of advanced airways have not reported positive findings, and the skill of intubation is frequently under attack. The NAEMSP has created a national database for tracking airway control maneuvers that is displayed in Figure 16.7.

Intubation by paramedics has come under scrutiny throughout the nation. A hypoxic brain injury from a missed intubation can result in millions of dollars of loss to an organization. The current scientific evidence indicates that intubation, especially in patients with a pulse or in pediatric patients, may be causing more harm than good. It is most likely the most important skill to measure and for which to collect data.

The collection of data on advanced airway control by EMS personnel should be monitored by the physician medical director and the quality-improvement team. The definition of an intubation attempt is considered to be anytime a laryngoscope blade passes the teeth or a nasal intubation enters the opening of the nose. The number of attempts and the length of time an attempt takes are important pieces of what is considered a successful intubation. If the patient desaturates during an intubation attempt, that could be considered an unsuccessful intubation and result in a bad outcome for the patient. This is particularly important for patients being intubated who have a pulse, in what the industry commonly calls rapid-sequence or facilitated intubation.

GEOGRAPHIC INFORMATION SYSTEMS

Special Project Data

In many communities the EMS agency may have or can identify a problem that is significantly impacting the health-care system. When a local issue has been identified, a specialized data-collection tool needs to be designed. Ideally, a multidisciplinary team should be assembled, and a template for data collection on a specific problem should be created. What makes this so important is that often those problems have solutions that may not be obvious until the data reveals them. Funding, budget adjustments, and grant money are often tied to such activities, and to sustain them real impact must be shown. Data-collection templates also help identify the population at risk and provide information to injury-prevention programs. Figure 16.8 is a representation of a local data-collection tool

for Clark County, Nevada. Pediatric drowning has long been a significant problem in the southwestern United States and provides an example of a local problem requiring data collection.

EMS managers often rely solely on the data from their PCR or quality-improvement operations. Within almost every large municipal or county organization, geographic information system (GIS) technology is being used for a variety of applications. This is a new field for EMS professionals, and the emerging technology is a powerful tool for EMS. The challenge facing EMS managers and leaders to meet response-time standards can be analyzed using GIS technology that empowers EMS personnel to make faster, more accurate decisions.

GIS is a proven technology widely used not only for planning but also for response, incident management, clinical research, and quality-improvement activities. GIS combines layers of information containing different data points to be manipulated as desired by the user. GIS is a computer-based system to capture, display, store, retrieve, manipulate, and analyze spatial information and its associated attributes. It combines spatial and tabulated information to produce maps for performing spatial analysis.

Spatial data requires three components that make up a GIS map: a vector, a raster, and tabular data. A **vector** is spatial data arranged by points, lines, and polygons. A **raster** is a method for the storage, processing, and display of nonvector spatial data such as imagery. Each given area is divided into rows and columns, which form a regular grid structure. Each cell must be rectangular in shape, although not necessarily square. Tabular data is information organized in a table format, like a database. Spatial data is tied to real-world coordinates, which allows overlaying of spatial data on different layers for proper display and analysis. GIS data also includes point data, line data, and area data. Point data identifies key locations such as cities, hospitals, fire stations, and schools. Line data includes pipelines, roads, rail lines, and power lines. Area data includes geopolitical boundaries, lakes, parks, and states.

GIS requires four elements—hardware, data, software, and people who know how to query the system. The hardware includes computers, data storage, data input devices, and data output. Data input converts existing geographic information on paper by digitizing it into a format that can be handled by a computer. Data must be collected in the correct format and should identify the source. Existing geographical information includes paper maps, aerial

NAEMSP AIRWAY MANAGEMENT REPORTING TEMPLATE

Patient demographic information:

Date: ____/____/_____   Dispatch Time: _____:_____ am / pm

EMS Service Name/No.:_____

Pt age (yr): _____   Patient sex: ❑ M ❑ F

1. Indication for invasive airway management (check one):
- ❑ Apnea or agonal respirations
- ❑ Airway reflex compromised
- ❑ Ventilatory effort compromised
- ❑ Injury/illness involving airway
- ❑ Adequate airway reflexes/vent effort, but potential for compromise
- ❑ Other _____

2. Was endotracheal intubation (ETI) attempted?
- ❑ Yes   ❑ No

3. If ETI not attempted – alternate method of airway support:
- ❑ Bag-Valve-Mask (BVM)   ❑ Combitube
- ❑ Needle Jet Ventilation   ❑ LMA
- ❑ Open Cricothyroidotomy   ❑ Other Cricothyroidotomy
- ❑ CPAP/BiPAP   ❑ Not Applicable (ETI Attempted)
- ❑ Other: _____

4-6. Patient subsets (Select Yes/No):

Is patient in cardiopulmonary arrest on intubation?   ❑ Yes ❑ No
Is patient a victim of trauma?   ❑ Yes ❑ No
Is patient *under* 18 years old?   ❑ Yes ❑ No

7-11. Vital signs prior to ETI attempt (leave blank if not obtained):

Pulse: ____ beats/min   Blood Pressure: ____ / ____ mmHg

Resp Rate: ____ breaths/min   SaO$_2$: ____ %

12-14. Glasgow Coma Score (GCS) before intubation:

Eye: ❑ none (1) ❑ pain (2) ❑ verbal (3) ❑ spontaneous (4)

Verbal: ❑ none (1) ❑ incomprehensible (2)
        ❑ inappropriate words (3)
        ❑ disoriented (4)   ❑ oriented (5)

Motor: ❑ no response (1)   ❑ extends to pain (2)
        ❑ flexes to pain (3)   ❑ withdraws from pain (4)
        ❑ localizes pain (5)   ❑ obeys commands (6)

15. Monitoring and treatment modalities concurrent with intubation (check *all* that apply):
- ❑ ECG monitor   ❑ Pulse-Oximetry
- ❑ IV access   ❑ C-spine immobilization
- ❑ CPR (chest compressions)   ❑ Gum Elastic Bougie
- ❑ BAAM   ❑ Endotrol Tube
- ❑ Other: _____

17. Level of training of each rescuer attempting intubation:

Rescuer	Level of Training (check one)
A[†]	❑ EMT-P ❑ EMT-I ❑ EMT-B ❑ Medic Student ❑ Nurse/PHRN ❑ Phys Asst ❑ MD/DO (attend) ❑ MD/DO (res) ❑ Other: _____
B[†]	❑ EMT-P ❑ EMT-I ❑ EMT-B ❑ Medic Student ❑ Nurse/PHRN ❑ Phys Asst ❑ MD/DO (attend) ❑ MD/DO (res) ❑ Other: _____
C[†]	❑ EMT-P ❑ EMT-I ❑ EMT-B ❑ Medic Student ❑ Nurse/PHRN ❑ Phys Asst ❑ MD/DO (attend) ❑ MD/DO (res) ❑ Other: _____

16-18. Provide information for each laryngoscopy attempt.
FOR ORAL ROUTE, EACH INSERTION OF BLADE (LARYNGOSCOPY) IS ONE "ATTEMPT."
FOR NASAL ROUTE, EACH PASS OF TUBE PAST NARES IS ONE "ATTEMPT."

Attempt	16. ETI Method	17. Who attempted?[†]	18. Was attempt successful?
#1	❑ OTI ❑ NTI ❑ Sedation ❑ RSI	❑ A ❑ B ❑ C	❑ Yes ❑ No
#2	❑ OTI ❑ NTI ❑ Sedation ❑ RSI	❑ A ❑ B ❑ C	❑ Yes ❑ No
#3	❑ OTI ❑ NTI ❑ Sedation ❑ RSI	❑ A ❑ B ❑ C	❑ Yes ❑ No
#4-24	❑ OTI ❑ NTI ❑ Sedation ❑ RSI	❑ A ❑ B ❑ C	❑ Yes ❑ No

Indicate drugs given to facilitate intubation:
- ❑ Midazolam ____ mg   ❑ Diazepam ____ mg
- ❑ Lidocaine ____ mg   ❑ Morphine ____ mg
- ❑ Etomidate ____ mg   ❑ Succinylcholine ____ mg
- ❑ Atropine ____ mg   ❑ Topical Spray
- ❑ Other – Specify: _____ - ____ mg
- ❑ Other – Specify: _____ - ____ mg

19-24. Endotracheal tube confirmation.

19. Auscultation	❑ Tracheal Placement	❑ Esophageal Placement	❑ Indeterminate	❑ Not Assessed	❑ Tube not placed.
20. Bulb Aspiration	❑ Tracheal Placement	❑ Esophageal Placement	❑ Indeterminate	❑ Not Assessed	❑ Tube not placed.
21. Syringe Aspiration	❑ Tracheal Placement	❑ Esophageal Placement	❑ Indeterminate	❑ Not Assessed	❑ Tube not placed.
22. Colorimetric ETCO$_2$	❑ Tracheal Placement	❑ Esophageal Placement	❑ Indeterminate	❑ Not Assessed	❑ Tube not placed.
23. Digital ETCO$_2$	❑ Tracheal Placement	❑ Esophageal Placement	❑ Indeterminate	❑ Not Assessed	❑ Tube not placed.
24. Waveform ETCO$_2$	❑ Tracheal Placement	❑ Esophageal Placement	❑ Indeterminate	❑ Not Assessed	❑ Tube not placed.
Other: _____	❑ Tracheal Placement	❑ Esophageal Placement	❑ Indeterminate	❑ Not Assessed	❑ Tube not placed.

25. Peak ETCO2 value: _____   ❑ Indeterminate

26. Was ETI successful for the overall encounter (on transfer of care to ED or helicopter)?
- ❑ Yes   ❑ No

27. Who determined the final placement (location) of ET tube?
- ❑ Rescuer performing intubation.
- ❑ Another rescuer on the same team.
- ❑ Receiving helicopter crew.
- ❑ Receiving hospital team.
- ❑ Other: _____

28-32. Vital signs after intubation attempt:

Pulse: ____ beats/min   Blood Pressure: ____ / ____ mmHg

Resp Rate: ____ breaths/min   SaO$_2$: ____ %

33. Critical complications encountered during airway management (Check *all* that apply):
- ❑ Failed intubation effort.
- ❑ Injury or trauma to patient from airway management effort.
- ❑ Adverse event from facilitating drugs.
- ❑ Esophageal intubation – delayed detection (after tube secured).
- ❑ Esophageal intubation – detected in ED.
- ❑ Tube dislodged during transport/patient care.
- ❑ Other: _____

34. If all intubation attempts FAILED, indicate suspected reasons for failed intubation (check all that apply):
- ❑ Inadequate patient relaxation   ❑ Orofacial trauma.
- ❑ Inability to expose vocal cords.   ❑ Secretions/blood/vomit.
- ❑ Difficult pt anatomy.   ❑ Unable to access pt.
- ❑ ETI attempted, but arrived at destination facility before accomplished.
- ❑ Not applicable – Successful field ETI   ❑ Other: _____

35. If all intubation attempts FAILED, indicate secondary (rescue) airway technique used (check all that apply):
- ❑ Bag-Valve-Mask (BVM) Ventilation   ❑ Needle/Jet Ventilation
- ❑ Combitube   ❑ Open Cricothyroidotomy
- ❑ Not applicable – Successful field ETI   ❑ Other: _____

36. Did secondary (rescue) airway result in satisfactory ventilation?
- ❑ Yes   ❑ No   ❑ Not applicable

37-38. Airway Management Times

Time of decision to intubate: _____:_____ am / pm
Time of successful intubation: _____:_____ am / pm
Time intubation abandoned: _____:_____ am / pm

Template Design by H. Wang, University of Pittsburgh, PA - May 23, 2003.

FIGURE 16.7 ■ NAEMSP Airway Management Reporting Template.
(Source: National Association of EMS Physicians. Reprinted with permission.)

SUBMERSION INCIDENT REPORT
FOR ALL DROWNING OR NEAR DROWNING INCIDENTS IN CLARK COUNTY
Fax completed form to SNHD EMS Office at: 702-383-1240
For questions call: 702-759-1050

CLARK COUNTY
HEALTH DISTRICT

BASIC INCIDENT INFORMATION
Date of Incident: _____ **Time of Incident**: _____
Your Agency's Incident Number: _____
Reporting Agency: _____
Incident Location City/Zip: _____

Type of Dwelling: ☐ House   ☐ Apartment/Condo
☐ Hotel/Motel   ☐ N/A   ☐ Other:

VICTIM INFORMATION
Age of Victim: _____   **Sex**: ☐ M   ☐ F

Victim's Race/Ethnicity: ☐ *Unknown*
☐ Hispanic   ☐ White   ☐ Black   ☐ American Indian
☐ Asian/PI   ☐ Other: _____

Victim Last Seen: ☐ *Unknown*
☐ Swimming   ☐ Playing Outside   ☐ Playing Inside
☐ Sleeping   ☐ Other: _____

Est. Time of Submersion: _____   ☐ *Unknown*

Type of Clothing Worn by Victim: ☐ *Unknown*
☐ Swim suit   ☐ Day clothing   ☐ Pajamas
☐ None   ☐ Other: _____

Personal Floatation Device (PFD) ☐ *Unknown*
☐ Yes   ☐ No
Appropriate PFD Type/Size ☐ *Unknown*
☐ Yes   ☐ No, why_____

WATER SOURCE INFORMATION
Site of Incident: ☐ *Unknown*
☐ Victim Residence   ☐ Relative Residence
☐ Neighbor Residence   ☐ Friend Residence
☐ Sitters/Daycare Provider   ☐ Public Pool
☐ County/City Park   ☐ Other: _____

Water Clarity: **Water Depth:**
☐ Clear   ☐ Cloudy   ☐ Under 18"(approx. depth ___)
☐ Muddy ☐ Green   ☐ 18" – 48"   ☐ Over 4'
☐ *Unknown*   ☐ *Unknown*

Water Type: ☐ *Unknown*
☐ Pool – in ground   ☐ Spa/Hot Tub   ☐ Bathtub
☐ Pool – above ground   ☐ Toilet   ☐ Bucket
☐ Child wading pool   ☐ Lake or pond   ☐ Stream/river
☐ Canal/irrigation ditch   ☐ Other: _____

Toys or other objects in water?   ☐ *Unknown*
☐ No ☐ Yes   If yes, describe: _____

Year pool/spa was built: _____ ☐ *Unknown* ☐ N/A

How long has current resident lived at this address?
_____   ☐ *Unknown* ☐ N/A

Form Completed by: _____

Contact Phone: _____

A: ADULT SUPERVISION
Age of Supervisor(s) at time of incident: ☐ *Unknown*
_____ ☐ Mother   _____ ☐ Father
_____ ☐ Sibling   _____ ☐ Babysitter/Childcare Provider
_____ ☐ Other (specify): _____

Supervisor activity immediately prior to incident:
_____   ☐ *Unknown*

Alcohol and/or drug use evident? ☐ Yes ☐ No   ☐ *Unknown*

B: BARRIER INFORMATION
Barriers Present Around Water: ☐ *Unknown* ☐ None
☐ Property Perimeter Fence   ☐ Isolation Pool Fence
  Fence Type: _____   *Fence Type:* _____
  Fence Height: _____   *Fence Height:* _____
☐ Self-closing/Self-latching gate
  Gate working properly?   ☐ *Yes* ☐ *No* ☐ *Unknown*
☐ Door/Window Alarm   ☐ Approved pool safety cover
☐ Pool Safety Net   ☐ Self-closing doors on house
☐ Perimeter pool alarm   ☐ In-Pool Alarm
☐ Turtle Alarm on child   ☐ Other: _____

Access to Pool by Victim: ☐ *Unknown*
☐ Direct Access/No Barriers
☐ Brought in to water area by other person
☐ Victim breached safety barrier(s): (choose all that apply)
  ☐ *Fence* ☐ *Pool Cover* ☐ *House Window*
  ☐ *Gate* ☐ *Pool Alarm* ☐ *House Door*
  ☐ *Other:* _____

Explain how victim got through barrier(s): _____
_____
_____

C: CLASSES/EMERGENCY PREPARATION
Was rescue equipment near water? (Shepherd's hook, life ring, etc…) ☐ Yes   ☐ No ☐ *Unknown*

Did bystanders attempt CPR?
☐ Yes ☐ No   ☐ *Unknown*

Did victim ever take swim lessons or water safety classes?
☐ Yes ☐ No   ☐ *Unknown*

Did supervisors ever take CPR?
☐ Yes ☐ No   ☐ *Unknown*

VICTIM CONDITION:
☐ Treated/released at scene
☐ Transported to hospital _____
☐ Deceased at scene

COMMENTS: _____
_____
_____
_____

FIGURE 16.8 ■ Clark County Drowning Sheet GIS.
(Source: Clark County Health District, Las Vegas)

- ♦ U.S. Geographical Survey data.
- ♦ FEMA data.
- ♦ U.S. Fire Administration.
- ♦ U.S. Census.
- ♦ Map information.
- ♦ Commercial GIS data providers.
- ♦ Fire-station locations.
- ♦ Tax assessor information.
- ♦ Political districts (city and county GIS departments).
- ♦ Hospitals.

FIGURE 16.9 ■ Common Sources of GIS Data.

photos, satellite images, addresses, and GPS locations. Satellites can produce an accurate picture of the terrain using Interferomic Synthetic Aperture Radar (IFSAR), available from GIS software. Figure 16.9 lists common sources of GIS data.

Data quality and analysis help EMS identify emergency call locations providing the emergency communications center with the closest unit to the call. GIS can help map the shortest distance to the call and to the most appropriate hospital. It can analyze incident volumes and trends for staffing and deployment purposes. As a power application for epidemiologist and homeland defense managers, GIS can correlate patient types and locations to identify disease outbreaks and the potential for people to be exposed to hazardous materials during a chemical release. GIS can show how disease is spreading and help apply resources to the correct location. For deploying EMS resources, GIS can match tax districts, funding, revenue projections from billing, and maps to help increase the efficiency of the resources.

Managing GIS

EMS managers periodically need to review and budget for upgrades in GIS. Routinely, as the amount of data going into the system increases and other EMS managers see the power behind GIS, the program will need more resources. Expect to budget for and manage increases in data-collection cost, computer memory, upgrades in video cards, larger data storage, color printers, and increased processor speeds. State-of-the-art GIS programs have high-end computers, a large server, and digitizing tablets for run reports and inventories; these high-end systems also may need laser printers and a larger printer called a *plotter*, used to print maps. The common GIS software packages on the market are ESRI ArcView, MARVLIS, MapInfo, Caliper, Intergraph, and Tactician.

EMS managers need to ensure the quality and accuracy of the data by examining the metadata; that is, data about the data, which focuses on content, quality, condition, and other characteristics of data. Standards from the National Standard for Spatial Data Accuracy (NSSDA) are in place for managers who oversee GIS data to ensure accuracy. The NSSDA implements a statistical and testing methodology for estimating the positional accuracy of points on maps and in digital geospatial data, with respect to geo-referenced ground positions of higher accuracy.

EMS managers need to ensure that GIS data can be transferred and will arrive correctly in another database or computer. Finally, data needs to be able to be projected correctly, so that the flat map accurately reflects the location of features on a round earth.

Automatic Vehicle Locators

Once an agency has integrated GIS into EMS operations, other related systems can be used to help make EMS operations more efficient. Automatic vehicle location (AVL) systems provide real-time tracking of vehicles by using GPS, wireless communication, and GIS mapping software to view the location of EMS vehicles. Having real-time locations provided means that EMS dispatching can be improved and response times shortened.

AVL technology can now do more than just locate a vehicle. Some AVL systems can interface with the computer-aided dispatch software and provide reports on drivers, vehicle use, and activity. Vehicle sensors also can record and transmit data on the use of lights and sirens, operate doors, and operate traffic-control devices for a given route of an emergency response. Panic or emergency traffic transmissions can be programmed into the AVL system to alert the communications center, management, and law enforcement if a crew is in trouble.

A common application of GIS is to use the technology and data to create a fluid deployment plan or assist the communications center with the dispatch of the unit closest to the call. When AVL is combined with GIS, response-time efficiency and standards of coverage can be adjusted to meet clinical and operational benchmarks. GIS can relate with previous call-volume statistics and predict seasonal adjustments to assist in placing vehicles in high-demand areas. Computer-aided dispatch software with GIS, call-history data, and traffic data can create deployment strategies for EMS units. One application combining this tech-

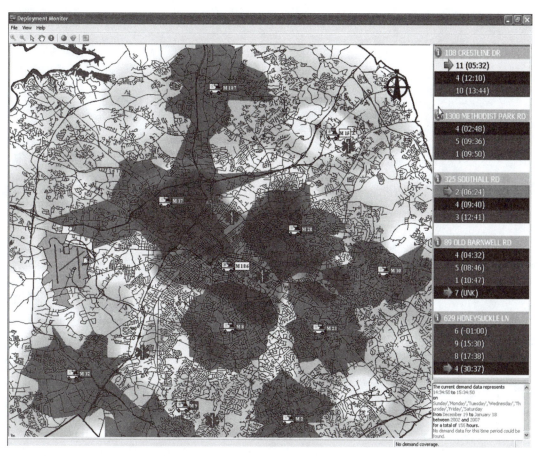

FIGURE 16.10 ■ Dynamic Service Area Generated with GIS and MARVLIS.
(Source: ESRI. Reprinted with permission.)

nology is offered in Figure 16.10. This figure shows a dynamic service area, the response area around each vehicle that represents the actual distance a vehicle can drive in a specified time period. GIS and computer-aided dispatch technology also can reroute vehicles with AVL around construction zones, areas with heavy traffic at specific times of the day, and other incidents.

GIS also can help display high-call-volume areas and areas that are not covered within the agency's response goals. This technology can reduce the number of move ups or cover assignments the EMS units make. Figure 16.11 shows a computer-aided dispatch (CAD) rerouting an EMS unit around an obstruction in the main route created by another EMS response to a vehicle accident in a high-volume geographic area.

■ RESEARCH OUTCOMES

Once data is collected, a key component of that data is defining in the research what an outcome is and how it is measured. EMS and the federal health-care system have several outcome definitions for various categories of patients. The most common research terms are "morbidity" and "mortality."

Mortality is a death. In most cases it is the death of the patient at some period identified in the EMS or health-care system. For example, a cardiac-arrest victim who dies within 24 hours of being admitted to the emergency department may be counted as a field death. The same patient who survives in a vegetative state for three months and then succumbs in the nursing home from an infection would be seen as a success in some prehospital systems, but in other systems this

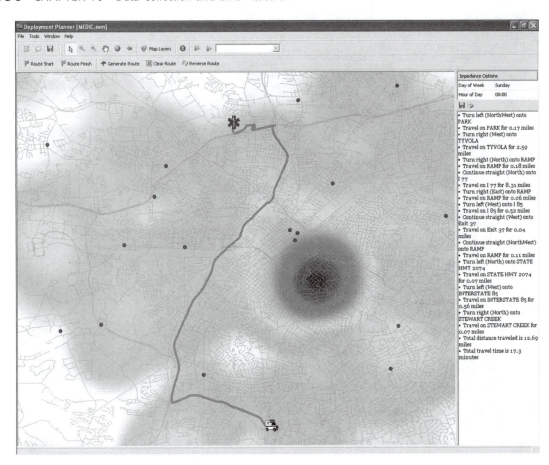

FIGURE 16.11 ■ GIS and AVL are being used to reroute a vehicle around another call or road closures. *(Source: ESRI. Reprinted with permission.)*

would be counted against the EMS system instead of the nursing facility. This varies not only in the health-care system but the legal system when a victim of a violent crime dies in a specific time frame, even if it is a death that was caused by a complication in the health-care system or an infection. Table 16.2 illustrates common cardiac-arrest definitions used in three different cities.

Morbidity is often seen as a disability or loss that does not result in death. Several neurological outcomes are used that relate to prehospital EMS outcomes. Morbidity and neurological outcome have typically been measured by the use of the Glasgow coma scale (GCS), but the GCS has given way to the disability rating scale (DRS), which is used now by most rehabilitation centers. Other scores such as the Frankel score, the American Spinal Cord Association's dermatomes, and ASIA score look at spinal-cord trauma and the amount of disability from a cord injury. The Barthel index rates

the neurological injuries of a patient based on disability. Table 16.3 shows how the outcomes of pediatric patients are characterized.

LITERATURE REVIEWS

When reviewing research that is involved with drugs or equipment, it is important to consider evidence-based practice. Evidence should come from a credible source and with predetermined criteria in human studies, if possible with sound methodology. Figure 16.12 lists the eight steps or levels involved with medical scientific research in the verification of evidence-based medicine. Once clinical evidence and research has been reviewed, drugs, medical procedures, and equipment are classified into seven different categories. The American Heart Association uses these terms to classify the effectiveness based on evidence-based medicine (Figure 16.13).

TABLE 16.2 ■ Examples of Variations in Standardized Definitions of Cardiac Arrest

County 1	*City 2*	*County 3*
Cardiac Arrest Inclusion	**Cardiac Arrest Inclusion**	**Cardiac Arrest Inclusion**
Any patient found without pulse and respirations.	Witnessed arrest only.	Patient found by ALS provider to be in V-fib or V-tach.
Cardiac Arrest Outcome	**Cardiac Arrest Outcome**	**Cardiac Arrest Outcome**
Arrived at the ED with a pulse and blood pressure.	Survived more than 24 hours in the hospital.	Survived to discharge neurologically intact.
Response Time	**Response Time**	**Response Time**
Public Safety Answering Point Time of call to arrival at scene.	First responder notification to arrival at scene.	Ambulance time of notification to time at patient's side.
Intubation Attempt	**Intubation Attempt**	**Intubation Attempt**
Scope passes teeth.	Tube passes cords.	(No definition)

In the absence of clinical research, medicine often relies on expert opinion, best practices in the industry, or the current practice plus or minus a related piece of research. The need for valid reliable data that is universally comparable is evident at every level of EMS system development and operation. Research is necessary for every activity from patient care to performance improvement. All of the components of the EMS research agenda for the future are dependent on data.

The success of validating our services will be dependent on data and information systems.

Specific tools have been used by researchers to collect data. Applying the correct tool to the correct action being taken by a researcher or EMS manager conducting research is important to ensure that the uniform protocols of medical research are met. Table 16.4 lists the tools used for various actions by a research effort.

TABLE 16.3 ■ Barthel Index on Pediatric Overall Performance

Score	Category	Description
1	Good	PCPC: Normal; normal age-appropriate activities. Medical and physical problems do not interfere with normal activity.
2	Mild disability	PCPC: Mild; minor chronic physical or medical problems present minor limitations but are compatible with normal life (e.g., asthma); preschool child has physical disability consistent with future independent functioning (e.g., single amputation) and is able to perform more than 75% of age-appropriate activities of daily living; school-age child is able to perform age-appropriate activities of daily living.
3	Moderate disability	PCPC: Moderate; medical and physical conditions are limiting as described below; preschool child cannot perform most age-appropriate activities of daily living; school-age child can perform most activities of daily living but is physically disabled (e.g., cannot participate in competitive physical activities).
4	Severe disability	PCPC: Severe; preschool child cannot perform most age-appropriate activities of daily living; school-age child is dependent on others for most activities of daily living.
5	Coma/vegetative state	PCPC: Coma/vegetative state.
6	Death	

Level 1	Large randomized clinical trials.
Level 2	Smaller randomized clinical trials.
Level 3	Prospective, controlled, nonrandomized cohort studies.
Level 4	Historic, nonrandomized cohort or case controlled studies.
Level 5	Case studies, no control group.
Level 6	Animal or mechanical studies.
Level 7	Extrapolations from existing data, theoretical analysis.
Level 8	Rational conjecture (common sense); common practice.

FIGURE 16.12 ■ National Institutes of Health (NIH) Steps in Clinical research.

Class I	Definite, excellent Level 1 evidence.
Class II	Acceptable and useful; no harm.
Class IIa	Good supportive evidence.
Class IIb	Fair supportive evidence.
Class IIg	Historical precedent or consensus.
Class III	Not acceptable; may be harmful.
Indeterminate	Insufficient data.

FIGURE 16.13 ■ American Heart Association Classification of Therapies.

RESEARCH TOOLS

Multi-voting

Teams and work groups use tools like brainstorming or a SWOT analysis to generate lists of process-related problems, potential solutions, and approaches or options to deal with an issue. Once this is done EMS research teams are sometimes unable to process the items on the list into a few manageable ideas that the organization can get its arms around and convince the field firefighters, EMTs, and paramedics of the action plan. When EMS managers recognize that more than one item has significant merit, multi-voting can be used to identify quickly the most important items on the list. Multi-voting is best suited for use in large groups that are reviewing long lists, and it is valuable when there is difficulty in reaching a consensus on the highest-priority items on a list. It is not used, however, when trying to reach a consensus on a single issue. Use the guidelines in Figure 16.14 to conduct a multi-voting exercise in a research/quality-improvement team or EMS stakeholder group.

Multi-voting is used to help quality-improvement teams focus on problem solving and identifying high-priority items in an efficient manner. Multi-voting allows each member to participate equally in the decision-making process; this works well in a union environment and fosters a collaborative approach with the labor group. This is particularly important in gaining acceptance and buy-in for future actions based on the decision.

Run Chart

When data about a work process is collected, the results can be displayed in a graph. A run chart is one type of graph that is used to see if work is performed in a consistent way (Figure 16.15). There are obvious changes as the work progresses over the course of time. A run chart can be prepared for any point of a

TABLE 16.4 ■ EMS Research Tool Selection Matrix

If you are working with →	Ideas	Teams or Groups	Numbers
You can use:			
Multi-voting	●	●	
Run chart			●
Histogram			●
Bar graphs	●	●	●
Column chart	●		●
Pie charts			●
Cause-and-effect diagram	●	●	
Flow charts	●	●	
Pareto diagram			●
Pictograms	●	●	

- Display the items under consideration on a flip chart, making sure to eliminate duplicate items.
- Number the items on the list to facilitate record keeping.
- Give each team member a number of votes equal to approximately one-half of the number of items on the list (e.g., 10 votes for a 20-item list).
- Have each team member vote for the items he believes are most important. Team members may cast all votes for one item, vote for several items, or vote for individual items until they use their allotted number of votes.
- Tally the votes.
- Select the four to six items that receive the highest number of votes. Discuss and rank order the items. If the team cannot establish the top four to six, remove the items having the fewest votes and then conduct another vote.

FIGURE 16.14 ◼ Guidelines for Conducting a Multi-voting Exercise.

1. Determine the problem or question to be studied.
2. Collect the appropriate type and adequate amount of data. (Note: Ask for assistance if you have questions about the sampling plan.)
3. Scale and label the horizontal, or x, axis to describe the process in the batch sequence or time period that was measured.
4. Scale and label the vertical, or y, axis for the characteristic or variable that is to be plotted.
5. Plot each data value in the sequence or chronological order that it was obtained.
6. Label the graph, including a description of the process and the sample size.

FIGURE 16.15 ◼ How to Construct a Run Chart.

work process that is to be measured and evaluated. In a run chart the data for a process measure are plotted either after several events of work are done or as work is completed over a period of time; for example, the number of IV starts per paramedic. This type of data is usually obtained in a sequence (over several months, for example). For a continuous process, data are usually obtained at set time intervals.

When interpreting a run chart, the following guidelines apply. A *trend* is a change in the process where values move in the same direction over time. EMS frequently sees trends that are often cyclic or seasonal. A *shift or run* is a process change in which the average or center line shifts. A *pattern* is any non-random result, such as a cycle that repeats over time. An *outlier* is a value that lies significantly outside the range of the rest of the data. These four cases are illustrated in Figure 16.16, which shows the analysis that can be identified with a run chart.

A run chart identifies questions about the work process and its performance; for example, are the results what would be expected? Can the shifts, trends,

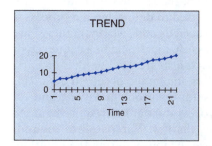

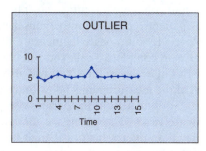

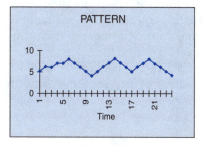

FIGURE 16.16 ◼ Analysis of Run Charts.
(*Source: Department of Transportation* EMS Quality Improvement Manual.)

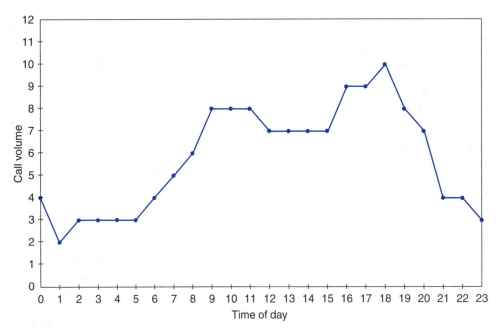

FIGURE 16.17 ■ Sample Analysis of Run Chart Identifying Response Volume.
(Source: Brady/Critical Care Paramedic. Reprinted with permission of Pearson Education, Upper Saddle River, NJ.)

or outliers be explained? The data also might reveal another common problem. Often a process will be free of trends, shifts, patterns, and outliers but is still unable to meet specifications. The solution to this problem is to identify improvements that will adjust the process to target or reduce the variability.

Run charts are used to determine if a process is performing as expected and whether there are changes in a process characteristic in a sequence or over time. Run charts also are used to identify early patterns and outliers among the observed data. This analysis can be useful for problem solving and for comparing to a process standard or requirement. As an example, a run chart was prepared to monitor ambulance response reliability during the month of January. Figure 16.17 illustrates that the number of late arrivals fluctuates considerably, particularly around January 6 and 26, when major snowstorms occurred. Also during this period the lowest number of late responses was five on January 11 and 12, while the highest number of late responses was 26 on January 7.

Histogram

It is important to collect and display data visually to allow field crews and the public, who may have limited statistical or research background, to interpret the data. One way to describe and evaluate performance is to display data in a chart called a *histogram*, a column graph where the height of the bars indicates the relative numbers or frequencies for given values of a variable. The values can be numeric, such as response time, or nonnumeric, such as days of the week. Histograms help to show how EMS crews are doing at the present time. In a histogram, data are grouped into defined intervals and displayed according to their frequency of occurrence in each interval. This method provides insights about performance and, in particular, the variation that normally occurs in work.

There are numerous situations where histograms can be used to show how much variation exists in work; for example, how much time it takes an EMS organization to complete a routine call, intubation success rates by patient type, and intravenous success rates per employee. If you repeatedly measure the length of time it takes to complete a call, you will observe that the time varies in each instance. You will also see, however, that all of the measurements fall within a certain range.

To construct a histogram, complete the steps in Figure 16.18. Histograms are conducted to visually display data over a historical time frame. Keep in mind that a histogram is a picture of the data distribution that includes a spread and a shape. This can provide clues about the variation that exists in the work per-

1. Count the number of observations in the data set.
2. Determine the range of the data. This is obtained by subtracting the smallest value from the largest value.
3. Decide the number of intervals, displayed as bars, to use. A good rule is 5 to 7 for 20 to 50 observations and 6 to 10 for more than 50 observations.
4. Divide the range by the number of intervals. Round the number to a whole number.
5. Select the boundaries for each interval so they are not overlapping.
6. Count the number of observations that fall within the boundaries of each interval.
7. Draw, scale, and label the horizontal (x-axis) and vertical (y-axis) axis lines of the chart. Label the x-axis for the intervals that cover the data range. Mark the vertical axis from zero until the highest frequency is included.
8. Draw vertical bars for each interval. The height of the bars equals the number of observations at that interval. The width of each of the bars should be the same.
9. Title the completed chart. The title should describe the nature of the observations being made that are summarized in the chart and the time frame in which the data was collected.

FIGURE 16.18 ▪ Steps to Construct a Histogram.

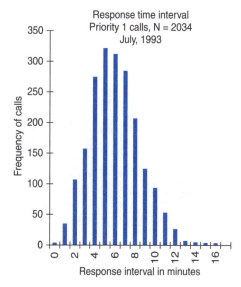

FIGURE 16.19 ▪ Histogram Sample.

formed. Distributions can have either a *positive skew* (tail of the distribution to the right) or *negative skew* (tail of the distribution to the left) direction from the center. By examining the spread and shape of a distribution, the extent of variation in a work process can be determined. This can provoke further discussions to identify the cause of variation and the measures needed either to control or to reduce it.

A foundation of a continuous-improvement effort is data collection. Data are used to better understand variation in a work process and determine how well standards are being met based on patient and other stakeholder expectations. A histogram is a useful tool to display these findings to identify current performance and show how work processes are improving over time.

In Figure 16.19 the histogram shows the response-interval performance of an EMS agency to emergencies during one month. This chart clearly illustrates how well the organization is meeting the needs of its patients.

Bar Graphs

A *bar graph* or *bar chart* is one of the simplest and most effective ways to display data. In a bar graph a bar is drawn for each category of data, allowing for a visual comparison of the results. For example, the informa-

tion that would need to indicate the types of EMS calls is presented well in a bar graph. The horizontal dimension gives the percent, while the vertical dimension shows the category labels. The bars are presented in numerical order starting with "undetermined after investigation" as the most frequent. Each bar also contains the number of EMS calls for that classification of injury or illness as additional information to the reader.

As a general rule the horizontal dimension in a bar graph is numeric, such as percentages or other numbers, while the vertical dimension shows the labels for the items in a category. It is not always necessary to include numbers in each bar, especially if there is an accompanying table or list, but they can be useful to readers unfamiliar with the data. If the numbers are omitted from the graph, a total number should be provided either in the title or a footnote.

Pie Charts

A pie chart is an effective way of showing how each component contributes to the whole. In a pie chart, each wedge represents the amount for a given category. The entire pie chart accounts for all the categories. For example, Figure 16.20 shows trauma injuries by type, and the percentages are included with each wedge label. Although the percentage numbers are not necessary, they aid in comparisons of the wedges. In developing pie charts one should follow these rules:

- Convert data to percentages.
- Keep the number of wedges to six or fewer.

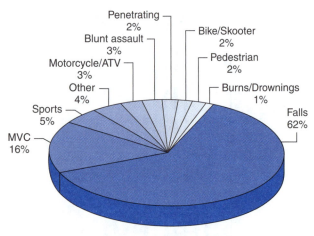

FIGURE 16.20 ■ Trauma Injuries Transported to Holy Family Hospital.

• If there are more than six categories, keep the most important five and group the rest into an "Other" category.
• Position the most important wedge starting at the 12-noon position.
• Maintain distinct color differences among the wedges.

While pie charts are popular, they are probably the least effective way of displaying results. For example, it may be difficult to compare wedges within a pie to determine rank. Similarly, it takes time and effort to compare several pie charts because they are separate figures.

Cause-and-Effect Diagram

Sometimes, a problem prevents a job from being completed as well as it should be. The problem may result from long-standing policies and procedures or because of a lack of adequate equipment or facilities. These problems can become more complicated to resolve when several people are working together to complete an assignment. A cause-and-effect diagram is used to show the causes of a problem. Since generally there is more than one cause to any problem, the diagram is used to further divide causes into groups or categories. This approach often uncovers the root causes of a problem, and when the root causes are identified, how much each cause contributes to a problem can be evaluated.

The following steps are used to construct a cause-and-effect diagram, sometimes called a "fish-bone diagram," as shown in Figure 16.21, because they resemble a fish skeleton when completed.

1. Develop a statement of the problem. Write it down on the right side of a piece of paper (the fish head). Draw a

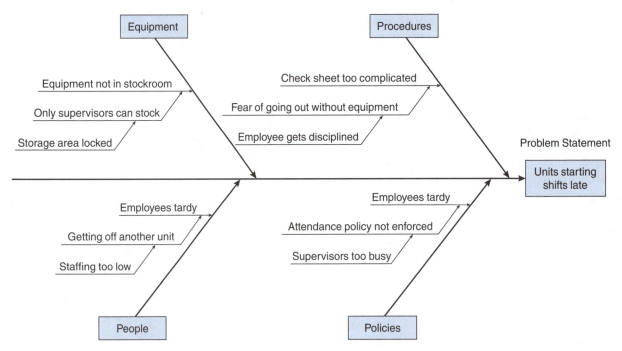

FIGURE 16.21 ■ Cause and Effect Diagram on Starting Shifts Late.
(Source: Brady/ Critical Care Paramedic. Reprinted with permission of Pearson Education, Upper Saddle River, NJ.)

central arrow across the middle of the page that points to the problem.

2. Brainstorm a list of probable causes of the problem. Write each of these down on another sheet of paper.

3. Review the list of causes, and identify the major categories. Write down the names of the categories as main branches (fish bones) off the central arrow.

4. Review the causes, and list each under the appropriate category. If necessary, revise or expand the list of categories.

5. Write down each cause as a small branch drawn off the main category branch for the category under which it falls.

Cause-and-effect diagrams can help illustrate clearly possible relationships among causes. They can be used to uncover the root causes of problems or specific problem steps or bottlenecks in a work process. By arranging possible causes into categories in a diagram, a better understanding of problems and the contributing factors develops. While a cause-and-effect diagram is an effective analysis tool, it only helps to identify possible causes or categories of problems. Even if everyone agrees on the items on the list, it is important to determine what is not known about each cause and how that information can be uncovered. If necessary, additional data must be collected and analyzed to identify and confirm actual causes.

Flowcharts

Everyday in EMS systems hundreds of tasks are completed in order to meet specific objectives. Much work flows among departments, offices, and other organizations. A flowchart illustrates the activities performed and the flow of resources and information in a process. Two types of flowcharts are particularly useful: high level and detailed. A high-level flowchart illustrates how major groups of related activities, often called *subprocesses*, interact in a process. Typically, four to seven subprocesses are shown in a flowchart. If you include only basic information, high-level flowcharts can readily show an entire process and its key subprocesses.

A detailed flowchart provides a wealth of information about activities at each step in a subprocess. An example of a detailed flowchart for two of the access and dispatch subprocesses is shown in Figure 16.22. It shows the sequence of the work and includes most or all of the steps, including rework steps that may be needed to overcome problems in the process. A quality-improvement team can increase the detail to show the individuals performing each activity or the time required to complete each activity. If necessary, the link

between various points in the subprocess and other high-level flowcharts of the process also can be shown.

Flowcharts are drawn using certain symbols as building blocks. Flow charts work well in teaching EMS providers to collect or report certain data elements. A square or rectangle identifies a step (task, activity) in the process. Activity blocks are the most common elements of a flowchart. They can be arranged in serial or parallel paths depending on how activities are actually performed. The name of the step is written inside, and a diamond identifies a decision or branch point in the process. Each path emerging from a decision block is labeled with one of the possible answers to a question that is posed at this point in the process. Decision blocks indicate conditional situations where the output of an activity needs to meet certain criteria before the process can continue. If the criteria are not met, a different set of activities follows. This is often called "rework" and is drawn as a feedback loop in the flowchart. An arrow indicates the sequence and direction of flow within the process. This is usually the transfer of an output of one activity to the next (where it becomes an input). A parallelogram identifies a material or information output or input from an activity. The name of the output (input) is written inside. A circle is used to indicate a continuation of the process flow elsewhere on the same page or on another page. The same label written on the connector symbol appears on another connector, where the process flow continues.

An EMS organization pursuing quality improvement is constantly looking for ways to improve the effectiveness and efficiency of its work. *Effectiveness* means producing the required results or output when needed. *Efficiency* means simply producing those results or outcomes the first time with minimum resources. In order to generate ideas on how to be more efficient and effective, it is helpful to define and document how activities are actually performed.

Flowcharts can be useful to identify activities in a process that reduce effectiveness and efficiency, as in Figure 16.22. For example, some activities may be redundant or repeated and others may be unnecessary. Activities may be performed in sequence, when they could be conducted at the same time to reduce the overall time for the process. Flowcharts can be used to identify conditions that cause delays and bottlenecks. This can bring focus to problems at various points within the process that need further evaluation and improvement.

ADA COUNTY EMS ADULT SELECTIVE SPINAL IMMOBILIZATION
ALGORITHM

IF A MAJOR TRAUMA PATIENT AND IS UNSTABLE, THEN IMMOBILIZE
AND DO NOT USE THIS PROTOCOL.
This Protocol is for patients>8 years of age

Determine Inclusionary Criteria:
• Falls from any height (greater than ground level) • Motor vehicle collisions • Any abrupt decelerating,
accelerating, or rotational force to head or spine • Direct trauma to the head or spine • Any significant
penetrating trauma to head, chest or abdomen
**Any other potential mechanism of injury deemed appropriate based on the patient's history or physical
exam.**

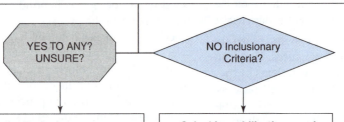

YES TO ANY?
UNSURE?

NO Inclusionary
Criteria?

**Assess for Exclusionary
Criteria:**

1. Patient Stability
   • Vital signs with in normal limits.
   • No signs of compromised perfusion
   • **UNSTABLE TRAUMA PATIENTS ARE
     GENERALLY NOT EXCLUDED IN THE
     FIELD**
2. Patient is a reliable historian
   • Conscious, alert, oriented to person,
     place, and time and situation.
   • No evidence of acute stress reaction
     or severs anxiety
   • No evidence or admission of intoxication
     or impairment by drug or alcohol use.
   • Patient is greater than eight (8) years old.
3. No Distracting injuries:
   • **Absence of major painful injuries that
     could distract the patient's ability to
     perceive pain.**
4. Normal neurological function in all extremities.
   • Sensory-Pain, pressure, and light touch
     are present
   • Numbness or tingling (paresthesia) are
     absent.
   • Motor-Strength is full and symmetrical.
5. Patient denies midline spine or neck pain.
6. Absence of spine or neck tenderness or
   deformity elicited on palpation and Axial Loading
7. (Only do if all other exclusionary findings are
   negative):
   Absence of spine or neck tenderness when patient
   moves head in the coronal, transverse and sagittal
   planes.
8. Able to complete all applicable assessments?

**Spinal immobilization may be
deferred.**

1. Immobilize at Paramedics discretion
2. Frequently reassess, If any signs and
   symptoms of spinal injury occur,
   immobilize.

Passes **ALL**
Exclusions?

Fails **ANY**
Exclusionary
Criteria?

IMMOBILIZE!

General Process

1. Routine patient care.
2. Potential for spinal injury
   determined. (i.e. Positive
   inclusion criteria.)
3. Spine stabilization is
   maintained untill need for
   immobilization is ruled-
   out.
4. Initial assessment is
   performed.
5. **If patient stable,**
   continue to assess for
   EXCLUSION CRITERIA

If the patient meets any
inclusion criteria, but fails
to meet ANY of the
applicable exclusion criteria,
or the assessment is
unable to be completed,
then the patient should be
immobilized.

IF THERE IS ANY DOUBT, IMMOBILIZE

FIGURE 16.22 ■ Flowchart for Field C-Spine Clearance.
(Source: ADA County EMS.)

1. Decide on the problem, the type of data needed, and the cause categories.
2. Collect or obtain data.
3. Order the causes or categories.
4. Calculate the cumulative total and the percentage of the total for each cause in a cumulative percent table.
5. Draw and label the horizontal (x-) axis, including an interval for each cause.
6. Draw, scale, and label the vertical (y-) axis on the left side of the diagram. Mark the y-axis from 0 through the cumulative total. Draw and label the vertical axis on the right side of the diagram. Mark this y-axis from 0 to 100% percent, corresponding to the cumulative total.
7. Draw the vertical bars for each cause, in the order of the highest to the lowest frequency (from left to right). The width of each bar should be the same.
8. Plot a point at the center of each bar equal to the cumulative totals, until the total adds up to 100%. Add a zero point at the left side of the first bar. Connect the points with straight lines.
9. Title the diagram to describe the nature of the observations and the time frame in which the data was collected.

FIGURE 16.23 ■ How to Construct a Pareto Diagram.

Pareto Diagram

There may be many causes for problems or conditions that adversely affect work processes. A Pareto dia-gram is a type of bar chart in which the bars representing each problem cause are arranged, or ranked, by their frequency in descending order. A Pareto diagram is useful in interpreting data and confirming the relationships that are suggested in cause-and-effect studies. This approach is based on the idea that 80% of the problem comes from 20% of the causes; the diagram is used to separate the "vital few" problem causes from the "trivial many." This aids in focusing on correcting or improving the vital few causes that contribute most to the problem. To create a Pareto diagram, follow the steps in Figure 16.23.

Figure 16.24 shows an example of a Pareto diagram. In this case an ambulance company wanted to identify the leading causes of incomplete data on run reports. Company managers found that about 35% (using the cumulative scale on the right) of the incomplete forms were missing the patient's zip code. Further, 35% of the missing data sheets showed that the ID of other responding agencies was missing. These two data elements accounted for 70% of all run reports with missing data. With this information, managers can work with field personnel to identify potential solutions. The benefit of the Pareto diagram analysis was to isolate the two major sources of missing

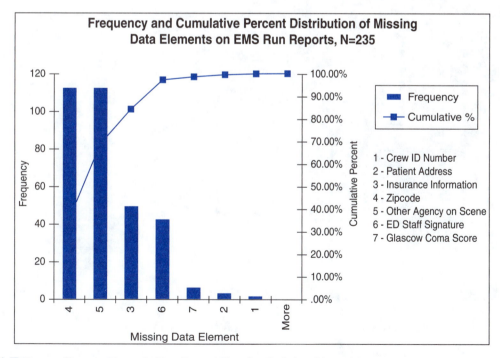

FIGURE 16.24 ■ Pareto Chart on Errors in Run Report Showing Relationship and Trends. *(Source: NHTSA and NFA.)*

data so that the most effective solutions can be pursued. Also, examining the specific data elements that are most likely to be missing may highlight specific or obvious solutions. For example, in the diagram in Figure 16.24, zip code data is missing because map books in the ambulance do not show zip codes. Simply adding a zip code map book could solve the problem.

Pictograms

The final type of chart takes advantage of pictures to display data. Data by geographical areas, such as counties, census tracts, or fire and EMS districts can be presented on maps showing the boundaries of the areas. Figure 16.25, for example, shows a large metro area with several EMS agencies. This map indicates the

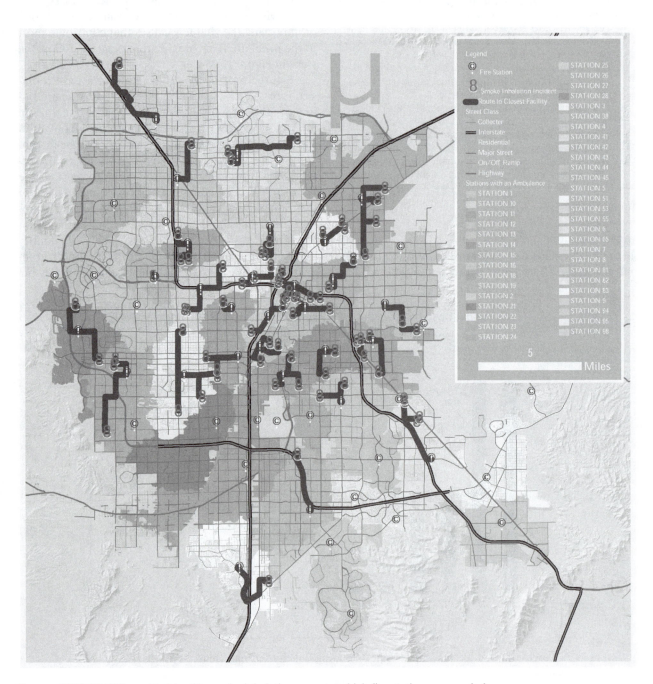

FIGURE 16.25 ■ GIS used to identify smoke inhalation cases to which fire stations responded.

number of smoke-inhalation patients in the Las Vegas Valley. It was used to help to determine which vehicles to place antidote therapy on. The key is that presentation in this manner is effective at presenting differences in regions. It can easily be seen that cases are clustered in certain areas.

ETHICS AND RESEARCH

Any research in EMS will typically involve human subjects. To the novice investigator it may not be apparent that strict ethical standards come into play when applying research to EMS activities. Once an agency has decided to investigate the science of a certain EMS practice, that study must be presented to an institutional review board (IRB) to be checked for safety and ethical constraints. IRBs are typically made up of people from the community with research, religious, sociological, and ethical training. The National Academy of Sciences maintains an IRB home page. Any research that is funded by the use of federal monies requires the researchers to follow the federal regulations for the Protection of Human Subjects. The NIH maintains a Web site with a tutorial on ethical requirements, and additional ethical information is available from the National Association of EMS Physicians.

The concern regarding informed consent for the use of new medications and procedures has provided a barrier to EMS research. Many agencies struggled to find a way to inform patients of the risk during an emergency situation. The FDA now provides for an "exception" from informed-consent requirements for emergency research. The regulations specify that a researcher must define the situations that constitute an emergency and what the requirements for informed consent will be. This process and the decision under these conditions need to be reviewed by an IRB. In most situations a public-education campaign on the research must be put out to the community, and people residing in the community should be given the option of securing a wristband or some other public-notification device that signals to EMS crews that this person does not want to participate in the research.

An IRB will have several preconceived issues with EMS research. Most IRBs have very little understanding of EMS. The perceived "standard of care" is not as well developed as it is for other medical professions, and an IRB may see the implementation of certain skills or procedures as a delay in the transportation. He will need to be educated on the role of EMS related to treating patients on the scene. Some subjects that should be included in studies will be excluded due to the barriers created in the prehospital environment.

CHAPTER REVIEW

Summary

Research is critical to EMS as an industry for several reasons: it validates what we do; it provides specific information to manage quality; it informs management on internal and external customers and groups of trends that affect EMS service delivery; and it provides a body of knowledge to define a profession. There are a variety of organizations that support EMS with data, including the National Highway Traffic Safety Administration (NHTSA), the National Institute of Standards and Technology (NIST), the Centers

for Disease Control and Prevention (CDC), the National Association of EMS Physicians (NAEMSP), and the International Association of Firefighters (IAFF) for starters. By participating in research and data collection, EMS organizations contribute to their future by validating their efforts. Consider the fact that the past 30 years of EMS have produced very little research that proves the efforts have been effective. If EMS is going to gain the respect that other allied-health professions enjoy, research will be key.

Research and Review Questions

1. Considering the story of the MAST study, how would you influence the use of the device today based on a review of the study?

2. What applications would GIS have in planning for a flu epidemic?

3. Research a local county for the GIS that is in use and the access and application a local EMS organization would have with it.
4. Research the CDC and state database for cardiac-arrest survival rates in a local county, and compare those with your agency's cardiac-arrest survival rates.
5. Contact a level 1 trauma hospital and determine the top five trauma injuries they have identified in their database over the past five years.
6. Obtain the EMS Airway Audit form from the National Association of EMS Physicians and establish a plan for implementing the data collection in your agency or community to track airway success.

References

American Heart Association. (2004). Cardiac arrest and cardiopulmonary resuscitation outcome report. *Circulation, 110,* 3385–3397.

American Heart Association, European Resuscitation Council, Heart and Stroke Association Foundation of Canada, & Australian Resuscitation Council. (1991). Recommended guidelines for uniform reporting of data from out-of-hospital cardiac arrest: the Utstein style. *Resuscitation, Sept* (22), 1–26.

American National Standards Institute, Information Technology. (1998). *Spatial data transfer standard (SDTS)* (ANSI-NCITS 320:1998). New York: Author.

Dawsen, D. E. (2006, July–September). National Emergency Medical Services Information System (NEMIS). *Prehospital Emergency Care, Vol. 10*(3), 60–64.

ESRI. (2005). *Mapping the future of public safety*. Redlands, CA: Author.

Institute of Medicine. (2006). *The future of emergency care in the United States health system*. Washington, DC: Author.

Jannig, J., & Sachs, G. (2003). Achieving excellence in the fire service. Upper Saddle River, NJ: Pearson Education.

Moore, L. (1999, Oct/Dec). Measuring quality and effectiveness. *Prehospital Emergency Care, 3*(4), 28–40.

National Highway Traffic Safety Administration. (2001). *National EMS research agenda* (DOT HS 809 674). Washington, DC: Author.

Sayre, M. R., White, L. J., Brown, L. H., McHenry, S. D. (2003, July). The national EMS research agenda executive summary. *Annals of Emergency Medicine, 42*(1), 159.

Swor, R., & Pirrallo, R. (Eds.). (2005). Improving quality in EMS (2nd ed.). Dubuque, IA: Kendall Hunt.

U.S. Fire Administration. (2004). *Fire data analysis handbook* (FA-266). Washington, DC: US Government Printing Office.

Wiederhold, R. (1992, April–June). Integrated information management system. *Prehospital Disaster Medicine, 7*(2), 14–15.

Yealy, D. M., & Scruggs, K. H. (1990). Study design and pretrial peer review in EMS research. *Prehospital and Disaster Medicine, 5*(2), 113–118.

Zaritsky, A., Nadkarni, V., Hazinski, M. F., Foltin, G., Quan, L., Wright, J., et al. (1995). Recommended guidelines for uniform reporting of pediatric advanced life support: The pediatric Utstein style. *Circulation, 1995*(92), 2006–2020.

Legal and Labor Relations

Accreditation Criteria

CAAS

106.04.01   There shall be a written procedure by which the employee can report alleged unfair policies or practices. Employees will be provided written copies of any grievance procedure utilized by the organization. A procedure must allow for input, investigation, recommendation, decision, execution, and feedback (pp. 420–424).

106.06   Disciplinary consequences and the events that will precipitate them must be clearly delineated (pp. 424–425).

106.11   The service shall have a written personal appearance standard for all employees. There shall be written standards delineating how employees should interact with the public, clients, and each other (pp. 428–430).

Objectives

After reading this chapter, the student should be able to:

17.1 Define the types of law applicable to EMS (pp. 410–411).

17.2 Identify and apply the federal legislation to EMS (pp. 411–415).

17.3 Understand and apply legislative mandates to EMS operations (pp. 411–412).

17.4 Design and understand the privacy compliance for an EMS operation that meets federal HIPAA regulations (pp. 412–413).

17.5 Identify case law affecting EMS operational systems (pp. 414–419).

17.6 Define due process, and apply the principles of progressive discipline to labor disputes involving EMS (pp. 420–425).

17.7 Recognize the Fair Labor Standards Act (FLSA) applications to EMS workers and fire-based EMS (pp. 417–419).

17.8 Define ethical behaviors and apply decision-making strategies when faced with ethical decisions (pp. 428–431).

17.9 Identify contemporary issues in EMS litigation (pp. 426–428).

Key Terms

administrative law (p. 411)
ambulance diversion
(p. 411)
arbitration (p. 421)
civil law (p. 410)
Consolidated Omnibus
Budget Reconciliation
Act (COBRA) (p. 411)
criminal law (p. 410)

deposition (p. 415)
discovery phase
(p. 415)
due process (p. 422)
Emergency Medical
Treatment and Active
Labor Act (EMTLA)
(p. 411)
ethics (p. 428)

Fair Labor Standards Act
(FLSA) (p. 417)
grievance (p. 422)
gross negligence (p. 412)
immunity laws (p. 417)
malpractice (p. 415)
progressive discipline
(p. 424)
qui tam (p. 413)

Point of View

During the early years of EMS system development, there were few concerns about medical-legal issues. Over the last decade, however, there have been increasing numbers of cases brought against prehospital providers (EMTs and paramedics) and against system administrators (municipalities and fire departments). As a result, those involved in EMS today must be informed and concerned about medical-legal issues surrounding the provision of emergency medical services.

Fire departments that provide EMS at any level (first responder, ALS, and/or transport services) must constantly monitor changes in local, state, and federal laws that may impact those services. Accurate documentation of services provided is essential.

Fire-department leaders and individual EMS providers must be knowledgeable about statutes that affect the provision of emergency medical services in their state, county, or city. Fire-department leaders should also initiate efforts to pass or protect laws,

ordinances, policies, procedures, and protocols to protect EMS system providers and EMS system administrators. Furthermore, a knowledgeable attorney should be involved in any issue that may have medical-legal ramifications.

In addition to medical-legal issues, there are EMS-related labor/management issues of which system leaders must be aware. One of the most common fire-based EMS legal issues deals with the federal Fair Labor Standards Act (FLSA). Many municipalities and EMS administrators have violated the act as it applies to single-role EMS employees working in a fire department or cross-trained multirole providers who are deployed for EMS only. In the following chapter, key cases are described with outcomes and landmark rulings with regard to FLSA violations.

Lori Moore, DrPH
Assistant to the General President
International Association of Firefighters

■ THE LAW

Medical-legal issues have become more and more significant to EMS managers and leaders. This means that EMS managers must know the law and the effects of certain types of law as they relate to EMS. It is important to identify specific types of law and what each term means. The National Fire Academy model EMS bachelor's degree program requires a course on EMS law as part of the core curriculum to heighten knowledge of legal issues.

The most important type of law to consider is **criminal law**, which has been established by legislatures to indicate public wrongs or crimes against the state. Criminal law prohibits actions that are harmful or destructive to society and identifies what is appropriate for the betterment of society. The federal, state, county, or city government prosecutes these actions or public wrongs; violations of criminal law can result in fines, imprisonment, probation, or restricted action.

Civil law is private law that is established between two recognizable parties, which may include corporations, partnerships, quasi-government structures, or other public entities. This type of law usually involves a plaintiff and a defendant. The plaintiff is the person or organization that has incurred a loss or been wronged, and a defendant is a person or organization that is accused of causing a loss or wronging the plaintiff. Civil law is based on torts or contracts, usually involving acts done by one person against another in a negligent or willful manner, causing an injury or loss. Many cases of civil law are based on common law or case law that has resulted from court cases that interpreted statutes or constitutional issues in previous disagreements. Sometimes these laws are overturned by higher courts or during a retrial when new information or case law is applied during an appeal process.

An area of civil law that pertains to the government's authority to enforce its rules, regulations, and statutes through criminal penalties, fines, seizures, or liens is known as **administrative law**. For example, a paramedic who has been granted a certification or license by the state but who violates protocol or compromises the administrative regulations of that certification can have administrative law applied to his situation. Any violation of the conditions of that certification or licensure would be conducted based on an administrative proceeding.

■ LEGAL APPLICATIONS

FEDERAL LEGISLATION AND EMS

The EMS manager and leadership should be aware of current legal statutes that affect the delivery of EMS or affect the EMS system. The first of those laws is the **Emergency Medical Treatment and Active Labor Act (EMTLA)**. This section of federal law was taken from an amendment to the Social Security Act passed in 1985 and is known as the **Consolidated Omnibus Budget Reconciliation Act (COBRA)**. The two terms are often used interchangeably. EMTLA covers any emergency medical condition and is designed to prevent unequal treatment of patients in need of emergency care. The law applies to every hospital and every patient. It requires that any individual that comes to a hospital be given a medical screening exam by qualified personnel, stabilized, and transferred to an appropriate facility if necessary. The hospital must maintain transfer records for up to five years.

When a hospital-owned ambulance or a nonhospital ambulance radios or phones a hospital, or an ambulance arrives on hospital property, there is a relationship between the embarked patient and the hospital that is covered by EMTLA. Many questions have arisen about what is hospital property. Recent rulings have specified that the hospital campus, including any sidewalks, parking lots, or driveways, qualifies as part of the hospital. In addition to the main campus, a patient or ambulance that is within 250 yards of the hospital or at any hospital clinic or facilities operating under the same Medicare ID number are covered. Initially, the law was generated to cover pregnant patients, but the final version encompassed all emergency medical conditions.

Several misconceptions about EMTLA have been placed on EMS by the hospital industry. In 1997 many EMS agencies stopped restocking from the hospital emergency departments. This was a reaction to the antikickback statute that covered how disposable supplies were restocked. In a letter dated November 25, 1997, the Inspector General responded, "Ambulance restocking arrangements with receiving hospitals do not necessarily violate anti-kickback statute. The key issues are whether or not the restocking is in return for referrals."

AMBULANCE DIVERSION

Hospital overcrowding is becoming a major problem in many communities around the country. As the baby-boomer generation begins to place a higher demand on the health-care system and the nursing shortage expands, this will increase the problem of **ambulance diversion**, a condition or event in which an ambulance is sent by radio or computer program to another hospital instead of to the closest facility or hospital of choice. Ambulance diversion has become a real problem in almost every EMS system. It is a detriment to patients and puts an undo strain on the EMS system when ambulances are not available for calls in their first-due district or high-priority staging location. Ambulance diversion frequently increases the time on task for an ambulance, increases response times, and often has some seasonal peaks. Being stuck in an ED causes low morale among paramedic and EMT crews and can have downstream effects on customer service. If the public is aware of ambulance-diversion issues, the patient may decide to bypass accessing EMS and take himself to the hospital, knowing he has a better chance at being admitted to that hospital once in the ED waiting room.

There are options for EMS diversion. The EMS crew can choose to ignore an order or can go to a secondary hospital with notification, if a paramedic deems the patient needs or wants to go only to that hospital. Like that of a walk-in emergency to the ED, a patient will not be diverted or refused treatment. Going to a more-distant hospital has potential negative consequences, such as a **gross negligence** complaint, a delay-of-care suit, an EMTLA violation, and negative public relations. *Arrington v. Wong* is a 2001 case that demonstrated the liability with ambulance diversion. It was established that ambulance diversion can occur only when the hospital lacks the resources (staff or facilities) necessary to treat an emergency patient.

Another key component of the EMTLA regulations is a provision that deals with the "dumping" of patients, when a patient is transferred by one hospital to another for insurance reasons or the lack of ability to pay. Under the EMTLA rules, a patient with an emergency medical condition that has not been stabilized cannot be transferred unless he has been offered stabilizing treatment and informed of the risk of transfer. The physician must sign a certification or document on the chart that the benefits of the transfer outweigh the effects or the risk of a transfer to the patient or an unborn child. If the hospital is without a physician, then a nurse or other health-care worker must consult with a physician for the transport.

Transfers must be done by qualified personnel. In some cases that may require a critical-care transport truck, nurse, or physician to accompany the patient. Hospitals with specialty capability (such as burn units, trauma centers, and pediatric specialty hospitals) that receive Medicare funding cannot refuse patients.

Hospital overcrowding has created a situation commonly known as *patient parking*. This occurs when a patient is prevented from being transferred from the ambulance gurney to a hospital bed or gurney, often when hospital personnel believe that they are not responsible for the patient. EMTLA states that a patient is presented to the hospital when the patient is on hospital property or with 250 feet of the hospital, including the parking lot. If the receiving hospital does not provide an initial screening exam and stabilizing treatment, it is in violation of the EMTLA law. The Centers for Medicare and Medicaid services do not recognize the distinction hospitals are making between EMS and the hospital.

PRIVACY LAW

The Health Insurance Portability and Accountability Act (HIPAA), a federal statute, was enacted in 1996. HIPAA has several key components that cover privacy, patient rights, and business and administrative requirements. HIPAA came about because of the increases in electronic filing from health-care providers and the need for congressional oversight to ensure privacy and standardized electronic claims. HIPAA requires the Department of Health and Human Services (DHHS) to oversee electronic reporting and to enforce the privacy of a patient's personal health information.

There are penalties for violating the confidential information of patients. The financial costs are divided between civil and criminal penalties. The civil penalties are $100 per violation and up to $25,000 per person, per year, for each violation. Criminal penalties can be assessed for up to $50,000 and one year in prison for obtaining and disclosing past medical history. Those penalties increase up to $100,000 and five years in prison for obtaining past medical history under false pretenses. An agency or individual can be fined up to $250,000 and given 10 years in prison for obtaining and disclosing past medical history with the intent to sell or use the information for commercial or personal gain or malicious harm. HIPAA also calls for a set of rules to be established for electronic billing.

In December 2001 the president signed into law the Administrative Simplification Compliance Act, which mandated that new Medicare claims and billing had to be conducted electronically after October 16, 2003. This was in response to HIPAA requiring electronic filing. This converted regulations into law based on the standards for electronic transactions, known as the transaction and code set (TCS). The TCS standards explain how to bill electronically for Medicare expenses.

There are waivers for electronic billing. If an EMS service has fewer than 25 full-time employees, it can be exempt from electronic transfer billing. However, this may delay working capital being transferred to an EMS agency because often electronic transfer can place funds in the agency's account in fewer than 30 days from the time the service charges are sent. The TCS rules require that billing or requests be submitted within 60 days of service. Also, the data, billing, and patient medical-history information must be secured in an area that is physically safe. Electronic billing

- ♦ Patient's name.
- ♦ Social Security number.
- ♦ Condition.
- ♦ Address.
- ♦ Age.
- ♦ Past, present, and future physical or mental health.
- ♦ Provisions of care to an individual.

FIGURE 17.1 ■ Protected Health Care Information Under HIPAA.

- ♦ Ensure that medical-history-protection policies and procedures are followed.
- ♦ Conduct protected medical history training.
- ♦ Handle patient requests for information or copies of PCR information.
- ♦ Handle and resolve patient complaints.
- ♦ Oversee system security.

FIGURE 17.2 ■ List of Roles for the Privacy Officer in EMS Organizations.

requires that the agency ensure privacy of patient-care information and that the information is transmitted in a secure manner.

The privacy rule in HIPAA requires that any agency or person that transmits health information in electronic form in connection with a covered transaction, which includes a provider of ambulance service or a first-responder agency, must have a privacy policy and protect confidential medical information. A covered transaction is a health-care claim, payment, premium payment, report of injury, or any other transaction as determined by DHHS. An EMS provider must protect and secure protected health information (PHI) as defined in Figure 17.1. This includes information in both written and oral form created or received by an EMS provider that is in the medical record.

HIPAA also calls for ambulance services and EMS agencies to provide a consent mechanism, a route for complaints, and a way to inform the patient of his rights. There are exemptions to the privacy rules. If a situation requires disclosure by law and involves issues of public health such as contagious diseases, abuse, neglect, or domestic violence, consent or authorization to transfer the information is not required. Several other situations also are exempt; for example, those requiring government oversight, law-enforcement activities, and organ- or tissue-donation requests do not need authorization.

It is not a violation to share information with certain "business associates"; for example, a billing company that subcontracts to fire and EMS agencies to collect payments for service. It also is not a violation when EMS first responders collect information and pass it off to the transporting service. Other types of business associations include those with legal consulting services, software vendors or tech support, and outside quality assurance and improvement contractors. Information also may be shared during a peer review. At no time is consent required for an EMS crew to share medical information among the health-care

team members who are attending to the patient. This is often a source of frustration between the hospital staff and EMS crews when it comes to patient follow-up or quality-improvement issues.

These services and the monitoring of privacy issues, electronic billing, and the TCS require a privacy officer, who may be the EMS chief officer or an EMS supervisor. If there is no EMS officer, this responsibility must fall on the fire chief or agency executive to ensure compliance with the rules, or the agency could forfeit the right to bill Medicare. In a society where a significant portion of the EMS response is covered under Medicare, few services can afford to be without that revenue stream. It is important for EMS agencies to have privacy policies in place and to budget for the implementation and continued operation of privacy protection in the agency. The roles and responsibilities of the person designated to be the privacy officer are listed in Figure 17.2. A job description for a privacy officer is shown in Figure 17.3. It is important for an EMS manager to determine who has access to protected health information and who is required to access it. Remember to budget for HIPAA compliance, including printing, legal consulting, and training hours for providers.

MEDICARE ABUSES

Medicare fraud has become rampant in the health-care industry. In 1990 the federal government began to crack down on health-care providers for overbilling and fraudulent billing practices. This type of lawsuit is called *qui tam* and is conducted under the False Claims Act of the federal Medicare law. It is an action brought by an informer under federal statute that establishes a penalty for the commission or omission of certain acts and provides that the same shall be recoverable in a civil action. Part of the money paid as the penalty for a false claim will go to the individual who blows the whistle; the remainder of the monies will go to the state or other institutions.

Position Title: EMS Privacy Officer
Immediate Supervisor: EMS Chief Officer [or Fire Chief or Executive Officer]

Duties and Responsibilities

- Chair and provide leadership for the creation and implementation of an organization-wide privacy program.
- Develop organization-wide privacy policies and procedures, including without limitation:
  - Notice of privacy practices.
  - Authorization forms.
  - Use and disclosure of protected health information.
  - Individual requests for access to protected health information.
  - Record-keeping and administrative requirements.

- Develop and chair the organization's [Privacy Oversight Committee] to assist in implementation of the organization's privacy program.
- Collaborate with other departments, such as legal counsel, corporate compliance, accounting, IT, and medical records to maintain organization compliance with federal and state laws regarding privacy, security, electronic transactions, and protection of information resources.
- Perform initial and periodic privacy and security-risk assessments and conduct ongoing monitoring activities in coordination with the organization's other compliance and audit functions.
- Oversee and direct initial and ongoing privacy training to members of the organization's workforce.
- Participate with legal counsel in the development, implementation, and ongoing monitoring of all business associate, responder, and chain-of-trust agreements.
- Report to the EMS Chief [or Fire Chief or Executive Officer] regarding the status of privacy compliance.
- Revise the privacy program as necessary to comply with changes in the law, regulations, professional ethics, and accreditation requirements and as necessary due to changes in patient/client mix, fire and EMS operations, and the overall health-care climate.
- Serve as liaison to regulatory and accrediting bodies for matters relating to privacy and security.
- Create a program permitting members of the organization's EMS workforce, patients, and members of the public to submit complaints regarding the organization's privacy policies, procedures, and practices, and ensure that all complaints are handled diligently and appropriately.
- Initiate and promote activities to foster privacy and security awareness and compliance within the organization for HIPAA compliance.
- Collaborate with legal counsel in handling any federal, state, and county government investigations of the organization regarding privacy or security.

Qualifications:

- Strong background in information security, including program analysis, development, and testing.
- Experience in health-industry compliance.
- Knowledge about information technology, medical records and other medical information, patient privacy and confidentiality, and release of information.
- Ability to communicate and work with many disciplines, such as management, physicians, information-systems specialist, health-information specialist, financial managers, state and federal agency officials, and patients/clients or other individuals upon whom the entity maintains or transmits individually identifiable health information.
- Ability to apply management and leadership skills to attain and maintain compliance in a cost-effective manner.

HIPAA Compliance: Is expected to protect the privacy and security of all protected health information (PHI) and electronic PHI (e-PHI) in accordance with department privacy and security policies, procedures, and practices, as required by federal [and state] law, and in accordance with general principles of professionalism as a health-care provider. Failure to comply with the department's policies and procedures regarding the privacy and security of PHI and e-PHI may result in disciplinary action up to and including termination of employment. May access PHI and e-PHI only to the extent that is necessary to complete job duties and may only share such information with those who have a need to know specific patient information to complete their job responsibilities related to treatment, payment, or other company operations. Is encouraged and expected to report, without the threat of retaliation, any concerns regarding the department's policies and procedures on patient privacy or security and any observed practices in violation of those policies to the designated privacy/information security officer. Is expected to actively participate in department privacy and security training and is required to communicate privacy policy information to coworkers, students, patients, and others in accordance with department policy.

FIGURE 17.3 ■ Sample Job Description of Privacy Officer.

In 1995 Mason County Fire District Five in Washington State was the target of a False Claims Act lawsuit brought by an ambulance-service owner who claimed the fire district billed for ALS services when only BLS services were provided. One year later the U.S. Department of Justice approved a settlement with the fire district, which agreed to pay three times the amount billed for over 700 Medicare claims. This case shows how important it is to ensure your billing practices are monitored for compliance and that EMS managers maintain their knowledge of billing practices.

LAWSUIT PROGRESSION

EMS managers and leadership, when faced with impending litigation, need to manage the process. Failure to do so can result in a greater loss to the organization. A lawsuit usually progresses in a predictable fashion.

In the prelitigation phase it is important for the EMS manager to make initial contact with insurance companies or the agency's legal staff and risk-management department. Most lawyers will ask for reports and names of witnesses to an incident. EMS managers should collect reports and prepare employees to be questioned by lawyers. A system should be in place that identifies and verifies that patient information may be released to the patient or his designee. This should be managed through a central location with appropriate releases and documentation of when and to whom information was released. It is important to stress whom employees should avoid and whom to talk to once it appears legal counsel is involved.

When an official complaint is filed, the insurance company, legal department, and risk managers need to be summoned. It is important that the agency's lawyer knows the EMS business. The EMS manager should help lead a strategic planning meeting that is made up of the risk manager, legal counsel, and EMS leadership. The **discovery phase** of a lawsuit requires bringing together all the information, both good and bad. This will require EMS managers and the employees involved to assist the legal team. During discovery EMS personnel are usually questioned by the victim's attorney in a deposition or a written interrogative. A **deposition** is a pretrial discovery procedure in which the testimony under oath of a witness is taken outside an open court. Interrogatives are a set or series of written questions served on one party in a proceeding for the purpose of a factual examination of a prospective witness. At no time should an EMS worker go to a deposition or answer interrogatives without a legal representative.

In some states a medical **malpractice** case must first attempt to go through an alternative dispute-resolution system. Some states have a medical screening committee that evaluates cases before they are allowed to proceed through the court system. Several different alternative dispute-resolution systems exist. Some involve mediation, facilitation, or direct negotiation. These represent an intermediate step between a trial and the discovery phase. They attempt to find common ground between the two parties and potentially identify settlement options. If an attempt at an alternative dispute resolution fails, then the case usually goes to trial. Some cases are heard by an administrative judge, and for cases that involve death or injury, lawyers usually call for a jury trial.

The time after the trial has been held is the most important phase of a lawsuit—it is the time that EMS leadership and risk managers sit down and ask themselves how to avoid the situation again.

■ STATE LAWS

ANTITRUST

In a landmark case in 1981—*Gold Cross v. City of Kansas City/Metropolitan Ambulance Service Trust*—a lawsuit defined the relationship between the federal government and the states when it comes to exclusive rights to provide ambulance service. Antitrust laws, commonly known as the Sherman Antitrust Act, were designed to enhance competition within industry. The Sherman Antitrust Act was amended in 1914 to prohibit discrimination in price between different purchasers of commodities of like grade and quality and to eliminate mergers that lessen competition or tend to create a monopoly. It assumes that more competition provides better service to the public. Also in 1914 the Federal Trade Commission Act was passed to restrict unfair methods of competition; it created the Federal Trade Commission to process antitrust legislation.

In response to poor quality of service from multiple providers, the Kansas City, Missouri, public-safety-improvement committee recommended a single provider for both nonemergency and emergency services. The city created a public entity called the Metropolitan Ambulance Service Trust (MAST), which was directed by the city to implement a public-utility model of ambulance transport, whereby a single provider is contracted to run both emergency and nonemergency calls. MAST first contracted with a private company with a state license to provide ambulance service and ultimately purchased all the provider equipment and relied on the contracted company to manage those resources. In 1981 Gold Cross Ambulance filed a lawsuit, claiming antitrust violations when the city implemented a publicly controlled single ambulance operator.

The Gold Cross lawsuit was based on a court case in Washington State in which the Central Pierce County Fire and Rescue District moved to take over ambulances from a private provider, Shepard Ambulance. In what was known as Parker Immunity, the state courts in *Shepard Ambulance, Inc., v. Pierce County Fire District 6*, doing business as Central Pierce Fire and Rescue, denied an injunction requested by Shepard Ambulance to continue to provide service or contract for services based on Washington State statutes. Parker Immunity required that the state legislatures authorize the challenge activity and that the actions must be done with the intent to displace competition. In the Washington State case, the state law allows for county fire departments to contract or directly provide services.

In *Gold Cross v. City of Kansas City/Metropolitan Ambulance Service Trust*, the district court ruled in favor of Kansas City and MAST, citing the state law; under Missouri law the city had the right to franchise ambulance service and regulate the provision of ambulance service on the basis of public need, rather than allowing unbridled competition. An appeal was made to the Missouri Supreme Court on the basis of due process and antitrust, and the Missouri Supreme Court ruled that due process was not violated and that Kansas City and MAST had acted in accordance with state law.

This case law was the legal precedent for fire and municipal services to be awarded the exclusive rights for emergency and nonemergency services and set the stage for government to hold private ambulances accountable for their performance or risk being replaced.

RIGHT TO PROVIDE SERVICE

Three key cases related to providing service and the rights of fire agencies to enter the ambulance service when an existing private provider is in place have been set as case law in California. The first, in May 1994, was brought to Federal District Court by American Medical Response (AMR) when the six fire districts in Sacramento County were routing 911 calls to their fire-based EMS units without authorization from Sacramento County. AMR filed on the basis of antitrust violations by having their market restricted when the fire agencies monopolized the 911 communications-center calls.

Several aspects of the California code worked in favor of the fire agencies. First, the California Health and Safety Code mandates that as of June 1, 1980, care be provided to cities by their fire agencies, as the minimum service level. The code also prohibits restricting cities and counties from providing that service. Cities and counties also have the right to increase service, and by cities' entering into an agreement with the county, the service can only be increased, not decreased. The district court stated there was no need to judge the case, letting the lower-court ruling stand, since the county had terminated its contract with AMR before the appeal was heard.

Upon the request of a city or fire district that contracted for or provided as of June 1, 1980, prehospital emergency medical services, a county shall enter into a written agreement with the city or fire district regarding the provision of prehospital emergency medical services for that city or fire district. Until such time that an agreement is reached, prehospital emergency medical service shall be continued at not less than the existing level and the administration of prehospital EMS by cities and fire district presently providing such service shall be retained by those cities and fire districts, except the level of prehospital emergency medical service may be reduced where the city council or the governing body of a fire district, pursuant to public hearing, determines that the reduction is necessary.

FIGURE 17.4 ■ Example of State Legislation Authorizing EMS Activities.

California cases continued to be litigated, based on the California Health and Safety Code, EMS Act. The right to provide EMS service comes from Section 1797.201 of the EMS act (Figure 17.4)

In August 1991 a controversy between the City of San Bernardino Fire Department and the County of San Bernardino developed over who had authority for the ambulance-service regulation. At the time the City of San Bernardino Fire Department and a private ambulance were providing paramedic service, and the private ambulance was providing transport in the city. The county established a protocol that stated that the first paramedic on scene was the one responsible for patient care. A poor working relationship between the private ambulance service and the fire-based units resulted in claims that the fire communications center was delaying the private-ambulance dispatch until a paramedic engine was on scene. The county filed suit, claiming authority over the city fire-paramedic services as the agency legally responsible for medical control, and won that aspect of the argument. However, the courts ruled that the County of San Bernardino had no authority to dispatch, regulate, or authorize providers within the cities. The County of San Bernardino went to the California Supreme Court and appealed the case.

The supreme court supported the judgment of the Court of Appeal insofar as it ruled that the City of San Bernardino Fire Department had the right to continue to administer its own prehospital emergency medical services. However, it reversed the Court of Appeal judgment, ruling that the city was not obligated to comply with the emergency service protocols and reversed the right of the city to provide general ambulance services or other types of services not provided as of June 1, 1980.

The court held that Health and Safety Code, §1797.201, which provides: "Upon the request of a city or fire district that contracted for or provided, as of June 1, 1980, prehospital emergency medical services [EMS], a county shall enter into a written agreement with a city or fire district regarding the provision of prehospital emergency medical services for that city or fire district," cannot be construed to terminate a city's or fire district's right to administer such services if it fails to request or enter into an agreement with a county by a certain date. Rather, until cities and fire districts reach agreements with counties, they are to retain administration of their prehospital EMS.

The court also held that the county, acting as the local EMS agency, did not exceed its authority under Health and Safety Code, §1798, subdivision (a), by subjecting the city to certain protocols governing the city's dispatch of EMS providers and their coordination at the emergency scene. The term "medical control," as used in Health and Safety Code, §1797.220, includes dispatch, patient-destination policies, patient-care guidelines, and quality-assurance requirements. The court further held that the city could not expand beyond the types of emergency medical services it provided as of June 1, 1980, and that the city could not exclude the county provider from furnishing such services. This established case law that restricts fire departments from removing another provider without negotiation.

IMMUNITY FROM LIABILITY

In the 1970s California became the first state to enact Good Samaritan legislation. Good Samaritan law reduces the liability of a would-be rescuer of a victim or injured person. Several conditions must be established to support a Good Samaritan defense in a medical-liability case. The rescuer cannot seek compensation from the victim, act recklessly, or intentionally do wrong. Cases of gross negligence have often violated Good Samaritan standards and resulted in monetary awards against volunteers and other EMS providers. Gross negligence is a severe violation of the standard that is expected of a reasonable provider and is essentially similar to recklessness and willful and wanton misconduct.

Almost every state now has Good Samaritan legislation, and many have some additional conditions placed in their legislation. Texas Good Samaritan laws initially covered all people except those who worked in emergency departments and was later rewritten to help reduce liabilities for physicians and emergency-department staff. Good Samaritan law requires the person to "act in good faith," and any maliciousness, recklessness, or neglect can be easily seen by a jury or administrative judge as grounds to deny Good Samaritan exemption if a lawsuit is filed.

In many states, legislatures have enacted **immunity laws** for fire and EMS providers. These laws are enacted to limit the liability or cap the award settlement of lawsuits filed against government-operated EMS and fire agencies. Often these laws have the same conditions of Good Samaritan laws, requiring the public-safety workers to act in good faith. Many immunity laws award caps. An example is the Nevada Revised Statute that limits liability to $50,000. This cap can be exceeded if it is found that the EMS workers were grossly negligent. In many cases the government entities settle with the plaintiffs and pay the maximum cap, assuming vicarious liability, meaning the agency responsible is liable for the actions of their employees. This includes the physician medical director of the EMS agency, which is operating under his license. The plaintiff's lawyers file on the individual EMS providers. Often the EMS providers will be sued, and the personal-liability clauses on their homeowner's policies will provide coverage for an individual. EMS managers should make sure that employees understand what protections are provided to them in the event that a lawsuit is filed on the agency as a result of the actions of an individual.

FAIR LABOR STANDARDS

In 1938 the **Fair Labor Standards Act (FLSA)** was passed to establish a minimum wage and protect employees from unpaid overtime. Today FLSA still impacts EMS operations, and there are several areas yet to be ruled on. Current FLSA issues, such as volunteering and the overtime rights of private employees in fire and EMS, are still in negotiation with government regulators. FLSA requires overtime pay for private-sector employees who work more than 40 hours per week or firefighters who work more than 53 hours per week. When it comes to EMS workers, there are conditions that must be met. In 1999 Congress passed what is known as the "7K exemption." This provision states that fire and law-enforcement employees can work over 40 hours without overtime pay. The 7K exemption defines an employee in fire-protection activities as one who has the legal authority and responsibility to engage in fire suppression.

Juan Vela v. City of Houston

Several lawsuits have been filed and ruled upon for employees, granting back wages to fire-based paramedics who did not have firefighting responsibility. In *Juan Vela v. City of Houston*, the Fifth Circuit Court ruled that the City of Houston had to pay 800 Houston Fire employees back wages. In December of 1999 a Ninth Circuit Court case involving the City of Los Angeles denied the city's motion that the city's dual-role cross-trained firefighters on an ambulance did not qualify for the 7K exemption. Additional standards were clarified by case law and an administrative hearing, requiring EMS workers—to be eligible for the 7K exemption—to be trained in fire suppression, have a legal authority to engage in fire suppression, be employed by a fire department, and engage in fire prevention and extinguishment of fire. The 2004 court cases in Pittsburgh resulted in the city's changing the work schedule and paying back wages of $150,000 to firefighters.

Quirk v. Baltimore County

In *Quirk v. Baltimore County* in 1995, the District Court for the State of Maryland found that the statutory language of the FLSA does not expressly bring EMS personnel within the terms of the Section 7(k) exemption. Indeed, the Congress specifically exempted employees involved in fire-protection and law-enforcement activities, but not those engaged in medical services generally. If Congress had truly wished to expand the reach of its exemption, it could have used more general language, such as "emergency-response activities." As a result the court had serious reservations about reading Section 7(k) expansively to exempt EMS personnel from the 40-hour overtime requirement.

In a decision mirroring its analysis in *West v. Anne Arundel County, Maryland*, the court ruled that in order for EMS personnel to fall within the scope of Section 7(K) of the Fair Labor Standards Act, they must satisfy the four-prong test contained in 29 CFR Section 553.210:

- *An organized fire department or fire-protection district employs the employee.* The court noted that the employees concede that the county had met its burden of proving that the employees are employed by an organized fire department.
- *The employee has been trained to the extent required by state and local ordinance.* The court noted that some of the employees have received instruction in firefighting techniques.
- *The employee has the legal authority and responsibility to engage in the prevention, control, or extinguishment of a fire of any type.* The court found that on rare occasions a few employees might have undertaken activities that could be considered fire-protection activities and that these employees were not disciplined as a result of their fire-protection activities.
- *The employee performs activities that are required for and directly concerned with the prevention, control, or extinguishment of fires, including such incidental nonfirefighting functions as the public agency's fire-protection activities.* The court noted that, while the employees asserted that the phrase "fire-protection activities" refers to the "prevention, control, and extinguishment of fires," they only occasionally fulfill this factor.

The court ruled that even if the four factors outlined above were met, the county must further demonstrate that the employees spend 80% or more of their time performing work that is incidental to or in conjunction with fire-protection activities. The court ruled that no one employee met this 80% requirement.

Horan v. King Co., Washington

In this case the court held that paramedics should be paid FLSA overtime based on the 40-hour-workweek standard, for the following reasons: first, work performance, and not simply training, should control whether paramedics are governed by Section 7(k) or the 40-hour overtime standard; next, if emergency medical service personnel spend more than 20% of their working time on activities unrelated to fires and law-enforcement emergencies, they should not be subjected to the higher firefighter and law-enforcement overtime standards under Section 7(K); and last, it cannot be said that paramedics are "regularly dispatched" to fires and law-enforcement scenes within the meaning of the Labor Department's regulation when such calls comprise a comparatively small percentage of their calls.

Edwards v. City of Memphis

In *Edwards v. City of Memphis*, pursuant to a settlement agreement reached between the parties, the plaintiffs and single-role EMS personnel were found to be entitled to overtime compensation for any work in excess of 40 hours per week. Furthermore, the City of Memphis was required to institute a cross-training program for EMS personnel, allowing them to become fully qualified in fire-suppression activities within a

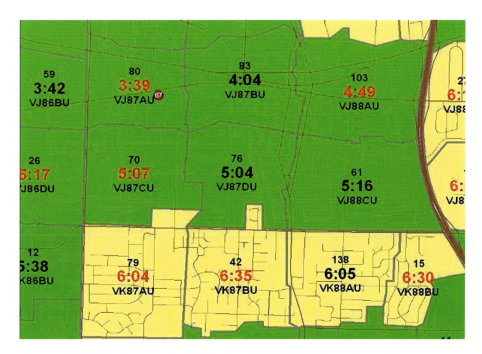

FIGURE 16.2 ■ Deployment Analysis Using GIS. This is information tracking response time compliance showing average response time (90% of responses under 6 minutes in green; tan map districts responses not met).
(Courtesy of Tualatin Valley Fire Rescue. Reprinted with permission.)

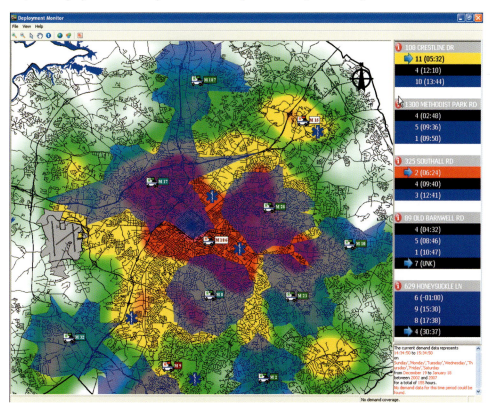

FIGURE 16.10 ■ Dynamic Service Area Generated with GIS and MARVLIS showing areas to which ambulances can respond in eight minutes or less.
(Source: ESRI. Reprinted with permission.)

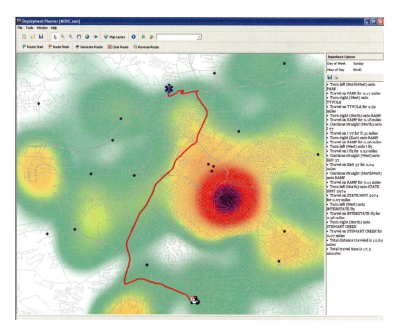

FIGURE 16.11 ■ GIS and AVL are being used to reroute a vehicle around another call or road closures.
(Source: ESRI. Reprinted with permission.)

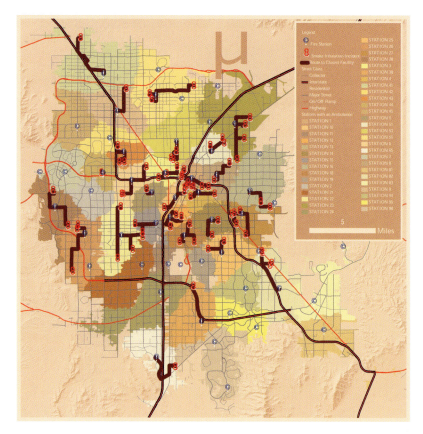

FIGURE 16.25 ■ GIS used to identify smoke inhalation cases to which fire stations responded.

three-year period with no loss in plaintiff employment. If any of the plaintiffs could not or did not successfully complete the cross-training agenda, such employees were given the right to transfer into a dispatch position or other positions within the city. Each plaintiff who was currently employed received $3,000, less applicable withholdings, and the opportunity to be cross-trained as a firefighter.

Christensen v. Harris County, Texas

Comp time, or banked time off in lieu of overtime pay, is not covered under FLSA. In a recent U.S. Supreme Court case—*Christensen v. Harris County, Texas*—the Court ruled that comp time belongs to the employee and an employer cannot specify how it is used. FLSA does allow for the employer to specify comp-time use or can cash out comp time. Most often how comp time is to be used is contractual in collective bargaining and the union contract.

FLSA also has influence on medical continuing education and promotional training. Under FLSA, unless your job title requires a certification, the employer does not have to pay for continuing education for the employee to keep that certification. For example, if the wages are firefighter wages and paramedic certification is considered premium pay, overtime expenses to maintain certification do not have to be compensated. Promotional training is not the employer's responsibility under FLSA, and overtime does not have to be paid for promotional training. However, if the job title and requirements are firefighter/paramedic, then the employer is responsible for the employee's certifications. Some state laws have attempted to establish labor laws targeted at EMS providers to ensure adequate pay for shift work. In 2006 the state of Missouri passed a minimum-wage law that included emergency workers; this law in effect requires agencies to pay employees of fire and EMS agencies overtime over 40 hours. Several agencies have taken actions to have emergency workers excluded.

■ LABOR RELATIONS

LANDMARK LEGAL CASES

EMS managers and leaders most often will be operating with one of three unions: the International Association of Firefighters (IAFF), the American Federation of State, County and Municipal Workers (AFSCME),

or in some organizations the Teamsters Union, all have local affiliates that represent EMS workers or firefighters. Most labor contracts with one of these organizations will require a manager to follow a specific process for disciplinary action and provide employees with certain rights. The IAFF has a tremendous amount of knowledge and experience in EMS, and many labor disputes involving EMS are mediated with resources from the international headquarters of the IAFF.

The history of labor law in the United States can be traced back to the 1920s. The union movement dates back to the 1926 Railway Labor Act, which gave employees the right to be represented by a union and to engage in union activity. It also established the basis for arbitration and mediation. In 1931 the Davis Bacon Act guaranteed a prevailing wage, which meant that government workers operating in union organizations would not have to compete with organizations offering lower wages. The Norris LaGuardia Act prevented court injunctions against legal unions unless they presented a clear and present danger to the organization. Real advances were made in 1935 with the passing of the Wagner Act, also known as the National Labor Relations Act, which established bargaining rights and collaborative approaches between labor and management. The Wagner Act established the National Labor Relations Board (NLRB) to monitor and establish procedures for elections. The NLRB also investigates complaints of unfair labor practices. Figure 17.5 lists what are considered unfair labor practices.

The Wagner Act was the beginning of the modern labor movement, with the establishment of the NLRB and the rights of workers to unionize and participate in collective bargaining. While the NLRB has no authority over government, state, city, or county workers, often labor contracts identify the NLRB as a source for conflict resolution between labor and management. Most fire and EMS organizations are mandated

---

- Interference with the efforts of employees to organize.
- Domination of the labor organization by the employer.
- Discrimination in the hiring or tenure of employees to discourage unions.
- Discrimination against or discharging of an employee for exercising the rights under this act.
- Refusal to bargain collectively with a representative of the employees.

FIGURE 17.5 ■ Unfair Labor Practices.

under state laws to participate in collective bargaining in return for the inability to strike. Most municipal governments participate in collective bargaining. In 1947 the Taft-Hartley Act placed additional controls on union activities.

COLLECTIVE BARGAINING

In public-safety collective bargaining there are some different approaches to the labor contract. In contrast to the private sector, where management signs off on a contract, most public-safety contracts require a government body of elected officials to sign off on or ratify a contract.

This process usually opens up the details of the contract to the media and to the public. In some jurisdictions the voters must ratify a new contract for EMS services.

Most collective-bargaining agreements in EMS have formal written agreements that often are called *clauses* or *articles*. Not all states have collective-bargaining laws (Figure 17.6). Routine clauses include federal or state constitutional statements, the composition of the bargaining unit, and reopening and severing the labor agreement. Union-security clauses define the membership and a provision for the recognition of the labor unit. Often this includes how a person becomes a union member and what the eligibility is for membership.

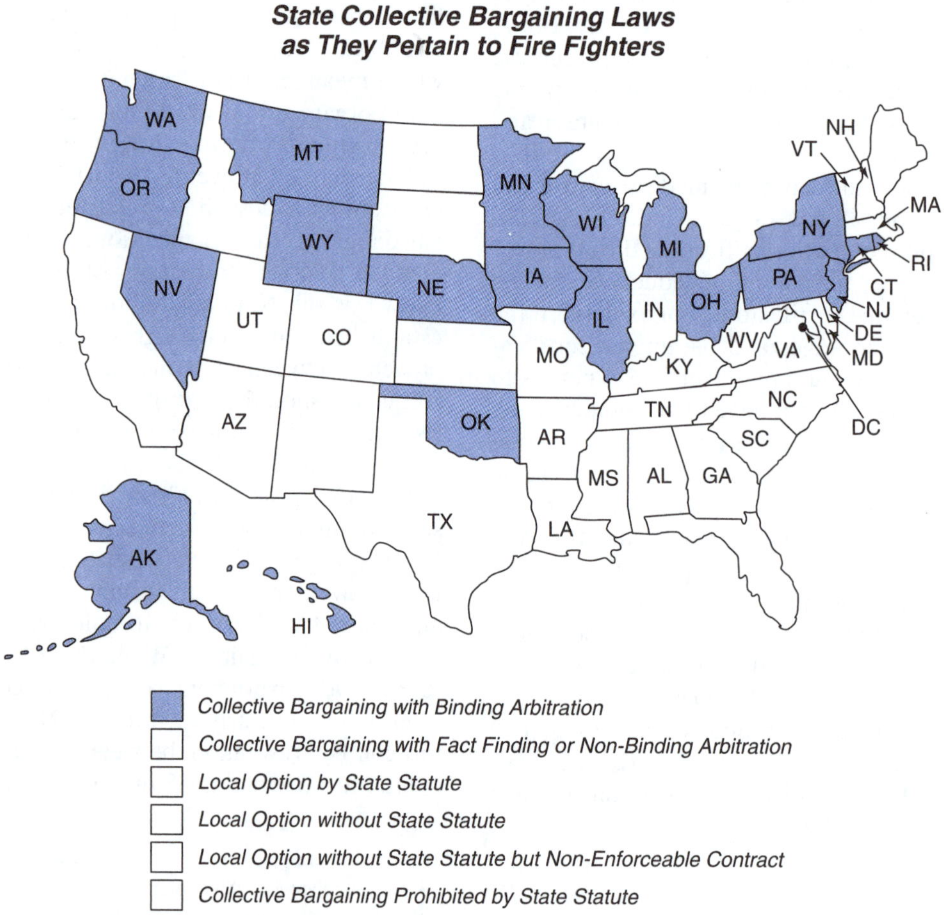

State Collective Bargaining Laws as They Pertain to Fire Fighters

- ▨ Collective Bargaining with Binding Arbitration
- ☐ Collective Bargaining with Fact Finding or Non-Binding Arbitration
- ☐ Local Option by State Statute
- ☐ Local Option without State Statute
- ☐ Local Option without State Statute but Non-Enforceable Contract
- ☐ Collective Bargaining Prohibited by State Statute

This map attempts to identify the similarities of state collective bargaining laws and how they pertain to fire fighters. It should be noted that these six categories are generalizations. As such, this map is not meant to imply that the states sharing the same category have identical collective bargaining laws. When alalyzing this map, bear in mind that there are no federal laws requiring an employer to bargain with representatives of your local. Because of this not only does each state operate differently with regards to collective bargaining, but in most instances, each municipality within each state operates differently with regards to how, if at all, they will negotiate with your local.

FIGURE 17.6 ■ State Collective Bargaining Laws as They Pertain to Firefighters. *(Source: IAFF Collective Bargaining. Reprinted with permission.)*

Some unions or local affiliates of the IAFF do not induct a person into the union until after he has cleared probation.

Management rights are the specific prerogatives retained by management. Often the right to hire, assign, or transfer employees is a management right. Management rights also extend to staffing levels, the workday, and the methods of supplying services to the public. Almost every contract includes a very detailed grievance article listing a step-by-step process of how discipline may be disputed and how differences or impasses are processed. Often a grievance article will detail an alternative dispute-resolution system and indicate the use of an arbitrator or mediator service. The conditions of employment specify wages, benefits, vacations, sick time, hours, and retirement contributions.

Collective bargaining creates the opportunity to negotiate how labor and employment processes are managed. It is a negotiation focused on solving problems and coming to reasonable agreements. Often it will focus on analyzing grievances and looking at options in such areas as the disciplinary system. Many times it focuses on analyzing grievances and looking at options in such areas as the disciplinary system. Contract negotiations should be based on facts. When proposing contractual language, a few general rules should be followed. The proposal should be written in clear, concise language; avoid long sentences, repetition, and lengthy paragraphs. Avoid using complicated legal terms and ambiguous wording, such as "in general," "whenever possible," and "except in emergencies." Proposals in a collective-bargaining negotiation should have documentation and examples to illustrate the problems to solve; for example, paramedic wages and how they may be affecting recruitment. Ensure before a proposal goes forward that it is proofread and checked for accuracy.

EMS managers and leaders should be prepared to make a full exchange of the proposal with the labor unit and agree to rules up front in the first meeting before any negotiations. Often there are ground rules that can be followed by both sides, which can help negotiations stay organized and move along. Figure 17.7 lists some common rules to follow.

Arbitration and Mediation

Mediation or facilitation occurs when labor and management cannot agree on a solution to a problem. Mediation is a process by which a neutral third party is

- Review collective-bargaining law and associated timetables.
- Determine a neutral meeting location and provide privacy for caucuses.
- Establish time frames for the meetings.
- Confirm the participants of each negotiating team.
- Ensure that everyone has equal status and that frank discussion can ensue.
- Establish recording criteria and how audio and written transcripts will be handled.
- Establish the right to caucuses and how to submit proposals and counterproposals.
- Establish procedures for a third party to be brought in to facilitate impasses.
- Decide on how agreed-upon items will be approved.
- Create a cooperative plan for dealing with the media.

FIGURE 17.7 ■ Common Negotiation Rules.

brought in to find common ground between the two groups. Mediators are often from federal mediation-and-conciliation services or a state-level agency with the same responsibilities. A mediator starts by talking to each party separately and then weighs the facts and statistical information. After the background information is collected, a mediator may try to persuade parties to work out a settlement. Fact finding is a method used to provide statistical information or background materials that may help each side reevaluate its position. Fact finding does not have the binding quality of arbitration.

Arbitration is a decision-making process by a neutral third party. Binding arbitration is often a clause placed in the contract or provided under state labor law that indicates a decision made by an arbitrator is final and binding on both parties (Figure 17.8). Arbitration is triggered after both parties reach an impasse and are unable to reach an agreement.

The two most important clauses in a contract negotiation that an EMS manager will be asked to administer or provide input on will be the discipline and grievance procedures. It is important to understand and create a system for how disciplinary disputes will be handled. Another important issue is how EMS pay incentives are created or whether EMS workers are paid as EMTs and paramedics according to a job description.

- Routine clauses.
- Union security.
- Management rights.
- Grievance procedures.
- Conditions of employment.

FIGURE 17.8 ■ Components of Collective Bargaining.

Due Process and EMS Performance

EMS operations at some point will need to discipline or discharge an employee from the organization. When employees fail to perform to medical standards, display improper conduct, or cause a loss to the organization, discipline follows. Sometimes the discipline is not appropriate for the action, or there is a discrepancy between what the labor group believes is appropriate and what management believes. Some situations, such as criminal acts, serious property damage, and lack of honesty, require immediate action. Poor performance often is due to a training issue or an underlying problem that requires an employee's assistance. When an official disagreement is placed in writing, it is called a **grievance**. Grievances proceed to discussions between management and labor, and if both sides stand firm, the grievance will be placed in an alternative dispute-resolution system, sent to an arbitrator, or referred to mediation. If all these processes fail to reach a mutual agreement, a grievance can be sent to district court to be ruled on by a judge.

When a disciplinary case is sent to an arbitrator or another dispute-resolution process by the union after discipline has been dispensed, it is often because management or the leadership did not follow procedures. Often these cases are solid in principle or identify actions that require discipline. However, managers who do not follow procedures, who do not adhere to what is required in union contracts, or who apply discipline in a biased manner violate employee rights. The procedural aspects are often called **due process**. When the process begins, the employee is entitled to due process, and the government cannot take away a person's life, liberty, or property without due process of law. Employees have a property right in their jobs. Due process involves the employee's being given notice of an investigation or pending or potential discipline. An agency should have a form that notifies an employee and records the events or actions that may result in discipline (Figure 17.9).

The employee has the right to be heard, and personnel regulations or union contracts define the time limits and how the process will proceed. In *Powell v. Mikulecky 891 F. 2d* 1454 (Tenth Cir. 1989), a discharged city firefighter sued individual city officials, alleging an inadequate pre-termination hearing, and won a settlement. The city terminated the firefighter after he had encouraged neighboring departments to sign a mutual-aid agreement. Often there are specific time limits, and if they are not met, the case is voided.

When an EMT or a paramedic is to receive discipline or potential termination before any outside or regulatory agency is involved, a proper investigation must be completed. Language in the contract with the labor group should include well-designed procedures for grievances (Figure 17.10).

Often investigative processes are outlined in union contracts, and EMS managers responsible for employees must be trained to follow procedures. An investigation is a follow-up, step-by-step inquiry on patient comments or observations. Information should be collected from supervisors who witness an event, employee witnesses who make oral complaints, and written complaints from inside and outside the organization. It is important for the EMS managers involved with the investigative process to remain objective and maintain strict confidentiality. If the issue is too close or has triggered emotions that will prevent a manager from nonbiased collection of the information, he is in danger of provoking a lawsuit unless he can put his personal feelings aside.

Investigations should be handled at the lowest level of management as possible, ideally between the employee and his firstline supervisor. When a complainant is outside the department, it is good practice to ask the person to place his complaint in writing as soon as possible. People's recollections tend to fade over time, and stories can change. A written statement serves to validate the seriousness of the complaint and attempts to keep people from embellishing the facts. Often a complainant will identify the real issues and help an EMS manager narrow down the problem. If at any point it appears there is a violation of the law, the investigation should be turned over to law enforcement. This may include drugs, theft, falsifying records, or embezzlement. If the initial interviews require more information, the inquiries can widen. A widened investigation may involve patients, citizens, other employees on shift, outside agencies, the medical director, and ED staff.

An EMS manager during the investigation should be a good listener. A series of open-ended questions should be asked as to who, what, when, why, and where and also how the complaint emerged. It is important that if a crew or team was involved with the complaint everyone be asked for individual statements and separated from each other to avoid collaboration on a story. It is important to collect any physical evidence, videotape, dispatch tapes, photographs, and other tangible items. The amount of surveillance videotape in public areas can be a valuable part of the investigation. In some cases the media, the public, or the complainant will make the information public and compromise any confidentiality. The EMS manager

---

Disciplinary Action Form

Employee Name _____ Position: _____
Assignment _____ Immediate Supervisor: _____

SECTION 1: INVESTIGATIVE INTERVIEW

This investigative interview concerns a problem that may lead to formal discipline. A summary of the problem is given below. You have the right to respond in person at this time or elect to respond in writing within 48 hours of the date/time signed by the interviewer below. Failure to respond either in person or in writing could result in disciplinary action being taken against you without your input. A decision regarding what, if any, disciplinary action may be imposed will be made after consideration of any response you may offer.

Purpose of investigative interview: ❏ CONDUCT   ❏ ATTENDANCE   ❏ PERFORMANCE

Interviewer Comments:
_____
_____
_____
_____

Employee Comments:
_____
_____
_____
_____

Union Representation: ❏ Present   ❏ Declined by Employee

Signature of Employee*: _____ Date/Time: _____
Signature of Interviewer: _____ Date/Time: _____
Signature of Witness: _____ Date/Time: _____

SECTION 2: DISCIPLINARY MEETING

Your response, if any, to the problem identified above has been considered. It has been decided to impose the action indicated below. Failure to correct a problem may result in more serious action being imposed against you at a later time.

Action to be taken: ❏ None   ❏ Counseling   ❏ Step 1   ❏ Step 2   ❏ Step 3 (# of shifts _____)   ❏ Step 4

Database Entry: Date ___/___/___   Admin Hearing Held? ❏ No   ❏ Yes ___/___/___

Effective Dates of Action: Active Date: ___/___/___   Inactive Date: ___/___/___

Union Representation: ❏ Present   ❏ Declined by Employee

Suggested Corrective Actions:
_____
_____

Signature of Employee*: _____ Date/Time: _____
(Acknowledging Receipt)

Signature of Interviewer: _____ Date/Time: _____
(Notify the next level of supervision when disciplinary action is imposed.)

Signature of Witness: _____ Date/Time: _____

*Disciplinary actions may be appealed. Consult the appropriate Labor Agreement for the proper procedures.

FIGURE 17.9 ■ Sample Employee Notification Form and Discipline Tracking Form.

Grievances or disputes which may be taken, including the interpretation of this Agreement, shall be settled in the following manner:

Step 1—The employee concerned may, in the presence of a representative of the Union, submit a grievance in writing to the employee's immediate supervisor. The supervisor shall attempt to adjust the grievance at that time and render a written decision within _____ working days.

Step 2—If the grievance is not settled at Step 1, the grievance shall be submitted to the Fire Chief within _____ working days who shall render a written decision within _____ working days after the receipt of the grievance.

Step 3—If the grievance is not settled at Step 2, the grievance shall be submitted within _____ working days to the (Personnel Director, City Manager, City Council, etc.) who shall render a written decision within _____ working days after the receipt of the grievance.

Step 4—If the grievance is not settled at Step 3, the grievance shall be submitted to arbitration by either of the parties upon notice to the other party.

Step 5—(Select the appropriate phrase[s].)

A. An impartial arbitrator shall be selected from a panel supplied by the American Arbitration Association or Federal Mediation and Conciliation Service upon the request of either party. The parties shall, within five (5) working days of receipt of the panel, make a selection of an arbitrator. In the event the parties cannot agree, the American Arbitration Association or the Federal Mediation and Conciliation service shall render a decision within thirty (30) working days after the case has been heard. The decision of the arbitrator will be final and binding upon both parties.

B. The Arbitration Board shall consist of a representative of the Employer, a representative of the Union, and a representative appointed by the (Director of the State Department of Labor and Industries, or Minister of Labour of the Provinces or other appropriate agency). A majority of the Board shall constitute a quorum.

C. The Arbitration Board shall consist of:
1. Appointee of the Employer.
2. Appointee of the Union.
3. The above two appointees shall select the third member of the Arbitration Board who shall be the Chairman. Failing to read an agreement within _____ working days as to who the Chairman shall be, the matter shall be referred to _____ who shall make the appointment of the Chairman of the Arbitration Board within _____ working days.

D. The majority of the Arbitration Board shall render a decision and failing to reach a majority, the decision of the Chairman shall prevail.

E. The findings of the Arbitration Board shall be final and binding upon both parties.

F. The arbitrator's expenses and compensation shall be borne equally by both parties.

G. If the Employer does not respond within the prescribed time limits, the grievance shall be settled in favor of the grievant.

H. The expenses and compensation of the arbitrators selected by the parties shall be borne by the respective parties, and the expenses and compensation of the Chairman of the Board of Arbitration shall be borne equally between the parties.

FIGURE 17.10 ■ IAFF Model Contract Language for Grievances.
(Source: IAFF. Reprinted with permission.)

should be sure to involve only employees who need to know or are required by law to have the information.

When faced with issues of performance or conduct in EMS-related activities, it is important for management to ask a series of questions designed to ensure due process has been followed by EMS managers who are disciplining employees. The employee should know what rule or expectation is involved with the incident and what the consequences are of not adhering to it. Rules should be applied in a consistent and predictable way. The facts should be collected and reported in a fair and systematic manner, and employees have the right to question the facts and present a defense. Did the employee have the opportunity to appeal the decision and was there **progressive discipline**, except in cases where a felony, violent crime, or theft has occurred? It is also important for EMS managers

to consider if there are any extenuating circumstances that influenced the employee's actions and were they considered when discipline was given.

Garrity Rule

The Garrity rule or Garrity warning is a protection that is utilized by many public-safety personnel each year (Figure 17.11). Garrity is an invocation that may be made by a public-safety employee being questioned regarding actions that may result in criminal prosecution. The Garrity rule goes by several different names including the Garrity right, the Garrity law, the Garrity advisement, and the Garrity warning. By invoking the Garrity rule, a public-safety worker is invoking his right not to incriminate himself. Any statements made after invoking Garrity may be used only for department-investigation purposes and not for criminal-prosecution

1. I am being questioned as part of an investigation by this agency into potential violations of department rules and regulations, or for my fitness for duty. This investigation concerns:

_____

_____

2. I have invoked my Miranda rights on the grounds that I might incriminate myself in a criminal matter.
3. I have been granted use of immunity. No answer given by me, nor evidence derived from the answer, may be used against me in any criminal proceeding, except for perjury or false swearing.
4. I understand that I must now answer questions specifically, directly, and narrowly related to the performance of my official duties or my fitness for office.
5. If I refuse to answer, I may be subject to discipline for that refusal, which can result in my dismissal from this agency.

6. Anything I say may be used against me in any subsequent department charges.
7. I have the right to consult with a representative of my collective-bargaining unit, or another representative of my choice, and have him or her present during the interview.

Assistant Prosecutor/Deputy Attorney General Authorizing:

_____

Signature:_____

Date: _____ Time: _____

Location: _____

Witnessed by: _____

Witnessed by: _____

FIGURE 17.11 ■ Garrity Warning for Employees Under Investigation.

purposes. The Garrity rule stems from the court case *Garrity v. New Jersey*, 385 U.S. 493 (1967), which was decided in 1966 by the U.S. Supreme Court. It was a traffic-ticket-fixing case, and officers were advised that they had to answer questions subjecting them to criminal prosecution or they would lose their jobs. The Court held that this was unconstitutional.

Technically, there are two prongs under the Garrity rights. First, if an employee is compelled to answer questions as a condition of employment, the employee's answers may not be used against the employee in a subsequent criminal prosecution. Second, the department becomes limited as to what it may ask. Such questions must be specifically, narrowly, and directly tailored to the employee's job. Thus, the basic thrust of the Garrity rights or Garrity rule is that a department member may be compelled to give statements under threat of discipline or discharge but those statements may not be used in the criminal prosecution of the individual officer. This means that the Garrity rule only protects an employee from criminal prosecution based upon statements he might make under threat of discipline or discharge. Also, the Garrity rule is not automatically triggered simply because questioning is taking place. The employee must announce that he wants the protections under Garrity. The statement shown in Figure 17.11 should be prepared in writing, and the employee should obtain a copy of it. If a written statement is being taken from an employee, the employee should insist that the Garrity warning actually be typed in the statement. Consult an attorney and

a union delegate for the laws regarding Garrity in your state before providing any statement.

Physician Prerogative

In situations that involve clinical performance and the delegated practice of a paramedic under the direction of the medical director, the physician may impose discipline that can result in a reduction in pay or a downgrade in or loss of certification. The physician has the right to eliminate, restrict, or suspend certifications or licensure under his delegated practice if after review of a paramedic's performance the physician loses confidence in the paramedic's ability to practice. This was upheld even in an organization with collective bargaining and a union contract. In *Hennepin CO v. Hennepin County Association of Paramedics*, the physician medical director withdrew a paramedic's certification under his licensure. The paramedic took the issue to union arbitration, but when the physician appealed to the district court, the court supported the physician, stating that a doctor cannot be forced to give medical control by way of collective bargaining.

EMS Pay Differentials

A common area in the contract that requires management's input in the collective-bargaining process is wages for EMS certifications. In fire-based EMS there is a system that commonly adds incentive pay onto regular salaries; in contrast, the private-ambulance and municipal-based ambulance services have salary schedules based on certifications. Incentive pay may

---

Clause 1:

(1) All bargaining unit personnel who obtain and maintain EMT-D certification shall receive $ ____ per pay period as incentive pay. The incentive pay will begin when proof of certification is furnished to and approved by the Fire Chief. An employee must have a current certification on file to be eligible for continuing EMT-D incentive pay. State guidelines will be used.

(2) All bargaining unit personnel who obtain and maintain EMS Instructor certification shall receive $ ____ per pay period as incentive pay. The incentive pay will begin when proof of certification is furnished to and approved by the Fire Chief. An employee must be an active instructor for the Fire Department and have a current certification on file to be eligible for continuing EMS instructor incentive pay. State guidelines will be used. Employees receiving this level of incentive pay shall not also draw EMT-D or EMT-P incentive pay.

(3) All bargaining unit personnel who obtain and maintain EMT-Paramedic (EMT-P) certification shall receive $ ____ per pay period as incentive pay. The incentive pay will begin when proof of certification is furnished to and approved by the Fire Chief. An employee must have a current certification on file to be eligible for continuing EMT-P incentive pay. State guidelines will be used. Employees receiving this level of incentive pay shall not also draw EMT-D incentive pay.

Clause 2:

Effective July 1, ____, Paramedics assigned to ambulance detail shall receive a wage differential (or premium) of ____% of the base wage for their rank. EMTs assigned to ambulance detail shall similarly receive a wage differential of ____% of the base wage for their rank.

Clause 3:

(1) Any employee who is assigned to an Advanced Life Support unit as a Paramedic shall be paid a differential of ____% of

his or her regular base rate. Effective January 1, ____, this premium pay shall be increased $ ____ bi-weekly. Only employees who have satisfactorily completed all required paramedic training shall be eligible for such assignment and pay differential.

(2) All employees in all titles will be required to become certified as an EMT and to maintain such certification as a term and condition of employment. However, any employee originally appointed to a title covered by this Agreement prior to January 1, ____, who is not certified as an EMT shall not be required to become certified. Such employees may choose to become certified. Once certified such employee must maintain his or her EMT certification as a term and condition of employment.

Any employee who was originally appointed prior to January 1, ____, in a title covered by this Agreement who is certified as an EMT must maintain his or her certification as a term and condition of employment.

Any employee who is assigned to an Advanced Life Support unit or a Basic Life Support unit as an EMT-Assigned (EMT-A) shall be paid a differential of ____% of his or her regular base rate. Effective January 1, ____, this premium shall be increased $____ bi-weekly.

Employees certified as an EMT but who are not assigned to an ambulance unit shall receive a differential of ____% of his or her regular base rate. Effective January 1, ____, this premium shall be increased $ ____ bi-weekly.

Only employees who have satisfactorily completed all required EMT training shall be eligible for such assignment and pay differential.

Any employee who is assigned to a unit as an EMT-I/D shall be paid a differential of ____% of his or her regular base rate.

FIGURE 17.12 ■ Model Contract Language for EMS Incentives.
(Source: IAFF. Reprinted with permission.)

be a stipend or a percentage of the base salary. In some states incentive pay is not calculated into retirement earnings so employers are likely to offer paramedic- and EMT-certification incentive pay because it does not require a contribution to a state or local pension plan. Figure 17.12 shows boilerplate contract language for personnel with EMS certifications.

■ LEGAL TRENDS IN EMS

EMS managers need to understand and routinely study the law as it relates to the ambulance and EMS industry. The number of lawsuits increased in 2006, when there was one lawyer for every 18 U.S. citizens,

with most of those lawyers in New York and California. The public will not tolerate incompetence from EMS. An aggressive proactive risk-management, employee-orientation program and reinforced values of the organization can stop trouble before it starts. It is important to conduct background checks on driving and for criminal records and to not tolerate misconduct, drugs, or alcohol on duty. Good screening and orientation can eliminate a lot of litigation. The trends in litigation are toward ambulance collisions, dispatching errors, civil rights, and medical malpractice.

CIVIL RIGHTS

A significant rise in civil rights violations and wrongful deaths has emerged in EMS. Most of these cases involve a patient's being restrained or handcuffed and not closely monitored by paramedics or law-enforcement officers. Many of these issues require the physician medical director's involvement, established protocols, and an understanding of the implications of the toxicity of stimulants and the development of a condition called *agitated delirium*. Many of the casualties of these situations involve a minority or protected class; if the community has a history of racial tension, what is an EMS call to assist law enforcement may result in EMS's being named along with law enforcement in a lawsuit over civil rights violations. Having a diversified workforce that is part of the community you are serving helps keep an open dialogue between EMS and the public.

FIRST AMENDMENT RIGHTS

Often the path management decides to pursue is not in agreement with the ideas of the crews on the street. Even more disruptive can be the one or more vocal opponents that turn business decisions into personal attacks or vendettas. EMS managers and leadership should know the law regarding First Amendment rights. The First Amendment protects freedom of speech, religion, and association. These are not unlimited rights, and only speech on matters of public concern is protected. If a public employee's speech is purely job related, it is not entitled to First Amendment protection. Legal counsel should be consulted before EMS management takes any disciplinary action against outspoken employees. In some cases the organization may set itself up for a First Amendment retaliation lawsuit. Often it is a disgruntled employee in the initial stage who begins complaining, and then further discipline leads to retaliation claims. This can bring in the International Association of Firefighters (IAFF) national representatives or ACLU lawyers.

The protections under the First Amendment also apply to brochures and other written materials. But if the agency, in its defense of its disciplinary actions, can prove the comments in the published material are made up or falsified, then slander or libel charges can be filed, with criminal and civil penalties possible.

COMMUNICATIONS LITIGATION

Medical-priority dispatch remains a common source of bad press and litigation for EMS agencies. The community now expects to be instructed on life-saving procedures and to receive prompt, appropriate resources when they call 911—this is now the standard of care. There is a mistaken assumption that medical-priority dispatch increases the liability to an agency, but most often litigation is the result of a lack of professionalism, burnout, and dispatcher abandonment by not providing prearrivals or hanging up on the caller. Lawsuits in Chicago—*Gant v. City of Chicago* and *Cooper v City of Chicago*—are just two examples of multimillion-dollar lawsuits for failure in the dispatch center that occur annually around the country. There have been no successful lawsuits against agencies that have implemented and managed a medical-priority dispatch.

VEHICLE INCIDENTS

Ambulance personnel responding to or returning from an EMS call are four times more likely to be involved in a crash versus people in their own personal vehicles. Annually, there are approximately 12,000 ambulance crashes and 120 deaths per year. Often the injuries caused by the incident are more serious than those that generated the original call. Intersections continue to be the major location of the majority of EMS vehicle incidents. If a lawsuit is filed and won, the average award is around $1 million.

EMS personnel often receive injuries that result in a workers' compensation lawsuit. Higher rates of injuries occur because of poor seatbelt use, existence of cabinets in the ambulance, unsecured equipment, and unsafe ambulance design. A priority dispatching system, driver certification, full-stop-at-red procedures, monitoring of seatbelt use, and securing of heavy items in the ambulances can reduce the damage and injury from an ambulance collision and potentially reduce litigation. The legal standard in most states requires emergency vehicles to be operated with due regard for the safety of all persons with no special exemptions for emergency driving.

MEDICAL MALPRACTICE

Negligence is the most common form of medical malpractice. An EMS worker has a duty to act within the jurisdiction the EMS agency responds to or is chartered to operate in; this may also include mutual-aid jurisdictions. A breach of that duty occurs when the worker does not observe a patient-care or operational standard, and that breach causes an injury. In other words, to be considered negligence, the injury must have been linked to the actions or lack of actions by EMS personnel.

Malfeasance occurs when an EMS worker performs an act that violates a standard of care or law. Nonfeasance occurs when an EMS worker fails to act; this is often called an act of omission. Contributory negligence occurs when the patient is said to have done something that contributed to his injury or death; for example, a person wrecking a vehicle after drinking could be found to be responsible for his own injuries. This has been used as a successful defense against medical-malpractice claims. Comparative negligence requires each agency, person, or participant that the claimant or plaintiff has filed on to pay for injuries or damages in the same proportion as his part in responsibility for the injury.

Many medical-malpractice lawsuits arise from inappropriate release or refusal to take a patient to the hospital. Minors, for example, lack the competency to consent to treatment, and authorization is required by a legal guardian. A patient who attempts to commit suicide is no longer considered mentally competent to refuse care, and treatment is required up to and including physical or chemical restraint of the patient.

■ ETHICS IN EMS

Ethics is a set of principles or a standard of conduct. The term comes from the Greek word *ethos,* which means character or custom. Commonly there are five ethical models: utilitarian, rights, fairness or justice, common good, and virtue.

- *Utilitarian.* An ethical approach that chooses the actions that will produce the greatest good for the greatest number. This is the concept of most emergency triage and modern medicine.
- *Rights.* An approach that operates on the concept that people have free will to choose their actions. An ethical decision made under the rights model will ask: does the action respect the moral rights of everyone involved? Any decision that violates a person's rights is considered wrong.
- *Fairness or justice.* This perspective is based on evaluating decision making on the fairness of an action and whether it treats everyone the same way or shows favoritism and discrimination.
- *Common good.* An approach based on the belief that society is made up of individuals and the ethical behavior of those individuals is linked to the community and bound by common interests and values.

- *Virtue.* An ethical model that uses certain ideals toward which people should strive and which provide for the full development of each individual's humanity. This model suggests that a person who has developed virtues will be naturally disposed to act in ways consistent with moral principles.

Ethical conduct requires doing the right thing, whether someone is watching or not. Ethical conduct involves beliefs, values, standards, and the principles of honor and morality. Ethics is concerned with how one should behave, and values determine how a person actually does behave. It is a guideline for human action, how people make decisions, and what motivates decision making. Some hot spots in ethics in the fire service and EMS are listed in Figure 17.13.

Ideally, an EMS organization that teaches ethics and reinforces moral principles should have few if any problems with legal issues. When people behave unethically, a tremendous amount of resources is wasted, trust and morale are destroyed, and management must do extensive damage control. As the Generation X and Y populations enter emergency services, the values and ethics of the traditional baby boomers will be in conflict with them. An EMS leader or manager will need to have a firm understanding of ethics and the common issues surrounding the workplace, EMS, the fire services, and medicine. Ethics should be incorporated into the behaviors modeled by EMS leaders and scheduled into basic EMS training. Ethical and social responsibility help reinforce our mission of community service. The challenge for EMS leadership is to sustain a culture that puts ethics and customer service in focus for the organization.

Conflicts of interest have many dimensions. In most agencies certain outside activities need to be addressed in conflict-of-interest statements, and civil service policies, union contracts, and government policies prohibit certain activities by employees. Embezzlement of public funds, bribery, and kickbacks on ambulance or apparatus contracts are all common forms of criminal

---

- ◆ Conflicts of interest.
- ◆ Appearance of fairness.
- ◆ Abuses of power.
- ◆ Sexual activity.
- ◆ Theft of government and personnel property.
- ◆ Drug abuse, abuse of vehicle, narcotics.
- ◆ Criminal activity off duty.

FIGURE 17.13 ■ Ethical Hot Spots in EMS.

I swear to fulfill, to the best of my ability and judgment, this covenant:

I will respect the hard-won scientific gains of those physicians in whose steps I walk, and gladly share such knowledge as is mine with those who are to follow.

I will apply, for the benefit of the sick, all measures [that] are required, avoiding those twin traps of overtreatment and therapeutic nihilism.

I will remember that there is art to medicine as well as science, and that warmth, sympathy, and understanding may outweigh the surgeon's knife or the chemist's drug.

I will not be ashamed to say "I know not," nor will I fail to call in my colleagues when the skills of another are needed for a patient's recovery.

I will respect the privacy of my patients, for their problems are not disclosed to me that the world may know. Most especially must I tread with care in matters of life and death. If it is given me to save a life, all thanks. But it may also be within my power to take a life;

this awesome responsibility must be faced with great humbleness and awareness of my own frailty. Above all, I must not play at God.

I will remember that I do not treat a fever chart, a cancerous growth, but a sick human being, whose illness may affect the person's family and economic stability. My responsibility includes these related problems, if I am to care adequately for the sick.

I will prevent disease whenever I can, for prevention is preferable to cure.

I will remember that I remain a member of society, with special obligations to all my fellow human beings, those sound of mind and body as well as the infirm.

If I do not violate this oath, may I enjoy life and art, respected while I live and remembered with affection thereafter. May I always act so as to preserve the finest traditions of my calling and may I long experience the joy of healing those who seek my help.

Written in 1964 by Louis Lasagna, academic dean of the School of Medicine at Tufts University, and used in many medical schools today.

FIGURE 17.14 ■ Hippocratic Oath—Modern Version.

offense that involve employees ignoring the law for their own self-interest. Kickbacks, or gifts for favorable treatment or purchases, are called *graft*. Most agencies have policies in place on the limits of gifts that may be given to and accepted by government employees. A trip paid for by a vendor or the accepting of expensive meals are just some examples and are often put to the test around the holidays. In some areas it is not uncommon for EMS workers to be offered tips or other gratuities for service.

Several other offenses have been reported in EMS agencies, including employees bird-dogging legal representation for personal-injury attorneys, running their private businesses from the station, and using EMS vehicles for personal business. Another area of potential problem is nepotism or the hiring of a family member. Some agencies allow relatives to be hired as long as that person cannot be in a position to directly supervise or be supervised by a relative.

Fairness or the appearance of fairness is important ethical behavior. It is important to conduct your EMS operations with a level playing field. Fairness is often a very personal issue and is subject to emotional influences. Judging a person too quickly can create distrust and damage relationships. The concept of emotional intelligence and delaying decisions or comments that provoke negative emotions can be taught to EMS providers. Abuse of power or position is another problem that creates ethical dilemmas.

Power is the capacity to ensure the outcomes a person wishes and to prevent those he does not want. Abuses of power seem to arise when a leader forgets the fact that he is given the power for one purpose only—to serve others.

A code of ethics is an important step in providing guidance to an organization's actions. In medicine the Hippocratic oath has long served as the foundation for ethical conduct (Figure 17.14). Although it has undergone several modifications since Hippocrates created it, the message remains the same and sets a vision for the practice of medicine.

Many organizations with EMS responsibilities use an ethics policy, a code of conduct, or an oath. Most professional organizations have at least one or even all. The National Association of EMTs has had an oath since 1978. Figure 17.15 is the NAEMT ethics code, offered as a statement of the organization's values and a guideline for ethical behavior in the course of the everyday duties of firefighters and EMTs.

An EMS manager can help create an ethical working environment by using education to teach ethical management solutions that promote accountability, integrity, and feedback. Management solutions should work to stop the inefficiencies and ineffectiveness that create an environment that allows for unethical behavior. The Josephson Institute on Ethics identifies six pillars of character that reinforce ethical behavior (Figure 17.16).

Be it pledged as an Emergency Medical Technician, I will honor the physical and judicial laws of God and man. I will follow that regimen which, according to my ability and judgment, I consider for the benefit of patients and abstain from whatever is deleterious and mischievous, nor shall I suggest any such counsel. Into whatever homes I enter, I will go into them for the benefit of only the sick and injured, never revealing what I see or hear in the lives of men unless required by law.

I shall also share my medical knowledge with those who may benefit from what I have learned. I will serve unselfishly and continuously in order to help make a better world for all mankind.

While I continue to keep this oath unviolated, may it be granted to me to enjoy life, and the practice of the art, respected by all men, in all times. Should I trespass or violate this oath, may the reverse be my lot.

So help me God.

Written by Charles B. Gillespie, M.D. Adopted by the National Association of Emergency Medical Technicians, 1978.

FIGURE 17.15 ■ Ethics Code Adopted by NAEMT. *(Source: National Association of Emergency Medical Technicians [NAEMT]. Reprinted with permission.)*

Much of the idea behind ethical behavior is that EMS workers are responsible to themselves, their families, their community, and the patient. It means being accountable for words, deeds, and actions and not blaming others but instead accepting responsibility as a public servant. Think ahead, and realize that all actions, words, and attitudes reflect choices, the consequences of which you are morally responsible for.

Having clear goals helps to create a balance between wants and needs. Gather facts before making an ethical decision—demand adequate information, involve stakeholders, and verify the credibility of information you are acting on. Consider the consequences, and filter your decision making through the Six Pillars in Figure 17.16.

Organizational values and ethics should be taught in orientation. New EMS employees should realize and be taught that they are responsible for

♦ Trustworthiness.
♦ Respect.
♦ Responsibility.
♦ Fairness.
♦ Caring.
♦ Citizenship.

FIGURE 17.16 ■ Josephson Institute Six Pillars of Character.

maintaining competency, being safe, and conducting themselves as a team in the community at large. When incidents occur they should take responsibility and accept corrective action. EMS personnel are responsible for serving and protecting their community. The ethical atmosphere in an organization should be felt every day.

When faced with a choice, EMS employees should ask several key questions: What is the impact? Is that impact safe, legal, within policy, and productive? What is the cost of that impact? What might be impacted? What might be compromised? How would my family feel if this decision were on the news?

A decision is making a choice, which will result in an action. All decisions involve a logical progression of thought to arrive at a choice. Decisions are based on standards and an ethic that involves morals and values that dictate the principles of conduct that result in action. Many decisions involve a moral temptation in which a person must choose between right and wrong. Others involve solving an ethical dilemma, a choice between a right and a right.

Ethical dilemmas have four models: individual versus community, truth versus loyalty, long-term versus short-term gains, and justice versus mercy. The first forces the individual to choose between something that benefits the individual and something that benefits the community. Often that choice can make the decision maker seem self-serving. Such decisions also can damage credibility when it is perceived that an EMS worker is not serving the community. An example of choosing the individual over the community occurs when a crew decides to leave its district for personal business, impacting its response time.

The truth-versus-loyalty ethical dilemma creates a problem between what is right and what protects the dignity or credibility of the organization and its leaders. Often in EMS, truth is placed in conflict with loyalty; for example, deployment issues under NFPA 1710 have placed management against the union in many communities.

Long-term versus short-term gains are another ethical dilemma. Consider negotiations between labor and management to create EMS field supervision. In the short term an EMS field supervisor is created because Paramedics that can contribute are not credentialed to be in a company-officer role. Five to ten years later, the lack of company-officer skills and authority within the organization has made the EMS programs less effective.

Lastly, the ethical dilemma of justice versus mercy produces two paths to implement discipline. Mercy can cause changes in the discipline process, when providers apologize or are harder on themselves than the system is. This can result in management deviating from a standard application of discipline.

A general rule when faced with problems in ethical decisions is to identify the consequences, operate under a set of principles, and treat people as you expect to be treated yourself. Realize that any decision will have an end, and abide by a set of rules that reflects caring and empathetic thinking.

CHAPTER REVIEW

Summary

The delivery of EMS does not always produce a positive outcome. When it does not, the legal process is employed to investigate. EMS managers need to know how to process and protect the organization. The boundaries within which EMS operates require a strong background in the legal aspects surrounding patient care. Labor disagreements are another common source of litigation. EMS managers and leadership must understand labor law, collective bargaining, and discipline with due process to avoid costly litigation. Legal issues involving EMS continue to arise, and a constant scanning of the work environment is required to avoid costly mistakes in the management of EMS operations. EMS law should be an ongoing part of an EMS manager's continuing education, and EMS leadership should look for expert legal advice with complex issues in EMS.

Research and Review Questions

1. List the steps in the discipline of a person for poor performance and conduct in your organization. How does it match with the requirements for due process and progressive discipline?
2. What are the qualifications for a dual-role cross-trained firefighter for eligibility for a 7k exemption under the Fair Labor Standards Act?
3. You are with a medium-size city fire department in the state of California. The chief decides he would like to take over ambulance transportation in the city. Can this be done? If so, how? Are there similar state laws in your jurisdiction that govern when the city, county, or fire district would be able to enter into the ambulance-transport business?
4. What are the four ethical decision-making perspectives in the chapter? With which perspective do you operate? By which perspective does your agency's leadership operate?
5. What are the most common causes of lawsuits in EMS?
6. A complaint was received by the EMS officer. List the steps for conducting an investigation.

References

Drobek, T. E., & Hoetmer, G. J. (1991). *Emergency management: Principles and practice for local government.* Washington, DC: International City/County Management Association.

Edwards, S. T. (2005). *Fire service personnel management.* Upper Saddle River, NJ: Pearson Education Inc.

Gold Cross Ambulance and Transfer and Standby Service, Inc. v. City of Kansas City, et al., 705 F.2D 005; 983-Trade Cas. (CCH) 65-339 (8th Cir. 1983).

Heimbach, L. J., & Wolfberg, D. M. (2006). Medical-legal concerns in EMS. In J. Berman & J. Krohmer (Eds.), *Principles of EMS systems* (3rd ed.). Boston: Jones & Bartlett and American College of Emergency Physicians.

International Association of Firefighters. (1998). *The collective bargaining process.* [Pamphlet]. Washington, DC: Author.

International Association of Firefighters. (1998). *Model contract language.* [Pamphlet]. Washington, DC: Author.

Lawsuits attempt to limit fire service EMS. (1996, May). *EMS Insider, 23*(5), 1–3.

Page, J. O., Wirth, S., & Wolfberg, J. (2005). *Ambulance service model personnel handbook.* Mechanisburg, PA: Page, Wolfberg & Wirth, LLC.

Schneid, T. D. (2001). *Legal liabilities in emergency medical services.* London: Taylor & Francis.

Schuler, R. S., & Huber, V. L. (1993). *Personnel and human resources management* (5th ed.). St. Paul, MN: West Publishing Company.

CHAPTER 18

EMS Management of Communications Centers

Accreditation Criteria

CAAS

Objectives

After reading this chapter, the student should be able to:

18.5 Evaluate and implement quality improvement/assurance programs (pp. 441–450).

18.6 Identify and implement training programs for emergency medical dispatch (pp. 450–451).

18.7 Recognize the legal case law related to communications centers (pp. 442–445).

Key Terms

AT&T language line (p. 441)

automated location identification (ALI) system (p. 440)

automatic number identification (ANI) system (p. 440)

automatic vehicle locator (AVL) (p. 445)

call prioritization (p. 445)

call-processing time (p. 441)

computer-aided dispatch (CAD) (p. 445)

continuing dispatch education (CDE) (p. 448)

control theory (p. 435)

cybernetics (p. 435)

emergency medical dispatch (EMD) (p. 433)

Medical Priority Dispatch System (MPDS) (p. 435)

multi-line telephone system (MLTS) (p. 440)

mobile data terminal (MDT) (p. 445)

National Academies of Emergency Dispatch (NAED) (p. 434)

National Crime Information Center (NCIC) (p. 434)

Point of View

The first step in getting help in any emergency is most often "making the right call." That usually means calling 911 and talking to a telecommunications operator. The more training these personnel have in emergency medical systems, especially in the area of emergency medical dispatching, the more efficient the delivery system. The priority is operational efficiency or "getting the right stuff to the right place in an effective period of time, as well as giving pre-arrival information to assist in keeping the patient or patients viable." After all, we can only do our "EMS magic" once we arrive on scene.

We are often told by responders that the original dispatch information is not what they find on scene. Well, maybe that's because we didn't ask the right questions to solicit the correct response. The initial dispatch information will cause responders to begin to put incident actions plans into place. This is usually based upon what they have been told in relation to their current training, protocols, and policies.

But what happens once we arrive on scene? Communications go a lot further than just our communications center. Are we speaking the same language as our ALS or BLS partners? Can we communicate with our mutual-aid resources, our medical delivery centers, and our trauma centers? And what if we are involved in a mass-casualty incident? Can we talk to the rest of the county, parish, regions, and state? Do we have the interoperability we thought we had? Effective communications are the backbone to your EMS system.

Mike Gavin
Director, Emergency Management
Poudre Fire Authority, Fort Collins, Colorado

■ EMERGENCY MEDICAL DISPATCH

The concept of **emergency medical dispatch (EMD),** or prioritizing medical dispatch, has been developing since the late 1970s. One of the goals and components of a comprehensive EMS system is to provide access to the most appropriate resources in response to requests for help. During the 1970s, it became apparent that the EMS dispatcher played a vital role in the EMS process. It became clear that most dispatchers had no uniform or consistent method of processing a medical call or making a decision on the deployment of resources to a response. Often there were major differences in the way that a dispatcher interpreted a call. This inconsistency led to overutilization of paramedic ambulances and underutilization of BLS units. In addition, matching the wrong ambulance to the wrong call and over- and underutilization of first responder agencies were common occurrences.

In 1979 the first emergency medical dispatch (EMD) program was started at the Salt Lake City Fire Department. After local success, it spread throughout Utah and into Colorado, Montana, and California. The system involved a dispatcher-specific training course built on the use of medically standardized, clinically based protocols. Those protocols were based on scripted interrogation questions that focused on identifying the level of response care required, the situations in which pre-arrival instructions are needed to collect information to relay to responding crews, and information about safety issues.

In 1983 the federal government endorsed the priority-dispatch model and issued a national standard curriculum, further identifying this process as the new standard of dispatch care. The generic name for protocols used in this document was the Emergency Medical Dispatch Protocol Reference System (EMDPRS). These no longer exist. The Salt Lake EMD/MPDS model (MPDS stands for medical priority dispatch system) went on to be adopted by the nonprofit **National Academies of Emergency Dispatch (NAED)** in 1988. In 1990 the Association of Public-Safety Communications Officials (APCO) issued a set of criteria-based guidelines and an associated training program for communications-center use.

While NAED and APCO use significantly different formats, they are considered the current standard-setting, nationally based organizations. Training programs grew to include comprehensive quality-assurance programs that have physician oversight, continuing dispatch training, protocol implementation and updating, and leadership and incident command training. EMD programs exist in many different types of communications centers and need to have support and administrative oversight of the clinical aspects by EMS leadership.

TYPES OF CENTERS

There are three primary classifications of dispatch centers: fire, law enforcement, and emergency medical communications. Fire communications centers, whose primary function is fire-related operations, require specialized training in fire-related areas of telecommunications. The training includes processing fire-alarm calls, dealing with private alarm providers, and assisting command officers in the implementation of the fire incident command systems. Fire communications centers also send resources and assist crews who are operating at hazardous-materials incidents and technical-rescue operations. Fire dispatchers need to understand the prioritization or arrangement of resources; for example, different resources are sent for automated fire alarms versus a car fire. Often the correct fire department units are sent according to what is designated by the computer-aided dispatch system. In manual systems SOPs are used to determine the correct units to send.

The primary work of law-enforcement communications centers is the call for police assistance. Operators are required to recognize situations; for example, traffic stops that put police officers at risk or into unsafe situations. Police communications operators must know how to protect and instruct via telephone the preservation of evidence and identification of witnesses. Police dispatchers must know how to manage radio traffic and resource direction during high-risk police operations; for example, police chases, hostage situations, or police standoffs. The police communications specialist is required to access the **National Crime Information Center (NCIC)** and to maintain certification to use this system.

Emergency medical-communications centers may be located at ambulance companies, fire departments, or medical-communications centers. Communications specialists in those centers deal with requests for assistance and require specialized training in questioning the callers, allocating emergency medical units, and providing the caller with instructions on how to provide aid to victims via the telephone. The EMD must be able to recognize the differences among the various types of prehospital resources needed on scenes. Most EMS communications centers employ some form of emergency medical-dispatch protocol as the basis for dispatch decision making and telephone care.

CERTIFICATION STANDARDS

In every communications center that handles public-safety calls, regardless of the type or agency, there are four common job tasks: questioning the caller, establishing communications, allocating resources, and managing resources. The communications center offers the customer, patient, and public a first contact with the EMS system. Long-standing ideas on the subject of job qualifications, job responsibilities, organizational structure, and operations control are being challenged by alternative approaches, such as fluid deployment, computer-aided dispatch, and emergency medical dispatch.

Three current options exist for implementing an EMD system. Priority Dispatch Corporation publishes the first option, the NAED EMD program, which is known as the **Medical Priority Dispatch System (MPDS)**. Priority Dispatch Corporation also produces a fire dispatch, police dispatch, and comprehensive quality-improvement program. The second option is APCO, which offers a program that is based on the NHTSA EMD training curriculum and is called Multipurpose Emergency Dispatch System (MEDS). MEDS is a 32-hour program that offers access to immediate dispatch, CPR instructions, DOT's *North American Emergency Response Guidebook* and offers the ability to access calls for case reviews. The third option is an EMD program and core-content material available free of charge from the NHTSA EMS division. The standardization, validation, legal, and administrative support are only available through APCO and NAED. Both of these sources conduct annual conferences, customized training, and continuing education.

Failures in communications can cascade through an entire system. For that reason EMS managers must make a collaborative effort to manage and support the efforts of the communications center, even if the center is not directly under the supervision of the EMS agency. The communications center is an integral part of the EMS management process, and the application of control theory is needed to manage it. **Control theory** is defined as the intervention and action necessary to promote appropriate behavior. The key term is *intervention*, since it distinguishes between managing and planning what resources are sent to the scene of a request for help.

To exert control EMS managers must communicate effectively with field crews and other employees. When confronted with a request for service, problems, or unique situations, EMS personnel must have reciprocal access to the manager and other resources. The communications center puts the systems in place to ensure that process. **Cybernetics** is the science of control, including the study of how control should be exercised and the flow of information necessary to achieve it. When applied to EMS operations, it is clear that to prevent adverse outcomes management needs effective communication in real time—as events are occurring.

ACCREDITATION

The National Academies of Emergency Dispatch is a nonprofit organization that has established a set of standards that recognize an accredited center of excellence.

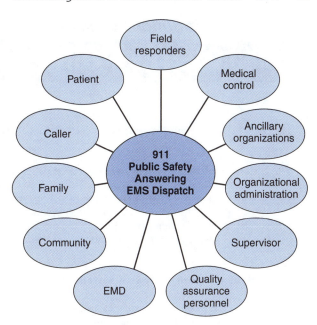

FIGURE 18.1 ■ Public Safety Answer Center Stakeholders.

This initially involves a self-study document that offers 20 points of action. The Academies have an organized College of Fellows, made up of industry experts and physicians, that can develop the NAED protocol, curriculum, and standards and evaluate communications centers. Academy accreditation assists stakeholders in successfully managing a communications center. Figure 18.1 lists the diversity of the stakeholders within a 911 communications center.

When you are preparing to submit or contribute to the accreditation of a 911 or medical communications center, a self-study guide is available from the National Academies of Emergency Dispatch. Figure 18.2 provides a road map to complete the accreditation self-study process. An EMS manager can assist in providing certain support components of this system to aid in the accreditation process, which must be formally documented and described.

ROLE OF COMMUNICATIONS CENTERS

Even without distractions or technical limitations, managers frequently do not have enough information to make good decisions. Because data transmission is limited due to the public's monitoring of the airwaves and serial records of mobile data transmissions, sensitive information may be delayed in getting to a manager. Two primary roles of a dispatch center are to serve as a message center and as the source of real-time control.

National Academy of Emergency Medical Dispatch

TWENTY POINTS OF ACCREDITATION

The Accreditation Self-Assessment Study must formally document and describe:

1. **Communication center overview and description.**

    a.   Document the total number of stations that are active (calltaking and dispatching) versus supervisory or standby (enter on line 9 of the application).

    b.   Include a floor plan showing the placement of each workstation.

    c.   List any current accreditations and the accrediting body.

2. **Medical Priority Dispatch System® (MPDS) version and licensing confirmation.**

    a.   Provide the following as applicable:
        i.   MPDS® version number
        ii. ProQA® version number
        iii. AQUA™ version number
        iv. ED-Q™ Scoring Standards version

    b.   Include documentation (or policy) stating that the most recent versions of the Protocols (ProQA and/or cardsets) and the Scoring Standards will be implemented within one year of their release.

3. **Current Academy EMD certification of all personnel authorized to process emergency calls.**

    a.   Provide a list of all EMDs, indicating their name, hire date, last certification date, next recertification date, and Academy EMD certification number.

4. **All EMD certification courses are conducted by Academy-certified instructors, and all case review is conducted by Academy-certified ED-Qs.**

    a.   If you have an in-house or contracted instructor, include their name, next recertification date, and certification number.

    b.   List all ED-Qs, indicating their name, next recertification date, and Academy ED-Q certification number.

5. **Full activity of quality improvement (QI) committee processes.**

    a.   Include copies of agendas and minutes of all Dispatch Review Committee (DRC) and Dispatch Steering Committee (DSC) meetings (minimum of three required in the six months immediately preceding the application).

    b.   List the names and titles of all committee members for:
        i.   Quality Improvement Unit
        ii. Dispatch Review Committee
        iii. Dispatch Steering Committee

    c.   List the objectives and tasks of each committee.

FIGURE 18.2 ■ National Academies of Emergency Dispatch (NAED) "Twenty Points of Accreditation Self-Assessment Study." *(Copyright NAED 2007. Reprinted with permission.)*

6. **NAED quality assurance and improvement methodology.**

   a. Attach a complete description of the methods used to evaluate EMD performance in using all elements of the MPDS correctly as outlined in the ED-Q Course Manual (consistent reviewing practices). The document should outline the following:

       i. How cases are randomly selected.

       ii. The minimum number of cases reviewed monthly.

       iii. Any special case review practices employed. This can include cases identified by the agency that warrant additional reviews. Examples are cardiac arrest, choking, and childbirth.

   b. Attach a detailed description of how EMD performance is checked, tabulated, and tracked.

   c. Include details and dates of when case review began and how scores were shared with each employee.

   d. Include details and dates of when shift and center scores were posted.

7. **Consistent case evaluation that meets or exceeds the Academy's minimum performance expectations.**

   a. Based on agency size, one of the following will apply:

       i. For agencies with less than 43,334 cases per year, the minimum is 25 cases per week.

       ii. For agencies with greater than 43,333 cases per year but less than 500,000, the minimum is 3% of the volume.

       iii. For agencies with greater than 500,000 cases per year, the minimum will vary from 1% to 3% based on the volume. (Use the Academy's calculator available on the ACE website. Provide a screen shot printout of the calculation and total).

   b. List the total number of emergency medical calls received by the center in the six months immediately prior to the accreditation application.

   c. List the total number of cases reviewed in the same time period.

8. **Historical Baseline QA data from initial implementation of structured Academy QA processes (first QI summary report, if available).**

   a. A Baseline QI Summary Report (or equivalent) that includes the following:

       i. Case Entry compliance

       ii. Key Question compliance

       iii. DLS compliance

           1. PDI compliance

           2. PAI compliance

       iv. Chief Complaint selection compliance

       v. Final Coding compliance

       vi. Total compliance score

   b. Determinant Drift Reports (or equivalent) for the center.

   c. If these are not available, please indicate on cover letter.

9. **Monthly average case evaluation compliance scores for the dispatch center for six months immediately preceding the accreditation application at or above accreditation levels.**

   a. Include a QI Summary report showing the agency has reached the following expected minimum performance levels for at least the three months preceding the application:

       • 95% Case Entry

       • 90% Key Questions

       • 90% PDIs

       • 95% PAIs

       • 95% Chief Complaint accuracy

       • 90% Final coding accuracy

       • 90% Total compliance scores

   b. Include a Center Drift Report showing that both risk and waste responses are 5% or less for the last three months prior to the application.

FIGURE 18.2 ■ Continued.

10. **Verification of correct case evaluation and QI techniques, validated through independent Academy review.**

  a. Include copies of 25 example case review audio files and completed case evaluation records for Academy assessment.
      i. 23 of these must be from the one-month period immediately preceding the application and must be selected purely at random, not cases specifically marked for feedback or other review.
      ii. State the process for random selection of these calls.
      iii. Include an additional 3 cases involving Pre-Arrival Instructions (the first Pre-Arrival case taken for each month in the three months immediately preceding this application).

11. **Implementation and/or maintenance of MPDS orientation and dispatch case feedback methodology for all field personnel.**

  a. Describe your MPDS field orientation process.
      i. Include copies of handouts, presentations, and any other materials used.
      ii. List the number of Field Responder Guides distributed, along with the dates these were given out.
  b. Describe your EMD case feedback methodology.
  c. Include a blank copy of the field feedback form utilized by your agency.
      i. Include documentation of the dates these were distributed to all field stations.

12. **Verification of local policies and procedures for implementation and maintenance of EMD. Include all policies relating to EMD practices, which must include the following:**

  a. Implementation and application of MPDS.
  b. Medical Director approval of all MPDS protocols, including those requiring local approval, to include:
      i. Obvious / Expected Death
      ii. OMEGA referrals (if applicable)
      iii. High-Risk Birth Situations
      iv. Protocol 33 Acuity levels (if applicable)
  c. Protocol compliance.
      i. Quality improvement
      ii. CDE requirements
      iii. Performance management and remediation
      iv. Customer service skills (how customer service scores are addressed by your agency)
      v. Language translation processes
      vi. Include a policy stating that all emergency medical calls are only processed by EMD-certified personnel, and that employees are removed from their calltaking duties if their certification is expired, suspended, or revoked.

13. **Copies of all documents pertaining to your Continuing Dispatch Education (CDE) Program.**

  a. Submit past six months' CDE schedules and topics.
  b. Submit EMD attendance records.
  c. Submit a CDE schedule draft for the next six months.

  ☐ Check this box if utilizing the *CDE EMD Advancement Series*.

FIGURE 18.2 ■ Continued.

14. **Secondary Emergency Notification of Dispatch (SEND) orientation.**

    a.   Include documentation of the distribution of SEND Protocol information to all police and fire dispatchers and to other agencies routinely forwarding emergency calls.
          i.   List others as appropriate.
    b.   Include documentation of agencies trained, copies of attendance records, and any training materials used for this process.

    ☐   Check this box if utilizing the *Special Procedures Briefing CD on SEND*.

15. **Established local response assignments for each MPDS Determinant Code.**

    a.   Include a description of the process for developing response configurations.
    b.   Include a list of all MPDS Determinant Codes and the response configuration locally assigned for each.
    c.   Include copies of the specific Dispatch Steering Committee (DSC) minutes with verification that all response configurations are approved.

16. **Maintenance and modification processes for local response assignments to MPDS Determinant Codes.**

    a.   Provide documentation about how MPDS local response assignments are regularly reviewed and how recommended changes are approved.

17. **Documentation for the call center's incidence (numbers) of all MPDS codes and levels.**

    a.   Each Chief Complaint (1–33).
    b.   Each individual Determinant Descriptor code (approximately 300).
    c.   Each Determinant Level (Ω, A, B, C, D, and E).

18. **Appointment and appropriate involvement of the Medical Director to provide oversight of the center's EMD activities.**

    a.   List the name, address, license number, and country/state/province (or equivalent) in which the Medical Director is licensed to practice.
    b.   Include a copy of the documentation appointing the Medical Director.
    c.   List the approved roles and responsibilities of the Medical Director within the dispatch system.

19. **Agreement to share nonconfidential EMD data with the Academy and others for the improvement of the MPDS and the enhancement of EMD in general.**

    a.   Include written verification, signed by the agency's senior executive, agreeing to the above requirement.
    b.   Include written verification, signed by the agency's senior executive, agreeing to submit the semiannual compliance summary reports to the Academy (submitted electronically through the Academy's website).

20. **Agreement to abide by the Academy's Code of Ethics and the standards set forth for an Accredited Center of Excellence.**

    a.   Include written verification, signed by the agency's senior executive, agreeing to the above requirement.
    b.   Provide verification and date of the prominent posting of the Code of Ethics and its location.

FIGURE 18.2 ■ Continued.

The communications center is a classic wheel system that is characteristic of a centralized management system in which managers deal with outside influences. Of all real-time operations, it is the communications center that is the initial point of contact with the external environment. Point of transfer of information and message-center data processing are key functions of the communications center. No one else in an EMS system has a hand on the pulse of the activity of the organization more than communications center personnel. Decisions by 911 dispatchers have more impact than any other factor on the system's response performance.

COMPONENTS OF THE COMMUNICATIONS SYSTEM

In 1968, 911 was designated the nationwide emergency number. Federal legislation and loans were made to telephone companies to upgrade local response numbers to 911. Since 1981 efforts have been under way to blanket the nation with a universally recognized, single-number system. Medicare requires that insurance, health-maintenance organizations, and managed-care recipients instruct their subscribers on the proper use of 911. This was done in response to delays created by subscribers having to call their health-plan number to get permission to call for emergency help. A public-opinion survey conducted for the American College of Emergency Physicians (ACEP) in 1992 found that nearly 50% of the adults in the United States still could not identify 911 as the emergency number or confused it with 411, the directory-assistance number. Fewer still, just one in five, had talked to their doctor about what to do in a medical emergency. The goal of every EMS agency should be to ensure that they are under the blanket of 911.

Enhanced 911 (E-911) is a system that provides dependable and precise information about the location of the emergency, incorporating the use of an **automatic number identification (ANI) system** that displays the phone number of the person calling the 911 center. Another system, the **automated location identification (ALI)**, displays the phone number, name, address, and appropriate agency to respond to the area where the caller using the phone is located. In many locations there are matrix systems in which the telephone is part of a **multi-line telephone system (MLTS)**. In some cases 911 calls made from telephones with MLTS may not precisely identify the location of the caller. For example, in a large building that has hundreds of phones registered to a single business client, the E-911 system may not validate the exact location and will require that additional information be solicited from the caller.

It is important that EMS leadership effectively market the 911 system. The nature of calls coming into the 911 system requires that a trained call taker respond directly to the caller with minimal delay, even when the caller may be incapacitated or in distress and unable to give the call center information. Additional information can be input by the communications center, such as the apartment or space number and phone location in places where a person lives with a pre-existing medical condition.

In the United States today, the cell phone is a tool used by many to report emergencies. In some areas a cell phone may be routed to a central 911 center that will require transferring of the call sometimes to several centers before the correct unit is dispatched. In the case of a cardiac arrest, for instance, when seconds can make the difference in the outcome, any delay is unacceptable. It is important for EMS agencies to publicize how a cell-phone call will be processed. The federal guidelines on wireless 911 calls will require vendors to use GPS to locate wireless 911 callers.

Several communities are now using triangulation of cell phones for this purpose. Consider the case of a person who witnesses a motor-vehicle crash on the way to Hoover Dam. He uses his cell phone to call 911, and the call is answered at a regional dispatch center that does not handle EMS requests in the city the caller is passing through. The call must then be transferred three times before it arrives at a center that can send the closest EMS resource. However, technology now can be used to pinpoint the location of cellular telephones and automatically route the call to the closest center.

DIVERSITY AND DISABILITY

Diversity in the United States poses several challenges to an EMS workforce that is traditionally not multilingual or well versed in languages other than English or Spanish. EMS and communications center leadership should provide for recruiting and training communications specialists who can speak local languages. Often under the stress of a medical emergency, people revert back to their native languages, and a dispatcher who can communicate with them will not only relieve some of their stress but will be able to gain important information that one who doesn't speak the language would not be able to obtain.

The city officials of Fremont, California, have identified over 30 languages in that community. EMS personnel could not be expected to learn and speak all of them. However, this situation is mitigated by the **AT&T language line**, an 800 service available to communications centers 24 hours a day. The language bank has linguists who can translate 160 different languages over the telephone. The use of the language bank requires the 911 communications specialist to be able to place a three-way call with the interpreter and the caller in order to get them the help they need.

Callers who are handicapped or impaired also may use special systems that the communications centers must be able to handle; for example, a system for the hearing impaired known as the telecommunications device for the deaf (TDD).

MANAGEMENT OF COMMUNICATIONS CENTERS

In most cases when a communications center services several agencies, it is called the *lead agency*. It is usually a government entity, such as a city or county, and is primarily responsible for the operations of the center. A user agency is any agency other than a lead agency but has a specific interest in EMS communications in the jurisdiction it serves. User agencies should have representation within the center's leadership and oversight functions.

An advisory committee, steering committee, or quality-improvement committee are excellent ways to implement oversight. These committees and their leadership should formulate a short-term and a long-term communications center plan. Leadership committees should consider meeting twice a month with an agenda of a minimum of eight action items; in addition, the committee has responsibility for addressing the items listed in Figure 18.3.

Any EMS communications plan should define and describe the relationships between agencies and

---

- ◆ Oversee and reduce administrative costs.
- ◆ Improve administrative services.
- ◆ Lower economic cost of service.
- ◆ Improve service benefits from the communications center.
- ◆ Review and update call-taking and dispatch procedures.
- ◆ Review, update, and implement technology.

FIGURE 18.3 ■ Tasks of the 911 Center Leadership Committee.

the EMS system. Part of the EMS communications center's strategic plan is to monitor and plan for replacement or reevaluation of communications equipment. This includes itemizing budgets and projecting replacement costs over the useful life span of the equipment.

Internal and external customer service should be incorporated into the management operations of a communications center. External customer service—the relationships the communications center has entered into outside the agency—need to be monitored regularly to ensure that external customers' needs are being met. It also is important to ensure that standards for devices that relate to unified communications, the collection of emergency information, and the transferring or handling of event information, as well as federal interoperability standards, are being met.

SELECTION AND EVALUATION

The management cycle should be applied to evaluations. One aspect of a 911 center's communications management cycle should be to calculate workload. Time studies measuring the call-processing time and frequency, overtime, and paperwork should be tracked. **Call-processing time** is the time it takes for a communications specialist to answer the call, verify the address, interrogate the caller, decide on resources needed, and initiate dispatch. Call-processing time starts when the call is received and ends when the dispatch or the alert of the responding resources is completed.

Categories should be established that represent the job task. To provide information on time spent on various tasks, communications tapes should be evaluated to validate workload, and tasks should be measured so that each aspect of the process can be managed; for example, call processing, administrative, training, and so on. Those categories then are matched with employee workload as shown in Figure 18.4. The workload is calculated by the number of emergency dispatchers on duty times the number of hours worked, divided by the work areas.

INDIVIDUAL PERFORMANCE

Once the task and percentage of each task per hour are calculated and averaged, the entire staff can establish averages. NAED's Medical Priority Dispatch System has established a performance checklist for individual communications specialists to rate their performance.

Type of Communication	Minutes	% Total
Routine traffic, radio	41.8	9.9
Routine traffic, telephone	7.6	1.8
Information request, radio	6.0	1.4
Information request, telephone	64.0	15.5
Emergency call traffic, radio	47.9	11.4
Emergency call traffic, telephone	82.1	19.5
Nonemergency call traffic, radio	10.1	2.4
Nonemergency call traffic, telephone	25.1	6.0
Relay to/from supervisor, radio	36.8*	8.8
Relay to/from supervisor, telephone	96.9**	23.3
TOTALS	**418.3**	**100.0**

Explanation of Terms:
– Routine traffic: Acknowledging status notification, etc.
– Information request: Noncall inquiries concerning status of other units, road information, or similar information.
– Emergency call traffic: Anything concerning acceptance, handling, or processing of an emergency call.
– Nonemergency call traffic: Anything concerning acceptance, handling, or processing of a routine transfer.
– Relay to/from supervisor: Any communication in which this dispatcher relays a message involving the on-duty field supervisor and a third party.

*At least four messages were clearly urgent, requiring immediate action to prevent unfavorable events. In two cases the delay in establishing contact with a supervisor was substantial.
**At least two messages were clearly urgent, requiring immediate action to prevent unfavorable events.

FIGURE 18.4 ■ Dispatcher Communication Time Analysis. *(Source: Prehospital and Disaster Medicine ©1992 Suter. Reprinted with permission.)*

Figure 18.5 displays the customer-service measurement and point-scoring system used to achieve a rating for communications personnel.

APCO offers a validated employee-evaluation program (EEP). This EEP is comprehensive, meets the Americans with Disabilities Act (ADA) and federal requirements, and is based on job-related, valid, criterion-based evidence. The test battery encompasses a personality profile, a distraction test, and a mental-abilities test.

CASE LAW

The 911 communications centers have long been the major source of litigation for public-safety agencies. Low pay, poor training, and mismanagement have resulted in millions of dollars of losses. In a 1991 journal article by the National Academies of Emergency Dispatch, 20 lawsuits against dispatch centers were reviewed. One unlucky agency had $50 million in asking

damages prior to the implementation of a medical priority-dispatch program. To date there have been no successful lawsuits against certified centers that use the Medical Priority Dispatch System. Several examples of case law and published studies have established that communications centers using a formal EMD system is now the expectation of the public.

Gant v. City of Chicago resulted in a multimillion-dollar award against the City of Chicago after the mother of a 19-year-old asthma patient called for EMS, which was only a block away. The phone rang 14 times on the first call and 26 times on the second call, and the patient subsequently died. A similar lawsuit—*Cooper v. City of Chicago*—resulted in a $3 million award when a man bleeding from an ulcer on his leg died after the 911 center refused to send EMS.

In *Ma v. City and County of San Francisco*, a legal opinion imposed liability on the city and county for failing to train dispatchers. In the San Francisco case, an asthma patient was kept on the phone for five minutes while short of breath and before an ambulance was dispatched. The ambulance was then sent to the wrong address.

In a 2003 case, *Eastburn v. Regional Fire Protection Authority*, the family of Felicia Kay Eastburn, who was then a three-year-old who had suffered an electric shock while bathing, called the 911 emergency dispatch center in Barstow, California, and the regional fire-protection district. The family alleged the 911 center failed to dispatch emergency personnel with emergency equipment and their daughter was denied early and prompt medical attention, resulting in a hypoxic brain injury. The trial failed to set forth any additional relevant facts that might support a finding of gross negligence or bad faith. The family made the additional allegation that the 911 dispatcher put them "on hold" during their telephone conversation, but such conduct would hardly amount to gross negligence or bad faith. Previous case law had defined gross negligence as "the want of even scant care or an extreme departure from the ordinary standard of conduct." The litigation was dismissed, and the 911 center was absolved of liability.

Why a sudden upsurge in litigation related to communications centers? The increase is due to management's resistance to priority-dispatch training and compliant protocol use, along with a mistaken belief that related protocols and processes increase liability. Low wages and high burnout rates are also contributing factors to litigation—a lack of professionalism in

Medical Dispatch Case Evaluation Record

Case #: _____ Date: _____ Time: _____ How obtained: **911 / E911 / Other**

Dispatcher(s): _____ Dispatcher ID: _____

Complaint description: _____ Shift: _____

Caller is: ☐ The patient *(1st party)* ☐ With patient *(2nd party)* ☐ Remote from patient *(3rd party)* ☐ Referring agency *(4th party)*

CASE ENTRY *(Primary Survey)*

	Yes	Obvious	No		Yes	Obvious	No	Insig
1. *Address* question asked?	☐	☐	☐	Address *verified?*	☐	☐	☐	
2. *Callback number* question asked?	☐	☐	☐	Callback number *verified?*	☐	☐	☐	
3. *Complaint description* question asked?	☐	☐	☐	Asked correctly?	☐	☐	☐	☐
3a. *Caller party* question asked?	☐	☐	☐	Asked correctly?	☐	☐	☐	☐
3b. *Patient count* question asked?	☐	☐	☐	Asked correctly?	☐	☐	☐	☐
3c. *Choking* question asked?	☐	☐	☐	Asked correctly?	☐	☐	☐	☐
4. *Age* question asked? _____	☐	☐	☐	Asked correctly?	☐	☐	☐	
4a. *Age* subquestion asked?	☐	☐	☐					
5. *Consciousness* question asked?	☐	☐	☐	Asked correctly?	☐	☐	☐	☐
6. *Breathing* question asked?	☐	☐	☐	Asked correctly?	☐	☐	☐	☐
6a. *Breathing* subquestion asked?	☐	☐	☐					
Gender of patient asked?	☐	☐	☐	Number of freelance questions asked: _____				

☐ Check if any questions were asked **out of order** (of questions that were asked) ** ECCS: Beginning _____ End _____

CHIEF COMPLAINT SELECTION

Chief Complaint Protocol selected: _____ ☐ Correct ☐ Incorrect Should have selected: _____

KEY QUESTIONS *(Secondary Survey)*

KQ asked?	Yes	Obvious	No	N/A	Insig	Asked incorrectly?	KQ asked?	Yes	Obvious	No	N/A	Insig	Asked incorrectly?
KQ _____	☐	☐	☐	☐	☐	☐	KQ _____	☐	☐	☐	☐	☐	☐
KQ _____	☐	☐	☐	☐	☐	☐	KQ _____	☐	☐	☐	☐	☐	☐
KQ _____	☐	☐	☐	☐	☐	☐	KQ _____	☐	☐	☐	☐	☐	☐
KQ _____	☐	☐	☐	☐	☐	☐	Category			All	Partial	None	Deduction
KQ _____	☐	☐	☐	☐	☐	☐	_____ Essential Info asked	☐		☐	☐		_____
KQ _____	☐	☐	☐	☐	☐	☐	_____ Essential Info asked	☐		☐	☐		_____
KQ _____	☐	☐	☐	☐	☐	☐	Number of freelance questions asked: _____						

☐ Check if any questions were asked **out of order** (of questions that were asked) ** ECCS: Beginning _____ End _____

DISPATCH LIFE SUPPORT INSTRUCTIONS *(Pre-Arrival & Post-Dispatch Instructions)*

	Yes	No	N/A		Yes	No	N/A
Appropriate to give Pre-Arrival Instructions?	☐	☐		*Appropriate* to give Post-Dispatch Instructions?	☐	☐	
Possible to give Pre-Arrival Instructions?	☐	☐	☐	*Possible* to give Post-Dispatch Instructions?	☐	☐	☐
(If yes) Were PAIs/PDIs given?	☐	☐	☐	*(If yes)* Were PDIs given?	☐	☐	☐
(If yes) Were they given correctly? **(C, M, D, J, A)** _____				*(If yes)* Were they given correctly? **(C, M, D, J, A)** _____			

[C]orrect **[M]**inor Mo**[D]**erate Ma**[J]**or **[A]**bsolute ** ECCS: Beginning _____ End _____

FINAL CODING

Determinant Code selected: _____ - _____ - _____ - _____ Determinant Code as reviewed: _____ - _____ - _____ - _____

TOTAL COMPLIANCE SCORE

Case Entry	_____
Chief Complaint Selection	_____
Key Questions	_____
Dispatch Life Support Instructions	_____
Final Coding	_____
Subtotal	_____ ÷ 5 = **TOTAL COMPLIANCE SCORE =**

Page 1 of 2 MPDS v11.0, v11.1, v11.2, v11.3 • SS8 • NAE 070405

FIGURE 18.5 ■ Scoring Sheet for 911 Communication Specialist.
(Copyright 2003 NAED, Salt Lake City, Utah. Reprinted with permission.)

COMMENTS

	N/A	Minor	Incorrect	Score
1. Displayed service attitude	–0	–3	–10	_____
2. Used correct volume/tone	–0	–3	–10	_____
3. Displayed compassion	–0	–3	–10	_____
4. Avoided gaps	–0	–3	–10	_____
5. Explained actions	–0	–3	–10	_____
6. Provided reassurance	–0	–3	–10	_____
7. Created expectations	–0		–10	_____
8. Used prohibited behavior	–0		–100	_____
9. Used calming techniques		N/A	No	
Case Entry		–0	–20	_____
Key Questions		–0	–20	_____
DLS Instructions		–0	–20	_____

If additional space is needed, attach a second sheet of paper.

Review date: _____ Reviewer: _____

Manager/supervisor: _____

TOTAL CUSTOMER SERVICE SCORE = [ ]

Calltaker: _____

SCORING CALCULATIONS

CASE ENTRY
100 points possible

25 points off if the address question was not asked
25 points off if the address was not verified
25 points off if the callback number question was not asked
25 points off if the callback number was not verified
33 points off if the age question was not asked
33 points off if the consciousness question was not asked
33 points off if the breathing question was not asked
20 points off if the age subquestion was not asked when appropriate
20 points off if the breathing subquestion was not asked when appropriate
20 points off for each question asked incorrectly
20 points off for each freelance question asked
10 points off if the questions were asked out of order (of questions that were asked)
10 points off if gender was not asked (if not obvious)

CHIEF COMPLAINT
100 points possible

33 points off if the complaint description question was not asked
20 points off if the complaint description question was asked incorrectly
10 points off for each appropriate subquestion (caller party, patient count, and choking) that was not asked
5 points off for each appropriate subquestion (caller party, patient count, and choking) that was asked incorrectly
67 points off if the calltaker chose an incorrect Chief Complaint Protocol

KEY QUESTIONS
100 points possible

Note: The value of each question is 100 points divided by the number of applicable questions.
Full value of the question off if the question was not asked at all
Half value of the question off if the question was asked incorrectly
10 points off if the questions were not asked in order (of questions that were asked)
20 points off for each freelance question asked
10 points off if some, but not all, required Essential Information was asked
20 points off if no required Essential Information was asked

DLS INSTRUCTIONS
100 points possible

Note: If PAIs are possible and appropriate, score PAIs and PDIs as a single DLS Instructions score and use the PAI section to score compliance. If PAIs are not possible, score only PDIs in the PDI section.
100 points off for **ABSOLUTE** deviation
50 points off for **MAJOR** deviation
25 points off for **MODERATE** deviation
10 points off for **MINOR** deviation
(See EMD-Q Scoring Standards for a complete description of DLS scoring calculations.)

FINAL CODING
100 points possible

100 points off if the calltaker should have shunted to another Chief Complaint Protocol, but did not
60 points off if the Determinant Level was incorrect
20 points off if the Determinant Descriptor was incorrect (the Chief Complaint and Determinant Level were both correct)
20 points off if a suffix was incorrect or was absent when appropriate

TOTAL COMPLIANCE SCORE

Total possible compliance score is 100%.
Add the points of all five scoring categories. Divide the sum by 5 to determine the total compliance score.

CUSTOMER SERVICE
100 points possible

For Customer Service Standards 1–7
10 points off if not applied
For Customer Service Standards 1–6
3 points off if applied with minor discrepancy
For Customer Service Standard 8
100 points off for use of any prohibited behavior

For Customer Service Standard 9 (Calming Techniques)
Case Entry 20 points off if not used and ECCS > 1
Key Questions 20 points off if not used and ECCS > 1
DLS Instructions 20 points off if not used and ECCS > 1

MPDS v11.0, v11.1, v11.2, v11.3 • SS8 • NAE 070405

FIGURE 18.5 ■ Continued

many dispatch centers allows for bad behavior to be tolerated and correct standards of conduct to be dismissed or ignored. Common sources that trigger legal action are lack of pre-arrival instructions and dispatcher hang-ups.

Priority dispatching has become, in legal terms, "the standard of care" recognized by the industry and the public. Priority-dispatch protocols and software are widely used to conduct **call prioritization** or the sorting of specific types of EMS calls based on the seriousness of the incident (Figure 18.6).

Emergency-medical-vehicle collisions (EMVCs) can be catastrophic events that provoke liability. EMVCs can cause a "wake effect," a vehicle collision that does not involve the EMS vehicle. This forces the EMS vehicle to discontinue the response and negatively affects response time. Call prioritization identifies calls that can be sent in a nonemergency fashion or can limit the number of vehicles being sent to an EMS response, thereby reducing the possibility of an EMVC (Figure 18.7). To date there have been no successful lawsuits against an agency that has properly implemented and managed an approved medical priority-dispatch system.

A computer-aided dispatch-and-triage system can safely identify cases requiring only BLS care. Although paramedic arrival may be delayed, most studies indicate no negative outcomes. One study found that using an EMS priority dispatch decreases the use of ALS on over 40% of the calls, increasing their availability for life-threatening calls. Figure 18.8 shows a sample card on stroke.

It has become a national standard to anticipate that when a person calls 911, the communications center should routinely provide pre-arrival instructions. Consequently, pre-arrival instruction should be a central topic for quality-improvement activities.

SYSTEM STATUS MANAGEMENT

The basis for system status and any deployment of vehicles, including fixed-base units, is **computer-aided dispatch (CAD).** The CAD is made up of hardware and software. The hardware includes screens, consoles, dialers, and buttons. The software manages information and recalls database resources to recommend vehicles to send to calls. Other software processes call information and provides a list of questions for the communications specialist to ask the caller to determine the nature of the medical emergency. The algorithm in the software then communicates with the other CAD software to recommend units based on the criticality of the call. This includes ALS versus BLS responses or the closest resources. CAD software can reference previous calls, duplicate calls, and hazards information.

Some CAD software systems interface with **automatic vehicle locator (AVL)** systems that track vehicles on a grid or map system. The CAD interfaces with the software that is prioritizing calls, then locates and sends the closest unit to the call. CAD software also has been developed to communicate with vehicles in the field via a **mobile data terminal (MDT)**. Response-time benchmarks, messages, and staffing can be downloaded to each unit and communicated back to the communications center. The CAD should be able to capture multiple data points and generate reports on EMD compliance, response times, and demographics of calls. The system also should reduce the

_____ Perform verbal skills in a clear and understandable manner, in the required language or languages established in the criteria as necessary to that dispatch provider agency.

_____ Perform alphanumeric transcription skills necessary to correctly record addresses, locations, and telephone numbers.

_____ Demonstrate an attitude of helpfulness and compassion toward the sick and injured patient and his/her caller advocate.

_____ Clearly guide callers in crisis though necessary interrogation procedures and the provision of telephone pre-arrival instructions.

_____ Efficiently and effectively organize multiple tasks and complicated situations and activities.

_____ Function within the team framework of public safety and EMS systems.

_____ Handle the levels of emotional stress clearly present in caller/patient crisis intervention, death and dying situations, call prioritization and triage, and multiple tasking.

_____ Elicit and assimilate call information to prioritize and properly consolidate (summarize) this information into a format used to inform the public safety responders.

_____ Demonstrate skilled use of the EMDPRS and appropriate compliance to interrogation questioning sequences and pre-arrival instruction.

_____ Demonstrate the ability to appropriately assign response configurations based on information gleaned from callers.

FIGURE 18.6 ■ Scoring Sheet for Emergency Dispatchers. *(Source: National Academies of Emergency Dispatch. Reprinted with permission.)*

CARDIAC/ RESPIRATORY ARREST: *ADULT*

KEY QUESTIONS

1. Conscipous?
2. Breathing?   NO = *Go to Card XX* NOW!!
3. Choked first?
4. Turning blue (cyanotic)?
5. Time down?
6. Age of patient?
7. Was Trauma involved?

INQUIRE OF CALLER

None required

DISPATCH PRIORITIES

Situation	Configuration	Mode	
Suspected cardiac or respiratory arrest	Closest EMTs Paramedics Ambulance	Urgent	

PRE-ARRIVAL INSTRUCTIONS

1. Get the phone next to the patient.
2. Possible obstructed airway - *GO TO CARD XX - CHOKING ADULT* as needed
3. ASK: "Does anybody there know CPR?"
4. Follow protocols below.

PROTOCOL - Patient Age 8 or Older

1. Establish unresponsiveness (gently shake and shout)
2. Call for additional HELP
3. MOUTH-TO-MOUTH

   a. Position patient on back on floor
   b. Strip the chest
   c. Kneel by patients side
   d. Open airway (head   chin)
   e. Check for breathing (look, listen, feel)
   f. Pinch the nose
   g. Completely cover patient's mouth with yours, give 2 full breaths into patient's lungs like blowing up a balloon.
   h. Did air enter lungs easily? Did chest rise?

      No- *Repeat sequence 1 time,* if still NO *then go to* CARD XX - CHOKING ADULT
   i. Is patient moving or breathing?
      YES- Stop. *Maintain/ monitor airway.*

4. Check for pulse (carotid, both sides)
   PULSE PRESENT - Stop

5. CHEST COMPRESSIONS

   a. Put heel of hand on center of patient's chest, right between the nipples.
   b. Put other hand on top of that.
   c. Push down firmly only on heels of hands. 1.5 to 2 inches.
   d. Do it 15 times. Count "1 and 2 and 3"
   e. Pinch the nose and lift the chin so the head lifts back.
   f. GIve 2 full breath like before.

6. Pump chest 15 times again (repeat step 5)
7. Keep doing patient starts breathing or help arrives.

IF PATIENT VOMITS:

1. Turn patient's head to side.
2. Sweep it out with your fingers, then continue

PANIC REPLY:

Listen to me. You need to calm down and do as I say to help your (friend, child, etc). Repeat this as necessary.

USEFUL INFORMATION

None Provided

3. Cardiac/ Respiratory Arrest

FIGURE 18.7 ■ Sample Cardiac/Respiratory Arrest Card.
(Source: National Academies of Emergency Dispatch. Reprinted with permission.)

amount of time necessary to complete the input of the data. Reports can be queried easily for the information listed in Figure 18.9 and should track each call.

Response time is a function of calculating the above sequences, and it varies from agency to agency. CAD software should interface with other computers used in the EMS system. The information should be able to be transferred to other centers electronically to avoid delays in resources being dispatched. This may include routing calls or requesting resources to and from other jurisdictions and quality-improvement

databases such as Codestat and the National Fire Incident Reporting System (NFIRS). The CAD should be able to capture the essential elements of the NHTSA minimum-data set. The CAD's communications model should allow for interoperability. Department of Homeland Security specifications call for radio systems to be interconnected between different bands and frequencies.

Retrieving data from the system is a challenge. Many EMS agencies collect mountains of data, but when asked if they can produce response-time statistics

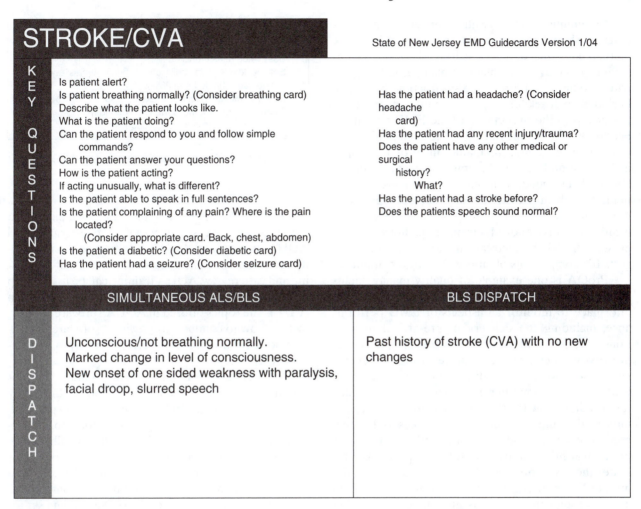

FIGURE 18.8 ■ Stroke Card from New Jersey EMS System.

or specific call demographics, they find it difficult, if not impossible. All agencies that come under the umbrella of the communications center need to be able to

- ◆ Response time.
- ◆ 911 system accessed.
- ◆ Call-processing time.
- ◆ Dispatch time.
- ◆ Turnout time.
- ◆ Travel time.
- ◆ Patient contact.
- ◆ En route to hospital.
- ◆ Arrival at hospital.
- ◆ Patient transfer.
- ◆ Unit in service.

FIGURE 18.9 ■ Data Collected at Communications Center.

have response reports, demographics, and the NHTSA data set generated.

QUALITY IMPROVEMENT PROCESSES

A vital part of any communications center is the quality-improvement process. This requires that the communications-center management and EMS leadership employ a process to capture feedback and review 3% of the yearly medically based calls. This feedback should be submitted by any party that interfaces with the communications center. This includes law enforcement, EMS crews, hospitals, and other communications centers. Feedback in writing, or a documented outcome of an event or cluster of call types, is an important scientific aspect of the research required to complete the quality-improvement process.

Communications specialists or emergency dispatchers should receive weekly or monthly feedback on their compliance to the specific EMD protocols, including interrogation, chief-complaint selection, dispatch coding, and pre-arrival-instruction provision. Periodic feedback displays a commitment by EMS leadership to the importance of the EMD operation. Feedback can be provided by the communications center management, EMS management or, in an optimal setting, an EMD quality improvement committee. Feedback also needs to include the outcome of the patients for which pre-arrivals or resources were sent. This committee should be composed of EMS, communications supervisors, EMD trainers, quality-assurance personnel, and the medical director. Table 18.1 suggests the components of an EMD QA/QI committee.

ProQA is one example of quality management-based automated protocol software that can be put into place to monitor all aspects of EMD performance, including the delivery of pre-arrival instructions. ProQA and other software programs use priority-dispatch protocols that record all the events and times involved in dispatching medical calls. ProQA allows communications managers to produce reports that allow for the communications center administration and EMS leadership to measure the correct provision and ultimate effectiveness of pre-arrival instructions. This software program integrates the EMD protocols from the National Academies of Emergency Dispatch and allows the EMD to move through the questions and share information with responders through the CAD system. ProQA provides a complete computerized record of the calls, the times, and every action taken.

◆ Average speed of answer (10 seconds or less).
◆ Abandoned-call rates (2% to 3%).
◆ Busy rates (less than 1%).
◆ Service levels (total calls minus busy signals and abandoned calls should be at 98%).
◆ First-call resolution (one call taker or dispatcher/no transfers).
◆ Queue waiting time (less than 60 seconds).
◆ On-hold waiting less than 15 seconds.

FIGURE 18.10 ■ Performance Indicators for 911 Centers.

ProQA also allows for data to be downloaded to a quality assurance case-review software program called AQUA. ProQA and AQUA help reduce the possibility of human error and reduce the time of call processing and case review. AQUA training can be used for up to eight hours of **continuing dispatch education (CDE)**. The topics offered in AQUA software training include how to conduct case reviews, software navigation, accreditation compliance, report analysis, and statistical reporting.

Data display from the automatic number identification (ANI) system should always be confirmed by the call taker or the 911 communications specialist. Communications centers also can conform to the National Performance Review Standards for call centers. EMS supervisors, communications supervisors, and/or quality improvement team leaders should review 5 to 10 calls per month for each frontline 911 communications operator or emergency dispatcher. A data set for the performance indicators for call centers is included in Figure 18.10.

The most effective QA/QI programs are used to identify good behaviors and showcase examples of excellent performance. An employee-rewards program can be tied to the communications center quality-improvement data. The NHTSA quality-improvement philosophy identifies the rewarding of excellent or above-average performance as a technique to improve substandard performance of a group. A field-feedback form should be used to encourage compliance and provide feedback to the center and the emergency dispatchers; an example of such a form is shown in Figure 18.11. It is difficult to have improvement without feedback, and the call takers and dispatcher only have the call for a few minutes and do not get to see the outcome. Failing to use a quality improvement program in the communications center will eventually weaken the center.

TABLE 18.1 ■ Suggested Membership in EMD QA/QI

- ◆ EMS supervisor (rotated annually if more than one).
- ◆ Communications center supervisor (rotated annually if more than one).
- ◆ Department EMS trainer.
- ◆ Department or agency dispatch center trainer.
- ◆ Paramedic or field EMT (rotated annually).
- ◆ Communications specialist or dispatcher (rotated annually).
- ◆ Quality assurance unit (QAU) leader.
- ◆ EMS medical director.

Field Feedback Report

NAEMD
National Academy of Emergency Medical Dispatch

Reported by: _____ Agency: _____

Date: _____ Time: _____ Run #: _____ Unit(s): _____

Dispatchers: _____ and _____

Response Team: _____ and _____

Problem Encountered: _____

_____

_____

_____

_____

_____

_____

_____

_____

_____

_____

Specific Protocol referred to: _____ #: _____

Operating procedure referred to: _____ #: _____

For QIU Use Only

Received at Quality Improvement Unit (Date): _____ By: _____

Investigation Outcome: _____

_____

_____

_____

_____

_____

_____

_____

_____

_____

_____

_____

_____

Case Review Completed (Date): _____ Compliance (%): _____ Correct Response Code: _____

Reported to: _____ at: _____

ED-Q's signature: _____ Date: _____

© 2007 NAEMD
This form may be photocopied and used with permission of the NAEMD

FIGURE 18.11 ■ Emergency Medical Dispatch Field Feedback Report.
(Source: National Academy of Emergency Dispatch. Reprinted with permission.)

EDUCATION AND TRAINING

Continuing Dispatch Education

Training is a process, not just an event. An ongoing continuing dispatch education (CDE) program is essential to ensure that the communications center changes with the environment it serves and strengthens the skills and abilities of the communications center personnel. NHTSA standards recommend that 911 communications centers have a minimum of one hour per month of continuing education that includes review of dispatch priorities, mock scenarios, techniques for pre-arrival instructions, and ride-alongs with EMS field crews.

Training can be conducted by using group training with workshops, seminars, and conferences related to emergency medical dispatch. The training calendar for the communications center should include a careful, detailed review of the specific protocols used, understanding of the concepts and rules involving problematic case types, and a variety of guest speakers to infuse new ideas and expand on areas of expertise that relate to the communications center. Group training also can encompass drills and exercises.

Teaching is the highest form of learning, and EMD personnel should participate in the EMS organization's speaker bureaus. Practical scenario training, which applies the EMD protocols to different situations, is likely to be encountered and should be ongoing and based on the feedback from the findings of the quality-improvement program (Figure 18.12).

Communications supervisors should be sent to outside seminars and conferences to bring new life to continuing dispatch education and the communications center. A CDE tracking process should be in place for each communications specialist to establish quality and ensure educational requirements are being met. In the event of litigation, this process also provides documentation that the employee was meeting and exceeding standards. Quality continuing education programs enhance the dispatcher's knowledge and build confidence.

Continuing education for a communications center should be linked to quality-assurance and quality-improvement activities. EMS leadership should track errors, retrain to correct errors, and track employee performance. A cumulative scoring system should be used to measure system performance, procedural errors, and individual compliance to protocols. Persistent performance problems after training and coaching need to receive disciplinary actions with due process. This allows management to maintain high levels of consistency within the system.

Some examples of CDE programs are the EMD Advancement Series from Priority Dispatch Corporation, using the National Academies of Emergency Dispatch protocols. The EMD Advancement Series is a computer-based program designed for continuing education for employees already certified in EMD and covers 33 protocols and topics. The APCO Institute offers online access to continuing dispatch education that appears in the APCO International monthly magazine *Public Safety Communications/APCO Bulletin*. Online and computer-assisted programs offer the benefit of self-paced and interactive multimedia. APCO provides a CDE program called *Training in a Box*, prepackaged training programs designed to be delivered in one to two hours. Each package has an instructor guide, a lesson outline, quizzes, PowerPoint® presentation, and handouts. APCO offers a variety of topics. Some topics of interest to EMS managers are presentations on large-fire scene operations, responder safety for fire and EMS, and EMD respiratory emergencies. Figure 18.13 is a sample policy on dispatch continuing education.

Additional Training

A weak area in an MCI is often the communications center. In some locations a communications specialist is incorporated into the planning and exercising of a disaster plan. Often the day-to-day medium-size incidents are not exercised or incorporated into a 911 communications center's training and continuing education. In cases where the center is a primary law-enforcement communications agency, often little or no ICS training is conducted with the personnel responsible for monitoring the radio during a multi-casualty or working-fire incident. Communications specialists

---

- ◆ Maintain and develop the communications operator's understanding of medical conditions.
- ◆ Improve skills in medical interrogation and protocol compliance.
- ◆ Improve the dispatcher's ability to correctly prioritize request for help.
- ◆ Maintain and improve skills in providing pre-arrival instructions.
- ◆ Maintain and improve skills of seldom-used procedures, call-outs, and emergency procedures.
- ◆ Provide opportunities for discussions, skill practice, and postincident analysis.
- ◆ Review and address specific issues found in the medical quality improvement process.

FIGURE 18.12 ■ Objectives for Continuing Dispatch Education (CDE) Program.

Continuing Education for EMD Dispatchers

EMD dispatchers at the Santa Barbara County Public Safety Dispatch Center will be required to complete 24 hours of continuing dispatch education (CDE) every two years as follows:

1. *Mandatory Continuing Education:* Will include attendance at a minimum of three tape-review sessions per year and continuous certification in cardiopulmonary resuscitation at the health-care provider level. Mandatory continuing education will total 16 hours of the 24 hours required every two years. Attendance at three or more tape-review sessions per year is highly encouraged.
2. *Optional Continuing Education:* Will be provided locally through the Santa Barbara County EMS Conference and through countywide base-hospital meetings. Optional opportunities for continuing education cannot exceed eight hours per two-year cycle.

FIGURE 18.13 ■ Sample Policy Statement on Continuing Education of 911 Communication Specialist.

should be trained on ICS 100 and ICS 200. Additional continuing-education topics on ICS for EMS and management of disasters should be incorporated into a comprehensive training program.

Case Study: Phoenix Fire

On a quarterly basis Phoenix Fire brings in their fire officers to run fire and emergency simulations in a command training center. On a quarterly basis members from the 911 communications center are brought in to participate in the training and provide real-time, CAD-assisted dispatching in a simulated environment. This type of integrated training teaches the 911 communications specialist to recognize and call out tactical benchmarks. The noise, confusion, and stress of an emergency scene make it difficult for incident commanders and division and group supervisors to call out benchmarks and tactical achievements on scene. With 911 communications personnel—who are isolated from the noise and confusion of the scene—prompting the benchmarks, better documentation is provided, and the communications personnel are able to assist in a logical progression of the event. Some examples of EMS tactical benchmarks include triage counts, transportation destinations, and extrication benchmarks that indicate access and removal of patient from transportation accidents.

RADIOS

Communications equipment needs a routine maintenance plan that includes evaluating the age and relia-

bility of equipment regularly. A maintenance plan should be developed and followed, and there should be a financial plan that forecasts repairs and maintenance needs. There also should be budget methodologies and revenue streams for new or replacement equipment. Most radio systems are pooled and have several components that need to be maintained. Federal grant monies are available for replacement equipment and upgrades to infrastructure involving communications systems.

Radio frequencies can be scanned, and the communications channels should be monitored to ensure that private users or cell-phone customers do not interfere with emergency communications. This can include the issue of prioritizing cell phones in an emergency. As EMS operations grow, the CAD radio system should have the capacity to assign tactical frequencies outside the main communications channels. On large-scale incidents many departments or services have the capability of using tactical frequencies. The training process and the ability to manage multiple frequencies to contact hospitals, mutual aid, and operational branches is accomplished through computer-switched systems such as 800- and 900-megahertz systems.

COMMUNICATIONS UNIT LEADER

In most large incidents the communications unit leader position will be staffed. Upon assignment the communications-unit leader should obtain a briefing from the logistics section chief or service-branch director. Afterward it is common to brief, organize, and staff the communications unit. Positions that may be included in the unit are the communications center manager and lead incident dispatcher.

In most large incidents, disasters, or disruptions in infrastructure, a message center is created, which requires a message-center manager. Adequate staff is assigned to answer phones and attend fax machines. The communications unit manager assesses the communications systems and frequencies in use and advises on communications capabilities and limitations.

It is important to develop and implement effective communications procedures (flow) internal and external to the incident and incident command post. The manager should continue to assess incident command post phone load and request additional lines as needed. He should prepare and implement the incident communications plan, obtain the current organizational chart,

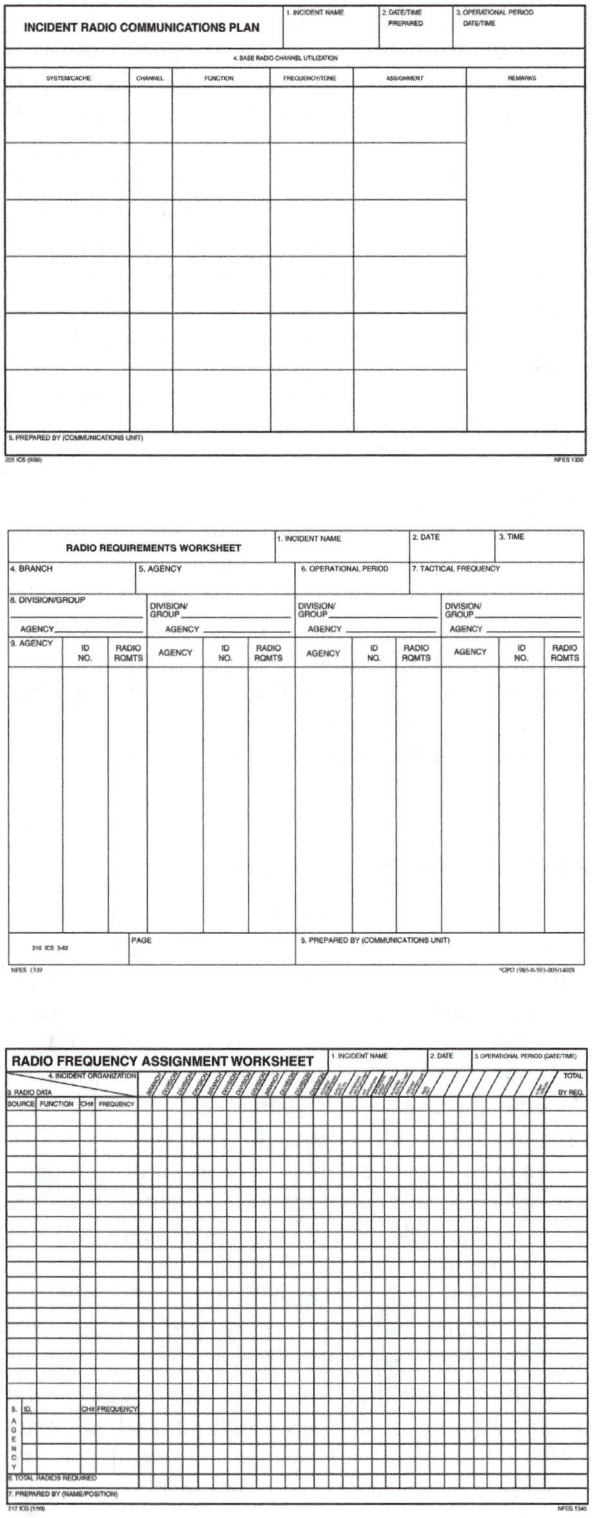

INCIDENT RADIO COMMUNICATIONS PLAN

RADIO REQUIREMENTS WORKSHEET

RADIO FREQUENCY ASSIGNMENT WORKSHEET

FIGURE 18.14 ■ IMS Communication Plan Worksheet.

determine the most hazardous tactical activity, and ensure adequate communications. He also should make communications assignments to all other operational elements, including volunteers, contract ambulances, and mutual-aid units.

Establish and post any specific procedures for use of incident command post communications equipment. The NIMS Form 205 is the national format for designating radio frequencies (Figure 18.14).

DISASTER MODE

In a disaster mode 911 centers should adopt an abbreviated procedure and only attempt to determine if the call is a medical emergency. A revised 911 EMS procedure is recommended for disasters: the dispatcher inquires whether the patient can be transported by other means and the caller is asked to cancel the ambulance if there is no longer a need. The focus should be to have the caller explain the need for an ambulance in order to assign a priority to the request. A suggested procedure for taking the call, modifying it, and processing it is shown in Figure 18.15.

■ THE FUTURE OF THE COMMUNICATIONS CENTER

In the future technology and computer systems will enhance the operations of the communications center. GIS software and trunked phone lines will be used to employ the reverse 911 systems, allowing public-safety agencies to notify residents in specific areas of dangerous conditions. Mapping software will route vehicles with pinpoint accuracy. Relational databases will access health-care databases and send real-time information to field ambulance crews.

In the future some system dispatch centers may be able to generate revenue. Consider the idea that a health maintenance organization may contract with a 911 communications center for the input of names and phone numbers for patients on antibiotics. The task would be to have a computer-assisted voice phone patients at home and remind them to take the full course of their medications as prescribed by their physicians. The potential savings from reduced antibiotic-resistant infections could provide for reimbursement by the HMO, perhaps even paying for additional personnel in the communications center.

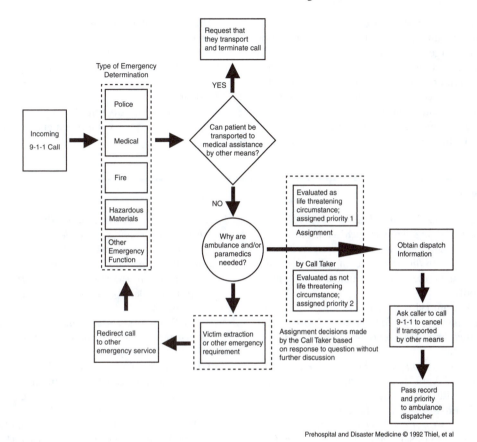

FIGURE 18.15 ■ Proposed Sequence of Steps for 911 Call Takers During Large-Scale Disasters. (*Source:* Prehospital and Disaster Medicine ©*1992 Suter. Reprinted with permission.*)

CHAPTER REVIEW

Summary

No one else has his fingers on the pulse of the EMS system to the degree that the 911 emergency dispatcher staffing the communications center does. Decisions made by these EMS professionals consistently have more impact than any other factors on system performance. By managing call processing and enhancing communications flow, EMS supervisors can exercise real-time control over each crew and the entire system and establish optimal performance. A call that starts badly due to ineffective communications center performance will translate to disruption in the field. Accreditation standards demand interaction among EMS managers and the communications center staff. The communications center is one of the most important components of an EMS system.

Research and Review Questions

1. What preparation is made to handle communications center calls that involve non-English-speaking patients or callers?

2. How does your 911 communications center measure workload?

3. How does your organization listen to and review calls and the voice communications of a 911 communications specialist or emergency dispatcher who directly answers 911 calls? Listen to a 911 tape provided by your communications center, and perform an audit on the call using a nationally recognized program to evaluate the communications center's performance.

4. Map your current procedure for core processes related to 911 calls. Include mapping how a call is handled and how a complaint to the 911 system is handled from initial contact to final resolution.

5. Review your agency's dispatch-center training. Design a CDE calendar for one year encompassing department-specific issues. State how those issues were identified.

6. Identify in your organization a plan for a communications-center quality-improvement program.

References

American Society for Testing and Materials (ASTM). *Standard practice for emergency medical dispatch* (F1258-90 ASTM). (1996). Retrieved April 20, 2006, from www.ct.gov/dps/lib/dps/office_of_education_and_data_management_files/telecommunicator_training_standards.pdf

Christen, H. T., & Maniscalco, P. M. (1998). *The EMS incident management system: EMS operations for mass casualty and high impact incidents*. Upper Saddle River, NJ: Prentice Hall.

Clawson, J. J. (1989). *Position paper: Emergency medical dispatch*. Retrieved April 28, 2006, from http://www.naefd.org/articles/positionpaper1.htm

Clawson, J. J., Martin, R. L., Langlo, T., & Maio, R. F. (1997). *The wake effect—Emergency vehicle-related collisions* (poster presentation). Salt Lake City, UT: National Academies of Emergency Dispatch.

International Fire Service Training Association. (2001). Telecommunicator (1st ed.). Stillwater, OK: Author.

National Highway Traffic Safety Administration. (1996). *Emergency medical dispatch: National standard curriculum.*, Washington, DC: Author.

National Performance Review. (1995). *Serving the American public: Best practices in telephone service*. Retrieved April 28, 2006, from govinfo.library.unt.edu/npr/library/review.html

Newkirk, W. L. (1984). *Managing emergency medical services principles and practices*. Reston, VA: Reston Publishing Company.

Sutter, R. E. (1992). Who calls the shots? EMS operations control and the role of the communications center. *Prehospital and Disaster Medicine, 7*, 395–399.

Thiel, C. C., Schneider, J. E., Hiatt, D., & Durkin, M. E. (1992). 9-1-1 EMS process in the Loma Prieta earthquake. *Prehospital and Disaster Medicine, 7*, 348–358.

EMS Special Operations

Accreditation Criteria

CAAS

CAAS does not offer accreditation criteria for the topic of this chapter.

Objectives

After reading this chapter, the student should be able to:

19.1 Explain the new mission of customer service in an all-hazards environment (pp. 460–461).
19.2 Describe the risk analysis process (pp. 400–468).
19.3 Describe the mitigation role of local community officials as it relates to EMS response (pp. 458–460).
19.4 Define an all-hazards approach and apply a checklist to preparing special events (pp. 460–464).
19.5 Given a medical evacuation of a patient by helicopter, identify the elements that affect or support local EMS special operations (pp. 481–482).
19.6 Identify existing documentation resources to use for special operations during a special event (pp. 470–474).
19.7 Identify the methods for developing and accessing state or statewide regional resource pools (pp. 465, 483–484).
19.8 Contrast the missions of tactical EMS (pp. 476–481).
19.9 Analyze the role of EMS in a special operations environment as it relates to the medical support of hazardous materials operations (pp. 482–483).
19.10 Explain the medical-legal issues involved in spectator care at mass gatherings (p. 467).
19.11 Discuss the process for implementing care systems at mass gatherings (pp. 462–476).

Key Terms

Point of View

EMS providers are prepared for the daily work and peak service-demand periods that face their jurisdictions. There are several models for determining the number of resources that are required for daily activities. However, most jurisdictions have special events or activities that require advanced planning and additional resources. Examples of these events are sporting events, major conventions, political visits, and special gatherings, which can occur in virtually any jurisdiction in the country. In addition, there is always the potential for a major disaster to happen anywhere in the world. And the expectation of the customers of that jurisdiction is the EMS agency should be able to respond to any emergency that occurs.

Planning for these events may become quite complex. In preparing event or disaster plans, responders will need to prepare to interface with federal, state, or regional agencies including the Secret Service, the FBI, the National Guard, and various police agencies and political entities, just to name a few. Failure to be able to work with all of the agencies will usually lead to a system failure that will affect patient care.

One such example occurred during a visit to Florida by a presidential candidate. The candidate landed at the local airport, closing the airport to all air traffic. A motorcade route was established, and the route was protected by various police agencies from the local, county, state, and federal governments. At about the same time, a citizen was experiencing chest pain and called 911 for assistance. ALS fire and EMS units were dispatched with a normal response time of three minutes. As the units approached the motorcade route, they were stopped by the police agencies and were not allowed to proceed across the intersection. The units were detained until the motorcade had crossed. They ended up with a 27-minute response time. They found their patient in cardiac arrest that resulted in a negative outcome. This event was noticed by the local press and reported, and the typical finger pointing followed. Interestingly enough, the candidate lost the close presidential election by only a few votes, in the county and state that determined the outcome for the country. Maybe more importantly, the local agencies lost the respect of their customers.

Jack McCart
Deputy Chief (Retired)
Boca Raton Fire, Florida

■ IMPORTANCE OF SPECIAL OPS

Often EMS is asked to provide care at events that require unique resources, system management, and control. EMS managers need to prepare their organizations to respond to such events by identifying potential **hazards**, determining potential resource needs, determining how those resources may be acquired, and developing a plan that enables the effective, specialized control of these events in concert with law enforcement and fire services resources. EMS usually is designed to meet the immediate everyday needs of a community.

Events like the 2005 hurricane Katrina, the Washington, D.C., anthrax incident, and the Columbine, Colorado, school shooting have shown that local EMS organizations can be easily overwhelmed. Events such as terrorism, mass-casualty vehicle collisions, storms, earthquakes, and technological emergencies can place an unusual demand upon the ability to provide continued background response to anticipated daily call volume. Only with thorough planning and preparation for these unique events can EMS continue to respond effectively to customers. Even if one of these incidents is unlikely in your community, your EMS agency may be asked to contribute people, equipment, and expertise to someone else's catastrophic event.

ROUTINE VERSUS SPECIAL OPS

When public or private EMS agencies are asked to provide, or are considering creating, a special operations capability within their organization, the capability should have a design with three common characteristics: purpose, resources, and structure. More and more, EMS missions are being extended due to the demands of the customers and the health-care community into an "all-hazards" mission.

EMS special operations should have a specific purpose and will require a tremendous commitment of personnel and money. Often crews may train for years before ever employing a skill or piece of equipment. The public has an expectation and EMS has a role in preventing loss to the community and ensuring the public trust with regard to these capabilities. Roles and responsibilities based on **risk assessment** and cost effectiveness contribute to the purpose of a special operations capability; for example, an EMS agency may see that its role is to qualify as a part of the federal response system with a FEMA task force. Another agency may identify a tactical medical program's purpose as a way to free up engine companies and ambulances from extended scene times that are involved with law enforcement events that impact EMS system-wide response times.

EMS resources are human, physical, and financial. Bringing resources to bear on a particular event to provide for an effective rescue or recovery of a victim requires the commitment of resources. When looking at resources, it is important to admit that any EMS provider possesses a finite amount of resources, and those resources have limited capabilities. However, due to the changing expectations and demands, EMS may be expected to apply these limited resources to an almost infinite list of possibilities. EMS systems providing routine medical and trauma care one moment may be thrust into mass-gathering events, civil disorders, technological emergencies, or disasters. Normally available resources may need to be supplemented.

The identification, acquisition, and deployment of resources are a major focus of this chapter. It may be impossible for an organization to have these specialized resources in-house. However, through an effective planning process, sources for specialized resources should be created or identified by EMS leadership. Agreements should be made prior to an emergency or special operation for the acquisition and deployment of resources. The sources for additional resources can be found at a local, regional, state, or federal level, and EMS managers must exercise and train on the requesting process, as well as at the field-level deployment of those resources. The proper order in which to access and notify when needed can prevent embarrassment and public backlash. Algorithms for notification and request along with a current phone list should be placed in command vehicles and communications centers.

A common characteristic of an EMS organization's response to a special operation is structure. The Incident Command System (ICS), pre-developed response plans, and the ability to form unified command structures help EMS bring control to situations that tax resources. Advance planning is essential to a coordinated and effective response to unique events. In addition to an agency's emergency operations plans, a variety of additional plans exist at the local, regional, state, and federal levels. Those plans identify hazards, potential resource requirements, sources of those resources, and a support structure that enables those resources to be deployed. The resources may be pre-deployed and organized into strike teams or task forces in anticipation of a need (such as an approaching hurricane or scheduled event), upon request or determination of actual need after the hazard strikes, or based upon the concept of time-phased forced deployment. The latter method projects resource needs based upon type of event, vulnerable populations, and pre-designated timelines. Therefore, time-phased forced deployment systems anticipate resource needs in advance and schedule their deployment to occur automatically so they arrive when the need occurs.

One key focus of this program with regard to resources is that of human resources. Not only does EMS need to identify what human resources are going to be required, but also the needs of those human resources themselves. EMS cannot reasonably expect human resources to operate effectively if managers do not ensure that the needs of personnel are met and that personnel have a reasonable certainty that their families are safe during extended operations at major emergencies and catastrophic disasters. This includes communications channels for keeping families informed.

MASS GATHERINGS

What does or does not constitute a special event or mass gathering is difficult to determine. The American College of Emergency Physicians defines a mass gathering as any collection of people greater than 1000 at one site or location. Other guidelines may be used to define it: a focus group discussing special events and mass gatherings has identified a special event as a non-routine activity within a community that brings together a large number of people. Emphasis is not placed on the total number of people attending but rather on the community's ability to respond to a large-scale emergency or disaster or the exceptional demands that the activity places on response services.

A community's special event requires additional planning, preparedness, and **mitigation** efforts of local emergency-response and public-safety agencies. A mass gathering can be a subset of a special event, such as an auto race at state fairgrounds. Mass gatherings are usually found at special events that attract large numbers of spectators or participants. Both special events and mass gatherings require planning and preferably an incident action plan. For example, an amusement park that attracts a large number of people is not considered a special event because large crowds are expected. A mass gathering does not imply that the event is a special event. Failure to prepare for all contingencies can lead to disastrous consequences.

Pre-Event Planning Matrix

The planning team should identify all of the major functions and responsibilities required by the event and assign appropriate agencies to manage each function or responsibility. Since responsibilities vary from jurisdiction to jurisdiction, it is most effective to assign responsibilities consistently to avoid duplication and promote efficient response to problems that may arise.

The pre-event planning matrix (Figure 19.1) is designed for you to choose the risks, hazards, or functions that are likely to be required by an event, and to assign each to a **primary agency** (P) or a secondary or support agency (S). The functions and responsibility assignments must be discussed and decided in the planning stages, not when an incident occurs. Additional room is provided in the matrix to add agencies or risks/functions as they may apply. Since responsibilities vary from jurisdiction to jurisdiction, certain risks or hazards are not consistently handled by one agency. This matrix is designed for you to choose the risks and hazards your agency is

accountable for handling. If more than one agency is tasked to respond to the risk or hazard, some overlap of responsibility may occur. One way to handle this is to place the numeral 1 in the primary agency position and the numeral 2 in the support agency's position.

Event Public Health and EMS Surveys

EMS managers faced with an event should conduct a pre-event public health and EMS survey for any venue intended for a mass-spectator event. Organizers should consult appropriate health authorities to ascertain the availability of running water (particularly for hand washing by food service and medical personnel); sufficient public toilets (with provision for pump out of portables and servicing as necessary during the event); adequate refrigeration for perishable food stuffs; recognized, approved suppliers of bulk food items delivered to the on-site food providers; sufficient number of covered containers for the storage of food and solid waste, including during the event; and appropriate storage and removal of liquid waste.

Public health inspectors should be available on site during the event to monitor public health compliance. Public health authorities on site should have legislated authority to enforce "cease operation" orders on on-site food providers who are in contravention of standards or who are otherwise operating contrary to the public interest. While this may seem a function of public health, it is important as EMS managers to ensure the health department contact has been made and is overseeing any food-related event because poor food handling techniques can quickly have a cascading event on the EMS and hospital system.

In most areas, vehicles other than ambulances can traverse the short distances between spectator areas and medical care facilities. Four-wheel-drive vehicles, for example, can be equipped with appropriate medical equipment (including, but not limited to, resuscitation equipment, trauma kit, and spine board) and serve as ambulances. In denser spectator areas, any vehicle can have access problems. You should consider using golf carts or off-road vehicle, either designed or modified to accept stretcher cases from these areas. For these reasons the ambulance network may have to consist of a mix of EMS providers on foot, golf-carts, four-wheel-drive vehicles, ambulance buses, and conventional ambulances to best facilitate patient transport requirements. Provide a magnetic-based beacon, portable radio, and appropriate marking for these vehicles. A communications network, designed to provide a coordinated response to requests

	County Agency	EMS	Emergency Management	FAA	FBI	Fire	Law Enforcement	Public Health	Public Works	State Agency	U.S. Secret Service	Utilities	
Crowd Control													
Demonstrations													
Dignitary Protection													
EMS													
Environmental Hazards													
Evacuation of Area													
Fire													
First Aid Stations													
Food Handling													
Food Waste													
HazMat													
Hostage w/o Terrorism													
Human Waste													
Kidnapping													
Lost Child													
Lost and Found													
Media Relations													
Motorcades													
Parking													
Permitting													

FIGURE 19.1 ■ Pre-Event Matrix Worksheet.

	County Agency	EMS	Emergency Management	FAA	FBI	Fire	Law Enforcement	Public Health	Public Works	State Agency	U.S. Secret Service	Utilities	
Potable Water													
Power Interruption													
Security/Governor													
Security/State Department													
Security													
Structural Collapse													
Terrorist act													
Terrorist Threat													
Ticketing													
Traffic Control													
Weapons of Mass Destruction													
Weather Hazards													

FIGURE 19.1 ■ Continued

for assistance, is essential. You may base the network on existing service networks, or event organizers may need to provide the network.

All-Hazards Influences

Our **all-hazards approach** can be influenced either directly or indirectly by a variety of statutory, regulatory, or planning requirements. Financial constraints of the local municipalities play a large role in determining the content and scope of resources available to bear on the planning and mitigation of hazards. Legislation, consensus standards, and administrative directives from government agencies create these influences. EMS leadership should be familiar with the influences on the EMS response to special operations. The influences are listed in Table 19.1.

Plan Components

A mass-gathering plan has the following characteristics:

- Hazards analysis.
- Vulnerability analysis.

TABLE 19.1 ■ Legal Influences on EMS Response to Special Operations

Type of Influence	Legislation
Statutory	◆ Robert T. Stafford Disaster Relief and Emergency Assistance Act ◆ Antiterrorism legislation ◆ Superfund Amendments and Reauthorization Act (SARA) ◆ State and local laws and ordinances
Regulatory	◆ Hazardous Waste Operations and Emergency Response, HAZWOPER regulation, 29 CFR 1910.120 ◆ Permit-Required Confined Spaces Final Rule (29 CFR 1910.146) ◆ Other applicable state and federal safety regulations
Consensus standards	◆ National Fire Protection Association (NFPA) standards ◆ National Association for Search & Rescue (NASAR) standards
Plans and planning	◆ National Response Plan (NRP) ◆ State Emergency Response Plan (ERP) ◆ City/County ERP ◆ Employer ERP ◆ Local Emergency Planning Committee (LEPC) plans ◆ Federal Radiological Emergency Response Plan (FRERP)

◆ Basic plan consideration (such as scope, activation, authorities, and concepts of operation).
◆ Resource identification.
◆ Support functions or mechanisms for resource deployment.
◆ Recovery activities.

Demands upon organizational resources are constantly increasing because of increased customer expectations. Because of these changing conditions, there is a requirement to assemble greater and greater resources to manage special operations and unique events that may occur in communities. The resources that are available individually, as well as the capabilities of those resources, are limited. EMS managers must identify in advance additional resources that can be brought to bear on virtually any type of event to which EMS may be called to assist.

There are multitudes of statutory, regulatory, and planning resources that can assist in this process at a local, then regional, state, and federal level. These also can be tremendous sources that can fill resource voids. However, if EMS fails to fill those voids during advance planning, EMS itself may have to provide those services when the need occurs.

Mutual-Aid Agreements

Most organizations have **mutual aid** or **automatic aid** agreements with neighboring communities for EMS resources. These agreements must be modified for EMS special operations. Mutual aid requires an agency's command officers to make specific requests for special-operations equipment (hazmat and technical rescue, for example). In many areas these resources will take time to assemble or will have to travel long distances to get to the scene. Automatic aid may be a condition to place in an inter-local or mutual-aid agreement so that the dispatch center without a command officer's input can send or request equipment based on the information from the call to the public-safety answering point. Agreements for specialty resources often require the cost recovery mechanism, worker compensation, replacement of equipment, and command issues to be worked out prior to the response.

DISASTER CONDITION

A disaster or emergency may overwhelm the capabilities of a state and its local governments, preventing them from providing a timely and effective response to meet the needs of the situation. For example, the occurrence of a large or catastrophic earthquake in a high-risk, high-population area will cause casualties, property loss, and disruption of normal life-support systems and will affect the economic, physical, and social infrastructure of the whole region.

A disaster or emergency has the potential to cause substantial health and medical problems, with hundreds or thousands of deaths and injuries, depending on factors such as time of occurrence, severity of impact, existing weather conditions, area demographics, and the nature of building construction. Deaths and injuries will occur principally from the collapse of buildings and other structures and from collateral events, such as fires and mudslides.

A disaster or emergency may cause significant damage particularly to the economic and physical infrastructure. An earthquake may trigger fires, floods, or other events that will multiply property losses and hinder the immediate emergency response effort. An earthquake or hurricane may cause significant damage to, or destroy, highway, airport, railway, marine, communications, water, waste disposal, electrical power, natural gas, and petroleum transmission systems.

Level of Care

Based on the profile of the incident, EMS managers need to categorize the level of care to be on standby at the incident. Considering the distance to, and accessibility of, the nearest hospital and its capability can help establish the level of care. Levels of care at special events or mass gatherings may include:

- Basic first-aid or pre-EMS certifications.
- Intermediate first aid plus IV therapy and oxygen.
- Advanced care and life support and early management of severe trauma.
- Site hospital full monitoring and ventilation. A surgical facility may be required depending on the level/type of event.
- Other level-of-care concerns including training staff, specialized rescue teams [Justin Sports Medicine for Professional Rodeo Cowboy Association (PRCA) rodeo events, paramedic standby for the International Boat Racing Association, extrication team for Championship Auto Racing Teams (CART) professional racing].
- Consulting medical personnel with experience of similar events to determine appropriate levels of care.

Ideally, event planners will pre-establish the coordination between venue medical services and those of the local community EMS responders (that is, establish how they will provide mutual aid if required). Further guidance on the establishment of medical care facilities and equipment requirements is available from local or regional disaster and health plans.

Planning Medical Care

Several issues need to be addressed in medical care planning at a special event. It is important to consider the factors that may affect patient volumes. The first is the setting at which the special event is being conducted.

- Indoor versus outdoor.
- Access routes to and from the site.
- Size of the crowd.
- Duration of the event.
- Extremes in temperature.
- Mobility of the crowd.
- Number and visibility of the aid stations and medical-treatment facilities.

FIGURE 19.2 ■ Event-Setting Considerations.

Figure 19.2 details a list of factors to be considered in an event setting. Outdoor events have been notorious for generating patients. The extent or duration of the incident can place demands on the crowd if they did not prepare themselves with water, food, or protection from extremes of climate.

The nature of the event helps to anticipate the type of patients that may be generated from a mass gathering or special event. For example, rock concerts can create problems from drugs, alcohol, or minor trauma. One study in southern California of 405 major concerts indicated rock concerts on an average generate 2.5 times more EMS requests than nonrock concerts. Altered mental status was almost five times greater at rock concerts versus nonrock concerts. In this same study the frequency of cardiac arrest increased at classical music concerts.

Offering medical assistance to an overdose victim at a rock concert poses a major difficulty in making appropriate field diagnoses when EMS crews do not know the drug. Even when victims are coherent, they may believe they took a particular drug when, in reality, they took a different drug or one that was adulterated. As is well documented, there is no quality control, or ethics, in the dealing of street drugs. To attempt to cope with on-site drug identification, the medical staff at a Watkins Glen, New York, rock concert with an estimated attendance of 600,000 established a mobile toxicology laboratory on site in a trailer adjacent to the hospital tent. This is the first time such a lab was included as part of the medical facility and EMS standby at a rock concert. The benefits of an onsite mobile toxicity lab include triage of drug overdoses for removal to hospitals, as well as reducing the number of people to be transported, which otherwise would be necessary, without toxicological diagnosis confirmation in the field.

Sporting events create major or minor trauma, mostly among the athletes, and environmental issues among the spectators created by heat or exhaustion. Athletes participating in extreme sports or competitions, such as marathons or the Ironman triathlon, may require close medical supervision and have complex physiological problems. Sporting events hosted by rival teams, especially playoff or championship games, that continue to serve alcohol through the entire event are more likely to generate patients.

Demonstrations can cause injuries, often respiratory issues when tear gas, pepper spray, or riot control agents are dispensed. High-profile events can trigger violence and cause minor injuries. They can create targets for terrorism, and EMS or rescue services may be called in to remove patients who have secured themselves to objects as part of the protest. It is important that any EMS standby for a politically sensitive event includes extrication equipment that will help cut chains, handcuffs, or other types of devices that protesters may use to secure themselves to fixed objects.

Patient volumes in most mass-gathering studies range from 0.12 to six per thousand spectators with cardiac arrest ranging from 0.3 to four per million spectators. Physician coverage at events also varies from events with attendance between 5000 and 50,000 attendees. A common staffing pattern is found with one paramedic/EMT or two paramedics for every 10,000 attendees and one ambulance for every 20,000 attendees. Basic first aid should be available to all attendees within four minutes and ALS within eight minutes with all patients who need definitive care being taken to definitive care within 30 minutes (Figure 19.3).

Logistics

Consider staffing a medical logistics position when planning a large event. Arrangements for facilities will be needed for medical personnel to operate in an environment in which the injured must make their way to that location. Logistics will need to make sure medical teams that patrol spectator areas are clearly identifiable. Logistics also will be responsible for alternative vehicles being available to transport spectators to the medical facility. Some specialty medical vehicles will be needed as appropriate to the terrain. Four-wheel-drive vehicles may be required for off-road areas and golf carts or similar vehicles required for high-density spectator areas. When an ambulance is not required, a "chauffeur system" can be provided to transport persons from the on-site medical facility to their own transport vehicle.

Logistics also will provide for how medical personnel will be notified of or summoned to spectators requiring assistance in public areas. The logistics unit must have a means of communication that will be available for EMS personnel to communicate with off-site medical personnel, event organizers, security, and other support personnel. Often this involves setting up a mobile command post to help establish interoperability with the agency.

EMS managers should monitor the sponsorship for conflicts between the event sponsor and any medical service operators. Often it helps that a level of on-site medical care be mandated by the regulatory agency for certain sizes and specific crowds given the nature of the event. A mix of medical personnel (first-aid providers, paramedics, nurses, doctors) can be used on site. Who will provide the personnel and how their costs will be funded should be identified. If the health service providers are not from the local area, they must be integrated with the local services. Are the selected personnel appropriately skilled to respond to anticipated medical problems at the event? They may require additional training and, at the very least, a briefing for the EMS operations at the event.

Logistics or planning should also provide maps for EMS crews and potential responders.

Medical Access to Venue

Consider risks associated with venue; for example, water in the vicinity or access problems for ambulances. Agreements must be reached among medical service providers on how medical teams will be able to locate individuals in need of attention. You should agree on the use of a common reference map or grid system so that everyone knows how to reach or rescue individuals in distress; for example, in crowded areas or through fixed seating. The mechanism of how patients

In the largest single-day event in the world, the Indianapolis Motor Speedway with over 400,000 people attending on race day, is a special operation. The infield hospital sees an average of 139 patients, or 0.35/1000, and first-aid stations treated an additional 400 attendees. The breakdown of patients seen by EMS at the facility and infield hospital included 16% intoxications, 15% lacerations, 11% preexisting conditions, and 8.5% heat illness.

FIGURE 19.3 ■ Case Example: Indianapolis 500.

will be transported on site should be worked out. A flying "V" pattern with security moving through a crowd protects the EMS crews and helps provide an area unobstructed by bystanders or a crowd. A dedicated access route, or emergency service lane, to allow rapid access to and from the venue for ambulances and other emergency vehicles should be established on all large incidents.

If the event itself poses a barrier to medical teams—for example, marathons or a parade—EMS should preposition and designate specific areas for crossing the event and ensure that law enforcement or security is stationed at the crossing. If spectators or patients need medical evacuation by helicopter, regulations should be followed and predesignated landing zones should be established on the venue map.

Medical Requirements

EMS should prepare for the most critical injury or illness foreseeable, such as cardiac arrest. Assessments should be made for the needs for a mobile team. Mobile teams may require prepacked medical kits. A determination should be made as to which mobile teams will provide care for the audience, any VIPs, and performers. The boundaries of what kind of care and services will be provided to the event should be defined; for example, which agency should provide services inside the venue and in the parking areas and which will provide services to the streets leading up to the event.

Medical Teams

When deploying medical teams, EMS managers need to consider the event size and location with regard to the medical infrastructure and available medical resources. EMS managers need to identify the number of teams needed for the event. For example, some older DOT and Emergency Management Institute recommendations call for one team per 10,000 people and one to two doctors per 50,000 people. Decisions should be made on who can see, treat, and discharge patients. An analysis of the event to identify peak periods or special circumstances requiring additional staff or the rotation of EMS units into the venue if necessary.

How the EMS and medical standby staff will be fed, watered, rested, and protected from the elements should be provisioned. Worker safety regulations that cover occupational health and safety (for example, protection from violence and crowd crushes) need to be enforced and covered in a site safety plan and as part of the briefing for the crews. Arrangements should be in place for the movement of medical teams onto and off the site, if the crews are to be rotated. Medical team members need to be appropriately dressed for the conditions and should be provided with cold weather gear, rain gear, camelbacks, or other items to protect against the elements. Medical team members should be easily identifiable. If the event is likely to draw foreign or ethnic populations, interpreters should be available. Medical teams must understand the command structure and their role within it and how to get more resources to the scene.

Mobile Teams

In tightly packed areas, particularly near stages and spectator areas, emergency medical responders on foot, bicycles, or golf carts may have the only access. Experience has shown that uniformed first responders on foot, circulating in dense spectator areas, are quite effective, and patrons will readily summon them in an emergency, even if the person requiring care is a stranger to them. Even if a clearly marked field hospital is visible, spectators are often unwilling to make the sometimes long trek to request assistance (since they may lose their seating position), particularly for a fellow spectator whom they may not know or for whom they fail to appreciate the seriousness of the emergency.

Identification of mobile teams, where ambulance or clinical uniforms are unsuitable, can be successfully accomplished by special "event uniforms." Mobile teams need to have communications equipment to keep EMS supervisors and the command post informed at all times. The Red Cross symbol is registered by the International Red Cross and its national societies and it should not be used as part of an "event uniform."

Medical-Aid Stations

Mobile teams will often move patients to the medical aid stations prior to the arrival of a transport vehicle. It is important to consider the establishment of medical aid posts at large events or events that may take a significant amount of time for an ambulance to arrive. Often intoxicated patients or minors need to be held and monitored until a competent adult arrives. Medical-aid stations have specific design, operational, and logistical requirements. Figure 19.4 lists the characteristics of a good medical-aid station.

Site Hospital

Depending on the nature of the event, a site or field hospital may be needed to provide care for the number of casualties anticipated. You also should make contingency plans in case of a major incident for which the resources of the field hospital may not be sufficient. Failure to plan for large numbers of casualties or severely injured patrons can result in long delays in providing medical treatment. It is important to provide a communications link between the site hospital and local hospitals. Often a special event is an ideal opportunity to deploy a **disaster medical assistance team (DMAT)**. For example the Halloween event on Maui every year draws thousands of spectators to the town of Lahaina. This local hospital is miles away with only one two-lane road. The Maui officials deploy the local DMAT for the event.

Site hospitals will require clean water, electricity for medical appliances, and adequate lighting in tent hospitals at night, which should, if possible, include a backup power system, washroom/rest facilities for the exclusive use of staff and patients, meals for medical staff, and tents for hospital use that have flooring as part of the structure to contain the service and to prevent ingress of water or insects.

Medical Equipment

The requirement for basic or advanced life support equipment depends on the type of event and the assessed risk of illness or injury. While standard lists of equipment will cover most requirements, you should review literature, post incident analysis from previous experiences, and current practices. Further equipment considerations include identifying requirements where the equipment should be mobile versus fixed and arrangements to resupply aid posts as required. Compatibility of on-site equipment with equipment used by ambulance and other health-care providers should be evaluated for inter-operability. A landline telephone should be established for ordering additional

- ◆ Provides easy ambulance access and egress.
- ◆ Is located within five minutes of all sections of the crowd.
- ◆ Has available a mode of transport to the crowd.
- ◆ Is clearly sign-posted from all directions.
- ◆ Is clearly identified.
- ◆ Is clearly marked on maps of layout.
- ◆ Is in a position known by security and other event personnel.
- ◆ Is stocked and staffed for the duration of the event and for spectator arrival and departure periods.
- ◆ Provides facilities for injured or sick patients to lie down.
- ◆ Ensures privacy in clinical areas.
- ◆ Provides some means of communication with the primary medical-control point, with venue control, and with mobile medical teams in the venue.
- ◆ Is located in as quiet a place as possible.

FIGURE 19.4 ■ Characteristics of a Medical Aid Station.

staff or supplies and for notifying hospitals of patient transfers with cellular telephones being used only as backup devices.

Documentation

Every special event will require documentation. A postincident analysis of the event and review of the number of requests for medical assistance that occurred at the event should be completed by EMS management. Tracking should be done of biological, chemical, and infectious disease exposures should they occur. Medical-legal issues, which must be addressed prior to the preparation of any documents, are as follows:

- ◆ Who has access to records?
- ◆ Who keeps the data and for how long?
- ◆ Who can give consent for treatment?

Every patient contact should be documented and recorded. Patient releases should be properly filled out and matched with prehospital care reports. HIPAA requirements must be followed and confidentiality preserved.

Ambulance Vehicles

Organizers should consult ambulance services to determine ambulance requirements for the event. Some considerations include prepositioned ambulances on site or to be called to the venue on an as-required basis. Providing a mix of basic and advanced life-saving ambulances at the event is more cost effective since the majority of requests for EMS are BLS in nature. Management must decide if ambulances are on site for participants (for example, at sporting events) or will they be available for injured spectators as well. If air ambulances are available or may be used, EMS managers at a special event site must secure a predetermined safety landing zone. Who will pay for the service and how it will be billed should be determined by the agency. In many jurisdictions, the promoter is made to provide the service or contract with fire and EMS agencies for staffing.

While conventional ambulances are appropriate for patient transfers to off-site medical facilities over good roads, such vehicles may be unsuitable for off-road use. Ad hoc roadways and cross-country terrain may require four-wheel-drive vehicles, particularly if grounds are saturated by recent rainfall. If four-wheel-drive ambulances are not available, consider evacuations by water rescue, lifeguard boats, or air units.

Other Medical Considerations

There are several further considerations for the clinical aspects of an EMS operation at a special event. Often at large-scale events, assaults occur, including sexual assaults. It is important to provide a separate facility and staff to counsel victims of sexual assault and also to collect evidence.

As seen with the Papal visit to Denver in 1995, a major EMS incident can be avoided by ensuring sufficient water supplies, especially if a delay in the event is anticipated or extremes in temperature are expected. Often outdoor events are complicated by water being available by vendors only. This places the lower-income attendee at risk if a person cannot afford the price. Emergency water supplies should be available to EMS workers to dispense in an emergency. Providing sprinkler systems or misting tents for crowds in hot, open areas, if they are suitable for the event, can help reduce the potential for mass casualties.

Provide welfare and information services (a helping and caring role). It is not uncommon for people to arrive at a special event or mass gathering and their medication is missing, stolen, or in luggage that did not arrive. A plan should be in place for assisting with forgotten or missing prescription medications; for example, having a medical control physician available to confirm prescriptions and phone in orders to a local pharmacy.

Sanitary spaces should be provided for a baby diaper-changing and caring facility. The disposal of contaminated material and clinical waste needs to be accounted for. How, and by whom, medical supplies will be obtained, including secure on-site storage of medications, should be part of the EMS preplan for a large-scale mass gathering.

Legal Issues

Some form of legislation usually governs or restricts public events or aspects of them. Some events that are extremely large or high impact require special state or local legislation. Local ordinances provide health and medical guidelines. Promoters should consider obtaining legal advice early in the planning stage. An event should cover a variety of liabilities that warrant consideration, including the following:

- Liability for injuries.
- Liability for acts or omissions.
- Liability for financial obligations incurred in responding to major emergencies occasioned by the event.
- Potential liability for the resultant effects of the event on normal emergency operations.

EMS managers should work with government officials to develop a permit process and require permits for parades, the sale and consumption of alcohol, and the sale of food items. Fire-safety inspections also are required, especially for any temporary structures that are set up like tents or portable buildings. Permission should be required if it will be necessary to close certain adjacent or peripheral roads or streets, especially if it will affect response times. A permit may be required for the mass gathering itself.

Most public-sector agencies have adopted a "user pays" policy for services provided at sporting and entertainment events. The purpose of this policy is to improve the allocation of statute resources in the general community by providing a means of charging for services deployed to plan for, and respond to, sporting and entertainment events. Event promoters should consult local and state authorities to determine relevant fee structures and charges for services provided, including payment of overtime costs for personnel. Promoters may be required to post a bond or provide liability insurance to cover the costs of response to emergencies, subsequent venue cleanup, traffic and crowd control, and other policing functions. The head of the planning

- Will staging the event require multiple venues?
- Is this kind of event normally conducted at a fixed facility?
- Will a fixed facility be used in ways that may not be considered normal for that facility?
- Is the event regularly conducted at a temporary venue?
- Is the event a "one-of-a-kind" project at a temporary venue?
- What services/utilities are available at the venue?
- What additional services and utilities will be required at the venue?
- Is there a need for backup services or utilities?
- What shelter facilities are available at the following locations:
  - Transport pickup and drop-off areas.
  - Spectator and official viewing areas.
  - Seated eating areas.
  - Pedestrian thoroughfares.
  - First-aid and medical centers.
  - Marshaling areas for competitors and officials.

In an urban setting or stadium venue, also find out:

- Could the adjacent streets on all sides be closed to other than emergency, service, and resident vehicles, creating a perimeter for access as well as a buffer zone?
- What is the duration of the event, and will it continue during the hours of darkness?
- Have you provided for the needs of people with disabilities?
- Does the date of the event conflict with other events to be conducted in the area?
- Will seasonal weather require any special contingency planning?
- Have you surveyed the proposed site (particularly outdoor sites) for inherent hazards associated with location, and have any been identified?

FIGURE 19.5 ■ Site Selection Assessment.

team must monitor progress made in satisfying all legal requirements throughout the planning stage of the event.

Site

You may need to consider a number of alternative venues for an event. EMS managers may be able to recommend appropriate venues based on health and safety considerations. Finding a suitable venue or set of venues can be difficult. Answering the questions in Figure 19.5 during the planning stage can aid in the selection of an appropriate event site.

As an EMS preplanning operation, venues should be evaluated to see if any aspects of the special event can be brought down by a severe storm or winds. If the site is adjacent to a waterway, and it is prone to flooding, any large group near water may require EMS or rescue resources to patrol on the water. If the

site layout is large enough and if the event triggers a mass-casualty incident, spaces should be identified for an on-site triage area to permit stabilizing medical treatment before critical patients are transported to local health-care facilities. Triage areas need to be accessible to ambulances to eliminate the need for carrying patients long distances. The site should allow for adequate crowd regulation, of which the most important aspect is for regimented seating areas and flow barriers. Are spectator overflow areas available to prevent crush loads should spectator turnout significantly exceed expectations (a common phenomenon at rock concerts)?

CONTINGENCY PLANS

Unfortunately, not every event runs smoothly. Often, incidents occur that are beyond the control of the planning team. Therefore, contingency plans for every event should be in place. An emergency response plan requires a comprehensive hazard and vulnerability analysis. Consultation among all parties who may respond to an emergency situation during the event is essential. Some important questions related to Incident Command System planning are listed in Figure 19.6.

But more than likely, every town or community probably already hosts a gathering of some sort that loads the local EMS resources to, or beyond, its limits. Examples could include community-sponsored events like fairs and festivals. Or your town might have been selected by an unsolicited group to hold a get-together, like a bike week or some other annual pilgrimage.

The key to organizationally surviving a population influx is in the process used to plan for increased service demands. One local agency cannot do this alone, considering that any event that produces patients may impact the region, including transportation routes and emergency resources. A community must approach the situation as a team, with a willingness to use all available resources to the fullest extent. The 1996 Olympic Games in Atlanta, Georgia, revealed how important principles and successful practices can be identified and applied to similar challenges in other communities. Without a strong commitment to customer service and public safety, the mission of any participating EMS agency could become sidetracked by the many influences exerted during the planning stages of the event.

- What weather conditions may require cancellation of the event?
- What weather conditions will postpone the event?
- How will storm warnings be monitored?
- What plans are in place for sudden, severe weather conditions, such as tornadoes? Will shelters be available?
- Who has the authority to make these decisions, and at what point do they exercise that authority?
- How is notification made of a cancellation or postponement?
- Are additional security personnel, including police, on standby or on call should an immediate increase in these services be required?
- Have you advised ambulance services and local hospitals of the nature of the event, provided an expected spectator profile, and estimated potential medical problems?
- Have you notified fire and rescue services of the nature of the event and identified the services that might be required?
- Have you identified the types of heavy equipment that could be required in a catastrophe event (for example, a grandstand collapse)? Have you made plans to obtain that equipment at any time, including during off-business hours?
- Have you advised counseling services of the nature of the event and identified the services that might be required?
- If the event is particularly dangerous, and deaths are a real possibility (for example, at automobile or power boat races or air shows), have you formulated plans to support any required coroner's investigation?
- Will a grid-type venue plan be available that is common to all emergency services, including access roads, pathways, major landmarks, spectators, performer and vendor areas? Will vendor locations or booths be numbered and be included on the venue plan?

FIGURE 19.6 ■ Factors Influencing ICS Planning at Events.

GENERAL CONSIDERATIONS

An EMS manager's goal is to develop and implement a plan that will maintain the present level of EMS in the community and provide an appropriate level of service for the scores of citizens and visitors who will be attending the event. EMS managers should build on previous experience from other events. A hazard analysis should be performed for each venue and other designated locations. Work hand-in-hand with law enforcement, the fire department, local security forces, and other event personnel to perform site surveys. This will educate others to your needs and enable them to incorporate life safety and fire and hazard prevention into the operation plans of each event. Information gathered in these surveys may reveal temporary construction, security efforts, and accessibility issues that indicate the need for strategic predeployment of personnel and equipment. Remember that EMS personnel will not operate totally independent from other public-safety services when responding to emergency situations.

EMS field operations personnel should be part of the staffing of any command centers that are established and should participate in mutual-aid groups. In certain instances, temporary facilities may need to be provided and staffed. These facilities may be inside high-security areas that require the sanitizing of all vehicles and personnel that enter the area. Additional ALS units may need to be obtained. Sources can be vendors or other EMS providers. Specially modified golf carts that are capable of providing ALS can service high-occupancy areas that have limited access for conventional ambulances yet require a rapid response. If circumstances indicate, these units actually could transport a patient to the hospital.

During the preevent planning phase, tabletop exercises and on-site simulations should be practiced up to the day operations begin. For events with multiple venues, a command center at each venue can enable some independent operation options. The use of an integrated incident management system will facilitate the coordinated operation of EMS, fire, and state and federal law-enforcement agencies and ensure that public safety is maintained at the highest level possible.

EMS managers need to ensure that minimum standards of coverage are met for special events. Many of these types of events, such as art fairs or community celebrations, can traverse an entire community, and the level of impact to the surrounding community should be anticipated with regard to access to the scene and exits to hospitals. EMS managers need to identify what activities will be taking place, where they will be, and how many people will be attending. Capacity figures for the venues need to be identified. Permitted occupancy load figures can be obtained from local building and fire officials. Consider past ticket sales and/or seats available to help guide your estimate. In cross-country events, such as bicycle races, the number of people attending and watching from along the route may be a true guess. EMS crews also may have difficulty accessing off-road areas and may need the use of alternative transportation or response vehicles like Segways, bike teams, or all terrain vehicles. Be sure to consider the required road and street closures for these types of events when planning response and transportation routes. Remember to factor in the weather and whether the vendor will provide water or allow water brought into the site.

Many cities and jurisdictions have an event permit process that helps mitigate problems for EMS managers. Certain fees are collected as part of the permit process, depending upon how many people will be attending. Some states require an EMS unit for every 8,000 persons in attendance. Several rules of thumb have proved valuable to cities that have hosted many mass gatherings. If alcohol is to be served or consumed, increase the number of EMS resources by 33%. If illicit drug use is anticipated, increase the number of EMS resources by 33%. The National Fire Protection Association (NFPA) Standard 101 Life Safety Code, state and local contemporary fire codes, and common sense can be used to guide the deployment of resources. Figure 19.7 shows an example of a permit process from the San Mateo, California, area. Event coordinators and EMS agencies must work together and coordinate through dispatch.

Mass-Gathering Crush Load

EMS managers should develop a comprehensive plan to provide medical support for all aspects of mass gatherings and anticipate crush load. Crush load is the assembly of persons inside or outside a facility that overwhelms the capacity of a given area, resulting in gridlock, limited access, and a surge in the crowd, which pins people against a fixed object, compromising life safety.

The first well-documented case of crush load occurred at a 1979 sold-out concert by The Who at the Cincinnati Coliseum, which resulted in 11 deaths. From 1992 through 2002, there were 232 deaths from crowd-safety failures at concerts and festivals around the world, and more than 66,000 people were injured. During a single week in 2003, a stampede to exit the E2 nightclub in Chicago left 21 dead by asphyxiation or heart attack; 100 others died either trying to escape or from injuries after they escaped a pyrotechnics fire at a Great White concert at The Station, a club in West Warwick, Rhode Island.

Risk of a mass-casualty event increases with festival seating. The National Fire Protection Administration's (NFPA) Life Safety Code®, NFPA 101® defines

festival seating as a form of audience/spectator accommodation in which no seating, other than a floor or ground surface, is provided for the audience/spectators gathered to observe a performance.

EMS Operations at Mass-Gathering Events

EMS managers should establish emergency medical units equipped with medical staff supported by a response vehicle and transport capability. Supplies and equipment to provide immediate emergency medical treatment to all should be staged or prepositioned at the scene. Mobile rescue and medical teams, the American Red Cross, and security services for maximum safety need to work in conjunction. A coordinated wedge with security or law enforcement in the lead is an important operational tactic to get to a patient in the middle of an event.

EMS personnel operating in this environment should have special training, equipment, policies, and procedures and have previously practiced the techniques needed to accomplish this mission. Their first exposure to these tactics should not be on the day of the event. EMS organizations should develop their plans using an established incident command system (ICS) compliant with the National Incident Management System (NIMS). These plans can be developed for mass-casualty events and terrorism incidents, building or structure collapse, bombing, hazardous materials spill, adverse weather, or any other disastrous incident.

The ICS 200 series forms offer the documentation tools necessary to collect, manage, and capture information critical to planning and managing the event. The ICS 202 is one example of form commonly used in incident management (Figure 19.8).

AMERICAN RED CROSS

Mass Care and Special Needs

In the last decade EMS has had to provide services to populations either displaced by man-made or natural events or mass gatherings. Critical factors specific to providing care at mass gatherings include accessibility, structural collapse, building construction, crowd capacity and control, evacuation capability of people and toxic gases and smoke, hazardous materials, pre-existing medical problems, safety, time, traffic patterns (both foot and vehicle), ventilation (both heat and air), water and food, weather, medical facilities,

County of Santa Clara
Emergency Medical Services Agency
SPECIAL EVENT REGISTRATION (EMS-909)

SECTION I: APPLICANT INFORMATION

Name of Applicant (Last, First, Middle)

Drivers License Number	State	Date of Birth	Phone Number

Business Address

City	State	Zip

Corporation/Organization Name (DBA)	Email Address	Fax

State of Incorporation	Tax Payer ID	Other Tax Payer ID

SECTION II: EVENT INFORMATION

Name of Event	Estimated Daily and Peak Attendance

Event Date(s)	Hours of Event(s)

Location of Event/Address (street address)

Thomas Brothers Map Reference	Event Sponsor	Type of Event

Brief Description of Event

☐ Attach a copy of the IAP and/or Medical Plan ☐ Attach a copy of the site plan/map

Risk Factors (alcohol, athletic events, large crowds, access issues, etc.)

SECTION III: EMS SERVICE PROVIDER INFORMATION

Name of Provider Agency	Contact Person/Phone/Email

Coverage Days/Times	Identify each resource that will be attached to the event below. Attach additional sheets as necessary.

Resource (ambulance, EMT, etc.)	Unit ID	F9 PGR	PTT ID	Staging Location	Hours of Coverage
Resource (ambulance, EMT, etc.)	Unit ID	F9 PGR	PTT ID	Staging Location	Hours of Coverage
Resource (ambulance, EMT, etc.)	Unit ID	F9 PGR	PTT ID	Staging Location	Hours of Coverage
Resource (ambulance, EMT, etc.)	Unit ID	F9 PGR	PTT ID	Staging Location	Hours of Coverage

Special Event Registration Form (EMS Form 909)
Page 1 of 2

FIGURE 19.7 ■ Sample Special Events Registration Form.
(Source: County of Santa Clara Emergency Medical Services Agency.)

SECTION IV: EMS Agency Use

Date Received	Duty Chief Assigned	Jurisdiction	Manager Review

Additional Considerations

Notifications

	Description	Responsible Party	Date
☐	Public Safety Jurisdiction		
☐	911- Ambulance Contractor		
☐	County Health Officer		
☐	Fire OA Coordinator		
☐	Law OA Coordinator		
☐	County Office of Emergency Services		
☐	County Communications		
☐			
☐			
☐			
☐			

*** NOTICE ***

A copy of this document must accompany the units assigned to the event or must
be immediately available through the service's dispatch center.

Special Event Registration Form (EMS Form 909)
Page 2 of 2

FIGURE 19.7 ■ Continued

1. Incident Name	2. Operational Period (Date / Time) From: To:	INCIDENT OBJECTIVES ICS 202-OS

3. Overall Incident Objective(s)

4. Objectives for specified Operational Period

5. Safety Message for specified Operational Period

Approved Site Safety Plan Located at:

6. Weather   See Attached Weather Sheet

7. Tides / Currents   See Attached Tide / Current Data

8. Time of Sunrise   Time of Sunset

9. Attachments (mark "X" if attached)

☐ Organization List (ICS 203-OS)	☐ Medical Plan (ICS 206-OS)	☐ Resource at Risk Summary (ICS 232-OS)
☐ Assignment List (ICS 204-OS)	☐ Incident Map(s)	☐ _____
☐ Communications List (ICS 205-OS)	☐ Traffic Plan	☐ _____

10. Prepared by: (Planning Section Chief)   Date / Time

INCIDENT OBJECTIVES	June 2000	ICS 202-OS

Electronic version: NOAA 1.0 June 1, 2000

FIGURE 19.8 ■ Sample ICS 202 Form.

INCIDENT OBJECTIVES (ICS FORM 202-OS)

Purpose. The Incident Objectives form describes the basic incident strategy, control objectives, and provides weather, tide and current information, and safety considerations for use during the next operational period. The Attachments list at the bottom of the form also serves as a table of contents for the Incident Action Plan.

Preparation. The Incident Objectives form is completed by the Planning Section following each formal Planning Meeting conducted in preparing the Incident Action Plan.

Distribution. The Incident Objectives form will be reproduced with the IAP and given to all supervisory personnel at the Section, Branch, Division/Group, and Unit levels. All completed original forms MUST be given to the Documentation Unit.

Item #	Item Title	Instructions
		NOTE: ICS form 202-OS, Incident Objectives, serves as part of the Incident Action Plan (IAP) (not complete until attachments are included).
1.	Incident Name	Enter the name assigned to the incident.
2.	Operational Period	Enter the time interval for which the form applies. Record the start and end date and time.
3.	Overall Incident Objective(s)	Enter clear, concise statements of the objectives for managing the response. These objectives usually apply for the duration of the incident.
4.	Objectives for specified Operational Period	Enter short, clear, concise statements of the objectives for the incident response for this operational period. Include alternatives.
5.	Safety Message for the specified Operational Period	Enter information such as known safety hazards and specific precautions to be observed during this operational period. If available, a safety message should be referenced and attached. At the bottom of this box, enter the location where approved Site Safety Plan is available for review.
6.	Weather	Attach a sheet with the observed and predicted weather.
7.	Tides/Currents	Attach a sheet with the predicted tide and current information for the specified operational period.
8.	Sunrise/Sunset	Enter predicted times for sunrise and/or sunset (local time, 24-hour clock) during the specified operational period.
9.	Attachments	Mark an "X" in boxes for forms attached to the IAP.
10.	Prepared By	Enter the name of the Planning Section Chief completing the form.
	Date/Time	Enter the Date (month, day, year) and Time (24-hour clock) the form was prepared.

FIGURE 19.8 ■ Continued

transportation for both team members and patients, and rescue.

In many cases EMS is required to provide mass care commonly thought of as food and shelter. In the federal response plan this is provided under ESF #6. The American Red Cross (ARC) is the lead agency for this function and is supported by agencies such as the Departments of Agriculture, Commerce, Defense, Health and Human Services, Housing and Urban Development, Transportation, and Veterans Affairs. In addition, FEMA, the General Services Administration (GSA), and the U.S. Postal Service also provide assistance to this vital function.

The following are some important concepts to understand with regard to the ARC and its services under ESF #6. The ARC is not a government agency. However, under a disaster or emergency declaration, it is empowered by legislation to provide the government's response to sheltering and feeding needs as if it were an actual branch of government. Since the ARC is not a government agency, EMS may access its resources without activation of the Federal Response Plan. However, in a major disaster, the ARC may require the use of government resources to accomplish its mission.

EMS managers should familiarize themselves with the services that the ARC can provide. On a routine basis, the organization responds to requests to assist with housing for fire victims, clothing, and food, as well as refreshments and rehabilitation assistance for EMS responders during protracted local operations. However, the real value of the ARC can be seen in times of disaster.

Basic Organization of the American Red Cross

The organizational structure of the ARC is simple. Local chapters report to a lead state chapter, which can in times of emergency coordinate directly with the ARC's national headquarters. In addition to this direct link to national resources, the local chapter also coordinates closely with local and state government officials. One of the ARC's main responsibilities in the Federal Response Plan is to provide shelter and basic needs for populations who have been impacted by an emergency or a disaster.

Shelter Operations

The ARC is responsible for the opening and management of shelters at the local level. This occurs when the local emergency manager and a representative of the local chapter select the appropriate shelters to be opened. Once the shelter selection is made, the ARC contacts its employees and volunteers to facilitate the shelter's opening and start up of operations. It is important to bear in mind the three primary roles of the ARC with regard to disaster mass care: shelter, food and clothing, and distribution of bulk materials.

Although it is a goal of the national headquarters of the ARC to have a nurse on location at each shelter, that resource is not always available. The roles of any nurses at a shelter are limited strictly to what are essentially advanced first-aid procedures. These procedures are published in the ARC's "Disaster Health Services Protocols," which are then adopted by a physician who serves as the local chapter's medical director. This delegated practice is very similar to that of EMS personnel. With such limits placed upon an ARC nurse, it is obvious that medical care at shelters will be strictly limited unless it is augmented by other resources. If these resources are not identified in advance, the burden will be placed directly upon the local EMS system provider.

MINIMIZING EMS SYSTEM BURDENS

The burden on EMS caused by sheltering operations can be minimized by EMS partnering with local agencies to develop a People with Special Needs (PSN) program. Specific contact with the local Office of Emergency Services (OES) can identify the process already in place. Cooperative planning by both EMS and OES can greatly reduce the planning burden on any one agency.

There are three significant burdens upon shelter medical personnel:

- Evacuees who did not bring or who run out of medication.
- Evacuees who require specialized medical care.
- Evacuees who have social and mental-health needs.

The situation can be managed very effectively by developing a PSN program by which those needs are identified in advance and resources are developed to meet those specific needs.

An effective PSN program allows for people with specific medical needs to register with local EMS or emergency management officials well in advance of

1. Those who can go to public shelters.
2. Those who have special assisted-care needs but no immediate medical-care needs (PSN shelters).
3. Those who need nursing assistance (nursing home shelters).
4. Those who require continuous medical care (hospital).

FIGURE 19.9 ■ Evacuation Classification of Patients with Special Needs.

- Task-force leader.
- ALS ambulance.
- Public-transportation bus.
- School bus.
- Staff cars.
- Wheelchair taxis or vans.

FIGURE 19.10 ■ People with Special Needs (PSN) Program Task Force.

the need to shelter. Once individuals have registered, the resources and needs that will be required in a disaster situation can be assessed better. EMS managers must be able to identify specialized transportation; for example, wheelchair transport vans or ambulances with onboard inverters.

Once the necessity for evacuation is identified (such as when a hurricane is approaching) and an evacuation order is issued, the 911 communications center or EMS agency can begin telephoning the registered persons to verify that they will need to be evacuated. During this process, the evacuees are categorized into one of the four categories listed in Figure 19.9. These verified evacuation needs then can be placed as "backlogged" calls in a 911 dispatch computer awaiting the actual evacuation order. Once the evacuation order is issued, the dispatchers simply can pull up the backlog calls and give them to a 911 communications operator who coordinates with task forces in the field. Based upon the categorization, the appropriate resource is sent to aid with evacuation and deliver the individuals to the appropriate type of shelter.

The actual process of evacuating people with special needs can be developed further to minimize EMS system impact even more by using a PSN Evacuation Task Force. In this case, only those requiring medical care during transport are transported by ALS ambulance. A suggested PSN program Task Force is listed in Figure 19.10. These Task Forces can be strategically located throughout the community to assist the registered PSN requests as well as normal evacuation assistance requests. Once the PSN dispatcher receives the list of evacuation requests, the information is transmitted to the appropriate task force leader, who then ensures that an appropriate resource is used. Generally, unless the evacuee requires medical care during transport, a bus being led by a staff car carries out the process.

One important consideration in the delivery of evacuees to a shelter is medical screening. Personnel should be directed to the registration area where ARC personnel can determine any medical needs that may not have been identified previously. Additionally cultural, ethnic, and racial considerations are important. Tension between ethnic groups or rival gangs can easily rise to violence in the cramped and stressful environment of a shelter. The provisions for children in a shelter should be considered carefully because the stress of caring for children during an evacuation can raise tensions in a shelter. Children and the elderly are two at-risk groups; provisions should be made for the supervision of these two groups in shelters to ensure criminals, thieves, and sexual predators are not allowed access to them.

It may be necessary to assign EMS resources to shelters to assist with emergency medical care. The need to do so should be determined by the size of the shelter and the potential for conditions that would not allow EMS units to arrive at a shelter should the need arrive. If the decision is made to place EMS in shelters, these personnel should be prepared to deal with any routine medical care needs that may be required. In addition, they should have routine medications available that can be distributed under protocol to persons who take such medication regularly.

Particular attention will need to be made to ensure public health is not threatened in the shelter populations. EMS assumes a role to help ensure clean drinking water and sanitary conditions and to make sure that the proper agencies report and respond to infectious diseases. EMS providers expect to provide initial screening for mental health emergencies. Collection and securing of patients who expire away from the general population may become EMS responsibility until coroner or mortician teams are available.

EMS SPECIAL TEAMS

Tactical Emergency Medical Support

Despite peaceful resolution always being the goal of a SWAT team and law enforcement, the unpredictable actions of criminals or terrorists can sometimes result in casualties. In preparation for such situations, many law-enforcement agencies have integrated highly specialized EMTs, paramedics, nurses, physician's assistants, and physicians into their SWAT teams. These medical providers then can maintain the wellness of the team's members and provide immediate medical care to anyone in need, whether they are law-enforcement officers, innocent bystanders, or suspects.

Provided by a number of organizations, **Tactical Emergency Medical Support (TEMS)** is the comprehensive out-of-hospital medical support of law enforcement's tactical teams during training and special operations. Provided by the DHS Federal Protective Service's Office of Protective Medicine, the goals of the **Counter Narcotics and Terrorism Programs (CONTOMS)** were to offer a nationally standardized curriculum, certification process, and quality improvement procedure to meet the needs of those EMTs, paramedics, and physicians who operate as part of a tactical/special operations team.

A common issue that EMS managers face is whether or not to allow EMS providers to be armed. Both positions require a significant amount of discussion and planning. One solution is to provide a less-than-lethal weapon to the EMS provider; for example, a TASER or beanbag gun. A law-enforcement officer is injured in one out of every 52 SWAT operations. In addition to the care of injured officers, the tactical medic often is responsible for rehabilitation and medical monitoring of personnel. Often police officers in a standoff or barricaded subject are in hot and heavy gear for hours, and their effectiveness can be jeopardized if they do not remain hydrated and monitored for fatigue and heat illnesses.

EMS managers should monitor the time on task or the entire event time an EMS unit is committed to a police standby. That time should have cost calculated and the hourly cost for the apparatus and crew totaled each year in addition to the number of times a unit had to cross a district or respond in a district where a unit was tied up on the police standby. The cost benefit of an on-call or assigned tactical medic can be measured against the cost per hour for the equipment to be on the call. If the unit's response times drop below the standards of coverage, compliance issues also may help justify a tactical EMS program.

Bike Teams and Alternative Response Programs

The use of bicycles is on the rise. There are over 300 EMS bike teams in the United States with most being in the top 200 cities. Bike medics have been incorporated into a variety of agencies including county and city fire agencies, villages, military bases, colleges, and hospital-based EMS. Experienced EMS bike-medic programs have been established in Las Vegas, Nevada; Troy, Ohio; and Toronto, Canada. Bike medics are cost effective and bridge the gap between medical personnel stationed in first-aid stations and ambulances. Bike units are more easily transported and can penetrate crowds and get to areas better than can golf carts, off-road vehicles, quads, and people on foot. Many agencies use their bike medics for special events or at locations that receive high pedestrian traffic. Figure 19.11 lists the common uses of bike-medic teams according to the International Police Mountain Bike Association (IPMBA).

There is no standard set of equipment that is recommended or required on bikes. Most EMS bike units use a rack-and-pannier system to mount specialty EMS bags. The type of bike and other equipment for bike medics has a variety of vendors.

The bike-medic training program is conducted by IPMBA and has certified over 20,000 cyclists (Figure 19.12). The course is called the IPMBA EMS Bicycle Operations certification course and includes bicycle-handling skills, night operations, bicycle maintenance, emergency maneuvers, fitness and nutrition, group riding, load placement, and scene safety. The IPMBA holds an annual conference that offers on-

---

- ◆ Special events.
- ◆ Rapid response in wilderness or off-road areas.
- ◆ Daily patrol in congested areas, such as tourist destinations.
- ◆ Patrol at large facilities like airports, depots, racetracks, shopping areas.
- ◆ Amusement parks/sports arenas/outdoor events.
- ◆ College and university campuses.
- ◆ Disaster response.
- ◆ Public-safety education.
- ◆ Mass casualties over a large area.

FIGURE 19.11 ■ List of Events for Bike Medic Programs.

FIGURE 19.12 ■ EMS Bike Teams Cover Special Operations.

bike and classroom activities. The IPMBA has a variety of resources to instruct agencies on financing, implementing, and operational issues for EMS bike teams.

■ PROTECTIVE OPERATIONS

ASSASSINATIONS

Protective operations are often referred to as dignitary protection. However, the term *dignitary*, by definition, refers to a person who holds a high rank or office and is, therefore, self-limiting in nature. Granted, those persons for whom EMS provides protective measures will more than likely be high-ranking officials. Virtually anyone may request a protective detail given a high enough potential threat. The EMS system should be prepared to provide those services if the threat is warranted. The protectee may be anyone—a political figure, media star, or a witness in a high-profile criminal case. The determining factors will be that of potential threat and importance of the protectee.

With the U.S. form of democratic government involving three branches (executive, judicial, and legislative), terrorists have come to realize that the assassination of any one individual may not obtain the desired results. These desired results include the breakdown of infrastructure, the instilling of anxiety and fear within the public, and the development of a lack in confidence in the government. This is not to say that an individual assassination attempt will not occur. What must be understood is that the assassination attempt may, in fact, be terrorism-based. And if so, the terrorist will be desirous of achieving maximum exposure through maximum fatalities. Therefore, any protective measures also should be considered potential terrorist events if the protectee is of a high enough rank and the detail involves mass-gathering events.

Causes of Assassinations

Revolutionary or political causes by groups or individuals fanatical in their desire to change the existing government or to establish a new government often resort to assassination to accomplish their goals. The targeted individuals represent the government that the assassin believes is the cause of repression, unfairness, persecution, and so on. The assassin hopes to overthrow the existing government by eliminating its representatives, as was seen in the 2005 bombing of the Madrid, Spain, transit system.

Economic causes are motivated by the belief that the victim is responsible for poor economic conditions affecting a nation, group of people, or the assassin directly. Ideological causes can trigger an assassin when the person or organization is convinced that the victim is endangering principles that the assassin believes to be of paramount importance. Ideological causes of assassination can stem from religious and social beliefs. The assassin hopes to change the existing system by eliminating key figures in the religious or the social systems, or the assassin may wish to draw attention to his or her group through the use of terror.

Personal causes are motivated by revenge, jealousy, hate, rage, or other strictly personal drives. Psychological causes—mental derangement, fanaticism, or emotional instability—are factors in most, if not all, actual attacks. Assassins almost always exhibit psychological problems even if the real reasons for the assassination are revolutionary, economic, and so on. It is not at all unusual to find more than one cause of assassination. Psychological causes remain the most common factor in assassination attempts.

A weapon at close range continues to be a method of attack, and there is little skill required. Handguns or knives are weapons that are often used. Usually some preattack intelligence gathering is necessary, if only to learn the victim's itinerary, and many times an investigation afterward will reveal that stalking has occurred.

Weapons at a distance do not require the victim to be at close range; the victim is usually more than 20 feet away. This method often involves a rifle but also may include a handgun. Distance requires more skill to succeed and usually a more sophisticated weapon. An attacker may miss due to the distance and strike civilians, which would require EMS to triage civilian injuries versus attending to the dignitary. Again, intelligence gathering is necessary to learn the victim's itinerary and to find the best attack site.

Explosives require even more sophistication and skill. Knowledge is needed to build, place, and detonate the bomb. Logistics and intelligence support is usually necessary to complete a bombing attack.

Kidnapping is not a form of assassination but still involves an attack. The immediate intention is not to kill the victim. The kidnap victim is more valuable alive than dead; however, death often follows when demands are not met. The protective detail must be eliminated immediately to successfully complete a kidnapping. Kidnapping is the most sophisticated of attack methods requiring attack team, intelligence team, safe haven, and logistical/surveillance team. Meticulous planning is imperative to a successful attack.

Counterassassination Methods

Law enforcement, EMS, and other emergency services cannot possibly guarantee a person's security. The terrorist has all the advantages. Protection is the elimination of surprise, taking into account any and all possibilities and planning for them. Law enforcement and public-safety agencies do this by eliminating the most dangerous first and working up to the more sophisticated. Figure 19.13 lists the common methods of dignitary protection.

PROTECTIVE DETAILS

Organization of Protective Details

The purpose of the protective detail is to safeguard the protectee from harm and from situations likely to endanger his or her person or liberty. Many times the protectee will find himself in a situation that is difficult to control. If the protectee wishes to be shielded from public view, the detail must honor his wishes. In order to protect an individual, a secure area must be established, which is known as the **concentric ring of safety**. Formation of the perimeter is dependent upon the threat, manpower, and resources available to a protective detail. The perimeter is directly around

---

- ◆ Protective formations for weapons at close range and at a distance.
- ◆ Surveillance detection for picking up of intelligence teams as they gather intelligence on the protectee and the protective security detail.
- ◆ Protective intelligence to determine if groups or individuals fit into any one of the assassination categories that would threaten the protectee.
- ◆ Security perimeters established to deny access to the protectee as much as is possible and to ensure that those who do have access do not pose a threat.
- ◆ Advance work that takes into account any potential danger points.
- ◆ Motorcades to move the protectee quickly and safely through potential danger points.
- ◆ Security agents responding to an attack if it presents itself.

FIGURE 19.13 ■ Types of Dignitary Protection.

the protectee. It should be staffed by protective detail agents only. Only those individuals with a personal need for access to the protectee are allowed inside (such as a protocol official, staff aid, or family member).

At each security perimeter, a checkpoint should be established to ensure proper access. Screening at the checkpoint can be as restrictive as requiring a person to provide positive identification, submit to a name check through police records or a visitor list, and undergo a search by a walk-through or hand-held metal detector. At each site the advance agent determines both where the security checkpoint should be and its screening process. He determines this based upon manpower available, the threat to the protectee, the number of persons or press expected at the function, and protocol considerations when diplomatic functions are planned.

It is important for everyone working the protectee's security detail to know whose work necessitates their involvement with the protectee. Permanent identification (lapel pins) are worn by members of the protective security detail. All forms of identification should be backed up with a formal guest list, and names checked off by a staff member of the security detail.

Security for the Protectee in Residence

The protectee should be most secure when in residence. There the protective detail has the greatest opportunity to control the environment through the use of the security ring, combined with agents posted, physical security hardware, electronic alarms, and a general familiarity with the area surrounding the residence. While the protectee is in residence, posting needs to be assigned so that all avenues of access to the protectee are secured by the detail. It is at the residence that the protective detail may be supplemented with physical barriers, alarm systems, and electronic equipment on a permanent basis. The systems deployed should be unobtrusive so as to allow the protectee privacy as well as security.

There should always be one member of the protective detail at the residence on a 24-hour basis to monitor alarms and ensure authorized personnel access while preventing unauthorized personnel access and unauthorized delivery of goods or services. This posting is called the command post or residence watch and should be maintained even when the protectee is away.

The Command Post

EMS will be managed by a team in the command post or residence that is a self-contained room in the protectee's residence, which serves as the perimeter ring. This room should be close enough to the protectee's living area to allow an agent or agents the ability to monitor effectively the residence and respond to an emergency on a moment's notice. Consideration should be given to placing the command post near an avenue most likely to bring danger, yet far enough away to provide some privacy to the protectee.

If the residence does not allow for a command post due to space considerations, then an alternate command post could be set up in the garage, a motor home, or a camper trailer located next to the residence. Wherever located, the command post also can act as a down room for detail members not standing post. The command post coordinates all communications and operations involving the protective detail.

The Safe Haven

Often EMS resources will be in another room in the residence within the perimeter of security; this is called the safe haven. This room provides temporary refuge to the protectee and family when additional time is needed for police or security personnel to respond to the residence under attack. When identifying a room in the residence as a safe haven, the factors listed in Figure 19.14 should be considered.

- *Accessibility.* The protected should not be exposed to additional danger while moving to the safe haven. If possible, this room should be the master bedroom.
- *Ability to secure.* Both the safe haven and its access route should be positioned such that they can be secured from all directions.
- *Ability to defend.* The safe haven should be within an established realistic perimeter that the protective detail can defend. It also should have a tactical advantage, such as high ground.
- *Ability to communicate.* The command post or headquarters must be accessible, in case of an emergency, to alert extra help.
- *Ability to escape.* If the perimeter ring is compromised, there must be an escape route.
- *Ability to hold.* With appropriate equipment and supplies, the safe haven should be able to hold for 15 to 60 minutes while under siege.

FIGURE 19.14 ■ Safe Haven Criteria.

The safe haven should contain emergency communications (telephone and radio), a fire extinguisher, water, and emergency lighting. The safe haven is considered to be the core of the ring.

Security Perimeters at an Office

The next location where a protectee should feel secure is at the office, which may be located in a government or privately owned building. To ensure proper security procedures are in effect at this location as well, create a command post and safe haven. In the perimeter area, there should be a post manned by members of the protective security detail to screen visitors and secure all avenues of access to the protectee. The staff of the protectee will screen visitors and coworkers and the only ones considered to have access to the perimeter.

If a garage exists under the office space, the protectee's motorcade should be secured in that area while the protectee is in the office. All persons having access to the building should be properly identified and screened before being allowed into the building. An identification system should be implemented for building employees to ensure unauthorized persons not be allowed inside. All suspicious persons, packages, or vehicles in the area should be reported. If possible, different entry and exit doors to the protectee's office should be used by the detail.

Hotel Security

When the protectee is traveling, the advance agent should ensure that the ring of security is established at hotels where the protectee is temporarily residing. The perimeter is the room, suite, or floor where the protectee is residing. The perimeter also should extend to the hotel rooms adjacent, above, and below the protectee's, in which staff or other detail members would reside. Access to the protectee's room should be granted to staff members, screened visitors, and cleared hotel personnel who are escorted by either a member of the staff or protective security. A temporary identification system must be employed, and the advance agent should obtain an approved list of hotel employees that has access to the protectee.

Depending on the circumstances and location of the protectee, there should be a post that allows the detail to screen visitors and prevent unauthorized access to the protectee's floor. The advance agent should

conduct a survey, prior to the protectee's arrival, to ensure all stairwells and elevators are controlled and all avenues of access are secured.

EMS PROTECTION LEVELS

Depending upon the threats levels involved, EMS may be requested to provide limited or continuous protective details. For example, the president of the United States, in addition to traveling with a physician and a nurse, is afforded 24-hour EMS protection. On the other hand, the vice president is generally provided EMS standby only when in public places or when threat levels are high. With these considerations, EMS may be asked to provide any variety of protective arrangements.

The methods used to protect the president of the United States (one of the most highly protected individuals in the world) can be applied in any protective detail. One of the first things that occurs when the president plans a visit to a community is that of the arrival of a Secret Service advance team. This team will begin to develop the protective plan and select agencies and personnel for involvement in the details. During this time, important items that must be addressed are security checks for all personnel who will come in close proximity to the president (such as EMS personnel) and the establishment of protective and community goals.

Being an elected official, the last thing that the president wants is poor publicity. Therefore, the Secret Service will do everything possible to ensure that local community concerns are addressed. Response routes, availability of resources, and impact of motorcades on traffic all can be identified. The most important concern for EMS to voice is that it must always maintain the established level of service to the community regardless of visiting dignitaries. Therefore, the system manager should take into account the potential impacts of the visit upon the ability of the service to provide that level of care. Questions that should be raised are:

- Can EMS still move emergency vehicles across motorcade routes, and at what point will that be prohibited?
- Will EMS need to relocate units to provide continuous service across routes?
- How many resources will be required?
- Does the protectee have medical staff traveling with her or him and at what level of training?

Your service may be requested to provide teams for the motorcade, each venue, and the residence depending upon the complexity of movements and the duration of the visit. Plans for the relief of these crews will need to be discussed with the Secret Service. Remember, during the advance work is your best and possibly only opportunity to address these issues. EMS must ensure that it has representation during the planning process. Again, make your community-response goals known early so that your concerns can be incorporated into the plan.

In addition to the field unit requirements, arrangements for hospital facilities will be made during the advance work. Primary and alternate facilities will be identified and potential methods of transport to those facilities will be developed. All personnel in the hospital emergency department will be cleared and an advance agent will be placed in the ED for the duration of the entire visit. Interruption of service in the ED generally is not anticipated unless threats are high or actual need for use occurs.

DIRECT CONTACT WITH THE PROTECTEE

The President

If an assassination attempt were to occur, the possibility of EMS personnel becoming directly involved with the treatment of the president is very remote. The first and foremost responsibility of the protecting agents is to get the president into a safe haven. This may be the limousine or other predefined location. The most likely role of the on-scene EMS personnel will be the treatment of any collateral injuries. However, EMS must remember to pay strict attention to the orders of the Secret Service to ensure their own personal safety.

The more likely direct contact with the president by EMS will be during other illnesses or injury. If such an incident should occur, the president obviously should receive as high a level of medical attention and confidentially as any other patient would receive. Remember that inquiring minds will want to know. The media—tabloids and others—will definitely seek out the EMS personnel for comment and any such comment should be well tempered and controlled.

Diplomats

If your agency has within its jurisdiction an embassy or consulate of another country, it is extremely im-

- Communicate with the Secret Service to let them know your goals early.
- Never sacrifice your level of service to the community.
- Make sure your personnel are well briefed and know the rules.
- When the protectee is near, it is "all business."

FIGURE 19.15 ■ Keys to Successful Involvement.

portant that your personnel understand the protocol associated with that facility. This also is true if a dignitary from a foreign country visits your community. Unlike the United States, many foreign countries have strict protocols with regards to personal contact with individuals based upon religious customs. Horror stories have arisen out of well-meaning EMS contact with dignitaries of other nations. In some countries, the simple touching of a male by a female, even in the context of medical care, is prohibited. Therefore, it is imperative that EMS personnel be aware of the protocols and customs associated with visitors, embassies, and consulates (Figure 19.15).

Remember also, an embassy or consulate is *not* sovereign territory of the United States, but it is considered the territory of the country it represents. Therefore, the laws of that country bind you when you enter. When in doubt, personnel should ask before doing anything. This reverts back to the very basic concept of "informed consent." However, the doctrine of implied consent may not be valid when you are actually standing in another country as you are in an embassy or consulate. Your closest U.S. State Department office can be a valuable source of information concerning these issues.

HELICOPTER TEAMS

The evacuation of a patient by helicopter is an EMS mission associated with high risks. Pressure to take or complete a mission, weather, nighttime flight, spatial disorientation resulting from lack of visual cues, and pilot training and inexperience were all identified as risk factors in the Safety Board's 1988 safety study of commercial EMS helicopter operations. The 2002 Air Medical Physician Association (AMPA) study cited additional risks, such as unprepared landing sites, complacency, and situational stress. Safely operating in such a high-risk environment calls for the systematic evaluation and management of these risks. EMS managers need to develop policies for when and how helicopters will be used by their agency.

EMS agencies should routinely benchmark helicopter response times in their jurisdiction. Company officers should call out when the helicopter is on the ground and when the helicopter lifts off the ground with the patient. Every evacuation of a patient by helicopter should be evaluated by the quality improvement team. Yearly safety training is required for helicopter landing and take-off procedures. Often this includes minimum clearance, slopes, and hand signals for landing and takeoff. Additional preparation by pre-designating land zones in safe areas that have pre-determined GPS coordinates can add an additional safety factor for helicopter operations in urban areas.

According to AMPA's study, an effective flight risk-evaluation program acknowledges and identifies threats, evaluates and prioritizes the risks, considers the probability that a risk will materialize, and prepares to avoid or manage those risks. EMS managers can help in this process by establishing procedures for dispatching, communicating, and landing helicopters.

Federal guidelines allow pilots responding to EMS operations to be notified of an assignment by a local 911 dispatch system or emergency hospital staff. Federal safety studies have identified that 911 dispatch personnel or hospital staff do not have expertise in or an understanding of the requirements of flight or landing procedures, particularly at night or in adverse conditions. When a pilot is dispatched by someone other than a flight dispatcher, the pilot would typically check the most accessible source of weather information available usually via computer, using sources that are not necessarily specific to aviation and begin the flight. The pilot would then have limited access to updated information.

■ OTHER SPECIAL OPERATIONS———

MEDICAL SUPPORT IN HAZMAT OPERATIONS

Every day millions of tons of hazardous materials are processed, transported, and used by business and industry. As an EMS manager, you can expect your crews to be involved in patient care or function in a support role to a hazardous materials team. An accidental release of these materials presents a potential danger to the public and the environment. Such an incident can be managed more expeditiously when the hazardous materials are specifically identified and characterized. Unfortunately, the contents of storage tanks and trucks may not be specifically or properly identified. Records or shipping papers may be inaccessible. Even with such information, an experienced person is needed to define the medical hazards and the gravity of the situation. The immediate need for information concerning a hazardous material during an incident is vital.

The EMT's employer will decide the level of care that the EMT will receive. Training is required for all responders who may be called to a hazardous materials incident. The National Fire Protection Association Standard 473 details the competencies for EMS providers at hazardous materials incidents and now requires EMS to have operational level certifications. An OSHA standard—29 CFR 1910.120—details the four levels of response for emergency responders:

- *First Responder Awareness.* The minimum training an EMT should receive, this level is designed to protect first responders. First responders are trained to recognize a problem and initiate a response from an agency with operational level training.
- *First Responder Operations.* This level is for first responders who are trained to start mitigation of releases or potential releases of hazardous materials. They require approximately eight hours of training, which includes training in the procedures for keeping hazardous materials from spreading and for protecting people, property, and the environment from exposure. First responders learn to wear and operate in level B protection and understand how to decontaminate patients who have been exposed to hazardous materials.
- *Hazardous Materials Technician.* This level of training is advanced and teaches a first responder to control a spill and operate in hazardous environments in chemical protective clothing and level A suits. About 60 to 80 hours of training are required.
- *Hazardous Materials Specialist.* This is the highest level of hazardous materials training, which provides advanced knowledge of monitoring devices, toxicology, and command and control of large-scale hazardous materials incidents. A minimum of 24 hours of additional training above technician level is required.

EMS managers need to make an investment in toxicology training for their medics. There are two programs readily available for paramedics or advanced life support providers. A three-day program through the University of Arizona called Advanced Hazmat Life Support covers many of the medical conditions that may be produced in a terrorist attack. The other program is the National Fire Academy's Advanced Life Support Response to Hazardous Materials Incidents. This course is two weeks long and contains significant reference to weapons of mass destruction. Both of these courses present opportunities for EMTs and paramedics to continue their education in accordance with regulatory and recertification agencies. The Fire Academy course instructs paramedics on how to create a toxmedic program from the ground up.

Toxmedic programs often incorporate separate pharmacology and protocols and require a strong background in chemistry and physiology for paramedics. A significant responsibility is to medically monitor fire department teams making entry. This includes preentry screening, monitoring, and post-incident rehabilitation.

USAR AND TECHNICAL RESCUE

There are nine classifications of technical-rescue incidents according to the National Fire Protection Association. A list of technical-rescue incidents is shown in Figure 19.16. As an EMS manager, it is important for you to establish interagency cooperation and train with technical-rescue teams before an incident. When EMS is summoned to a technical-rescue incident, the minimum qualifications require rescuers to be certified at the BLS level and have rescue-awareness training. All personnel at the site need to understand that they must operate under an incident-management system. Consensus standards identify that EMTs need to be trained to the awareness level if responding to rescue incidents under the NFPA 1670: Standard on Operations and Training for Technical Search and Rescue Incidents.

Awareness-level training represents the minimum capabilities of a responder who, in the course of his duties, could be called upon to respond to or be first on scene of a technical-rescue incident. Awareness-level operations could involve search, rescue, and recovery of victims in technical-rescue scenarios. At this level, the EMS manager and field crew's responsibility generally is not considered a rescuer but rather support personnel for the operations. The EMS crew conducting rescue operations will need to get specialized training and become familiar with the FEMA USAR medical plan (Figure 19.17).

Operational-level training is designed for responders who will have the capability of hazard recognition, equipment use, and techniques necessary to conduct a technical rescue. Rescue operations are usually supervised by a rescuer certified at the technician level.

The federal system for **Urban Search and Rescue (USAR)** relies on local fire and EMS organizations to organize into 36-person FEMA USAR teams. These teams require federally mandated training and a specialized equipment list. Within each FEMA USAR team is a medical group designed to care for the team and any potential victims. While it may not be economical to support a FEMA USAR with your organization, EMS agencies can contribute to regional, state, or other teams. The medical specialist course is offered in several locations and is required as part of the mandated training for FEMA EMS operations at an USAR event. The medical specialist course involves training to care for victims of crush syndrome.

STRUCTURAL COLLAPSE

In an operational setting during a large-scale event such as an earthquake, you or one of your crews may be first on scene at a structural collapse where you will need to evaluate the existing and potential conditions. As an EMS manager, you should know where to locate the resources available to respond to a structural collapse. It is important as a first-on-scene emergency responder to maintain site control and scene safety. It also is important to understand the types of construction and how the components will act. There

---

- ◆ Water rescue.
- ◆ Rope rescue.
- ◆ Rescue from confined spaces.
- ◆ Wilderness search and rescue.
- ◆ Trench rescue.
- ◆ Vehicle and machinery rescue.
- ◆ Dive search and rescue.
- ◆ Collapse rescue.
- ◆ Any other rescue operations requiring specialized training.

FIGURE 19.16 ■ Technical Rescue Incidents.

MEDICAL PLAN	INCIDENT	REPORTING UNIT **ESF-9**	FORM **US&R—015** 3/96

DISASTER #:	OPS PERIOD:	DATE/TIME PREPARED:	PREPARED BY:

A   GENERAL EVENT INFORMATION

Event Type: Date/Time of Event:

Location: Travel Time: Time Change:

Situation Assessment:

Lifelines Affected: [ ] Water   [ ] Electricity   [ ] Gas [ ] Sanitation   [ ] Telephone

[ ] Cellular System   [ ] Roadways   [ ] Airports   [ ] Railroad

Probability of Recurrence:

B   LOCAL AREA CONDITIONS

Weather Conditions: Avg Temp: Day —   Night —   Sunrise —   Sunset —

Precip. —   Humidity —   THI/WC Factor —

Forcast (3 day):

Wind Speed/Direction:

Terrain:

Access/Egress:

Endemic Threats: Disease:

(incl for canine)   Insects:

Animals:

Botanicals:

Technical Hazards: [ ] Chemical Storage   [ ] Biomedical   [ ] Radioactive   [ ] Other

Site	Material	ID Number	Fire/Expl Hazard	Health Hazard	Mitigation

C   LOCAL RESOURCES

Medical/EMS POC: Phone #:   Contact Method:

Veterinary POC: Phone #:   Contact Method:

DoD Medcical POC: Phone #:   Contact Method:

DFO ESF-8 Rep: Phone #:   Contact Method:

DFO ESF-9 Rep: Phone #:   Contact Method:

Facilities/#   [ ] Emerg Med —   [ ] Trauma Cntr —   [ ] Burn Cntr —

[ ] HBO —   [ ] Peds —   [ ] Vet —

Name	Location	Capab/Assessment	Travel Time	POC	Comm Method

EMS Transport   [ ] ALS Units   [ ] BLS Units   [ ] Aircraft/Type

Name	Location	Capab/Destination	Response Time	POC	Comm Method

Notes:

FIGURE 19.17 ■ Urban Search and Rescue (USR) Medical Plan.

484

MEDICAL PLAN	INCIDENT	REPORTING UNIT **ESF-9**	FORM **US&R—015** 3/96
DISASTER #:	OPS PERIOD:	DATE/TIME PREPARED:	PREPARED BY:

D MEDICAL ANCILLARY INFORMATION

Casualty Estimates: [ ] Dead: [ ] Injured: [ ] Homeless:

Injury Profile/# : [ ] Trauma   [ ] Burn   [ ] Crush   [ ] HazMat

[ ] Victim Age Range   [ ] Baseline Med Problems   [ ] Antic. Length Entrapment

EMS Triage Tags (type/in use?)

Casualty Collection Points:

Transfer Procedures:

Processing of Deceased: Coroner POC:

  Comm Method: Forms:

  Morgue Locations:

  Procedures:

Medical Agencies/Teams in Area:

Medical Resupply Resources:

Resource	Name	Location	Procedure	POC	Comm Method

Political/Religious Medical Considerations:

E. EVACUATION PROCEDURES FOR INJURED/ILL TF PERSONNEL

Contact: Phone #:   Radio Freq:

Tested: [ ]   Date/Hour: By:

Medevac Locations:

  Procedures:

Mode of Transport: Destination:

Route of Travel:

TF Member Accompanying: Notified: [ ] TF Leader [ ] Spons. Org.

F. TASK FORCE HEALTH MAINTENANCE

Med Cache Requirements:

Rehydration (water consumpt./person/hour):   quarts

Stress Assessment:

Uniform Adjustments: [ ] Heat   [ ] Cold   [ ] Other Protection   [ ] Work Cycle

Base of Ops Issues: [ ] Shelter   [ ] Water Source   [ ] Sanitation

[ ] Fresh Food Source   [ ] Safe Food Prep   [ ] Quiet Rest Area   [ ] Wash/Hygiene Area

[ ] Canine Facilities   [ ] Animal/Insect Control [ ] Weather Impact Minimized

In Transit Considerations:

Notes:

G. SUMMARY OF RECOMMENDATIONS/PLANNED MEDICAL ACTIVITIES

Name/Title (print):	Date/Time:
Signature:	Addendum Attached: [ ] Yes [ ] No

FEMA US&R INCIDENT SUPPORT TEAM

FIGURE 19.17 ■ Continued

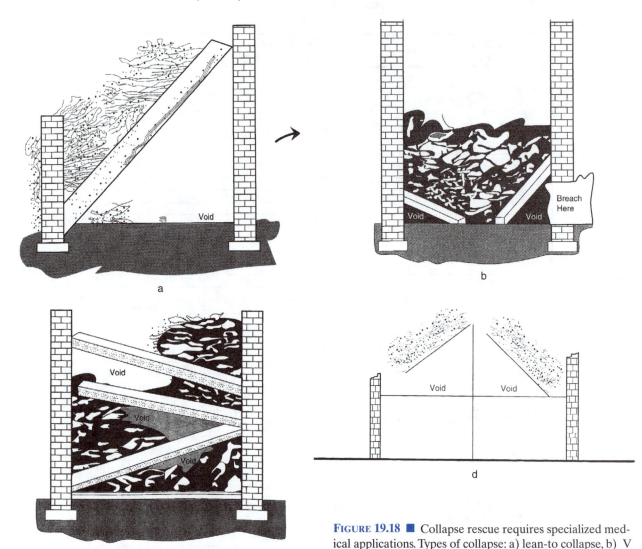

FIGURE 19.18 ■ Collapse rescue requires specialized medical applications. Types of collapse: a) lean-to collapse, b) V collapse, c) pancake collapse, d) tent collapse.

are several types of structural collapse patterns, as shown in Figure 19.18.

For each of these types of collapses, it is important for the EMS manager to understand the FEMA agency search-and-rescue marking system and structure marking system so as to recognize hazards and to know which structures have been searched. The FEMA search-and-rescue marking for a damaged building is indicated by an "X." This system was used extensively in New Orleans after Katrina (Figures 19.19 and 19.20). The first slash of the X is made when the team enters the building. The second slash completing the X is made when the team leaves the building. The top portion of the X then is dated, and the time that search team exited is marked. The right side of the X is for marking personal hazards. The bottom part of the X is for the number of live or dead victims found in the structure. The left side of the X is marked with the identification of who searched the structure.

The building marking system from FEMA USAR teams is slightly different. A building marking is a two-foot by two-foot box, spray painted in international orange. A simple clear box means the structure is safe to enter. An arrow close to the box will indicate the safest route into the structure. A box with a diagonal line through it means the building has significant damage. The X inside the box means that the building is not safe to enter. "HM" painted next to the box indicates hazardous materials.

FIGURE 19.19 ■ A Large-Scale Disaster Requires Managers to Have Planned Ahead.
(Source: FEMA: Greg Henshall.)

Certain techniques are useful to remove readily accessible victims from structural collapse. EMS agencies with fire or single-service backgrounds may be asked to participant in organized search-and-rescue operations. A FEMA-qualified USAR team requires a significant amount of logistics. FEMA has classified smaller teams for light- and medium-rescue assignments. USAR teams are scattered throughout the United States and are often composed of personnel from multiple departments.

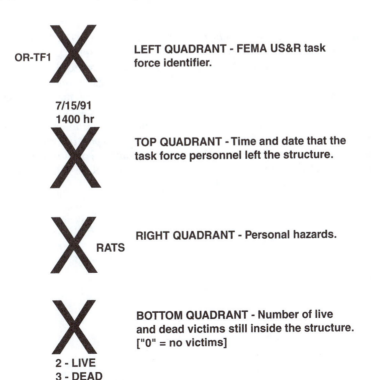

FIGURE 19.20 ■ USAR Markings Indicating Search Results in Damaged Buildings.

CHAPTER REVIEW

Summary

EMS is constantly being challenged to provide services in environments as part of a special-operations team. The all-hazards approach to emergency services is now creating an expectation that EMS can deliver services to patients during events that require technical rescue, hazardous materials response, and law enforcement scenarios. Preparation for such incidents involves added responsibilities for more training and interagency relationships. More specialty teams will be needed in the future, and as it becomes necessary to regionalize these resources, EMS managers will be required to design and implement services within these units and establish procedures for interfacing with special operations. Today's EMS leadership must have some knowledge of EMS special operations. Planning is the key to success in special events: the better the plan, the less chance for problems.

Research and Review Questions

1. What at the four phases of managing a special operation?
2. What are the four levels of hazardous materials training?
3. Explain the process and training needs for a tactical EMS operation.
4. How would an EMS agency respond to accommodate a patient with special needs during an evacuation emergency, power failure, or heat wave?
5. An outdoor concert is coming to your community. What steps can you take to ensure EMS assist in providing a safe environment? Does the preparation change if the concert is a bluegrass or a rock concert?

References

Christen, H. T., and Maniscalco, P. M. (1998). *The EMS incident management system: Operations for mass casualty and high impact incidents.* Upper Saddle River, NJ: Prentice Hall.

Ciottone, G. R. (2006). *Disaster medicine.* San Diego, CA: Elsevier.

De Lorenzo, R. A., and Porter, R. S. (1999). *Tactical emergency care: Military and operational out-of-hospital medicine.* Upper Saddle River, NJ: Brady-Prentice Hall.

Dewitt, I. (2001). *Tactical medicine: An introductory to law enforcement emergency care.* Boulder, CO: Paladin Press.

Downey, R., Jr. (1992). *The rescue company.* Tulsa, OK: PennWell Books.

Hnatow, D. A., & Gordon, D. J. (1991). Medical planning for mass gatherings: A retrospective review of the San Antonio papal mass. *Prehospital and Disaster Medicine 6*(4), 443–450.

National Transportation Safety Board. (2006). *Special investigation report on emergency medical services (EMS) operations: Aviation special investigation report* (NTSB/SIR-06/01). Washington, DC: Author.

Glossary

501(C)(3) organizations a provision of the U.S. Internal Revenue Code [26 U.S.C. § 501(c)] that lists 28 types of nonprofit organizations exempt from some federal income taxes. Sections 503 through 505 list the requirements for attaining such exemptions. Many states reference Section 501(c) for definitions of organizations exempt from state taxation as well.

Accidental Death and Disability: The Neglected Disease of Modern Society a National Academy of Sciences "white paper" that outlines the problems related to traumatic injuries, deaths, and disability cost.

active learning a process whereby learners are actively engaged in the learning process, rather than passively absorbing lectures.

activity-based cost system a cost method that assigns costs to activities and cost objects based on the consumption of resources.

administrative law an area of civil law that pertains to the government's authority to enforce its rules, regulations, and statues through criminal penalties, fines, seizures, or liens.

administrative skills skills that involve planning, negotiating, and coaching.

affective domain a type of teaching and learning that addresses interests, attitudes, opinions, appreciations, values, and emotions.

algorithmic protocol a series of steps based on recognized or objective facts or conditions.

all-hazards approach the development of plans, procedures, and resources that can be assembled in a modular fashion and brought to bear upon either a crisis or a scheduled event for the purpose of reducing medical risks to the EMS system's customers and response personnel.

ambulance a vehicle for emergency medical care that provides all of the following: a patient compartment to accommodate an EMT or paramedic and two litter patients so positioned that the primary patient can be given intensive life support during transit; equipment and supplies for emergency care at the scene as well as during transport; two-way radio communication; and, when necessary, equipment for light rescue/extrication procedures.

ambulance diversion a condition or event in which an ambulance is sent by radio or computer programming to a hospital other than the one of choice or the one closest to the emergency scene.

Ambulance Manufacturers Division (AMD) a division of the National Truck Equipment Association (NTEA); for the last 25 years has represented a majority of ambulance manufacturers in the United States.

ambulance specification a construction document for the design and building of an ambulance; should meet the federal KKK standards for ambulances to be eligible for the Star of Life.

ambulance strike team a group of five ambulances of the same type with common communications and a leader. *Also called* EMS strike team.

American College of Emergency Physicians (ACEP) a membership organization that supports quality emergency medical care and promotes the interest of emergency medicine physicians.

Applied Strategic Planning Model created by Dr. J. William Pfeiffer, this National Fire Academy planning model involves a nine-step process.

arbitration a decision-making process by a neutral third party; binding arbitration is often a clause placed in the contract or provided under state labor law that indicates a decision made by an arbitrator is final and binding to both parties.

area command command established either to oversee management of separate ICS organizations at multiple incidents or of multiple ICS organizations at a very large incident.

articulation the term used for transferring one degree in its entirety to another college or university as part of a higher degree.

assessment center process a series of exercises with which an employee is given the opportunity to demonstrate skills to a group of trained observers.

assisting agency an agency directly contributing tactical or service resources to another agency.

asynchronous learning learning that occurs at various times without an instructor being present, often by way of Web-based programs, self study, or CD-ROM learning programs.

AT&T language line an 800-number service available to communication centers 24 hours a day; its language bank has 160 different languages and linguists to translate over the telephone.

attributable risk the amount that a particular type of injury is associated with a particular exposure in a population.

automated location identification (ALI) system technology that locates a landline telephone or a cellular transmission.

automated resource management systems (ARMS) a secure system that maintains an inventory of typed federal, state, local, and privately owned resources.

automatic aid aid for which the dispatch center without a command officer's input can send or request equipment based on the information from the call to the public safety answering center.

automatic number identification (ANI) system technology that identifies the location of a caller to 911.

automatic vehicle locator (AVL) a CAD software system that tracks vehicles on a grid or map system.

average cost curve a formula that is equal to total cost divided by the number of goods or services produced (Q, or quantity). It also is equal to the sum of average variable costs divided by Q plus average fixed costs (total fixed costs divided by Q).

avoidance a risk-management technique that eliminates or does not expose the agency to operations or procedures that would result in a loss.

baby boomer a person born between 1946 and 1964.

balanced score card (BSC) a term that refers to the principle that what gets measured gets done and that financial performance is not the only measure of EMS success.

bathtub curve a graph used to represent the failure rate of a product during its life cycle; namely, the product experiences early "infant mortality" failures when first introduced, then exhibits random failures with constant failure rate during its "useful life," and finally experiences "wear out" failures as the product exceeds its design lifetime.

behavior modification approach an approach that minimizes the influence of thoughts and feelings and stresses the learning theory that people do what they do because they find it rewarding.

benchmarking a process with which management evaluates various aspects of the organization in relation to best practices, usually with the aim of increasing some aspect of performance; may be a one-off event, but often treated as a continuous process in which organizations continually seek to challenge their practices.

brainstorming a method of shared problem solving in which all members of a group spontaneously contribute ideas; a process undertaken by a person to solve a problem by rapidly generating a variety of possible solutions.

branches organizational level having functional or geographical responsibility for major aspects of incident operations.

brand a recognized physical logo, label, or representation of a person, organization, or product that identifies a distinct set of principles or values.

branding a process that establishes direction, leadership, clarity of purpose, inspiration, and energy for an agency's most important asset.

business plan an accountability program for the current budget year; sets the stage for multi-year budgets and operational targets.

call avoidance a business practice that focuses on solving a customer's problem on first contact. *Also called* first-call resolution.

call prioritization the sorting of a specific type of EMS call based on the seriousness of the incident.

call processing time the time it takes for a communication specialist to answer a call and then verify the address, decide on resources, and initiate the dispatch.

capital equipment cost the cost of equipment used to manufacture a product, provide a service, or use to sell, store, and deliver merchandise that will not be sold in the normal course of business, but will be used and worn out or consumed; any single asset that has an acquisition cost of $5,000 or more and a useful life of more than one year, whether purchased outright, acquired through a capital lease, or through donation; certain constructed or fabricated items and certain component parts.

capitated contract a contract in which an HMO gives an annual lump sum payment to the provider to cover an estimated number of patients under their plan.

carve-out language wording in a contract for special circumstances billing with an HMO or other insurance entity that allows for reimbursement when a prudent layperson calls for an ambulance.

Centers for Medicare and Medicaid Services (CMS) a federal agency within the U.S. Department of Health and Human Services (DHHS) that administers the Medicare program and works in partnership with state governments to administer Medicaid, the State Children's Health Insurance Program (SCHIP), and health insurance portability standards. Other responsibilities include the administrative simplification standards for the Health Insurance Portability and Accountability Act of 1996 (HIPAA). *Previously known as* Health Care Financing Administration (HCFA).

certified ambulance manager an internal management training program provided through the American Ambulance Association; has an ongoing offering for ambulance industry managers to develop manager and leadership skills.

chief fire officer designation (CFOD) the highest designation and credential for a public safety official; involves a credential review and documentation of experience.

chief medical officer designation (CMOD)   a credential designed to apply career development activities and education from the National Fire Academy to EMS providers who may or may not be affiliated with fire-based EMS organizations. The Center for Public Safety Excellence has taken up the credentialing of the chief EMS officer designation.

civil law   laws established between two recognizable parties, which may include corporations, partnerships, quasi-government structures, or other public entities; based on torts or contracts, usually involving acts done by one person against another in a negligent or willful manner causing injury or loss.

clinical research   research that involves the study of direct patient care activities.

coaching   refers to managers fostering skill development, imparting knowledge, and instilling values and behaviors that will help subordinates achieve organizational goals and prepare them for more challenging assignments or performance.

collective bargaining   a process often mandated that outlines what can or will be negotiated between an employer and a labor group; governed by federal and state statutory law, administrative agency regulations, and judicial decisions.

command staff   positions established to assign responsibility for key activities not specifically identified in the general staff functional elements; may include the public information officer (PIO), safety officer (SO), and the liaison officer (LNO), in additional to various others, as required and assigned by incident command.

Commission on Accreditation of Allied Health Education Programs (CAAHEP)   the largest programmatic accreditor in the health science field; in collaboration with its committees on accreditation, it reviews and accredits over 2000 educational programs in 20 health science occupations.

commitment planning   a planning technique used to search for and find common ground through a labor management partnership or by stakeholder consensus-building.

Committee on Accreditation of Educational Programs for EMS Professions (CoAEMSP)   under the auspices of CAAHEP, an organization committed to continuously improving the quality of EMS education through accreditation and recognition services.

communicable disease   a disease that can spread from one person to another.

communications protocols   definitions of the method and timing of communications by way of on-line medical control.

computer-aided dispatch (CAD)   the basic system status and any deployment of vehicles including fixed-base units.

computer-based testing (CBT)   a method of administering tests in which the responses are electronically recorded, assessed, or both; makes use of a computer or an equivalent electronic device. *Also called* e-exam, computerized testing, and computer-administered testing.

concentric ring of safety   the formation of a protective perimeter dependent upon the threat, manpower, and resources available to a protective detail.

Consolidated Omnibus Budget Reconciliation Act (COBRA)   a federal law with a subsection that regulated public access to emergency care and mandated emergency care or evaluation of non-insured people; defines when a person is on hospital property and the parameters for a person to be transferred to another hospital.

constant staffing   a plan that uses one person for each budgeted position on each unit and fills vacations and sick time with current staff on overtime.

continuing dispatch education (CDE)   ongoing training for dispatch personnel integral to continuous quality improvement.

Continuing Education Coordinating Board for Emergency Medical Services (CECBEMS)   an organization that approves submissions for continuing education credits for EMS education programs on a national basis.

continuous quality improvement (CQI)   a management approach to the continuous study and improvement of the processes of providing health-care services to meet the needs of patients and other persons; focuses on making an entire system's outcomes better by constantly adjusting and improving the system itself, instead of searching out and getting rid of "bad apples" (outliers).

control theory   the intervention and action necessary to promote appropriate behavior.

core elements   the major areas of prehospital instruction and/or performance identified in the National EMS Training Blueprint.

corrective maintenance   a maintenance program that depends upon each user identifying specific problems and reporting them to the maintenance manager for corrective action.

cost per mile   a cost calculated by taking the specific vehicle cost including the fuel, insurance, maintenance, and indirect cost to support that vehicle in a budget cycle or calendar year and dividing it by the annual mileage accumulated on that vehicle.

cost shifting   a method that directs the cost of service to other providers or health-care agencies.

Counter Narcotics and Terrorism Programs (CONTOMS)   programs designed to meet the need for specialized medical training that supports law enforcement operations; the first federal programs to address the complex response issues of a terrorist or ultra-violence incident that might confront state and local emergency responders.

crash injury management   a first responder program for law enforcement officers created by NHSTA in 1970 that was 40 hours of advanced first aid training.

criminal law   a type of law established by legislatures to indicate public wrongs or crimes against the state.

critical indicator   clearly defined measurements that compare various input and process characteristics.

critical path method (CPM)   a graphic view of projects; predicts the time required to complete a project by showing which activities are critical to maintaining the program or construction of the physical plant.

critical to quality (CTQ)   refers to identified protocol requirements, customer expectations, and the process used for delivering EMS.

critical vehicle-failure (CVF) rate   the average frequency with which something mechanical fails on a fleet vehicle.

crush load   the assembly of persons inside or outside a facility who overwhelm the capacity of a given area, resulting in gridlock, limited access, and a surge in the crowd, which pins people against a fixed object, compromising life and safety.

culture   a pattern of behavior that helps remove uncertainty for employees and provides a context for doing the organization's work.

customer   in the broadest sense, the recipient of the service provided or the purchaser of the services.

customer complaint   any indication that a service or product does not meet the customer or public's expectations.

cybernetics   the science of control; the study of how control should be exercised and the flow of information necessary to achieve it.

data   the facts or figures. Singular *datum*.

Degrees at a Distance Program (DDP)   a program sponsored by the National Fire Academy that includes a significant Web-based and online component for emergency service personnel; awards a bachelor's degree.

delegated practice   a certain standard of emergency patient care, established by medical oversight, which is carried out by prehospital providers in the field.

Delphi technique   a method for obtaining forecasts from a panel of independent experts over two or more rounds.

deposition   a pretrial discovery procedure whereby the testimony under oath of a witness is taken outside of an open court.

descriptive protocols   narratives that detail a condition or situation and usually contain educational material.

designated infection-control officer   an officer appointed by an ambulance service, EMS first response, or fire service; receives notification from health-care facilities of exposures to infectious diseases dangerous to public health and notifying the indicated care provider(s) of an exposure to an infectious disease dangerous to public or employee health.

destination examination   a check for contract compliance and minor defects or defects subject to correction by the manufacturer; occurs when the vehicle arrives at its destination.

destination protocols   definitions of the appropriate receiving facility for a patient with any given medical condition.

Develop a Curriculum (DACUM)   a process for occupational analysis; involves the local men and women in the organization with the reputation for being top performers at their jobs.

direct services   the services provided to the public; includes patient care activities, public education programs, preventive maintenance programs, and special events.

direct transmission   occurs when a pathogen is transmitted directly from an infected individual to one who had not yet been infected.

disaster medical assistance team (DMAT)   a group of professional and volunteer medical personnel organized to provide essential emergency medical care and patient evacuation during times of natural or man-made disaster, or in times of national security emergencies; augments and aids local jurisdictions overwhelmed in managing patient care.

disaster mortuary operational response team (DMORT)   a federal level response team designed to provide mortuary assistance in the case of a mass fatality or cemetery-related incident. These teams work under the local jurisdictional authorities such as coroner/medical examiners, law enforcement, and emergency managers.

discovery phase   the portion of a lawsuit that requires the parties to provide information, both good and bad, concerning the case.

disease management coordinator   a person in the hierarchy of the health insurance industry who, for each disease process, looks at how to reduce medical costs and provide cost-effective activities such as injury prevention.

distributive learning   a portion of the contact hours of a course or an organized teaching/learning event that occurs outside of the physical presence of the teacher.

division   a common designation that represents geographic responsibilities of an incident; a tactical level management position in the command organization of NIMS.

DOT national standardized curricula   standards and guidelines complete with objectives and lesson plans to be used for instruction to the various levels or certifications in EMS.

due process   standards for the fair treatment of citizens by federal, state, and local governments.

E code   an information code used in the International Classification of Diseases (ICD-9) system that identifies the external cause of an injury.

economy of scale   an increase in efficiency of production as the number of goods or services being produced increases.

educational approach   an approach that attempts to initiate behavioral changes by informing a target group about potential hazards and persuading them to adopt safer behavior; based on the belief that individuals will do the right thing if they understand why and know how to carry it out.

educational intervention   application of a methodology, lesson plan, or technique that addresses an objective; enforcement activities and enactment of laws.

educational research research that examines the methods used to prepare, remediate, and provide recertification activities to EMS providers.

effectiveness conformity to requirements; the degree to which a service is performed in the correct and desired manner.

Emergency Management Assistance Compact (EMAC) a congressionally ratified organization that provides form and structure to interstate mutual aid.

emergency medical dispatch (EMD) protocols that identify the level of care needed, to collect information to relay to responding crews, and to gain information about safety issues.

Emergency Medical Services Systems Act of 1973 (Public Law 93–154) federal legislation that established that EMS should be composed of 15 identified components and that provided federal funds for EMS systems.

Emergency Medical Treatment and Active Labor Act (EMTLA) a U.S. Act of Congress passed in 1986 as part of the Consolidated Omnibus Budget Reconciliation Act; requires hospitals and ambulance services to provide care to anyone needing emergency treatment regardless of citizenship, legal status or ability to pay. *Also known as* 42 U.S.C. § 1395dd, EMTALA.

emergency operations center a secure location that coordinates support functions and provides resources support for the management of an incident.

Emergency Planning and Community Right-to-Know Act a law meant to encourage and support emergency planning efforts at the state and local levels and to provide the public and local governments with information concerning potential chemical hazards present in their communities. *Also known as* Right-to-Know Act of 1986 or SARA Title III, the Superfund Amendment Reauthorization Act.

emergency support function (ESF) consolidates multiple agencies that perform similar functions into a single, cohesive unit as defined by the National Response Plan.

emotional intelligence (EI) an emerging science that looks at how to control emotions to avoid a negative response and recognize triggers.

Employees Assistance Program (EAP) a voluntary or management mandated confidential counseling, referral, and information service that provides a broad range of services to employees and their immediate families; may include mental health services, drug rehabilitation, and financial counseling services.

employee forecasting an EMS management activity that estimates the number and type of personnel needed to meet organizational objectives.

EMS Agenda for the Future a consensus document for a national EMS strategic plan facilitated by the Department of Transportation (DOT) and the National Highway Traffic Safety Administration (NHTSA).

EMS complaint any indication that patient care or EMS service did not meet the customer's need or the customer's standard of care.

EMS education portfolio typically contains examples of educational products (lesson plans, tests, slide presentations, games, and so on) generated by an instructor; allows an instructor to present a representative body of work for review and comment.

EMS fellowship typically a two-year residency program to guide physicians in development of the necessary experience in operations, training, research, and administration in order to contribute to EMS systems in a variety of settings.

EMS for Children (EMS-C) a federally funded program under the U.S. Health Resources and Services Administration to improve the way EMS serves children in emergency situations.

EMS strike team See *ambulance strike team.*

EMS task force a group of different resources with common communications and a designated leader.

EMS Week a week of celebration to recognize the efforts of EMS in the United States; held annually in May.

EMT-Basic an emergency care provider who has the knowledge and skill of the EMS First Responder, but is also qualified to function as the minimum staff for an ambulance.

EMT-Intermediate an emergency care provider who has the knowledge and skills of the EMT-Basic, but can also perform essential advanced techniques and administer a limited number of medications.

EMT-Paramedic an emergency care provider who has the competencies of an EMT-Intermediate, but also has enhanced assessment skills and can administer additional interventions and medications.

enforcement attempts efforts to reduce dangerous behaviors through legislation and enforcement of that legislation.

enterprise account an account often set up in government budgets to contain money within a specific operation.

environmental interventions application of changes to the environment or a product design so as to automatically protect everyone.

epidemiology the study of distribution and determinants of health-related states and events in specified populations and the application of this study to control health problems.

E-series a certification test for technicians who work on ambulances; a test administered by the Emergency Vehicle Technician Certification Commission.

ethics a set of principles or a standard of conduct.

evidence-based medicine a practice that indicates which prehospital interventions and actions produce a positive outcome for the patient.

executive fire officer (EFO) a designation granted by the National Fire Academy for completion of its four-year professional development program.

executive summary a part of a business plan that is generally written at the end of the development process; identifies the key issues related to the plan.

experimental design defines a population, applies a treatment, intervention, or change, and then measures the effects.

exposure contact with a pathogen or disease-causing organism, toxic substance, or harmful environment.

external customer represents anyone who calls 911 and experiences service from any of the following: fire, EMS, and communication center staff, hospital staff, nursing home staff, law enforcement agencies, neighboring fire and EMS agencies, contracted services, and elected officials.

failure mode and effect analysis (FMEA) a way of looking into the future and determining where the potential for failure might exist.

Fair Labor Standards Act (FLSA) a law passed in 1936 by Congress to regulate working hours, establish a minimum wage, and protect non-exempt employees from unpaid overtime.

Federal Interagency Committee on EMS (FICEMS) coordinates various federal agencies involved in EMS, including the Department of Health and Human Services (DHHS), the Department of Homeland Security (DHS), and the National Highway Traffic Safety Administration (NHTSA).

festival seating seating or accommodation in which no seating other than a floor or ground surface is provided for the audience/spectators gathered to observe a performance.

field data displayed in a specific place in the database.

Fire and Emergency Services Higher Education (FESHE) an initiative chartered to create a model framework for EMS and fire academic degrees.

Fire Department Strategic Planning Model a 12-step process that merges the Fire Academy model and the Bryson nonprofit model.

FIRESCOPE a system for coordinating firefighting resources in Southern California. *An acronym for* Firefighting Resources of California Organized for Potential Emergencies.

First Responder an emergency care provider who is at the EMS entry level or lowest certification level in some states, and who uses a limited amount of equipment to perform initial assessment and intervention. *Also called* EMS first responder.

fixed costs costs that remain unchanged in total for a given time period despite wide changes in the related level of total activity or volume.

fixed expenses those aspects of the service that continue to be the same regardless of the level of activity, such as salaries, depreciation, preventive vehicle maintenance, and training.

functional analysis an evaluation that begins with the organization's mission and the identification of those functions that enable the organization to achieve its mission.

functional services services provided within the EMS agency, such as yearly physicals, employees assistances program (EAP), training, uniforms, office support, and utilities.

Garrity Rights a U.S. Supreme Court decision [Garrity v. New Jersey, 385 U.S. 493 (1967)] that protects public employees against being compelled to make a choice between self-incrimination and loss of job.

general staff incident management personnel who represent the major functional elements of the ICS, including the operations section chief, planning section chief, logistics section chief, and finance/administration section chief.

generation X people born between 1965 and 1980.

government budget a budget assigned to a federal, state, city, or county organization.

graft the acquisition of money, position, and so on by dishonest or unjust means, as by actual theft or by taking advantage of a public office or any position of trust or employment to obtain fees, perquisites, profits on contracts, legislation, pay for work not done or service not performed; illegal or unfair practice for profit or personal advantage.

grievance an official disagreement placed in writing by a bargaining unit member.

gross negligence a severe violation of a standard that is expected or that a reasonable provider would provide; essentially similar to recklessness and willful and wanton misconduct.

gross receipts tax a tax taken from the sales of local businesses.

group a common designation that represents a functional (job) responsibility at an incident; a tactical level management position in the command organization of NIMS.

Haddon Matrix a method of systematically addressing injury prevention; developed in 1968 by William Haddon, Jr., a public health physician with the New York State Health Department.

Hawthorne Effect a phenomenon thought to occur when people who are aware they are being observed during a research study temporarily change their behavior or performance.

hazard a situation or event that can cause a loss.

hazard analysis identification of hazards that exist within a community.

Health Care Financing Administration (HCFA) See *Centers for Medicare and Medicaid Services (CMS)*.

Healthy People 2010 a national initiative to collect the national agendas of National Training Initiative (NTI) agencies to identify injury prevention targets.

Highway Safety Act of 1966 federal legislation that resulted from a "white paper" investigation by the National Academy of Sciences into highway traffic safety deaths.

hygiene factors factors that when present can keep employees from being dissatisfied, but they do not cause satisfaction with the work.

ICD-9 codes   See *International Classification of Diseases (ICD-9)*.

identity   the physical manifestation of a business brand.

image   the perceived sum of the entire organization, its objectives, and its plans; encompasses the organization's products, services, management style, and communications activities.

immediate supervision   on-line supervision; involves medical direction by a physician to a field provider.

immunity lawsuits   federal laws enacted to ensure privacy of personal medical information and to set standards for the disclosure of medical information.

impact fees   one-time fees often assessed on new construction, new business, or activities in order to establish infrastructure in the district or community.

implementation fidelity   measures how closely the design of a program equals the actual implementation of a program.

incident action plan (IAP)   the overall incident objectives and strategies established by incident command or unified command.

incident commander (IC)   the command position in the organization of NIMS; individual responsible for establishing incident management objectives and strategies.

incident management   the use of best business practices in planning, directing, organizing, coordinating, communicating, delegating, and evaluating emergency incidents.

incident management team (IMT)   typically a multi-discipline entity comprised of the command and general staff members of an incident organization assembled to manage an incident.

incremental planning approach   a master strategic plan that focuses on incremental initiatives that are flexible, reversible, and bite size and requires innovation, pilot projects, and experimenting.

indexing   part of a top-down approach in a demand analysis; uses ratio of workload to employee based on increases in call volume or other EMS activities such as injury prevention and pubic education.

indirect costs   costs of those resources necessary for logistics or infrastructure, but that cannot be traced directly to a specific product or service provided by the EMS agency.

indirect transmission   transmission of a disease that occurs when an inanimate object serves as a temporary reservoir for the infectious agent.

infect   spread illness as a result of a pathogen that has entered the body.

infectious   likely to spread illness as a result of invasion of the body by a disease organism such as bacteria, virus, fungi, or parasite.

injury severity score (ISS)   a mathematical coding applied to victims of trauma identifying the location on the body, type, and severity of an injury with expected survivability.

inspirational leadership   a leadership style for complex, unpredictable environments that have unstable and conflicting goals; requires a charismatic leader and new approaches to service delivery.

integrated information management system (IIMS)   a name commonly used to describe the system that collects data for EMS.

integrative budgeting system   a line-item budget modification that uses three computerized categories: personnel, operations, and capital outlays.

intern   an emergency care provider who is new to the EMS system and participates in an orientation and supervised practice program. *Also called* preceptee.

internal customer   an individual, entity, or organization involved in, or with, the operation of the EMS system.

International Classification of Diseases (ICD-9)   the world classification of medical conditions, diseases, and injuries that allows medical providers to bill government insurances, including Medicare and Medicaid. *Also called* ICD-9 codes.

internship   the planned orientation program that helps to introduce and integrate the intern into the EMS system or agency.

interpersonal skills   the ability to contribute to an organization by way of interaction with others; the ability to understand the feelings, attitudes, and motives of others from what they say and do.

job card   a report filled out upon repair or maintenance of any fleet vehicle; includes the vehicle number, odometer reading, hours and cost of labor used, and any miscellaneous expenses around the event, such as example towing.

job description   a document that provides information about the task, duties, and responsibilities of a position in an organization.

job specification   a list of knowledge, skills, and abilities an individual must have in order to succeed in the position.

job task analysis   a systematic process of determining the skills, abilities, and knowledge required for performing jobs in an organization.

Kaizen   a program that encourages continuous and gradual improvement through "evolution" rather than "revolution."

key driver   quantifiable measurements that reflect the critical success factors of an EMS organization.

knowledge, skills, and abilities (KSAs)   usually referred to in the context of requirements for a job or position.

leader emergence   a process by which peers recognize the leadership potential of a group member and allow that person to influence their futures.

leadership   the process of guiding others toward goal accomplishment; an individual's ability to get people to follow or take action.

LEADS project   See *Longitudinal EMT Attributes and Demographic Study*.

line-item budget a simple budget that includes each item or service on a separate line; a budget that lists specific items or services by division, department, unit, and occasionally into smaller units.

local emergency planning committees (LEPCs) groups at the state and local levels that increase the public's awareness of hazardous chemicals within their communities.

Longitudinal Emergency Medical Technician Attributes & Demographics Study (LEADS) a project that began in 1998 led by a team of researchers from the National Registry of EMTs that looks at longitudinal data for work activities, working conditions, and job satisfaction.

Malcolm Baldrige National Quality Award established by the U.S. Department of Commerce based on criteria used by organizations to evaluate their overall quality and performance.

malpractice improper care, conduct, or treatment by a physician, hospital, or other provider of health care.

management the rational assessment of a situation, the development of goals and strategies, and the design, organization, direction, and control of the activities required to attain the goals.

managerial skills the skills of analysis, planning, problem solving, and communicating (especially listening).

marginal cost cost of training, supplies, equipment, fuel, and accelerated maintenance and vehicle replacement schedules.

mass-casualty incident incident resulting from man-made or natural causes, resulting in illness or injuries that exceed or overwhelm the EMS and hospital capabilities of a locality, jurisdiction, or region.

mass gathering any collection of greater than 1000 people at one site or location.

master plans either 10- or 20-year plans; for EMS operations, it is often difficult to plan more than 10 years out due to the constant changes in the health-care system.

mean the arithmetic average.

medical authority the legal or regulatory agency with legal oversight; often state law, county ordinance, or some other legal body granting authority for the EMS system to operate.

medical necessity a necessity established when the patient's condition is such that transfer by any means other than an ambulance would endanger the health of the patient.

Medical Priority Dispatch System (MPDS) originally developed by Dr. Jeff Clawson and now controlled by the National Academies of Emergency Dispatch; teaches 911 operators how to handle difficult callers, identify the correct chief complaint, assign appropriate resources, effectively communicate between responders and callers, and provide life-saving support; the most advanced and comprehensive EMD system available.

mentor a trusted counselor or guide.

mentoring the offering of advice, information, or guidance by a person who has useful experiences, skills, or expertise for another individual's personal and professional growth.

mission statement a statement that addresses what an organization is going to do and for whom the organization will do it.

mitigation refers to the proactive, protective measures used to reduce the vulnerabilities to or potential for loss.

mixed method a research method that utilizes both qualitative and quantitative data.

mobile data terminal (MDT) a vehicle-based computer or terminal that enables communications with vehicles in the field and the CAD system.

morality refers to the ability to distinguish between right and wrong.

morbidity a disability or loss that does not result in a death.

mortality a death.

motivating factors when present, these factors will cause workers to be satisfied with the job, but when absent they do not cause dissatisfaction.

multi-line telephone system (MLTS) a matrix system using a computer to send multiple calls over a single line; a system common in locations where phones use a one-digit number to get an outside line.

mutual aid assistance provided by one agency to another and in return the other agency can expect help when needed; requires an agency's command officers to make a specific request for assistance from a neighboring jurisdiction.

Mutual Aid System Task Force (MASTF) a multi-agency team plan for an interstate mutual aid system.

Myers Briggs Type Indicator (MBTI) process that identifies normal personality differences among people.

National Academies of Emergency Dispatch (NAED) a non-profit standard-setting organization promoting safe and effective emergency dispatch services worldwide.

National Association of EMS Physicians (NAEMSP) an organization of physicians and other professionals partnering to provide leadership and foster excellence in prehospital emergency medical services.

National Center for Injury Prevention and Control (NCIPC) a branch of the CDC that provides services through three other branches: unintentional injury, violence prevention, and acute care; rehabilitation research; and disability prevention.

national crime information computers (NCIC) a computerized index of criminal justice information (such as criminal record history information, fugitives, stolen properties, missing persons) available to federal, state, and local law enforcement and other criminal justice agencies; operational 24 hours a day, 365 days a year.

National Fire Academy (NFA) a facility in Emmitsburg, Maryland, operated by the U.S. Fire Administration that provides fire and emergency services training.

National Fire Service Intrastate Mutual Aid System (IMAS) a system meant to support the creation of formalized, comprehensive, and exercised intrastate mutual aid plans

National Highway Traffic Safety Administration (NHTSA) a division of the U.S. Department of Transportation (DOT) with oversight for the nation's EMS system development and training.

National Institute for Automotive Service Excellence (ASE) the organization that certifies mechanics and reviews service providers to ensure that standards for repair and maintenance are completed to industry and manufacturer's standards.

National Institute of Occupational Safety and Health (NIOSH) under the management of the CDC, this agency is charged with supporting OSHA with research and scientific analysis of occupational injury.

National Registry of Emergency Medical Technicians (NREMT) a private organization that provides national certification, data collection, and testing of EMTs and paramedics.

National Response Plan (NRP) a plan that contains emergency support functions for a national response to man-made and natural disasters.

National Standard Curriculum (NSC) created and adopted by NHTSA and the National Association of State EMS directors for each of the four levels of EMS providers.

National Training Institute (NTI) a national coalition formed to build essential skills for injury prevention professionals.

National Wildfire Coordinating Group (NWCG) a federal agency that was chartered to coordinate fire management programs of the various participating federal and state agencies.

N code an information code used in the International Classification of Diseases (ICD-9) system, identifying the nature of an injury.

near miss an unintentional, unsafe act that could have resulted in an injury, fatality, or property damage.

Needle Stick Prevention Act Public Law 106–430 was passed in 2000 and requires OSHA to update bloodborne pathogen standards; requires that employees participate in the evaluation and selection of devices.

objective measurable statements that are consistent with the system's or agency's mission, vision, and key drivers.

objective structured clinical examination (OSCE) a program based on planned clinical encounters in which an EMT or paramedic interviews, examines, informs, or otherwise interacts with a standardized patient.

observation a common-sense approach to developing a set of standards for job performance.

on-line medical control real-time (via radio or telephone) direction of prehospital providers in the delivery of emergency medical care.

operational plans short-term plans that generally range from three to five years.

oval mapping technique a method used to help prioritize objectives; requires that EMS managers apply a collaborative approach to help set the goals of the organization that will be used in the strategic plan.

oxygen system test a test performed by pressurizing to 150 psi with dry air or nitrogen and assessing the ability of the ambulance to hold that pressure for four hours.

pathogen an infectious biological agent, especially a living microorganism such as a bacterium or fungus, that causes disease or illness to a person.

Patient Self-Determination Act (PSDA) an act that recognizes a patient's right to self-determination including the right to refuse life-saving treatment.

patient transportation group supervisor the supervisor who obtains all required transportation and causes the patients to be transported to the appropriate hospitals.

PDCA cycle plan, do, check, and act.

performance a term that refers to an employee's accomplishments in an assigned task.

performance-based budget a budget that uses statements of missions, goals, and objectives to explain why the money is being spent; a way to allocate resources to achieve specific objectives based on program goals and measured results.

performance-based reimbursement a reimbursement based on the EMS providing care and transport based on medical necessity and evidence-based medicine.

performance measure an action performed by an organization that can be measured on the basis of a standard or benchmark.

persuasion approach an approach that subscribes to the idea that people are motivated after a careful argument and the triggering of their motivational hot buttons.

physician certification statement refers to an HCFA rule that requires a signature or statement signed by a physician or physician assistant stating that the ambulance transport was medically necessary.

position analysis questionnaire one technique for producing standards and designing positions within an organization.

power the ability to get work done.

preceptor an EMS provider who serves as a clinical instructor to new employees and students, assisting with the transition into the field environment.

predetermined response matrix an agreed upon plan for emergency response agencies involving the deployment of resources in advance of an expected event.

pre-EMS emergency care provided by laypeople, bystanders, and others with an occupational requirement to be trained in first aid.

prehospital care report (PCR) an EMS report; a written report on the care provided by EMS personnel.

presumptive legislation a law that assumes the employee would have contracted a disease only in the work environment.

preventive maintenance (PM) a proactive maintenance program to prevent a malfunction or failure, not just correct it after it occurs.

primary agency the agency with the primary responsibility or qualified resources to manage an event.

primary EMS instructor a member of the educational team who is the main educator in charge of a group of students who are attending a course; often referred to as the lead instructor or instructor of record at an agency or institution.

process management the application of knowledge, skills, tools, techniques, and systems to define, visualize, measure, control, report, and improve processes; improvement of work activities and work flow across functional or department boundaries.

professional development skills required for maintaining a specific career path; skills offered through continuing education, including the more general skills related to personal development.

program-based budget a budget that links the system costs with results. *Also called* planning-programming-budgeting.

Program Evaluation and Review Technique (PERT) a way to organize activities that have more random completion times; generally used for complex strategic projects that may take years to develop and bring online.

progressive discipline discipline that involves an increasing level of severity for actions that violate the organizations rules, standard operating procedures, or standard of conduct.

prospective supervision off-line or indirect supervision that provides standards, policies, procedures, and treatment protocols for the medical direction of EMS providers.

protected medical history the information in written or oral form created or received by an EMS provider; the patient's medical record.

protégé a person seeking help and guidance from a mentor.

protocols a set of written rules that are to be followed by providers such as EMTs or nurses; stricter than guidelines, medical protocols often carry the weight of law.

public information, education, and relations (PIER) a program that was developed through the National Highway Traffic Safety Administration (NHTSA) to help EMS providers understand their role in identifying, implementing, and evaluating injury prevention strategies within their communities.

public service announcement (PSA) noncommercial advertisements typically on radio or television, aimed at the public good; media advertisements meant to modify public attitudes by raising awareness about specific issues. *Also called* community service announcements (CSAs).

qualitative research research that often is anecdotal; composed of soft data, which does not have enough numbers or objective measurements to be statistically significant.

quality a character, characteristic, or property of anything that makes it good or bad, commendable or reprehensible; the degree of excellence that a thing possesses.

quality assurance (QA) retrospective review or inspection of services or processes that is intended to identify problems; in health care, the activities and programs intended to provide adequate confidence that the quality of patient care will satisfy stated or implied requirements or needs.

quality control (QC) a process through which actual performance is measured and compared with goals, and the difference is acted upon; the use of operational techniques and statistical methods to measure and predict quality.

quality improvement (QI) the sum of all activities undertaken to continuously examine and improve products and services; the attainment or process of attaining a new level of performance or quality that is superior to any previous level.

quality indicators a set of performance measures that together define quality for an industry or organization.

quality of care the degree to which health services for individuals and populations increase the likelihood of desired health outcomes and are consistent with current professional knowledge.

quality planning the successful design or redesign of a system to perform to the quality standards expected by patients or other stakeholders.

quantitative research research based on objective and measurable factors; an approach to classify features, count them, and construct more complex statistical models in an attempt to explain what is observed.

queuing theory the study of waiting lines and their consequences.

qui tam a type of lawsuit involving the false claims act of the federal Medicare law.

range the difference between the lowest and highest value in a distribution.

raster a method for the storage, processing, and display of spatial data.

rating errors errors in judgment that occur when one individual evaluates another.

rational strategic planning a planning approach that typically takes the following steps: plan, act, and evaluate.

recognition primed decision making (RPDM) a retrieval process that draws on past experiences; recalling an incident and responding in a successful way based on what the outcomes were from previous experiences.

regulatory inspection mandated periodical inspections conducted by the government authority that licenses ambulances.

relational database the sharing of data between two different collection sources or databases.

reliability the degree to which interviews, tests, and selection procedures yield comparable data over the period of time; the degree to which a measure is free from random errors; refers to the consistency of the results or measurements.

rescue/extrication group supervisor the supervisor responsible for managing the rescue of entrapped victims.

retrospective supervision off-line or indirect medical direction; includes medical audit of a specific response or overall system quality management.

risk analysis a determination of the potential for an actual release from a facility, based upon historical release data and onsite work practices; an analysis of hazards and vulnerabilities.

risk assessment the evaluation of people, places, or things likely to incur a loss should a hazard actually occur if no protective action is taken to minimize or eliminate the loss.

risk management any activity that involves the evaluation of or comparison of risks and the development, selection, and implementation of control measures that change, reduce, or eliminate the probability or the consequences of harmful actions.

risk priority number (RPN) a benchmark in which the automotive industry allies three parameters for vehicle failure, using severity, occurrence, and detection to give each one a score between 1 and 10 (not critical to extremely critical).

road test a part of an acceptance criteria for new vehicles to be driven a total of 150 miles, 75 miles of that total on highways at a speed of 70 mph, 30 miles on city streets at 30 mph, 15 miles on gravel or dirt roads at 35 mph, and 5 miles on cross-country operations that are muddy or open field areas.

Robert Wood Johnson Foundation a privately funded health care foundation that historically has granted money for EMS system development and initiation of new directions in prehospital care.

role modeling a process in which an individual identifies with and assumes the values and behaviors of another person, which ultimately results in behavior modification that is usually permanent.

scientific management a method in management theory that determines changes to improve labor productivity.

seat time a term used for an instructional method in which the student is actually in the classroom and is usually measured in hours under a system called the *Carnegie unit.*

secondary EMS instructor an individual who can assist with lab and clinical instruction; a level of instructor by which the individual is eased into instruction under the guidance of a primary EMS instructor.

sentinel event an undesirable event or phenomenon that triggers the need for further analysis and investigation; a monitored event that signifies the potential for a significant system or provider deficiency, which should be examined or investigated by the appropriate regulatory or EMS manager or appropriate regulatory or contracting entity.

service audit an analysis of the EMS system for strengths and weaknesses; an assessment of all of the services being delivered by an emergency services.

silent generation people born before 1946.

simulation a computer model that helps to predict employee needs based on data collected in computer-aided dispatching software or run-volume statistics.

situational awareness the ability to generate actionable knowledge through the use of timely and accurate information.

Six Sigma a way to manage an EMS organization by putting the patient or customer first, using facts and data to find better solutions to service.

social influence approach an approach that uses social pressures to conform and advocates a campaign that influences the entire community's collective behavior.

Society for Academic Emergency Medicine (SAEM) an organization dedicated to the improvement of care of the acutely ill or injured patient by improving research and education.

spatial data requires three components that make up a GIS map: a vector, raster, and tabular data.

speaker's bureau consists of employees, generally paramedics and EMTs, who have expertise or are comfortable speaking on certain subjects.

special protocols rules developed to address special circumstances that might be encountered within the EMS system.

specialty care transport (SCT) units or interfacility transports that require levels of care beyond the scope of the paramedic, such as a care from a critical-care nurse or neonatal specialty nurse.

stakeholders individuals or organizations who receive EMS services and have some interest in the operation of the EMS organization, such as the patient's family, the community in which the EMS system operates, government officials, the patient's insurer/third-party payor, and health-care providers.

standardized patient an individual who is scripted and rehearsed to portray an actual patient with a specific set of symptoms or clinical findings.

standing orders rules that are more specific than treatment protocols; usually included in a protocol where delay in treatment might have a harmful effect or outcome.

Star of Life a blue, six-pointed star outlined with a white border, which features the rod of Asclepius in the center. Originally designed and governed by the U.S. National Highway Traffic Safety Administration (NHTSA) (under the U.S. Department of Transportation (DOT). Traditionally, in the United States the logo was used as a stamp of authentication or certification for ambulances, paramedics, or other EMS personnel.

state emergency response commission (SERC) every state governor has established a SERC as a requirement of the Superfund Amendments and Reauthorization Act of 1986 (SARA), Title III.

strategic map enables EMS managers to define and communicate the cause-and-effect relationship and their EMS organization's value to the community.

Strategic National Stockpile (SNS) a program maintained by the Centers for Disease Control and Prevention (CDC); maintains large quantities of medicine and medical supplies to protect the U.S. public if there is a public health emergency (such as a terrorist attack, flu outbreak, earthquake) severe enough to cause local supplies to run out.

strategic planning refers to the entire set of processes and behaviors that an organization uses to identify, prioritize, focus, and schedule action in order for it to remain viable in the future.

succession planning a process that recruits employees, develops their skills and abilities, and prepares them for advancement, all while retaining them to ensure a return on the organization's training investment.

supporting agency the agency that has resources or expertise to support the primary agency.

SWOT analysis a strategic planning tool used to evaluate a project or a business venture; the acronym stands for Strengths, Weaknesses, Opportunities, and Threats.

synchronous education learning that occurs at a specific time when the teacher communicates remotely using chat rooms or bulletin boards.

tactical alert a state of preparedness necessary to cope with civil disturbances, violence, or a situation that could threaten EMS responders and the public.

Tactical Emergency Medical Support (TEMS) a specialized team of EMS responders trained to operate with law enforcement to provide medical support to SWAT operations and other high-risk law enforcement scenarios.

technical skills skills gained from training and certification related to a trade or profession.

theory X an early theory of human relations, based on the belief that employees dislike work and managers must therefore coerce, control, and direct them.

theory Y a theory of human relations that assumes employees accept work as natural and managers do not need to use exclusively external controls to manage them.

third-party payee generally referred to as anyone responsible for the ambulance other than Medicare, Medicaid, or the self-paying patient.

three Es of prevention refers to the standard for application of management principles to injury prevention and includes engineering, education, and enforcement.

total quality management (TQM) a continuous quality improvement management system directed from the top, but empowering employees and focusing on systemic, not individual, employee problems.

treatment protocols guidelines that define the scope of prehospital intervention that will be practiced by providers.

treatment unit leader the leader who establishes an area where multiple patients can be collected and treated.

trend a change in the process where values move in the same direction over time.

trend analysis a top-down approach to demand analysis; focuses on past hiring.

unified command a command structure that enables agencies with different legal, geographic, and functional responsibilities to coordinate, plan, and interact effectively.

unit-hour utilization a measurement of workload for EMS agencies that was developed from a process called *queuing theory*.

upbilling a misrepresentation of provided services by billing for a more expensive service.

urban search and rescue (USAR) a "multi-hazard" discipline; involves the location, rescue (extrication), and initial medical stabilization of victims trapped in confined spaces.

Urban Search and Rescue Team (USAR) consists of two 31-person teams, four canines, and a comprehensive equipment cache. Members work in four areas of specialization: search, to find victims trapped after a disaster; rescue, which includes safely digging victims out of collapsed concrete and metal; technical, structural specialists who make rescues safe for the rescuers; and medical, which cares for the victims before and after a rescue.

Utstein criteria gold standard for measuring cardiac arrest survival since 1991.

validity a term that refers to what a test measures and how well the test measures it; the extent and accuracy to which research measures what it is designed to measure.

values statements statements that communicate the leader's beliefs, values, and outcomes as they relate to the organization's mission and vision statement; should be positive and provide a focus to the organization.

variable cost a cost that changes in total proportion to changes in the related level of total activity or volume.

variable expenses expenses that change with activity and include unscheduled vehicle maintenance, station repair, medical supplies, linen, fuel, office supplies and reports, accounting and legal expenses.

vector spatial data arranged by points, lines, and polygons.

vehicle checkouts daily or shift-start assessments made on the mechanical aspects of the emergency vehicle, including fluids, belts, and tires.

veterinary medical assistance team (VMAT) team designed to assist the local veterinary community with the care of animals, veterinary oversight, and advice concerning animal-related issues and public health during a disaster or following a request from an appropriate agency.

vision statement a statement that provides a futuristic look at, and broad guidance for, the EMS agency or system; helps to make sure everyone is going in the same direction.

vulnerability analysis the identification and quantification of those persons or segments of the population who are vulnerable to certain hazards.

water spray test a test that subjects the vehicle to a water spray at 25 psi for 15 minutes to identify any evidence of a leak.

Web-based Injury Statistics Query and Reporting System (WISQARS™) an interactive database system that provides customized reports of injury-related data and can query information by state and cause of death and injury.

Wedworth-Townsend Act law that established the state legal framework for the establishment of paramedic programs in the State of California in 1971. *Also known as* Wedworth-Townsend Pilot Paramedic Act (SB 772).

Weingarten Rights refers to the right of employees to be assisted without prejudice by an advisor who may be an attorney or union official.

workplace violence any violent act, including physical assaults and threats of assault, directed toward persons at work or on duty.

work profiling a job analysis technique based on questionnaires.

zero-based budget a technique of planning and decision making that reverses the process of traditional budgeting; a form of budgeting that requires EMS managers to ask what if a program is eliminated rather than assuming that a basis exists for a budget item for the program.

Index